Some Fundamental Constants[a]

Quantity	Symbol	Value[b]
Atomic mass unit	u	$1.660\ 540\ 2(10) \times 10^{-27}$ kg
		$931.494\ 32(2\ 8)$ MeV/c^2
Avogadro's number	N_A	$6.022\ 136\ 7(36) \times 10^{23}$ particles/mol
Bohr magneton	$\mu_B = \dfrac{e\hbar}{2m_e}$	$9.274\ 015\ 4(31) \times 10^{-24}$ J/T
Bohr radius	$a_0 = \dfrac{\hbar^2}{m_e e^2 k_e}$	$5.291\ 772\ 49\ (24) \times 10^{-11}$ m
Boltzmann's constant	$k_B = R/N_A$	$1.380\ 658\ (12) \times 10^{-23}$ J/K
Compton wavelength	$\lambda_C = \dfrac{h}{m_e c}$	$2.426\ 310\ 58(2\ 2) \times 10^{-12}$ m
Coulomb constant	$k_e = \dfrac{1}{4\pi\epsilon_0}$	$8.987\ 551\ 787 \times 10^9$ N·m^2/C^2 (exact)
Deuteron mass	m_d	$3.343\ 586\ 0(20) \times 10^{-27}$ kg
		$2.013\ 553\ 214\ (24)$ u
Electron mass	m_e	$9.109\ 389\ 7(54) \times 10^{-31}$ kg
		$5.485\ 799\ 03(1\ 3) \times 10^{-4}$ u
		$0.510\ 999\ 06(1\ 5)$ MeV/c^2
Electron-volt	eV	$1.602\ 177\ 33(4\ 9) \times 10^{-19}$ J
Elementary change	e	$1.602\ 177\ 33(4\ 9) \times 10^{-19}$ C
Gas constant	R	$8.314\ 510\ (70)$ J/K·mol
Gravitational constant	G	$6.672\ 59(8\ 5) \times 10^{-11}$ N·m^2/kg^2
Hydrogen ground state energy	$E_1 = -\dfrac{e^2 k_e}{2a_0}$	$-13.605\ 698\ (40)$ eV
Josephson frequency–voltage ratio	$2e/h$	$4.835\ 976\ 7(14) \times 10^{14}$ Hz/V
Magnetic flux quantum	$\Phi_0 = \dfrac{h}{2e}$	$2.067\ 834\ 61(6\ 1) \times 10^{-15}$ T·m^2
Neutron mass	m_n	$1.674\ 928\ 6(10) \times 10^{-27}$ kg
		$1.008\ 664\ 904\ (14)$ u
		$939.565\ 63(2\ 8)$ MeV/c^2
Nuclear magneton	$\mu_n = \dfrac{e\hbar}{2m_p}$	$5.050\ 786\ 6(17) \times 10^{-27}$ J/T
Permeability of free space	μ_0	$4\pi \times 10^{-7}$ T·m/A (exact)
Permittivity of free space	$\epsilon_0 = 1/\mu_0 c^2$	$8.854\ 187\ 817 \times 10^{-12}$ C^2/N·m^2 (exact)
Planck's constant	h	$6.626\ 075\ (40) \times 10^{-34}$ J·s
	$\hbar = h/2\pi$	$1.054\ 572\ 66(6\ 3) \times 10^{-34}$ J·s
Proton mass	m_p	$1.672\ 623\ (10) \times 10^{-27}$ kg
		$1.007\ 276\ 470\ (12)$ u
		$938.272\ 3(28)$ MeV/c^2
Rydberg constant	R_H	$1.097\ 373\ 153\ 4(13) \times 10^7$ m^{-1}
Speed of light in vacuum	c	$2.997\ 924\ 58 \times 10^8$ m/s (exact)

[a] These constants are the values recommended in 1986 by CODATA, based on a least-squares adjustment of data from different measurements. For a more complete list, see E. R. Cohen and B. N. Taylor, *Rev. Mod. Phys.* 59:1121, 1987.

[b] The numbers in parentheses for the values above represent the uncertainties of the last two digits.

Solar System Data

Body	Mass (kg)	Mean Radius (m)	Period (s)	Distance from the Sun (m)
Mercury	3.18×10^{23}	2.43×10^{6}	7.60×10^{6}	5.79×10^{10}
Venus	4.88×10^{24}	6.06×10^{6}	1.94×10^{7}	1.08×10^{11}
Earth	5.98×10^{24}	6.37×10^{6}	3.156×10^{7}	1.496×10^{11}
Mars	6.42×10^{23}	3.37×10^{6}	5.94×10^{7}	2.28×10^{11}
Jupiter	1.90×10^{27}	6.99×10^{7}	3.74×10^{8}	7.78×10^{11}
Saturn	5.68×10^{26}	5.85×10^{7}	9.35×10^{8}	1.43×10^{12}
Uranus	8.68×10^{25}	2.33×10^{7}	2.64×10^{9}	2.87×10^{12}
Neptune	1.03×10^{26}	2.21×10^{7}	5.22×10^{9}	4.50×10^{12}
Pluto	$\approx 1.4 \times 10^{22}$	$\approx 1.5 \times 10^{6}$	7.82×10^{9}	5.91×10^{12}
Moon	7.36×10^{22}	1.74×10^{6}	—	—
Sun	1.991×10^{30}	6.96×10^{8}	—	—

Physical Data Often Used[a]

Average Earth–Moon distance	3.84×10^{8} m
Average Earth–Sun distance	1.496×10^{11} m
Average radius of the Earth	6.37×10^{6} m
Density of air (20°C and 1 atm)	1.20 kg/m^3
Density of water (20°C and 1 atm)	1.00×10^{3} kg/m^3
Free-fall acceleration	9.80 m/s^2
Mass of the Earth	5.98×10^{24} kg
Mass of the Moon	7.36×10^{22} kg
Mass of the Sun	1.99×10^{30} kg
Standard atmospheric pressure	1.013×10^{5} Pa

[a] These are the values of the constants as used in the text.

Some Prefixes for Powers of Ten

Power	Prefix	Abbreviation	Power	Prefix	Abbreviation
10^{-24}	yocto	y	10^{1}	deka	da
10^{-21}	zepto	z	10^{2}	hecto	h
10^{-18}	atto	a	10^{3}	kilo	k
10^{-15}	femto	f	10^{6}	mega	M
10^{-12}	pico	p	10^{9}	giga	G
10^{-9}	nano	n	10^{12}	tera	T
10^{-6}	micro	μ	10^{15}	peta	P
10^{-3}	milli	m	10^{18}	exa	E
10^{-2}	centi	c	10^{21}	zetta	Z
10^{-1}	deci	d	10^{24}	yotta	Y

Principles *of* Physics

A Calculus-Based Text

third edition

VOLUME 1

Principles *of* Physics

A Calculus-Based Text

THIRD EDITION

VOLUME 1

Raymond A. Serway

James Madison University

John W. Jewett, Jr.

California State Polytechnic University—Pomona

HARCOURT COLLEGE PUBLISHERS

Fort Worth   Philadelphia   San Diego   New York   Orlando   Austin
San Antonio   Toronto   Montreal   London   Sydney   Tokyo

Publisher: Emily Barrosse
Publisher: John Vondeling
Marketing Strategist: Kathleen S. McLellan
Developmental Editor: Ed Dodd
Project Editor: Bonnie Boehme
Production Manager: Charlene Catlett Squibb
Art Director and Text Designer: Carol Bleistine

Cover Image and Credit: Competitor on luge course./© *Tony Stone Images*

Frontmatter Images and Credits: Title page: Downhill skier *(Jean Y. Ruszniewski/Stone)*; p. vii: Space shuttle *(NASA)*; p. vii: Drag racing *(George Lepp/Stone)*; p. viii: Vince Carter dunking *(Jed Jacobsohn/Allsport)*; p. viii: Golf club hitting ball *(Courtesy of Michael Hans/Photo Researchers, Inc.)*; p. ix: Rolling cylinder *(Henry Leap and Jim Lehman)*; p. ix: Pendulum *(© Bob Emmott, Photographer)*; p. x: Ultrasound image of a human fetus *(U.H.B. Trust/Stone)*; p. x: Plane breaking the sound barrier *(Courtesy of U.S. Navy. Photo by Ensign John Gay)*; p. xi: Twin Falls on the Island of Kauai, Hawaii *(Bruce Byers/ FPG)*; p. xii: Rock climbers *(Scott Markewitz/FPG)*; p. xiv: Edwin Aldrin *(Courtesy of NASA)*; p. xv: Roller coaster *(Frank Cezus/FPG International)*; p. xviii: Windmills *(Billy Hustace/Stone)*; p. xix: Glass shattering *(© 1992 Ben Rose/The Image Bank)*; p. xxiii: Sky surfer *(Jump Run Productions/Image Bank)*; p. xxiv: Stonehenge *(© John Serafin, Peter Arnold, Inc.)*; p. xxv: Wine bottle defying gravity *(Charles D. Winters)*.

PRINCIPLES OF PHYSICS, Third Edition, Volume 1
ISBN: 0-03-033599-X
Library of Congress Catalog Card Number: 00-110798

Address for domestic orders:
Harcourt College Publishers, 6277 Sea Harbor Drive, Orlando, FL 32887-6777
1-800-782-4479
e-mail collegesales@harcourt.com

Address for international orders:
International Customer Service, Harcourt, Inc.
6277 Sea Harbor Drive, Orlando FL 32887-6777
(407) 345-3800
Fax (407) 345-4060
e-mail hbintl@harcourt.com

Address for editorial correspondence:
Harcourt College Publishers, Public Ledger Building, Suite 1250,
150 S. Independence Mall West, Philadelphia, PA 19106-3412

Web Site Address
http://www.harcourtcollege.com

Printed in the United States of America

0123456789 048 10 987654321

Contents Overview

v

Contents

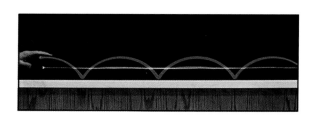

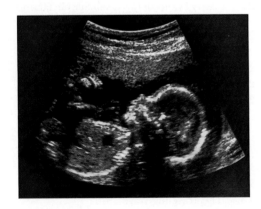

Preface

Principles of Physics is designed for a one-year introductory calculus-based physics course for engineering and science students and for premed students taking a rigorous physics course. This third edition contains many new pedagogical features—most notably, a contextual approach to enhance motivation, an increased emphasis on avoiding misconceptions, and a problem-solving strategy that uses a modeling approach. Based on comments from users of the second edition and reviewers' suggestions, a major effort was made to improve organization, clarity of presentation, precision of language, and accuracy throughout.

This project was conceived because of well-known problems in teaching the introductory calculus-based physics course. The course content (and hence the size of textbooks) continues to grow, while the number of contact hours with students has either dropped or remained unchanged. Furthermore, traditional one-year courses cover little if any 20th-century physics.

In preparing this textbook, we were motivated by the spreading interest in reforming this course, primarily through the efforts of the Introductory University Physics Project (IUPP) sponsored by the American Association of Physics Teachers and the American Institute of Physics. The primary goals and guidelines of this project are to

- Reduce course content following the "less may be more" theme;
- Incorporate contemporary physics naturally into the course;
- Organize the course in the context of one or more "story lines";
- Treat all students equitably.

Recognizing a need for a textbook that could meet these guidelines several years ago, we studied the various proposed IUPP models and the many reports from IUPP committees. Eventually, one of us (RAS) became actively involved in the review and planning of one specific model, initially developed at the U.S. Air Force Academy, entitled "A Particles Approach to Introductory Physics." Part of the summer of 1990 was spent at the Academy working with Colonel James Head and Lt. Col. Rolf Enger, the primary authors of the Particles model, and other members of that department. This most useful collaboration was the starting point of this project.

The coauthor (JWJ) became involved with the IUPP model called "Physics in Context," developed by John Rigden (American Institute of Physics), David Griffiths (Oregon State University), and Lawrence Coleman (University of Arkansas at Little Rock). This involvement led to the contextual overlay that is used in this book and described in detail later in the Preface.

The combined IUPP approach in this book has the following features:

- It is an evolutionary approach (rather than a revolutionary approach), which should meet the current demands of the physics community.

- It deletes many topics in classical physics (such as alternating current circuits and optical instruments) and places less emphasis on rigid body motion, optics, and thermodynamics.
- Some topics in 20th-century physics, such as special relativity, energy quantization, and the Bohr model of the hydrogen atom, are introduced early in the textbook.
- A deliberate attempt is made to show the unity of physics.
- As a motivational tool, the textbook connects physics principles to interesting social issues, natural phenomena, and technological advances.

OBJECTIVES

This introductory physics textbook has two main objectives: to provide the student with a clear and logical presentation of the basic concepts and principles of physics, and to strengthen an understanding of the concepts and principles through a broad range of interesting applications to the real world. To meet these objectives, we have emphasized sound physical arguments and problem-solving methodology. At the same time, we have attempted to motivate the student through practical examples that demonstrate the role of physics in other disciplines, including engineering, chemistry, and medicine.

CHANGES IN THE THIRD EDITION

A number of changes and improvements have been made in the third edition of this text. Many of these are in response to current trends in science education and to comments and suggestions provided by the reviewers of the manuscript and instructors using the first two editions. The following represent the major changes in the third edition:

Content   While the overall content of the textbook is similar to that of the second edition, several changes were implemented. A global approach to energy and energy transfer is introduced in Chapter 6 and has been incorporated throughout the book. A discussion of molar specific heats of gases has been added to Chapter 17. Also in Chapter 17 is the first law of thermodynamics expressed as $\Delta U = Q + W$, rather than the common expression that appears in many physics textbooks, $\Delta U = Q - W$. This form follows naturally from the global approach to energy introduced in Chapter 6 and is consistent with the form of the law that most chemistry books use. The use of this form of the first law follows a recommendation made by a committee appointed by the American Physical Society. Finally, many sections have been streamlined, deleted, or combined with other sections to allow for a more balanced presentation.

Organization   We have incorporated a "context overlay" scheme into the textbook, in response to the "Physics in Context" approach in the IUPP. This new feature adds interesting applications of the material covered in the third edition to real issues. We have developed this feature to be flexible, so that the instructor who does not wish to follow the contextual approach can simply ignore the additional contextual features without sacrificing complete coverage of the existing material. We feel, though, that the benefits students will gain from this approach will be many.

The context overlay organization divides the text into eight sections, or "Contexts," after Chapter 1, as follows:

Context Number	Context	Physics Topics	Chapters
1	Mission to Mars	Classical mechanics	2–11
2	Earthquakes	Vibrations and waves	12–14
3	Search for the *Titanic*	Fluids	15
4	Global Warming	Thermodynamics	16–18
5	Lightning	Electricity	19–21
6	Magnetic Levitation Vehicles	Magnetism	22–23
7	Lasers	Optics	24–27
8	The Cosmic Connection	Modern physics	28–31

Each Context begins with an introduction, leading to a "central question" that motivates study within the Context. The final section of each chapter is a "Context Connection," which discusses how the material in the chapter relates to the Context and the central question. The final chapter in each Context is followed by a "Context Conclusion." Each conclusion uses the principles learned in the context to respond fully to the central question. Each chapter, as well as the Context Conclusions, includes problems related to the context material.

Pitfall Prevention   These new features are placed in the margins of the text and address common student misconceptions and situations in which students often follow unproductive paths. Over 200 Pitfall Preventions are provided to help students avoid common mistakes and misunderstandings.

Quick Quizzes   Several Quick Quiz questions are included in each chapter to provide students opportunities to test their understanding of the physical concepts presented. The questions require students to make decisions on the basis of sound reasoning. Some of them help students overcome common misconceptions. Answers to all Quick Quiz questions are found at the end of each chapter.

Modeling   A modeling approach, based on four types of models commonly used by physicists, is introduced to help students understand they are solving problems that approximate reality. They must then learn how to test the validity of the model. This approach also helps students see the unity in physics, as a large fraction of problems can be solved with a small number of models. A general problem-solving strategy using the modeling approach is introduced in Chapter 1.

Alternative Representations   Emphasis is placed on alternative representations of the information, including mental, pictorial, graphical, tabular, and mathematical representations. Many problems are easier to solve if the information is presented in alternative ways, to reach the many different methods students use to learn.

Line-by-Line Revision   The text has been carefully edited to improve clarity of presentation and precision of language. We hope that the result is a book both accurate and enjoyable to read.

Problems   In an effort to improve clarity and quality, the end-of-chapter problems were substantially revised. Approximately 40% of the problems (about 575) are new to this edition, and most of these new problems are at the intermediate level (as identified by blue problem numbers). Many problems require students to make order-of-magnitude calculations. All problems have been carefully edited and reworded where necessary. Solutions to approximately 20% of the end-of-chapter problems are included in the *Student Solutions Manual and Study Guide*. Boxed numbers identify these problems. A smaller subset of solutions will be posted on the World Wide Web **(http://www.harcourtcollege.com/physics)** and will be accessible to students and instructors using *Principles of Physics*. The web icon identifies these problems. See the next section for a complete description of other features of the problem set.

Web Notes   Useful World Wide Web addresses are provided as marginal notes (indicated by a ▭ **WEB** ▭ icon) to encourage students to explore extensions of the material beyond what is covered in the text. In particular, the Contexts allow for rich opportunities for further explorations on the Web.

Biomedical Applications   For biology and premed students, ▨ icons point the way to various practical and interesting applications of physical principles to biology and medicine.

TEXT FEATURES

Most instructors would agree that the textbook selected for a course should be the student's primary guide for understanding and learning the subject matter. Furthermore, the textbook should be easily accessible as well as styled and written to facilitate instruction and learning. With these points in mind, we have included many pedagogical features that are intended to enhance the textbook's usefulness to both students and instructors. These features are as follows:

Style   To facilitate rapid comprehension, we have attempted to write the book in a clear, logical, and engaging style. The somewhat informal and relaxed writing style is intended to increase reading enjoyment. New terms are carefully defined, and we have tried to avoid the use of jargon.

Previews   Most chapters begin with a brief preview that includes a discussion of the particular chapter's objectives and content.

Important Statements and Equations   Most important statements and definitions are set in boldface type or are highlighted with a background screen for added emphasis and ease of review. Similarly, important equations are highlighted with a tan background screen to facilitate location.

Problem-Solving Hints   We have included general strategies for solving the types of problems featured both in the examples and in the end-of-chapter problems. This feature helps students identify necessary steps in solving problems and eliminate any uncertainty they might have. Problem-solving strategies are highlighted with a light blue-gray screen for emphasis and ease of location.

Marginal Notes   Comments and notes appearing in the margin can be used to locate important statements, equations, and concepts in the text.

Illustrations and Tables   The readability and effectiveness of the text material and worked examples are enhanced by the large number of figures, diagrams, photographs, and tables. Full color adds clarity to the artwork and makes illustrations as realistic as possible. For example, vectors are color coded, and curves in graphs are drawn in color. The three-dimensional appearance of many illustrations has been improved in this third edition. The color photographs have been carefully selected, and their accompanying captions have been written to serve as an added instructional tool.

Mathematical Level We have introduced calculus gradually, keeping in mind that students often take introductory courses in calculus and physics concurrently. Most steps are shown when basic equations are developed, and reference is often made to mathematical appendices at the end of the textbook. Vector products are discussed in detail later in the text, where they are needed in physical applications. The dot product is introduced in Chapter 6, which addresses work and energy; the cross product is introduced in Chapter 10, which deals with rotational dynamics.

Worked Examples A large number of worked examples of varying difficulty are presented to promote students' understanding of concepts. In many cases, the examples serve as models for solving the end-of-chapter problems. Because of the increased emphasis on understanding physical concepts, many examples are conceptual in nature and are labeled as such. The examples are set off in boxes, and the answers to examples with numerical solutions are highlighted with a tan screen.

Worked Example Exercises Many of the worked examples are followed immediately by exercises with answers. These exercises are intended to promote interactivity between the student and the textbook and to immediately reinforce the student's understanding of concepts and problem-solving techniques. The exercises represent extensions of the worked examples.

Questions Questions requiring verbal responses are provided at the end of each chapter. Over 500 questions are included in this edition. Some questions provide the student with a means of self-testing the concepts presented in the chapter. Others could serve as a basis for initiating classroom discussions. Answers to selected questions are included in the *Student Solutions Manual and Study Guide.*

Significant Figures Significant figures in both worked examples and end-of-chapter problems have been handled with care. Most numerical examples and problems are worked out to either two or three significant figures, depending on the accuracy of the data provided.

Problems The end-of-chapter problems are more numerous in this edition and more varied (in all, over 1 800 problems are given throughout the text). For the convenience of both the student and the instructor, about two thirds of the problems are keyed to specific sections of the chapter, including Context Connection sections. The remaining problems, labeled "Additional Problems," are not keyed to specific sections. An 🦓 icon identifies problems dealing with applications to the life sciences and medicine. One or more problems in each chapter ask students to make an order-of-magnitude calculation based on their own estimated data. Other types of problems are described in more detail below. Answers to odd-numbered problems are provided at the end of the book.

Usually, the problems within a given section are presented so that the straight-forward problems (those with black problem numbers) appear first; these straight-forward problems are followed by those of increasing difficulty. For ease of identification, the numbers of intermediate-level problems are printed in blue, and those of challenging problems are printed in magenta.

Solutions to approximately 20% of the problems in each chapter are in the *Student Solutions Manual and Study Guide.* Among these, selected problems are identified with web icons and have their solutions posted on the World Wide Web at **http://www.harcourtcollege.com/physics.**

Review Problems   Many chapters include review problems requiring the student to relate concepts covered in the chapter to those discussed in previous chapters. These problems could be used by students in preparing for tests and by instructors for special assignments and classroom discussions.

Paired Problems   As an aid for students learning to solve problems symbolically, paired numerical and symbolic problems are included in Chapters 1 through 4 and 16 through 21. Paired problems are identified by a common tan background screen.

Computer- and Calculator-Based Problems   Most chapters include one or more problems whose solution requires the use of a computer or graphing calculator. Modeling of physical phenomena enables students to obtain graphical representations of variables and to perform numerical analyses.

Units   The international system of units (SI) is used throughout the text. The British engineering system of units (conventional system) is used only to a limited extent in the chapters on mechanics and thermodynamics.

Summaries   Each chapter contains a summary that reviews the important concepts and equations discussed in that chapter.

Appendices and Endpapers   Several appendices are provided at the end of the textbook. Most of the appendix material represents a review of mathematical concepts and techniques used in the text, including scientific notation, algebra, geometry, trigonometry, differential calculus, and integral calculus. Reference to these appendices is made throughout the text. Most mathematical review sections in the appendices include worked examples and exercises with answers. In addition to the mathematical reviews, the appendices contain tables of physical data, conversion factors, atomic masses, and the SI units of physical quantities, as well as a periodic table of the elements and a list of Nobel prize recipients. Other useful information, including fundamental constants and physical data, planetary data, a list of standard prefixes, mathematical symbols, the Greek alphabet, and standard abbreviations of units of measure, appears on the endpapers.

ANCILLARIES

The ancillary package has been updated substantially and streamlined in response to suggestions from users of the second edition. The most essential changes in the student package are a *Student Solutions Manual and Study Guide* with a tighter focus on problem-solving, the *Student Tools CD-ROM,* and the *Saunders Core Concepts in Physics CD-ROM* developed by Archipelago Productions. Instructors will find increased support for their teaching efforts with new electronic materials.

Student Ancillaries

Student Solutions Manual and Study Guide   by John R. Gordon, Ralph McGrew, and Raymond A. Serway. This two-volume manual features detailed solutions to approximately 20% of the end-of-chapter problems from the textbook. Boxed numbers identify those problems in the textbook whose complete solutions are found in the manual. The manual also features a list of important equations and concepts, as well as answers to selected end-of-chapter questions.

Student Tools CD-ROM This CD-ROM contains tools designed to enhance the learning of physical concepts and train students to become better problem-solvers. It includes a textbook version of the highly acclaimed Interactive Physics™ software by MSC Working Knowledge, more than 100 Interactive Physics™ simulations keyed to appropriate worked examples and selected end-of-chapter problems (as identified by the 🖳 icon), and support for working those end-of-chapter problems that require the use of computers.

Saunders Core Concepts in Physics CD-ROM This CD-ROM package developed by Archipelago Productions applies the power of multimedia to the introductory physics course, offering full-motion animation and video, engaging interactive graphics, clear and concise text, and guiding narration. *Saunders Core Concepts in Physics CD-ROM* focuses on those concepts students usually find most difficult in the course, drawing from topics in mechanics, thermodynamics, electric fields, magnetic fields, and optics. The animations and graphics are presented to aid the student in developing accurate conceptual models of difficult topics—topics often too complex to be explained in words or chalkboard illustrations. The CD-ROM also presents step-by-step explorations of problem-solving strategies and provides animations of problems in order to promote conceptual understanding and sharpen problem-solving skills. Textbook topics further explored on the CD-ROM are identified by marginal 💿 icons that give the appropriate module and screen number(s). Students should look to the CD-ROM for help in understanding these topics.

Student Web Site Students will have access to an abundance of material at **http://www.harcourtcollege.com/physics.** The Web Site features special topic essays by guest authors, practice problems with answers, and optional topics that accompany selected chapters of the textbook. Also included are selected solutions from the *Student Solutions Manual and Study Guide* and a glossary that includes more than 300 physics terms. Students also can take practice quizzes in our Practice Exercises and Testing area.

Physics Laboratory Manual, Second Edition by David Loyd. Updated and redesigned, this manual supplements the learning of basic physical principles while introducing laboratory procedures and equipment. Each chapter includes a pre-laboratory assignment, objectives, an equipment list, the theory behind the experiment, step-by-step experimental procedures, and questions. A laboratory report form is provided for each experiment so that students can record data and make calculations. Students are encouraged to apply statistical analysis to their data so they can develop the ability to judge the validity of their results.

So You Want to Learn Physics: A Preparatory Course with Calculus by Rodney Cole. This introductory-level book is useful to those students who need additional preparation before or during a calculus-based course in physics. The friendly, straightforward style makes it easier to understand how mathematics is used in the context of physics.

Life Science Applications for Physics   by Jerry Faughn. This supplement provides examples, readings, and problems from the biological sciences as they relate to physics. Topics include "Friction in Human Joints," "Physics of the Human Circulatory System," "Physics of the Nervous System," and "Ultrasound and Its Applications." This supplement is useful in those courses taken by a significant number of premed students.

Instructor's Ancillaries

Instructor's Manual with Solutions   by Ralph McGrew, Jeffery Saul, and Charles Teague. This manual consists of complete, worked solutions to all the problems in the textbook. The solutions to problems new to the third edition are marked for easy identification by the instructor. New to this edition of the manual are suggestions on how to teach difficult topics and help students overcome common misconceptions. These suggestions are based on recent research in physics education.

Instructor's Web Site   The instructor's area at **http://www.harcourtcollege.com/physics** includes a listing of overhead transparencies; a guide to relevant experiments in David Loyd's *Physics Laboratory Manual, Second Edition;* a correlation guide between sections in *Principles of Physics* and modules in the *Saunders Core Concepts in Physics CD-ROM;* supplemental problems with answers; optional topics to accompany selected chapters of the textbook; and a syllabus generator.

Instructor's Resource CD-ROM   This CD-ROM accompanying the third edition of *Principles of Physics* has been created to provide instructors with an exciting new tool for classroom presentation. The CD-ROM contains a collection of graphics files of line art from the textbook. These files can be opened directly or can be imported into a variety of presentation packages. The labels for each piece of art have been enlarged and boldfaced to facilitate classroom viewing. The CD-ROM also contains electronic files of the *Instructor's Manual* and *Test Bank*.

Instructor Options for Online Homework

WebAssign: A Web-Based Homework System   WebAssign is a Web-based homework delivery, collection, grading, and recording service developed at North Carolina State University. Instructors who sign up for WebAssign can assign frequent homework to their students, using questions and problems taken directly from *Principles of Physics*. WebAssign gives students immediate feedback on their homework and helps them to master information and skills, leading to greater competence and better grades. WebAssign can free instructors from the drudgery of grading homework and recording scores, allowing them to devote more time to meeting with students and preparing classroom presentations.

WebAssign is being used in scores of educational institutions by tens of thousands of students and hundreds of instructors. Most of the numerical problems that can be assigned have different values, so that each student has a unique problem to solve. This feature motivates independent thinking within the context of collaborative learning.

Details about and a demonstration of WebAssign are available at **http://wasnet01ws.physics.ncsu.edu/info/.** For more information about ordering this service, contact WebAssign at **webassign@ncsu.edu**

CAPA: A Computer-Assisted Personalized Approach   CAPA is a network system for learning, teaching, assessment, and administration. It provides students with personalized problem sets, quizzes, and examinations consisting of qualitative conceptual problems and quantitative problems. CAPA was developed through a collaborative effort of the Physics–Astronomy, Computer Science, and Chemistry Departments at Michigan State University. Students are given instant feedback and relevant hints via the Internet and may correct errors without penalty before an assignment's due date. The system records each student's participation and performance on assignments, quizzes, and examinations; records are available online to both the individual student and his or her instructor. For more information, visit the CAPA Web site at: **http://capa4.lite.msu.edu/homepage/**

Homework Service   With this service, instructors can reduce their grading workload by assigning thought-provoking homework problems using the World Wide Web. Instructors browse problem banks, select those they wish to assign to their students, and then let the Homework Service take over the delivery and grading. This system was developed and is maintained by Fred Moore at the University of Texas **(moore@physics.utexas.edu).** Students download their unique problems, submit their answers, and obtain immediate feedback; if students' answers are incorrect, they can resubmit them. This rapid grading feature facilitates effective learning. After the due date of their assignments, students can obtain the solutions to their problems. Minimal online connect time is required. The Homework Service uses algorithm-based problems: This means that each student solves sets of problems different from those given to other students. Details about and a demonstration of this service are available at **http://hw.ph.utexas.edu/hw.html**

Printed Test Bank   by Edward Adelson. Contains approximately 2,000 multiple-choice questions. It is provided for the instructor who does not have access to a computer.

Computerized Test Bank   Available in Windows™ and Macintosh® formats, the *Computerized Test Bank* contains more than 2,000 multiple-choice questions, representing every chapter of the text. The *Test Bank* enables the instructor to create many unique tests by allowing the editing of questions and the addition of new questions. The software program solves all problems and prints each answer on a separate grading key. All questions have been reviewed for accuracy.

Overhead Transparency Acetates   This collection of transparencies consists of 200 full-color figures from the text and features large print for easy viewing in the classroom.

Instructor's Manual for Physics Laboratory Manual   by David Loyd. Each chapter contains a discussion of the experiment, teaching hints, answers to selected questions, and a post-laboratory quiz with short-answer and essay questions. It also includes a list of the suppliers of scientific equipment and a summary of the equipment needed for each of the laboratory experiments in the manual.

Harcourt College Publishers may provide complementary instructional aids and supplements or supplement packages to those adopters qualified under our adoption policy. Please contact your sales representative for more information. If as an adopter or potential user you receive supplements you do not need, please return them to your sales representative or send them to

Attn: Returns Department
Troy Warehouse
465 South Lincoln Drive
Troy, MO 63379

TEACHING OPTIONS

Although some topics found in traditional textbooks have been omitted from this textbook, instructors may find that the current text still contains more material than can be covered in a two-semester sequence. For this reason, we would like to offer the following suggestions. If you wish to place more emphasis on contemporary topics in physics, you should consider omitting parts or all of Chapters 15, 16, 17, 18, 24, 25, and 26. On the other hand, if you wish to follow a more traditional approach that places more emphasis on classical physics, you could omit Chapters 9, 11, 28, 29, 30, and 31. Either approach can be used without any loss in continuity. Other teaching options would fall somewhere between these two extremes by choosing to omit some or all of the following sections, which can be considered optional:

3.6	Relative Velocity
7.7	Energy Diagrams and Stability of Equilibrium
9.9	General Relativity
10.11	Rolling of Rigid Bodies
12.6	Damped Oscillations
12.7	Forced Oscillations
14.7	Nonsinusoidal Wave Patterns
15.8	Other Applications of Fluid Dynamics
16.6	Distribution of Molecular Speeds
17.7	Molar Specific Heats of Ideal Gases
17.8	Adiabatic Processes for an Ideal Gas
17.9	Molar Specific Heats and the Equipartition of Energy
20.10	Capacitors with Dielectrics
22.11	Magnetism in Matter
26.5	Lens Aberrations
27.9	Diffraction of X-Rays by Crystals
28.13	Tunneling Through a Potential Energy Barrier

ACKNOWLEDGMENTS

The third edition of this textbook was prepared with the guidance and assistance of many professors who reviewed part or all of the manuscript, the pre-revision text, or both. We wish to acknowledge the following scholars and express our sincere appreciation for their suggestions, criticisms, and encouragement:

Yildirim M. Aktas, *University of North Carolina—Charlotte*
Alfonso M. Albano, *Bryn Mawr College*
Michael Bass, *University of Central Florida*
James Carolan, *University of British Columbia*
Kapila Clara Castoldi, *Oakland University*
Michael Dennin, *University of California, Irvine*
Madi Dogariu, *University of Central Florida*
William Fairbank, *Colorado State University*
Marco Fatuzzo, *University of Arizona*
Patrick Gleeson, *Delaware State University*
Christopher M. Gould, *University of Southern California*
James D. Gruber, *Harrisburg Area Community College*
John B. Gruber, *San Jose State University*
Gail Hanson, *Indiana University*
Dieter H. Hartmann, *Clemson University*
Michael J. Hones, *Villanova University*
Roger M. Mabe, *United States Naval Academy*

Thomas P. Marvin, *Southern Oregon University*
Martin S. Mason, *College of the Desert*
Wesley N. Mathews, Jr., *Georgetown University*
Ken Mendelson, *Marquette University*
Allen Miller, *Syracuse University*
John W. Norbury, *University of Wisconsin—Milwaukee*
Romulo Ochoa, *The College of New Jersey*
Melvyn Oremland, *Pace University*
Steven J. Pollock, *University of Colorado—Boulder*
Rex D. Ramsier, *The University of Akron*
Charles R. Rhyner, *University of Wisconsin—Green Bay*
Dennis Rioux, *University of Wisconsin—Oshkosh*
Gregory D. Severn, *University of San Diego*
Shirvel Stanislaus, *Valparaiso University*
Randall Tagg, *University of Colorado at Denver*
Robert Watkins, *University of Virginia*

This book was carefully checked for accuracy by Edward Gibson (California State University, Sacramento), Chris Vuille (Embry–Riddle Aeronautical University), and Ronald Jodoin (Rochester Institute of Technology).

We thank the following people for their suggestions and assistance during the preparation of earlier editions of this textbook:

Edward Adelson, *Ohio State University*
Subash Antani, *Edgewood College*
Harry Bingham, *University of California, Berkeley*
Anthony Buffa, *California Polytechnic State University, San Luis Obispo*
Ralph V. Chamberlin, *Arizona State University*
Gary G. DeLeo, *Lehigh University*
Alan J. DeWeerd, *Creighton University*
Gordon Emslie, *University of Alabama at Huntsville*
Donald Erbsloe, *United States Air Force Academy*
Philip Fraundorf, *University of Missouri—St. Louis*
Todd Hann, *United States Military Academy*
Gerald Hart, *Moorhead State University*
Richard W. Henry, *Bucknell University*
Laurent Hodges, *Iowa State University*
Joey Huston, *Michigan State University*
Herb Jaeger, *Miami University*
David Judd, *Broward Community College*
Thomas H. Keil, *Worcester Polytechnic Institute*

V. Gordon Lind, *Utah State University*
David Markowitz, *University of Connecticut*
John W. McClory, *United States Military Academy*
L. C. McIntyre, Jr., *University of Arizona*
Alan S. Meltzer, *Rensselaer Polytechnic Institute*
Roy Middleton, *University of Pennsylvania*
Clement J. Moses, *Utica College of Syracuse University*
Anthony Novaco, *Lafayette College*
Desmond Penny, *Southern Utah University*
Prabha Ramakrishnan, *North Carolina State University*
Rogers Redding, *University of North Texas*
Perry Rice, *Miami University*
Janet E. Seger, *Creighton University*
Antony Simpson, *Dalhousie University*
Harold Slusher, *University of Texas at El Paso*
J. Clinton Sprott, *University of Wisconsin at Madison*
Cecil Thompson, *University of Texas at Arlington*
Chris Vuille, *Embry–Riddle Aeronautical University*
James Whitmore, *Pennsylvania State University*

We are indebted to the developers of the IUPP models, "A Particles Approach to Introductory Physics" and "Physics in Context," upon which much of the pedagogical approach in this textbook is based.

Ralph McGrew coordinated the end-of-chapter problems. Problems new to this edition were written by Michael Browne, Michael Hones, Robert Forsythe, John Jewett, Ralph McGrew, Laurent Hodges, Boris Korsunsky, Richard Cohen, John DiNardo, Ronald Bieniek, and Raymond Serway. Robert Beichner and John Gerty contributed ideas for problems. Students Eric Peterman, Karl Payne, and Alexander Coto made corrections in problems from the previous edition, as did instructors Vasili Haralambous, Frank Hayes, Eugene Mosca, David Aspnes, and Erika Hermon.

We are grateful to John R. Gordon and Ralph McGrew for writing the *Student Solutions Manual and Study Guide,* and we thank Michael Rudmin for its attractive layout. Ralph McGrew, Jeffery Saul, and Charles Teague have prepared an excellent *Instructor's Manual,* and we thank them (and Michael Rudmin, again, for his excellent work on its layout). During the development of this text, the authors benefited from many useful discussions with colleagues and other physics instructors, including Robert Bauman, William Beston, Don Chodrow, Jerry Faughn, John R. Gordon, Kevin Giovanetti, Dick Jacobs, Harvey Leff, Clem Moses, Dorn Peterson, Joseph Rudmin, and Gerald Taylor. Special thanks and recognition go to the professional staff at Harcourt College Publishers—in particular, Ed Dodd, Frank Messina, Bonnie Boehme, Carol Bleistine, and Kathleen S. McLellan. We are most appreciative of the proofreading by Margaret Mary Anderson, the final copy editing by Linda Davoli, the excellent artwork produced by Rolin Graphics, and the dedicated photo research efforts of Dena Digilio-Betz. We sincerely appreciate the wisdom and enthusiasm of our publisher and good friend John Vondeling, who continues to publish high-quality instructional products for science education.

Finally, we are deeply indebted to our wives and children for their love, support, and long-term sacrifices.

Raymond A. Serway
Leesburg, Virginia

John W. Jewett, Jr.
Pomona, California

To the Student

It is appropriate to offer some words of advice that should benefit you, the student. Before doing so, we assume you have read the Preface, which describes the various features of the text that will help you through the course.

HOW TO STUDY

Very often instructors are asked, "How should I study physics and prepare for examinations?" There is no simple answer to this question, but we would like to offer some suggestions based on our own experiences in learning and teaching over the years.

First and foremost, maintain a positive attitude toward the subject matter, keeping in mind that physics is the most fundamental of all natural sciences. Other science courses that follow will use the same physical principles, so it is important that you understand and are able to apply the various concepts and theories discussed in the text.

The Contexts in the text will help you understand how the physical principles relate to real issues, phenomena, and applications. Be sure to read the Context Introductions, Context Connection sections in each chapter, and Context Conclusions. These will be most helpful in motivating your study of physics.

CONCEPTS AND PRINCIPLES

It is essential that you understand the basic concepts and principles before attempting to solve assigned problems. You can best accomplish this goal by carefully reading the textbook before you attend your lecture on the covered material. When reading the text, you should jot down those points that are not clear to you. We've purposely left wide margins in the text to give you space for doing this. Also be sure to make a diligent attempt at answering the questions in the Quick Quizzes as you come to them in your reading. We have worked hard to prepare questions that help you judge for yourself how well you understand the material. Pay careful attention to the many Pitfall Preventions throughout the text. These will help you avoid misconceptions, mistakes, and misunderstandings as well as maximize the efficiency of your time by minimizing adventures along fruitless paths. During class, take careful notes and ask questions about those ideas that are unclear to you. Keep in mind that few people are able to absorb the full meaning of scientific material after only one reading. Several readings of the text and your notes may be necessary. Your lectures and laboratory work supplement reading of the textbook and should clarify some of the more difficult material. You should minimize your memorization of material. Successful memorization of passages from the text, equations, and derivations does not necessarily indicate that you understand the material. Your understanding of the material will be enhanced through a combination of efficient study habits, discussions with other students and with instructors, and your ability to solve the problems presented in the textbook. Ask questions whenever you feel clarification of a concept is necessary.

STUDY SCHEDULE

It is important for you to set up a regular study schedule, preferably a daily one. Make sure you read the syllabus for the course and adhere to the schedule set by your instructor. The lectures will be much more meaningful if you read the corresponding textual material before attending them. As a general rule, you should devote about two hours of study time for every hour you are in class. If you are having trouble with the course, seek the advice of the instructor or other students who have taken the course. You may find it necessary to seek further instruction from experienced students. Very often, instructors offer review sessions in addition to regular class periods. It is important that you avoid the practice of delaying study until a day or two before an exam. More often than not, this approach has disastrous results. Rather than undertake an all-night study session, briefly review the basic concepts and equations and get a good night's rest. If you feel you need additional help in understanding the concepts, in preparing for exams, or in problem-solving, we suggest that you acquire a copy of the *Student Solutions Manual and Study Guide* that accompanies this textbook; this manual should be available at your college bookstore.

USE THE FEATURES

You should make full use of the various features of the text discussed in the preface. For example, marginal notes are useful for locating and describing important equations and concepts, and **boldfaced** type indicates important statements and definitions. Many useful tables are contained in the Appendices, but most are incorporated in the text where they are most often referenced. Appendix B is a convenient review of mathematical techniques.

Answers to odd-numbered problems are given at the end of the textbook, answers to Quick Quizzes are located at the end of each chapter, and answers to selected end-of-chapter questions are provided in the *Student Solutions Manual and Study Guide*. The exercises (with answers) that follow some worked examples represent extensions of those examples; in most of these exercises, you are expected to perform a simple calculation. Their purpose is to test your problem-solving skills as you read through the text. Problem-Solving Strategies and Hints are included in selected chapters throughout the text and give you additional information about how you should solve problems. The Table of Contents provides an overview of the entire text, while the Index enables you to locate specific material quickly. Footnotes sometimes are used to supplement the text or to cite other references on the subject discussed.

After reading a chapter, you should be able to define any new quantities introduced in that chapter and to discuss the principles and assumptions used to arrive at certain key relations. The chapter summaries and the review sections of the *Student Solutions Manual and Study Guide* should help you in this regard. In some cases, it may be necessary for you to refer to the index of the text to locate certain topics. You should be able to correctly associate with each physical quantity the symbol used to represent that quantity and the unit in which the quantity is specified. Furthermore, you should be able to express each important relation in a concise and accurate prose statement.

PROBLEM-SOLVING

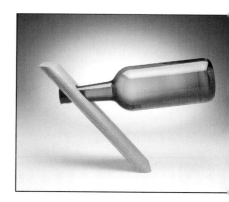

R. P. Feynman, Nobel laureate in physics, once said, "You do not know anything until you have practiced." In keeping with this statement, we strongly advise that you develop the skills necessary to solve a wide range of problems. Your ability to solve problems will be one of the main tests of your knowledge of physics; therefore, you should try to solve as many problems as possible. It is essential that you understand basic concepts and principles before attempting to solve problems. It is good practice to try to find alternative solutions to the same problem. For example, you can solve problems in mechanics using Newton's laws, but very often an alternative method that draws on energy considerations is more direct. You should not deceive yourself into thinking you understand a problem merely because you have seen it solved in class. You must be able to solve the problem and similar problems on your own.

The approach to solving problems should be carefully planned. A systematic plan is especially important when a problem involves several concepts. First, read the problem several times until you are confident you understand what is being asked. Look for any key words that will help you interpret the problem and perhaps allow you to make certain assumptions. Your ability to interpret a question properly is an integral part of problem-solving. Second, you should acquire the habit of writing down the information given in a problem and those quantities that need to be found; for example, you might construct a table listing both the quantities given and the quantities to be found. This procedure is sometimes used in the worked examples of the textbook. Finally, after you have decided on the method you feel is appropriate for a given problem, proceed with your solution. General problem-solving strategies of this type are included in the text and are highlighted with a light blue-gray screen. We have also developed a general problem-solving strategy, making use of models, to help guide you through complex problems. This strategy is located at the end of Chapter 1. If you follow the steps of this procedure, you will find it easier to come up with a solution and also gain more from your efforts.

Often, students fail to recognize the limitations of certain equations or physical laws in a particular situation. It is very important that you understand and remember the assumptions underlying a particular theory or formalism. For example, certain equations in kinematics apply only to a particle moving with constant acceleration. These equations are not valid for describing motion whose acceleration is not constant, such as the motion of an object connected to a spring or the motion of an object through a fluid.

EXPERIMENTS

Physics is a science based on experimental observations. In view of this fact, we recommend that you try to supplement the text by performing various types of "hands-on" experiments, either at home or in the laboratory. For example, the common Slinky™ toy is excellent for studying traveling waves; a ball swinging on the end of a long string can be used to investigate pendulum motion; various masses attached to the end of a vertical spring or rubber band can be used to determine their elastic nature; an old pair of Polaroid sunglasses and some discarded lenses and a magnifying glass are the components of various experiments in optics; and the approximate measure of the acceleration due to gravity can be determined simply by measuring with a stopwatch the time it takes for a ball to drop from a known height. The list of such experiments is endless. When physical models are not available, be imaginative and try to develop models of your own.

NEW MEDIA

We strongly encourage you to use one or more of the following multimedia products that accompany this textbook. It is far easier to understand physics if you see it in action, and these new materials will enable you to become a part of that action.

Student Tools CD-ROM You may purchase the dual-platform (Windows™- and Macintosh®-compatible) *Student Tools CD-ROM*. In addition to support for working those end-of-chapter problems that require use of a computer, this CD-ROM contains a textbook version of the Interactive Physics™ program by MSC Working Knowledge. Interactive Physics™ simulations are keyed to the following figures, worked examples, Quick Quizzes, and end-of-chapter problems (identified in the text with the 🖳 icon).

Saunders Core Concepts in Physics CD-ROM   Alternatively or in addition, you may purchase the *Saunders Core Concepts in Physics CD-ROM* developed by Archipelago Productions. This CD-ROM provides a complete multimedia presentation of selected topics in mechanics, thermodynamics, electromagnetism, and optics. It contains more than 350 movies—both animated and live video—that bring to life laboratory demonstrations, "real-world" examples, graphic models, and step-by-step explanations of essential mathematics. Those CD-ROM modules that supplement the material in *Principles of Physics* are identified in the margin of the text by the 🔵 icon.

It is our sincere hope that you too will find physics an exciting and enjoyable experience and that you will profit from this experience, regardless of your chosen profession. Welcome to the exciting world of physics!

The scientist does not study nature because it is useful; he studies it because he delights in it, and he delights in it because it is beautiful. If nature were not beautiful, it would not be worth knowing, and if nature were not worth knowing, life would not be worth living.

Henri Poincaré

List of Life Science Applications/Problems

About the Authors

RAYMOND A. SERWAY received his doctorate at Illinois Institute of Technology and is Professor Emeritus at James Madison University. In 1990, he received the Madison Scholar Award at James Madison University, where he taught for 17 years. Dr. Serway began his teaching career at Clarkson University, where he conducted research and taught from 1967 to 1980. He was the recipient of the Distinguished Teaching Award at Clarkson University in 1977 and of the Alumni Achievement Award from Utica College in 1985. As Guest Scientist at the IBM Research Laboratory in Zurich, Switzerland, he worked with K. Alex Müller, 1987 Nobel Prize recipient. Dr. Serway also was a visiting scientist at Argonne National Laboratory, where he collaborated with his mentor and friend, Sam Marshall. In addition to earlier editions of this textbook, Dr. Serway is the co-author of *Physics for Scientists and Engineers, Fifth Edition; College Physics, Fifth Edition;* and *Modern Physics, Second Edition.* He also is the author of the high-school textbook *Physics,* published by Holt, Rinehart, & Winston. In addition, Dr. Serway has published more than 40 research papers in the field of condensed matter physics and has given more than 70 presentations at professional meetings. Dr. Serway and his wife Elizabeth enjoy traveling, golfing, and spending quality time with their four children and five grandchildren.

JOHN W. JEWETT, JR., earned his doctorate at Ohio State University, specializing in optical and magnetic properties of condensed matter. He is currently Professor of Physics at California State Polytechnic University—Pomona. Throughout his teaching career, Dr. Jewett has been active in promoting science education. In addition to receiving four National Science Foundation grants, he helped found and direct the Southern California Area Modern Physics Institute (SCAMPI). He also is the director of Science IMPACT (Institute for Modern Pedagogy and Creative Teaching), which works with teachers and schools to develop effective science curricula. Dr. Jewett's honors include four Meritorious Performance and Professional Promise awards, selection as Outstanding Professor at California State Polytechnic University for 1991–1992, and the Excellence in Undergraduate Physics Teaching Award from the American Association of Physics Teachers (AAPT) in 1998. He has given many presentations both domestically and abroad, including multiple presentations at national meetings of the AAPT. Dr. Jewett is the author of *The World of Physics—Mysteries, Magic, and Myth,* which provides many connections between physics and everyday experiences. In addition to his work as the co-author for the third edition of *Principles of Physics,* he is also contributing author for *Physics for Scientists and Engineers, Fifth Edition,* and co-author of *Global Issues,* a four-volume set of instruction manuals in integrated science for high school. Dr. Jewett enjoys playing piano, traveling, and collecting antique quack medical devices, as well as spending time with his wife Lisa and their children.

Pedagogical Color Chart

Mechanics

Displacement and position vectors

Linear (**v**) and angular (ω) velocity vectors

Velocity component vectors

Force vectors (**F**)

Force component vectors

Acceleration vectors (**a**)

Acceleration component vectors

Linear (**p**) and angular (**L**) momentum vectors

Torque vectors (τ)

Linear or rotational motion directions

Springs

Pulleys

Electricity and Magnetism

Electric fields

Magnetic fields

Positive charges

Negative charges

Resistors

Batteries and other dc power supplies

Switches

Capacitors

Inductors (coils)

Voltmeters

Ammeters

Galvanometers

ac Generators

Ground symbol

Light and Optics

Light rays

Lenses and prisms

Mirrors

Objects

Images

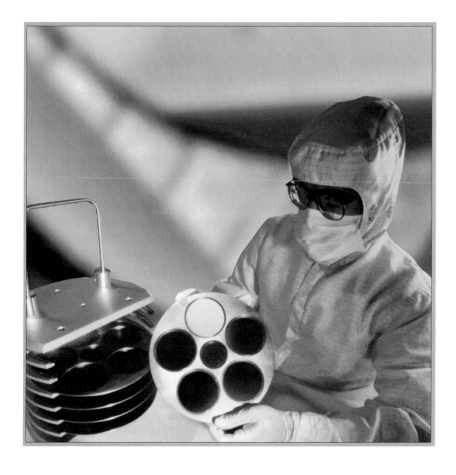

A technician operates machinery used to produce gallium arsenide circuit chips, whose operation is based on the principles of physics. *(Courtesy of TRW)*

An Invitation to Physics

Physics, the most fundamental physical science, is concerned with the basic principles of the universe. It is the foundation on which engineering, technology, and the other sciences—astronomy, biology, chemistry, and geology—are based. The beauty of physics lies in the simplicity of its fundamental theories and in the manner in which just a small number of basic concepts, equations, and assumptions can alter and expand our view of the world around us.

Classical physics, developed prior to 1900, includes the theories, concepts, laws, and experiments in classical mechanics, thermodynamics, and electromagnetism. For example, Galileo Galilei (1564–1642) made significant contributions to classical mechanics through his work on the laws of motion with constant acceleration. In the same era, Johannes Kepler (1571–1630) used astronomical observations to develop empirical laws for the motions of planetary bodies.

The most important contributions to classical mechanics, however, were provided by Isaac Newton (1642–1727), who developed classical mechanics as a systematic theory and was one of the originators of the calculus as a mathematical tool. Although major developments in classical physics continued in the 18th century, thermodynamics and electromagnetism were not developed until the latter part of the 19th century, principally because the apparatus for controlled experiments was either too crude or unavailable until then. Although many electric and magnetic phenomena had been studied earlier, the work of James Clerk Maxwell (1831–1879) provided a unified theory of electromagnetism. In this text we shall treat the various disciplines of classical physics in separate sections; however, we will see that the disciplines of mechanics and electromagnetism are basic to all the branches of physics.

A major revolution in physics, usually referred to as *modern physics*, began near the end of the 19th century. Modern physics developed mainly because many physical phenomena could not be explained by classical physics. The two most important developments in this modern era were the theories of relativity and quantum mechanics. Einstein's theory of relativity completely revolutionized the traditional concepts of space, time, and energy. Einstein's theory correctly describes the motion of objects moving at speeds comparable to the speed of light. The theory of relativity also shows that the speed of light is the upper limit of the speed of an object and that mass and energy are related. Quantum mechanics was formulated by a number of distinguished scientists to provide descriptions of physical phenomena at the atomic level.

Scientists continually work at improving our understanding of fundamental laws, and new discoveries are made every day. In many research areas a great deal of overlap exists among physics, chemistry, and biology. Evidence for this overlap is seen in the names of some subspecialties in science—biophysics, biochemistry, chemical physics, biotechnology, and so on. Numerous technological advances in recent times are the result of the efforts of many scientists, engineers, and technicians. Some of the most notable developments in the latter half of the 20th century are (1) space missions to the moon and other planets, (2) microcircuitry and high-speed computers, (3) sophisticated imaging techniques used in scientific research and medicine, and (4) several remarkable accomplishments in genetic engineering. The impact of such developments and discoveries on our society has indeed been great, and future discoveries and developments will very likely be exciting, challenging, and of great benefit to humanity.

In order to investigate the impact of physics on developments in our society, we will use a *contextual* approach to the study of the content in this textbook. The book is divided into eight *Contexts*, which relate the physics to social issues, natural phenomena, or technological applications, as outlined here:

Chapters	Context
2–11	Mission to Mars
12–14	Earthquakes
15	Search for the *Titanic*
16–18	Global Warming
19–21	Lightning
22–23	Magnetic Levitation Vehicles
24–27	Lasers
28–31	The Cosmic Connection

The Contexts provide a storyline for each section of the text, which will help provide relevance and motivation for studying the material.

The Hubble Space Telescope in the final stages of construction before launch. (*Lockheed Missiles and Space Co., Inc.*)

Each Context begins with a *central question,* which forms the focus for the study of the physics in the Context. The final section of each chapter is a "Context Connection," in which the material in the chapter is explored with the central question in mind. At the end of each Context, a Context Conclusion brings together all of the principles necessary to respond as fully as possible to the central question.

In the first chapter, we investigate some of the mathematical fundamentals and problem-solving strategies that we will use in our study of physics. The first Context, *Mission to Mars,* is introduced in Chapter 2, where the principles of classical mechanics are applied to the problem of transferring a spacecraft from Earth to Mars.

chapter 1

These controls in the cockpit of a commercial aircraft assist the pilot in maintaining control over the *velocity* of the aircraft — how fast it is traveling and in what direction it is traveling — allowing it to land safely. Quantities that are defined by both a magnitude and a direction, such as velocity, are called *vectors.*

(Mark Wagner/Stone)

Introduction and Vectors

The goal of physics is to provide a quantitative understanding of certain basic phenomena that occur in our Universe. Physics is a science based on experimental observations and mathematical analyses. The main objective behind such experiments and analyses is to develop theories that explain the phenomenon being studied and to relate those theories to other established theories. Fortunately, it is possible to explain the behavior of various physical systems using relatively few fundamental laws. Analytical procedures require the expression of those laws in the language of mathematics, the tool that provides a bridge between theory and experiment. In this chapter we shall discuss a few mathematical concepts and techniques that will be used throughout the text.

1.1 • STANDARDS OF LENGTH, MASS, AND TIME

If we measure a certain quantity and wish to describe it to someone, a unit for the quantity must be specified and defined. For example, it would be meaningless for a visitor from another planet to talk to us about a length of 8 "glitches" if we did not

• *Definition of the meter*

• *Definition of the kilogram*

• *Definition of the second*

know the meaning of the unit glitch. On the other hand, if someone familiar with our system of measurement reports that a wall is 2.0 meters high and our unit of length is defined to be 1.0 meter, we then know that the height of the wall is twice our fundamental unit of length. Likewise, if we are told that a person has a mass of 75 kilograms and our unit of mass is defined as 1.0 kilogram, then that person has a mass 75 times larger than our fundamental unit of mass. An international committee has agreed on a system of definitions and standards to describe fundamental physical quantities. It is called the **SI system** (Système International) of units. Its units of length, mass, and time are the meter, kilogram, and second, respectively.

Length

In A.D. 1120, King Henry I of England decreed that the standard of length in his country would be the yard, and that the yard would be precisely equal to the distance from the tip of his nose to the end of his outstretched arm. Similarly, the original standard for the foot adopted by the French was the length of the royal foot of King Louis XIV. This standard prevailed until 1799, when the legal standard of length in France became the **meter,** defined as one ten-millionth of the distance from the equator to the North Pole.

Many other systems have been developed in addition to those just discussed, but the advantages of the French system have caused it to prevail in most countries and in scientific circles everywhere. Until 1960, the length of the meter was defined as the distance between two lines on a specific bar of platinum–iridium alloy stored under controlled conditions. This standard was abandoned for several reasons, a principal one being that the limited accuracy with which the separation between the lines can be determined does not meet the current requirements of science and technology. The definition of the meter was modified to be equal to 1 650 763.73 wavelengths of orange–red light emitted from a krypton-86 lamp. In October 1983, the meter was redefined to be the **distance traveled by light in a vacuum during a time of 1/299 792 458 second.** This arises from the establishment of the speed of light in a vacuum as exactly 299 792 458 meters per second.

Mass

Mass represents a measure of the resistance of an object to changes in its motion. The SI unit of mass, the **kilogram,** is defined as the **mass of a specific platinum–iridium alloy cylinder kept at the International Bureau of Weights and Measures at Sèvres, France.** At this point, we should add a word of caution. Many beginning students of physics tend to confuse the physical quantities called *weight* and *mass*. For the present we shall not discuss the distinction between them; they will be clearly defined in later chapters. For now you should note that they are distinctly different quantities.

Time

Before 1960, the standard of time was defined in terms of the average length of a solar day in the year 1900. (A solar day is the time interval between successive appearances of the Sun at the highest point it reaches in the sky each day.) The basic unit of time, the **second,** was defined to be $(1/60)(1/60)(1/24) = 1/86\ 400$ of the average solar day. In 1967 the second was redefined to take advantage of the great precision obtainable with a device known as an atomic clock, which uses the

characteristic frequency of the cesium-133 atom as the "reference clock." The second is now defined as **9 192 631 770 times the period of oscillation of radiation from the cesium atom.** It is possible today to purchase clocks and watches that receive radio signals from the atomic clock in Colorado, which the clock or watch uses to continuously reset itself to the correct time.

WEB

For more information on Stonehenge, visit **www.stonehenge.org.uk**

Approximate Values for Length, Mass, and Time

Approximate values of various lengths, masses, and time intervals are presented in Tables 1.1, 1.2, and 1.3, respectively. Note the wide range of values for these quan-

WEB

Visit the National Institute of Standards and Technology at **www.NIST.gov**

TABLE 1.1	Approximate Values of Some Measured Lengths
	Length (m)
Distance from Earth to most remote quasar known	1.4×10^{26}
Distance from Earth to most remote normal galaxies known	4×10^{25}
Distance from Earth to nearest large galaxy (M 31, the Andromeda galaxy)	2×10^{22}
Distance from Sun to nearest star (Proxima Centauri)	4×10^{16}
One lightyear	9.46×10^{15}
Mean orbit radius of Earth	1.5×10^{11}
Mean distance from Earth to Moon	3.8×10^{8}
Distance from equator to North Pole	1×10^{7}
Mean radius of Earth	6.4×10^{6}
Typical altitude of orbiting Earth satellite	2×10^{5}
Length of a football field	9.1×10^{1}
Length of a housefly	5×10^{-3}
Size of smallest dust particles	1×10^{-4}
Size of cells of most living organisms	1×10^{-5}
Diameter of a hydrogen atom	1×10^{-10}
Diameter of a uranium nucleus	1.4×10^{-14}
Diameter of a proton	1×10^{-15}

PITFALL PREVENTION 1.3
Reasonable values

The generation of intuition about typical values of quantities suggested here is critical. An important step in solving problems is to think about your result at the end of a problem and determine if it seems reasonable. If you are calculating the mass of a housefly and arrive at a value of 100 kg, this is *unreasonable*—there is an error somewhere. If you are calculating the length of a spacecraft on a launch pad and end up with a value of 10 cm, this is *unreasonable*—look for a mistake.

TABLE 1.2

**Masses of
Various Objects
(Approximate Values)**

	Mass (kg)
Visible universe	10^{52}
Milky Way galaxy	10^{42}
Sun	2×10^{30}
Earth	6×10^{24}
Moon	7×10^{22}
Shark	3×10^{2}
Human	7×10^{1}
Frog	1×10^{-1}
Mosquito	1×10^{-5}
Bacterium	1×10^{-15}
Hydrogen atom	1.67×10^{-27}
Electron	9.11×10^{-31}

TABLE 1.3 Approximate Values of Some Time Intervals

	Interval (s)
Age of the Universe	5×10^{17}
Age of the Earth	1.3×10^{17}
Time since the fall of the Roman Empire	5×10^{12}
Average age of a college student	6.3×10^{8}
One year	3.2×10^{7}
One day (time for one revolution of Earth about its axis)	8.6×10^{4}
Time between normal heartbeats	8×10^{-1}
Period of audible sound waves	1×10^{-3}
Period of typical radio waves	1×10^{-6}
Period of vibration of an atom in a solid	1×10^{-13}
Period of visible light waves	2×10^{-15}
Duration of a nuclear collision	1×10^{-22}
Time for light to cross a proton	3.3×10^{-24}

TABLE 1.4

**Some Prefixes for Powers
of Ten**

Power	Prefix	Abbreviation
10^{-24}	yocto	y
10^{-21}	zepto	z
10^{-18}	atto	a
10^{-15}	femto	f
10^{-12}	pico	p
10^{-9}	nano	n
10^{-6}	micro	μ
10^{-3}	milli	m
10^{-2}	centi	c
10^{-1}	deci	d
10^{3}	kilo	k
10^{6}	mega	M
10^{9}	giga	G
10^{12}	tera	T
10^{15}	peta	P
10^{18}	exa	E
10^{21}	zetta	Z
10^{24}	yotta	Y

tities.* You should study the tables and begin to generate an intuition for what is meant by a mass of 100 kilograms, for example, or by a time interval of 3.2×10^{7} seconds.

Systems of units commonly used in science, commerce, manufacturing, and everyday life are the (1) SI system, in which the units of length, mass, and time are the meter (m), kilogram (kg), and second (s), respectively, and (2) the British engineering system (sometimes called the conventional system), in which the units of length, mass, and time are the foot (ft), slug, and second, respectively. Throughout most of this text we shall use SI units because they are almost universally accepted in science and industry. We will make limited use of conventional units in the study of classical mechanics.

Some of the most frequently used prefixes for the powers of ten and their abbreviations are listed in Table 1.4. For example, 10^{-3} m is equivalent to 1 millimeter (mm), and 10^{3} m is 1 kilometer (km). Likewise, 1 kg is 10^{3} grams (g), and 1 megavolt (MV) is 10^{6} volts (V).

1.2 • DENSITY AND ATOMIC MASS

The variables length, time, and mass are examples of *fundamental quantities*. A much larger list of variables contains *derived quantities*. These are quantities that can be expressed as a combination of fundamental quantities. Common examples are *area*, which is a combination of two lengths, and *speed*, which is a combination of a length and a time interval.

In this section, we will investigate another derived quantity, **density.** The density ρ (Greek letter rho; a table of the letters in the Greek alphabet is provided at the back of the book) of any substance is defined as its *mass per unit volume:*

---

* If you are unfamiliar with the use of powers of ten (scientific notation), you should review Appendix B.1.

$$\rho \equiv \frac{m}{V} \qquad\qquad [1.1]$$

which is a combination of mass and three lengths. For example, aluminum has a density of 2.70×10^3 kg/m^3, and lead has a density of 11.3×10^3 kg/m^3. An extreme difference in density can be imagined by thinking about holding a 10-centimeter (cm) cube of Styrofoam in one hand and a 10-cm cube of lead in the other. A list of densities for various substances is given in Table 1.5.

All ordinary matter consists of atoms, and each atom is made up of electrons and a nucleus. Practically all of the mass of an atom is contained in the nucleus, which consists of protons and neutrons. Thus, we can understand one reason why the densities of the elements differ — different elements have different numbers of protons and neutrons.

The numbers of protons and neutrons are related to the **atomic mass** of an element, which is defined as the mass of a single atom of the element measured in **atomic mass units** u, where 1 u = $1.660\ 540\ 2 \times 10^{-27}$ kg. The atomic mass of lead is 207 u and that of aluminum is 27.0 u. However, the ratio of atomic masses, 207 u/27.0 u = 7.67, does not correspond to the ratio of densities, $(11.3 \times 10^3$ kg/m$^3)/(2.70 \times 10^3$ kg/m$^3) = 4.19$. The discrepancy is due to the difference in atomic spacings and atomic arrangements in the crystal structures of the two elements.

TABLE 1.5	
Densities of Various Substances	
Substance	Density ρ (kg/m^3)
Platinum	21.45×10^3
Gold	19.3×10^3
Uranium	18.7×10^3
Lead	11.3×10^3
Copper	8.93×10^3
Iron	7.86×10^3
Aluminum	2.70×10^3
Magnesium	1.75×10^3
Water	1.00×10^3
Air at atmospheric pressure	$0.001\ 2 \times 10^3$

Example 1.1 How Many Atoms in the Cube?

A solid cube of aluminum (density 2.70 g/cm^3) has a volume of 0.200 cm^3. It is known that 27.0 g of aluminum contains 6.02×10^{23} atoms. How many aluminum atoms are contained in the cube?

Solution Because density equals mass per unit volume, the mass of the cube is

$$m = \rho V = (2.70 \text{ g/cm}^3)(0.200 \text{ cm}^3) = 0.540 \text{ g}$$

To solve this problem, we will set up a ratio based on the fact that the mass of a sample of material is proportional to the number of atoms contained in the sample. This technique of solving by ratios is very powerful and should be studied and understood so that it can be applied in future problem solving. Let us express our proportionality as $m = kN$, where m is the mass of the sample, N is the number of atoms in the sam-

ple, and k is an unknown proportionality constant. We write this relationship twice, once for the actual sample of aluminum in the problem and once for a 27.0-g sample, and then we divide the first equation by the second:

$$m_{\text{sample}} = kN_{\text{sample}} \longrightarrow \frac{m_{\text{sample}}}{m_{27.0\text{ g}}} = \frac{kN_{\text{sample}}}{kN_{27.0\text{ g}}}$$
$$m_{27.0\text{ g}} = kN_{27.0\text{ g}}$$

Notice that the unknown proportionality constant k cancels, so we do not need to know its value. We now substitute the values:

$$\frac{0.540 \text{ g}}{27.0 \text{ g}} = \frac{N_{\text{sample}}}{6.02 \times 10^{23} \text{ atoms}}$$

$$N_{\text{sample}} = \frac{(0.540 \text{ g})(6.02 \times 10^{23}\text{atoms})}{27.0 \text{ g}}$$

$$= 1.20 \times 10^{22} \text{ atoms}$$

1.3 • DIMENSIONAL ANALYSIS

The word *dimension* has a special meaning in physics. It denotes the physical nature of a quantity. Whether a distance is measured in units of feet or meters or furlongs, it is a distance. We say its dimension is *length*.

PITFALL PREVENTION 1.4

Setting up ratios

When using ratios to solve a problem, keep in mind that *ratios come from equations.* Students often just say, "the ratio of this to this is equal to the ratio of that to that" without thinking about the actual mathematical relationships. For example, a student may say, "the ratio of the areas of two circles is equal to the ratio of their radii." This ignores the fact that the area depends on the *square* of the radius. By using the technique of dividing equations in Example 1.1, these errors can be avoided.

The symbols used in this book to specify the dimensions* of length, mass, and time are L, M, and T, respectively. We shall often use square brackets [ ] to denote the dimensions of a physical quantity. For example, in this notation the dimensions of velocity v are written $[v] = L/T$, and the dimensions of area A are $[A] = L^2$. The dimensions of area, volume, velocity, and acceleration are listed in Table 1.6, along with their units in the two common systems. The dimensions of other quantities, such as force and energy, will be described as they are introduced in the text.

In many situations, you may be faced with having to derive or check a specific equation. Although you may have forgotten the details of the derivation, a useful and powerful procedure called *dimensional analysis* can be used as a consistency check, to assist in the derivation, or to check your final expression. Dimensional analysis makes use of the fact that **dimensions can be treated as algebraic quantities.** For example, quantities can be added or subtracted only if they have the same dimensions. Furthermore, the terms on both sides of an equation must have the same dimensions. By following these simple rules, you can use dimensional analysis to help determine whether an expression has the correct form, because the relationship can be correct only if the dimensions on the two sides of the equation are the same.

To illustrate this procedure, suppose you wish to derive an expression for the position x of a car at a time t if the car starts from rest at $t = 0$ and moves with constant acceleration a. In Chapter 2 we shall find that the correct expression for this special case is $x = \frac{1}{2} at^2$. Let us check the validity of this expression from a dimensional analysis approach.

The quantity x on the left side has the dimension of length. For the equation to be dimensionally correct, the quantity on the right side must also have the dimension of length. We can perform a dimensional check by substituting the basic dimensions for acceleration, L/T^2, and time, T, into the equation $x = \frac{1}{2}at^2$. That is, the dimensional form of the equation $x = \frac{1}{2}at^2$ can be written as

$$L = \frac{L}{\cancel{T}^2} \cdot \cancel{T}^2 = L$$

The units of time cancel as shown, leaving the unit of length. Notice that the number $\frac{1}{2}$ in the equation has no units, so it does not enter into the dimensional analysis.

TABLE 1.6	Units of Area, Volume, Velocity, and Acceleration			
System	**Area (L^2)**	**Volume (L^3)**	**Velocity (L/T)**	**Acceleration (L/T^2)**
SI	m^2	m^3	m/s	m/s^2
British engineering	ft^2	ft^3	ft/s	ft/s^2

* The *dimensions* of a variable will be symbolized by a capitalized, nonitalic letter, such as, in the case of length, L. The *symbol* for the variable itself will be italicized, such as L for the length of an object, or t for time.

Example 1.2 Analysis of an Equation

Show that the expression $v_f = v_i + at$ is dimensionally correct, where v_f and v_i represent speeds at two instants of time, a is acceleration, and t is an instant of time.

Solution The dimensions of the speeds are

$$[v_f] = [v_i] = \frac{L}{T}$$

and the dimensions of acceleration are L/T^2. Thus, the dimensions of at are

$$[at] = \frac{L}{T^2} \cdot T = \frac{L}{T}$$

and the expression is dimensionally correct. On the other hand, if the expression were given as $v_f = v_i + at^2$, it would be dimensionally *incorrect*. Try it and see!

1.4 • CONVERSION OF UNITS

Sometimes it is necessary to convert units from one system to another, or to convert within a system, for example, from kilometers to meters. Equalities between SI and conventional units of length are as follows:

$$1 \text{ mile} = 1\,609 \text{ m} = 1.609 \text{ km} \qquad 1 \text{ ft} = 0.304\,8 \text{ m} = 30.48 \text{ cm}$$

$$1 \text{ m} = 39.37 \text{ in.} = 3.281 \text{ ft} \qquad 1 \text{ in.} = 0.025\,4 \text{ m} = 2.54 \text{ cm}$$

A more complete list of equalities can be found in Appendix A.

Units can be treated as algebraic quantities that can cancel each other. To perform a conversion, a quantity can be multiplied by a **conversion factor,** which is a fraction equal to 1, with numerator and denominator having different units, to provide the desired units in the final result. For example, suppose we wish to convert 15.0 in. to centimeters. Because 1 in. = 2.54 cm, we multiply by a conversion factor that is the appropriate ratio of these equal quantities, and find that

$$15.0 \text{ in.} = (15.0 \text{ in.}) \left(\frac{2.54 \text{ cm}}{1 \text{ in.}} \right) = 38.1 \text{ cm}$$

PITFALL PREVENTION 1.5
Always include units

When performing calculations, make it a habit to include the units on every quantity and carry the units through the entire calculation. Avoid the temptation to drop the units during the calculation steps and then apply the appropriate unit to the number that results for an answer. By including the units in every step, you can detect errors if the units for the answer are incorrect.

(*Left*) Conversion of miles to kilometers. (*Right*) This automobile speedometer gives speed readings in both miles per hour and kilometers per hour. You should confirm the equality for a few readings on the dial. (*Paul Silverman, Fundamental Photographs*)

where the ratio in parentheses is equal to 1. Notice that we choose to put the unit of an inch in the denominator and it cancels with the unit in the original quantity. The remaining unit is the centimeter, which is our desired result.

Example 1.3   The Density of a Cube

The mass of a solid cube is 856 g, and each edge has a length of 5.35 cm. Determine the density ρ of the cube in SI units.

Solution   We will convert the mass and length to SI units before calculating the density. Because 1 kg = 1 000 g and 1 m = 100 cm, the mass m and volume V in SI units are

$$m = (856 \text{ g}) \left(\frac{1.00 \text{ kg}}{1.00 \times 10^3 \text{ g}} \right) = 0.856 \text{ kg}$$

$$L = (5.35 \text{ cm}) \left(\frac{1.00 \text{ m}}{1.00 \times 10^2 \text{ cm}} \right) = 5.35 \times 10^{-2} \text{ m}$$

Now, the volume of the cube is

$$V = L^3 = (5.35 \times 10^{-2} \text{ m})^3 = (5.35)^3 \times 10^{-6} \text{ m}^3$$
$$= 1.53 \times 10^{-4} \text{ m}^3$$

Therefore the density of the cube is

$$\rho = \frac{m}{V} = \frac{0.856 \text{ kg}}{1.53 \times 10^{-4} \text{ m}^3} = \boxed{5.59 \times 10^3 \text{ kg/m}^3}$$

1.5   •   ORDER-OF-MAGNITUDE CALCULATIONS

It is often useful to compute an approximate answer to a given physical problem even when little information is available. This answer can then be used to determine whether a more precise calculation is necessary. Such an approximation is usually based on certain assumptions, which must be modified if greater precision is needed. Thus, we will sometimes refer to an *order of magnitude* of a certain quantity as the power of ten of the number that describes that quantity. Usually, when an order-of-magnitude calculation is made, the results are reliable to within about a factor of 10. If a quantity increases in value by three orders of magnitude, this means that its value increases by a factor of $10^3 = 1\ 000$. We use the symbol ~ for "is on the order of." Thus,

$$0.008\ 6 \sim 10^{-2} \qquad 0.002\ 1 \sim 10^{-3} \qquad 700 \sim 10^3$$

The spirit of attempting order-of-magnitude calculations, sometimes referred to as "guesstimates" or "ball-park figures," is captured by the following quotation: "Make an estimate before every calculation, try a simple physical argument . . . before every derivation, guess the answer to every puzzle."*

Example 1.4   The Number of Atoms in a Solid

Estimate the number of atoms in 1 cm^3 of a solid.

Solution   From Table 1.1 we note that the diameter d of an atom is about 10^{-10} m. Let us assume that the atoms in the solid are spheres of this diameter. Then the volume of each sphere is about 10^{-30} m^3 (more precisely, volume = $4\pi r^3/3 = \pi d^3/6$, where $r = d/2$). Therefore, because

1 cm^3 = 10^{-6} m^3, the number of atoms in the solid is on the order of $10^{-6}/10^{-30} = 10^{24}$ atoms.

A more precise calculation would require knowledge of the density of the solid and the mass of each atom. However, our estimate agrees with the more precise calculation to within a factor of 10.

---

* E. Taylor and J. A. Wheeler, *Spacetime Physics,* San Francisco, W. H. Freeman, 1966, p. 60.

Example 1.5   How Much Gas Do We Use?

Estimate the number of gallons of gasoline used by all U.S. cars each year.

Solution   Because there are about 280 million people in the United States, an estimate of the number of cars in the country is 7×10^7 (assuming one car and four people per family). We can also estimate that the average distance traveled per year is 1×10^4 miles. If we assume gasoline consumption of 0.05 gal/mi (equivalent to 20 miles per gallon), each car uses about 5×10^2 gal/year. Multiplying this by the total number of cars in the United States gives an estimated total consumption of about 10^{11} gal, which corresponds to a yearly consumer expenditure on the order of 10^2 billion dollars. This is probably a low estimate because we haven't accounted for commercial consumption.

1.6 • SIGNIFICANT FIGURES

When certain quantities are measured, the measured values are known only to within the limits of the experimental uncertainty. The value of the uncertainty can depend on various factors, such as the quality of the apparatus, the skill of the experimenter, and the number of measurements performed. The number of **significant figures** in a measurement can be used to express something about the uncertainty.

As an example of significant figures, consider the population of New York state, as reported in a published road atlas: 18 044 505. Notice that this number reports the population *to the level of one individual*. We would describe this number as having eight significant figures. Can the population really be this accurate? First of all, is the census process accurate enough to measure the population to one individual? By the time this number was actually published, had the number of births and immigrations into the state balanced the number of deaths and emigrations out of the state, so that the change in the population is exactly zero? This same number for the population is reported in the 1994 edition of the road atlas *and* the 1998 edition—could it be a reasonable expectation that the population did not change by even one person in four years?

The claim that the population is measured and known to the level of one individual is unjustified. We would describe this by saying that *there are too many significant figures in the measurement.* To account for the inherent uncertainty in the census-taking process and the inevitable changes in population by the time the number is read in the road atlas, it might be better to report the population as something like 18.0 million. This number has three significant figures, rather than the eight significant figures in the published population. The fact that the identical populations are reported in the 1994 and 1998 editions suggests that these measurements might come from the census performed every ten years. Thus, for this number to be somewhat valid for a ten-year cycle of editions, it might be better to report the population as 2×10^7, which has only one significant figure.

Let us look at a more scientific example. Suppose we are asked in a laboratory experiment to measure the area of a rectangular plate using a meter stick as a measuring instrument. Let us assume that the accuracy to which we can measure a particular dimension of the plate is ± 0.1 cm. If the length of the plate is measured to be 16.3 cm, we can claim only that its length lies somewhere between 16.2 cm and 16.4 cm. In this case, we say that the measured value has three significant figures. Likewise, if its width is measured to be 4.5 cm, the actual value lies between 4.4 cm and 4.6 cm. This measured value has only two significant figures. Note that the significant figures include the first estimated digit. Thus, we could write the measured values as 16.3 ± 0.1 cm and 4.5 ± 0.1 cm.

Suppose now that we would like to find the area of the plate by multiplying the two measured values. If we were to claim that the area is (16.3 cm)(4.5 cm) = 73.35 cm², our answer would be unjustifiable because it contains four significant figures, which is greater than the number of significant figures in either of the measured lengths. The following is a good rule of thumb to use in determining the number of significant figures that can be claimed:

> When multiplying several quantities, the number of significant figures in the final answer is the same as the number of significant figures in the quantity having the lowest number of significant figures. The same rule applies to division.

Applying this rule to the previous multiplication example, we see that the answer for the area can have only two significant figures because the length of 4.5 cm has only two significant figures. Thus, all we can claim is that the area is 73 cm², realizing that the value can range between (16.2 cm)(4.4 cm) = 71 cm² and (16.4 cm)(4.6 cm) = 75 cm².

Zeros may or may not be significant figures. Those used to position the decimal point in such numbers as 0.03 and 0.0075 are not significant. Thus there are one and two significant figures, respectively, in these two values. When the positioning of zeros comes after other digits, however, there is the possibility of misinterpretation. For example, suppose the mass of an object is given as 1 500 g. This value is ambiguous because we do not know whether the last two zeros are being used to locate the decimal point or whether they represent significant figures in the measurement. To remove this ambiguity, it is common to use scientific notation to indicate the number of significant figures. In this case, we would express the mass as 1.5×10^3 g if the measured value has two significant figures, 1.50×10^3 g if it has three significant figures, and 1.500×10^3 g if it has four. Likewise, 0.000 150 should be expressed in scientific notation as 1.5×10^{-4} if it has two significant figures or as 1.50×10^{-4} if it has three significant figures. The three zeros between the decimal point and the digit 1 in the number 0.000 150 are not counted as significant figures because they are present only to locate the decimal point. In general, a **significant figure** in a measurement is a reliably known digit (other than a zero used to locate the decimal point) or the first estimated digit.

For addition and subtraction, the number of decimal places must be considered when you are determining how many significant figures to report.

> When numbers are added or subtracted, the number of decimal places in the result should equal the smallest number of decimal places of any term in the sum.

For example, if we wish to compute 123 + 5.35, the answer is 128 and not 128.35. If we compute the sum 1.000 1 + 0.000 3 = 1.000 4, the result has the correct number of decimal places; consequently it has five significant figures even though one of the terms in the sum, 0.000 3, has only one significant figure. Likewise, if we perform the subtraction 1.002 − 0.998 = 0.004, the result has only one significant figure even though one term has four significant figures and the other has three. In this book, **most of the numerical examples and end-of-chapter problems will yield answers having three significant figures.**

If the number of significant figures in the result of an addition or subtraction must be reduced, a general rule for rounding off numbers states that the last digit retained is to be increased by 1 if the last digit dropped is greater than 5. If the last digit dropped is less than 5, the last digit retained remains as it is. If the last digit dropped is equal to 5, the remaining digit should be rounded to the nearest even number. (This helps avoid accumulation of errors in long arithmetic processes.)

Example 1.6   Installing a Carpet

A carpet is to be installed in a room whose length is measured to be 12.71 m (four significant figures) and whose width is measured to be 3.46 m (three significant figures). Find the area of the room.

Solution   If you multiply 12.71 m by 3.46 m on your calculator, you will obtain an answer of 43.976 6 m². How many of these numbers should you claim? Our rule of thumb for multiplication tells us that you can claim only the number of significant figures in the quantity with the smallest number of significant figures. In this example, that number is three (in the width 3.46 m), so we should express our final answer as 44.0 m².

1.7 • COORDINATE SYSTEMS

Many aspects of physics deal in some way or another with locations in space. For example, the mathematical description of the motion of an object requires a method for specifying the position of the object. Thus, we first discuss how to describe the position of a point in space. This is done by means of coordinates in a graphical representation. A point on a line can be located with one coordinate; a point in a plane is located with two coordinates; and three coordinates are required to locate a point in space.

 See the *Core Concepts in Physics CD-ROM,* Screen 2.2

A coordinate system used to specify locations in space consists of:

- A fixed reference point *O*, called the origin
- A set of specified axes or directions with an appropriate scale and labels on the axes
- Instructions that tell us how to label a point in space relative to the origin and axes

One convenient coordinate system that we will use frequently is the *cartesian coordinate system,* sometimes called the *rectangular coordinate system.* Such a system in two dimensions is illustrated in Figure 1.1. An arbitrary point in this system is labeled with the coordinates (x, y). Positive x is taken to the right of the origin, and positive y is upward from the origin. Negative x is to the left of the origin, and negative y is downward from the origin. For example, the point *P*, which has coordinates (5, 3), may be reached by going first 5 m to the right of the origin and then 3 m above the origin (*or* by going 3 m above the origin and then 5 m to the right). Similarly, the point *Q* has coordinates $(-3, 4)$, which correspond to going 3 m to the left of the origin and 4 m above the origin.

Sometimes it is more convenient to represent a point in a plane by its *plane polar coordinates* (r, θ), as in Figure 1.2a. In this coordinate system, r is the length of the line from the origin to the point, and θ is the angle between that line and a fixed axis, usually the positive x axis, with θ measured counterclockwise. From the right triangle in Figure 1.2b, we find $\sin \theta = y/r$ and $\cos \theta = x/r$. (A review of trigonometric functions is given in Appendix B.4.) Therefore, starting with plane polar coordinates, one can obtain the cartesian coordinates through the equations

$$x = r \cos \theta \qquad\qquad [1.2]$$

$$y = r \sin \theta \qquad\qquad [1.3]$$

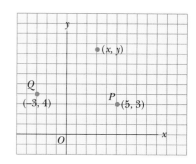

Figure 1.1

Designation of points in a cartesian coordinate system. Each square in the *xy* plane is 1 m on a side. Every point is labeled with coordinates (x, y).

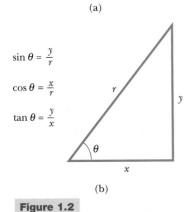

(a)

$$\sin \theta = \frac{y}{r}$$

$$\cos \theta = \frac{x}{r}$$

$$\tan \theta = \frac{y}{x}$$

(b)

Figure 1.2

(a) The plane polar coordinates of a point are represented by the distance r and the angle θ, where θ is measured in a clockwise direction from the positive x axis. (b) The right triangle used to relate (x, y) to (r, θ).

  See Screen 2.3

• *Definition of displacement along a line*

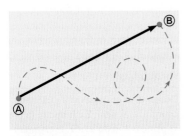

Figure 1.3

As a particle moves from Ⓐ to Ⓑ along an arbitrary path represented by the broken line, its displacement is a vector quantity shown by the arrow drawn from Ⓐ to Ⓑ.

Furthermore, it follows that

$$\tan \theta = \frac{y}{x} \qquad [1.4]$$

and

$$r = \sqrt{x^2 + y^2} \qquad [1.5]$$

You should note that these expressions relating the coordinates (x, y) to the coordinates (r, θ) apply only when θ is defined as in Figure 1.2a, where positive θ is an angle measured *counterclockwise* from the positive x axis. Other choices are made in navigation and astronomy. If the reference axis for the polar angle θ is chosen to be other than the positive x axis, or the sense of increasing θ is chosen differently, then the corresponding expressions relating the two sets of coordinates will change.

1.8 • VECTORS AND SCALARS

Each of the physical quantities that we shall encounter in this text can be placed in one of two categories: It is either a scalar or a vector. A **scalar** is a quantity that is completely specified by a positive or negative number with appropriate units. On the other hand, a **vector** is a physical quantity that must be specified by both magnitude and direction.

The number of apples in a basket is an example of a scalar quantity. If you are told there are 38 apples in the basket, this completely specifies the information; no specification of direction is required. Other examples of scalars are temperature, volume, mass, and time intervals. The rules of ordinary arithmetic are used to manipulate scalar quantities—they can be freely added and subtracted (assuming that they have the same units!), multiplied and divided.

Force is an example of a vector quantity. To completely describe the force on an object, we must specify both the direction of the applied force and the magnitude of the force.

Another simple example of a vector quantity is the **displacement** of a particle, defined as its *change in position*. Suppose the particle moves from some point Ⓐ to a point Ⓑ along a straight path, as in Figure 1.3. This displacement can be represented by drawing an arrow from Ⓐ to Ⓑ, where the arrowhead represents the direction of the displacement and the length of the arrow represents the magnitude of the displacement. If the particle travels along some other path from Ⓐ to Ⓑ, such as the broken line in Figure 1.3, its displacement is still the vector from Ⓐ to Ⓑ. The vector displacement along any indirect path from Ⓐ to Ⓑ is defined as being equivalent to the displacement represented by the direct path from Ⓐ to Ⓑ. The magnitude of the displacement is the shortest distance between the end points. Thus, the **displacement of a particle is completely known if its initial and final coordinates are known.** The path need not be specified. In other words, the **displacement is independent of the path,** if the end points of the path are fixed.

It is important to note that the **distance** traveled by a particle is distinctly different from its displacement. The distance traveled (a scalar quantity) is the length of the path, which in general can be much greater than the magnitude of the displacement (see Fig. 1.3).

If the particle moves along the x axis from position x_i to position x_f, as in Figure 1.4, its displacement is given by $x_f - x_i$. (The indices i and f refer to the initial

(a) The number of apples in the basket is one example of a scalar quantity. Can you think of other examples? *(Superstock)* (b) Jennifer pointing in the right direction. *(Photo by Raymond A. Serway)*

and final values.) We use the Greek letter delta (Δ) to denote the *change* in a quantity. Therefore, we define the change in the position of the particle (the displacement) as

$$\Delta x \equiv x_f - x_i \qquad\qquad [1.6]$$

From this definition we see that Δx is positive if x_f is greater than x_i and negative if x_f is less than x_i. For example, if a particle changes its position from $x_i = -5$ m to $x_f = 3$ m, its displacement is $\Delta x = 8$ m.

Many physical quantities in addition to displacement are vectors. They include velocity, acceleration, force, and momentum, all of which will be defined in later chapters. In this text we will use boldface letters, such as **A**, to represent vectors. Another common notation for vectors with which you should be familiar is the use of an arrow over the letter: $\vec{A}$. This notation is useful in handwriting on paper or a chalkboard, where using bold letters is inconvenient.

The magnitude of the vector **A** is written A or, alternatively, $|\mathbf{A}|$. The magnitude of a vector is always positive and carries the units of the quantity that the vector represents, such as meters for displacement or meters per second for velocity. Vectors combine according to special rules, which will be discussed in Sections 1.9 and 1.10.

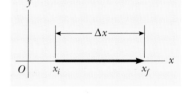

Figure 1.4

A particle moving along the x axis from x_i to x_f undergoes a displacement $\Delta x = x_f - x_i$.

THINKING PHYSICS 1.1

Consider your commute to work or school in the morning. Which is larger, the distance you travel or the magnitude of the displacement vector?

Reasoning Unless you have a very unusual commute, the distance traveled *must* be larger than the magnitude of the displacement vector. The distance includes all of the twists and turns you make in following the roads from home to work or school. On the other hand, the magnitude of the displacement vector is the length of a

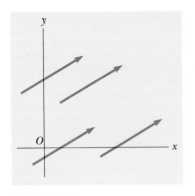

 See Screen 2.4

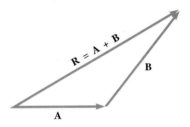

Figure 1.6

When vector **B** is added to vector **A**, the resultant **R** is the vector that runs from the tail of **A** to the tip of **B**.

PITFALL PREVENTION 1.6
Vector addition versus scalar addition

Keep in mind that **A** + **B** = **C** is very different from $A + B = C$. The first is a vector sum, which must be handled carefully, such as with the graphical method described here. The second is a simple algebraic addition of numbers that is handled with the normal rules of arithmetic.

straight line from your home to work or school. This is often described informally as "the distance as the crow flies." The only way that the distance could be the same as the magnitude of the vector is if your commute is a perfect straight line, which is highly unlikely! The distance could *never* be less than the magnitude of the displacement vector, because the shortest distance between two points is a straight line.

1.9 • SOME PROPERTIES OF VECTORS

Equality of Two Vectors Two vectors **A** and **B** are defined to be equal if they have the same units, the same magnitude, and the same direction. That is, **A** = **B** only if $A = B$ *and* **A** and **B** point in the same direction. For example, all the vectors in Figure 1.5 are equal even though they have different starting points. This property allows us to translate a vector parallel to itself in a diagram without affecting the vector.

Addition When two or more vectors are added together, they must *all* have the same units. For example, it would be meaningless to add a velocity vector to a displacement vector because they are different physical quantities. Scalars obey the same rule. For example, it would be meaningless to add time intervals and temperatures.

The rules for vector sums are conveniently described using geometry. To add vector **B** to vector **A**, first draw a diagram of vector **A** on graph paper, with its magnitude represented by a convenient scale, and then draw vector **B** to the same scale with its tail starting from the tip of **A**, as in Figure 1.6. The *resultant vector* **R** = **A** + **B** is the vector drawn from the tail of **A** to the tip of **B**. If these vectors are displacements, **R** is the single displacement that has the same effect as the displacements **A** and **B** performed one after the other.

This process is known as the *triangle method of addition*, because the three vectors can be geometrically modeled as the sides of a triangle. An alternative graphical procedure for adding two vectors, known as the *parallelogram rule of addition*, is shown in Figure 1.7a. In this construction, the tails of the two vectors **A** and **B** are together, and the resultant vector **R** is the diagonal of a parallelogram formed with **A** and **B** as its sides.

When vectors are added, the sum is independent of the order of the addition. This can be seen for two vectors from the geometric construction in Figure 1.7b

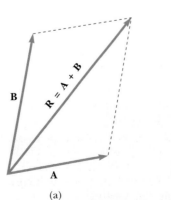

(a)

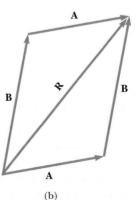

(b)

Figure 1.7

(a) In this construction, the resultant **R** is the diagonal of a parallelogram having sides **A** and **B**.
(b) This construction shows that **A** + **B** = **B** + **A**—vector addition is commutative.

and is known as the **commutative law of addition:**

$$\mathbf{A} + \mathbf{B} = \mathbf{B} + \mathbf{A} \qquad [1.7]$$

If three or more vectors are added, their sum is independent of the way in which they are grouped. A geometric demonstration of this for three vectors is given in Figure 1.8. This is called the **associative law of addition:**

$$\mathbf{A} + (\mathbf{B} + \mathbf{C}) = (\mathbf{A} + \mathbf{B}) + \mathbf{C} \qquad [1.8]$$

Geometric constructions can also be used to add more than three vectors. This is shown in Figure 1.9 for the case of four vectors. The resultant vector sum $\mathbf{R} = \mathbf{A} + \mathbf{B} + \mathbf{C} + \mathbf{D}$ is the *vector that closes the polygon formed by the vectors being added.* In other words, $\mathbf{R}$ is the *vector drawn from the tail of the first vector to the tip of the last vector.* Again, the order of the summation is unimportant.

Thus we conclude that a vector is a quantity that has both magnitude and direction and also obeys the laws of vector addition described in Figures 1.6 to 1.9.

Negative of a Vector The negative of the vector $\mathbf{A}$ is defined as the vector that, when added to $\mathbf{A}$, gives zero for the vector sum. That is, $\mathbf{A} + (-\mathbf{A}) = 0$. The vectors $\mathbf{A}$ and $-\mathbf{A}$ have the same magnitude but opposite directions.

Subtraction of Vectors The operation of vector subtraction makes use of the definition of the negative of a vector. We define the operation $\mathbf{A} - \mathbf{B}$ as vector $-\mathbf{B}$ added to vector $\mathbf{A}$:

$$\mathbf{A} - \mathbf{B} = \mathbf{A} + (-\mathbf{B}) \qquad [1.9]$$

A diagram for subtracting two vectors is shown in Figure 1.10.

Multiplication of a Vector by a Scalar If a vector $\mathbf{A}$ is multiplied by a positive scalar quantity s, the product $s\mathbf{A}$ is a vector that has the same direction as $\mathbf{A}$ and magnitude sA. If s is a negative scalar quantity, the vector $s\mathbf{A}$ is directed opposite to $\mathbf{A}$. For example, the vector $5\mathbf{A}$ is five times greater in magnitude than $\mathbf{A}$ and has the same direction as $\mathbf{A}$. On the other hand, the vector $-\frac{1}{3}\mathbf{A}$ has one third the magnitude of $\mathbf{A}$ and points in the direction opposite $\mathbf{A}$ (because of the negative sign).

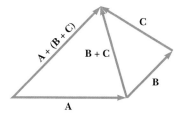
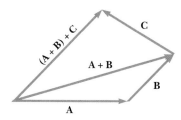

Figure 1.8

Geometric constructions for verifying the associative law of addition.

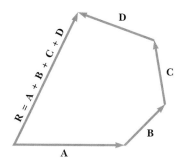

Figure 1.9

Geometric construction for summing four vectors. The resultant vector $\mathbf{R}$ closes the polygon and points from the tail of the first vector to the tip of the final vector.

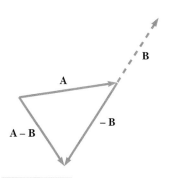

Figure 1.10

This construction shows how to subtract vector $\mathbf{B}$ from vector $\mathbf{A}$: Add the vector $-\mathbf{B}$ to vector $\mathbf{A}$. The vector $-\mathbf{B}$ is equal in magnitude and opposite to the vector $\mathbf{B}$.

Multiplication of Two Vectors   Two vectors **A** and **B** can be multiplied in two different ways to produce either a scalar or a vector quantity. The **scalar product** (or dot product) $\mathbf{A} \cdot \mathbf{B}$ is a scalar quantity equal to $AB \cos \theta$, where θ is the angle between **A** and **B**. The **vector product** (or cross product) $\mathbf{A} \times \mathbf{B}$ is a vector quantity whose magnitude is equal to $AB \sin \theta$. We shall discuss these products more fully in Chapters 6 and 10, where they are first used.

Quick Quiz 1.1

The magnitudes of two vectors **A** and **B** are $A = 8$ units and $B = 3$ units. What are the largest and smallest possible values for the magnitude of the resultant vector $\mathbf{R} = \mathbf{A} + \mathbf{B}$?

Quick Quiz 1.2

If vector **B** is added to vector **A**, under what circumstances does the resultant vector $\mathbf{A} + \mathbf{B}$ have magnitude $A + B$? Under what circumstances is the resultant vector equal to zero?

See Screen 2.5

PITFALL PREVENTION 1.7

Component vectors versus components

The vectors $\mathbf{A}_x$ and $\mathbf{A}_y$ are the *component vectors* of **A**. These should not be confused with the scalars A_x and A_y, which we shall always refer to as the *components* of **A**.

PITFALL PREVENTION 1.8

x Components

Equation 1.10 for the x and y components of a vector associates the cosine of the angle with the x component and the sine of the angle with the y component. This occurs *solely* because we chose to measure the angle with respect to the x axis, so don't memorize these equations. Invariably, you will face a problem in the future in which the angle is measured with respect to the y axis, and the equations will be incorrect. It is much better to always think about which side of the triangle containing the components is adjacent to the angle and which side is opposite, and then to assign the sine and cosine accordingly.

1.10 • COMPONENTS OF A VECTOR AND UNIT VECTORS

The geometric method of adding vectors is not the recommended procedure in situations where great precision is required, or in three-dimensional problems, because we are forced to draw them on two-dimensional paper. In this section we describe a method of adding vectors that makes use of the *projections* of a vector along the axes of a rectangular coordinate system.

Consider a vector **A** lying in the xy plane and making an arbitrary angle θ with the positive x axis, as in Figure 1.11a. The vector **A** can be represented by its rectangular **components,** A_x and A_y. The component A_x represents the projection of **A** along the x axis, and A_y represents the projection of **A** along the y axis. The components of a vector, which are scalar quantities, can be positive or negative. For example, in Figure 1.11a, A_x and A_y are both positive. The absolute values of the components are the magnitudes of the associated **component vectors** $\mathbf{A}_x$ and $\mathbf{A}_y$.

Figure 1.11b shows the component vectors again, but with the y component vector shifted so that it is added vectorially to the x component vector. This diagram shows us two important features. First, a vector is equal to the sum of its component vectors. Thus, the combination of the component vectors is a valid substitute for the actual vector. The second feature is that the vector and its component vectors form a right triangle. Thus, we can let the triangle be a model for the vector and can use right triangle trigonometry to analyze the vector. The legs of the triangle are of lengths proportional to the components (depending on what scale factor you have chosen), and the hypotenuse is of a length proportional to the magnitude of the vector.

From Figure 1.11b and the definition of the sine and cosine of an angle, we see that $\cos \theta = A_x/A$ and $\sin \theta = A_y/A$. Hence, the components of **A** are given by

$$A_x = A \cos \theta \qquad \text{and} \qquad A_y = A \sin \theta \qquad \text{[1.10]}$$

It is important to note that when using these component equations, θ must be measured counterclockwise from the positive x axis. From our triangle, it follows that the magnitude of **A** and its direction are related to its components through the Pythagorean theorem and the definition of the tangent function:

$$A = \sqrt{A_x^{\,2} + A_y^{\,2}} \qquad \text{[1.11]}$$

$$\tan \theta = \frac{A_y}{A_x} \qquad \text{[1.12]}$$

To solve for θ, we can write $\theta = \tan^{-1}(A_y/A_x)$, which is read "$\theta$ equals the angle the tangent of which is the ratio A_y/A_x." *Note that the signs of the components A_x and A_y depend on the angle θ.* For example, if $\theta = 120°$, A_x is negative and A_y is positive. On the other hand, if $\theta = 225°$, both A_x and A_y are negative. Figure 1.12 summarizes the signs of the components when **A** lies in the various quadrants.

If you choose reference axes or an angle other than those shown in Figure 1.11, the components of the vector must be modified accordingly. In many applications it is more convenient to express the components of a vector in a coordinate system having axes that are not horizontal and vertical but are still perpendicular to each other. Suppose a vector **B** makes an angle θ' with the x' axis defined in Figure 1.13. The components of **B** along these axes are given by $B_{x'} = B \cos \theta'$ and $B_{y'} = B \sin \theta'$, as in Equation 1.10. The magnitude and direction of **B** are obtained from expressions equivalent to Equations 1.11 and 1.12. Thus, we can express the components of a vector in *any* coordinate system that is convenient for a particular situation.

Vector quantities are often expressed in terms of unit vectors. **A unit vector is a dimensionless vector with a magnitude of one and is used to specify a given direction.** Unit vectors have no other physical significance. They are used simply as a bookkeeping convenience in describing a direction in space. We will use the symbols **i**, **j**, and **k** to represent unit vectors pointing in the x, y, and z directions, respectively. Thus, the unit vectors **i**, **j**, and **k** form a set of mutually perpendicular vectors as shown in Figure 1.14a, where the magnitude of each unit vector equals one; that is, $|\mathbf{i}| = |\mathbf{j}| = |\mathbf{k}| = 1$.

Consider a vector **A** lying in the xy plane, as in Figure 1.14b. The product of the component A_x and the unit vector **i** is the component vector $A_x\mathbf{i}$ parallel to the x axis with magnitude A_x. Likewise, $A_y\mathbf{j}$ is a component vector of magnitude A_y parallel to the y axis. When using the unit form of a vector, we are simply multiplying a vector (the unit vector) by a scalar (the component). Thus, the unit-vector notation for the vector **A** is written

$$\mathbf{A} = A_x\mathbf{i} + A_y\mathbf{j} \qquad \text{[1.13]}$$

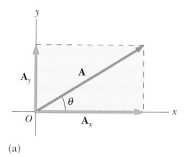

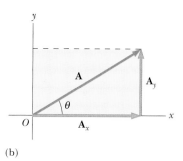

Figure 1.11

(a) A vector **A** lying in the xy plane can be represented by its component vectors $\mathbf{A}_x$ and $\mathbf{A}_y$. (b) The y component vector $\mathbf{A}_y$ can be moved to the right so that it adds to $\mathbf{A}_x$. The vector sum of the component vectors is **A**. These three vectors form a right triangle.

PITFALL PREVENTION 1.9

Tangents on calculators

 Generally, the inverse tangent function on calculators provides an angle between $-90°$ and $+90°$. As a consequence, if the vector you are studying lies in the second or third quadrant, the angle measured from the positive x axis will be the angle your calculator returns plus 180°.

	y	
A_x negative		A_x positive
A_y positive		A_y positive
A_x negative		A_x positive
A_y negative		A_y negative

Figure 1.12

The signs of the components of a vector **A** depend on the quadrant in which the vector is located.

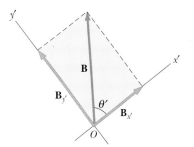

Figure 1.13

The components of vector **B** in a coordinate system that is tilted.

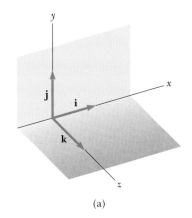

(a)

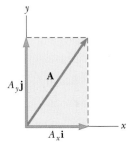

Figure 1.14

(a) The unit vectors, **i**, **j**, and **k** are directed along the x, y, and z axes, respectively. (b) A vector **A** lying in the xy plane has component vectors $A_x\mathbf{i}$ and $A_y\mathbf{j}$ where A_x and A_y are the components of **A**.

Now suppose we wish to add vector **B** to vector **A**, where **B** has components B_x and B_y. The procedure for performing this sum is to simply add the x and y components separately. The resultant vector $\mathbf{R} = \mathbf{A} + \mathbf{B}$ is therefore

$$\mathbf{R} = (A_x + B_x)\mathbf{i} + (A_y + B_y)\mathbf{j} \qquad [1.14]$$

Thus, the components of the resultant vector are given by

$$R_x = A_x + B_x$$

$$R_y = A_y + B_y \qquad [1.15]$$

The magnitude of **R** and the angle it makes with the x axis can then be obtained from its components using the relationships

$$R = \sqrt{R_x^2 + R_y^2} = \sqrt{(A_x + B_x)^2 + (A_y + B_y)^2} \qquad [1.16]$$

$$\tan \theta = \frac{R_y}{R_x} = \frac{A_y + B_y}{A_x + B_x} \qquad [1.17]$$

The procedure just described for adding two vectors **A** and **B** using the component method can be checked using a diagram like Figure 1.15.

The extension of these methods to three-dimensional vectors is straightforward. If **A** and **B** both have x, y, and z components, we express them in the form

$$\mathbf{A} = A_x\mathbf{i} + A_y\mathbf{j} + A_z\mathbf{k}$$

$$\mathbf{B} = B_x\mathbf{i} + B_y\mathbf{j} + B_z\mathbf{k}$$

The sum of **A** and **B** is

$$\mathbf{R} = \mathbf{A} + \mathbf{B} = (A_x + B_x)\mathbf{i} + (A_y + B_y)\mathbf{j} + (A_z + B_z)\mathbf{k} \qquad [1.18]$$

The same procedure can be used to add three or more vectors.

If a vector **R** has x, y, and z components, the magnitude of the vector is

$$R = \sqrt{R_x^2 + R_y^2 + R_z^2}$$

The angle θ_x that **R** makes with the x axis is given by

$$\cos \theta_x = \frac{R_x}{R}$$

with similar expressions for the angles with respect to the y and z axes.

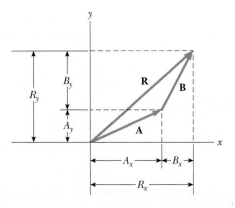

Figure 1.15

A geometric construction showing the relation between the components of the resultant **R** of two vectors and the individual components.

THINKING PHYSICS 1.2

You may have asked someone directions to a destination in a city and been told something like, "walk 3 blocks east and then 5 blocks south." If so, are you experienced with vector components?

Reasoning Yes, you are! Although you may not have thought of vector component language when you heard these directions, this is exactly what the directions represent. The perpendicular streets of the city reflect an *xy* coordinate system—we can assign the *x* axis to the east-west streets, and the *y* axis to the north-south streets. Thus, the comment of the person giving you directions can be translated as, "Undergo a displacement vector that has an *x* component of + 3 blocks and a *y* component of − 5 blocks." You would arrive at the same destination by undergoing the *y* component first, followed by the *x* component, demonstrating the commutative law of addition.

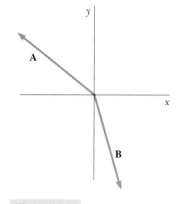

Figure 1.16

(Quick Quiz 1.3)

Quick **Quiz 1.3**

Figure 1.16 shows two vectors lying in the *xy* plane. Determine the signs of (a) the *x* components of **A** and **B**, (b) the *y* components of **A** and **B**, and (c) the *x* and *y* components of **A** + **B**.

Quick **Quiz 1.4**

If one component of a vector is not zero, can the magnitude of the vector be zero? Explain.

Quick **Quiz 1.5**

If **A** + **B** = 0, what can you say about the components of the two vectors?

Quick **Quiz 1.6**

Under what circumstances can the component of a vector be equal to the magnitude of the vector?

Example 1.7   The Sum of Two Vectors

Find the sum of two vectors **A** and **B** lying in the *xy* plane and given by

$$\mathbf{A} = 2.00\mathbf{i} + 3.00\mathbf{j} \quad \text{and} \quad \mathbf{B} = 5.00\mathbf{i} - 4.00\mathbf{j}$$

Solution   It might be helpful for you to draw a diagram of the vectors to clarify what they look like on the *xy* plane. Using the rule given by Equation 1.14, we solve this problem mathematically as follows. Note that $A_x = 2.00$, $A_y = 3.00$, $B_x = 5.00$, and $B_y = -4.00$. Therefore, the resultant vector **R** is

$$\mathbf{R} = \mathbf{A} + \mathbf{B} = (2.00 + 5.00)\mathbf{i} + (3.00 - 4.00)\mathbf{j}$$
$$= 7.00\mathbf{i} - 1.00\mathbf{j}$$

or

$$R_x = 7.00, \; R_y = -1.00$$

The magnitude of **R** is

$$R = \sqrt{R_x^{\,2} + R_y^{\,2}} = \sqrt{(7.00)^2 + (-1.00)^2} = \sqrt{50.0} = 7.07$$

EXERCISE   Find the angle θ that the resultant vector **R** makes with the positive *x* axis.

Answer   352°

Example 1.8   The Resultant Displacement

A particle undergoes three consecutive displacements: $\Delta \mathbf{r}_1 =$ (1.50$\mathbf{i}$ + 3.00$\mathbf{j}$ − 1.20$\mathbf{k}$) cm, $\Delta \mathbf{r}_2 =$ (2.30$\mathbf{i}$ − 1.40$\mathbf{j}$ − 3.60$\mathbf{k}$) cm, and $\Delta \mathbf{r}_3 =$ (−1.30$\mathbf{i}$ + 1.50$\mathbf{j}$) cm. Find the components of the resultant displacement and its magnitude.

Solution   We use Equation 1.18 for three vectors:

$\mathbf{R} = \Delta\mathbf{r}_1 + \Delta\mathbf{r}_2 + \Delta\mathbf{r}_3 = (1.50 + 2.30 - 1.30)\mathbf{i}$ cm

$\quad + (3.00 - 1.40 + 1.50)\mathbf{j}$ cm $+ (-1.20 - 3.60 + 0)\mathbf{k}$ cm

$\quad = (2.50\mathbf{i} + 3.10\mathbf{j} - 4.80\mathbf{k})$ cm

That is, the resultant displacement has components $R_x =$ 2.50 cm, $R_y =$ 3.10 cm, and $R_z = -$ 4.80 cm. Its magnitude is

$$R = \sqrt{R_x^2 + R_y^2 + R_z^2}$$
$$= \sqrt{(2.50 \text{ cm})^2 + (3.10 \text{ cm})^2 + (-4.80 \text{ cm})^2} = \boxed{6.24 \text{ cm}}$$

Example 1.9   Taking a Hike

A hiker begins a two-day trip by first walking 25.0 km due southeast from her car. She stops and sets up her tent for the night. On the second day she walks 40.0 km in a direction 60.0° north of east, at which point she discovers a forest ranger's tower.

(a) Determine the components of the hiker's displacements in the first and second days.

Solution   If we denote the displacement vectors on the first and second days by **A** and **B**, respectively, and use the car as the origin of coordinates, we obtain the vectors shown in the diagram in Figure 1.17. Notice that the resultant vector **R** can be drawn in the diagram to provide you with an approximation of the final result of the two hikes.

Displacement **A** has a magnitude of 25.0 km and is 45.0° southeast. Its components are

$A_x = A \cos(-45.0°) = (25.0 \text{ km})(0.707) = \boxed{17.7 \text{ km}}$

$A_y = A \sin(-45.0°) = (25.0 \text{ km})(-0.707) = \boxed{-17.7 \text{ km}}$

The positive value of A_x indicates that the *x* coordinate increased in this displacement. The negative value of A_y indicates that the *y* coordinate decreased in this displacement. Notice in the diagram of Figure 1.17 that vector **A** lies in the fourth quadrant, consistent with the signs of the components we calculated.

The second displacement **B** has a magnitude of 40.0 km and is 60.0° north of east. Its components are

$B_x = B \cos 60.0° = (40.0 \text{ km})(0.500) = \boxed{20.0 \text{ km}}$

$B_y = B \sin 60.0° = (40.0 \text{ km})(0.866) = \boxed{34.6 \text{ km}}$

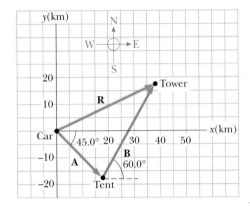

Figure 1.17

(Example 1.9) The total displacement of the hiker is the vector **R** = **A** + **B**.

(b) Determine the components of the hiker's total displacement for the trip.

Solution   The resultant displacement vector for the trip, **R** = **A** + **B**, has components given by

$R_x = A_x + B_x = 17.7 \text{ km} + 20.0 \text{ km} = \boxed{37.7 \text{ km}}$

$R_y = A_y + B_y = -17.7 \text{ km} + 34.6 \text{ km} = \boxed{16.9 \text{ km}}$

In unit-vector form, we can write the total displacement as

$$\mathbf{R} = (37.7\mathbf{i} + 16.9\mathbf{j}) \text{ km}$$

EXERCISE   Determine the magnitude and direction of the total displacement.

Answer   41.3 km, 24.1° north of east from the car.

1.11 • MODELING, ALTERNATIVE REPRESENTATIONS, AND PROBLEM-SOLVING STRATEGY

Most courses in general physics require the student to learn the skills of problem solving, and examinations usually include problems that test such skills. This section describes some useful ideas that will enable you to enhance your understanding of physical concepts, increase your accuracy in solving problems, eliminate initial panic or lack of direction in approaching a problem, and organize your work.

One of the primary problem-solving methods in physics is to form an appropriate **model** of the problem. **A model is a simplified substitute for the real problem that allows us to solve the problem in a relatively simple way.** As long as the predictions of the model agree to our satisfaction with the actual behavior of the real system, the model is valid. If the predictions do not agree, then the model must be refined or replaced with another model. The power of modeling is in its ability to reduce a wide variety of very complex problems to a limited number of classes of problems that can be approached in similar ways.

• *Models*

In science, a model is very different from, for example, an architect's scale model of a proposed building, which appears as a smaller version of what it represents. A scientific model is a theoretical construct and may have no visual similarity to the physical problem. A simple application of modeling is presented in Example 1.10, and we shall encounter many more examples of models as the text progresses.

Models are needed because the actual operation of the Universe is extremely complicated. Suppose, for example, we are asked to solve a problem about the Earth's motion around the Sun. The Earth is very complicated, with many processes occurring simultaneously. These include weather processes, seismic activity, and ocean movements, as well as the multitude of processes involving human activity. Trying to maintain knowledge and understanding of all these processes is an impossible task.

The modeling approach recognizes that none of these processes affects the motion of the Earth around the Sun to a measurable degree. Thus, all of these details are ignored. In addition, as we shall find in a later chapter, the size of the Earth does not affect the gravitational force between the Earth and the Sun—only the masses of the Earth and Sun and the distance between them determine this force. In a simplified model, the Earth is imagined to be a particle, an object with mass but zero size. This replacement of an extended object by a particle is called the **particle model,** which is used extensively in physics. By analyzing the motion of a particle with the mass of the Earth in orbit around the Sun, we find that the predictions of the motion of the particle are in excellent agreement with the actual motion of the Earth.

The two primary conditions for using the particle model are as follows:

• The size of the actual object is of no consequence in the analysis of its motion
• Any internal processes occurring in the object are of no consequence in the analysis of its motion

Both of these conditions are in action in modeling the Earth as a particle—its radius is not a factor in determining its motion, and internal processes such as thunderstorms, earthquakes, and manufacturing processes can be ignored.

Four categories of models used in this book will help us to understand and solve physics problems. The first category is the **geometric model.** In this model,

• *Geometric model*

we form a geometric construction that represents the real situation. We then set aside the real problem and perform an analysis of the geometric construction. Consider a popular problem in elementary trigonometry, as in the following example.

Example 1.10 Finding the Height of a Tree

You wish to find the height of a tree, but cannot measure it directly. You stand 50.0 m from the tree, and determine that a line of sight from the ground to the top of the tree makes an angle of 25.0° with the ground. How tall is the tree?

Solution Figure 1.18 shows the tree and a right triangle corresponding to the information in the problem superimposed over it. (We assume that the tree is exactly perpendicular to a perfectly flat ground.) In the triangle, we know the length of the horizontal leg and the angle between the hypotenuse and the horizontal leg. We can find the height of the tree by calculating the length of the vertical leg. We do this with the tangent function:

$$\tan \theta = \frac{opposite\ side}{adjacent\ side} = \frac{h}{50.0\ \text{m}}$$

$$h = (50.0\ \text{m})\tan \theta = (50.0\ \text{m})\tan 25.0° = \boxed{23.3\ \text{m}}$$

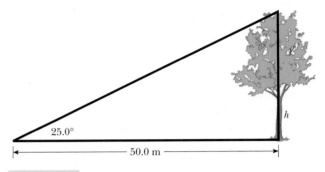

Figure 1.18

(Example 1.10) The height of a tree can be found by measuring the distance from the tree and the angle of sight to the top above the ground. This problem is a simple example of geometrically *modeling* the actual problem.

You may have performed a problem very similar to Example 1.10, but never thought about the notion of modeling. From the modeling approach, however, once we draw the triangle in Figure 1.18, the triangle is a geometric model of the real problem—it is a *substitute*. Until we reach the end of the problem, we no longer imagine the problem to be about a tree, but to be about a triangle. We use trigonometry to find the vertical leg of the triangle, leading to a value of 23.3 m. Because this leg *represents* the height of the tree, we can now return to the original problem and claim that the height of the tree is 23.3 m.

Other examples of geometric models include modeling the Earth as a perfect sphere, a pizza as a perfect disk, a meter stick as a long rod with no thickness, and an electric wire as a long straight cylinder.

Example 1.10 is an example of the modeling approach in which the modeling is so subtle that you may not even notice it. In fact, you should be able to solve this problem with no knowledge of modeling. More complicated physics problems that you will encounter, however, may appear to be very difficult using a "traditional" approach to problem solving, but can be simplified profoundly with a modeling approach.

- *Simplification model*

The particle model is an example of the second category of models, which we will call **simplification models.** In a simplification model, details that are not significant in determining the outcome of the problem are ignored. When we study rotation in Chapter 10, objects will be modeled as *rigid bodies*. All of the molecules in a rigid body maintain their exact positions with respect to one another. This is a simplification model, and we adopt it because a spinning rock is much easier to analyze than a spinning block of gelatin, which is *not* a rigid body. Other simplification models will assume that quantities such as friction forces either remain constant or are proportional to some power of the object's speed. If the predictions of the model agree with experimental measurements on the real situation, the model is valid. If there are discrepancies between the predictions of the model and the experimental results, we may have to modify the model and analyze the problem again.

The third category is that of **analysis models,** which are problems we have solved before. An important technique in problem solving is to cast a new problem into a form similar to one we have already solved. We will see our first analysis models in Chapter 2.

The fourth category of models is **structural models.** These are generally used to understand the behavior of a system that is far different in scale from our macroscopic world—either much smaller or much larger—so that we cannot interact with it directly. As an example, the notion of a hydrogen atom as an electron in a circular orbit around a proton is a structural model of the atom. We will discuss this model and structural models in general in Chapter 11.

Intimately related to the notion of modeling is that of forming alternative **representations** of the problem. **A representation is a method of viewing or presenting the information related to the problem.** Scientists must be able to communicate complex ideas to individuals without scientific backgrounds. The best representation to use in conveying the information successfully will vary from one individual to the next. Some will be convinced by a well-drawn graph, and others will require a picture. Physicists are often persuaded to agree with a point of view by examining an equation, but nonphysicists may not be convinced by this mathematical representation of the information.

- *Representations*

A word problem, such as those at the ends of the chapters in this book, is one representation of a problem. In the "real world" that you will enter after graduation, the initial representation of a problem may be just an existing situation, such as the effects of global warming, or a patient in danger of dying. You may have to identify the important data and information, and then cast the situation into an equivalent word problem!

As a familiar example of alternative representations, consider how you might represent a song. As you listen to a song, you form a *mental representation* of the song. This representation includes any emotional responses you have to the song, any memories the song evokes, any analysis of the chord structure you might be performing, and so on. The song also has an *audio representation*. This is related to the actual physical variations in pressure in the air as the song is played. Your ear–brain combination is sensitive to this representation and converts the audio representation into a mental representation. Another representation is the *digital representation,* referring to the pattern of pits in the compact disc for the song. This representation requires a compact disc player to interpret the representation and convert it to an audio representation. Yet another representation of the song is the sheet music for the song—the *written representation*. This representation can be interpreted only by a trained musician and is subject to variation among different musicians. By using the voice or a musical instrument, the written representation can be converted to an audio representation.

All four representations convey the information about the song, but in very different ways. This is what we try to do in solving physics problems—think about the information in the problem in several different ways to help us understand and solve it. Several types of representations can be of help in this endeavor:

- **Mental representation**—From the description of the problem, imagine a scene that describes what is happening in the word problem, and let time

Figure 1.19

A pictorial representation of a pop foul being hit by a baseball player.

progress so that you understand the situation and can predict what changes will occur in the situation. This step is critical in approaching *every* problem.

• **Pictorial representation**—Drawing a picture of the situation described in the word problem can be of great assistance in understanding the problem. In Example 1.10, the pictorial representation in Figure 1.18 allows us to identify the triangle as a geometric model of the problem. In architecture, a blueprint is a pictorial representation of a proposed building.

Generally, a pictorial representation describes *what you would see* if you were observing the problem. For example, Figure 1.19 shows a pictorial representation of a baseball player hitting a short pop foul. Any coordinate axes included in your pictorial representation will be in two dimensions: x and y axes.

• **Simplified pictorial representation**—It is often useful to redraw the pictorial representation without complicating details by applying a simplification model. This is similar to the discussion of the particle model described earlier. In a pictorial representation of the Earth in orbit around the Sun, you might draw the Earth and the Sun as spheres, with possibly some attempt to draw continents to identify which sphere is the Earth. In the simplified pictorial representation, the Earth and the Sun would be drawn simply as dots, representing particles. Figure 1.20 shows a simplified pictorial representation corresponding to the pictorial representation of the baseball trajectory in Figure 1.19. The notations v_x and v_y refer to the components of the velocity vector for the baseball. We shall use such simplified pictorial representations throughout the book.

• **Graphical representation**—In some problems, drawing a graph that describes the situation can be very helpful. In mechanics, for example, position–time graphs can be of great assistance. Similarly, in thermodynamics, pressure–volume graphs are essential to understanding. Figure 1.21 shows a graphical representation of the position as a function of time of a block on the end of a vertical spring as it oscillates up and down. Such a graph is helpful for understanding simple harmonic motion, which we study in Chapter 12.

This is different from a pictorial representation, which is also a two-dimensional display of information, but whose axes, if any, represent *length* coordinates. In a graphical representation, the axes may represent any two related variables. For example, a graphical representation may have axes that are tem-

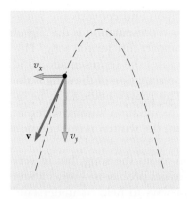

Figure 1.20

A simplified pictorial representation for the situation shown in Figure 1.19.

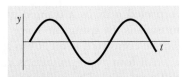

Figure 1.21

A graphical representation of the position as a function of time of a block hanging from a spring and oscillating.

perature and time. Thus, in comparison to a pictorial representation, a graphical representation is generally *not* something you would see when observing the problem.

- **Tabular representation**—It is sometimes helpful to organize the information in tabular form to help make it clearer. For example, some students find that making tables of known quantities and unknown quantities is helpful. The periodic table is an extremely useful tabular representation of information in chemistry and physics.
- **Mathematical representation**—This is often the ultimate goal in solving a problem. You want to move from the information contained in the word problem, through various representations of the problem that allow you to understand what is happening, to one or more equations that represent the situation in the problem and can be solved mathematically for the desired result.

As we study physics in this textbook, we shall use modeling and alternative representations in numerical examples to help you learn these powerful techniques.

An important way to become a skilled problem solver is to adopt a problem-solving strategy. Many chapters in this text include a section labeled "Problem-Solving Strategies" that should help you through the rough spots. The following steps are commonly used to develop a general problem-solving strategy for numerical problems:

1. Read the problem carefully at least twice. Be sure you understand the nature of the problem before proceeding further. Imagine a movie, running in your mind, of what happens in the problem. This step allows you to set up the mental representation of the problem.
2. Draw a suitable diagram with appropriate labels and coordinate axes if needed. This provides the pictorial representation. Simplify the problem by drawing a simplified pictorial representation, which eliminates the unnecessary details. Use a simplification model to remove additional unnecessary details if the conditions for the model are satisfied. If appropriate, generate a graphical representation. If you find it helpful, generate a tabular representation. If it helps you solve the problem, identify a useful geometric model from the diagrams.
3. If the problem can be cast into a form similar to one you have previously analyzed, identify an analysis model for the problem. From the model, identify the basic physical principle or principles that are involved, listing the knowns and unknowns. Select a basic relationship or derive an equation that can be used to find the unknown, and symbolically solve the equation for the unknown. This step allows you to establish a mathematical representation of the problem.
4. Substitute the given values in the problem, along with the appropriate units, into the equation(s). Obtain a numerical value with units for the unknown.
5. Think about your final result. You can have confidence in your result if questions such as the following can be properly answered: Do the units match? Is the answer of a reasonable order of magnitude? Is the positive or negative sign proper and meaningful? Is the result consistent with your initial mental representation of the problem?

Although the problem-solving strategy we just described may look quite complicated, it may not be necessary to perform all of these steps for a given problem. Examples in the early chapters of this text focus on how to apply these steps explicitly to help you become an effective problem solver. Once you have developed an organized system for examining problems and extracting relevant information, you will become a more confident problem solver in physics as well as in other areas.

SUMMARY

Mechanical quantities can be expressed in terms of three fundamental quantities—**length, mass,** and **time**—which in the SI system have the units **meters** (m), **kilograms** (kg), and **seconds** (s), respectively. It is often useful to use the method of **dimensional analysis** to check equations and to assist in deriving expressions.

The **density** of a substance is defined as its mass per unit volume.

Vectors are quantities that have both magnitude and direction and obey the vector law of addition. **Scalars** are quantities that add algebraically.

Two vectors **A** and **B** can be added using either the triangle method or the parallelogram rule. In the triangle method (see Fig. 1.6), the vector **R** = **A** + **B** runs from the tail of **A** to the tip of **B**. In the parallelogram method (see Fig. 1.7a), **R** is the diagonal of a parallelogram having **A** and **B** as its sides.

The x component A_x of the vector **A** is equal to its projection along the x axis of a coordinate system, where $A_x = A \cos \theta$, and θ is the angle **A** makes with the x axis. Likewise, the y component A_y of **A** is its projection along the y axis, where $A_y = A \sin \theta$.

If a vector **A** has an x component equal to A_x and a y component equal to A_y, the vector can be expressed in unit-vector form as **A** = A_x**i** + A_y**j**. In this notation, **i** is a unit vector in the positive x direction and **j** is a unit vector in the positive y direction. Because **i** and **j** are unit vectors, $|\mathbf{i}| = |\mathbf{j}| = 1$. In three dimensions, a vector can be expressed as **A** = A_x**i** + A_y**j** + A_z**k**, where **k** is a unit vector in the z direction.

The resultant of two or more vectors can be found by resolving all vectors into their x, y, and z components and adding their components:

$$\mathbf{R} = \mathbf{A} + \mathbf{B} = (A_x + B_x)\mathbf{i} + (A_y + B_y)\mathbf{j} + (A_z + B_z)\mathbf{k} \quad [1.18]$$

Problem-solving skills and physical understanding can be improved by **modeling** the problem and by constructing different **representations** of the problem. Models helpful in solving problems include **geometric, simplification,** and **analysis models.** Scientists use **structural models** to understand systems larger or smaller in scale than those with which we normally have direct experience. Helpful representations include the **mental, pictorial, simplified pictorial, graphical, tabular,** and **mathematical representations.**

QUESTIONS

1. Suppose that the three fundamental standards of the metric system were length, *density,* and time rather than length, *mass,* and time. The standard of density in this system is to be defined as that of water. What considerations about water would need to be addressed to make sure that the standard unit of density is as accurate as possible?

2. What types of natural phenomena could serve as alternative time standards?

3. The height of a horse is sometimes given in units of "hands." Why is this a poor standard of length?

4. Express the following quantities using the prefixes given in Table 1.4: (a) 3×10^{-4} m, (b) 5×10^{-5} s, (c) 72×10^2 g.

5. As one moves upward in atomic number in the periodic table, the atomic masses of the elements increase. Because the atoms are becoming more and more massive, why

doesn't the *density* of elemental materials increase in the same way?

6. Suppose that two quantities A and B have different dimensions. Determine which of the following arithmetic operations *could* be physically meaningful: (a) $A + B$, (b) A/B, (c) $B - A$, (d) AB.

7. What accuracy is implied in an order-of-magnitude calculation?

8. Apply an order-of-magnitude calculation to an everyday situation you might encounter. For example, how far do you walk or drive each day?

9. Estimate your age in seconds.

10. Which of the following are vectors and which are not: force, temperature, the volume of water in a can, the ratings of a TV show, the height of a building, the velocity of a sports car, the age of the Universe?

11. A vector **A** lies in the *xy* plane. For what orientations of **A** will both of its components be negative? For what orientations will its components have opposite signs?

12. A book is moved once around the perimeter of a tabletop with the dimensions 1.0 m × 2.0 m. If the book ends up at its initial position, what is its displacement? What is the distance traveled?

13. While traveling along a straight interstate highway you notice that the mile marker reads 260. You travel until you reach the 150-mile marker and then retrace your path to the 175-mile marker. What is the magnitude of your resultant displacement from the 260-mile marker?

14. If the component of vector **A** along the direction of vector **B** is zero, what can you conclude about the two vectors?

15. If **A** = **B,** what can you conclude about the components of **A** and **B**?

16. Is it possible to add a vector quantity to a scalar quantity? Explain.

17. A roller coaster travels 135 ft at an angle of 40.0° above the horizontal. How far does it move horizontally and vertically?

18. The resolution of vectors into components is equivalent to replacing the original vector with the sum of two vectors, whose sum is the same as the original vector. An infinite number of pairs of vectors will satisfy this condition; we choose a pair with one vector parallel to the *x* axis and the second parallel to the *y* axis. What difficulties would be introduced by defining components relative to axes that are not perpendicular—for example, the *x* axis and a *y* axis oriented at 45° to the *x* axis?

19. Identify the type of model (geometric, simplification, structural) represented by each of the following: (a) In its orbit around the Sun, the Earth is treated as a particle. (b) The distance the Earth travels around the Sun is calculated as 2π times the Earth–Sun distance. (c) The atomic structure of a solid material is imagined to consist of small masses (atoms) connected to neighboring masses by springs. (d) As an object is dropped, air resistance is ignored. (e) The volume of water in a bottle is estimated by calculating the volume of a cylinder. (f) A baseball is hit by a bat. In studying the motion of the baseball, any distortion of the ball while it is in contact with the bat is not considered. (g) In the early 20th century, the atom was proposed to consist of electrons in orbit around a very small but massive nucleus.

PROBLEMS

1, 2, 3 = straightforward, intermediate, challenging   □ = full solution available in the *Student Solutions Manual and Study Guide*

web = solution posted at **http://www.harcourtcollege.com/physics/**   ▯ = computer useful in solving problem

 = Interactive Physics   ▮ = paired numerical/symbolic problems   ▬ = life science application

---

Note: Consult the endpapers, appendices, and tables in the text whenever necessary in solving problems. For this chapter Appendix B.3 may be particularly useful. Answers to odd-numbered problems are in the back of the book.

Section 1.2   Density and Atomic Mass

1. The standard kilogram is a platinum–iridium cylinder 39.0 mm in height and 39.0 mm in diameter. What is the density of the material?

2. The mass of the planet Saturn is 5.64×10^{26} kg, and its radius is 6.00×10^7 m. Calculate its average density.

3. A major motor company displays a die-cast model of its first automobile, made from 9.35 kg of iron. To celebrate its 100th year in business, a worker will recast the model in gold from the original dies. What mass of gold is needed to make the new model?

4. What mass of a material with density ρ is required to make a hollow spherical shell having inner radius r_1 and outer radius r_2?

5. On your wedding day your lover gives you a gold ring of mass 3.80 g. Fifty years later its mass is 3.35 g. On the average, how many atoms were abraded from the ring during each second of your marriage? The atomic mass of gold is 197 u.

6. A structural I-beam is made of steel. A view of its cross-section and its dimensions are shown in Figure P1.6. The

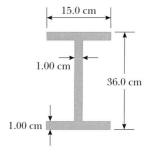

Figure P1.6

density of steel is 7.56×10^3 kg/m^3. (a) What is the mass of a section 1.50 m long? (b) Assume that the atoms are predominantly iron, with atomic mass 55.9 u. How many atoms are in this section?

7. For many electronic applications, such as in computer chips, it is desirable to make components as small as possible to keep the temperature of the components low and increase the speed of the device. Thin metallic coatings (films) can be used instead of wires to make electrical connections. Gold is especially useful because it does not oxidize readily. Its atomic mass is 197 u. A gold film can be no thinner than the size of a gold atom. Calculate the minimum coating thickness, assuming that a gold atom occupies a cubical volume in the film that is equal to the volume it occupies in a large piece of metal. This geometric model yields a result of the correct order of magnitude.

Section 1.3   Dimensional Analysis

8. Which of the following equations are dimensionally correct? (a) $v_f = v_i + ax$; (b) $y = (2 \text{ m}) \cos(kx)$, where $k = 2 \text{ m}^{-1}$.

9. The position of a particle when moving under uniform acceleration is some function of the elapsed time and the acceleration. Suppose we write this position as $x = ka^m t^n$, where k is a dimensionless constant. Show by dimensional analysis that this expression is satisfied if $m = 1$ and $n = 2$. Can this analysis give the value of k?

10. Figure P1.10 shows a *frustrum of a cone*. Of the following mensuration (geometrical) equations, which describes (a) the total circumference of the flat circular faces, (b) the volume, and (c) the area of the curved surface? (i) $\pi(r_1 + r_2)[h^2 + (r_1 - r_2)^2]^{1/2}$, (ii) $2\pi(r_1 + r_2)$, (iii) $\pi h(r_1^2 + r_1 r_2 + r_2^2)$

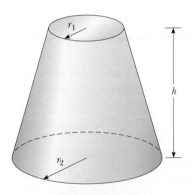

Figure P1.10

11. The consumption of natural gas by a company satisfies the empirical equation $V = 1.50\ t + 0.008\ 00\ t^2$, where V is the volume in millions of cubic feet and t is the time in months. Express this equation in units of cubic feet and

seconds. Put the proper units on the coefficients. Assume a month is 30.0 days.

Section 1.4   Conversion of Units

12. Suppose your hair grows at the rate of 1/32 in. per day. Find the rate at which it grows in nanometers per second. Because the distance between atoms in a molecule is on the order of 0.1 nm, your answer suggests how rapidly layers of atoms are assembled in this protein synthesis.

13. Assume that it takes 7.00 min to fill a 30.0-gal gasoline tank. (a) Calculate the rate at which the tank is filled in gallons per second. (b) Calculate the rate at which the tank is filled in cubic meters per second. (c) Determine the time, in hours, required to fill a 1-m^3 volume at the same rate. (1 U.S. gal = 231 in.^{3})

14. An ore loader moves 1 200 tons/h from a mine to the surface. Convert this to lb/s, using 1 ton = 2 000 lb.

15. A *section* of land has an area of 1 square mile and contains 640 acres. Determine the number of square meters in 1 acre.

16. At the time of this book's printing, the U. S. national debt is about \$6 trillion. (a) If payments were made at the rate of \$1 000 per second, how many years would it take to pay off the debt, assuming no interest were charged? (b) A dollar bill is about 15.5 cm long. If six trillion dollar bills were laid end to end around the Earth's equator, how many times would they encircle the Earth? Take the radius of the Earth at the equator to be 6 378 km. (*Note*: Before doing any of these calculations, try to guess at the answers. You may be very surprised.)

17. One gallon of paint (volume = 3.78×10^{-3} m^3) covers an area of 25.0 m^2. What is the thickness of the paint on the wall?

18. A hydrogen atom has a diameter of approximately 1.06×10^{-10} m, as defined by the diameter of the spherical electron cloud around the nucleus. The hydrogen nucleus has a diameter of approximately 2.40×10^{-15} m. (a) For a scale model, represent the diameter of the hydrogen atom by the length of an American football field (100 yards = 300 feet), and determine the diameter of the nucleus in millimeters. (b) The atom is how many times larger in volume than its nucleus?

19. One cubic meter (1.00 m^3) of aluminum has a mass of 2.70×10^3 kg, and 1.00 m^3 of iron has a mass of 7.86×10^3 kg. Find the radius of a solid aluminum sphere that will balance a solid iron sphere of radius 2.00 cm on an equal-arm balance.

20. Let ρ_{Al} represent the density of aluminum and ρ_{Fe} that of iron. Find the radius of a solid aluminum sphere that balances a solid iron sphere of radius r_{Fe} on an equal-arm balance.

Section 1.5 Order-of-Magnitude Calculations

21. Find the order of magnitude of the number of Ping-Pong
web balls that would fit into a typical-size room (without being
crushed). In your solution, state the quantities you measure or estimate and the values you take for them.

22. An automobile tire is rated to last for 50 000 miles. To an
order of magnitude, through how many revolutions will it
turn? In your solution, state the quantities you measure or
estimate and the values you take for them.

23. Compute the order of magnitude of the mass of a bathtub
half full of water. Compute the order of magnitude of the
mass of a bathtub half full of pennies. In your solution, list
the quantities you take as data and the value you measure
or estimate for each.

24. Soft drinks are commonly sold in aluminum cans. To an
order of magnitude, how many such containers are
thrown away or recycled each year by U.S. consumers?
How many tons of aluminum does this represent? In your
solution, state the quantities you measure or estimate and
the values you take for them.

Section 1.6 Significant Figures

25. Carry out the following arithmetic operations: (a) the sum
of the measured values 756, 37.2, 0.83, and 2.5; (b) the
product $0.003\ 2 \times 356.3$; (c) the product $5.620 \times \pi$.

26. The *tropical year,* the time from vernal equinox to vernal equinox, is the basis for our calendar. It contains
365.242 199 days. Find the number of seconds in a tropical year.

27. How many significant figures are in the following
numbers: (a) 78.9 ± 0.2, (b) 3.788×10^9, (c) 2.46×10^{-6},
(d) 0.005 3?

28. The radius of a solid sphere is measured to be $(6.50 \pm
0.20)$ cm, and its mass is measured to be (1.85 ± 0.02) kg.
Determine the density of the sphere in kilograms per cubic meter and the uncertainty in the density.

29. A sidewalk is to be constructed around a swimming pool
that measures (10.0 ± 0.1) m by (17.0 ± 0.1) m. If the
sidewalk is to measure (1.00 ± 0.01) m wide by $(9.0 \pm
0.1)$ cm thick, what volume of concrete is needed, and
what is the approximate uncertainty of this volume?

Section 1.7 Coordinate Systems

30. The polar coordinates of a point are $r = 5.50$ m and $\theta =
240°$. What are the cartesian coordinates of this point?

31. A fly lands on one wall of a room. The lower left-hand corner of the wall is selected as the origin of a two-dimensional cartesian coordinate system. If the fly is located at
the point having coordinates (2.00, 1.00) m, (a) how far is
it from the lower left-hand corner of the room? (b) what is
its location in polar coordinates?

32. Two points in the xy plane have cartesian coordinates
$(2.00, -4.00)$ m and $(-3.00, 3.00)$ m. Determine (a) the
distance between these points and (b) their polar coordinates.

33. If the polar coordinates of the point (x, y) are (r, θ), determine the polar coordinates for the points: (a) $(-x, y)$, (b)
$(-2x, -2y)$, and (c) $(3x, -3y)$.

Section 1.8 Vectors and Scalars

Section 1.9 Some Properties of Vectors

34. Each of the displacement vectors **A** and **B** shown in Figure
P1.34 has a magnitude of 3.00 m. Find graphically (a) **A** +
B, (b) **A** − **B**, (c) **B** − **A**, (d) **A** − 2**B**. Report all angles
counterclockwise from the positive x axis.

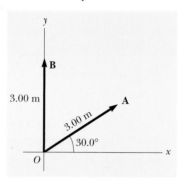

Figure P1.34   Problems 34 and 47.

35. A skater glides along a circular path of radius 5.00 m. If he
web coasts around one half of the circle, find (a) the magnitude of the displacement vector and (b) how far the person skated. (c) What is the magnitude of the displacement
if he skates all the way around the circle?

36. A plane flies from base camp to lake A, a distance of 280
km at a direction of 20.0° north of east. After dropping off
supplies, it flies to lake B, which is 190 km at 30.0° west of
north from lake A. Graphically determine the distance
and direction from lake B to the base camp.

37. A roller coaster car moves 200 ft horizontally and then rises
135 ft at an angle of 30.0° above the horizontal. It then travels 135 ft at an angle of 40.0° downward. What is its displacement from its starting point? Use graphical techniques.

Section 1.10 Components of a Vector
and Unit Vectors

38. Find the horizontal and vertical components of the 100-m
displacement of a superhero who flies from the top of a
tall building following the path shown in Fig. P1.38.

39. A vector has an x component of -25.0 units and a y component of 40.0 units. Find the magnitude and direction of
this vector.

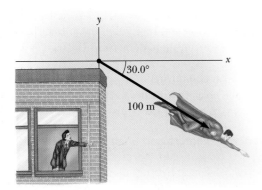

Figure P1.38

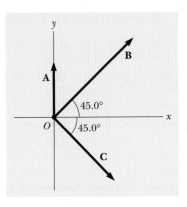

Figure P1.43

40. Given the vectors $\mathbf{A} = 2.00\mathbf{i} + 6.00\mathbf{j}$ and $\mathbf{B} = 3.00\mathbf{i} - 2.00\mathbf{j}$, (a) draw the vector sum $\mathbf{C} = \mathbf{A} + \mathbf{B}$ and the vector difference $\mathbf{D} = \mathbf{A} - \mathbf{B}$. (b) Calculate $\mathbf{C}$ and $\mathbf{D}$, first in terms of unit vectors and then in terms of polar coordinates, with angles measured with respect to the $+x$ axis.

41. A person going for a walk follows the path shown in Figure P1.41. The total trip consists of four straight-line paths. At the end of the walk, what is the person's resultant displacement measured from the starting point?

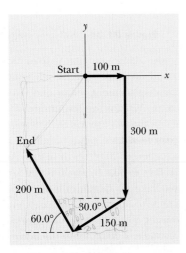

Figure P1.41

42. Vector $\mathbf{A}$ has x and y components of -8.70 cm and 15.0 cm, respectively; vector $\mathbf{B}$ has x and y components of 13.2 cm and -6.60 cm, respectively. If $\mathbf{A} - \mathbf{B} + 3\mathbf{C} = 0$, what are the components of $\mathbf{C}$?

43. Three displacement vectors of a croquet ball are shown in
web Figure P1.43, where $|\mathbf{A}| = 20.0$ units, $|\mathbf{B}| = 40.0$ units, and $|\mathbf{C}| = 30.0$ units. Find (a) the resultant in unit-vector notation and (b) the magnitude and direction of the resultant displacement.

44. Consider the displacement vectors $\mathbf{A} = (3\mathbf{i} + 3\mathbf{j})$ m, $\mathbf{B} = (\mathbf{i} - 4\mathbf{j})$ m, and $\mathbf{C} = (-2\mathbf{i} + 5\mathbf{j})$ m. Use the component method to determine (a) the magnitude and direction of the vector $\mathbf{D} = \mathbf{A} + \mathbf{B} + \mathbf{C}$, and (b) the magnitude and direction of $\mathbf{E} = -\mathbf{A} - \mathbf{B} + \mathbf{C}$.

45. Consider two vectors $\mathbf{A} = 3\mathbf{i} - 2\mathbf{j}$ and $\mathbf{B} = -\mathbf{i} - 4\mathbf{j}$. Calculate (a) $\mathbf{A} + \mathbf{B}$, (b) $\mathbf{A} - \mathbf{B}$, (c) $|\mathbf{A} + \mathbf{B}|$, (d) $|\mathbf{A} - \mathbf{B}|$, (e) the directions of $\mathbf{A} + \mathbf{B}$ and $\mathbf{A} - \mathbf{B}$.

46. The helicopter view in Figure P1.46 shows two people pulling on a stubborn mule. Find (a) the single force that is equivalent to the two forces shown, and (b) the force that a third person would have to exert on the mule to make the resultant force equal to zero. The forces are measured in units of newtons.

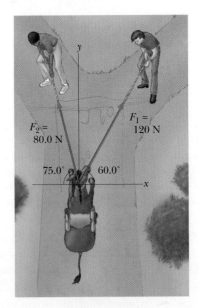

Figure P1.46

47. Use the component method to add the vectors $\mathbf{A}$ and $\mathbf{B}$ shown in Figure P1.34. Express the resultant $\mathbf{A} + \mathbf{B}$ in unit-vector notation.

48. A billiard ball undergoes two displacements. The first has a magnitude of 150 cm and makes an angle of 120° with the positive *x* axis. The *resultant* displacement has a magnitude of 140 cm and is directed at an angle of 35.0° to the positive *x* axis. Find the magnitude and direction of the second displacement.

49. The vector **A** has *x*, *y*, and *z* components of 8.00, 12.0, and −4.00 units, respectively. (a) Write a vector expression for **A** in unit-vector notation. (b) Obtain a unit-vector expression for a vector **B** one-fourth the length of **A** pointing in the same direction as **A**. (c) Obtain a unit-vector expression for a vector **C** three times the length of **A** pointing in the direction opposite the direction of **A**.

50. In an assembly operation illustrated in Figure P1.50, a robot moves an object first straight upward and then also to the east, around an arc forming one quarter of a circle of radius 4.80 cm that lies in an east–west vertical plane. The robot then moves the object upward and to the north, through a quarter of a circle of radius 3.70 cm that lies in a north–south vertical plane. Find (a) the magnitude of the total displacement of the object, and (b) the angle the total displacement makes with the vertical.

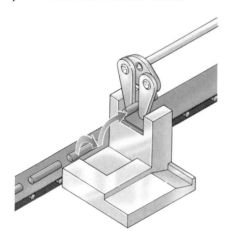

Figure P1.50

51. Vector **B** has *x*, *y*, and *z* components of 4.00, 6.00, and 3.00 units, respectively. Calculate the magnitude of **B** and the angles that **B** makes with the coordinate axes.

52. If **A** = (6.00**i** − 8.00**j**) units, **B** = (−8.00**i** + 3.00**j**) units, and **C** = (26.0**i** + 19.0**j**) units, determine *a* and *b* such that *a***A** + *b***B** + **C** = 0.

53. Vectors **A** and **B** have equal magnitudes of 5.00. If the sum of **A** and **B** is the vector 6.00**j**, determine the angle between **A** and **B**.

Section 1.11 Modeling, Alternative Representations, and Problem-Solving Strategy

54. A surveyor measures the distance across a straight river by the following method: Starting directly across from a tree on the opposite bank, she walks 100 m along the riverbank to establish a baseline. Then she sights across to the tree. The angle from her baseline to the tree is 35.0°. How wide is the river?

55. A crystalline solid consists of atoms stacked up in a repeating lattice structure. Consider a crystal as shown in Figure P1.55a. The atoms reside at the corners of cubes of side *L* = 0.200 nm. One piece of evidence for the regular arrangement of atoms comes from the flat surfaces along which a crystal separates, or cleaves, when it is broken. Suppose this crystal cleaves along a face diagonal, as shown in Figure P1.55b. Calculate the spacing *d* between two adjacent atomic planes that separate when the crystal cleaves.

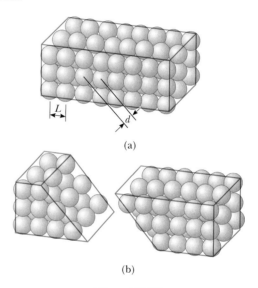

(a)

(b)

Figure P1.55

Additional Problems

56. When data are known to three significant figures, we write 6.379 m = 6.38 m and 6.374 m = 6.37 m. When a number ends in 5, we write 6.375 m = 6.38 m. We could equally well write 6.375 m = 6.37 m, "rounding down" instead of "rounding up," because we would change the number 6.375 by equal increments in both cases, but we choose the convention of rounding to an even number. Now consider an order-of-magnitude estimate, in which factors rather than increments are important. We write 500 m $\sim$ 10^3 m because 500 differs from 100 by a factor of 5 but it differs from 1000 only by a factor of 2. We write 437 m $\sim$ 10^3 m and 305 m $\sim$ 10^2 m. What length differs from 100 m and from 1000 m by equal factors, so that we could equally well choose to represent its order of magnitude either as $\sim 10^2$ m or as $\sim 10^3$ m?

57. The basic function of the carburetor of an automobile is to "atomize" the gasoline and mix it with air to promote rapid combustion. As an example, assume that 30.0 cm³ of gasoline is atomized into *N* spherical droplets, each with a radius of 2.00×10^{-5} m. What is the total surface area of these *N* spherical droplets?

58. In physics it is important to use mathematical approximations. Demonstrate for yourself that for small angles ($< 20°$)

$$\tan \alpha \approx \sin \alpha \approx \alpha = \pi \alpha'/180°$$

where α is in radians and α' is in degrees. Use a calculator to find the largest angle for which $\tan \alpha$ may be approximated by $\sin \alpha$ if the error is to be less than 10.0%.

59. There are nearly $\pi \times 10^7$ s in one year. Find the percentage error in this approximation, where "percentage error" is defined as

$$\frac{|\text{assumed value} - \text{true value}|}{\text{true value}} \times 100\%$$

60. A child loves to watch as you refill a transparent plastic bottle with shampoo. Every horizontal cross-section of the bottle is a circle, but the diameters of the circles have different values, so that the bottle is much wider in some places than in others. You pour in bright green shampoo with a constant volume flow rate of 16.5 cm³/s. At what rate is the fluid level in the bottle rising (a) at a point where the diameter of the bottle is 6.30 cm and (b) at a point where the diameter is 1.35 cm?

61. The distance from the Sun to the nearest star is 4×10^{16} m. The Milky Way galaxy is roughly a disk of diameter $\sim 10^{21}$ m and thickness $\sim 10^{19}$ m. Find the order of magnitude of the number of stars in the Milky Way. Assume the distance between the Sun and our nearest neighbor is typical.

62. An air-traffic controller notices two aircraft on his radar screen. The first is at altitude 800 m, horizontal distance 19.2 km, and 25.0° south of west. The second aircraft is at altitude 1 100 m, horizontal distance 17.6 km, and 20.0° south of west. What is the distance between the two aircraft? (Place the *x* axis west, the *y* axis south, and the *z* axis vertical.)

63. Two vectors **A** and **B** have precisely equal magnitudes. For the magnitude of **A** + **B** to be larger than the magnitude of **A** − **B** by the factor *n*, what must be the angle between them?

64. A pirate has buried his treasure on an island with five trees, located at the following points: (30.0 m, − 20.0 m), (60.0 m, 80.0 m), (− 10.0 m, − 10.0 m), (40.0 m, − 30.0 m), and (− 70.0 m, 60.0 m), all measured relative to some origin, as in Figure P1.64. His ship's log instructs you to start at tree A and move toward tree B, but to cover only one-half the distance between A and B. Then move toward tree C, covering one-third the distance between your cur-

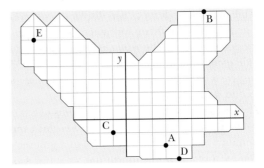

Figure P1.64

rent location and C. Next move toward D, covering one-fourth the distance between where you are and D. Finally move toward E, covering one-fifth the distance between you and E, stop, and dig. (a) Assume you have correctly determined the order in which the pirate labeled the trees as A, B, C, D, and E, as shown in the figure. What are the coordinates of the point where his treasure is buried? (b) Now rearrange the order of the trees [e.g., B(30 m, − 20 m), A(60 m, 80 m), E(− 10 m, − 10 m), C(40 m, − 30 m), and D(− 70 m, 60 m)], and repeat the calculation to show that the answer does not depend on the order in which the trees are labeled.

65. Consider a game in which *N* children position themselves at equal distances around the circumference of a circle. At the center of the circle is a rubber tire. Each child holds a rope attached to the tire and, at a signal, pulls on his rope. All children exert forces of the same magnitude *F*. In the case *N* = 2, it is easy to see that the net force on the tire will be zero, because the two oppositely directed force vectors add to zero. Similarly, if *N* = 4, 6, or any even integer, the resultant force on the tire must be zero, because the forces exerted by each pair of oppositely positioned children will cancel. When an odd number of children are around the circle, it is not so obvious that the total force on the central tire will be zero. (a) Calculate the net force on the tire in the case *N* = 3, by adding the components of the three force vectors. Choose the *x* axis to lie along one of the ropes. (b) Determine the net force for the general case where *N* is any integer, odd or even, greater than one. Proceed as follows: Assume that the total force is not zero. Then it must point in some particular direction. Let every child move one position clockwise. Give a reason that the total force must then have a direction turned clockwise by 360°/*N*. Argue that the total force must nevertheless be the same as before. Explain that the contradiction proves that the magnitude of the force is zero. This problem illustrates a widely use-

ful technique of proving a result "by symmetry"—by using a bit of the mathematics of *group theory*. The particular situation is actually encountered in physics and chemistry when an array of electric charges (ions) exerts electric forces on an atom at a central position in a molecule or in a crystal.

66. A rectangular parallelepiped has dimensions a, b, and c, as in Figure P1.66. (a) Obtain a vector expression for the face diagonal vector $\mathbf{R}_1$. What is the magnitude of this vector? (b) Obtain a vector expression for the body diagonal vector $\mathbf{R}_2$. Note that $\mathbf{R}_1$, $c\mathbf{k}$, and $\mathbf{R}_2$ form a right triangle, and prove that the magnitude of $\mathbf{R}_2$ is $\sqrt{a^2 + b^2 + c^2}$.

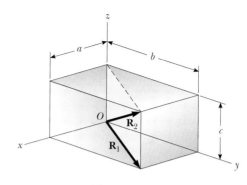

Figure P1.66

ANSWERS TO QUICK QUIZZES

1.1 The largest possible magnitude of the resultant occurs when the two vectors are in the same direction. In this case, the magnitude of the resultant is the sum of the magnitudes of **A** and **B**: $R = A + B = 11$ units. The smallest possible magnitude of the resultant occurs when the two vectors are in opposite directions, and the magnitude is the absolute value of the difference of the magnitudes of **A** and **B**: $R = |A - B| = 5$ units.

1.2 The resultant has magnitude $A + B$ when **A** is oriented in the same direction as **B**. The resultant vector is zero only when the two vectors **A** and **B** have the same magnitude and are oriented in opposite directions.

1.3 (a) Because **A** is oriented toward the left, its x component is negative. Because **B** is oriented toward the right, its x component is positive. (b) Because **A** is oriented upward, its y component is positive. Because **B** is oriented down-

ward, its y component is negative. (c) The resultant vector $\mathbf{R} = \mathbf{A} + \mathbf{B}$ lies in the third quadrant, so both x and y components are negative.

1.4 No. The magnitude of a vector **A** is equal to $\sqrt{A_x^2 + A_y^2 + A_z^2}$. Therefore, if any component is nonzero, A cannot be zero.

1.5 The fact that $\mathbf{A} + \mathbf{B} = 0$ tells you that $\mathbf{A} = -\mathbf{B}$. Therefore, the components of the two vectors must have equal magnitudes and opposite signs: $A_x = -B_x$, $A_y = -B_y$, and $A_z = -B_z$.

1.6 The only circumstances under which this can happen is if the vector is oriented along the positive direction for one of the three axes: x, y, or z. Then the magnitude of the vector is the same as the component associated with its direction and the other components are zero.

THE WIZARD OF ID   **By Parker and Hart**

By permission of John Hart and Field Enterprises, Inc.

Mission to Mars

I n our opening Context, we shall investigate the physics necessary to send a spacecraft from Earth to Mars. If the two planets were sitting still in space, millions of kilometers apart, this would be a difficult enough proposition, but keep in mind that we are launching the spacecraft from a moving object, the Earth, and are aiming at a moving target, Mars. Despite these apparent difficulties, we can use the principles of physics to plan a successful mission.

Travel in space began in the early 1960s, with the launch of human-occupied spacecraft in both the United States and the Soviet Union. The first human to ride into space was Yuri Gagarin, who made a one-orbit trip in 1961 in the

The roving robot in the Mars Pathfinder mission explores the area of the Martian surface near the landing spot. *(NASA)*

Space shuttles have been operating since the 1980s, delivering satellites to orbit, carrying astronauts on repair missions for the Hubble Space Telescope, and providing appropriate environmental conditions for scientific experimentation. *(NASA)*

Soviet spacecraft Vostok. Competition between the two countries resulted in a "space race," which led to the successful landing of American astronauts on the Moon in 1969.

In the 1970s, the Viking Project landed spacecraft on Mars to analyze the soil for signs of life. These tests were inconclusive.

American efforts in the 1980s focused on the development and implementation of the space shuttle system, a reusable space transportation system.

The United States returned to Mars in the 1990s, with Mars Global Surveyor, designed to perform careful mapping of the Martian surface, and Mars Pathfinder, which landed on Mars and deployed a roving robot to analyze rocks and soil. In 1999, Mars Polar Lander was launched to land near the polar ice cap and search for water. Mars Surveyor 2001 is scheduled for launch in early 2001. Not all trips have been successful—Mars Climate Orbiter was lost in 1999 due to communication errors between the builder of the spacecraft and the mission control team.

Many individuals dream of one day establishing colonies on Mars. This is far in the future—we are still learning much about Mars today and have yet taken only a handful of trips to the planet. Travel to

Mars is still not an everyday occurrence, although we learn more from each mission. In this Context, we address the central question,

How can we undertake a successful transfer of a spacecraft from Earth to Mars?

Motion in One Dimension

Dynamics is the study of the motion of an object and the relationship
of this motion to such physical concepts as force and mass. Before
beginning our study of dynamics, however, it is important to be able
to *describe* motion using the concepts of space and time without regard to the causes of the motion. This portion of mechanics is called *kinematics*. In
this chapter we shall consider motion along a straight line, that is, one-dimensional motion. Chapter 3 extends our discussion to two-dimensional motion.

From everyday experience, we recognize that motion represents continuous
change in the position of an object. If you are driving from New York to Los Angeles, your position on the Earth's surface is changing. If you travel at an average
speed of 50 miles per hour (mi/h), how long will it take to make this trip? This is
the kind of question we can answer with the material in this chapter.

The movement of an object through space (translation) may be accompanied
by the rotation or vibration of the object. Such motions can be quite complex. It
is often possible to simplify matters, however, by temporarily ignoring rotation
and internal motions of the moving object. This is the simplification model that
we call the particle model, discussed in Chapter 1. In many situations, an *object*
can be treated as a *particle* if the only motion being considered is translation
through space. We will use the particle model extensively in this and the next few
chapters.

2.1 • AVERAGE VELOCITY

 See the *Core Concepts in Physics CD-ROM*, Screen 3.2

We begin our study of kinematics with the notion of average velocity. You may be familiar with a similar notion, average speed, from experiences with driving. If you drive your car 100 miles according to your odometer, and it takes 2.0 hours to do this, your average speed is (100 mi)/(2.0 h) = 50 mi/h. For a particle moving through a distance d in a time interval Δt, the average speed $\bar{v}$ is mathematically defined as

$$\bar{v} = \frac{d}{\Delta t} \qquad\qquad [2.1]$$

Speed is not a vector, so there is no direction associated with average speed.

Average velocity may be a little less familiar to you, due to the vector nature of the velocity. Let us start by imagining the motion of a particle, which, through the particle model, can represent the motion of many types of objects. We shall restrict our study at this point to one-dimensional motion along the x axis.

The motion of a particle is completely specified if the position of the particle in space is known at all times. Consider a car moving back and forth along the x axis, and imagine that we take data on the position of the car every 10 seconds. Figure 2.1a is a *pictorial* representation of this one-dimensional motion that shows the positions of the car at 10-second intervals. The six data points we have recorded are represented by the letters Ⓐ through Ⓕ. Table 2.1 is a *tabular* representation of the motion. It lists the data as entries for position at each time. The black dots in Figure 2.1b show a *graphical* representation of the motion. Such a plot is often called a **position–time graph.** The curved line in Figure 2.1b cannot be unambiguously drawn through our six data points because we have no information about what happened between the data points. The curved line is, however, a *possible* graphical representation of the position of the car at all instants of time during the 50 seconds.

If a particle is moving during a time interval $\Delta t = t_f - t_i$, the displacement of the particle is described as $\Delta \mathbf{x} = \mathbf{x}_f - \mathbf{x}_i = (x_f - x_i)\mathbf{i}$. (Recall that displacement is defined as the change in the position of the particle, which is equal to its final position value minus its initial position value.) Because we are considering only one-dimensional motion in this chapter, we shall drop the vector notation at this point and pick it up again in Chapter 3. The direction of a vector in this chapter will be indicated by means of a positive or negative sign.

The **average velocity** $\bar{v}_x$ of the particle is defined as the ratio of its displacement Δx to the time interval Δt:

$$\bar{v}_x = \frac{\Delta x}{\Delta t} = \frac{x_f - x_i}{t_f - t_i} \qquad\qquad [2.2]$$

where the subscript x indicates motion along the x axis. From this definition we see that average velocity has the dimensions of length divided by time—m/s in SI units and ft/s in conventional units. The average velocity is *independent* of the path taken between the initial and final points. This independence is a major difference from the average speed discussed at the beginning of this section. The average velocity is independent of path because it is proportional to the displacement Δx, which depends only on the initial and final coordinates of the particle. Average speed (a scalar) is found by dividing the *distance* traveled by the time interval, whereas average velocity (a vector) is the *displacement* divided by the time interval. Thus, average velocity gives us no details of the motion—it only gives us the result

TABLE 2.1		
Positions of the Car at Various Times		
Position	*t* (s)	*x* (m)
Ⓐ	0	30
Ⓑ	10	52
Ⓒ	20	38
Ⓓ	30	0
Ⓔ	40	− 37
Ⓕ	50	− 53

PITFALL PREVENTION 2.1

Average speed and average velocity

 The magnitude of the average velocity is *not* the average speed. Consider a particle moving from the origin to $x = 10$ m and then back to the origin in a time interval of 4.0 s. The magnitude of the average velocity is zero, because the particle ends the time interval at the same position at which it started—the displacement is zero. The average speed, however, is the total distance divided by the time interval: 20 m/4.0 s = 5.0 m/s.

On this basketball court, players run back and forth for the entire game. The average speed of the players over the duration of the game is nonzero. The average velocity of the players over the duration of the game is approximately zero because they keep returning to the same point over and over again. *(Ken White/Allsport)*

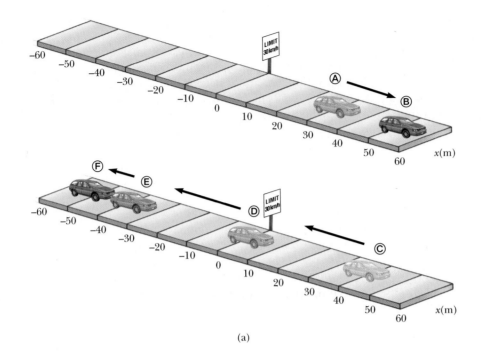

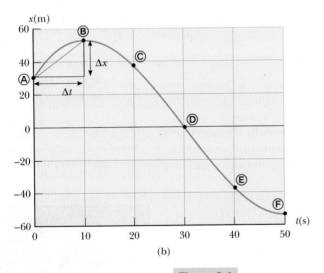

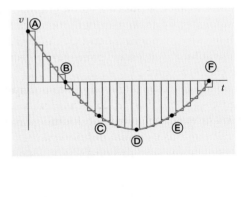

(a)

(b)

(c)

Figure 2.1

(a) A pictorial representation of the motion of a car. The positions of the car at six instants of time are shown and labeled. (b) A graphical representation, known as a position–time graph, of the motion of the car in part (a). The average velocity $\bar{v}_x$ in the interval $t = 0$ to $t = 10$ s is obtained from the slope of the straight line connecting the points Ⓐ and Ⓑ. (c) A velocity–time graph of the motion of the car in part (a).

PITFALL PREVENTION 2.2

A new way of averaging

You are likely familiar with calculating the average of a set of values by adding the values and dividing by the number of values. If you are faced with a problem in which the values of velocity are given at several instants, resist the temptation to add the values and divide by the number of values—this is *not* a valid method for finding the average velocity. Equation 2.2 is the prescription for finding the average velocity.

of the motion. Finally, note that the average velocity in one dimension can be positive or negative, depending on the sign of the displacement. (The time interval Δt is always positive.) If the x coordinate of the particle increases during the time interval (i.e., if $x_f > x_i$), then Δx is positive and $\bar{v}_x$ is positive. This corresponds to an average velocity in the positive x direction. On the other hand, if the coordinate decreases over time ($x_f < x_i$), Δx is negative; hence, $\bar{v}_x$ is negative. This corresponds to an average velocity in the negative x direction.

The average velocity can also be interpreted geometrically, as seen in the graphical representation in Figure 2.1b. A straight line can be drawn between any two points on the curve. Figure 2.1b shows such a line drawn between points Ⓐ and Ⓑ. Using a geometric model, this line forms the hypotenuse of a right triangle of height Δx and base Δt. The slope of the hypotenuse is the ratio $\Delta x/\Delta t$. Therefore, we see that the **average velocity of the particle during the time interval t_i to t_f is equal to the slope of the straight line joining the initial and final points on the position–time graph.** For example, the average velocity of the car between points Ⓐ and Ⓑ is $\bar{v}_x = (52 \text{ m} - 30 \text{ m})/(10 \text{ s} - 0) = 2.2 \text{ m/s}$.

We can also identify a geometric interpretation for the total displacement during the time interval. Figure 2.1c shows the velocity–time graphical representation of the motion in Figures 2.1a and 2.1b. The time interval has been divided into small increments of duration Δt_n. During each of these increments, if we model the velocity as constant during the short increment, the displacement of the particle is given by $\Delta x_n = v_n \Delta t_n$. Geometrically, the product on the right side of this expression represents the area of a thin rectangle associated with each time increment in Figure 2.1c—the height of the rectangle (from the time axis) is v_n and the width is Δt_n. The total displacement of the particle will be the sum of the displacements during each of the increments:

$$\Delta x \approx \sum_n \Delta x_n = \sum_n v_n \Delta t_n$$

This is an approximation because we have modeled the velocity as constant in each increment, which is not the case. The term on the right represents the total area of all the thin rectangles. Now let us take the limit of this expression as the time increments shrink to zero, in which case the approximation becomes exact:

$$\Delta x = \lim_{\Delta t_n \to 0} \sum_n \Delta x_n = \lim_{\Delta t_n \to 0} \sum_n v_n \Delta t_n$$

In this limit, the sum of the areas of all of the very thin rectangles becomes equal to the area under the curve. Thus, **the displacement of a particle during the time interval t_i to t_f is equal to the area under the curve between the initial and final points on the velocity–time graph.**

Example 2.1 Calculate the Average Velocity

A particle moving along the x axis is located at $x_i = 12$ m at $t_i = 1$ s and at $x_f = 4$ m at $t_f = 3$ s. Find its displacement and average velocity during this time interval.

Solution First establish the mental representation—imagine the particle moving along the axis. Based on the information in the problem, which way is it moving? You may find it useful to draw a pictorial representation, but for this simple example, we will go straight to the mathematical representation. The displacement is

$$\Delta x = x_f - x_i = 4 \text{ m} - 12 \text{ m} = \boxed{-8 \text{ m}}$$

The average velocity is, according to Equation 2.2,

$$\bar{v}_x = \frac{\Delta x}{\Delta t} = \frac{4 \text{ m} - 12 \text{ m}}{3 \text{ s} - 1 \text{ s}} = \boxed{-4 \text{ m/s}}$$

Because the displacement is negative for this time interval, we conclude that the particle has moved to the left, toward decreasing values of x. Is this consistent with your mental representation? Keep in mind that it may not have *always* been moving to the left. We only have information about its location at two points in time. After $t_i = 1$ s, it could have moved to the right, turned around, and ended up farther to the left than its original position by the time $t_f = 3$ s. To be completely confident that we know the motion of the particle, we would need to have information about its location at *every* instant of time.

Example 2.2   A jogger runs in a straight line, with a magnitude of average velocity of 5.00 m/s for 4.00 min, and then with a magnitude of average velocity of 4.00 m/s for 3.00 min. (a) What is the magnitude of the final displacement from her initial position? (b) What is the magnitude of her average velocity during this entire time interval of 7.00 min?

Solution   (a) The fact that this problem involves a jogger is not important—we model the jogger as a particle. We have data for two separate portions of the motion, so we use these data to find the displacement for each portion, using Equation 2.2:

$$\bar{v}_x = \frac{\Delta x}{\Delta t} \quad\longrightarrow\quad \Delta x = \bar{v}_x\,\Delta t$$

$$\Delta x_{\text{portion 1}} = (5.00\ \text{m/s})(4.00\ \text{min})\left(\frac{60\ \text{s}}{1\ \text{min}}\right)$$

$$= 1.20 \times 10^3\ \text{m}$$

$$\Delta x_{\text{portion 2}} = (4.00\ \text{m/s})(3.00\ \text{min})\left(\frac{60\ \text{s}}{1\ \text{min}}\right)$$

$$= 7.20 \times 10^2\ \text{m}$$

We add these two displacement to find the total displacement of 1.92×10^3 m.

(b) We now have the data we need to find the average velocity for the entire time interval using Equation 2.2:

$$\bar{v}_x = \frac{\Delta x}{\Delta t} = \frac{1.92 \times 10^3\ \text{m}}{7.00\ \text{min}}\left(\frac{1\ \text{min}}{60\ \text{s}}\right) = 4.57\ \text{m/s}$$

Notice, as indicated in Pitfall Prevention 2.2, that the average velocity is *not* calculated as the simple arithmetic mean of the two velocities given in the problem.

2.2 • INSTANTANEOUS VELOCITY

See Screen 3.3

We would like to be able to define the velocity of a particle at a particular instant in time, rather than during a finite interval of time. For example, you may drive your car through a displacement of magnitude 40 miles and it may take exactly 1 hour to do this, from 1:00:00 PM to 2:00:00 PM. Then your average velocity is of magnitude 40 mi/h for the 1-h interval. But how fast were you going at the particular *instant* of time 1:20:00 PM? It is likely that your velocity varied during the trip, due to hills, traffic lights, slow drivers, and the like, so that there was not a single velocity maintained during the entire hour of travel. The velocity of a particle at any instant of time is called the **instantaneous velocity.**

Consider again the motion of the car shown in Figure 2.1a. Figure 2.2a is the graphical representation again, with two blue lines representing average velocities over very different time intervals. The dark blue line represents the average velocity we calculated earlier over the interval from Ⓐ to Ⓑ. The light blue line represents the average velocity over the much longer interval Ⓐ to Ⓕ. How well does either of these represent the instantaneous velocity at point Ⓐ? In Figure 2.1a, the car begins to move to the right, which we identify as a positive velocity. The average velocity from Ⓐ to Ⓕ is *negative* (because the slope of the line from Ⓐ to Ⓕ is negative), so this clearly is not an accurate representation of the instantaneous velocity at Ⓐ. The average velocity from interval Ⓐ to Ⓑ is *positive*, so this velocity at least has the right sign.

In Figure 2.2b, we show the result of drawing the lines representing the average velocity of the car as point Ⓑ is brought closer and closer to point Ⓐ. As this occurs, the slope of the blue line approaches that of the green line, which is the line drawn tangent to the curve at point Ⓐ. As Ⓑ approaches Ⓐ, the time interval that includes point Ⓐ becomes infinitesimally small. Thus, the average velocity over this interval as the interval shrinks to zero can be interpreted as the instantaneous velocity at point Ⓐ. Furthermore, the slope of the line tangent to the curve

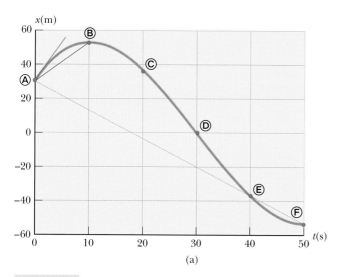

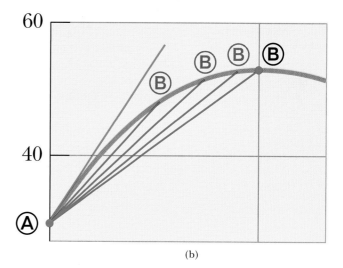

(a)

(b)

Figure 2.2

(a) Position–time graph for the motion of the car in Figure 2.1. (b) An enlargement of the upper left-hand corner of the graph in part (a) shows how the blue line between positions Ⓐ and Ⓑ approaches the green tangent line as point Ⓑ is moved closer to point Ⓐ.

at Ⓐ is the instantaneous velocity at the time t_A. In other words, **the instantaneous velocity v_x equals the limiting value of the ratio $\Delta x/\Delta t$ as Δt approaches zero:***

$$v_x \equiv \lim_{\Delta t \to 0} \frac{\Delta x}{\Delta t}$$

In the calculus notation, this limit is called the *derivative* of x with respect to t, written dx/dt:

$$v_x \equiv \lim_{\Delta t \to 0} \frac{\Delta x}{\Delta t} = \frac{dx}{dt} \qquad [2.3]$$

The instantaneous velocity can be positive, negative, or zero. When the slope of the position–time graph is positive, such as at point Ⓐ in Figure 2.3, v_x is positive. At point Ⓒ, v_x is negative because the slope is negative. Finally, the instantaneous velocity is zero at the peak Ⓑ (the turning point), where the slope is zero. From here on, we shall usually use the word *velocity* to designate instantaneous velocity.

The instantaneous speed of a particle is defined as the magnitude of the instantaneous velocity vector. Hence, by definition, *speed* can never be negative.

Quick Quiz 2.1

Are members of the highway patrol more interested in your average speed or your instantaneous speed as you drive?

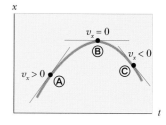

Figure 2.3

In the position–time graph shown, the velocity is positive at Ⓐ, where the slope of the tangent line is positive; the velocity is zero at Ⓑ, where the slope of the tangent line is zero; and the velocity is negative at Ⓒ, where the slope of the tangent line is negative.

---

* Note that the displacement Δx also approaches zero as Δt approaches zero. As Δx and Δt become smaller and smaller, however, the ratio $\Delta x/\Delta t$ approaches a value equal to the *true* slope of the line tangent to the x versus t curve.

PITFALL PREVENTION 2.5

Instantaneous speed and instantaneous velocity

In Pitfall Prevention 2.1, we argued that the magnitude of the average velocity is not the average speed. Notice the difference when discussing instantaneous values. The magnitude of the instantaneous velocity *is* the instantaneous speed. In an infinitesimal time interval, the magnitude of the displacement is equal to the distance traveled by the particle.

Quick Quiz 2.2

Under what conditions is the magnitude of the average velocity of a particle moving in one dimension smaller than the average speed over some time interval?

If you are familiar with the calculus, you should recognize that specific rules exist for taking the derivatives of functions. These rules, which are listed in Appendix B.6, enable us to evaluate derivatives quickly.

Suppose x is proportional to some power of t, such as

$$x = At^n$$

where A and n are constants. (This is a very common functional form.) The derivative of x with respect to t is

$$\frac{dx}{dt} = nAt^{n-1}$$

For example, if $x = 5t^3$ we see that $dx/dt = 3(5)t^{3-1} = 15t^2$.

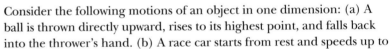

THINKING PHYSICS 2.1

Consider the following motions of an object in one dimension: (a) A ball is thrown directly upward, rises to its highest point, and falls back into the thrower's hand. (b) A race car starts from rest and speeds up to 100 m/s along a straight line. (c) A *Voyager* spacecraft drifts through empty space at constant velocity. Are there any instants of time in the motion of these objects at which the instantaneous velocity at the instant and the average velocity over the entire interval are the same? If so, identify the point(s).

Reasoning (a) The average velocity over the entire interval for the thrown ball is zero—the ball returns to the starting point at the end of the time interval. There is one point at which the instantaneous velocity is zero—at the top of the motion. (b) The average velocity for the motion of the race car cannot be evaluated unambiguously with the information given, but its magnitude must be some value between 0 and 100 m/s. Because the magnitude of the instantaneous velocity of the car will have every value between 0 and 100 m/s at some time during the interval, there must be some instant at which the instantaneous velocity is equal to the average velocity over the entire interval. (c) Because the instantaneous velocity of the spacecraft is constant, its instantaneous velocity at *any* time and its average velocity over *any* time interval are the same.

THINKING PHYSICS 2.2

Does an automobile speedometer measure average or instantaneous speed?

Reasoning Based on the fact that changes in the speed of the automobile are reflected in changes in the speedometer reading, we might be tempted to say that a speedometer measures instantaneous speed. This would not be quite correct, however. The reading on a speedometer is related to the rotation of the wheels. Often, a rotating magnet, whose rotation is related to that of the wheels, is used in a speedometer. The reading is based on the *time interval* between rotations of the magnet. Thus, the reading is an average speed over this interval. This interval is generally very short, so that the average speed over the interval is a good approximation to the instantaneous speed at any instant within the interval. As long as the changes in speed are gradual, this approximation is very good. If the automobile brakes to a quick stop, however, the speedometer may not be able to keep up with the rapid changes in speed, and the measured average speed over the time interval may be different from the instantaneous speed at the end of the interval.

Example 2.3   The Limiting Process

The position of a particle moving along the x axis varies in time according to the expression* $x = 3t^2$, where x is in meters and t is in seconds. Find the velocity in terms of t at any time.

Solution   The position–time graphical representation for this motion is shown in Figure 2.4. We can compute the velocity at any time t by using the definition of the instantaneous velocity. If the initial coordinate of the particle at time t is $x_i = 3t^2$, the coordinate at a later time $t + \Delta t$ is

$$x_f = 3(t + \Delta t)^2 = 3[t^2 + 2t\,\Delta t + (\Delta t)^2]$$
$$= 3t^2 + 6t\,\Delta t + 3(\Delta t)^2$$

Therefore, the displacement in the time interval Δt is

$$\Delta x = x_f - x_i = (3t^2 + 6t\,\Delta t + 3(\Delta t)^2) - (3t^2)$$
$$= 6t\,\Delta t + 3(\Delta t)^2$$

The average velocity in this time interval is

$$\bar{v}_x = \frac{\Delta x}{\Delta t} = \frac{6t\,\Delta t + 3(\Delta t)^2}{\Delta t} = 6t + 3\,\Delta t$$

To find the instantaneous velocity, we take the limit of this expression as Δt approaches zero. In doing so, we see that the term $3\,\Delta t$ goes to zero; therefore,

$$v_x = \lim_{\Delta t \to 0} \frac{\Delta x}{\Delta t} = 6t$$

Notice that this expression gives us the velocity at *any* general time t. It tells us that v_x is increasing linearly in time. It is then a straightforward matter to find the velocity at some

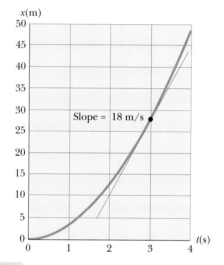

Figure 2.4

(Example 2.3) Position–time graph for a particle having an x coordinate that varies in time according to $x = 3t^2$. Note that the instantaneous velocity at $t = 3.0$ s is obtained from the slope of the blue line tangent to the curve at this point.

specific time from the expression $v_x = 6t$ by substituting the value of the time. For example, at $t = 3.0$ s, the velocity is $v_x = 6(3) = 18$ m/s. Again, this can be checked from the slope at $t = 3.0$ s (the blue line in Figure 2.4).

We can also find v_x by taking the first derivative of x with respect to time, as in Equation 2.3. In this example, $x = 3t^2$, and we see that $v_x = dx/dt = 6t$, in agreement with our result from taking the limit explicitly.

---

* Simply to make it easier to read, we write the equation as $x = 3t^2$ rather than as $x = (3.00 \text{ m/s}^2)\, t^{2.00}$. When an equation summarizes measurements, consider its coefficients to have as many significant digits as other data quoted in a problem. Also consider its coefficients to have the units required for dimensional consistency. When we start our clocks at $t = 0$, we usually do not mean to limit precision to a single digit. Consider any zero value in this book to have as many significant figures as you need.

Example 2.4   Average and Instantaneous Velocity

A particle moves along the *x* axis. Its *x* coordinate varies with time according to the expression $x = -4t + 2t^2$, where *x* is in meters and *t* is in seconds. The position–time graph for this motion is shown in Figure 2.5.

   (a) Determine the displacement of the particle in the time intervals $t = 0$ to $t = 1$ s and $t = 1$ s to $t = 3$ s.

Reasoning   The Thinking Physics examples in this book have a response called "Reasoning." Some numerical examples, such as this one, will also have a Reasoning section to emphasize the fact that you should make a habit of *thinking* about a problem and its solution before diving into the equations. This problem provides a graphical representation of the motion in Figure 2.5. In your mental representation, note that the particle moves in the negative *x* direction for the first second of motion, stops instantaneously at $t = 1$ s, and then heads back in the positive *x* direction for $t > 1$ s. Remember that this is a one-dimensional problem, so the curve in Figure 2.5 does not represent the path the particle follows through space—be sure not to confuse a graphical representation with a pictorial representation of the motion in space (see Figure 2.1 for a comparison). In your mental representation, you should imagine the particle moving to the left and then to the right, with all of the motion taking place along a single line.

Solution   In the first time interval (Ⓐ to Ⓑ) we set $t_i = 0$ and $t_f = 1$ s. Because $x = -4t + 2t^2$, the displacement during the first time interval is

$$\Delta x_{AB} = x_f - x_i = -4(1) + 2(1)^2 - [-4(0) + 2(0)^2]$$
$$= -2 \text{ m}$$

Likewise, in the second time interval (Ⓑ to Ⓓ) we can set $t_i = 1$ s and $t_f = 3$ s. Therefore, the displacement in this interval is

$$\Delta x_{BD} = x_f - x_i = -4(3) + 2(3)^2 - [-4(1) + 2(1)^2]$$
$$= 8 \text{ m}$$

These displacements can also be read directly from the position–time graph (see Fig. 2.5).

   (b) Calculate the average velocity in the time intervals $t = 0$ to $t = 1$ s and $t = 1$ s to $t = 3$ s.

Solution   In the first time interval, $\Delta t = t_f - t_i = 1$ s. Therefore, using Equation 2.2 and the result from (a) gives

$$\bar{v}_x = \frac{\Delta x_{AB}}{\Delta t} = \frac{-2 \text{ m}}{1 \text{ s}} = -2 \text{ m/s}$$

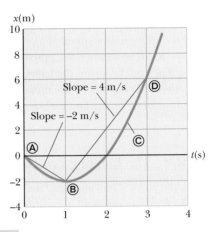

Figure 2.5

(Example 2.4) Position–time graph for a particle having an *x* coordinate that varies in time according to $x = -4t + 2t^2$.

Likewise, in the second time interval, $\Delta t = 2$ s; therefore,

$$\bar{v}_x = \frac{\Delta x_{BD}}{\Delta t} = \frac{8 \text{ m}}{2 \text{ s}} = 4 \text{ m/s}$$

These values agree with the slopes of the lines joining these points in Figure 2.5.

   (c) Find the instantaneous velocity of the particle at $t = 2.5$ s (point Ⓒ).

Solution   We can find the instantaneous velocity at any time *t* by taking the first derivative of *x* with respect to *t*:

$$v_x = \frac{dx}{dt} = \frac{d}{dt}(-4t + 2t^2) = -4 + 4t$$

Thus, at $t = 2.5$ s, we find that

$$v_x = -4 + 4(2.5) = 6 \text{ m/s}$$

   We can also obtain this result by measuring the slope of the position–time graph at $t = 2.5$ s. Do you see any symmetry in the motion? For example, are there points at which the speed is the same? Is the velocity the same at these points?

2.3 • ANALYSIS MODELS — THE PARTICLE UNDER CONSTANT VELOCITY

As mentioned in Section 1.11, the third category of models used in this book is that of **analysis models.** Such models help us analyze the situation in a physics problem and guide us toward the solution. **An analysis model is a problem we have solved before—it is a description of either (1) the behavior of some physical entity or (2) the interaction between that entity and the environment.** When you encounter a new problem, you should identify the fundamental details of the problem and attempt to recognize which of the types of problems you have already solved might be used as a model for the new problem. For example, suppose an automobile is moving along a straight freeway at a constant speed. Is it important that it is an automobile? Is it important that it is a freeway? If the answers to both questions are "no," then we model the situation as a particle under constant velocity, which we will discuss in this section.

This method is somewhat similar to the common practice in the legal profession of finding "legal precedents." If a previously resolved case can be found that is very similar legally to the present one, it is offered as a model, and an argument is made in court to link them logically. The finding in the previous case can then be used to sway the finding in the present case. We will do something similar in physics. For a given problem, we search for a "physics precedent"—a model with which we are already familiar and that can be applied to the present problem.

We shall generate analysis models based on four fundamental simplification models. The first simplification model is the particle model discussed in Chapter 1. We will look at a particle under various behaviors and environmental interactions. Further analysis models are introduced in later chapters with simplification models of a *system*, a *rigid body*, and a *wave*. Once we have introduced these analysis models, we shall see that they appear again and again later in the book in different situations.

Let us use Equation 2.2 to build our first analysis model for solving problems. We imagine a particle moving with a constant velocity. This model can be applied in *any* situation in which an entity that can be modeled as a particle is moving with constant velocity. This situation occurs frequently, so this is an important model.

If the velocity of a particle is constant, its instantaneous velocity at any instant during a time interval is the same as the average velocity over the interval, $v_x = \bar{v}_x$. Thus, we start with Equation 2.2 to generate an equation to be used in the mathematical representation of this situation:

$$v_x = \bar{v}_x = \frac{\Delta x}{\Delta t} \quad \longrightarrow \quad \Delta x = x_f - x_i = v_x \Delta t$$

$$\longrightarrow \quad x_f = x_i + v_x \Delta t$$

This equation tells us that the position of the particle is given by the sum of its original position x_i at time $t = 0$ plus the displacement $v_x \Delta t$ that occurs during the time interval Δt. In practice, we usually choose the time at the beginning of the interval to be $t_i = 0$ and the time at the end of the interval to be $t_f = t$, so our equation becomes

$$x_f = x_i + v_x t \qquad \text{(for constant } v_x \text{)} \qquad \qquad \textbf{[2.4]}$$

• *A particle under constant velocity*

This is the primary equation used in the model of a particle under constant velocity—it can be applied to particles or objects if they can be modeled as particles.

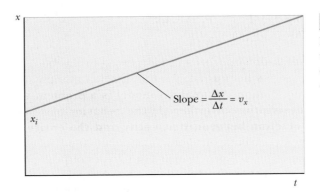

Figure 2.6

Position–time graph for a particle under constant velocity. The value of the constant velocity is the slope of the line.

Figure 2.6 is a graphical representation of the particle under constant velocity. On this position–time graph, the slope of the line representing the motion is constant and equal to the magnitude of the velocity. This is consistent with the mathematical representation, Equation 2.4, which is the equation of a straight line. The slope of the straight line is v_x and the y intercept is x_i in both representations.

Example 2.5 Modeling a Runner as a Particle

A runner has her track coach determine her velocity while she runs at a constant rate. The coach starts the stopwatch at the moment the runner passes and stops it after the runner has passed another point 20 m away. The time interval indicated on the stopwatch is 4.4 s.

(a) What is the runner's velocity?

Solution We model the runner as a particle, as we did in Example 2.2 because the size of the runner and the movement of arms and legs are unnecessary details. This, in combination with the fact that the velocity is stated as being constant, allows us to use Equation 2.4 to find the velocity:

$$v_x = \frac{x_f - x_i}{\Delta t} = \frac{20 \text{ m} - 0}{4.4 \text{ s}} = \boxed{4.5 \text{ m/s}}$$

(b) What is the position of the runner after 10 s has passed?

Solution In this part of the problem, we use Equation 2.4 to find the position of the particle at the time $t = 10$ s. Using the velocity found in part (a),

$$x_f = x_i + v_x t = 0 + (4.5 \text{ m/s})(10 \text{ s}) = \boxed{45 \text{ m}}$$

The mathematical manipulations for the particle under constant velocity stem from Equation 2.2 and its descendent, Equation 2.4. These equations can be used to solve for any variable in the equations that happens to be unknown if the other variables are known. For example, in part (b) of Example 2.5, we find the position when the velocity and the time are known. Similarly, if we know the velocity and the final position, we could use Equation 2.4 to find the time at which the runner is at this position. We shall present more examples of a particle under constant velocity in Chapter 3.

A particle under constant velocity moves with a constant speed along a straight line. Now consider a particle moving with a constant speed along a curved path. The primary equation for this situation is Equation 2.1, with the average speed replaced by the constant speed. As an example, imagine a particle moving at a constant speed in a circular path. If the speed is 5.00 m/s, and the radius of the path is 10.0 m, we can calculate the time required to complete one trip around the circle:

$$v = \bar{v} = \frac{d}{\Delta t} \quad \longrightarrow \quad \Delta t = \frac{d}{v} = \frac{2\pi r}{v} = \frac{2\pi (10.0 \text{ m})}{5.00 \text{ m/s}} = 12.6 \text{ s}$$

2.4 • ACCELERATION

When the velocity of a particle changes with time, the particle is said to be *accelerating.* For example, the speed of a car increases when you "step on the gas"; the car slows down when you apply the brakes; it changes direction when you turn the wheel—all of these are accelerations. We will need a precise definition of acceleration for our studies of motion.

Suppose a particle moving along the x axis has a velocity v_{xi} at time t_i and a velocity v_{xf} at time t_f. The **average acceleration** $\bar{a}_x$ of a particle in the time interval $\Delta t = t_f - t_i$ is defined as the ratio $\Delta v_x/\Delta t$, where $\Delta v_x = v_{xf} - v_{xi}$ is the *change* in velocity in this time interval:

$$\bar{a}_x \equiv \frac{v_{xf} - v_{xi}}{t_f - t_i} = \frac{\Delta v_x}{\Delta t} \qquad \text{[2.5]}$$

In drag racing, acceleration is a highly desired quantity. In a distance of 1/4 mi, speeds of over 320 mi/h are reached, with the entire distance being covered in under 5 s. *(George Lepp/Stone)*

• *Definition of average acceleration*

Thus, **acceleration is a measure of how rapidly the velocity is changing.** Acceleration is a vector quantity having dimensions of length divided by (time)^2, or L/T^2. Some of the common units of acceleration are meters per second per second (m/s^2) and feet per second per second (ft/s^2). For example, an acceleration of 2 m/s^2 means that the velocity changes by 2 m/s during each second of time that passes.

In some situations, the value of the average acceleration may be different for different time intervals. It is therefore useful to define the **instantaneous acceleration** as the limit of the average acceleration as Δt approaches zero, analogous to the definition of instantaneous velocity discussed in Section 2.2:

$$a_x \equiv \lim_{\Delta t \to 0} \frac{\Delta v_x}{\Delta t} = \frac{dv_x}{dt} \qquad \text{[2.6]}$$

• *The instantaneous acceleration is the derivative of the velocity with respect to time*

That is, the **instantaneous acceleration** equals the derivative of the velocity with respect to time, which by definition is the slope of the velocity–time graph. Note that if a_x is positive, the acceleration is in the positive x direction, whereas negative a_x implies acceleration in the negative x direction. This does not necessarily mean that the particle is *moving* in the negative x direction, a point which we shall address in more detail shortly. From now on, we use the term *acceleration* to mean instantaneous acceleration.

Because $v_x = dx/dt$, the acceleration can also be written

$$a_x = \frac{dv_x}{dt} = \frac{d}{dt}\left(\frac{dx}{dt}\right) = \frac{d^2 x}{dt^2} \qquad \text{[2.7]}$$

that is, the acceleration equals the *second derivative* of the position with respect to time.

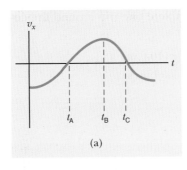

(a)

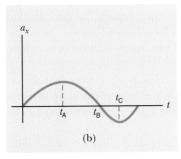

(b)

Figure 2.7

The instantaneous acceleration can be obtained from the velocity–time graph (a). At each instant, the acceleration in the a_x versus t graph (b) equals the slope of the line tangent to the v_x versus t curve.

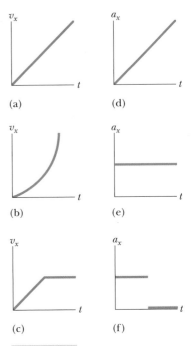

(a)   (d)

(b)   (e)

(c)   (f)

Figure 2.8

(Quick Quiz 2.3) Parts (a), (b), and (c) are velocity–time graphs of objects in one-dimensional motion. The possible acceleration–time graphs of each object are shown in scrambled order in parts (d), (e), and (f).

Figure 2.7 shows how the acceleration–time curve can be derived from the velocity–time curve. In these diagrams, the acceleration of a particle at any time is simply the slope of the velocity–time graph at that time. Positive values of the acceleration correspond to those points (between t_A and t_B) where the velocity in the positive x direction is increasing in magnitude (the particle is speeding up), or those points (between $t = 0$ and t_A) where the velocity in the negative x direction is decreasing in magnitude (the particle is slowing down). The acceleration reaches a maximum at time t_A, when the slope of the velocity–time graph is a maximum. The acceleration then goes to zero at time t_B, when the velocity is a maximum (i.e., when the velocity is momentarily not changing and the slope of the v versus t graph is zero). Finally, the acceleration is negative when the velocity in the positive x direction is decreasing in magnitude (between t_B and t_C) or when the velocity in the negative direction is increasing in magnitude (after t_C).

Quick Quiz 2.3

Using Figure 2.8, match each of the velocity–time graphical representations on the left with the acceleration–time graphical representation on the right that best describes the motion.

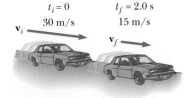

$t_i = 0$ $t_f = 2.0$ s
30 m/s   15 m/s

$\mathbf{v}_i$ $\mathbf{v}_f$

Figure 2.9

The velocity of the car decreases from 30 m/s to 15 m/s in a time interval of 2.0 s.

As an example of the computation of acceleration, consider the pictorial representation of the motion of a car in Figure 2.9. In this case, the velocity of the car has changed from an initial value of 30 m/s to a final value of 15 m/s in a time interval of 2.0 s. The average acceleration during this time interval is

$$\overline{a}_x \equiv \frac{15 \text{ m/s} - 30 \text{ m/s}}{2.0 \text{ s}} = -7.5 \text{ m/s}^2$$

The negative sign in this example indicates that the acceleration vector is in the negative *x* direction (to the left in Figure 2.9). For the case of motion in a straight line, the direction of the velocity of an object and the direction of its acceleration are related as follows. **When the object's velocity and acceleration are in the same direction, the object is speeding up in that direction. On the other hand, when the object's velocity and acceleration are in opposite directions, the speed of the object decreases in time.**

To help with this discussion of the signs of velocity and acceleration, let us take a peek ahead to Chapter 4, where we shall relate the acceleration of an object to the *force* on the object. We will save the details until that later discussion, but for now, let us borrow the notion that **force is proportional to acceleration:**

$$\mathbf{F} \propto \mathbf{a}$$

This proportionality indicates that acceleration is caused by force. What's more, as indicated by the vector notation in the proportionality, force and acceleration are in the same direction. Thus, let us think about the signs of velocity and acceleration by forming a mental representation in which a force is applied to the object to cause the acceleration. Again consider the case in which the velocity and acceleration are in the same direction. This is equivalent to an object moving in a given direction, and experiencing a force that pulls on it in the same direction. It is clear in this case that the object speeds up! If the velocity and acceleration are in opposite directions, the object moves one way and a force pulls in the opposite direction. In this case, the object slows down! It is very useful to equate the direction of the acceleration in these situations to the direction of a force because it is easier from our everyday experience to think about what effect a force will have on an object than to think only in terms of the direction of the acceleration.

PITFALL PREVENTION 2.6

Negative acceleration

Keep in mind that *negative acceleration does not necessarily mean that an object is slowing down.* If the acceleration is negative, and the velocity is negative, the object is speeding up!

PITFALL PREVENTION 2.7

Deceleration

The word *deceleration* has a common popular connotation as *slowing down.* When combined with the misconception in Pitfall Prevention 2.6 that negative acceleration means slowing down, the situation can be further confused by the use of the word *deceleration.* We will not use this word in this text.

Quick Quiz 2.4

If a car is traveling eastward and slowing down, what is the direction of the force on the car? Explain.

Quick Quiz 2.5

Can a particle with constant acceleration ever stop and stay stopped?

Example 2.6 Average and Instantaneous Acceleration

The velocity of a particle moving along the *x* axis varies in time according to the expression $v_x = (40 - 5t^2)$, where *t* is in seconds.

(a) Find the average acceleration in the time interval $t = 0$ to $t = 2.0$ s.

Solution Build your mental representation from the mathematical expression given for the velocity. For example, which way is the particle moving at $t = 0$? How does the velocity change in the first few seconds—does it move faster or slower? The velocity–time graphical representation for this function is given in Figure 2.10. The velocities at $t_i = t_A = 0$ and $t_f = t_B = 2.0$ s are found by substituting these values of t into the expression given for the velocity:

$$v_{xA} = 40 - 5t_A^2 \text{ m/s} = 40 - 5(0)^2 \text{ m/s} = 40 \text{ m/s}$$

$$v_{xB} = 40 - 5t_B^2 \text{ m/s} = 40 - 5(2.0)^2 \text{ m/s} = 20 \text{ m/s}$$

Therefore, the average acceleration in the specified time interval $\Delta t = t_B - t_A$ is

$$\overline{a}_x \equiv \frac{20 \text{ m/s} - 40 \text{ m/s}}{2.0 \text{ s}} = \boxed{-10 \text{ m/s}^2}$$

The negative sign is consistent with the fact that the slope of the line joining the initial and final points on the velocity–time graph is negative.

(b) Determine the acceleration at $t = 2.0$ s.

Solution   Because this question refers to a specific instant of time, it is asking for an instantaneous acceleration. The velocity at time t is $v_{xi} = 40 - 5t^2$, and the velocity at time $t + \Delta t$ is

$$v_{xf} = 40 - 5(t + \Delta t)^2 = 40 - 5t^2 - 10t\,\Delta t - 5(\Delta t)^2$$

Therefore, the change in velocity over the time interval Δt is

$$\Delta v_x = v_{xf} - v_{xi} = -10t\,\Delta t - 5(\Delta t)^2$$

Dividing this expression by Δt and taking the limit of the result as Δt approaches zero gives the acceleration at *any* time t:

$$a_x \equiv \lim_{\Delta t \to 0} \frac{\Delta v_x}{\Delta t} = \lim_{\Delta t \to 0} (-10t - 5\Delta t) = -10t$$

Therefore, at $t = 2.0$ s we find that

$$a_x = (-10)(2.0) \text{ m/s}^2 = \boxed{-20 \text{ m/s}^2}$$

This result can also be obtained by measuring the slope of the velocity–time graph at $t = 2.0$ s (see Fig. 2.10) or by taking the derivative of the velocity expression.

EXERCISE   When struck by a club, a golf ball initially at rest acquires a speed of 31.0 m/s. If the ball is in contact with the club for 1.17 ms, what is the magnitude of the average acceleration of the ball?

Answer 2.65×10^4 m/s^2

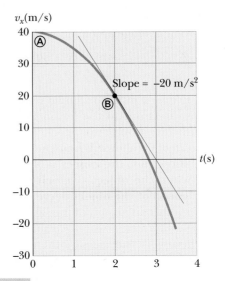

Figure 2.10

(Example 2.6) The velocity–time graph for a particle moving along the x axis according to the relation $v_x = (40 - 5t^2)$. The acceleration at $t = 2.0$ s is obtained from the slope of the blue tangent line at that time.

EXERCISE   A particle moves with a velocity given by the expression $v = 4t^2$. (a) What is the acceleration of the particle at $t = 0.25$ s? (b) Is the particle speeding up or slowing down at this instant? Why?

Answer   (a) 2.0 m/s^2 (b) Speeding up, because both the velocity and acceleration are in the positive direction.

EXERCISE   A particle moves with a velocity $v = -4.0$ m/s at time $t = 2.0$ s. At $t = 3.0$ s, its velocity is -1.0 m/s. The acceleration during this time interval is constant. (a) What is the sign of the acceleration during this interval? (b) What is the magnitude of the acceleration during this interval? (c) Is the particle speeding up or slowing during this interval? Why?

Answer   (a) Positive (b) 3.0 m/s^2 (c) Slowing down, because the acceleration is in a direction opposite to the velocity. This result is consistent with the fact that the magnitude of the velocity is smaller at the end of the time interval than at the beginning.

2.5 • MOTION DIAGRAMS

The concepts of velocity and acceleration are often confused with each other, but in fact they are quite different quantities. It is instructive to make use of the specialized pictorial representation called a **motion diagram** to describe the velocity and acceleration vectors while an object is in motion.

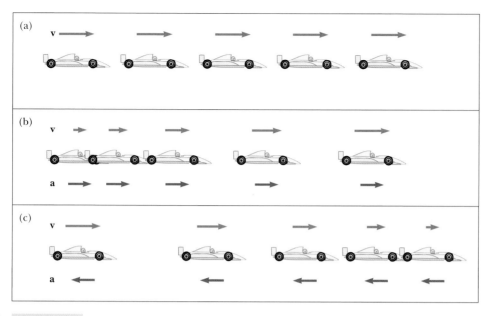

Figure 2.11

(a) Motion diagram for a car moving at constant velocity. (b) Motion diagram for a car whose constant acceleration is in the direction of its velocity. The velocity vector at each instant is indicated by a red arrow, and the constant acceleration vector by a violet arrow. (c) Motion diagram for a car whose constant acceleration is in the direction *opposite* the velocity at each instant.

A *stroboscopic photograph* of a moving object shows several images of the object, taken as the strobe light flashes at a constant rate. Figure 2.11 represents three sets of strobe photographs of cars moving along a straight roadway in a single direction, from left to right. The time intervals between flashes of the stroboscope are equal in each part of the diagram. To distinguish between the two vector quantities, we use red for velocity vectors and violet for acceleration vectors in Figure 2.11. The vectors are sketched at several instants during the motion of the object. Let us describe the motion of the car in each diagram.

In Figure 2.11a, the images of the car are equally spaced—the car moves through the same displacement in each time interval. Thus, the car moves with *constant positive velocity* and has *zero acceleration*. We could model the car as a particle and describe it as a particle under constant velocity.

In Figure 2.11b, the images of the car become farther apart as time progresses. In this case, the velocity vector increases in time because the car's displacement between adjacent positions increases as time progresses. Thus, the car is moving with a *positive velocity* and a *positive acceleration*. The velocity and acceleration are in the same direction. In terms of our earlier force discussion, imagine a force pulling on the car in the same direction it is moving—it speeds up.

In Figure 2.11c, we interpret the car as slowing down as it moves to the right because its displacement between adjacent positions decreases as time progresses. In this case, the car moves initially to the right with a *positive velocity* and a *negative acceleration*. The velocity vector decreases in time and eventually reaches zero. (This type of motion is exhibited by a car that skids to a stop after applying its brakes.) From this diagram we see that the acceleration and velocity vectors are *not* in the same direction. The velocity and acceleration are in opposite directions.

In terms of our earlier force discussion, imagine a force pulling on the car opposite to the direction it is moving—it slows down.

The violet acceleration vectors in Figures 2.11b and 2.11c are all of the same length. Thus, these diagrams represent a motion with constant acceleration. This important type of motion is discussed in the next section.

2.6 • THE PARTICLE UNDER CONSTANT ACCELERATION

 See Screen 3.4

If the acceleration of a particle varies in time, the motion may be complex and difficult to analyze. A very common and simple type of one-dimensional motion occurs when the acceleration is constant, such as for the motion of the cars in Figures 2.11b and 2.11c. In this case, the average acceleration over any time interval equals the instantaneous acceleration at any instant of time within the interval. Consequently, the velocity increases or decreases at the same rate throughout the motion. The particle under constant acceleration forms the basis of a common analysis model that we can apply to appropriate problems and is often used to model situations such as falling objects and braking cars.

If we replace $\overline{a}_x$ with the constant a_x in Equation 2.5, we find that

$$a_x = \frac{v_{xf} - v_{xi}}{t_f - t_i}$$

For convenience, let $t_i = 0$ and t_f be any arbitrary time t. With this notation, we can express the acceleration as

$$a_x = \frac{v_{xf} - v_{xi}}{t}$$

or, solving for v_{xf},

$$v_{xf} = v_{xi} + a_x t \qquad \text{(for constant } a_x\text{)} \qquad [2.8]$$

This expression enables us to predict the velocity at *any* time t if the initial velocity and constant acceleration are known. It is the first of four equations that can be used to solve problems in which a particle is under constant acceleration. A graphical representation of position versus time for this motion is shown in Figure 2.12a. The velocity–time graph shown in Figure 2.12b is a straight line, the slope of which is the constant acceleration a_x. This is consistent with the fact that $a_x = dv_x/dt$ is a constant. From this graph and from Equation 2.8, we see that the velocity at any time t is the sum of the initial velocity v_{xi} and the change in velocity $a_x t$ due to the acceleration. The graph of acceleration versus time (Fig. 2.12c) is a straight line with a slope of zero, because the acceleration is constant. If the acceleration were negative, the slope of Figure 2.12b would be negative and the horizontal line in Figure 2.12c would be below the time axis.

We can generate another equation for constant acceleration by recalling a result from Section 2.1, that the displacement of a particle is the area under the curve on a velocity–time graph. Because the velocity varies linearly with time (Fig. 2.12b), the area under the curve is the sum of a rectangular area (under the horizontal dashed line in Fig. 2.12b) and a triangular area (from the horizontal dashed line upward to the curve). Thus,

$$\Delta x = v_{xi}\,\Delta t + \tfrac{1}{2}(v_{xf} - v_{xi})\Delta t$$

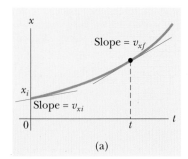

(a)

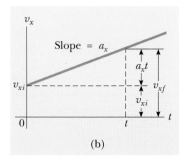

(b)

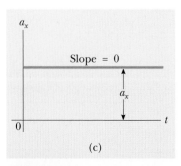

(c)

Figure 2.12

A particle moving along the x axis with constant acceleration a_x; (a) the position–time graph, (b) the velocity–time graph, and (c) the acceleration–time graph.

This can be simplified as follows:

$$\Delta x = (v_{xi} + \tfrac{1}{2}v_{xf} - \tfrac{1}{2}v_{xi})\Delta t = \tfrac{1}{2}(v_{xi} + v_{xf})\Delta t$$

In general, from Equation 2.2, the displacement for a time interval is

$$\Delta x = \bar{v}_x \Delta t$$

Comparing these last two equations, we find that the average velocity in any time interval is the arithmetic mean of the initial velocity v_{xi} and the final velocity v_{xf}:

$$\bar{v}_x = \tfrac{1}{2}(v_{xi} + v_{xf}) \qquad \text{(for constant } a_x) \qquad \text{[2.9]}$$

● *Average velocity for a particle under constant acceleration*

Remember that this expression is valid only when the acceleration is constant, that is, when the velocity varies linearly with time.

We now use Equations 2.2 and 2.9 to obtain the position as a function of time. Again we choose $t_i = 0$, at which time the initial position is x_i. This gives

$$\Delta x = \bar{v}_x \Delta t = \tfrac{1}{2}(v_{xi} + v_{xf})t$$

$$x_f = x_i + \tfrac{1}{2}(v_{xi} + v_{xf})t \qquad \text{(for constant } a_x) \qquad \text{[2.10]}$$

● *Position as a function of velocity and time*

We can obtain another useful expression for the position by substituting Equation 2.8 for v_{xf} in Equation 2.10:

$$x_f = x_i + \tfrac{1}{2}[v_{xi} + (v_{xi} + a_x t)]t$$

$$x_f = x_i + v_{xi}t + \tfrac{1}{2}a_x t^2 \qquad \text{(for constant } a_x) \qquad \text{[2.11]}$$

● *Position as a function of time for a particle under constant acceleration*

Note that the position at any time t is the sum of the initial position x_i, the displacement $v_{xi}t$ that would result if the velocity remained constant at the initial velocity, and the displacement $\tfrac{1}{2}a_x t^2$ due to the fact that the particle is accelerating. Consider again the position–time graph for motion under constant acceleration shown in Figure 2.12a. The curve representing Equation 2.11 is a parabola, as shown by the t^2 dependence in the equation. The slope of the tangent to this curve at $t = 0$ equals the initial velocity v_{xi}, and the slope of the tangent line at any time t equals the velocity at that time.

Finally, we can obtain an expression that does not contain the time by substituting the value of t from Equation 2.8 into Equation 2.10. This gives

$$x_f = x_i + \tfrac{1}{2}(v_{xi} + v_{xf})\left(\frac{v_{xf} - v_{xi}}{a_x}\right) = x_i + \frac{v_{xf}^2 - v_{xi}^2}{2a_x}$$

$$v_{xf}^2 = v_{xi}^2 + 2a_x(x_f - x_i) \qquad \text{(for constant } a_x) \qquad \text{[2.12]}$$

● *Velocity as a function of position for a particle under constant acceleration*

This is *not* an independent equation, because it arises from combining Equations 2.8 and 2.10. It is useful, however, for those problems in which a value for the time is not involved.

If motion occurs in which the constant value of the acceleration is *zero*, then

$$\left.\begin{array}{l} v_{xf} = v_{xi} \\ x_f = x_i + v_{xi}t \end{array}\right\} \qquad \text{when } a_x = 0$$

That is, when the acceleration is zero, the velocity remains constant and the position changes linearly with time. In this case, the particle under constant *acceleration* becomes the particle under constant *velocity* (Equation 2.4).

TABLE 2.2	Kinematic Equations for Motion in a Straight Line Under Constant Acceleration
Equation	**Information Given by Equation**
$v_{xf} = v_{xi} + a_x t$	Velocity as a function of time
$x_f = x_i + \frac{1}{2}(v_{xf} + v_{xi})t$	Position as a function of velocity and time
$x_f = x_i + v_{xi}t + \frac{1}{2}a_x t^2$	Position as a function of time
$v_{xf}^2 = v_{xi}^2 + 2a_x(x_f - x_i)$	Velocity as a function of position

Note: Motion is along the *x* axis. At $t = 0$, the position of the particle is x_i and its velocity is v_{xi}.

Equations 2.8, 2.10, 2.11, and 2.12 are four kinematic expressions that may be used to solve any problem in one-dimensional motion of a particle (or an object that can be modeled as a particle) with constant acceleration. Keep in mind that these relationships were derived from the definitions of velocity and acceleration, together with some simple algebraic manipulations and the requirement that the acceleration be constant. It is often convenient to choose the initial position of the particle as the origin of the motion, so that $x_i = 0$ at $t = 0$. We will see cases, however, in which we must choose the value of x_i to be something other than zero.

The four kinematic equations for the particle under constant acceleration are listed in Table 2.2 for convenience. The choice of which kinematic equation or equations you should use in a given situation depends on what is known beforehand. Sometimes it is necessary to use two of these equations to solve for two unknowns, such as the position and velocity at some instant. You should recognize that the quantities that vary during the motion are velocity v_{xf}, position x_f, and time *t*. The other quantities—x_i, v_{xi}, and a_x—are *parameters* of the motion and remain constant.

Solving the exercises and problems will give you a great deal of practice in the use of these equations. You will often discover that more than one method can be used to obtain a solution.

PROBLEM-SOLVING STRATEGY | **Particle Under Constant Acceleration**

The following procedure is recommended for solving problems that involve an object undergoing a constant acceleration:

1. Think about what is going on physically in the problem. Establish the mental representation.
2. Simplify the problem as much as possible. Confirm that the problem involves either a particle or an object that can be modeled as a particle and that it is moving with a constant acceleration.
3. Construct an appropriate pictorial representation, such as a motion diagram, or a graphical representation.
4. Make sure all the units in the problem are consistent. That is, if positions are measured in meters, be sure that velocities have units of m/s and accelerations have units of m/s^2.
5. Choose a coordinate system to be used throughout the problem.

6. Set up the mathematical representation. Choose an instant to call the "initial" time $t = 0$ and another the "final" time t. Let your choice be guided by what you know about the particle and what you want to know about it. The initial instant need not be when the particle starts to move and the final instant will only rarely be when the particle stops moving. Identify all the quantities given in the problem and a separate list of those to be determined. A tabular representation of these quantities may be helpful to you. Select from the list of kinematic equations the one or ones that will enable you to determine the unknowns. Solve the equations.

7. Once you have determined your result, check to see if your answers are consistent with the mental and pictorial representations and that your results are realistic.

Example 2.7   Accelerating an Electron

An electron in the cathode-ray tube of a television set enters a region in which it accelerates uniformly in a straight line from a speed of 3.00×10^4 m/s to a speed of 5.00×10^6 m/s in a distance of 2.00 cm. For what length of time is the electron accelerating?

Reasoning   For this example, we shall identify the individual steps in the Problem-Solving Strategy. In subsequent examples, you should identify the portions of the solution that correspond to each step. For step 1, think about the electron moving through space. Note that it is moving faster at the end of the interval than before, so imagine it speeding up as it covers the 2.00-cm displacement. In step 2, ignore the fact that this is an electron and that it is in a television. The electron is easily modeled as a particle and the phrase "accelerates uniformly" tells us that this is a particle under constant acceleration. In step 3, all of the parts of Figure 2.12 represent the motion of the particle as a function of time, although you may want to graph velocity versus position because no time is given in the problem. In step 4, note that all units are metric, although we must convert 2.00 cm to meters to put all units in SI. In step 5, we make the simple choice of the x axis lying along the straight line mentioned in the text of the problem.

Solution   We are now ready to move on to step 6, in which we develop the mathematical representation of the problem. Notice that no acceleration is given in the problem and that

the time interval is requested, which provides a hint that we should use an equation that does not involve acceleration. We can find the time at which the particle is at the end of the 2.00-cm distance from Equation 2.10:

$$x_f = x_i + \tfrac{1}{2}(v_{xi} + v_{xf})t \quad \longrightarrow \quad t = \frac{2(x_f - x_i)}{v_{xi} + v_{xf}}$$

$$t = \frac{2(0.0200 \text{ m})}{3.00 \times 10^4 \text{ m/s} + 5.00 \times 10^6 \text{ m/s}}$$

$$= 7.95 \times 10^{-9} \text{ s}$$

Finally, we check if the answer is reasonable (step 7). The final speed is on the order of 10^6 m/s. Let us estimate the time required to move 1 cm at this speed:

$$\Delta t = \frac{\Delta x}{v} \approx \frac{0.01 \text{ m}}{10^6 \text{ m/s}} \approx 10^{-8} \text{ s} = 10 \times 10^{-9} \text{ s}$$

This is the same order of magnitude as our answer, so this provides confidence that our answer is reasonable.

EXERCISE   A hockey puck sliding on a frozen lake comes to rest after traveling 200 m. Its initial speed was 3.00 m/s. (a) What was its acceleration if it is assumed to have been constant? (b) How long was it in motion? (c) What was its speed after traveling 150 m?

Answer   (a) -2.25×10^{-2} m/s^2 (b) 133 s (c) 1.50 m/s

Example 2.8   Watch Out for the Speed Limit!

A car traveling at a constant velocity of magnitude 45.0 m/s passes a trooper hidden behind a billboard. One second after the speeding car passes the billboard, the trooper sets out

from the billboard to catch it, accelerating at a constant rate of 3.00 m/s^2. How long does it take her to overtake the speeding car?

Reasoning   We will point out again in this example steps in the Problem-Solving Strategy. A pictorial representation (step 3) of the situation is shown in Figure 2.13. Establish the mental representation (step 1) of this situation for yourself; in the following solution, we will go straight to the mathematical representation. As you become more proficient at solving physics problems, a quick thought about the mental representation may be enough to allow you to skip pictorial representations and go right to the mathematics. Let us model the speeding car as a particle under constant velocity and the trooper as a particle under constant acceleration (step 2). We shall ignore the fact that these are cars—we will imagine the speeder and the trooper as point particles undergoing the motion described in the problem.

Solution   Note that all units are in the same system (step 4). To solve this problem algebraically, we will write an expression for the position of each vehicle as a function of time. It is convenient to choose the origin at the position of the billboard (step 5) and take $t_B = 0$ as the time the trooper begins moving (step 6). At that instant, the speeding car has already traveled a distance of 45.0 m, because it has traveled at a constant speed of $v_x = 45.0$ m/s for 1.00 s—it is at point Ⓑ in Figure 2.13. Thus, the initial position of the speeding car is $x_i = x_B = 45.0$ m. We do *not* choose $t = 0$ as the time at which the car passes the trooper (point Ⓐ in Figure 2.13), because then the acceleration of the trooper is not constant during the problem. Her acceleration is $a_x = 0$ for the first second and then 3.00 m/s² after that. Thus, we could not model the trooper as a particle under constant acceleration with this choice.

Because the car moves with constant velocity, its acceleration is zero, and applying Equation 2.4 gives us

$$x_f = x_B + v_x t \longrightarrow x_{car} = 45.0 \text{ m} + (45.0 \text{ m/s})t$$

Note that at $t = 0$, this expression gives the car's correct initial position, $x_{car} = 45.0$ m.

For the trooper, who starts from the origin at $t = 0$, we have $x_i = 0$, $v_{xi} = 0$, and $a_x = 3.00$ m/s². Hence, from Equation 2.11 for a particle under constant acceleration, the position of the trooper as a function of time is

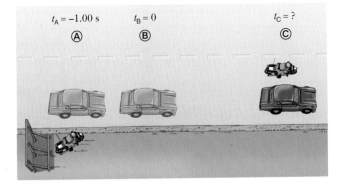

$$v_{x\,car} = 45.0 \text{ m/s}$$
$$a_{x\,car} = 0$$
$$a_{x\,trooper} = 3.00 \text{ m/s}^2$$

$t_A = -1.00$ s $t_B = 0$ $t_C = ?$

Figure 2.13

(Example 2.8) A speeding car passes a hidden trooper. The trooper catches up to the car at point Ⓒ.

$$x_f = x_i + v_{xi}t + \tfrac{1}{2}a_x t^2 \longrightarrow x_{trooper} = \tfrac{1}{2}a_x t^2$$
$$= \tfrac{1}{2}(3.00 \text{ m/s}^2)t^2$$

The trooper overtakes the car at the instant that $x_{trooper} = x_{car}$, which is at position Ⓒ in Figure 2.13:

$$\tfrac{1}{2}(3.00 \text{ m/s}^2)t^2 = 45.0 \text{ m} + (45.0 \text{ m/s})t$$

This gives the quadratic equation (dropping the units)

$$1.50t^2 - 45.0t - 45.0 = 0$$

whose positive solution is $t = 31.0$ s.

From your everyday experience, is this value reasonable (step 7)? (For help in solving quadratic equations, see Appendix B.2.)

EXERCISE   In the preceding solution, the problem was solved in the mathematical representation. This problem can also be solved using a graphical representation. On the *same* graph, plot the position-versus-time function for each vehicle and from the intersection of the two curves determine the time at which the trooper overtakes the speeding car.

2.7 • FREELY FALLING OBJECTS

It is well known that all objects, when dropped, fall toward the Earth with nearly constant acceleration. Legend has it that Galileo Galilei first discovered this fact by observing that two different weights dropped simultaneously from the Leaning Tower of Pisa hit the ground at approximately the same time. (Air resistance plays a role in the falling of an object, but, for now, we shall model falling objects as if they are falling through a vacuum—a simplification model.) Although there is some doubt that this particular experiment was actually carried out, it is well established that Galileo did perform many systematic experiments on objects moving on inclined planes. Through careful measurements of distances and time intervals, he was able to show that the displacement from an origin of an object starting from rest is proportional to the square of the time the object is in motion. This observation is consistent with one of the kinematic equations we derived for a particle under constant acceleration (Eq. 2.11, with $v_{xi} = 0$). Galileo's achievements in mechanics paved the way for Newton in his development of the laws of motion.

You might want to try the following experiment. Drop a coin and a crumpled-up piece of paper simultaneously from the same height. Do they hit the floor at the same time? In the idealized case of no air resistance, the two objects experience the same motion and hit the floor at the same time. In a real (nonideal) experiment, air resistance cannot be ignored, and there will be a small time difference between their arrivals at the floor. In the idealized case, however, where air resistance *is* ignored, such motion is referred to as *free-fall*. If this same experiment could be conducted in a good vacuum, where air friction is truly negligible, the paper and coin would fall with the same acceleration, regardless of the shape or weight of the paper, even if the paper were still flat. This point is illustrated very convincingly in Figure 2.14, which is a photograph of an apple and a feather falling in a vacuum. On August 2, 1971, such an experiment was conducted on the Moon by astronaut David Scott. He simultaneously released a geologist's hammer and a falcon's feather, and in unison they fell to the lunar surface. This demonstration surely would have pleased Galileo!

We shall denote the magnitude of the free-fall acceleration with the symbol g, representing a vector acceleration $\mathbf{g}$. At the surface of the Earth, g is approximately 9.80 m/s², or 980 cm/s², or 32 ft/s². Unless stated otherwise, we shall use the value 9.80 m/s² when doing calculations. Furthermore, we shall assume that the vector $\mathbf{g}$ is directed downward toward the center of the Earth.

WEB

The following site contains images of Galileo's experimental notes on motion: **www.imss.fi.it/ms72**

- *Free-fall acceleration*
  $\mathbf{g} = 9.80 \ m/s^2 \ down$

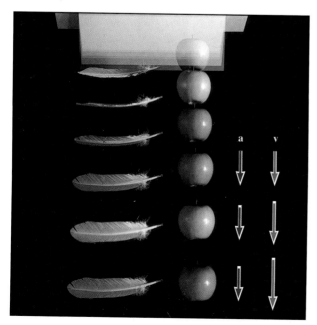

Figure 2.14

An apple and a feather, released from rest in a 4-ft vacuum chamber, fall at the same rate, regardless of their masses. Ignoring air resistance, all objects fall to the Earth with the same acceleration of magnitude 9.80 m/s², as indicated by the violet arrows in this multiflash photograph. The velocity of the two objects increases linearly with time, as indicated by the series of red arrows. *(© 1993 James Sugar/Black Star)*

When we use the expression *freely falling object,* we do not necessarily mean an object dropped from rest. A freely falling object is an object moving freely under the influence of gravity alone, regardless of its initial motion. Thus, objects thrown upward or downward and those released from rest are all freely falling objects once they are released!

It is important to emphasize that any freely falling object experiences an acceleration directed *downward,* as shown in the strobe photograph of a falling apple and a falling feather (see Fig. 2.14). If we ignore air resistance (the apple and feather are falling in a vacuum chamber in Figure 2.14), then a freely falling object (released from rest or projected with a vertical velocity) moves in one dimension with constant acceleration. Therefore, we can model the object as a particle under constant acceleration.

In previous examples in this chapter, the particles were truly undergoing constant acceleration, as stated in the problem. Thus, it may have been difficult to understand the need for modeling. Now, we can begin to see the need for modeling—we really are *modeling* a real falling object with an analysis model. Notice that we are (1) ignoring air resistance and (2) assuming that the free-fall acceleration is constant. Thus, the model of a particle under constant acceleration is a *replacement* for the real problem, which could be more complicated. If air resistance and any variation in *g* are small, however, the model should make predictions that agree closely with the real situation.

The equations developed in Section 2.6 for objects moving with constant acceleration can be applied to the falling object. The only necessary modification that we need to make in these equations for freely falling objects is to note that the motion is in the vertical direction, so we will use *y* instead of *x*, and that the acceleration is downward and of magnitude 9.80 m/s². Thus, for a freely falling object we commonly take $a_y = -g = -9.80$ m/s², where the negative sign indicates that the acceleration of the object is downward. The choice of negative for the downward direction is arbitrary, but common.

Quick Quiz 2.6

A ball is thrown upward. While the ball is in free-fall, (a) what happens to its speed? (b) Does its acceleration increase, decrease, or remain constant?

THINKING PHYSICS 2.3

A sky diver steps out of a stationary helicopter. A few seconds later, another sky diver steps out, so that they both fall along the same vertical line. Ignore air resistance, so that both sky divers fall with the same acceleration and let us model the sky divers as particles under constant acceleration. Does the vertical separation distance between them stay the same? Does the difference in their speeds stay the same?

Reasoning At any given instant of time, the speeds of the jumpers are definitely different, because one had a head start over the other. In any time interval, however, each jumper increases his or her speed by the same amount, because they have the same acceleration. Thus, the difference in speeds remains the same. The first jumper will always be moving with a higher speed than the second. In a given time interval, then, the first jumper will have a larger displacement than the second. Thus, the separation distance between them increases.

Example 2.9 Try to Catch the Dollar

Emily challenges David to catch a dollar bill as follows. She holds the bill vertically, as in Figure 2.15, with the center of the bill between David's index finger and thumb. David must catch the bill after Emily releases it without moving his hand downward. The reaction time of most people is at best about 0.2 s. Who would you bet on?

Reasoning Place your bets on Emily. There is a time delay between the instant Emily releases the bill and the time David reacts and closes his fingers. We model the bill as a particle. When released, the bill will probably flutter downward to the floor due to the effects of the air, but for the very early part of its motion, we will assume that it can be modeled as a particle falling through a vacuum. Because the bill is in free-fall and undergoes a downward acceleration of magnitude 9.80 m/s^2, in 0.2 s it falls a distance of $y = \frac{1}{2}gt^2 \approx 0.2$ m = 20 cm. This distance is about twice the distance between the center of the bill and its top edge (≈ 8 cm). Thus, David will be unsuccessful. You might want to try this "trick" on one of your friends.

Figure 2.15

(Example 2.9) *(George Semple)*

Example 2.10 Not a Bad Throw for a Rookie!

A stone is thrown at point Ⓐ from the top of a building with an initial velocity of 20.0 m/s straight upward. The building is 50.0 m high, and the stone just misses the edge of the roof on its way down, as in the pictorial representation of Figure 2.16. Determine (a) the time at which the stone reaches its maximum height, (b) the maximum height above the rooftop, (c) the time at which the stone returns to the level of the thrower, (d) the velocity of the stone at this instant, (e) the velocity and position of the stone at $t = 5.00$ s, and (f) the position of the stone at $t = 6.00$ s. How does the model fail in this last part of the problem?

Solution Think about the mental representation—the stone rises upward, slowing down. It stops momentarily (point Ⓑ) and then begins to fall downward again. During the entire motion, it is accelerating downward because the gravitational force is always pulling downward on it. Ignoring air resistance, we model the stone as a particle under constant acceleration.

(a) To begin the mathematical representation, we find the time at which the stone reaches the maximum height, point Ⓑ. We use the vertical modification of Equation 2.8, noting

that $v_{yf} = 0$ at the maximum height:

$$v_{yf} = v_{yi} + a_y t \quad \longrightarrow \quad 0 = 20.0 \text{ m/s} + (-9.80 \text{ m/s}^2)t_B$$

$$t_B = \frac{20.0 \text{ m/s}}{9.80 \text{ m/s}^2} = 2.04 \text{ s}$$

(b) This value of time can be substituted into Equation 2.11, to give the maximum height measured from the position of the thrower:

$$y_{max} = y_B = y_A + v_{yA}t + \tfrac{1}{2}a_y t_A{}^2$$

$$= 0 + (20.0 \text{ m/s})(2.04 \text{ s}) + \tfrac{1}{2}(-9.80 \text{ m/s}^2)(2.04 \text{ s})^2$$

$$= 20.4 \text{ m}$$

(c) When the stone is back at the height of the thrower, the y coordinate is zero. From Equation 2.11, letting $y_f = y_C = 0$, we obtain the expression

$$y_C = y_A + v_{yA}t + \tfrac{1}{2}a_y t_C{}^2 \quad \longrightarrow \quad 0 = 20.0t_C - 4.90t_C{}^2$$

This is a quadratic equation and has two solutions for t_C. The equation can be factored to give

$$t_C(20.0 - 4.90t_C) = 0$$

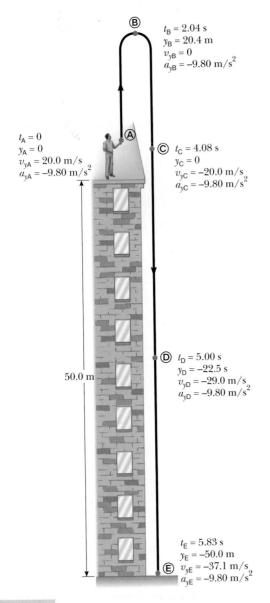

Ⓑ $t_B = 2.04$ s
$y_B = 20.4$ m
$v_{yB} = 0$
$a_{yB} = -9.80$ m/s^2

$t_A = 0$
$y_A = 0$
$v_{yA} = 20.0$ m/s
$a_{yA} = -9.80$ m/s^2

Ⓐ

Ⓒ $t_C = 4.08$ s
$y_C = 0$
$v_{yC} = -20.0$ m/s
$a_{yC} = -9.80$ m/s^2

50.0 m

Ⓓ $t_D = 5.00$ s
$y_D = -22.5$ s
$v_{yD} = -29.0$ m/s
$a_{yD} = -9.80$ m/s^2

$t_E = 5.83$ s
$y_E = -50.0$ m
Ⓔ $v_{yE} = -37.1$ m/s
$a_{yE} = -9.80$ m/s^2

Figure 2.16

(Example 2.10) Position, velocity, and acceleration at various instants of time for a freely falling particle initially thrown upward with a velocity $v_y = 20.0$ m/s.

One solution is $t_C = 0$, corresponding to the time the stone starts its motion. The other solution—the one we are after—is $t_C = 4.08$ s. Note that this is twice the value for t_B. The fall from Ⓑ to Ⓒ is the reverse of the rise from Ⓐ to Ⓑ and the stone takes exactly the same time to undergo each part of the motion.

(d) The value for t_C found in part (c) can be inserted into Equation 2.8 to give

$$v_{yC} = v_{yA} + a_y t_C$$
$$= 20.0 \text{ m/s} + (-9.80 \text{ m/s}^2)(4.08 \text{ s})$$
$$= -20.0 \text{ m/s}$$

Note that the velocity of the stone when it arrives back at its original height is equal in magnitude to its initial velocity but opposite in direction. This, along with the equal time intervals noted at the end of part (c), indicates that the motion to this point is symmetric.

(e) From Equation 2.8, the velocity at Ⓓ after 5.00 s is

$$v_{yD} = v_{yA} + a_y t_D$$
$$= 20.0 \text{ m/s} + (-9.80 \text{ m/s}^2)(5.00 \text{ s})$$
$$= -29.0 \text{ m/s}$$

We can use Equation 2.11 to find the position of the stone at $t = 5.00$ s:

$$y_D = y_A + v_{yA}t + \tfrac{1}{2}a_y t_D^2$$
$$= 0 + (20.0 \text{ m/s})(5.00 \text{ s}) + \tfrac{1}{2}(-9.80 \text{ m/s}^2)(5.00 \text{ s})^2$$
$$= -22.5 \text{ m}$$

(f) We use Equation 2.11 again to find the position of the stone at $t = 6.00$ s:

$$y = 0 + (20.0 \text{ m/s})(6.00 \text{ s}) + \tfrac{1}{2}(-9.80 \text{ m/s}^2)(6.00 \text{ s})^2$$
$$= -56.4 \text{ m}$$

The failure of the model lies in the fact that the building is only 50.0 m high, so that the stone cannot be at a position 6.4 m below ground. Our model did not include the fact that the ground exists at $y = -50.0$ m, so the mathematical representation gives us an answer that is not consistent with our expectations in this case.

EXERCISE Find (a) the velocity of the stone just before it hits the ground at Ⓔ and (b) the total time the stone is in the air.

Answer (a) -37.1 m/s (b) 5.83 s

context connection
2.8 • LIFTOFF ACCELERATION

We now have our first opportunity to address a Context in a closing section, as we will do in each remaining chapter. Our present Context is *Mission to Mars,* and our central question is, *How can we undertake a successful transfer of a spacecraft from Earth to Mars?*

The first step on our mission to Mars is to lift the spacecraft from the surface of the Earth and place it in orbit. Before the rocket engines are fired and the launch system begins to move, the velocity and acceleration of the spacecraft are both zero. When the engines fire, an upward force is applied to the launch system, resulting in an upward acceleration of the launch system and the attached spacecraft.

The acceleration of the launch system is not uniform. There are two primary reasons for this. The first is that the force from the rocket engines may not be constant, as adjustments are made to keep the launch system moving according to the preflight plan. The second reason is that the mass of the launch system decreases as the fuel is expended. As we shall learn in Chapter 4, the acceleration of an object is related to its mass. As the launch system loses mass, it becomes easier for the engines to accelerate it, and its acceleration increases.

Early launch systems, such as those for the Apollo, Gemini, and Mercury missions, were rather harsh on astronauts, with launch accelerations as high as 78 m/s^2. This can be expressed as $8g$, or eight times the free-fall acceleration. As a result, astronauts felt pressed into their flight seats with a force many times their weight! The space shuttle launch system is much easier on astronauts, with a maximum acceleration of about $3g$ at the time of solid rocket booster separation.

If you watch a shuttle launch on television, you can measure the time interval between the beginning of motion of the shuttle and the passing of the bottom of the shuttle by the top of the launch tower. From this time, and knowledge of the height of the launch tower, you can calculate the acceleration and speed of the shuttle, as in the following example.

WEB

For information on Mars exploration, visit
cmex=www.arc.nasa.gov

Example 2.11   Space Shuttle Acceleration

During a shuttle launch (Fig. 2.17), the bottom of the launch system clears the launch tower at a time of 6.0 s after liftoff. The height of the launch tower from the initial position of the bottom of the launch system is 75 m. Determine (a) the acceleration, assumed constant, of the launch system during these 6.0 s; and (b), the velocity of the launch system as the bottom clears the tower.

Reasoning   The shuttle is a large object, but we shall model it as a particle. The position of this particle is chosen at the bottom of the launch system because it is the bottom about which we have information. Let us assume that the acceleration is constant so that we can model the bottom of the launch system as a particle under constant acceleration and use the kinematic equations. If the acceleration is not truly

constant during this time interval, what we shall calculate is the *average* acceleration during the interval.

Solution   (a) Using Equation 2.11 and realizing that the initial velocity is zero (because the launch system begins from rest) and defining the initial position of the bottom of the launch system as the origin, we have

$$y_f = y_i + v_{yi}t + \tfrac{1}{2}a_yt^2 = 0 + 0 + \tfrac{1}{2}a_yt^2$$

$$\longrightarrow \quad a_y = \frac{2y_f}{t^2} = \frac{2(75 \text{ m})}{(6.0 \text{ s})^2} = \boxed{4.2 \text{ m/s}^2}$$

(b) We use the kinematic equation for velocity, Equation 2.8:

$$v_{yf} = v_{yi} + a_yt = 0 + (4.2 \text{ m/s}^2)(6.0 \text{ s}) = \boxed{25 \text{ m/s}}$$

This velocity, of course, is in the upward direction.

Notice that the acceleration of the shuttle in Example 2.11 is only about 0.4*g*. As mentioned earlier, as the shuttle moves toward solid rocket booster separation, at about 120 seconds into the launch, the acceleration rises to about 3*g*.

As we investigate two-dimensional motion in the next chapter, we shall look at the next step in our mission after our spacecraft is launched from the surface of the Earth. We shall consider the details of its circular parking orbit around the Earth as it waits for the proper time to be launched to Mars.

SUMMARY

The **average speed** of a particle during some time interval is equal to the ratio of the distance *d* traveled by the particle and the time interval Δt:

$$\overline{v} = \frac{d}{\Delta t} \qquad [2.1]$$

The **average velocity** of a particle moving in one dimension during some time interval is equal to the ratio of the displacement Δx and the time interval Δt:

$$\overline{v}_x \equiv \frac{\Delta x}{\Delta t} \qquad [2.2]$$

The **instantaneous velocity** of a particle is defined as the limit of the ratio $\Delta x/\Delta t$ as Δt approaches zero:

$$v_x \equiv \lim_{\Delta t \to 0} \frac{\Delta x}{\Delta t} = \frac{dx}{dt} \qquad [2.3]$$

The **speed** of a particle is defined as the magnitude of the instantaneous velocity vector.

If the velocity v_x is constant, the preceding equations can be modified and used to solve problems describing the motion of a particle under constant velocity:

$$v_x = \frac{\Delta x}{\Delta t} \qquad [2.4]$$

$$x_f = x_i + v_x t$$

The **average acceleration** of a particle moving in one dimension during some time interval is defined as the ratio of the change in its velocity Δv_x and the time interval Δt:

$$\overline{a}_x \equiv \frac{\Delta v_x}{\Delta t} \qquad [2.5]$$

The **instantaneous acceleration** is equal to the limit of the ratio $\Delta v_x/\Delta t$ as $\Delta t \to 0$. By definition, this equals the derivative of v_x with respect to t, or the time rate of change of the velocity:

$$a_x \equiv \lim_{\Delta t \to 0} \frac{\Delta v_x}{\Delta t} = \frac{dv_x}{dt} \qquad [2.6]$$

The slope of the tangent to the x-versus-t curve at any instant gives the instantaneous velocity of the particle.

The slope of the tangent to the v-versus-t curve gives the instantaneous acceleration of the particle.

The **equations of kinematics** for a particle moving along the x axis with uniform acceleration a_x (constant in magnitude and direction) are

$$v_{xf} = v_{xi} + a_x t \qquad [2.8]$$

$$x_f = x_i + \tfrac{1}{2}(v_{xi} + v_{xf})t \qquad [2.10]$$

$$x_f = x_i + v_{xi}t + \tfrac{1}{2}a_x t^2 \qquad [2.11]$$

$$v_{xf}{}^2 = v_{xi}{}^2 + 2a_x(x_f - x_i) \qquad [2.12]$$

These equations can be used to solve problems describing the motion of a particle under constant acceleration.

An object falling freely experiences an acceleration directed toward the center of the Earth. If air friction is ignored, and if the altitude of the motion is small compared with the Earth's radius, then one can assume that the magnitude of the free-fall acceleration g is constant over the range of motion, where g is equal to 9.80 m/s^2, or 32 ft/s^2. Assuming y to be positive upward, the acceleration is given by $-g$, and the equations of kinematics for an object in free-fall are the same as those already given, with the substitutions $x \to y$ and $a_y \to -g$.

QUESTIONS

1. The average velocity for a particle moving in one dimension has a positive value. Is it possible for the instantaneous velocity of the particle to have been negative at any time in the interval? Suppose the particle started at the origin, $x = 0$. If the average velocity is positive, could the particle ever have been in the $-x$ region?

2. If the average velocity of an object is zero in some time interval, what can you say about the displacement of the object for that interval?

3. Can the instantaneous velocity of an object during a time interval ever be greater in magnitude than the average velocity over the entire interval? Can it ever be less?

4. Consider the following combinations of signs and values for velocity and acceleration of a particle with respect to a one-dimensional x axis:

	Velocity	Acceleration
a.	Positive	Positive
b.	Positive	Negative
c.	Positive	Zero
d.	Negative	Positive
e.	Negative	Negative
f.	Negative	Zero
g.	Zero	Positive
h.	Zero	Negative

Describe what a particle is doing in each case, and give a real-life example for an automobile on an east-west one-dimensional axis, with east considered the positive direction.

5. Can the equations of kinematics (Eqs. 2.8 through 2.12) be used in a situation where the acceleration varies in time? Can they be used when the acceleration is zero?

6. A student at the top of a building of height h throws one ball upward with an initial speed v_i and then throws a second ball downward with the same initial speed. How do the final velocities of the balls compare when they reach the ground?

7. Two cars are moving in the same direction in parallel lanes along a highway. At some instant, the velocity of car A exceeds the velocity of car B. Does this mean that the acceleration of A is greater than that of B? Explain.

8. Car A traveling south from New York to Miami has a speed of 25 m/s. Car B traveling west from New York to Chicago also has a speed of 25 m/s. Are their velocities equal? Explain.

9. You drop a ball from a window on an upper floor of a building. It strikes the ground with velocity v. You now repeat the drop, but you have a friend down on the street, who throws another ball upward at velocity v. Your friend throws the ball upward at exactly the same time that you drop yours from the window. At some location, the balls pass each other. Is this location at the halfway point between window and ground, above this point, or below this point?

10. Galileo experimented with balls rolling down inclined planes in order to reduce the acceleration along the plane and thus reduce the rate of descent of the balls. Suppose the angle that the incline makes with the horizontal is θ. How would you expect the acceleration along the plane to decrease as θ decreases? What specific trigonometric dependence on θ would you expect for the acceleration?

11. A ball rolls in a straight line along the horizontal direction. Using motion diagrams (or multiflash photographs) as in Figure 2.11, describe the velocity and acceleration of

B.C.   By John Hart

By permission of John Hart and Field Enterprises Inc.

Figure Q2.13   See Question 13.

the ball for each of the following situations: (a) The ball moves to the right at a constant speed. (b) The ball moves from right to left and continually slows down. (c) The ball moves from right to left and continually speeds up. (d) The ball moves to the right, first speeding up at a constant rate and then slowing down at a constant rate.

12. A plant growing rapidly next to a building doubles in height each week. At the end of the 25th day, the plant reaches the height of the building. At what time was the plant one fourth the height of the building?

13. A pebble is dropped into a water well, and the splash is heard 16 s later, as illustrated in the cartoon strip. What is the approximate distance from the rim of the well to the water's surface?

14. The motion of the Earth's crustal plates is described by a model referred to as *plate tectonic motion*. Measurements indicate that coastal portions of southern California have northward plate tectonic speeds of 2.5 cm per year. Estimate the time it would take for this motion to carry southern California to Alaska.

15. A heavy object falls from a height h under the influence of gravity. It is released at $t = 0$ and strikes the ground at time t. Ignore air resistance. (a) When the object is at height $0.5h$, is the time earlier than $0.5t$, later than $0.5t$ or equal to $0.5t$? (b) When the time is $0.5t$, is the height of the object larger than $0.5h$, smaller than $0.5h$, or equal to $0.5h$?

PROBLEMS

1, 2, 3 = straightforward, intermediate, challenging   □ = full solution available in the *Student Solutions Manual and Study Guide*

web = solution posted at **http://www.harcourtcollege.com/physics/**   💻 = computer useful in solving problem

🖼 = Interactive Physics   ▪ = paired numerical/symbolic problems   ▨ = life science application

Section 2.1   Average Velocity

1. The position of a pinewood derby car is observed at various times; the results are summarized in the following table. Find the average velocity of the car for (a) the first second, (b) the last 3 s, and (c) the entire period of observation.

t (s)	0	1.0	2.0	3.0	4.0	5.0
x (m)	0	2.3	9.2	20.7	36.8	57.5

2. A motorist drives north for 35.0 min at 85.0 km/h and then stops for 15.0 min. He next continues north, traveling 130 km in 2.00 h. (a) What is his total displacement? (b) What is his average velocity?

3. The displacement versus time for a certain particle moving along the x axis is shown in Figure P2.3. Find the average velocity in the time intervals (a) 0 to 2 s, (b) 0 to 4 s, (c) 2 s to 4 s, (d) 4 s to 7 s, (e) 0 to 8 s.

4. A particle moves according to the equation $x = 10t^2$, where x is in meters and t is in seconds. (a) Find the aver-

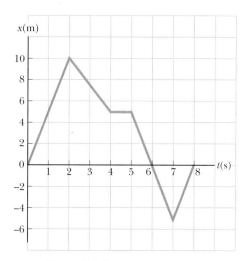

Figure P2.3 Problems 3 and 9.

age velocity for the time interval from 2.00 s to 3.00 s. (b) Find the average velocity for the time interval from 2.00 s to 2.10 s.

5. A person first walks at a constant speed v_1 along a straight line from *A* to *B* and then back along the line from *B* to *A* at a constant speed v_2. What are (a) her average speed over the entire trip and (b) her average velocity over the entire trip?

Section 2.2 Instantaneous Velocity

6. The position of a particle moving along the *x* axis varies in time according to the expression $x = 3t^2$, where *x* is in meters and *t* is in seconds. Evaluate its position (a) at $t = 3.00$ s and (b) at 3.00 s $+ \Delta t$. (c) Evaluate the limit of $\Delta x/\Delta t$ as Δt approaches zero, to find the velocity at $t = 3.00$ s.

7. A position–time graph for a particle moving along the *x*
web axis is shown in Figure P2.7. (a) Find the average velocity

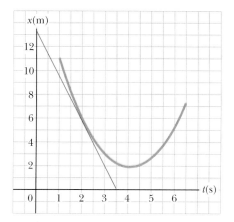

Figure P2.7

in the time interval $t = 1.5$ s to $t = 4.0$ s. (b) Determine the instantaneous velocity at $t = 2.0$ s by measuring the slope of the tangent line shown in the graph. (c) At what value of *t* is the velocity zero?

8. (a) Use the data in Problem 1 to construct a smooth graph of position versus time. (b) By constructing tangents to the $x(t)$ curve, find the instantaneous velocity of the car at several instants. (c) Plot the instantaneous velocity versus time and, from this, determine the average acceleration of the car. (d) What was the initial velocity of the car?

9. Find the instantaneous velocity of the particle described in Figure P2.3 at the following times: (a) $t = 1.0$ s, (b) $t = 3.0$ s, (c) $t = 4.5$ s, and (d) $t = 7.5$ s.

Section 2.3 Analysis Models — The Particle Under Constant Velocity

10. A truck tractor pulls two trailers at a constant speed of 100 km/h. It takes 0.600 s for the big rig to completely pass onto a bridge 400 m long. For what duration of time is all or part of the truck-trailer combination on the bridge?

11. A hare and a tortoise compete in a race over a course 1.00 km long. The tortoise crawls straight and steadily at its maximum speed of 0.200 m/s toward the finish line. The hare runs at its maximum speed of 8.00 m/s toward the goal for 0.800 km and then stops to tease the tortoise. How close to the goal can the hare let the tortoise approach before resuming the race, which the tortoise wins in a photo finish? Assume that, when moving, both animals move steadily at their respective maximum speeds.

Section 2.4 Acceleration

12. A 50.0-g superball traveling at 25.0 m/s bounces off a brick wall and rebounds at 22.0 m/s. A high-speed camera records this event. If the ball is in contact with the wall for 3.50 ms, what is the magnitude of the average acceleration of the ball during this time interval? (*Note:* 1 ms $= 10^{-3}$ s.)

13. Figure P2.13 shows a graph of v_x versus *t* for the motion of a motorcyclist as he starts from rest and moves along the road in a straight line. (a) Find the average acceleration for the time interval $t_0 = 0$ to $t_1 = 6.0$ s. (b) Estimate the time at which the acceleration has its greatest positive value and the value of the acceleration at that instant. (c) When is the acceleration zero? (d) Estimate the maximum negative value of the acceleration and the time at which it occurs.

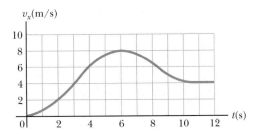

Figure P2.13

14. An object moves along the x axis according to the equation $x(t) = (3.00t^2 - 2.00t + 3.00)$ m. Determine (a) the average speed between $t = 2.00$ s and $t = 3.00$ s, (b) the instantaneous speed at $t = 2.00$ s and at $t = 3.00$ s, (c) the average acceleration between $t = 2.00$ s and $t = 3.00$ s, and (d) the instantaneous acceleration at $t = 2.00$ s and $t = 3.00$ s.

15. A particle moves along the x axis according to the equation $x = 2.00 + 3.00t - t^2$, where x is in meters and t is in seconds. At $t = 3.00$ s, find (a) the position of the particle, (b) its velocity, and (c) its acceleration.

16. A student drives a moped along a straight road as described by the velocity-versus-time graph in Figure P2.16. Sketch this graph in the middle of a sheet of graph paper. (a) Directly above your graph, sketch a graph of position versus time, aligning the time coordinates of the two graphs. (b) Sketch a graph of the acceleration versus time directly below the v_x-t graph, again aligning the time coordinates. On each graph, show the numerical values of x and a_x for all points of inflection. (c) What is the acceleration at $t = 6$ s? (d) Find the position (relative to the starting point) at $t = 6$ s. (e) What is the moped's final position at $t = 9$ s?

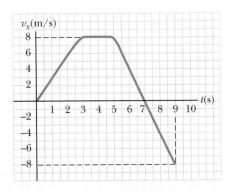

Figure P2.16

17. A particle starts from rest and accelerates as shown in Figure P2.17. Determine: (a) the particle's speed at $t = 10$ s and at $t = 20$ s and (b) the distance traveled in the first 20 s.

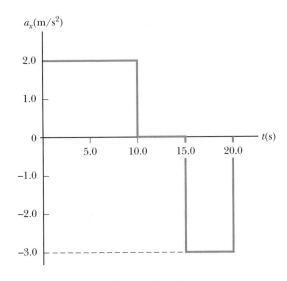

Figure P2.17

Section 2.5 Motion Diagrams

18. Draw motion diagrams for (a) an object moving to the right at constant speed, (b) an object moving to the right and speeding up at a constant rate, (c) an object moving to the right and slowing down at a constant rate, (d) an object moving to the left and speeding up at a constant rate, and (e) an object moving to the left and slowing down at a constant rate. (f) How would your drawings change if the changes in speed were not uniform; that is, if the speed were not changing at a constant rate?

Section 2.6 The Particle Under Constant Acceleration

19. A speedboat moving at 30.0 m/s approaches a no-wake buoy marker 100 m ahead. The pilot slows the boat with a constant acceleration of -3.50 m/s^2 by reducing the throttle. (a) How long does it take the boat to reach the buoy? (b) What is the velocity of the boat when it reaches the buoy?

20. A certain automobile manufacturer claims that its superdeluxe sports car will accelerate from rest to a speed of 42.0 m/s in 8.00 s. Under the (improbable) assumption that the acceleration is constant, (a) determine the acceleration of the car. (b) Find the distance the car travels in the first 8.00 s. (c) What is the speed of the car 10.0 s after it begins its motion if it continues to move with the same acceleration?

21. An object moving with uniform acceleration has a velocity of 12.0 cm/s in the positive x direction when its x coordinate is 3.00 cm. If its x coordinate 2.00 s later is -5.00 cm, what is its acceleration?

22. The minimum distance required to stop a car moving at 35.0 mi/h is 40.0 ft. What is the minimum stopping dis-

Figure P2.29   (*Left*) Col. John Stapp on the rocket sled. (*Courtesy of the U.S. Air Force*)
(*Right*) Col. Stapp's face is contorted by the stress of rapid negative acceleration. (*Photri, Inc.*)

tance for the same car moving at 70.0 mi/h, assuming the same rate of acceleration?

23. The initial velocity of an object is 5.20 m/s. What is its velocity after 2.50 s (a) if it accelerates uniformly at 3.00 m/s^2 and (b) if it accelerates uniformly at -3.00 m/s^2?

24. A particle moves along the x-axis. Its position is given by the equation $x = 2.0 + 3.0t - 4.0t^2$ with x in meters and t in seconds. Determine (a) its position at the instant when it changes direction and (b) its velocity when it returns to the position it had at $t = 0$.

25. A jet plane comes in for a landing with a speed of 100 m/s and can accelerate at a maximum rate of -5.00 m/s^2 as it comes to rest. (a) From the instant the plane touches the runway, what is the minimum time it needs before it can come to rest? (b) Can this plane land at a small tropical island airport where the runway is 0.800 km long?

26. A car is approaching a hill at 30.0 m/s when its engine suddenly fails just at the bottom of the hill. The car moves with a constant acceleration of -2.00 m/s^2 while coasting up the hill. (a) Write equations for the position along the slope and for the velocity as functions of time, taking $x = 0$ at the bottom of the hill, where $v_i = 30.0$ m/s. (b) Determine the maximum distance the car travels up the hill.

27. The driver of a car slams on the brakes when he sees a tree blocking the road. The car slows uniformly with an acceleration of -5.60 m/s^2 for 4.20 s, making straight skid marks 62.4 m long, ending at the tree. With what speed does the car then strike the tree?

28. *Help! One of our equations is missing!* We describe constant-acceleration motion with the variables v_{xf}, x_f, and t, and parameters v_{xi}, a_x, and x_i. Of the equations in Table 2.2, the first does not involve x_f or x_i. The second does not contain a_x, the third omits v_{xf}, and the last leaves out t. So to complete the set there should be an equation *not* involving v_{xi}. Derive it from the others. Use it to solve Problem 27 in one step.

29. For many years the world's land speed record was held by Colonel John P. Stapp, USAF. On March 19, 1954, he rode a rocket-propelled sled that moved down a track at 632 mi/h. He and the sled were safely brought to rest in 1.40 s. (Fig. P2.29) Determine (a) the negative acceleration he experienced and (b) the distance he traveled during this negative acceleration.

30. In a linear accelerator, an electron is accelerated to 1.00% of the speed of light in 40.0 m before it coasts 60.0 m to a target. (a) What is the electron's acceleration during the first 40.0 m? (b) How long does the total flight take?

31. A truck on a straight road starts from rest, accelerating at 2.00 m/s^2 until it reaches a speed of 20.0 m/s. Then the truck travels for 20.0 s at constant speed until the brakes are applied, stopping the truck in a uniform manner in an additional 5.00 s. (a) How long is the truck in motion? (b) What is the average velocity of the truck for the motion described?

32. A ball starts from rest and accelerates at 0.500 m/s^2 while moving down an inclined plane 9.00 m long. When it reaches the bottom, the ball rolls up another plane, where, after moving 15.0 m, it comes to rest. (a) What is the speed of the ball at the bottom of the first plane? (b) How long does it take to roll down the first plane? (c) What is the acceleration along the second plane? (d) What is the ball's speed 8.00 m along the second plane?

33. Speedy Sue, driving at 30.0 m/s, enters a one-lane tunnel. She then observes a slow-moving van 155 m ahead traveling at 5.00 m/s. Sue applies her brakes but can accelerate only at -2.00 m/s^2 because the road is wet. Will there be a collision? If so, determine how far into the tunnel and at what time the collision occurs. If not, determine the distance of closest approach between Sue's car and the van.

Section 2.7 Freely Falling Objects

> *Note:* In all problems in this section, ignore the effects of air resistance.

34. A golf ball is released from rest from the top of a very tall building. Calculate (a) the position and (b) the velocity of the ball after 1.00 s, 2.00 s, and 3.00 s.

35. A student throws a set of keys vertically upward to her
web sorority sister, who is in a window 4.00 m above. The keys are caught 1.50 s later by the sister's outstretched hand. (a) With what initial velocity were the keys thrown? (b) What was the velocity of the keys just before they were caught?

36. A ball is thrown directly downward, with an initial speed of 8.00 m/s from a height of 30.0 m. After what time interval does the ball strike the ground?

37. A baseball is hit so that it travels straight upward after being struck by the bat. A fan observes that it takes 3.00 s for the ball to reach its maximum height. Find (a) its initial velocity and (b) the height reached by the ball.

38. *Every morning at seven o'clock*
There's twenty terriers drilling on the rock.
The boss comes around and he says, "Keep still
And bear down heavy on the cast-iron drill
And drill, ye terriers, drill." And drill, ye terriers, drill.
It's work all day for sugar in your tea
Down beyond the railway. And drill, ye terriers, drill.
The foreman's name was John McAnn.
By God, he was a blamed mean man.
One day a premature blast went off
And a mile in the air went big Jim Goff. And drill . . .
Then when next payday came around
Jim Goff a dollar short was found.
When he asked what for, came this reply:
"You were docked for the time you were up in the sky." And
drill . . .
—American folksong
What was Goff's hourly wage? State the assumptions you make in computing it.

39. A daring ranch hand sitting on a tree limb wishes to drop
web vertically onto a horse galloping under the tree. The constant speed of the horse is 10.0 m/s, and the distance from the limb to the level of the saddle is 3.00 m. (a) What must be the horizontal distance between the saddle and limb when the ranch hand makes his move? (b) How long is he in the air?

40. A boy stands at the edge of a bridge 20.0 m above a river and throws a stone straight down with a speed of 12.0 m/s. He throws another pebble straight upward with the same speed so that it misses the edge of the bridge on the way back down and falls into the river. For each stone find (a) the velocity as it reaches the water and (b) the average velocity while it is in flight.

41. A ball is thrown vertically upward from the ground with an initial speed of 15.0 m/s. (a) How long does it take the ball to reach its maximum altitude? (b) What is its maximum altitude? (c) Determine the velocity and acceleration of the ball at $t = 2.00$ s.

42. The height of a helicopter above the ground is given by $h = 3.00t^3$, where h is in meters and t is in seconds. After 2.00 s, the helicopter pilot lets go of a small mailbag. How long after its release does the mailbag reach the ground?

Section 2.8 Context Connection — Liftoff Acceleration

43. Jules Verne in 1865 suggested sending people to the Moon by firing a space capsule from a 220-m-long cannon with a final velocity of 10.97 km/s. What would have been the unrealistically large acceleration experienced by the space travelers during launch? Compare your answer with the free-fall acceleration 9.80 m/s².

Additional Problems

44. Figure P2.44 represents part of the performance data of a car owned by a proud physics student. (a) Calculate from the graph the total distance traveled. (b) What distance does the car travel between the times $t = 10$ s and $t = 40$ s? (c) Draw a graph of its acceleration versus time between $t = 0$ and $t = 50$ s. (d) Write an equation for x as a function of time for each phase of the motion, represented by (i) $0a$, (ii) ab, (iii) bc. (e) What is the average velocity of the car between $t = 0$ and $t = 50$ s?

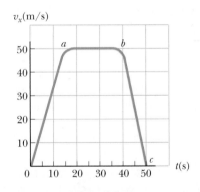

Figure P2.44

45. A test rocket is fired vertically upward from a well. A catapult gives it an initial velocity of 80.0 m/s at ground level. Its engines then fire and give it an upward acceleration of 4.00 m/s² until it reaches an altitude of 1000 m. At that point its engines fail and the rocket goes into free-fall, with an acceleration of -9.80 m/s². (a) How long is the rocket in motion above the ground? (b) What is its maximum altitude? (c) What is its velocity just before it collides with the Earth? (*Hint:* Consider the motion while the engine is operating separate from the free-fall motion.)

46. A motorist drives along a straight road at a constant speed of 15.0 m/s. Just as she passes a parked motorcycle police officer, the officer starts to accelerate at 2.00 m/s^2 to overtake her. Assuming the officer maintains this acceleration, (a) determine the time it takes the police officer to reach the motorist. Find (b) the speed and (c) the total displacement of the officer as he overtakes the motorist.

47. Setting new world records in a 100-m race, Maggie and Judy cross the finish line in a dead heat, both taking 10.2 s. Accelerating uniformly, Maggie took 2.00 s and Judy 3.00 s to attain maximum speed, which they maintained for the rest of the race. (a) What was the acceleration of each sprinter? (b) What were their respective maximum speeds? (c) Which sprinter was ahead at the 6.00-s mark and by how much?

48. A commuter train travels between two downtown stations. Because the stations are only 1.00 km apart, the train never reaches its maximum possible cruising speed. During rush hour the engineer minimizes the travel time t between the two stations by accelerating at $a_1 = 0.100$ m/s^2 for a time t_1 and then immediately braking with acceleration $a_2 = -0.500$ m/s^2 for a time t_2. Find the minimum time of travel t and the time t_1.

49. A hard rubber ball, released at chest height, falls to the pavement and bounces back to nearly the same height. When it is in contact with the pavement, the lower side of the ball is temporarily flattened. Suppose that the maximum depth of the dent is on the order of 1 cm. Compute an order-of-magnitude estimate for the maximum acceleration of the ball while it is in contact with the pavement. State your assumptions, the quantities you estimate, and the values you estimate for them.

50. At NASA's John H. Glenn research center in Cleveland, Ohio, free-fall research is performed by dropping experiment packages from the top of an evacuated shaft 145 m high. Free-fall imitates the so-called microgravity environment of a satellite in orbit. (a) What is the maximum time for free-fall if a package falls the entire 145 m? (b) Actual NASA specifications allow for a 5.18-s free-fall time. How far do the packages drop and (c) what is their speed at 5.18 s? (d) What constant acceleration is required to stop a package in the distance remaining in the shaft after its 5.18-s fall?

51. An inquisitive physics student and mountain climber climbs a 50.0-m cliff that overhangs a calm pool of water. He throws two stones vertically downward, 1.00 s apart, and observes that they cause a single splash. The first stone has an initial speed of 2.00 m/s. (a) How long after release of the first stone do the two stones hit the water? (b) What was the initial velocity of the second stone? (c) What is the velocity of each stone at the instant the two hit the water?

52. A rock is dropped from rest into a well. (a) If the sound of the splash is heard 2.40 s later, how far below the top of the well is the surface of the water? The speed of sound in air (at the ambient temperature) is 336 m/s. (b) If the travel time for the sound is ignored, what percentage error is introduced when the depth of the well is calculated?

53. To protect his food from hungry bears, a boy scout raises his food pack with a rope that is thrown over a tree limb at height h above his hands. He walks away from the vertical rope with constant velocity v_{boy}, holding the free end of the rope in his hands (Fig. P2.53). (a) Show that the speed v of the food pack is given by $x(x^2 + h^2)^{-1/2} v_{boy}$, where x is the distance he has walked away from the vertical rope. (b) Show that the acceleration a of the food pack is $h^2(x^2 + h^2)^{-3/2} v_{boy}^2$. (c) What values do the acceleration and velocity v have shortly after he leaves the point under the pack ($x = 0$)? (d) What values do the pack's velocity and acceleration approach as the distance x continues to increase?

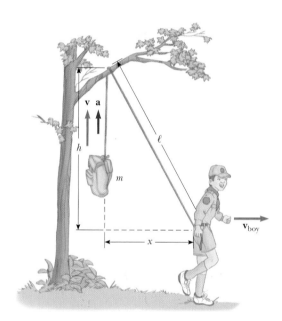

Figure P2.53 Problems 53 and 54.

54. In Problem 53, let the height h equal 6.00 m and the speed v_{boy} equal 2.00 m/s. Assume that the food pack starts from rest. (a) Tabulate and graph the speed–time graph. (b) Tabulate and graph the acceleration–time graph. (Let the range of time be from 0 s to 5.00 s and the time intervals be 0.500 s.)

55. Liz rushes down onto a subway platform to find her train already departing. She stops and watches the cars go by. Each car is 8.60 m long. The first moves past her in 1.50 s and the second in 1.10 s. Find the constant acceleration of the train.

56. Astronauts on a distant planet toss a rock into the air. With the aid of a camera that takes pictures at a steady rate, they record the height of the rock as a function of time as given in the following table. (a) Find the average velocity of the rock in the time interval between each measurement and the next. (b) Using these average velocities to

TABLE P2.56		Height of a Rock versus Time	
Time (s)	Height (m)	Time (s)	Height (m)
0.00	5.00	2.75	7.62
0.25	5.75	3.00	7.25
0.50	6.40	3.25	6.77
0.75	6.94	3.50	6.20
1.00	7.38	3.75	5.52
1.25	7.72	4.00	4.73
1.50	7.96	4.25	3.85
1.75	8.10	4.50	2.86
2.00	8.13	4.75	1.77
2.25	8.07	5.00	0.58
2.50	7.90		

approximate instantaneous velocities at the midpoints of the time intervals, make a graph of velocity as a function of time. Does the rock move with constant acceleration? If so, plot a straight line of best fit on the graph and calculate its slope to find the acceleration.

57. Two objects, *A* and *B*, are connected by a rigid rod that has a length *L*. The objects slide along perpendicular guide rails, as shown in Figure P2.57. If *A* slides to the left with a constant speed *v*, find the speed of *B* when $\alpha = 60.0°$.

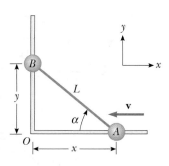

Figure P2.57

ANSWERS TO QUICK QUIZZES

2.1 A member of the highway patrol is interested in your instantaneous speed. Regardless of your speeds at all other times, if your instantaneous speed at the instant that it is measured is higher than the speed limit, you may receive a speeding ticket.

2.2 If the particle moves along a line without changing direction, the displacement and distance over any time interval are the same. As a result, the magnitude of the average velocity and the average speed are the same. If the particle reverses direction, however, the displacement would be less than the distance. In turn, the magnitude of the average velocity would be smaller than the average speed.

2.3 Graph (a) has a constant slope, indicating a constant acceleration; this is represented by graph (e). Graph (b) represents a speed that is increasing constantly but not at a uniform rate. Thus, the acceleration must be increasing, and the graph that best indicates this is (d). Graph (c) depicts a velocity that first increases at a constant rate, indicating constant acceleration. Then the velocity stops increasing and becomes constant, indicating zero acceleration. The best match to this is graph (f).

2.4 If the car is slowing down, the direction of its acceleration is opposite to its direction of motion. Thus, the acceleration and, therefore, the force are directed to the west.

2.5 No. For an accelerating particle to stop at all, the velocity and acceleration must have opposite signs, so that the speed is decreasing. Given that this is the case, the particle will eventually come to rest. If the acceleration remains constant, however, the particle must begin to move again, opposite to the direction of its original motion. If the particle comes to rest and then stays at rest, the acceleration has become zero at the moment the motion stops. This is the case for a braking car—the acceleration is negative and goes to zero as the car comes to rest.

2.6 (a) As it travels upward, its speed decreases by 9.80 m/s during each second of its motion. When it reaches the peak of its motion, its speed becomes zero. As the ball moves downward, its speed increases by 9.80 m/s each second. (b) The acceleration of the ball remains constant while it is in the air. The magnitude of its acceleration is the free-fall acceleration, $g = 9.80$ m/s².

A welder cuts holes through a
heavy metal construction beam
with a hot torch. The sparks generated in the process all follow a
path that is described by a curve
called a *parabola.*

(©The Telegraph Colour Library/FPG)

Motion in Two Dimensions

n this chapter we shall study the kinematics of an object that can be modeled
as a particle moving in a plane. This is two-dimensional motion. Some common examples of motion in a plane are the motions of satellites in orbit
around the Earth, projectiles, such as a thrown baseball, and the motion of
electrons in uniform electric fields. We shall also study a particle in uniform circular motion and discuss various aspects of particles moving in curved paths.

3.1 • THE POSITION, VELOCITY, AND ACCELERATION VECTORS

In Chapter 2 we found that the motion of a particle moving along a straight line is
completely specified if its position is known as a function of time. Now let us extend this idea to motion in the *xy* plane. We will find equations for position and velocity that are the same as those in Chapter 2, except for their vector nature.

We begin by describing the position of a particle with a **position vector r,**
drawn from the origin of a reference frame to the location of the particle in the *xy*

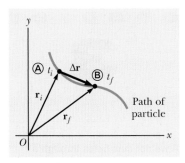

Figure 3.1

A particle moving in the *xy* plane is located with the position vector **r** drawn from the origin to the particle. The displacement of the particle as it moves from Ⓐ to Ⓑ in the time interval $\Delta t = t_f - t_i$ is equal to the vector $\Delta \mathbf{r} = \mathbf{r}_f - \mathbf{r}_i$.

plane, as in Figure 3.1. At time t_i, the particle is at the point Ⓐ, and at some later time t_f, the particle is at Ⓑ, where the subscripts *i* and *f* refer to initial and final values. As the particle moves from Ⓐ to Ⓑ in the time interval $\Delta t = t_f - t_i$, the position vector changes from $\mathbf{r}_i$ to $\mathbf{r}_f$. As we learned in Chapter 2, the displacement of a particle is the difference between its final position and initial position:

$$\Delta \mathbf{r} \equiv \mathbf{r}_f - \mathbf{r}_i \qquad [3.1]$$

The direction of $\Delta \mathbf{r}$ is indicated in Figure 3.1.

The **average velocity** $\bar{\mathbf{v}}$ of the particle during the time interval Δt is defined as the ratio of the displacement to the time interval:

$$\bar{\mathbf{v}} \equiv \frac{\Delta \mathbf{r}}{\Delta t} \qquad [3.2]$$

Because displacement is a vector quantity and the time interval is a scalar quantity, we conclude that the average velocity is a *vector* quantity directed along $\Delta \mathbf{r}$. The average velocity between points Ⓐ and Ⓑ is *independent of the path* between the two points. This is because the average velocity is proportional to the displacement, which in turn depends only on the initial and final position vectors and not on the path taken between those two points. As with one-dimensional motion, if a particle starts its motion at some point and returns to this point via any path, its average velocity is zero for this trip because its displacement is zero.

Consider again the motion of a particle between two points in the *xy* plane, as shown in Figure 3.2. As the time intervals over which we observe the motion become smaller and smaller, the direction of the displacement approaches that of the line tangent to the path at the point Ⓐ.

The **instantaneous velocity v** is defined as the limit of the average velocity $\Delta \mathbf{r}/\Delta t$ as Δt approaches zero:

$$\mathbf{v} \equiv \lim_{\Delta t \to 0} \frac{\Delta \mathbf{r}}{\Delta t} = \frac{d\mathbf{r}}{dt} \qquad [3.3]$$

- *Definition of instantaneous velocity*

That is, the instantaneous velocity equals the derivative of the position vector with respect to time. The direction of the instantaneous velocity vector at any point in a particle's path is along a line that is tangent to the path at that point and in the direction of motion. The magnitude of the instantaneous velocity is called the *speed*.

As a particle moves from point Ⓐ to point Ⓑ along some path as in Figure 3.3, its instantaneous velocity changes from $\mathbf{v}_i$ at time t_i to $\mathbf{v}_f$ at time t_f.

The **average acceleration** $\bar{\mathbf{a}}$ of a particle over a time interval is defined as the ratio of the change in the instantaneous velocity $\Delta \mathbf{v}$ to the time interval Δt:

$$\bar{\mathbf{a}} \equiv \frac{\mathbf{v}_f - \mathbf{v}_i}{t_f - t_i} = \frac{\Delta \mathbf{v}}{\Delta t} \qquad [3.4]$$

PITFALL PREVENTION 3.1

Vector addition

The vector addition that was discussed in Chapter 1 involved displacement vectors. Because we are familiar with movements through space due to our everyday experience, the addition of displacement vectors can be understood easily. The notion of vector addition can be applied to *any* type of vector quantity. Figure 3.3, for example, shows the addition of velocity vectors using the tip-to-tail approach.

Because the average acceleration is the ratio of a vector quantity $\Delta \mathbf{v}$ and a scalar quantity Δt, we conclude that $\bar{\mathbf{a}}$ is a vector quantity directed along $\Delta \mathbf{v}$. As indicated in Figure 3.3, the direction of $\Delta \mathbf{v}$ is found by adding the vector $-\mathbf{v}_i$ (the negative of $\mathbf{v}_i$) to the vector $\mathbf{v}_f$, because by definition $\Delta \mathbf{v} \equiv \mathbf{v}_f - \mathbf{v}_i$.

The **instantaneous acceleration a** is defined as the limiting value of the ratio $\Delta\mathbf{v}/\Delta t$ as Δt approaches zero:

$$\mathbf{a} \equiv \lim_{\Delta t \to 0} \frac{\Delta\mathbf{v}}{\Delta t} = \frac{d\mathbf{v}}{dt} \qquad [3.5]$$

• *Definition of instantaneous acceleration*

That is, the instantaneous acceleration equals the derivative of the velocity vector with respect to time.

It is important to recognize that various changes can occur that represent a particle undergoing an acceleration. First, the magnitude of the velocity vector (the speed) may change with time as in straight-line (one-dimensional) motion. Second, the direction of the velocity vector may change with time as its magnitude remains constant. Finally, both the magnitude and the direction of the velocity vector may change.

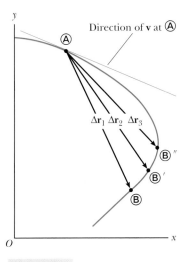

Figure 3.2

As a particle moves between two points, its average velocity is in the direction of the displacement vector $\Delta\mathbf{r}$. As the end point of the path is moved from Ⓑ to Ⓑ′ to Ⓑ″, the respective displacements and corresponding time intervals become smaller and smaller. In the limit that the end point approaches Ⓐ, Δt approaches zero and the direction of $\Delta\mathbf{r}$ approaches that of the line tangent to the curve at Ⓐ. By definition, the instantaneous velocity at Ⓐ is in the direction of this tangent line.

Quick Quiz 3.1

(a) Can an object accelerate if its speed is constant? (b) Can an object accelerate if its velocity is constant?

Quick Quiz 3.2

(a) Can a particle have a constant velocity and a varying speed? (b) Can a particle have a constant speed and a varying velocity? Give examples of a positive answer and explain a negative answer.

THINKING PHYSICS 3.1

The gas pedal in an automobile is commonly called the *accelerator*. Are there any other controls that could also be considered accelerators?

Reasoning   The gas pedal is called the accelerator because the common usage of the word *acceleration* refers to an *increase in speed*. The scientific definition, however, is that acceleration occurs *whenever the velocity changes in any way*. Thus, the *brake pedal* can also be considered an accelerator because it causes the car to slow down. The *steering wheel* is also an accelerator because it changes the direction of the velocity vector.

Figure 3.3

A particle moves from position Ⓐ to position Ⓑ. Its velocity vector changes from $\mathbf{v}_i$ at time t_i to $\mathbf{v}_f$ at time t_f. The vector addition diagrams at the upper right show two ways of determining the vector $\Delta\mathbf{v}$ from the initial and final velocities.

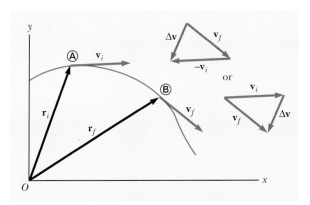

3.2 • TWO-DIMENSIONAL MOTION WITH CONSTANT ACCELERATION

Let us consider two-dimensional motion during which the magnitude and direction of the acceleration remain unchanged. In this situation, we shall investigate motion as a two-dimensional version of the analysis in Section 2.6.

The motion of a particle can be determined if its position vector $\mathbf{r}$ is known at all times. The position vector for a particle moving in the xy plane can be written

$$\mathbf{r} = x\mathbf{i} + y\mathbf{j} \qquad [3.6]$$

where x, y, and $\mathbf{r}$ change with time as the particle moves. If the position vector is known, the velocity of the particle can be obtained from Equations 3.3 and 3.6:

$$\mathbf{v} = \frac{d\mathbf{r}}{dt} = \frac{dx}{dt}\mathbf{i} + \frac{dy}{dt}\mathbf{j} = v_x\mathbf{i} + v_y\mathbf{j} \qquad [3.7]$$

Because we are assuming that $\mathbf{a}$ is constant in this discussion, its components a_x and a_y are also constants. Therefore, we can apply the equations of kinematics to the x and y components of the velocity vector separately. Substituting $v_x = v_{xf} = v_{xi} + a_x t$ and $v_y = v_{yf} = v_{yi} + a_y t$ into Equation 3.7 gives

$$\mathbf{v}_f = (v_{xi} + a_x t)\mathbf{i} + (v_{yi} + a_y t)\mathbf{j}$$
$$= (v_{xi}\mathbf{i} + v_{yi}\mathbf{j}) + (a_x\mathbf{i} + a_y\mathbf{j})t$$

$$\mathbf{v}_f = \mathbf{v}_i + \mathbf{a}t \qquad [3.8]$$

- *Velocity vector as a function of time for constant acceleration*

This result states that the velocity $\mathbf{v}_f$ of a particle at some time t equals the vector sum of its initial velocity $\mathbf{v}_i$ and the additional velocity $\mathbf{a}t$ acquired at time t as a result of its constant acceleration. This is the same as Equation 2.8, except for its vector nature.

Similarly, from Equation 2.11 we know that the x and y coordinates of a particle moving with constant acceleration are

$$x_f = x_i + v_{xi}t + \tfrac{1}{2}a_x t^2 \qquad \text{and} \qquad y_f = y_i + v_{yi}t + \tfrac{1}{2}a_y t^2$$

Substituting these expressions into Equation 3.6 gives

$$\mathbf{r}_f = (x_i + v_{xi}t + \tfrac{1}{2}a_x t^2)\mathbf{i} + (y_i + v_{yi}t + \tfrac{1}{2}a_y t^2)\mathbf{j}$$
$$= (x_i\mathbf{i} + y_i\mathbf{j}) + (v_{xi}\mathbf{i} + v_{yi}\mathbf{j})t + \tfrac{1}{2}(a_x\mathbf{i} + a_y\mathbf{j})t^2$$

$$\mathbf{r}_f = \mathbf{r}_i + \mathbf{v}_i t + \tfrac{1}{2}\mathbf{a}t^2 \qquad [3.9]$$

- *Position vector as a function of time for constant acceleration*

This equation implies that the final position vector $\mathbf{r}_f$ is the vector sum of the initial position vector $\mathbf{r}_i$ plus a displacement $\mathbf{v}_i t$, arising from the initial velocity of the particle, and a displacement $\tfrac{1}{2}\mathbf{a}t^2$, resulting from the uniform acceleration of the particle. It is the same as Equation 2.11, except for its vector nature.

Pictorial representations of Equations 3.8 and 3.9 are shown in Figures 3.4a and 3.4b. Note from Figure 3.4b that $\mathbf{r}_f$ is generally not along the direction of $\mathbf{v}_i$ or $\mathbf{a}$, because the relationship between these quantities is a vector expression. For the same reason, from Figure 3.4a we see that $\mathbf{v}_f$ is generally not along the direction of $\mathbf{v}_i$ or $\mathbf{a}$. Finally, if we compare the two figures, we see that $\mathbf{v}_f$ and $\mathbf{r}_f$ are not in the same direction.

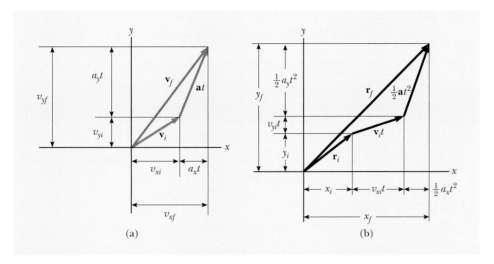

(a)　　　　　(b)

Figure 3.4

Vector representations and components of (a) the velocity and (b) the position of a particle under constant acceleration **a**.

Because Equations 3.8 and 3.9 are *vector* expressions, we may also write their x and y component equations:

$$\mathbf{v}_f = \mathbf{v}_i + \mathbf{a}t \longrightarrow \begin{cases} v_{xf} = v_{xi} + a_x t \\ v_{yf} = v_{yi} + a_y t \end{cases}$$

$$\mathbf{r}_f = \mathbf{r}_i + \mathbf{v}_i t + \tfrac{1}{2}\mathbf{a}t^2 \longrightarrow \begin{cases} x_f = x_i + v_{xi} t + \tfrac{1}{2}a_x t^2 \\ y_f = y_i + v_{yi} t + \tfrac{1}{2}a_y t^2 \end{cases}$$

These components are illustrated in Figure 3.4. In other words, **two-dimensional motion having constant acceleration is equivalent to two *independent* motions in the x and y directions having constant accelerations a_x and a_y.** Motion in the x direction does not affect motion in the y direction, and vice-versa. Thus, there is no new model for a particle under two-dimensional constant acceleration—the appropriate model is just the one-dimensional particle under constant acceleration applied twice, in the x and y directions separately!

Example 3.1　Motion in a Plane

A particle moves through the origin of an xy coordinate system at $t = 0$ with initial velocity $\mathbf{v}_i = (20\mathbf{i} - 15\mathbf{j})$ m/s. The particle moves in the xy plane with an acceleration $\mathbf{a} = 4.0\mathbf{i}$ m/s^2. (a) Determine the components of velocity as a function of time and the total velocity vector at any time.

Reasoning　Establish the mental representation—think about what the particle is doing. From the given information, we see that the particle starts off moving to the right and downward and accelerates only toward the right—what will the particle do under these conditions? It may help if you draw a pictorial representation. Because the ac-

celeration is only in the x direction, the moving particle can be modeled as one under constant acceleration in the x direction, and one under constant velocity in the y direction.

Solution　(a) With $v_{xi} = 20$ m/s and $a_x = 4.0$ m/s^2, the equations of kinematics give us for the x direction,

$$v_{xf} = v_{xi} + a_x t = (20 + 4.0t)$$

Also, with $v_{yi} = -15$ m/s and $a_y = 0$,

$$v_{yf} = v_{yi} = -15 \text{ m/s}$$

Therefore, using these results and noting that the velocity vector **v** has two components, we find

$$\mathbf{v}_f = v_{xf}\mathbf{i} + v_{yf}\mathbf{j} = [(20 + 4.0t)\mathbf{i} - 15\mathbf{j}]$$

Note that only the *x* component varies in time, reflecting the fact that acceleration occurs only in the *x* direction.

(b) Calculate the velocity and speed of the particle at *t* = 5.0 s.

Solution   Taking *t* = 5.0 s, the velocity expression from (a) gives

$$\mathbf{v}_f = \{[20 + 4(5.0)]\mathbf{i} - 15\mathbf{j}\} \text{ m/s} = (40\mathbf{i} - 15\mathbf{j}) \text{ m/s}$$

That is, at *t* = 5.0 s, v_{xf} = 40 m/s and v_{yf} = − 15 m/s. Knowing these two components for this two-dimensional motion, we know the numerical value of the velocity vector. To deter-mine the angle θ that $\mathbf{v}_f$ makes with the *x* axis, we use the fact that $\tan \theta = v_{yf}/v_{xf}$, or

$$\theta = \tan^{-1}\left(\frac{v_{yf}}{v_{xf}}\right) = \tan^{-1}\left(\frac{-15 \text{ m/s}}{40 \text{ m/s}}\right) = -21°$$

The speed is the magnitude of $\mathbf{v}_f$:

$$v_f = |\mathbf{v}_f| = \sqrt{v_{xf}^2 + v_{yf}^2} = \sqrt{(40)^2 + (-15)^2} \text{ m/s} = 43 \text{ m/s}$$

In examining our result, we find that $v_f > v_i$. Does this make sense to you? (Remember step 5—**Think**—in the general problem-solving strategy in Chapter 1.)

EXERCISE   Determine the *x* and *y* coordinates of the particle at any time *t* and the position vector at this time.

Answer $x_f = 20t + 2.0t^2$ $y_f = -15t$
$\mathbf{r}_f = (20t + 2.0t^2)\mathbf{i} - 15t\mathbf{j}$

3.3 • PROJECTILE MOTION

See the *Core Concepts in Physics CD-ROM*, Screen 3.5

• *Assumptions of projectile motion*

Anyone who has observed a baseball in motion (or, for that matter, any object thrown into the air) has observed projectile motion. The ball moves in a curved path when thrown at some angle with respect to the Earth's surface. This very common form of motion is surprisingly simple to analyze if the following two assumptions are made in building a model for these types of problems: (1) the free-fall acceleration *g* is constant over the range of motion and is directed downward,* and (2) the effect of air resistance is negligible.† With these assumptions, the path of a projectile, called its *trajectory*, is *always* a parabola. **We shall use a simplification model based on these assumptions throughout this chapter.**

If we choose our reference frame such that the *y* direction is vertical and positive upward, then $a_y = -g$ (as in one-dimensional free-fall) and $a_x = 0$ (because the only possible horizontal acceleration is due to air resistance, and it is ignored). Furthermore, let us assume that at *t* = 0, the projectile leaves the origin (point Ⓐ, $x_i = y_i = 0$) with speed v_i, as in Figure 3.5. If the vector $\mathbf{v}_i$ makes an angle θ_i with the horizontal, we can identify a right triangle in the diagram as a geometric model, and from the definitions of the cosine and sine functions we have

$$\cos \theta_i = \frac{v_{xi}}{v_i} \quad \text{and} \quad \sin \theta_i = \frac{v_{yi}}{v_i}$$

---

* This approximation is reasonable as long as the range of motion is small compared with the radius of the Earth (6.4×10^6 m). In effect, this approximation is equivalent to assuming that the Earth is flat within the range of motion considered, and that the maximum height of the object is small compared to the radius of the Earth.

† This approximation is often *not* justified, especially at high velocities. In addition, the spin of a projectile, such as a baseball, can give rise to some very interesting effects associated with aerodynamic forces (for example, a curve ball thrown by a pitcher).

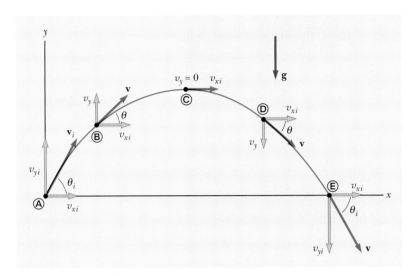

Figure 3.5

The parabolic path of a projectile that leaves the origin (point Ⓐ) with a velocity $\mathbf{v}_i$. The velocity vector $\mathbf{v}$ changes with time in both magnitude and direction. The change in the velocity vector is the result of acceleration in the negative y direction. The x component of velocity remains constant in time because no acceleration occurs in the horizontal direction. The y component of velocity is zero at the peak of the path (point Ⓒ).

Therefore, the initial x and y components of velocity are

$$v_{xi} = v_i \cos \theta_i \qquad \text{and} \qquad v_{yi} = v_i \sin \theta_i$$

Substituting these expressions into Equations 3.8 and 3.9 with $a_x = 0$ and $a_y = -g$ gives the velocity components and position coordinates for the projectile at any time t:

$$v_{xf} = v_{xi} = v_i \cos \theta_i = \text{constant} \qquad [3.10] \qquad \bullet \quad \textit{Horizontal velocity component}$$

$$v_{yf} = v_{yi} - gt = v_i \sin \theta_i - gt \qquad [3.11] \qquad \bullet \quad \textit{Vertical velocity component}$$

$$x_f = x_i + v_{xi}t = (v_i \cos \theta_i)\,t \qquad [3.12] \qquad \bullet \quad \textit{Horizontal position component}$$

$$y_f = y_i + v_{yi}t - \tfrac{1}{2}gt^2 = (v_i \sin \theta_i)\,t - \tfrac{1}{2}gt^2 \qquad [3.13] \qquad \bullet \quad \textit{Vertical position component}$$

From Equation 3.10 we see that v_{xf} remains constant in time and is equal to v_{xi}; there is no horizontal component of acceleration. Thus, we model the horizontal motion as a particle moving with constant velocity. For the y motion, note that the equations for v_{yf} and y_f are similar to Equations 2.8 and 2.11 for freely falling objects. Thus, we can apply the model of a particle moving with constant acceleration to the y component. In fact, *all* of the equations of kinematics developed in Chapter 2 are applicable to projectile motion.

If we solve for t in Equation 3.12 and substitute this expression for t into Equation 3.13, we find that

$$y_f = (\tan \theta_i)\,x_f - \left(\frac{g}{2v_i^2 \cos^2 \theta_i} \right) x_f^2 \qquad [3.14]$$

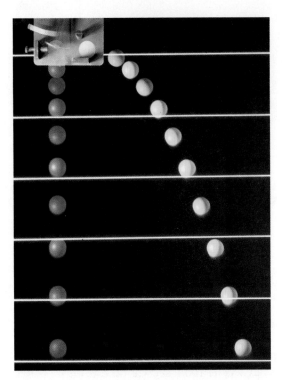

This strobe photograph of two balls released simultaneously illustrates both free fall (red ball) and projectile motion (yellow ball). The yellow ball was projected horizontally, while the red ball was released from rest. Can you explain why both balls reach the floor simultaneously? *(Richard Megna, Fundamental Photographs)*

which is valid for angles in the range $0 < \theta_i < \pi/2$. This expression is of the form $y = ax - bx^2$, which is the equation of a parabola that passes through the origin. Thus, we have proven that the trajectory of a projectile can be geometrically modeled as a parabola. The trajectory is *completely* specified if v_i and θ_i are known.

The vector expression for the position of the projectile as a function of time follows directly from Equation 3.9, with $\mathbf{a} = \mathbf{g}$:

• *Vector expression for the position of a projectile*

$$\mathbf{r}_f = \mathbf{r}_i + \mathbf{v}_i t + \tfrac{1}{2}\mathbf{g}t^2$$

This equation gives the same information as the combination of Equations 3.12 and 3.13 and is plotted in Figure 3.6. Note that this expression for $\mathbf{r}_f$ is consistent with Equation 3.13, because the expression for $\mathbf{r}_f$ is a vector equation and $\mathbf{a} = \mathbf{g} = -g\mathbf{j}$ when the upward direction is taken to be positive.

The position of a particle can be considered the superposition of its original position $\mathbf{r}_i$, the term $\mathbf{v}_i t$, which would be the displacement if no acceleration were present, and the term $\tfrac{1}{2}\mathbf{g}t^2$, which arises from the acceleration due to gravity. In other words, if no gravitational acceleration occurred, the particle would continue to move along a straight path in the direction of $\mathbf{v}_i$.

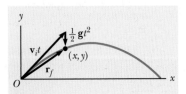

Figure 3.6

The position vector $\mathbf{r}$ of a projectile whose initial velocity at the origin is $\mathbf{v}_i$. The vector $\mathbf{v}_i t$ would be the position vector of the projectile if gravity were absent. To this is added the vector $\tfrac{1}{2}\mathbf{g}t^2$ due to the downward gravitational acceleration of the projectile.

Horizontal Range and Maximum Height of a Projectile

Let us assume that a projectile is launched over flat ground from the origin at $t = 0$ with a positive v_y component, as in Figure 3.7. There are two special points that are interesting to analyze: the peak point Ⓐ, which has cartesian coordinates ($R/2$, h), and the landing point ⓑ, having coordinates (R, 0). The distance R is called the *horizontal range* of the projectile, and h is its *maximum height*. Because of the symmetry of the trajectory, the projectile is at the maximum height h when its x position is half of the range R. Let us find h and R in terms of v_i, θ_i, and g.

We can determine h by noting that at the peak, $v_{yA} = 0$. Therefore, Equation 3.11 can be used to determine the time t_1 it takes to reach the peak:

$$t_1 = \frac{v_i \sin \theta_i}{g}$$

Substituting this expression for t_1 into Equation 3.13 and replacing y_f with h gives h in terms of v_i and θ_i:

$$h = (v_i \sin \theta_i) \frac{v_i \sin \theta_i}{g} - \frac{1}{2}g \left(\frac{v_i \sin \theta_i}{g} \right)^2$$

$$h = \frac{v_i^2 \sin^2 \theta_i}{2g} \qquad \qquad \textbf{[3.15]}$$

Notice from the mathematical representation how you could increase the peak height h: You could launch the projectile with a larger initial velocity, at a higher angle, or at a location with lower free-fall acceleration, such as on the Moon. Is this consistent with your mental representation of this situation?

The range R is the horizontal distance traveled in twice the time it takes to reach the peak, that is, in a time $2t_1$. (This can be seen by setting $y_f = 0$ in Equation 3.13 and solving the quadratic for t. One solution of this quadratic is $t = 0$, and the second is $t = 2t_1$.) Using Equation 3.12 and noting that $x_f = R$ at $t = 2t_1$, we find that

$$R = (v_i \cos \theta_i) 2t_1 = (v_i \cos \theta_i) \frac{2v_i \sin \theta_i}{g} = \frac{2v_i^2 \sin \theta_i \cos \theta_i}{g}$$

Because $\sin 2\theta = 2 \sin \theta \cos \theta$, R can be written in the more compact form

$$R = \frac{v_i^2 \sin 2\theta_i}{g} \qquad \qquad \textbf{[3.16]}$$

Notice from the mathematical expression how you could increase the range R: You could launch the projectile with a larger initial velocity or at a location with lower free-fall acceleration, such as on the Moon. Is this consistent with your mental representation of this situation?

The range also depends on the angle of the initial velocity vector. The maximum possible value of R from Equation 3.16 is given by $R_{max} = v_i^2/g$. This result follows from the fact that the maximum value of $\sin 2\theta_i$ is unity, which occurs when $2\theta_i = 90°$. Therefore, R is a maximum when $\theta_i = 45°$.

Figure 3.8 illustrates various trajectories for a projectile of a given initial speed. As you can see, the range is a maximum for $\theta_i = 45°$. In addition, for any θ_i other than $45°$, a point with coordinates (R, 0) can be reached by using either one of two complementary values of θ_i, such as $75°$ and $15°$. Of course, the maximum height and the time of flight will be different for these two values of θ_i.

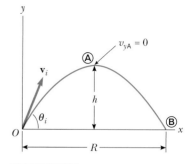

Figure 3.7

A projectile launched from the origin at $t = 0$ with an initial velocity $\mathbf{v}_i$. The maximum height of the projectile is h, and its horizontal range is R. At ⓐ, the peak of the trajectory, the projectile has coordinates $\left(\dfrac{R}{2}, h \right)$.

PITFALL PREVENTION 3.2
The height and range equations

Keep in mind that Equations 3.15 and 3.16 are useful for calculating h and R only for a symmetric path, as shown in Figure 3.7. If the path is not symmetric, *do not use these equations*. The general expressions given by Equations 3.10 through 3.13 are the *more important* results because they give the coordinates and velocity components of the projectile at *any* time t for *any* trajectory.

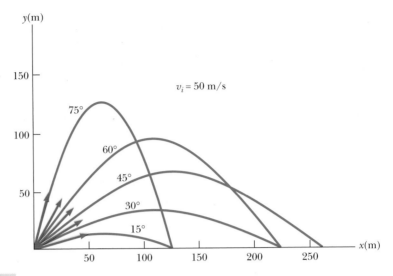

Figure 3.8

A projectile launched from the origin with an initial speed of 50 m/s at various angles of projection. Note that complementary values of θ_i will result in the same value of R.

PITFALL PREVENTION 3.3

Acceleration at the highest point

As discussed in Pitfall Prevention 2.10 in Chapter 2, many people have trouble with the very topmost point of the trajectory of a projectile, claiming that the acceleration of the ball at the highest point is zero. This interpretation arises from a confusion between zero vertical velocity and zero acceleration. Even though the ball has momentarily come to rest at the highest point (vertically—in two dimensions, it is still moving horizontally), it is still accelerating because the gravitational force is still acting. If the ball were to experience zero acceleration at the highest point, then the velocity would not change—the ball would continue to move horizontally once it reached the highest point! This doesn't happen because the acceleration is indeed not equal to zero.

PROBLEM-SOLVING STRATEGY | **Projectile Motion**

We suggest that you use the following approach when solving projectile motion problems:

1. Select a coordinate system.
2. Resolve the initial velocity vector into x and y components.
3. Treat the horizontal motion and the vertical motion independently.
4. Analyze the horizontal motion of the projectile as a particle under constant velocity.
5. Analyze the vertical motion of the projectile as a particle under constant acceleration.

Quick Quiz 3.3

Suppose you run at constant speed and wish to throw a ball so that you can catch it as it comes back down. In what direction should you throw the ball?

Quick Quiz 3.4

As a projectile moves in its parabolic path, is there any point along its path where the velocity and acceleration vectors are (a) perpendicular to each other? (b) parallel to each other?

THINKING PHYSICS 3.2

A home run is hit in a baseball game. The ball is hit from home plate into the stands, along a parabolic path. What is the acceleration of the ball (a) while it is rising, (b) at the highest point of the trajectory, and (c) while it is descending after reaching the highest point? Ignore air resistance.

Reasoning The answers to all three parts are the same—the acceleration is that due to gravity, $a_y = -9.80$ m/s^2, because the gravitational force is pulling downward on the ball during the entire motion. During the rising part of the trajectory, the downward acceleration results in the decreasing positive values of the vertical component of the velocity of the ball. During the falling part of the trajectory, the downward acceleration results in the increasing negative values of the vertical component of the velocity.

Example 3.2 That's Quite an Arm

A stone is thrown from the top of a building at an angle of 30.0° to the horizontal and with an initial speed of 20.0 m/s, as in Figure 3.9. If the height of the building is 45.0 m, (a) how long is the stone "in flight"?

Reasoning Looking at the pictorial representation in Figure 3.9, it is clear that this is not a symmetric trajectory. Thus, we cannot use Equations 3.15 and 3.16. We use the more general approach described by the Problem-Solving Strategy and represented by Equations 3.10 to 3.13.

Solution The initial x and y components of the velocity are

$$v_{xi} = v_i \cos \theta_i = (20.0 \text{ m/s})(\cos 30.0°) = 17.3 \text{ m/s}$$

$$v_{yi} = v_i \sin \theta_i = (20.0 \text{ m/s})(\sin 30.0°) = 10.0 \text{ m/s}$$

To find t, we use the vertical motion, in which we model the particle as one moving with constant acceleration. We use Equation 3.13 with $y_f = -45.0$ m and $v_{yi} = 10.0$ m/s (we have chosen the top of the building as the origin, as in Figure 3.9):

$$y_f = y_i + v_{yi}t - \tfrac{1}{2}gt^2$$

$$-45.0 \text{ m} = 0 + (10.0 \text{ m/s})t - \tfrac{1}{2}(9.80 \text{ m/s}^2)t^2$$

Solving the quadratic equation for t gives, for the positive root, $t = 4.22$ s. Does the negative root have any physical meaning? (Can you think of another way of finding t from the information given?)

(b) What is the speed of the stone just before it strikes the ground?

Solution The y component of the velocity just before the stone strikes the ground can be obtained using Equation 3.11, with $t = 4.22$ s:

$$v_{yf} = v_{yi} - gt$$

$$= 10.0 \text{ m/s} - (9.80 \text{ m/s}^2)(4.22 \text{ s}) = -31.4 \text{ m/s}$$

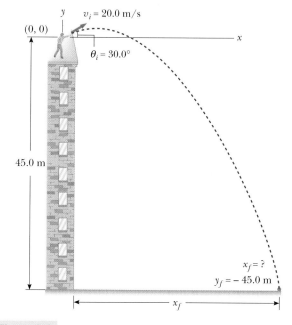

Figure 3.9

(Example 3.2)

In the horizontal direction, the appropriate model is the particle under constant velocity. Because $v_{xf} = v_{xi} = 17.3$ m/s, the speed as the stone strikes the ground is

$$v_f = \sqrt{v_{xf}^2 + v_{yf}^2} = \sqrt{(17.3)^2 + (-31.4)^2} \text{ m/s} = 35.9 \text{ m/s}$$

EXERCISE Where does the stone strike the ground?

Answer 73.0 m from the base of the building

Example 3.3 The Stranded Explorers

An Alaskan rescue plane drops a package of emergency rations to a stranded party of explorers, as shown in the pictorial representation in Figure 3.10. If the plane is traveling horizontally at 40.0 m/s at a height of 100 m above the ground, where does the package strike the ground relative to the point at which it is released?

Reasoning We ignore air resistance, so we model this as a particle in two-dimensional free-fall, which, as we have seen, is modeled by a combination of a particle moving with constant velocity in the x direction and a particle moving with constant acceleration in the y direction. The coordinate system for this problem is selected as shown in Figure 3.10, with the positive x direction to the right and the positive y direction upward.

Solution Consider first the horizontal motion of the package. From Equation 3.12, the position is given by $x_f = x_i + v_{xi}t$. The initial x component of the package velocity is the same as that of the plane when the package is released, 40.0 m/s. We define the initial position $x_i = 0$ right under the plane at the instant the package is released. Thus,

$$x_f = x_i + v_{xi}t = 0 + (40.0 \text{ m/s})t$$

If we know t, the time at which the package strikes the ground, we can determine x_f, the distance traveled by the package in the horizontal direction. At present, however, we have no information about t. To find t, we turn to the equations for the vertical motion of the package, modeling the particle as being under constant acceleration. We know that at the instant the package hits the ground, its y coordinate is -100 m. We also know that the initial component of velocity v_{yi} of the package in the vertical direction is zero, because the package was released with only a horizontal component of velocity. From Equation 3.13, we have

$$y_f = y_i + v_{yi}t - \tfrac{1}{2}gt^2$$

$$-100 \text{ m} = 0 + 0 - \tfrac{1}{2}(9.80 \text{ m/s}^2)t^2$$

$$t^2 = 20.4 \text{ s}^2$$

$$t = 4.52 \text{ s}$$

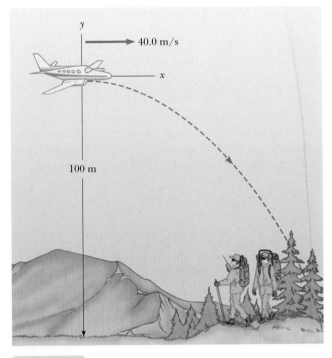

y

→ 40.0 m/s

x

100 m

Figure 3.10

(Example 3.3)

This value for the time at which the package strikes the ground is substituted into the equation for the x coordinate to give us

$$x_f = (40.0 \text{ m/s})(4.52 \text{ s}) = \boxed{181 \text{ m}}$$

The package hits the ground 181 m to the right of the point at which it was dropped in Figure 3.10.

EXERCISE What are the horizontal and vertical components of the velocity of the package just before it hits the ground?

Answer $v_{xf} = 40.0$ m/s; $v_{yf} = -44.3$ m/s

Example 3.4 Javelin Throwing at the Olympics

An athlete throws a javelin a distance of 80.0 m at the Olympics held at the equator, where $g = 9.78$ m/s^2. Four years later, the Olympics are held at the North Pole, where $g = 9.83$ m/s^2. Assuming that the thrower provides the javelin with exactly the same initial velocity as he did at the equator, how far does the javelin travel at the North Pole?

Reasoning In the absence of any information about how the javelin is affected by moving through the air, we adopt the free-fall model for the javelin. Track and field events are normally held on flat fields. Thus, we surmise that the javelin returns to the same vertical position from which it was thrown and, therefore, the trajectory is symmetric. This al-

A javelin can be thrown over a very long distance by a world class athlete. *(Tony Duffy/Allsport)*

---

lows us to use Equations 3.15 and 3.16 to analyze the motion. The difference in range is due to the difference in the free-fall acceleration at the two locations. This difference is due to the rotational motion of the Earth.

Solution   We use Equation 3.16 to express the range of the particle at each of the two locations:

$$R_{equator} = \frac{v_i^2 \sin 2\theta_i}{g_{equator}}$$

$$R_{North\ Pole} = \frac{v_i^2 \sin 2\theta_i}{g_{North\ Pole}}$$

We divide the first equation by the second to establish a relationship between the ratio of the ranges and the ratio of the free-fall accelerations. Because the problem states that the same initial velocity is provided to the javelin at both locations, v_i and θ_i are the same in the numerator and denomina-

tor of the ratio, which gives us:

$$\frac{R_{equator}}{R_{North\ Pole}} = \frac{\left(\dfrac{v_i^2 \sin 2\theta_i}{g_{equator}}\right)}{\left(\dfrac{v_i^2 \sin 2\theta_i}{g_{North\ Pole}}\right)} = \frac{g_{North\ Pole}}{g_{equator}}$$

We can now solve this equation for the range at the North Pole and substitute the numerical values:

$$R_{North\ Pole} = \frac{g_{equator}}{g_{North\ Pole}} R_{equator} = \frac{9.78\ m/s^2}{9.83\ m/s^2}\ (80.0\ m)$$

$$= 79.6\ m$$

Notice one of the advantages of this powerful technique of setting up ratios—we do not need to know the magnitude (v_i) nor the direction (θ_i) of the initial velocity. As long as they are the same at both locations, they cancel in the ratio.

EXERCISE   Suppose that the javelin thrower wished to match his 80.0-m throw at the equator in the Olympics held at the North Pole. By what percentage must he increase the magnitude of the initial velocity of the javelin to throw it 80.0 m at the North Pole?

Answer   0.26%

EXERCISE   It has been said that in his youth George Washington threw a silver dollar across a river. Assuming that the river was 75 m wide, (a) what *minimum initial speed* was necessary to throw the coin across the river, and (b) how long was the coin in flight?

Answer   (a) 27 m/s (b) 3.9 s

3.4 • THE PARTICLE IN UNIFORM CIRCULAR MOTION

See Screen 3.6

Figure 3.11a shows a car moving in a circular path with *constant speed v*. Such motion is called **uniform circular motion** and serves as the basis for a new group of problems that we can solve.

It is often surprising to students to find that **even though an object moves at a constant speed in a circular path, it still has an acceleration.** To see why, consider the defining equation for average acceleration, $\bar{\mathbf{a}} = \Delta\mathbf{v}/\Delta t$ (Eq. 3.4). The acceleration depends on *the change in the velocity vector*. Because velocity is a vector quantity, an acceleration can be produced in two ways, as mentioned in Section 3.1: by a change in the *magnitude* of the velocity or by a change in the *direction* of the velocity. The latter situation is occurring for an object moving with constant speed in a circular path. The velocity vector is always tangent to the path of the object and perpendicular to the radius of the circular path. We now show that the acceleration vector in uniform circular motion is always perpendicular to the path and always points toward the center of the circle. An acceleration of this nature is

PITFALL PREVENTION 3.4

Acceleration of a particle in uniform circular motion

Many students have trouble with the notion of a particle moving in a circular path at constant speed and yet having an acceleration. This is due to the everyday interpretation of acceleration as meaning *speeding up* or *slowing down*. But remember that acceleration is defined as a change in the *velocity*, not a change in the *speed*. In circular motion, the velocity vector is changing in direction, so there is indeed an acceleration.

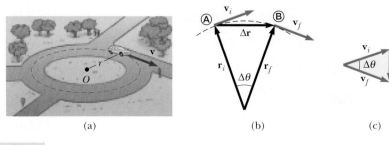

(a) (b) (c)

Figure 3.11

(a) A car moving along a circular path at constant speed is in uniform circular motion. (b) As the particle moves from Ⓐ to Ⓑ, its velocity vector changes from $\mathbf{v}_i$ to $\mathbf{v}_f$. (c) The construction for determining the direction of the change in velocity $\Delta\mathbf{v}$, which is toward the center of the circle for small $\Delta\theta$.

called a **centripetal acceleration** (*centripetal* means *center seeking*), and its magnitude is

- *Magnitude of centripetal acceleration*

$$a_c = \frac{v^2}{r} \qquad [3.17]$$

where r is the radius of the circle. The subscript on the acceleration symbol reminds us that the acceleration is centripetal.

Let us first argue conceptually that the acceleration must be perpendicular to the path followed by the particle. If this were not true, there would be a component of the acceleration parallel to the path and, therefore, parallel to the velocity vector. Such an acceleration component would lead to a change in the speed of the object, which we model as a particle, along the path. But this is inconsistent with our setup of the problem—the particle moves with constant speed along the path. Thus, for *uniform* circular motion, the acceleration vector can only have a component perpendicular to the path, which is toward the center of the circle.

To derive Equation 3.17, consider the pictorial representation of the position and velocity vectors in Figure 3.11b. In addition, the figure shows the vector representing the change in position, $\Delta\mathbf{r}$. The particle follows a circular path, part of which is shown by the dotted curve. The particle is at Ⓐ at time t_i, and its velocity at that time is $\mathbf{v}_i$; it is at Ⓑ at some later time t_f, and its velocity at that time is $\mathbf{v}_f$. Let us also assume that $\mathbf{v}_i$ and $\mathbf{v}_f$ differ only in direction; their magnitudes are the same (i.e., $v_i = v_f = v$, because it is *uniform* circular motion). To calculate the acceleration of the particle, let us begin with the defining equation for average acceleration (Eq. 3.4):

$$\overline{\mathbf{a}} = \frac{\mathbf{v}_f - \mathbf{v}_i}{t_f - t_i} = \frac{\Delta\mathbf{v}}{\Delta t}$$

In Figure 3.11c, the velocity vectors in Figure 3.11b have been redrawn tail to tail. The vector $\Delta\mathbf{v}$ connects the tips of the vectors, representing the vector addition, $\mathbf{v}_f = \mathbf{v}_i + \Delta\mathbf{v}$. In Figures 3.11b and 3.11c, we can identify triangles that can serve as geometric models to help us analyze the motion. The angle $\Delta\theta$ between the two position vectors in Figure 3.11b is the same as the angle between the velocity vectors in Figure 3.11c, because the velocity vector $\mathbf{v}$ is always perpendicular to the position vector $\mathbf{r}$. Thus, the two triangles are *similar*. (Two triangles are similar if the angle between any two sides is the same for both triangles and if the ratio of the lengths of these sides is the same.) This enables us to write a relationship between the lengths of the sides for the two triangles:

$$\frac{|\Delta \mathbf{v}|}{v} = \frac{|\Delta \mathbf{r}|}{r}$$

where $v = v_i = v_f$ and $r = r_i = r_f$. This equation can be solved for $|\Delta \mathbf{v}|$ and the expression so obtained can be substituted into $\overline{\mathbf{a}} = \Delta \mathbf{v}/\Delta t$ (Eq. 3.4) to give the magnitude of the average acceleration over the time interval for the particle to move from Ⓐ to Ⓑ:

$$|\overline{\mathbf{a}}| = \frac{v}{r}\frac{|\Delta \mathbf{r}|}{\Delta t}$$

Now imagine that points Ⓐ and Ⓑ in Figure 3.11b become extremely close together. As Ⓐ and Ⓑ approach each other, Δt approaches zero, and the ratio $|\Delta \mathbf{r}|/\Delta t$ approaches the speed v. In addition, the average acceleration becomes the instantaneous acceleration at point Ⓐ. Hence, in the limit $\Delta t \rightarrow 0$, the magnitude of the acceleration is

$$a_c = \frac{v^2}{r}$$

Thus, in uniform circular motion, the acceleration is directed inward toward the center of the circle and has magnitude v^2/r.

In many situations it is convenient to describe the motion of a particle moving with constant speed in a circle of radius r in terms of the **period** T, which is defined as the time required for one complete revolution. In the time T the particle moves a distance of $2\pi r$, which is equal to the circumference of the particle's circular path. Therefore, because its speed is equal to the circumference of the circular path divided by the period, or $v = 2\pi r/T$, it follows that

$$T \equiv \frac{2\pi r}{v} \qquad\qquad [3.18]$$

The particle in uniform circular motion is a very common physical situation and is useful as an analysis model for problem solving. If a particle, or an object that can be modeled as a particle, moves in a circular path at a uniform speed, we can use the results of the preceding discussion, with the important equations represented by Equation 3.17 for the centripetal acceleration and Equation 3.18 for the period.

PITFALL PREVENTION 3.5

Centripetal acceleration is not constant

We derived the magnitude of the centripetal acceleration vector and found it to be constant for uniform circular motion. But *the centripetal acceleration vector is not constant.* It always points toward the center of the circle, but continuously changes direction as the particle moves around the circular path.

● *Period of a particle in uniform circular motion*

THINKING PHYSICS 3.3

An airplane travels from Los Angeles to Sydney, Australia. After cruising altitude is reached, the instruments on the plane indicate that the ground speed holds rock-steady at 700 km/h and that the heading of the airplane does not change. Is the velocity of the airplane constant during the flight?

Reasoning The velocity is not constant, due to the curvature of the Earth. Even though the speed does not change, and the heading is always toward Sydney (is this actually true?), the airplane travels around a significant portion of the circumference of the Earth. Thus, the direction of the velocity vector does indeed change. We could extend this by imagining that the airplane passes over Sydney and continues (assuming it has enough fuel!) around the Earth until it arrives at Los Angeles again. It is impossible for an airplane to have a constant velocity (relative to the Universe, not to the Earth's surface) and return to its starting point.

Example 3.5 **The Centripetal Acceleration of the Earth**

What is the centripetal acceleration of the Earth as it moves in its orbit around the Sun?

Solution We shall model the Earth as a particle and approximate the Earth's orbit as circular (it's actually slightly elliptical, as we discuss in Chapter 11). Although we don't know the orbital speed of the Earth, with the help of Equation 3.18, we can recast Equation 3.17 in terms of the period

of the Earth's orbit, which we know is one year:

$$a_c = \frac{v^2}{r} = \frac{\left(\dfrac{2\pi r}{T}\right)^2}{r} = \frac{4\pi^2 r}{T^2}$$

$$= \frac{4\pi^2(1.5 \times 10^{11} \text{ m})}{(1 \text{ yr})^2}\left(\frac{1 \text{ yr}}{3.15 \times 10^7 \text{ s}}\right)^2$$

$$= \boxed{6.0 \times 10^{-3} \text{ m/s}^2}$$

See Screen 3.6

3.5 • TANGENTIAL AND RADIAL ACCELERATION

Let us consider the motion of a particle along a curved path where the velocity changes both in direction and in magnitude, as described in Figure 3.12. In this situation, the velocity vector is always tangent to the path; however, the acceleration vector **a** is at some angle to the path. At each of three points Ⓐ, Ⓑ, and Ⓒ in Figure 3.12, we draw dotted circles that form geometric models of circular paths for a portion of the actual path at each point. The radius of the model circles is equal to the radius of curvature of the path at each point.

As the particle moves along the curved path in Figure 3.12, the direction of the total acceleration vector **a** changes from point to point. This vector can be resolved into two components, based on an origin at the center of the model circle: a radial component a_r along the radius of the model circle, and a tangential component a_t perpendicular to this radius. The *total* acceleration vector **a** can be written as the vector sum of the component vectors:

• *Total acceleration*

$$\mathbf{a} = \mathbf{a}_r + \mathbf{a}_t \qquad \text{[3.19]}$$

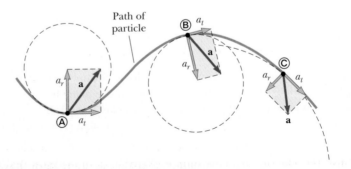

Figure 3.12

The motion of a particle along an arbitrary curved path lying in the *xy* plane. If the velocity vector **v** (always tangent to the path) changes in direction and magnitude, the acceleration vector **a** has a tangential component a_t and a radial component a_r.

The tangential acceleration arises from the change in the speed of the particle, and has a magnitude given by

$$a_t = \frac{d|\mathbf{v}|}{dt}$$

[3.20]   • *Tangential acceleration*

The radial acceleration is due to the change in direction of the velocity vector and is given by

$$a_r = -a_c = -\frac{v^2}{r}$$

• *Centripetal acceleration*

where r is the radius of curvature of the path at the point in question, which is the radius of the model circle. We recognize the radial component of the acceleration as the centripetal acceleration discussed in Section 3.4. The negative sign indicates that the direction of the centripetal acceleration is toward the center of the model circle, opposite the direction of the radial unit vector $\hat{\mathbf{r}}$, which always points away from the center of the circle.

Because $\mathbf{a}_r$ and $\mathbf{a}_t$ are perpendicular component vectors of $\mathbf{a}$, it follows that $a = \sqrt{a_r^2 + a_t^2}$. At a given speed, a_r is large when the radius of curvature is small (as at points Ⓐ and Ⓑ in Fig. 3.12) and small when r is large (such as at point Ⓒ). The direction of $\mathbf{a}_t$ is either in the same direction as $\mathbf{v}$ (if v is increasing) or opposite $\mathbf{v}$ (if v is decreasing).

In the case of uniform circular motion, where v is constant, $a_t = 0$ and the acceleration is always radial, as described in Section 3.4. In other words, uniform circular motion is a special case of motion along a curved path. Furthermore, if the direction of $\mathbf{v}$ doesn't change, then no radial acceleration occurs and the motion is one-dimensional ($a_r = 0$, but a_t may not be zero).

3.6 • RELATIVE VELOCITY

In Section 1.7, we discussed the need for a fixed reference point as the origin of a coordinate system used to locate the position of a point. We have made observations of position, velocity, and acceleration of a particle with respect to this reference point. Now imagine that we have two observers making measurements of a particle located in space, and one of the observers moves with respect to the other at constant velocity. Each observer can define a coordinate system with an origin fixed with respect to him or her. The origins of the two coordinate systems are in motion with respect to each other. In this section, we explore how we relate the measurements of one observer to that of the other.

 See Screen 3.7

As an example, consider two cars, a red one and a blue one, moving on a highway in the same direction, both with speeds of 60 mi/h, as in Figure 3.13. If we identify the red car as a particle to be observed, an observer on the side of the road measures a speed for this car of 60 mi/h. Now consider an observer riding in the blue car. This observer looks out the window and sees that the red car is always in the same position with respect to the blue car. Thus, this observer measures a speed for the red car of *zero*. This simple example demonstrates that speed measurements differ in different frames of reference. Both observers look at the same particle (the red car) and arrive at different values for its speed. Both are correct; the difference in their measurements is due to the relative velocity of their frames of reference.

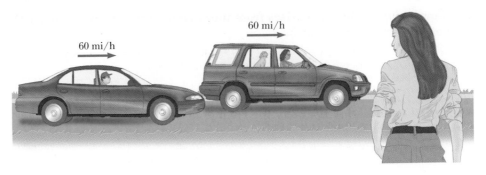

60 mi/h

60 mi/h

Two observers measure the speed of the red car. Observer O is standing on the ground beside the highway. Observer O' is in the blue car.

Let us now generate a mathematical representation that will allow us to calculate one observer's measurements from the other's. Consider a particle located at point P in an xy plane, as shown in Figure 3.14. Imagine that the motion of this particle is being observed by two observers. Observer O is in reference frame S. Observer O' is in reference frame S', which moves with velocity $\mathbf{v}_{O'O}$ with respect to S, where the subscripts indicate the following: The first subscript describes what is being observed, and the second describes who is doing the observing. Thus, $\mathbf{v}_{O'O}$ is the velocity of observer O' as measured by observer O. At $t = 0$, the origins of the reference frames coincide. Thus, modeling the origin of S' as a particle under constant velocity, the origins of the two reference frames are separated by a displacement $\mathbf{v}_{O'O}t$ at time t. This displacement is shown in Figure 3.14. Also shown in the figure are the position vectors $\mathbf{r}_{PO}$ and $\mathbf{r}_{PO'}$ for point P from each of the two origins. These are the position vectors that the two observers would use to describe the location of point P, using the same subscript notation. From the diagram, we see that these three vectors form a vector addition triangle:

$$\mathbf{r}_{PO} = \mathbf{r}_{PO'} + \mathbf{v}_{O'O}t$$

Notice the order of subscripts in this expression. The subscripts on the left side are the same as the first and last subscripts on the right. The second and third subscripts on the right are both O'. These subscripts are helpful in analyzing these types of situations. On the left, we are looking at the position vector that points directly to P from O, as described by the subscripts. On the right, the same point P is

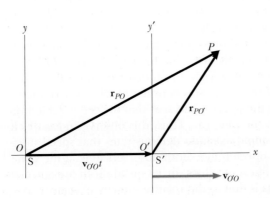

Position vectors for an event occurring at point P for two observers. Observer O' is moving to the right at speed $v_{O'O}$ with respect to observer O.

located by first going to P from O' and then describing where O' is relative to O, again as suggested by the subscripts.

Let us now differentiate this expression with respect to time to find an expression for the velocity of a particle located at point P:

$$\frac{d}{dt}(\mathbf{r}_{PO}) = \frac{d}{dt}(\mathbf{r}_{PO'} + \mathbf{v}_{O'O}t) \quad \longrightarrow \quad \mathbf{v}_{PO} = \mathbf{v}_{PO'} + \mathbf{v}_{O'O} \qquad \textbf{[3.21]}$$

This expression relates the velocity of the particle as measured by O to that measured by O' and the relative velocity of the two reference frames.

In the one-dimensional case, this equation reduces to

$$v_{PO} = v_{PO'} + v_{O'O}$$

Often, this equation is expressed in terms of the observer O', as follows:

$$v_{PO'} = v_{PO} - v_{O'O} \qquad \textbf{[3.22]}$$

• *Relative velocity in one dimension*

and is called the **relative velocity**—the velocity of a particle as measured by a moving observer (moving with respect to another observer). In our car example, observer O is standing on the side of the road. Observer O' is in the blue car. Both observers are measuring the speed of the red car, which is located at point P. Thus,

$$v_{PO} = 60 \text{ mi/h}$$

$$v_{O'O} = 60 \text{ mi/h}$$

and, using Equation 3.22,

$$v_{PO'} = v_{PO} - v_{O'O} = 60 \text{ mi/h} - 60 \text{ mi/h} = 0$$

The result of our calculation agrees with our previous intuitive discussion.

This equation will be used in Chapter 9, when we discuss special relativity. We shall find that this simple expression is valid for low-speed particles but is no longer valid when the particle or observers are moving at speeds close to the speed of light.

Example 3.6   A Boat Crossing a River

A boat heading due north crosses a wide river with a speed of 10.0 km/h relative to the water. The river has a current such that the water moves with uniform speed of 5.00 km/h due east relative to the ground. (a) What is the velocity of the boat relative to a stationary observer on the side of the river?

Reasoning   It is often useful to use subscripts other than O and P that make it easy to identify the observers and the object being observed. Observer O is standing on the side of the river. Because he is at rest with respect to the Earth, we will use the subscript E for this observer. Let us identify an imaginary observer O' at rest in the water, floating with the current. Because he is at rest with respect to the water, we will use the subscript w for this observer. Both observers are looking at the boat, denoted by the subscript b. We can identify the velocity of the boat relative to the water as $\mathbf{v}_{bw} = 10.0\mathbf{j}$ km/h. The velocity of the water relative to the Earth is that of the current in the river, $\mathbf{v}_{wE} = 5.00\mathbf{i}$ km/h.

Solution   We are looking for the velocity of the boat relative to the Earth, so, from Equation 3.21,

$$\mathbf{v}_{bE} = \mathbf{v}_{bw} + \mathbf{v}_{wE} = (10.0\mathbf{j} + 5.00\mathbf{i}) \text{ km/h}$$

This vector addition is shown in Figure 3.15a. The speed of the boat relative to the observer on shore is found from the

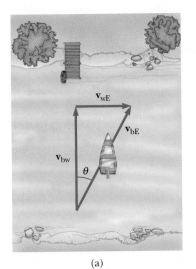

(a)

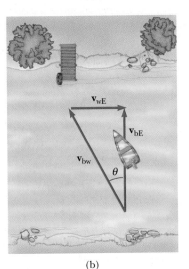

(b)

Figure 3.15

(Example 3.6)

Pythagorean theorem:

$$v_{bE} = \sqrt{v_{bw}^2 + v_{wE}^2} = \sqrt{(10.0)^2 + (5.00)^2} \text{ km/h}$$

$$= \boxed{11.2 \text{ km/h}}$$

The direction of the velocity vector can be found with the inverse tangent function:

$$\theta = \tan^{-1}\left(\frac{v_{wE}}{v_{bw}}\right) = \tan^{-1}\left(\frac{5.00}{10.0}\right) = \boxed{26.6°}$$

(b) At what angle should the boat be headed if it is to travel directly north across the river, and what is the speed of the boat relative to the Earth?

Solution We now want $\mathbf{v}_{bE}$ to be pointed due north, as shown in Figure 3.15b. From the vector triangle,

$$\theta = \sin^{-1}\left(\frac{v_{wE}}{v_{bw}}\right) = \sin^{-1}\left(\frac{5.00 \text{ km/h}}{10.0 \text{ km/h}}\right) = \boxed{30.0°}$$

The speed of the boat relative to the Earth is

$$v_{bE} = \sqrt{v_{bw}^2 - v_{wE}^2} = \sqrt{(10.0)^2 - (5.00)^2} \text{ km/h}$$

$$= \boxed{8.66 \text{ km/h}}$$

context connection

3.7 • CIRCULAR ORBITS

In Chapter 2, we considered some information about the liftoff phase of our spacecraft on its way to Mars. Now let us look at the next phase of its motion, in which the spacecraft is placed in a parking orbit around the Earth. We shall assume that the parking orbit is circular so that the spacecraft can be modeled as a particle in uniform circular motion.

The spacecraft is actually in free-fall, but it never moves closer to the Earth. This is a difficult concept to accept, but it might help to remember a discussion from Section 3.3 in which it was mentioned that if there were no gravitational acceleration, the particle would continue to move along a straight path in the direc-

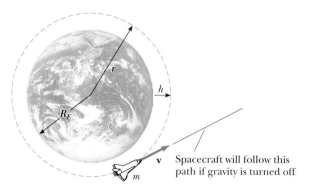

Figure 3.16

If the Earth's gravity were to be turned off, a satellite initially in orbit would follow the straight line path shown.

v Spacecraft will follow this path if gravity is turned off

tion of its original velocity. The gravitational force causes it to "fall" below the straight path. In the case of the spacecraft, if gravity were to be turned off, the straight line path would take the spacecraft tangentially off its orbit into space, as in Figure 3.16. The gravitational force causes the spacecraft to "fall" below this straight line path. With a spacecraft in orbit, the speed of the spacecraft is designed so that this falling below the straight line path keeps the spacecraft at a fixed height *h* above the Earth's surface.

We can use the model of a particle in uniform circular motion to analyze the orbit of a spacecraft around the Earth, as in the following example.

WEB

For a *New York Times* retrospective on *Sputnik*, the first Earth satellite, visit **http://www.nytimes.com/ partners/aol/special/sputnik/**

Example 3.7 A Spacecraft in Orbit

A spacecraft is in orbit at a height of 200 km above the surface of the Earth. The period of its orbit is 88.2 min.

(a) What is the centripetal acceleration of the spacecraft?

Solution It is not stated explicitly that the orbit is circular, but we make that assumption based on the fact that only one height above the Earth's surface is given. We shall use the equation derived in Example 3.5 relating the centripetal acceleration to the period. We first recognize that the radius of the circular orbit of the spacecraft is the height above the Earth's surface *plus* the radius of the Earth. Thus,

$$a_c = \frac{4\pi^2 r}{T^2}$$

$$= \frac{4\pi^2 (2.00 \times 10^5 \text{ m} + 6.37 \times 10^6 \text{ m})}{(88.2 \text{ min})^2} \left(\frac{1 \text{ min}}{60 \text{ s}}\right)^2$$

$$= 9.26 \text{ m/s}^2$$

Notice that this is slightly less than the free-fall acceleration at the surface of the Earth. This should not be a surprise, because the spacecraft is simply *falling* as it orbits the Earth, so its centripetal acceleration *should* be equal to the free-fall acceleration. The value is less than that at the surface due to the decrease in the free-fall acceleration as one moves higher off the surface of the Earth.

(b) What is the speed of the spacecraft in its orbit?

Solution We use Equation 3.18 to find the speed:

$$v = \frac{2\pi r}{T} = \frac{2\pi (2.00 \times 10^5 \text{ m} + 6.37 \times 10^6 \text{ m})}{(88.2 \text{ min})} \left(\frac{1 \text{ min}}{60 \text{ s}}\right)$$

$$= 7.80 \times 10^3 \text{ m/s}$$

This is equivalent to over 17,000 mi/h.

The motion of our spacecraft in orbit, with the rocket engine turned off, is completely governed by gravity, resulting in a circular orbit around the Earth. We must have control over the motion of our spacecraft beyond that provided by gravity. In the next chapter, we shall see how to use the rocket engine to change our velocity from an initial vector to a final desired vector.

SUMMARY

If a particle moves with *constant* acceleration **a** and has velocity $\mathbf{v}_i$ and position $\mathbf{r}_i$ at $t = 0$, its velocity and position vectors at some later time t are

$$\mathbf{v}_f = \mathbf{v}_i + \mathbf{a}t \qquad [3.8]$$

$$\mathbf{r}_f = \mathbf{r}_i + \mathbf{v}_i t + \tfrac{1}{2}\mathbf{a}t^2 \qquad [3.9]$$

For two-dimensional motion in the xy plane under constant acceleration, these vector expressions are equivalent to two component expressions, one for the motion along x and one for the motion along y.

Projectile motion is a special case of two-dimensional motion under constant acceleration, where $a_x = 0$ and $a_y = -g$. In this case, the horizontal components of Equations 3.8 and 3.9 reduce to those of a particle under constant velocity:

$$v_{xf} = v_{xi} = \text{constant} \qquad [3.10]$$

$$x_f = x_i + v_{xi}t \qquad [3.12]$$

and the vertical components of Equations 3.8 and 3.9 are those of a particle under constant acceleration:

$$v_{yf} = v_{yi} - gt \qquad [3.11]$$

$$y_f = y_i + v_{yi}t - \tfrac{1}{2}gt^2 \qquad [3.13]$$

where $v_{xi} = v_i \cos \theta_i$, $v_{yi} = v_i \sin \theta_i$, v_i is the initial speed of the projectile, and θ_i is the angle $\mathbf{v}_i$ makes with the positive x axis.

A particle moving in a circle of radius r with constant speed v undergoes a **centripetal acceleration** because the direction of **v** changes in time. The magnitude of this acceleration is

$$a_c = \frac{v^2}{r} \qquad [3.17]$$

and its direction is always toward the center of the circle.

If a particle moves along a curved path in such a way that the magnitude and direction of **v** change in time, the particle has an acceleration vector that can be described by two components: (1) a radial component $a_r = -a_c$ arising from the change in direction of **v**, and (2) a tangential component a_t arising from the change in magnitude of **v**.

If an observer O' is moving with velocity $\mathbf{v}_{O'O}$ with respect to observer O, their measurements of the velocity of a particle located at point P are related according to

$$\mathbf{v}_{PO} = \mathbf{v}_{PO'} + \mathbf{v}_{O'O} \qquad [3.21]$$

The velocity $\mathbf{v}_{PO'}$ is called the **relative velocity**—the velocity of a particle as measured by a moving observer (moving at constant velocity with respect to another observer).

QUESTIONS

1. If you know the position vectors of a particle at two points along its path and also know the time it took to move from one point to the other, can you determine the particle's instantaneous velocity? Its average velocity? Explain.

2. Explain whether the following particles have an acceleration: (a) a particle moving in a straight line with constant speed and (b) a particle moving around a curve with constant speed.

3. Correct the following statement: "The racing car rounds the turn at a constant velocity of 90 miles per hour."

4. A spacecraft drifts through space at a constant velocity. Suddenly a gas leak in the side of the spacecraft causes a constant acceleration of the spacecraft in a direction perpendicular to the initial velocity. The orientation of the spacecraft does not change, so that the acceleration remains perpendicular to the original direction of the velocity. What is the shape of the path followed by the spacecraft in this situation?

5. A ball is projected horizontally from the top of a building. One second later another ball is projected horizontally from the same point with the same velocity. At what point in the motion will the balls be closest to each other? Will the first ball always be traveling faster than the second ball? What will be the time difference between when the balls hit the ground? Can the horizontal projection velocity of the second ball be changed so that the balls arrive at the ground at the same time?

6. At the end of a pendulum's arc, its velocity is zero. Is its acceleration also zero at that point?

7. If a rock is dropped from the top of a sailboat's mast, will it hit the deck at the same point whether the boat is at rest or in motion at constant velocity?

8. Two projectiles are thrown with the same magnitude of initial velocity, one at an angle θ with respect to the level ground and the other at angle $90° - \theta$. Both projectiles will strike the ground at the same distance from the projection point. Will both projectiles be in the air for the same time interval?

9. A projectile is launched at some angle to the horizontal with some initial speed v_i, and air resistance is ignored. Is

the projectile a freely falling body? What is its acceleration in the vertical direction? What is its acceleration in the horizontal direction?

10. State which of the following quantities, if any, remain constant as a projectile moves through its parabolic trajectory: (a) speed, (b) acceleration, (c) horizontal component of velocity, (d) vertical component of velocity.

11. The maximum range of a projectile occurs when it is launched at an angle of 45.0° with the horizontal, if air resistance is ignored. If air resistance is not ignored, will the optimum angle be greater or less than 45.0°? Explain.

12. An object moves in a circular path with constant speed v. (a) Is the velocity of the object constant? (b) Is its acceleration constant? Explain.

13. A projectile is launched on the Earth with some initial velocity. Another projectile is launched on the Moon with the *same* initial velocity. Ignoring air resistance, which projectile has the greater range? Which reaches the greater altitude? (Note that the free-fall acceleration on the Moon is about 1.6 m/s².)

14. A coin on a table is given an initial horizontal velocity such that it ultimately leaves the end of the table and hits the floor. At the instant the coin leaves the end of the table, a ball is released from the same height and falls to the floor. Explain why the two objects hit the floor simultaneously, even though the coin has an initial velocity.

15. Describe how a driver can steer a car traveling at constant speed so that (a) the acceleration is zero or (b) the magnitude of the acceleration remains constant.

16. An ice skater executes a figure eight, consisting of two equal, tangent circular paths. During the first loop she increases her speed uniformly, and during the second loop she moves at a constant speed. Make a sketch of her acceleration vector at several points along the path of motion.

17. Construct motion diagrams showing the velocity and acceleration of a projectile at several points along its path if (a) the projectile is launched horizontally and (b) the projectile is launched at an angle θ with the horizontal.

18. A baseball is thrown such that its initial x and y components of velocity are known. Ignoring air resistance, describe how you would calculate, at the instant the ball reaches the top of its trajectory, (a) its coordinates, (b) its velocity, and (c) its acceleration. How would these results change if air resistance were taken into account?

PROBLEMS

1, 2, 3 = straightforward, intermediate, challenging   □ = full solution available in the *Student Solutions Manual and Study Guide*

web = solution posted at **http://www.harcourtcollege.com/physics/**   💻 = computer useful in solving problem

📓 = Interactive Physics   ▓ = paired numerical/symbolic problems   ◣ = life science application

Section 3.1   The Position, Velocity, and Acceleration Vectors

1. A motorist drives south at 20.0 m/s for 3.00 min, then turns west and travels at 25.0 m/s for 2.00 min, and finally travels northwest at 30.0 m/s for 1.00 min. For this 6.00-min trip, find (a) the total vector displacement, (b) the average speed, and (c) the average velocity. Let the positive x axis point east.

2. Suppose that the position vector for a particle is given as a function of time by $\mathbf{r}(t) = x(t)\mathbf{i} + y(t)\mathbf{j}$, with $x(t) = at + b$ and $y(t) = ct^2 + d$, where $a = 1.00$ m/s, $b = 1.00$ m, $c = 0.125$ m/s², and $d = 1.00$ m. (a) Calculate the average velocity during the time interval from $t = 2.00$ s to $t = 4.00$ s. (b) Determine the velocity and the speed at $t = 2.00$ s.

3. A golf ball is hit off a tee at the edge of a cliff. Its x and y coordinates as functions of time are given by the following expressions:

$$x = (18.0 \text{ m/s})t$$

and

$$y = (4.00 \text{ m/s})t - (4.90 \text{ m/s}^2)t^2$$

(a) Write a vector expression for the ball's position as a function of time, using the unit vectors $\mathbf{i}$ and $\mathbf{j}$. By taking derivatives, obtain expressions for (b) the velocity vector $\mathbf{v}$ as a function of time and (c) the acceleration vector as a function of time. Next use unit-vector notation to write expressions for (d) the position, (e) the velocity, and (f) the acceleration of the golf ball, all at $t = 3.00$ s.

Section 3.2   Two-Dimensional Motion with Constant Acceleration

4. At $t = 0$, a particle moving in the xy plane with constant acceleration has a velocity of $\mathbf{v}_i = (3.00\mathbf{i} - 2.00\mathbf{j})$ m/s and is at the origin. At $t = 3.00$ s, the particle's velocity is $\mathbf{v}_f = (9.00\mathbf{i} + 7.00\mathbf{j})$ m/s. Find (a) the acceleration of the particle and (b) its coordinates at any time t.

5. A fish swimming in a horizontal plane has velocity $\mathbf{v}_i = (4.00\mathbf{i} + 1.00\mathbf{j})$ m/s at a point in the ocean where the displacement from a certain rock is $\mathbf{r}_i = (10.0\mathbf{i} - 4.00\mathbf{j})$ m. After the fish swims with constant acceleration for 20.0 s, its velocity is $\mathbf{v}_f = (20.0\mathbf{i} - 5.00\mathbf{j})$ m/s. (a) What are the components of the acceleration? (b) What is the direction of the acceleration with respect to unit vector $\mathbf{i}$? (c) If the fish maintains constant acceleration, where is it at $t = 25.0$ s, and in what direction is it moving?

6. A particle initially located at the origin has an acceleration of $\mathbf{a} = 3.00\mathbf{j}$ m/s^2 and an initial velocity of $\mathbf{v}_i = 5.00\mathbf{i}$ m/s. Find (a) the vector position and velocity at any time t and (b) the coordinates and speed of the particle at $t = 2.00$ s.

7. It is not possible to see very small objects, such as viruses, using an ordinary light microscope. An electron microscope can view such objects using an electron beam instead of a light beam. Electron microscopy has proved invaluable for investigations of viruses, cell membranes and subcellular structures, bacterial surfaces, visual receptors, chloroplasts, and the contractile properties of muscles. The "lenses" of an electron microscope consist of electric and magnetic fields that control the electron beam. As an example of the manipulation of an electron beam, consider an electron traveling away from the origin along the x axis in the xy plane with initial velocity $\mathbf{v}_i = v_i\mathbf{i}$. As it passes through the region $x = 0$ to $x = d$, the electron experiences acceleration $\mathbf{a} = a_x\mathbf{i} + a_y\mathbf{j}$, where a_x and a_y are constants. Assume that $v_i = 1.80 \times 10^7$ m/s, $a_x = 8.00 \times 10^{14}$ m/s^2, and $a_y = 1.60 \times 10^{15}$ m/s^2. Determine at $x = d = 0.010\ 0$ m (a) the position of the electron, (b) the velocity of the electron, (c) the speed of the electron, and (d) the direction of travel of the electron (i.e., the angle between its velocity and the x axis).

8. The determined coyote is out once more in pursuit of the elusive roadrunner. The coyote wears a pair of Acme jet-powered roller skates, which provide a constant horizontal acceleration of 15.0 m/s^2 (Fig. P3.8). The coyote starts at rest 70.0 m from the brink of a cliff at the instant the roadrunner zips past him in the direction of the cliff. (a) If the roadrunner moves with constant speed, determine the minimum speed he must have in order to reach the cliff

Coyoté Roadrunner
Stupidus Delightus
BEEP
BEEP

Figure P3.8

before the coyote. At the edge of the cliff, the roadrunner escapes by making a sudden turn, and the coyote continues straight ahead. His skates remain horizontal and continue to operate while he is in flight, so that the coyote's acceleration is $(15.0\ \mathbf{i} - 9.80\mathbf{j})$ m/s^2. (b) If the cliff is 100 m above the flat floor of a canyon, determine where the coyote lands in the canyon. (c) Determine the components of the coyote's impact velocity.

Section 3.3 Projectile Motion

Note: Ignore air resistance in all problems and take $g = 9.80$ m/s^2 at the Earth's surface.

9. In a local bar, a customer slides an empty beer mug down the counter for a refill. The bartender is momentarily distracted and does not see the mug, which slides off the counter and strikes the floor 1.40 m from the base of the counter. If the height of the counter is 0.860 m, (a) with what velocity did the mug leave the counter, and (b) what was the direction of the mug's velocity just before it hit the floor?

10. In a local bar, a customer slides an empty beer mug down the counter for a refill. The bartender is momentarily distracted and does not see the mug, which slides off the counter and strikes the floor at distance d from the base of the counter. If the height of the counter is h, (a) with what velocity did the mug leave the counter, and (b) what was the direction of the mug's velocity just before it hit the floor?

11. A tennis player standing 12.6 m from the net hits the ball at 3.00° above the horizontal. To clear the net, the ball must rise at least 0.330 m. If the ball just clears the net at the apex of its trajectory, how fast was the ball moving when it left the racquet?

12. An astronaut on a strange planet finds that she can jump a maximum horizontal distance of 15.0 m if her initial speed is 3.00 m/s. What is the free-fall acceleration on the planet?

13. A ball is tossed from an upper-story window of a building. The ball is given an initial velocity of 8.00 m/s at an angle of 20.0° below the horizontal. It strikes the ground 3.00 s later. (a) How far horizontally from the base of the building does the ball strike the ground? (b) Find the height from which the ball was thrown. (c) How long does it take the ball to reach a point 10.0 m below the level of launching?

14. A cannon with a muzzle speed of 1 000 m/s is used to start an avalanche on a mountain slope. The target is 2 000 m from the cannon horizontally and 800 m above the cannon. At what angle, above the horizontal, should the cannon be fired?

15. A place-kicker must kick a football from a point 36.0 m (about 40 yards) from the goal, and half the crowd hopes the ball will clear the crossbar, which is 3.05 m high. When kicked, the ball leaves the ground with a speed of 20.0 m/s at an angle of 53.0° to the horizontal. (a) By how much does the ball clear or fall short of clearing the crossbar? (b) Does the ball approach the crossbar while still rising or while falling?

16. One strategy in a snowball fight is to throw a snowball at a high angle over level ground. While your opponent is watching the first one, a second snowball is thrown at a low angle timed to arrive before or at the same time as the first one. Assume both snowballs are thrown with a speed of 25.0 m/s. The first one is thrown at an angle of 70.0° with respect to the horizontal. (a) At what angle should the second snowball be thrown to arrive at the same point as the first? (b) How many seconds later should the second snowball be thrown after the first to arrive at the same time?

17. The speed of a projectile when it reaches its maximum height is one half of its speed when it is at half its maximum height. What is the initial projection angle of the projectile?

18. The small archerfish (length 20–25 cm) lives in brackish waters of southeast Asia. This aptly named creature captures its prey by shooting a stream of water drops at an insect, either flying or at rest. After the bug falls into the water, the fish gobbles it up. The archerfish has high accuracy at distances of 1.2 m to 1.5 m and sometimes makes hits at distances up to 3.5 m. A groove in the roof of its mouth, along with a curled tongue, forms a tube that enables the fish to impart high velocity to the water in its mouth when it suddenly closes its gill flaps. Suppose the archerfish shoots at a target 2.00 m away, at an angle of 30.0° above the horizontal. With what speed must the water stream be launched if it is not to drop more than 3.00 cm vertically on its path to the target?

19. A firefighter at a distance d from a burning building directs a stream of water from a fire hose at angle θ_i above the horizontal as in Figure P3.19. If the initial speed of the stream is v_i, at what height h does the water strike the building?

20. A soccer player kicks a rock horizontally off a 40.0-m high cliff into a pool of water. If the player hears the sound of the splash 3.00 s later, what was the initial speed given to the rock? Assume the speed of sound in air to be 343 m/s.

21. A playground is on the flat roof of a city school, 6.00 m above the street below. The vertical wall of the building is 7.00 m high, forming a 1.00-m-high railing around the playground. A ball falls to the street below, and a passerby returns it by launching it at an angle of 53.0° above the horizontal at a point 24.0 m from the base of the building

Figure P3.19 *(Frederick McKinney/FPG International)*

Figure P3.22 *(Left, Jed Jacobsohn/Allsport; right, Bill Lee/Dembinsky Photo Associates)*

wall. The ball takes 2.20 s to reach a point vertically above the wall. Find (a) the speed at which the ball was launched, (b) the vertical distance by which the ball clears the wall, and (c) the distance from the wall to the point on the roof where the ball lands.

22. A basketball star covers 2.80 m horizontally in a jump to dunk the ball (Fig. P3.22). His motion through space can be modeled as that of a particle at his *center of mass*, which we will define in Chapter 8. His center of mass is at elevation 1.02 m when he leaves the floor. It reaches a maximum height of 1.85 m above the floor and is at elevation 0.900 m when he touches down again. Determine (a) his time of flight (his "hang time"), (b) his horizontal and (c) vertical velocity components at the instant of takeoff, and (d) his takeoff angle. (e) For comparison, determine the hang time of a white-tailed deer making a jump with center-of-mass elevations $y_i = 1.20$ m, $y_{max} = 2.50$ m, $y_f = 0.700$ m.

23. A fireworks rocket explodes at height h, the peak of its vertical trajectory. It throws out burning fragments in all directions, but all at the same speed v. The bits of ash fall to the ground without air resistance. Find the smallest angle that the final velocity of an impacting fragment makes with the horizontal.

Section 3.4   The Particle in Uniform Circular Motion

24. The orbit of the Moon about the Earth is approximately circular, with a mean radius of 3.84×10^8 m. It takes 27.3 days for the Moon to complete one revolution about the Earth. Find (a) the mean orbital speed of the moon and (b) its centripetal acceleration.

25. The athlete shown in Figure P3.25 rotates a 1.00-kg discus along a circular path of radius 1.06 m. The maximum speed of the discus is 20.0 m/s. Determine the magnitude of the maximum radial acceleration of the discus.

Figure P3.25 *(Allsport)*

26. From information on the endsheets of this book, compute the radial acceleration of a point on the surface of the Earth at the equator, due to the rotation of the Earth about its axis.

27. A tire 0.500 m in radius rotates at a constant rate of 200 rev/min. Find the speed and acceleration of a small

stone lodged in the tread of the tire (on its outer edge). (*Hint:* In one revolution, the stone travels a distance equal to the circumference of its path, $2\pi r$.)

28. As their booster rockets separate, space shuttle astronauts typically feel accelerations up to 3*g*, where $g = 9.80$ m/s². In their training, astronauts ride in a device where they experience such an acceleration as a centripetal acceleration. Specifically, the astronaut is fastened securely at the end of a mechanical arm which then turns at constant speed in a horizontal circle. Determine the rotation rate, in revolutions per second, required to give an astronaut a centripetal acceleration of 3.00*g* while in circular motion with radius 9.45 m.

Section 3.5 Tangential and Radial Acceleration

29. A train slows down as it rounds a sharp horizontal turn, slowing from 90.0 km/h to 50.0 km/h in the 15.0 s that it takes to round the bend. The radius of the curve is 150 m. Compute the acceleration at the moment the train speed reaches 50.0 km/h. Assume it continues to slow down at this time at the same rate.

30. A point on a rotating turntable 20.0 cm from the center accelerates from rest to a final speed of 0.700 m/s in 1.75 s. At $t = 1.25$ s, find the magnitude and direction of (a) the radial acceleration, (b) the tangential acceleration, and (c) the total acceleration of the point.

31. Figure P3.31 represents the total acceleration of a particle moving clockwise in a circle of radius 2.50 m at a certain time. At this instant, find (a) the radial acceleration, (b) the speed of the particle, and (c) its tangential acceleration.

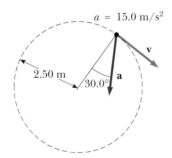

Figure P3.31

32. A ball swings in a vertical circle at the end of a rope 1.50 m long. When the ball is 36.9° past the lowest point on its way up, its total acceleration is $(-22.5\mathbf{i} + 20.2\mathbf{j})$ m/s². At that instant, (a) sketch a vector diagram showing the components of its acceleration, (b) determine the magnitude of its radial acceleration, and (c) determine the speed and velocity of the ball.

Section 3.6 Relative Velocity

33. A river has a steady speed of 0.500 m/s. A student swims upstream a distance of 1.00 km and swims back to the starting point. If the student can swim at a speed of 1.20 m/s in still water, how long does the trip take? Compare this with the time the trip would take if the water were still.

34. How long does it take an automobile traveling in the left lane at 60.0 km/h to pull alongside a car traveling in the right lane at 40.0 km/h if the cars' front bumpers are initially 100 m apart?

35. The pilot of an airplane notes that the compass indicates a heading due west. The airplane's speed relative to the air is 150 km/h. If there is a wind of 30.0 km/h toward the north, find the velocity of the airplane relative to the ground.

36. A child in danger of drowning in a river is being carried downstream by a current that has a speed of 2.50 km/h. The child is 0.600 km from shore and 0.800 km upstream of a boat landing when a rescue boat sets out. (a) If the boat proceeds at its maximum speed of 20.0 km/h relative to the water, what heading relative to the shore should the pilot take? (b) What angle does the boat velocity make with the shore? (c) How long does it take the boat to reach the child?

37. A science student is riding on a flatcar of a train traveling along a straight horizontal track at a constant speed of 10.0 m/s. The student throws a ball into the air along a path that he judges to make an initial angle of 60.0° with the horizontal and to be in line with the track. The student's professor, who is standing on the ground nearby, observes the ball to rise vertically. How high does she see the ball rise?

38. Two swimmers, Alan and Beth, start together at the same point on the bank of a wide stream that flows with a speed v. Both move at the same speed c ($c > v$), relative to the water. Alan swims downstream a distance L and then upstream the same distance. Beth swims so that her motion relative to the Earth is perpendicular to the banks of the stream. She swims the distance L and then back the same distance, so that both swimmers return to the starting point. Which swimmer returns first? (*Note:* First guess the answer.)

Section 3.7 Context Connection — Circular Orbits

39. The astronaut orbiting the Earth in Figure P3.39 (see next page) is preparing to dock with a Westar VI satellite. The satellite is in a circular orbit 600 km above the Earth's surface, where the free-fall acceleration is 8.21 m/s². Take the radius of the Earth as 6400 km. Determine the speed of the satellite and the time required to complete one orbit around the Earth.

Figure P3.39 *(Courtesy of NASA)*

40. An astronaut on the surface of the Moon fires a cannon to launch an experiment package, which leaves the barrel moving horizontally. (a) What must be the muzzle speed of the probe so that it travels completely around the Moon and returns to its original location? (b) How long does this trip around the Moon take? Assume that the free-fall acceleration on the Moon is one-sixth that on the Earth.

Additional Problems

41. When baseball players throw the ball in from the outfield, they usually allow it to take one bounce before it reaches the infield, on the theory that the ball arrives sooner that way. Suppose that the angle at which a bounced ball leaves the ground is the same as the angle at which the outfielder threw it, as in Figure P3.41, but that the ball's speed after the bounce is one half of what it was before the bounce. (a) Assuming the ball is always thrown with the same initial speed, at what angle θ should the fielder throw the ball to make it go the same distance D with one bounce (blue path) as a ball thrown upward at 45.0° with no bounce (green path)? (b) Determine the ratio of the times for the one-bounce and no-bounce throws.

42. A ball on the end of a string is whirled around in a horizontal circle of radius 0.300 m. The plane of the circle is 1.20 m above the ground. The string breaks and the ball lands 2.00 m (horizontally) away from the point on the ground directly beneath the ball's location when the string breaks. Find the radial acceleration of the ball during its circular motion.

43. A projectile is launched up an incline (incline angle ϕ) with an initial speed v_i at an angle θ_i with respect to the horizontal ($\theta_i > \phi$), as shown in Figure P3.43. (a) Show that the projectile travels a distance d up the incline, where

$$d = \frac{2v_i^2 \cos \theta_i \sin(\theta_i - \phi)}{g \cos^2 \phi}.$$

(b) For what value of θ_i is d a maximum, and what is that maximum value?

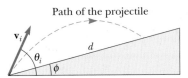

Figure P3.43

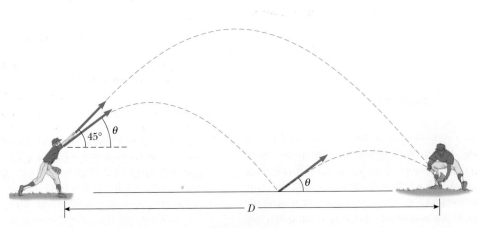

Figure P3.41

44. A fisherman sets out upstream from Metaline Falls on the Pend Oreille River in northwestern Washington State. His small boat, powered by an outboard motor, travels at a constant speed v in still water. The water flows at a lower constant speed v_w. After traveling upstream for 2.00 km, his ice chest falls out of the boat. He notices that the chest is missing only after he has gone upstream for another 15.0 minutes. At that point he turns around and heads back downstream, all the time traveling at the same speed relative to the water. He catches up with the floating ice chest just as it is about to go over the falls at his starting point. How fast is the river flowing? Solve this problem in two ways. (a) First, use the Earth as a reference frame. (With respect to the Earth, the boat travels upstream at speed $v - v_w$ and downstream at $v + v_w$.) (b) Second, use the water as the reference frame. (This is a simpler and more elegant approach. It has important application in many more complicated problems, such as calculating the motion of rockets and Earth satellites and analyzing the scattering of subatomic particles from massive targets.)

45. A home run is hit in such a way that the baseball just clears the top row of bleachers, 21.0 m high, located 130 m from home plate. The ball is hit at an angle of 35.0° to the horizontal, and air resistance is negligible. Find (a) the initial speed of the ball, (b) the time it takes the ball to reach the cheap seats, and (c) the velocity components and the speed of the ball when it passes over the top row. (Assume the ball is hit at a height of 1.00 m above the ground.)

46. A small massive ball is suspended at the bottom end of a light cord 1.00 m long, with the top end of the cord fixed. The ball is set swinging around in a vertical circle. Let θ represent the instantaneous angle the cord makes with the vertical. When the ball is in the two positions where the cord is horizontal, $\theta = 90°$ and $\theta = 270°$, its speed is 5.00 m/s. (a) Find the magnitude of the ball's radial acceleration and tangential acceleration for these positions. (b) Draw vector diagrams to determine the direction of the total acceleration for these two positions. (c) Calculate the magnitude and direction of the total acceleration.

47. A basketball player who is 2.00 m tall is standing on the floor 10.0 m from the basket, as in Figure P3.47. If he shoots the ball at a 40.0° angle with the horizontal, at what initial speed must he throw so that it goes through the hoop without striking the backboard? The basket height is 3.05 m.

48. A ball is thrown with an initial speed v_i at an angle θ_i with the horizontal. The horizontal range of the ball is R, and the ball reaches a maximum height $R/6$. In terms of R and g, find (a) the time the ball is in motion, (b) the ball's speed at the peak of its path, (c) the initial vertical component of its velocity, (d) its initial speed, and (e) the angle

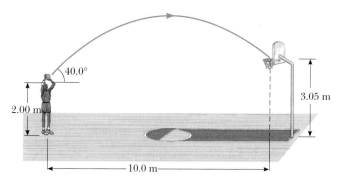

Figure P3.47

θ_i. Suppose the ball is thrown at the same initial speed found in (d), but at the angle appropriate for reaching the greatest height that it can. (f) Find this height. Suppose the ball is thrown at the same initial speed but at the angle for greatest possible range. (g) Find this maximum horizontal range.

49. A person standing at the top of a hemispherical rock of radius R kicks a ball (initially at rest on the top of the rock) to give it horizontal velocity $\mathbf{v}_i$ as in Figure P3.49. (a) What must be its minimum initial speed if the ball is never to hit the rock after it is kicked? (b) With this initial speed, how far from the base of the rock does the ball hit the ground?

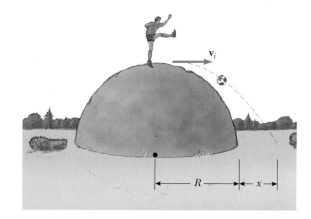

Figure P3.49

50. Your grandfather is copilot of a bomber, flying horizontally over level terrain, with a speed of 275 m/s relative to the ground, at an altitude of 3 000 m. (a) The bombardier releases one bomb. How far will it travel horizontally between its release and its impact on the ground? Ignore the effects of air resistance. (b) The bombardier is inexperienced and does not call, "Bombs away!" Consequently, the pilot maintains the plane's original course, altitude, and speed. Where will the plane be when the bomb hits the ground? (c) At what angle from the vertical was the telescopic bomb sight set so that the bomb hits the target seen in the sight at the time of release?

51. A quarterback throws a football toward a receiver with an initial speed of 20.0 m/s, at an angle of 30.0° above the horizontal. At that instant, the receiver is 20.0 m from the quarterback. In what direction and with what constant speed should the receiver run in order to catch the football at the level at which it was thrown?

52. A truck loaded with cannonball watermelons stops suddenly to avoid running over the edge of a washed-out bridge (Fig. P3.52). The quick stop causes a number of melons to fly off the truck. One melon rolls over the edge with an initial speed $v_i = 10.0$ m/s in the horizontal direction. A cross-section of the bank has the shape of the bottom half of a parabola with its vertex at the edge of the road, and described by the equation $y^2 = 16x$, where x and y are measured in meters. What are the x and y coordinates of the melon when it splatters on the bank?

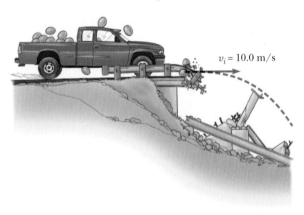

$v_i = 10.0$ m/s

Figure P3.52

53. A car is parked on a steep incline overlooking the ocean. The incline makes an angle of 37.0° below the horizontal. The negligent driver leaves the car in neutral, and the parking brakes are defective. The car rolls from rest down the incline with a constant acceleration of 4.00 m/s², traveling 50.0 m to the edge of a vertical cliff. The cliff is 30.0 m above the ocean. Find (a) the speed of the car when it reaches the edge of the cliff and the time it takes to get there, (b) the velocity of the car when it lands in the ocean, (c) the total time the car is in motion, and (d) the position of the car when it lands in the ocean, relative to the base of the cliff.

54. A catapult launches a rocket at an angle of 53.0° above the horizontal with an initial speed of 100 m/s. The rocket engine immediately starts a burn and for 3.00 s the rocket moves along its initial line of motion with an acceleration of 30.0 m/s². Then its engine fails, and the rocket proceeds to move in free fall. Find (a) the maximum altitude reached by the rocket, (b) its total time of flight, and (c) its horizontal range.

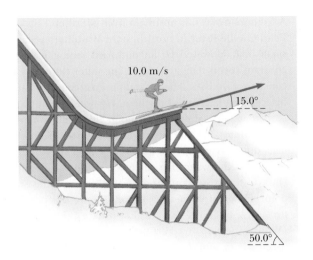

10.0 m/s

15.0°

50.0°

Figure P3.55

55. A skier leaves the ramp of a ski jump with a velocity of 10.0 m/s, 15.0° above the horizontal, as in Figure P3.55. The slope is inclined at 50.0°, and air resistance is negligible. Find (a) the distance from the ramp to where the jumper lands and (b) the velocity components just before the landing. (How do you think the results might be affected if air resistance were included? Note that jumpers lean forward in the shape of an airfoil, with their hands at their sides, to increase their distance. Why does this work?)

56. In a television picture tube (a cathode-ray tube) electrons are emitted with velocity $\mathbf{v}_i$ from a source at the origin of coordinates. The initial velocities of different electrons make different angles θ with the x axis. As they move a distance D along the x axis, the electrons are acted on by a constant electric field, giving each a constant acceleration $\mathbf{a}$ in the x direction. At $x = D$ the electrons pass through a circular aperture, oriented perpendicular to the x axis. At the aperture, the velocity imparted to the electrons by the electric field is much larger than $\mathbf{v}_i$ in magnitude. Show that the velocity vectors of electrons going through the aperture radiate from a certain point on the x axis, which is not the origin. Determine the location of this point. This point is called a *virtual source* and is important in determining where the electron beam hits the screen of the tube.

57. Do not hurt yourself; do not strike your hand against anything. Within these limitations, describe what you do to give your hand a large acceleration. Compute an order-of-magnitude estimate of this acceleration, stating the quantities you measure or estimate and their values.

58. An enemy ship is on the east side of a mountain island, as shown in Figure P3.58. The enemy ship has maneuvered to within 2 500 m of the 1 800-m-high mountain peak and only can shoot projectiles with an initial speed of 250 m/s. If the western shoreline is horizontally 300 m from the peak, what are the distances from the western shore at which a ship can be safe from the bombardment of the enemy ship?

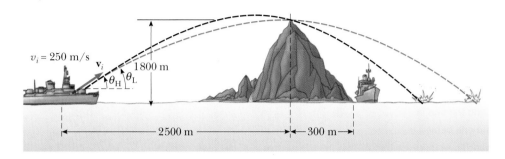

Figure P3.58

ANSWERS TO QUICK QUIZZES

3.1 (a) Yes. An object moving along a curved path at constant speed is accelerating because the direction of the velocity vector is changing. (b) No. If the velocity is constant, there is no change in the velocity and the acceleration is zero.

3.2 (a) No. If the velocity is constant, the magnitude of the instantaneous velocity vector, which is the speed, is constant. (b) Yes. An electron traveling around a curved path at a constant speed has a varying velocity because the direction of the velocity vector is changing.

3.3 You should simply throw it straight up in the air. Because the ball is moving along with you, it will follow a parabolic trajectory with a horizontal component of velocity that is the same as yours.

3.4 (a) At the top of its flight, **v** is horizontal and **a** is vertical. This is the only point at which the velocity and acceleration vectors are perpendicular. (b) If the object is thrown straight up or down, then **v** and **a** will be parallel throughout the downward motion. Otherwise, the velocity and acceleration vectors are never parallel.

A locomotive pulls on the cars of
a long train. The acceleration of
a given car depends on forces
between that car and the one in
front and the one behind.

(Bruce Hands/Stone)

The Laws of Motion

n the preceding two chapters on kinematics, we described the motion of particles based on the definitions of position, velocity, and acceleration. Aside from our discussion of gravity for objects in free-fall, we did not address what causes an object to move as it does. We would like to be able to answer general questions related to the causes of motion, such as "What mechanism causes changes in motion?" and "Why do some objects accelerate at higher rates than others?" In this first chapter on *dynamics*, we shall describe the change in motion of particles, using the concepts of force and mass. We then discuss the three fundamental laws of motion, which are based on experimental observations and were formulated about three centuries ago by Sir Isaac Newton.

4.1 • THE CONCEPT OF FORCE

Everyone has a basic understanding of the concept of force as a result of everyday experiences. When you push or pull an object, you exert a force on it. You exert a force when you throw or kick a ball. In these examples, the word *force* is associated with the result of muscular activity and with some change in the state of motion of an object. Forces do not always cause an object to move, however. For example, as you sit reading this book, the gravitational force acts on your body, and yet you remain stationary. You can push on a heavy block of stone and yet fail to move it.

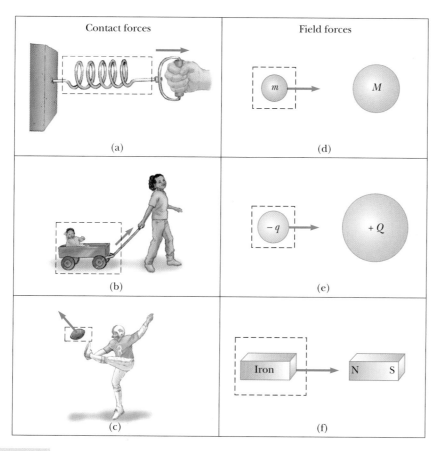

Contact forces	Field forces
(a)	(d)
(b)	(e)
(c)	(f)

Figure 4.1

Some examples of forces applied to various objects. In each case a force is exerted on the particle or object within the boxed area. The environment external to the boxed area provides this force.

This chapter is concerned with the relation between the force on an object and the change in motion of that object. If you pull on a spring, as in Figure 4.1a, the spring stretches. If the spring is calibrated, the distance it stretches can be used to measure the strength of the force. If a child pulls on a wagon, as in Figure 4.1b, the wagon moves. When a football is kicked, as in Figure 4.1c, it is both deformed and set in motion. These are all examples of a class of forces called *contact forces.* That is, they represent the result of physical contact between two objects.

• *Contact forces*

Another class of forces, which do not involve physical contact between two objects but act through empty space, are known as *field forces.* The gravitational force between two objects that causes the free-fall acceleration described in Chapters 2 and 3 is an example of this class of force, and is illustrated in Figure 4.1d. This gravitational force keeps objects bound to the Earth and gives rise to what we commonly call the *weight* of an object. The planets of our solar system are bound under the action of gravitational forces. Another common example of a field force is the electric force that one electric charge exerts on another electric charge, as in Figure 4.1e. These charges might be an electron and proton forming a hydrogen atom. A third example of a field force is the force that a bar magnet exerts on a piece of iron, as shown in Figure 4.1f.

• *Field forces*

The distinction between contact forces and field forces is not as sharp as you may have been led to believe by the preceding discussion. At the atomic level, all

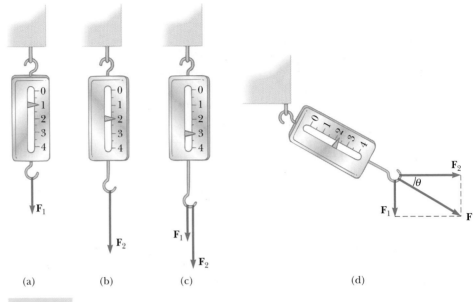

Figure 4.2

The vector nature of a force is tested with a spring scale. (a) A downward vertical force $\mathbf{F}_1$ elongates the spring 1.00 cm. (b) A downward vertical force $\mathbf{F}_2$ elongates the spring 2.00 cm. (c) When $\mathbf{F}_1$ and $\mathbf{F}_2$ are applied simultaneously, the spring elongates by 3.00 cm. (d) When $\mathbf{F}_1$ is downward and $\mathbf{F}_2$ is horizontal, the combination of the two forces elongates the spring $\sqrt{(1.00)^2 + (2.00)^2} = \sqrt{5.00}$ cm.

the forces classified as contact forces turn out to be due to electric (field) forces of the type illustrated in Figure 4.1e. Nevertheless, in understanding macroscopic phenomena, it is convenient to use both classifications of forces.

It is sometimes convenient to use the linear deformation of a spring to measure force, as in the case of a common spring scale. Suppose a force is applied vertically to a spring that has a fixed upper end, as in Figure 4.2a. We can calibrate the spring by defining the unit force $\mathbf{F}_1$ as the force that produces an elongation of 1.00 cm. If a force $\mathbf{F}_2$, applied as in Figure 4.2b, produces an elongation of 2.00 cm, the magnitude of $\mathbf{F}_2$ is 2.00 units. If the two forces $\mathbf{F}_1$ and $\mathbf{F}_2$ are applied simultaneously, as in Figure 4.2c, the elongation of the spring is 3.00 cm because the forces are applied in the same direction and their magnitudes add. If the two forces $\mathbf{F}_1$ and $\mathbf{F}_2$ are applied in perpendicular directions, as in Figure 4.2d, the elongation is $\sqrt{(1.00)^2 + (2.00)^2}$ cm $= \sqrt{5.00}$ cm $= 2.24$ cm. The single force $\mathbf{F}$ that would produce this same elongation is the vector sum of $\mathbf{F}_1$ and $\mathbf{F}_2$, as described in Figure 4.2d. That is, $|\mathbf{F}| = \sqrt{F_1{}^2 + F_2{}^2} = 2.24$ units, and its direction is $\theta = \tan^{-1}(-0.500) = -26.6°$. **Because forces have been experimentally verified to behave as vectors, you must use the rules of vector addition to obtain the net force on an object.**

4.2 • NEWTON'S FIRST LAW

See the *Core Concepts in Physics CD-ROM*, Screen 4.2

Before about 1600, scientists felt that the natural state of matter was the state of rest, because moving objects on the Earth tended to come to rest if a force on them was not maintained. (Today, of course, we recognize that friction can cause a moving object to come to rest.) Galileo took a different approach to motion and the natural state of matter. He devised thought experiments, such as that of an object moving on a frictionless surface, and concluded that it is not the nature of an object to stop once set in motion; rather, it is an object's nature to resist changes in its motion. In his words, "Any velocity, once imparted to a moving body, will be rigidly maintained as long as the external causes of retardation are removed."

This new approach to motion was later formalized by Newton in a statement that has come to be known as **Newton's first law of motion:**

> In the absence of external forces, an object at rest remains at rest and an object in motion continues in motion with a constant velocity (that is, with a constant speed in a straight line).

In simpler terms, we can say that **when no force acts on a body, its acceleration is zero.** Thus, an object has a tendency to maintain its original state of motion in the absence of a force. This tendency is called **inertia,** and Newton's first law is sometimes called the **law of inertia.** It is not a common experience to observe Newton's first law, because it is difficult to completely eliminate all forces on an object, especially the force of friction.

An example of almost uniform motion on a nearly frictionless plane is the motion of a light disk on a layer of air, as in Figure 4.3. If the disk is given an initial velocity, it coasts a great distance before coming to rest. This idea is used in the game of air hockey, in which the disk makes many collisions with the walls of the air hockey table before coming to rest.

Consider a spacecraft traveling in space, far removed from any planets or other matter. The spacecraft requires some propulsion system to change its velocity. If the propulsion system is turned off when the spacecraft reaches a velocity **v**, however, the spacecraft "coasts" in space with that velocity, and the astronauts enjoy a "free ride" (i.e., no propulsion system is required to keep them moving at the velocity **v**).

Finally, recall our discussion in Chapter 2 about the proportionality between force and acceleration:

$$\mathbf{F} \propto \mathbf{a}$$

Newton's first law tells us that the velocity does not change if no force acts on an object—the object maintains its state of motion. The preceding proportionality tells us that if a force *does* act, then a change does occur in the motion, measured by the acceleration. This notion will form the basis of Newton's second law, and we shall provide more details on this concept shortly.

Inertial Frames

Newton's first law defines a special set of reference frames called inertial frames. **An inertial frame of reference is one in which Newton's first law is valid.** For example, suppose you are on a perfectly smoothly moving train, observing a stationary air hockey puck on a perfectly level operating air hockey table. If the train were moving at a constant velocity, you would see the puck remain at rest, obeying Newton's first law. If the train were to accelerate in the forward direction, however, you would see the puck accelerate toward the back of the train—the same thing happens when something on your dashboard slides off when you step on the gas pedal. The train moving at constant velocity represents an inertial frame. The puck at rest remains at rest, and Newton's first law is obeyed. The accelerating train is not an inertial frame. According to you as the observer on the train, there

Isaac Newton (1642–1727)

Newton, an English physicist and mathematician, was one of the most brilliant scientists in history. Before the age of 30, he formulated the basic concepts and laws of mechanics, discovered the law of universal gravitation, and invented the mathematical methods of calculus. As a consequence of his theories, Newton was able to explain the motions of the planets, the ebb and flow of the tides, and many special features of the motions of the Moon and the Earth. He also interpreted many fundamental observations concerning the nature of light. His contributions to physical theories dominated scientific thought for two centuries and remain important today. *(Giraudon/Art Resource)*

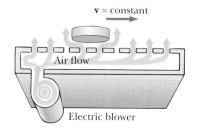

Figure 4.3

Air hockey takes advantage of Newton's first law to make the game more exciting. A disk moving on a column of air is an approximation of uniform motion, that is, motion in which the acceleration is zero.

PITFALL PREVENTION 4.1
Newton's first law

Newton's first law does *not* say what happens for an object with *zero net force*, that is, multiple forces that cancel; it says what happens *in the absence of a force*. This subtle but important difference allows us to define force as that which causes a change in the motion. The description of an object under the effect of forces that balance is covered by Newton's second law.

is no force on the puck, yet it accelerates from rest toward the back of the train, violating Newton's first law.

Any reference frame that moves with constant velocity with respect to an inertial frame is itself an inertial frame. The Earth is not an inertial frame because of its orbital motion about the Sun and rotational motion about its own axis, both of which represent centripetal accelerations. As the Earth travels in its nearly circular orbit about the Sun, it experiences a centripetal acceleration of about 6.0×10^{-3} m/s^2 toward the Sun, as found in Example 3.5. In addition, because the Earth rotates about its own axis once every 24 h, a point on the equator experiences an additional centripetal acceleration of 3.4×10^{-2} m/s^2 toward the center of the Earth. These accelerations are small compared with g, however, and can often be ignored—this is a simplification model. In most situations we shall model a reference frame on or near the Earth's surface as an inertial frame.

Quick Quiz 4.1

Is it possible to have motion in the absence of a force? Is it possible to have force in the absence of motion?

Quick Quiz 4.2

(a) If a single force acts on an object, does the object accelerate? (b) If an object experiences an acceleration, is a force acting on it? (c) If an object experiences no acceleration, is no force acting on it?

See Screen 4.3

4.3 • INERTIAL MASS

• *Inertial mass*

Consider playing catch with either a basketball or a bowling ball. If both are coming toward you with the same velocity, which would be easier to catch? Which one would be harder to throw? The bowling ball has more mass than the basketball, so it will be harder to catch and to throw. When you either catch or throw the ball, you are changing its motion. Newton's first law tells us that if we make no attempt to change the motion of an object (i.e., if we apply no force on it), then its motion remains uniform. What if we *do* attempt to change its motion by applying a force on the object, as we do with the basketball and bowling ball? We find that the object resists this change. We quantify this resistance as the *mass* of the object, and, in particular, we call this the inertial mass. **Inertial mass is the measure of an object's resistance to a change in motion in response to an external force.** The bowling ball has more inertial mass than the basketball, so it is harder to change the motion of the bowling ball by catching or throwing it. Inertial mass is different in definition from gravitational mass, which we discuss later in this chapter, but has the same value, so we call them both simply *mass*.

A quantitative measurement of mass can be made by comparing the accelerations that a given force produces on different objects. Suppose a force acting on an object of mass m_1 produces an acceleration $\mathbf{a}_1$, and the *same force* acting on an object of mass m_2 produces an acceleration $\mathbf{a}_2$. The ratio of the two masses is defined as the *inverse* ratio of the magnitudes of the accelerations produced by the same force:

$$\frac{m_1}{m_2} \equiv \frac{a_2}{a_1}$$ [4.1]

WEB

For a "virtual museum of Sir Isaac Newton and the history of science," visit **www.newton.org.uk**

If one mass is standard and known—say, 1 kg—the mass of an unknown object can be obtained from acceleration measurements. For example, if the standard 1-kg object undergoes an acceleration of 3 m/s^2 under the influence of some force, an object experiencing an acceleration of 1.5 m/s^2 under the action of the same force must have a mass of 2 kg.

Mass is an inherent property of an object, independent of the object's surroundings and of the method used to measure it. It is an experimental fact that mass is a scalar quantity. As a scalar quantity, mass obeys the rules of ordinary arithmetic. That is, several masses can be combined in simple numerical fashion—if you combine a 3-kg mass with a 5-kg mass, the total mass is 8 kg. This can be verified experimentally by comparing the acceleration of each object produced by a known force with the acceleration of the combined system using the same force.

Mass should not be confused with weight. **Mass and weight are two different quantities.** What we call the weight of an object is equal to the magnitude of the gravitational force exerted by the planet on which the object resides. For example, a person who weighs 180 lb on Earth weighs only about 30 lb on the Moon. On the other hand, the mass of an object is the same everywhere. A given object exhibits a fixed amount of resistance to changes in motion regardless of its location. An object of mass 2 kg on Earth also has a mass of 2 kg on the Moon.

• *Mass and weight are different quantities*

4.4 • NEWTON'S SECOND LAW – THE PARTICLE UNDER A NET FORCE

   See Screen 4.4

Newton's first law explains what happens to an object when no force acts on it: It either remains at rest or moves in a straight line with constant speed. This allows us to define an inertial frame of reference. It also allows us to identify force as that which changes motion. Newton's second law answers the question of what happens to an object that has a nonzero net force acting on it, based on our discussion of mass in the preceding section.

Imagine you are pushing a block of ice across a frictionless horizontal surface. When you exert some horizontal force **F**, the block moves with some acceleration **a**. Experiments show that if you apply a force twice as large to the same object, the acceleration doubles. If you increase the applied force to 3**F**, the original acceleration is tripled, and so on. From such observations, we conclude that **the acceleration of an object is directly proportional to the net force acting on it.** We alluded to this proportionality in our discussion of acceleration in Chapter 2. We are now ready to extend that discussion.

These observations and those in Section 4.3 relating mass and acceleration are summarized in **Newton's second law:**

PITFALL PREVENTION 4.2

Force is the cause of changes in motion

Be sure that you are clear on the role of force. Many times, students make the mistake of thinking that force is the cause of motion. But we can have motion in the absence of forces, as described in Newton's first law. Be sure to understand that force is the cause of *changes* in motion.

The acceleration of an object is directly proportional to the net force acting on it and inversely proportional to its mass.

We write this as

$$\mathbf{a} \propto \frac{\Sigma \mathbf{F}}{m}$$

where $\Sigma\mathbf{F}$ is the **net force,** which is the vector sum of *all* forces acting on the object of mass *m*. If the object consists of a system of individual elements, the net force is the vector sum of all forces *external* to the system. Any *internal* forces—that is, forces between elements of the system—are not included because they do not affect the motion of the entire system. The net force is also sometimes called the *resultant force,* the *sum of the forces,* the *total force,* or the *unbalanced force.*

Newton's second law in mathematical form is a statement of this relationship that makes the preceding proportionality an equality:*

• *Newton's second law*

$$\sum \mathbf{F} = m\mathbf{a} \qquad\qquad \text{[4.2]}$$

Note that Equation 4.2 is a *vector* expression and hence is equivalent to the following three component equations:

$$\sum F_x = ma_x \qquad \sum F_y = ma_y \qquad \sum F_z = ma_z \qquad \text{[4.3]}$$

Quick Quiz 4.3

If an object experiences a net force in the *x* direction, does it *move* in the *x* direction?

Newton's second law introduces us to a new analysis model—the particle under a net force. If a particle, or an object that can be modeled as a particle, is under the influence of a net force, then Equation 4.2, the mathematical statement of Newton's second law, can be used to describe its motion. The acceleration is constant if the net force is constant. Thus, the particle under a net force may have its motion predicted as a particle under constant acceleration. The net force does not have to be constant, however. We shall investigate situations in this chapter and the next involving both constant and varying forces.

Unit of Force

The SI unit of force is the **newton,** which is defined as the force that, when acting on a 1-kg mass, produces an acceleration of 1 m/s^2.

From this definition and Newton's second law, we see that the newton can be expressed in terms of the fundamental units of mass, length, and time:

$$1\ \text{N} \equiv 1\ \text{kg} \cdot \text{m/s}^2 \qquad\qquad \text{[4.4]}$$

The units of mass, acceleration, and force are summarized in Table 4.1. Most of the calculations we shall make in our study of mechanics will be in SI units. Equalities between units in the SI and British engineering systems are given in Appendix A.

TABLE 4.1	Units of Mass, Acceleration, and Force		
System of Units	**Mass** **(M)**	**Acceleration** **(L/T^2)**	**Force** **(ML/T^2)**
SI	kg	m/s^2	N = kg·m/s^2
British engineering	slug	ft/s^2	lb = slug·ft/s^2

---

* Equation 4.2 is valid only when the speed of the object is much less than the speed of light. We will treat the relativistic situation in Chapter 9.

THINKING PHYSICS 4.1

In a train, such as that in the opening photograph of this chapter, the cars are connected by *couplers*. The couplers between the cars exert forces on the cars as the train is pulled by the locomotive in the front. Imagine that the train is speeding up in the forward direction. As you imagine moving from the locomotive to the caboose, does the force exerted by the couplers *increase, decrease,* or *stay the same?* What if the engineer applies the brakes? How does the force vary from locomotive to caboose in this case? (Assume that the only brakes applied are those on the engine.)

Reasoning   The force *decreases* from the front of the train to the back. The coupler between the locomotive and the first car must apply enough force to accelerate all of the remaining cars. As we move back along the train, each coupler is accelerating less mass behind it. The last coupler only has to accelerate the caboose, so it exerts the smallest force. If the brakes are applied, the force decreases from front to back of the train also. The first coupler, at the back of the locomotive, must apply a large force to slow down all of the remaining cars. The final coupler must only apply a force large enough to slow down the mass of the caboose.

Example 4.1   An Accelerating Hockey Puck

A 0.30-kg hockey puck slides on the horizontal frictionless surface of an ice rink. It is struck simultaneously by two different hockey sticks. The two constant forces that act on the puck as a result of the hockey sticks are parallel to the ice surface and are shown in the pictorial representation in Figure 4.4. The force $\mathbf{F}_1$ has a magnitude of 5.0 N, and $\mathbf{F}_2$ has a magnitude of 8.0 N. Determine the acceleration of the puck while it is in contact with the two sticks.

Solution   The puck is modeled as a particle under a net force. We first find the components of the net force. The component of the net force in the *x* direction is

$$\sum F_x = F_{1x} + F_{2x} = F_1 \cos 20° + F_2 \cos 60°$$
$$= (5.0 \text{ N})(0.940) + (8.0 \text{ N})(0.500) = 8.7 \text{ N}$$

The component of the net force in the *y* direction is

$$\sum F_y = F_{1y} + F_{2y} = -F_1 \sin 20° + F_2 \sin 60°$$
$$= -(5.0 \text{ N})(0.342) + (8.0 \text{ N})(0.866) = 5.2 \text{ N}$$

Now we use Newton's second law in component form to find the *x* and *y* components of acceleration:

$$a_x = \frac{\sum F_x}{m} = \frac{8.7 \text{ N}}{0.30 \text{ kg}} = 29 \text{ m/s}^2$$

$$a_y = \frac{\sum F_y}{m} = \frac{5.2 \text{ N}}{0.30 \text{ kg}} = 17 \text{ m/s}^2$$

The acceleration has a magnitude of

$$a = \sqrt{(29)^2 + (17)^2} \text{ m/s}^2 = \boxed{34 \text{ m/s}^2}$$

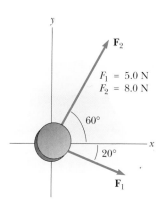

Figure 4.4

(Example 4.1) A hockey puck moving on a frictionless surface accelerates in the direction of the net force, $\Sigma\mathbf{F} = \mathbf{F}_1 + \mathbf{F}_2$.

and its direction is

$$\theta = \tan^{-1} \frac{a_y}{a_x} = \tan^{-1} \frac{17 \text{ m/s}^2}{29 \text{ m/s}^2} = \boxed{30°}$$

relative to the positive *x* axis.

EXERCISE   Determine the components of a third force that, when applied to the puck along with the two in Figure 4.4, causes the puck to have zero acceleration.

Answer $F_x = -8.7 \text{ N}, F_y = -5.2 \text{ N}$

EXERCISE A 6.0-kg object undergoes an acceleration of 2.0 m/s². (a) What is the magnitude of the net force acting on the object? (b) If this same force is applied to a 4.0-kg object, what acceleration does it produce?

Answer (a) 12 N (b) 3.0 m/s²

EXERCISE A 1.80×10^3-kg car is traveling in a straight line with a speed of 25.0 m/s. What is the magnitude of the constant horizontal force needed to bring the car to rest in a distance of 80.0 m?

Answer 7.03×10^3 N

PITFALL PREVENTION 4.5

Weight "of an object"

We are familiar with the common phrase, the "weight of an object." Keep in mind that, despite this everyday phrase, weight is not a property of an object. It is a measure of the gravitational force between the object and the Earth. Thus, weight is a property of a *system* of items—the object and the Earth. As the system changes, for example, by lifting the object well off the surface, the weight changes, due to the decreasing value of *g*. We could also move the object into a new system, consisting of the object and Mars, for example. Then the weight will be very different from that on the Earth because of the different value of *g* on Mars.

4.5 • THE GRAVITATIONAL FORCE AND WEIGHT

We are well aware that all objects are attracted to the Earth. The force exerted by the Earth on an object is the **gravitational force** $\mathbf{F}_g$. This force is directed toward the center of the Earth.* The magnitude of the gravitational force is called the **weight** F_g of the object.

We have seen in Chapters 2 and 3 that a freely falling object experiences an acceleration $\mathbf{g}$ directed toward the center of the Earth. A freely falling object has only one force on it, the gravitational force, so the net force on the object is equal to the gravitational force:

$$\sum \mathbf{F} = \mathbf{F}_g$$

Because the acceleration of a freely falling object is equal to the free-fall acceleration $\mathbf{g}$, it follows that

$$\sum \mathbf{F} = m\mathbf{a} \longrightarrow \mathbf{F}_g = m\mathbf{g}$$

or, in magnitude,

$$F_g = mg \tag{4.5}$$

Astronaut Edwin E. Aldrin, Jr., walking on the Moon after the *Apollo 11* lunar landing. The life-support system strapped to his back weighed 300 lb on Earth. During his training, a 50-lb mock-up system was used to simulate the reduction of gravity on the Moon. Although this effectively simulated the reduced weight of the unit on the Moon, it did not correctly represent the unchanging mass. The system was easier to accelerate during the training on Earth by jumping or twisting than it was on the Moon. *(Courtesy of NASA)*

---

* This statement represents a simplification model in that it ignores the fact that the mass distribution of the Earth is not perfectly spherical.

Because it depends on g, weight varies with location, as we mentioned in Section 4.3. Objects weigh less at higher altitudes than at sea level because g decreases with increasing distance from the center of the Earth. Hence, weight, unlike mass, is not an inherent property of an object. For example, if an object has a mass of 70 kg, then its weight in a location where $g = 9.80$ m/s² is $mg = 686$ N. At the top of a mountain where $g = 9.76$ m/s², the object's weight would be 683 N. Thus, if you want to lose weight without going on a diet, climb a mountain or weigh yourself at 30 000 ft during an airplane flight.

Because $F_g = mg$, we can compare the masses of two objects by measuring their weights with a spring scale. At a given location (so that g is fixed) the ratio of the weights of two objects equals the ratio of their masses.

Equation 4.5 quantifies the gravitational force on the object, but notice that this equation does not require the object to be moving. Even for a stationary object, or an object on which several forces act, Equation 4.5 can be used to calculate the magnitude of the gravitational force. This results in a subtle shift in the interpretation of m in the equation. The mass m in Equation 4.5 is playing the role of determining the strength of the gravitational attraction between the object and the Earth. This is a completely different role from that previously described for mass, that of measuring the resistance to changes in motion in response to an external force. Thus, we call m in this type of equation the **gravitational mass.** Despite this quantity being different from inertial mass, it is one of the experimental conclusions in Newtonian dynamics that gravitational mass and inertial mass have the same value.

PITFALL PREVENTION 4.6
Kilogram is not a unit of weight

You may have seen the "conversion" 1 kg = 2.2 lb. Be sure not to misinterpret this common statement. Despite popular statements of weights expressed in kilograms, the kilogram is not a unit of *weight*, it is a unit of *mass*. On the Earth's surface, an object with a mass of 1 kg will have a weight of 2.2 lbs, but this will not be true for the object on the Moon. The conversion statement is not an equality (even though you may see it written that way); it is an *equivalence* that is only valid on the surface of the Earth.

Quick Quiz 4.4

Suppose you are talking by interplanetary telephone to your friend, who lives on the Moon. He tells you that he has just won a piece of gold weighing one newton in a contest. Excitedly, you tell him that you entered the Earth version of the same contest and also won a newton of gold! Who's richer?

Quick Quiz 4.5

A baseball of mass m is thrown upward with some initial speed. If air resistance is ignored, what is the magnitude of the force on the ball (a) when it reaches half its maximum height and (b) when it reaches its peak?

PITFALL PREVENTION 4.7
Differentiate between *g* and g.

Be sure not to confuse the italicized symbol g that we use for the magnitude of the free-fall acceleration with the symbol g that is used for grams.

THINKING PHYSICS 4.2

In the absence of air friction, it is claimed that all objects fall with the same acceleration. A heavier object is pulled to the Earth with more force than a light object. Why does the heavier object not fall faster?

Reasoning It is indeed true that the heavier object is pulled with a larger force. The *strength* of the force is determined by the gravitational mass of the object. The *resistance* to the force and, therefore, to the change in motion of the object, is represented by the inertial mass. Thus, if an object has twice as much mass as another, it is pulled to the Earth with twice the force, but it also exhibits twice the resistance to having its motion changed. These factors cancel, so that the change in motion, the acceleration, is the same for all objects, regardless of mass.

See Screen 4.5

- *A statement of Newton's third law*

4.6 • NEWTON'S THIRD LAW

Newton's third law conveys the notion that forces are always interactions between two objects: **If two objects interact, the force $\mathbf{F}_{12}$ exerted by object 1 on object 2 is equal in magnitude but opposite in direction to the force $\mathbf{F}_{21}$ exerted by object 2 on object 1:**

$$\mathbf{F}_{12} = -\mathbf{F}_{21} \qquad [4.6]$$

When it is important to designate forces as interactions between two objects, we will use this subscript notation, where $\mathbf{F}_{ab}$ means "the force exerted *by* a *on* b." The third law, illustrated in Figure 4.5a, is equivalent to stating that **forces always occur in pairs,** or that **a single isolated force cannot exist.** The force that object 1 exerts on object 2 may be called the *action force* and the force of object 2 on object 1 the *reaction force*. In reality, either force can be labeled the action or reaction force. **The action force is equal in magnitude to the reaction force and opposite in direction. In all cases, the action and reaction forces act on different objects and must be of the same type.** For example, the force acting on a freely falling projectile is the gravitational force exerted by the Earth on the projectile $\mathbf{F}_g = \mathbf{F}_{Ep}$ (E = Earth, p = projectile), and the magnitude of this force is *mg*. The reaction to this force is the gravitational force exerted by the projectile on the Earth $\mathbf{F}_{pE} = -\mathbf{F}_{Ep}$. The reaction force $\mathbf{F}_{pE}$ must accelerate the Earth toward the projectile just as the action force $\mathbf{F}_{Ep}$ accelerates the projectile toward the Earth. Because the Earth has such a large mass, however, its acceleration due to this reaction force is negligibly small.

Another example of Newton's third law in action is shown in Figure 4.5b. The force $\mathbf{F}_{hn}$ exerted by the hammer on the nail (the action) is equal in magnitude and opposite the force $\mathbf{F}_{nh}$ exerted by the nail on the hammer (the reaction). This latter force stops the forward motion of the hammer when it strikes the nail.

You directly experience the third law if you slam your fist against a wall or kick a football with your bare foot. You can feel the force back on your fist or your foot. You should be able to identify the action and reaction forces in these cases.

PITFALL PREVENTION 4.8
Newton's third law

 This is such an important and often misunderstood notion that it will be repeated here in a Pitfall Prevention. Newton's third law action and reaction forces act on *different* objects. Two forces acting on the same object, even if they are equal in magnitude and opposite in direction, *cannot* be an action–reaction pair.

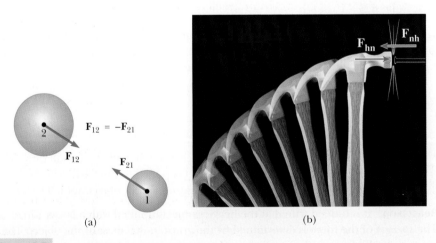

Figure 4.5

Newton's third law. (a) The force $\mathbf{F}_{12}$ exerted by object 1 on object 2 is equal in magnitude and opposite in direction to the force $\mathbf{F}_{21}$ exerted by object 2 on object 1. (b) The force $\mathbf{F}_{hn}$ exerted by the hammer on the nail is equal in magnitude and opposite in direction to the force $\mathbf{F}_{nh}$ exerted by the nail on the hammer. (*John Gillmoure, The Stock Market*)

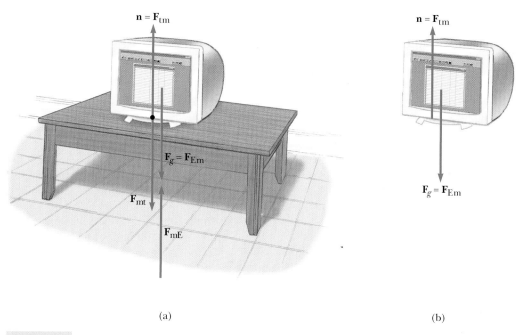

(a)   (b)

Figure 4.6

(a) When a computer monitor is sitting on a table, several forces are acting. (b) The free-body diagram for the monitor. The forces acting on the monitor are the normal force $\mathbf{n} = \mathbf{F}_{tm}$ and the gravitational force $\mathbf{F}_g = \mathbf{F}_{Em}$. These are the only forces acting on the monitor (notice "m" as the second subscript on these forces, but no other forces) and the only forces that should appear in a free-body diagram for the monitor. The other forces in part (a) are as follows. The reaction force to $\mathbf{n}$ is the force $\mathbf{F}_{mt}$ of the monitor on the table. The reaction to $\mathbf{F}_g$ is the force $\mathbf{F}_{mE}$ of the monitor on the Earth.

The Earth exerts a gravitational force $\mathbf{F}_g$ on any object. If the object is a computer monitor at rest on a table, as in the pictorial representation in Figure 4.6a, the reaction force to $\mathbf{F}_g = \mathbf{F}_{Em}$ is the force exerted by the monitor on the Earth $\mathbf{F}_{mE} = -\mathbf{F}_{Em}$. The monitor does not accelerate because it is held up by the table. The table exerts on the monitor an upward force $\mathbf{n} = \mathbf{F}_{tm}$, called the **normal force.*** This force prevents the monitor from falling through the table; it can have any value needed, up to the point of breaking the table. From Newton's second law, we see that, because the monitor has zero acceleration, it follows that $\Sigma\mathbf{F} = \mathbf{n} - m\mathbf{g} = 0$, or $n = mg$. The normal force balances the gravitational force on the monitor, so that the net force on the monitor is zero. The reaction to $\mathbf{n}$ is the force exerted by the monitor downward on the table, $\mathbf{F}_{mt} = -\mathbf{F}_{tm}$.

Note that the forces acting on the monitor are $\mathbf{F}_g$ and $\mathbf{n}$, as shown in Figure 4.6b. The two reaction forces $\mathbf{F}_{mE}$ and $\mathbf{F}_{mt}$ are exerted on objects other than the monitor. Remember, the two forces in an action–reaction pair always act on two different objects.

Figure 4.6 illustrates an extremely important difference between a pictorial representation and a simplified pictorial representation for solving problems involving forces. Figure 4.6a shows many of the forces in the situation—those on the monitor, one on the table, and one on the Earth. Figure 4.6b, by contrast, shows only the forces on *one object,* the monitor. This is a critical pictorial representation

• *Normal force*

PITFALL PREVENTION 4.9

_n_ does not always equal _mg_

In the situation shown in Figure 4.6, we find that $n = mg$. There are many situations in which the normal force has the same magnitude as the gravitational force, but *do not* adopt this as a general rule—this is a common student pitfall. If the problem involves an object on an incline, if there are applied forces with vertical components, or if there is a vertical acceleration of the system, then $n \neq mg$. *Always* apply Newton's second law to find the relationship between n and mg.

*The word *normal* is used because the direction of $\mathbf{n}$ is always *perpendicular* to the surface.

- *A free-body diagram is a critical step when applying Newton's laws*

called a **free-body diagram.** When analyzing a particle under a net force, we are interested in the net force on one object, an object of mass *m*, which we will model as a particle. Thus, a free-body diagram helps us to isolate only those forces on the object and eliminate the other forces from our analysis. The free-body diagram can be simplified further, if you wish, by representing the object, such as the monitor in this case, as a particle, by simply drawing a dot.

Quick Quiz 4.6

If a sports car collides head-on with a massive truck, which vehicle experiences the greater force? Which vehicle experiences the greater acceleration?

Quick Quiz 4.7

What force causes an automobile to move? A propeller-driven airplane? A rocket? A rowboat?

Quick Quiz 4.8

A large man and a small boy stand facing each other on frictionless ice. They put their hands together and push toward each other, so that they move away from each other. Who exerts the larger force? Who experiences the large acceleration? Who moves away with the higher velocity? Who moves over a longer distance while their hands are in contact?

THINKING PHYSICS 4.3

A horse pulls on a sled with a horizontal force, causing it to accelerate as in Figure 4.7a. Newton's third law says that the sled exerts a force of equal magnitude and opposite direction on the horse. In view of this, how can the sled accelerate—don't these forces cancel?

Reasoning When applying Newton's third law, it is important to remember that the forces involved act on different objects. Notice that the force exerted by the horse acts *on the sled*, whereas the force exerted by the sled acts *on the horse.* Because these forces act on different objects, they cannot cancel.

The horizontal forces exerted on the *sled* alone are the forward force **F** exerted by the horse and the backward force of friction f_{sled} between sled and surface (Fig. 4.7b). When **F** exceeds f_{sled}, the sled accelerates to the right.

The horizontal forces exerted on the *horse* alone are the forward friction force f_{horse} from the ground and the backward force **F** exerted by the sled (Fig. 4.7c). The resultant of these two forces causes the horse to accelerate. When f_{horse} exceeds **F**, the horse accelerates to the right.

The two forces described in the text of the problem act on different objects, so they cannot cancel. If we consider the horse and sled as a system, the two forces described are *internal* to the system and, therefore, cannot affect the motion of the system.

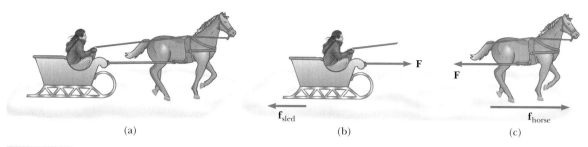

(a)   (b)   (c)

Figure 4.7

(Thinking Physics 4.3)

4.7 • APPLICATIONS OF NEWTON'S LAWS

See Screen 4.6

In this section we present some simple applications of Newton's laws to objects that are either in equilibrium (**a** = 0) or accelerating under the action of constant external forces. We shall assume that the objects behave as particles so that we need not worry about rotational motion or other complications. In this section we also apply some additional simplification models. We ignore the effects of friction for those problems involving motion. This is equivalent to stating that the surfaces are *frictionless*. We usually ignore the masses of any ropes or strings involved. In this approximation, the magnitude of the force exerted at any point along a string is the same at all points along the string. In problem statements, the terms *light* and *of negligible mass* are used to indicate that a mass is to be ignored when you work the problem. These two terms are synonymous in this context.

When we apply Newton's laws to an object, we shall be interested only in those external forces that act *on the object*. For example, in Figure 4.6 the only external forces acting on the monitor are **n** and $\mathbf{F}_g$. The reactions to these forces, $\mathbf{F}_{mt}$ and $\mathbf{F}_{mE}$, act on the table and on the Earth, respectively, and do not appear in Newton's second law as applied to the monitor. The importance of drawing a proper free-body diagram to ensure that you are considering the correct forces cannot be stressed enough.

When an object such as a block is being pulled by a rope or string attached to it, the rope exerts a force on the object. The magnitude of this force is called the **tension** in the rope. Its direction is along the rope, away from the object.

Consider a crate being pulled to the right on a frictionless, horizontal surface, as in Figure 4.8a. Suppose you are asked to find the acceleration of the crate and the force the floor exerts on it. First, note that the horizontal force being applied to the crate acts through the rope. The force the rope exerts on the crate is denoted by the symbol **T** and its magnitude is the tension in the rope.

Because we are interested only in the motion of the crate, we must be able to *identify any and all external forces acting on it*. These are illustrated in the free-body diagram in Figure 4.8b. In addition to the force **T**, the free-body diagram for the crate includes the gravitational force $\mathbf{F}_g$ and the normal force **n** exerted by the floor on the crate. The *reactions* to the forces we have listed—namely, the force exerted by the crate on the rope, the force exerted by the crate on the Earth, and the force exerted by the crate on the floor—are not included in the free-body diagram because they act on *other* objects and not on the crate.

Now let us apply Newton's second law to the crate. First we must choose an appropriate coordinate system. In this case it is convenient to use the coordinate

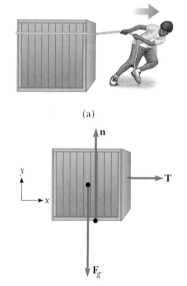

(a)

(b)

Figure 4.8

(a) A crate being pulled to the right on a frictionless surface. (b) The free-body diagram that represents the external forces on the crate.

Rock climbers depend on the tension forces in ropes for their safety. *(Scott Markewitz/FPG)*

system shown in Figure 4.8b, with the *x* axis horizontal and the *y* axis vertical. We can apply Newton's second law in the *x* direction, *y* direction, or both, depending on what we are asked to find in the problem. In addition, we may be able to use the equations of motion for the particle under constant acceleration that we discussed in Chapter 2. You should use these equations only when the acceleration is constant, however, which is the case if the net force is constant. For example, if the force **T** in Figure 4.8 is constant, then the acceleration in the *x* direction is also constant, because $\mathbf{a} = \mathbf{T}/m$. Hence, if we need to find the position or the velocity of the crate at some instant of time, we can use the equations of motion with constant acceleration.

The Particle in Equilibrium

Objects that are either at rest or moving with constant velocity are said to be in **equilibrium.** From Newton's second law with $\mathbf{a} = 0$, this condition of equilibrium can be expressed as

$$\sum \mathbf{F} = 0 \qquad [4.7]$$

This statement signifies that the vector sum of all the forces (the net force) acting on an object in equilibrium is zero.* If a particle is subject to forces but exhibits an acceleration of zero, we use Equation 4.7 to analyze the situation, as we shall see in some following examples.

Usually, the problems we encounter in our study of equilibrium are easier to solve if we work with Equation 4.7 in terms of the components of the external forces acting on an object. By this we mean that, in a two-dimensional problem, the sum of all the external forces in the *x* and *y* directions must separately equal zero; that is,

$$\sum F_x = 0 \qquad \sum F_y = 0 \qquad [4.8]$$

The extension of Equations 4.8 to a three-dimensional situation can be made by adding a third component equation, $\sum F_z = 0$.

In a given situation, we may have balanced forces on an object in one direction, but unbalanced forces in the other. Thus, for a given problem, we may need to model the object as a particle in equilibrium for one component and a particle under a net force for the other.

Quick Quiz 4.9

Consider the two situations shown in Figure 4.9, in which no acceleration occurs. In both cases, all individuals pull with a force of magnitude *F* on a rope attached to a spring scale. Is the reading on the spring scale in part (i) of the figure (a) greater than, (b) less than, or (c) equal to the reading in part (ii)?

The Accelerating Particle

In a situation in which a nonzero net force is acting on an object, the object is accelerating, and we use Newton's second law in order to determine the features of the motion:

$$\sum \mathbf{F} = m\mathbf{a}$$

---

* This is only one condition of equilibrium for an object. A second condition of equilibrium is a statement of rotational equilibrium. This condition will be discussed in Chapter 10.

In practice, this equation is broken into components, so that two (or three) equations can be handled independently. The representative suggestions and problems that follow should help you to solve problems of this kind.

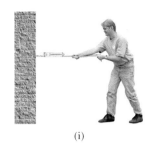

(i)

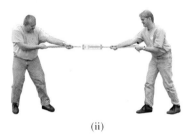

(ii)

Figure 4.9

(Quick Quiz 4.9) (i) An individual pulls with a force of magnitude *F* on a spring scale attached to a wall. (ii) Two individuals pull with forces of magnitude *F* in opposite directions on a spring scale attached between two ropes.

PROBLEM-SOLVING STRATEGY | **Particle Under a Net Force**

The following procedure is recommended when dealing with problems involving the application of Newton's second law for a particle under a net force:

1. **Think**—establish your mental representation of the situation. Draw a diagram of the system.
2. Isolate the object whose motion is being analyzed. Draw a free-body diagram for this object, showing all external forces acting on it. For systems containing more than one object, draw a *separate* diagram for each object.
3. Establish convenient coordinate axes for each object and find the components of the forces along those axes. Apply Newton's second law, $\Sigma \mathbf{F} = m\mathbf{a}$, in the *x* and *y* directions for each object. If the object is in equilibrium in either direction, set the right hand side of Newton's second law equal to zero.
4. Solve the component equations for the unknowns. Remember that, in order to obtain a complete solution, you must have as many independent equations as you have unknowns.
5. If appropriate to the situation (the net force is constant), and necessary to respond to the problem, use the equations of kinematics from Chapter 2 to find the unknowns.

We now embark on a series of examples that demonstrate how to solve problems involving a particle under a net force. You should read and study these examples very carefully.

PITFALL PREVENTION 4.12
Again, *m*a is not a force.

Because of the frequent occurrence of this misconception, it bears repeating that the product *m*a is *not* a force. Avoid the temptation to include an "*m*a force" in your free-body diagram.

THINKING PHYSICS 4.4

You have most likely had the experience of standing in an elevator that accelerates upward as it leaves to move toward a higher floor. In this case, you *feel* heavier. If you are standing on a bathroom scale at the time, it *measures* a force larger than your weight. Thus, you have tactile and measured evidence that may lead you to believe that you are heavier in this situation. *Are* you heavier?

Reasoning No, you are not—the gravitational force on you is unchanged. To provide the upward acceleration, the floor or the bathroom scale must apply an upward force larger than the gravitational force. It is this larger force that you feel, which you interpret as feeling heavier. A bathroom scale measures this force, so its reading increases.

Example 4.2 A Traffic Light at Rest

A traffic light weighing 122 N hangs from a cable tied to two other cables fastened to a support, as in Figure 4.10a. The upper cables make angles of 37.0° and 53.0° with the horizontal. These upper cables are not as strong as the vertical cable, and will break if the tension in them exceeds 100 N. Does the traffic light remain in this situation, or will one of the cables break?

Reasoning Let us assume that the cables do not break so that no acceleration of any sort occurs in any direction. Thus, we use the model of a particle in equilibrium for both *x* and *y* components. We shall construct two free-body diagrams. The first of these is for the traffic light, shown in Figure 4.10b; the second is for the knot that holds the three cables together, as in Figure 4.10c. The knot is a convenient point to choose because all the forces in which we are interested act through this point. Because the acceleration of the system is zero, we can use the equilibrium conditions that the net force on the light is zero, and the net force on the knot is zero.

Solution Considering Figure 4.10b, we apply the equilibrium condition in the *y* direction, $\sum F_y = 0 \rightarrow T_3 - F_g = 0$. This leads to $T_3 = F_g = 122$ N. Thus, the force $\mathbf{T}_3$ exerted by the vertical cable balances the weight of the light.

Considering the knot next, we choose the coordinate axes as shown in Figure 4.10c and resolve the forces into their *x* and *y* components, as shown in the following table:

Force	*x* component	*y* component
$\mathbf{T}_1$	$- T_1 \cos 37.0°$	$T_1 \sin 37.0°$
$\mathbf{T}_2$	$T_2 \cos 53.0°$	$T_2 \sin 53.0°$
$\mathbf{T}_3$	0	$- 122$ N

Equations 4.8 give us:

(1) $\quad \sum F_x = T_2 \cos 53.0° - T_1 \cos 37.0° = 0$

(2) $\quad \sum F_y = T_1 \sin 37.0° + T_2 \sin 53.0° - 122 \text{ N} = 0$

We solve (1) for T_2 in terms of T_1 to give

$$T_2 = T_1 \left(\frac{\cos 37.0°}{\cos 53.0°} \right) = 1.33 T_1$$

This value for T_2 is substituted into (2) to give

$$T_1 \sin 37.0° + (1.33 T_1)(\sin 53.0°) - 122 \text{ N} = 0$$

$$T_1 = \boxed{73.4 \text{ N}}$$

Then, we calculate T_2:

$$T_2 = 1.33 T_1 = \boxed{97.4 \text{ N}}$$

Both of these values are less than 100 N (just barely for T_2!), so the cables do not break.

EXERCISE If the angles of the cables can be adjusted, in what situation does $T_1 = T_2$?

Answer When the angles the cables make with the horizontal are equal.

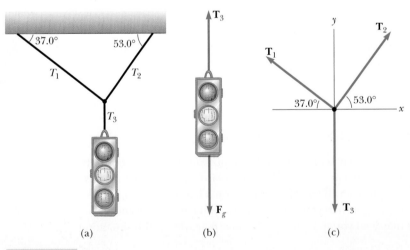

(a) (b) (c)

Figure 4.10

(Example 4.2) (a) A traffic light suspended by cables. (b) The free-body diagram for the traffic light. (c) The free-body diagram for the knot in the cable.

Example 4.3 A Sled on Frictionless Snow

A child on a sled is released on a frictionless, inclined hill of angle θ, as in Figure 4.11a. (a) Determine the acceleration of the sled after it is released.

Reasoning   We identify the system of the sled and the child as our object of interest. We model the system as a particle of mass m. Newton's second law can be used to determine the acceleration of the particle. First, we construct the free-body diagram for the particle as in Figure 4.11b. The only forces on the particle are the normal force **n** acting perpendicularly to the incline and the gravitational force $m\mathbf{g}$ acting vertically downward. For problems of this type involving inclines, it is convenient to choose the coordinate axes with x *along* the incline and y *perpendicular* to it. Then, we replace $m\mathbf{g}$ by a combination of a component vector of magnitude $mg \sin \theta$ along the *positive x* axis (down the incline) and one of magnitude $mg \cos \theta$ in the *negative y* direction.

Solution   Applying Newton's second law in component form to the particle and noting that $a_y = 0$ gives

$$(1) \quad \sum F_x = mg \sin \theta = ma_x$$

$$(2) \quad \sum F_y = n - mg \cos \theta = 0$$

From (1) we see that the acceleration along the incline is provided by the component of the gravitational force parallel to the incline, which gives us

$$(3) \quad a_x = \boxed{g \sin \theta}$$

Note that the acceleration given by (3) is *independent* of the mass of the particle—it depends only on the angle of inclination and on g. From (2) we conclude that the component of the gravitational force perpendicular to the incline is *balanced* by the normal force; that is, $n = mg \cos \theta$. (Notice, as pointed out in Pitfall Prevention 4.9, that n does not equal mg in this case.)

Special Cases   When $\theta = 90°$, (3) gives us $a_x = g$ and (2) gives us $n = 0$. This case corresponds to the particle in free-fall. (For our choice of coordinate system, positive x is in the downward direction when $\theta = 90°$, which is why the acceleration is $+ g$ rather than $- g$.) When $\theta = 0°$, $a_x = 0$ and $n = mg$ (its maximum value). This corresponds to the situation in which the particle is on a level surface and not accelerating.

This technique of looking at special cases of limiting situations is often useful in checking an answer. In this situation, if the angle θ goes to 90°, we know intuitively that the object should be falling parallel to the surface of the incline. The fact that (3) mathematically reduces to $a_x = g$ when $\theta = 90°$ gives us confidence in our answer. It doesn't prove that the answer is correct, but if the acceleration does not reduce to g, this would tell us that the answer is incorrect.

(b) Suppose the sled is released from rest at the top of the hill, and the distance from the front of the sled to the bottom of the hill is d. How long does it take the front of the sled to reach the bottom, and what is its speed just as it arrives at that point?

Solution   In part (a), we found $a_x = g \sin \theta$, which is constant. Hence, we can model the system as a particle under constant acceleration for the motion parallel to the incline. We use Equation 2.11, $x_f = x_i + v_{xi}t + \frac{1}{2}a_xt^2$, to describe the position of the front edge of the sled. We define the initial position as $x_i = 0$ and the final position as $x_f = d$. Because the sled starts sliding from rest, $v_{xi} = 0$. With these values, Equation 2.11 becomes simply $d = \frac{1}{2}a_xt^2$, or

$$(4) \quad t = \sqrt{\frac{2d}{a_x}} = \boxed{\sqrt{\frac{2d}{g \sin \theta}}}$$

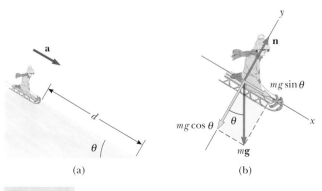

(a)

30.0°

Figure 4.11

(Example 4.3) (a) A child on a sled sliding down a frictionless incline. (b) The free-body diagram for the system.

Figure 4.12

A child holding a sled at rest on a frictionless hill.

This answers the first question as to the length of time to reach the bottom. Now, to determine the speed when the sled arrives at the bottom, we use Equation 2.12, $v_{xf}^2 = v_{xi}^2 + 2a_x(x_f - x_i)$ with $v_{xi} = 0$, and we find that $v_{xf}^2 = 2a_xd$, or

$$(5) \quad v_{xf} = \sqrt{2a_xd} = \boxed{\sqrt{2gd \sin \theta}}$$

As with the acceleration parallel to the incline, t and v_{xf} are *independent* of the mass of the sled and child.

EXERCISE   A child holds a sled weighing 77.0 N at rest on a frictionless snow-covered incline at 30.0° as in Figure 4.12. Find (a) the magnitude of the force the child must exert on the rope, and (b) the magnitude of the force the incline exerts on the sled.

Answer   (a) 38.5 N (b) 66.7 N

Example 4.4   The Atwood Machine

When two objects with unequal masses are hung vertically over a light, frictionless pulley as in Figure 4.13a, the arrangement is called an *Atwood machine*. The device is sometimes used in the laboratory to measure the free-fall acceleration. Calculate the magnitude of the acceleration of the two objects and the tension in the string.

Reasoning   Think about the mental representation suggested by Figure 4.13a—as one object moves upward, the other object moves downward. Because the objects are connected by an inextensible string, they must have the same magnitude of acceleration. The objects in the Atwood machine are subject to the gravitational force as well as to the forces exerted by the strings connected to them. We model the objects as particles under a net force. The free-body diagrams for the two objects are shown in Figure 4.13b. Two forces act on each object: the upward force **T** exerted by the string and the downward gravitational force. In a problem such as this in which the pulley is modeled as massless and frictionless, the tension in the string on both sides of the pulley is the same. If the pulley has mass or is subject to a friction force, the tensions on either side are not the same and the situation requires the techniques of Chapter 10.

In these types of problems, involving strings that pass over pulleys, we must be careful about the sign convention. Notice that if m_1 goes up, then m_2 goes down. Thus, m_1 going up and m_2 going down should be represented equivalently as far as a sign convention is concerned. We can do this by defining our sign convention with up as positive for m_1 and down as positive for m_2, as shown in Figure 4.13a.

With this sign convention, the magnitude of the net force exerted on m_1 is $T - m_1g$, while the magnitude of the net force exerted on m_2 is $m_2g - T$. We have chosen the signs of the forces to be consistent with the choices of the positive direction for each object.

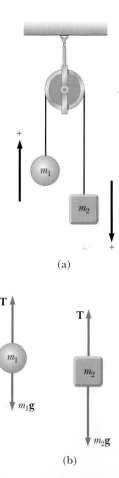

(a)

(b)

Figure 4.13

(Example 4.4) The Atwood machine. (a) Two objects connected by a light string over a frictionless pulley. (b) The free-body diagrams for m_1 and m_2.

Solution When Newton's second law is applied to m_1, we find

$$(1) \quad \sum F_y = T - m_1g = m_1a$$

Similarly, for m_2 we find

$$(2) \quad \sum F_y = m_2g - T = m_2a$$

When (2) is added to (1), T cancels and we have

$$-m_1g + m_2g = m_1a + m_2a$$

Solving this for the acceleration a,

$$(3) \quad a = \left(\frac{m_2 - m_1}{m_1 + m_2} \right) g$$

If $m_2 > m_1$, the acceleration given by (3) is positive — m_1 goes up and m_2 goes down. Is this consistent with your mental representation? If $m_1 > m_2$, the acceleration is negative and the masses move in the opposite direction.

If (3) is substituted into (1), we find

$$(4) \quad T = \left(\frac{2m_1m_2}{m_1 + m_2} \right) g$$

Special Cases When $m_1 = m_2$, (3) and (4) give us $a = 0$ and $T = m_1g = m_2g$, as we would intuitively expect for the balanced case. Also, if $m_2 \gg m_1$, $a \approx g$ (a freely falling object) and $T \approx 0$. For such a large mass m_2, we would expect m_1 to have little effect, so that m_2 is simply falling. Thus, our results are consistent with our intuitive predictions in both of these limiting situations.

EXERCISE Find the acceleration and tension of an Atwood machine in which $m_1 = 2.00$ kg and $m_2 = 4.00$ kg.

Answer $a = 3.27$ m/s^2, $T = 26.1$ N

Example 4.5 One Block Pushes Another

Two blocks of masses m_1 and m_2, with $m_1 > m_2$, are placed in contact with each other on a frictionless, horizontal surface, as in Figure 4.14a. A constant horizontal force **F** is applied to m_1 as shown. (a) Find the magnitude of the acceleration of the system of two blocks.

Reasoning and Solution Both blocks must experience the *same* acceleration because they are in contact with each other and remain in contact with each other. We model the system of both blocks as a particle under a net force. Because **F** is the only horizontal force exerted on the system, we have

$$\sum F_x(\text{system}) = F = (m_1 + m_2)a$$

$$(1) \quad a = \frac{F}{m_1 + m_2}$$

(b) Determine the magnitude of the contact force between the two blocks.

Reasoning and Solution The contact force is internal to the system of two blocks. Thus, we cannot find this force by modeling the whole system as a single particle. We now need to treat each of the two blocks individually as a particle under a net force. We first construct a free-body diagram for each block, as shown in Figures 4.14b and 4.14c, where the contact force is denoted by **P**. From Figure 4.14c we see that the only horizontal force acting on m_2 is the contact force

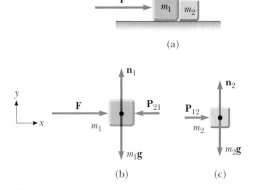

Figure 4.14

(Example 4.5)

$\mathbf{P}_{12}$ (the force exerted by m_1 on m_2), which is directed to the right. Applying Newton's second law to m_2 gives

$$(2) \quad \sum F_x = P_{12} = m_2a$$

Substituting the value of the acceleration a given by (1) into (2) gives

$$(3) \quad P_{12} = m_2a = \left(\frac{m_2}{m_1 + m_2} \right) F$$

From this result, we see that the contact force P_{12} is *less* than the applied force F. This is consistent with the fact that the force required to accelerate m_2 alone must be less than the force required to produce the same acceleration for the system of two blocks. Compare this with the forces in the couplers in the train of Thinking Physics 4.1.

It is instructive to check this expression for P_{12} by considering the forces acting on m_1, shown in Figure 4.14b. The horizontal forces acting on m_1 are the applied force **F** to the right and the contact force **P**$_{21}$ to the left (the force exerted by m_2 on m_1). From Newton's third law, **P**$_{21}$ is the reaction to

P_{12}, so $P_{21} = P_{12}$. Applying Newton's second law to m_1 gives

$$(4) \quad \sum F_x = F - P_{21} = F - P_{12} = m_1 a$$

Solving for P_{12} and substituting the value of a from (1) into (4) gives

$$P_{12} = F - m_1 a = F - m_1 \left(\frac{F}{m_1 + m_2} \right) = \left(\frac{m_2}{m_1 + m_2} \right) F$$

This agrees with (3), as it must.

Quick Quiz 4.10

Imagine that the force **F** in Example 4.5 is applied toward the left on the right-hand mass m_2. Is the magnitude of the force **P**$_{12}$ now (a) greater than, (b) less than, or (c) equal to the magnitude of the force **P**$_{12}$ in Example 4.5?

Example 4.6 Weighing a Fish in an Elevator

A person weighs a fish on a spring scale attached to the ceiling of an elevator, as shown in Figure 4.15. Show that if the elevator accelerates, the spring scale reads a weight different from the true weight of the fish.

Reasoning An observer on the accelerating elevator is not in an inertial frame. We need to analyze this situation in an inertial frame, so let us imagine observing it from the stationary ground. We model the fish as a particle under a net force. The external forces acting on the fish are the downward gravitational force **F**$_g$ and the upward force **T** exerted on it by the hook hanging from the bottom of the scale. (It might be more fruitful in your mental representation to imagine that the hook is a string connecting the fish to the scale.) Because the tension is the same everywhere in the hook supporting the fish, the hook pulls downward with a force of magnitude T on the spring scale. Thus, the tension T in the hook is also the reading of the spring scale.

If the elevator is either at rest or moves at constant velocity, the fish is not accelerating and is a particle in equilibrium, which gives us $\sum F_y = T - mg = 0 \rightarrow T = mg$. If the elevator accelerates either up or down, however, the tension is no longer equal to the weight of the fish because $T - mg$ does not equal zero.

Solution If the elevator accelerates with an acceleration **a** relative to an observer in an inertial frame outside the elevator, Newton's second law applied to the fish in the vertical di-

rection gives us:

$$\sum F_y = T - mg = ma_y$$

which leads to

$$(1) \quad T = mg + ma_y$$

Thus, we conclude from (1) that the scale reading T is greater than the weight mg if **a** is upward as in Figure 4.15a. Furthermore, we see that T is less than mg if **a** is downward as in Figure 4.15b. For example, if the weight of the fish is 40.0 N, and **a** is upward with $a_y = 2.00 \text{ m/s}^2$, then the scale reading is

$$T = mg + ma_y = mg\left(1 + \frac{a_y}{g}\right)$$
$$= (40.0 \text{ N})\left(1 + \frac{2.00 \text{ m/s}^2}{9.80 \text{ m/s}^2}\right) = \boxed{48.2 \text{ N}}$$

If $a_y = -2.00 \text{ m/s}^2$, so that **a** is downward, then

$$T = mg\left(1 + \frac{a_y}{g}\right) = (40.0 \text{ N})\left(1 + \frac{-2.00 \text{ m/s}^2}{9.80 \text{ m/s}^2}\right) = \boxed{31.8 \text{ N}}$$

Hence, if you buy a fish in an elevator, make sure the fish is weighed while the elevator is at rest or accelerating downward!

Special Case If the cable breaks, the elevator falls freely so that $a_y = -g$, and from (1) we see that the tension T is zero; that is, the fish appears to be weightless.

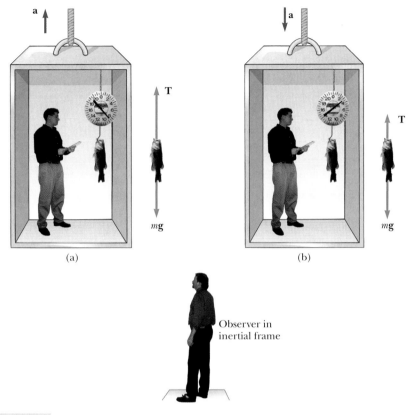

Figure 4.15

(Example 4.6) (a) When the elevator accelerates *upward*, the spring scale reads a value *greater* than the true weight of the fish. (b) When the elevator accelerates *downward*, the spring scale reads a value *less* than the true weight.

context connection

4.8 • CONTROLLING THE SPACECRAFT IN EMPTY SPACE

Our entire trip from Earth to Mars will be governed by the laws of dynamics. We shall look at the details of the planned trip in the Context Conclusion, following Chapter 11. In this section, we present an example of a *midcourse correction*. As you read this discussion, note that all three of Newton's laws are involved.

Suppose we are in a situation such as that shown in Figure 4.16, such that our spacecraft is on the wrong trajectory toward our destination. Our actual trajectory

Figure 4.16

Our spacecraft is on a trajectory different from the desired trajectory. How shall we fire the engines to put the spacecraft on the correct trajectory?

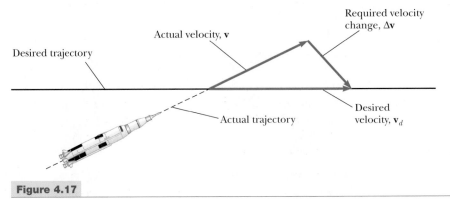

Figure 4.17

We set up the three velocity vectors required to solve the problem posed by Figure 4.16—the actual velocity, the desired velocity, and the change in velocity.

is different from the desired trajectory. Newton's *first* law tells us that we will continue on our incorrect trajectory in the absence of a force. How can we return our spacecraft to the correct trajectory? In what direction should we fire the rocket engines? We can answer these questions by first setting up a pictorial representation showing the velocity vectors, as in Figure 4.17.

The vector **Δv** for the change in velocity in Figure 4.17 indicates the direction of the acceleration that we must impart to the rocket, because $\mathbf{a} = \Delta\mathbf{v}/\Delta t$. Thus, we must fire the rockets in exactly the opposite direction at the time we cross the desired trajectory, as shown in Figure 4.18. (We imagine that there are thrusters on the side of the spacecraft that allow it to rotate in space to the position shown.) This is an application of Newton's *third* law—the rocket exerts on the exhaust gases a force $\mathbf{F}_{rg}$ (the force *by* the rocket *on* the gases) in one direction, and the gases exert a force $\mathbf{F}_{gr} = -\mathbf{F}_{rg}$ in the opposite direction on the rocket. We must also fire the rocket for just the right amount of time to give us the desired change in velocity. We can find the required engine burn time from Newton's *second* law, because we know the desired change in velocity and we assume that we know the

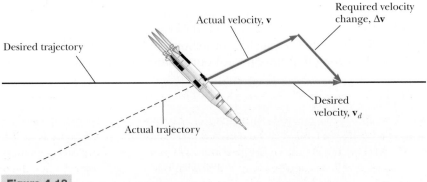

Figure 4.18

We must fire the engines in the direction opposite to the change in velocity vector so that, by Newton's third law, the rocket experiences a force in the direction of this vector.

mass m of the rocket. We write Newton's second law for components along the direction of $\Delta\mathbf{v}$ in Figure 4.18:

$$F_{gr} = ma = m\frac{\Delta v}{\Delta t} \longrightarrow \Delta t = \frac{m\Delta v}{F_{gr}}$$

As long as the midcourse correction is fairly minor, the required burn time Δt can be approximated with this equation because we know the mass m of the rocket and the force F_{gr} from the engine, and we found the required Δv from Figure 4.18.

For larger corrections, two major complications arise, which we have left out of this simple discussion and which may require a more refined model:

- The mass of the rocket changes for longer burns, so that $\Sigma\mathbf{F} = m\mathbf{a}$ is not the appropriate form of Newton's second law. (The more appropriate form involves momentum, discussed in Chapter 8.) This may require the use of calculus.
- If the time Δt for the engine burn is relatively long, by the time the desired velocity is reached, the rocket may be on a parallel trajectory above the desired trajectory in Figure 4.16. Thus, the timing of the initiation of the engine burn must be handled carefully.

We shall not concern ourselves with these complications because we are restricting ourselves to a relatively simple approach to our contextual problem.

In the next chapter, we investigate more details about the gravitational force. This is the only force on our spacecraft once the rocket engines have been shut down, so it is an important factor in our analysis of our trip to Mars.

SUMMARY

Newton's first law states that, if no force is exerted on an object, an object at rest remains at rest, and an object in uniform motion in a straight line maintains that motion.

Newton's first law defines an **inertial frame of reference,** which is a frame in which Newton's first law is valid.

Newton's second law states that the acceleration of an object is directly proportional to the net force acting on the object and inversely proportional to the object's mass. Thus, the net force on an object equals the product of the mass of the object and its acceleration,

$$\sum \mathbf{F} = m\mathbf{a} \qquad [4.2]$$

The **weight** of an object is equal to the product of its mass (a scalar quantity) and the magnitude of the free-fall acceleration, or

$$F_g = mg \qquad [4.5]$$

If the acceleration of an object is zero, then the particle is in equilibrium, with the appropriate equations being

$$\sum F_x = 0 \qquad \sum F_y = 0 \qquad [4.8]$$

Newton's third law states that if two objects interact, the force exerted on object 1 by object 2 is equal in magnitude but opposite in direction to the force exerted on object 2 by object 1. Thus, an isolated force cannot exist in nature.

QUESTIONS

1. Draw a free-body diagram for each of the following objects: (a) a projectile in motion in the presence of air resistance, (b) a rocket leaving the launch pad with its engines operating, (c) an athlete running along a horizontal track.

2. In the motion picture *It Happened One Night* (Columbia Pictures, 1934), Clark Gable is standing inside a stationary bus in front of Claudette Colbert, who is seated. The bus suddenly starts moving forward and Clark falls into Claudette's lap. Why did this happen?

3. A passenger sitting in the rear of a bus claims that he was injured when the driver slammed on the brakes, causing a suitcase to come flying toward the passenger from the front of the bus. If you were the judge in this case, what disposition would you make? Why?

4. As you sit in a chair, the chair pushes up on you with a normal force. This force is equal to your weight and in the opposite direction. Is this force the Newton's third law reaction to your weight?

5. A space explorer is in a spacecraft moving through space, far from any planet or star. She notices a large rock, taken as a specimen from an alien planet, floating around the cabin of the spacecraft. Should she push it gently toward a storage compartment or kick it toward the compartment? Why?

6. The observer in the accelerating elevator of Example 4.6 would claim that the "weight" of the fish is *T*, the scale reading. This is obviously wrong. Why does this observation differ from that of a person outside the elevator, at rest with respect to the Earth?

7. Identify the action–reaction pairs in the following situations: a man takes a step; a snowball hits a girl in the back; a baseball player catches a ball; a gust of wind strikes a window.

8. While a football is in flight, what forces act on it? What are the action–reaction pairs while the football is being kicked and while it is in flight?

9. A rubber ball is dropped onto the floor. What force causes the ball to bounce back into the air?

10. What is wrong with the statement "Because the car is at rest, no forces are acting on it"? How would you correct this sentence?

11. If gold were sold by weight, would you rather buy it in Denver or in Death Valley? If it were sold by mass, at which of the two locations would you prefer to buy it? Why?

12. A weight lifter stands on a bathroom scale. He pumps a barbell up and down. What happens to the reading on the bathroom scale as this is done? Suppose he is strong enough to actually *throw* the barbell upward. How does the reading on the scale vary now?

13. The mayor of a city decides to fire some city employees because they will not remove the sags from the cables that support the city traffic lights. If you were a lawyer, what defense would you give on behalf of the employees? Who do you think would win the case in court?

14. In a tug-of-war between two athletes, each athlete pulls on the rope with a force of 200 N. What is the tension in the rope? If the rope does not move, what force does each athlete exert against the ground?

15. Suppose a truck loaded with sand accelerates at 0.5 m/s^2 on a highway. If the driving force on the truck remains constant, what happens to the truck's acceleration if it leaks sand at a constant rate through a hole in its bottom?

16. Can an object exert a force on itself? Argue for your answer.

PROBLEMS

1, 2, 3 = straightforward, intermediate, challenging   □ = full solution available in the *Student Solutions Manual and Study Guide*

web = solution posted at **http://www.harcourtcollege.com/physics/**   🖥 = computer useful in solving problem

📲 = Interactive Physics   ■ = paired numerical/symbolic problems   🔖 = life science application

---

Section 4.3   Inertial Mass

1. A force **F** applied to an object of mass m_1 produces an acceleration of 3.00 m/s^2 . The same force applied to a second object of mass m_2 produces an acceleration of 1.00 m/s^2. (a) What is the value of the ratio m_1/m_2? (b) If m_1 and m_2 are combined, find their acceleration under the action of the force **F**.

2. (a) A car with a mass of 850 kg is moving to the right with a constant speed of 1.44 m/s. What is the total force on the car? (b) What is the total force on the car if it is moving to the left?

Section 4.4   Newton's Second Law — The Particle Under a Net Force

3. An object of mass 3.00 kg undergoes an acceleration given by $\mathbf{a} = (2.00\mathbf{i} + 5.00\mathbf{j}) \text{ m/s}^2$. Find the vector expression for the net force acting on it and the magnitude of the net force.

4. The largest-caliber antiaircraft gun operated by the German air force during World War II was the 12.8-cm Flak 40. This weapon fired a 25.8-kg shell with a muzzle speed of 880 m/s. What propulsive force was necessary to attain the muzzle speed within the 6.00-m barrel? (Assume the shell moves horizontally with constant acceleration and neglect friction.)

5. Two forces, $\mathbf{F}_1 = (-6\mathbf{i} - 4\mathbf{j})$ N and $\mathbf{F}_2 = (-3\mathbf{i} + 7\mathbf{j})$ N, act on a particle of mass 2.00 kg that is initially at rest at coordinates $(-2.00 \text{ m}, +4.00 \text{ m})$. (a) What are the components of the particle's velocity at $t = 10.0$ s? (b) In what direction is the particle moving at $t = 10.0$ s? (c) What displacement does the particle undergo during the first 10.0 s? (d) What are the coordinates of the particle at $t = 10.0$ s?

6. Two people pull as hard as they can on horizontal ropes attached to a boat that has a mass of 200 kg. If they pull in the same direction, the boat has an acceleration of 1.52 m/s² to the right. If they pull in opposite directions, the boat has an acceleration of 0.518 m/s² to the left. What is the force exerted by each person on the boat? (Disregard any other horizontal forces on the boat.)

7. Two forces $\mathbf{F}_1$ and $\mathbf{F}_2$ act on a 5.00-kg object. If $F_1 =$ 20.0 N and $F_2 = 15.0$ N, find the accelerations in (a) and (b) of Figure P4.7.

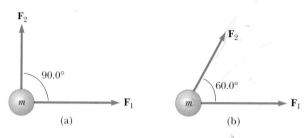

Figure P4.7

8. Three forces, given by $\mathbf{F}_1 = (-2.00\mathbf{i} + 2.00\mathbf{j})$ N, $\mathbf{F}_2 = (5.00\mathbf{i} - 3.00\mathbf{j})$ N, and $\mathbf{F}_3 = (-45.0\mathbf{i})$ N, act on an object to give it an acceleration of magnitude 3.75 m/s². (a) What is the direction of the acceleration? (b) What is the mass of the object? (c) If the object is initially at rest, what is its speed after 10.0 s? (d) What are the velocity components of the object after 10.0 s?

9. A 3.00-kg object is moving in a plane, with its x and y coordinates given by $x = 5t^2 - 1$ and $y = 3t^3 + 2$, where x and y are in meters and t is in seconds. Find the magnitude of the net force acting on this object at $t = 2.00$ s.

Section 4.5 The Gravitational Force and Weight

10. A woman weighs 120 lb. Determine (a) her weight in newtons and (b) her mass in kilograms.

11. If a man weighs 900 N on Earth, what would he weigh on Jupiter, where the acceleration due to gravity is 25.9 m/s²?

12. The gravitational force on a baseball is $-F_g\mathbf{j}$. A pitcher throws the baseball with velocity $v\,\mathbf{i}$ by uniformly accelerating it straight forward horizontally for a time t. If the ball starts from rest, (a) through what distance does it accelerate before its release? (b) What force does the pitcher exert on the ball?

13. An electron of mass 9.11×10^{-31} kg has an initial speed of 3.00×10^5 m/s. It travels in a straight line, and its speed increases to 7.00×10^5 m/s in a distance of 5.00 cm. Assuming its acceleration is constant, (a) determine the force exerted on the electron and (b) compare this force with the weight of the electron, which we ignored.

14. Define one pound as the weight of an object of mass 0.453 592 37 kg at a location where the acceleration of gravity is 32.174 0 ft/s². Express the pound as one quantity with one SI unit.

15. The distinction between mass and weight was discovered after Jean Richer transported pendulum clocks from Paris to French Guiana in 1671. He found that they ran slower there quite systematically. The effect was reversed when the clocks returned to Paris. How much weight would you personally lose in traveling from Paris, where $g = 9.809\ 5$ m/s², to Cayenne, where $g = 9.780\ 8$ m/s²? (We will consider how the free-fall acceleration influences the period of a pendulum in Section 12.4.)

16. Besides its weight, a 2.80-kg object is subjected to one other constant force. The object starts from rest and in 1.20 s undergoes a displacement of $(4.20\ \text{m})\mathbf{i} - (3.30\ \text{m})\mathbf{j}$, where the direction of $\mathbf{j}$ is the upward vertical direction. Determine the other force.

Section 4.6 Newton's Third Law

17. You stand on the seat of a chair and hop off. (a) During the time you are in flight down to the floor, the Earth is lurching up toward you with an acceleration of what order of magnitude? In your solution explain your logic. Model the Earth as a perfectly solid object. (b) The Earth moves up through a distance of what order of magnitude?

18. The average speed of a nitrogen molecule in air is about 6.70×10^2 m/s, and its mass is 4.68×10^{-26} kg. (a) If it takes 3.00×10^{-13} s for a nitrogen molecule to hit a wall and rebound with the same speed but moving in the opposite direction, what is the average acceleration of the molecule during this time interval? (b) What average force does the molecule exert on the wall?

19. A 15.0-lb block rests on the floor. (a) What force does the floor exert on the block? (b) If a rope is tied to the block and run vertically over a pulley, and the other end is attached to a free-hanging 10.0-lb weight, what is the force of the floor on the 15.0-lb block? (c) If we replace the 10.0-lb weight in part (b) with a 20.0-lb weight, what is the force exerted by the floor on the 15.0-lb block?

Section 4.7 Applications of Newton's Laws

20. A bag of cement of weight 325 N hangs from three wires as suggested in Figure P4.20. Two of the wires make angles $\theta_1 = 60.0°$ and $\theta_2 = 25.0°$ with the horizontal. If the system is in equilibrium, find the tensions T_1, T_2, and T_3 in the wires.

21. A bag of cement of weight F_g hangs from three wires as shown in Figure P4.20. Two of the wires make angles θ_1

and θ_2 with the horizontal. If the system is in equilibrium, show that the tension in the left-hand wire is

$$T_1 = F_g \cos \theta_2 / \sin(\theta_1 + \theta_2)$$

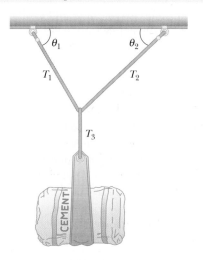

Figure P4.20   *Problems 20 and 21.*

22. You are a judge in a children's kite-flying contest, and two children will win prizes for the kites that pull most strongly and least strongly on their strings. To measure string tensions, you borrow a weight hanger, some slotted weights, and a protractor from your physics teacher, and use the following protocol, illustrated in Figure P4.22: Wait for a child to get her kite well controlled, hook the hanger onto the kite string about 30 cm from her hand, pile on weight until that section of string is horizontal, record the mass required, and record the angle between the horizontal and the string running up to the kite. (a) Explain how this method works. As you construct your explanation, imagine that the children's parents ask you about your method, that they might make false assumptions about your ability without concrete evidence, and that your explanation is an

opportunity to give them confidence in your evaluation technique. (b) Find the string tension if the mass is 132 g and the angle of the kite string is 46.3°.

23. A 1.00-kg object is observed to accelerate at 10.0 m/s^2 in a
web direction 30.0° north of east (Fig. P4.23). The force $\mathbf{F}_2$ acting on the object has a magnitude of 5.00 N and is directed north. Determine the magnitude and direction of the force $\mathbf{F}_1$ acting on the object.

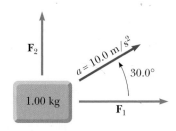

Figure P4.23

24. A simple accelerometer is constructed by suspending an object of mass m from a string of length L that is tied to the roof of a car. As the car accelerates the string makes a constant angle of θ with the vertical. (a) Assuming that the string mass is negligible compared to m, derive an expression for the car's acceleration in terms of θ and show that it is independent of the mass m and the length L. (b) Determine the acceleration of the car when $\theta = 23.0°$.

25. The systems shown in Figure P4.25 are in equilibrium. If the spring scales are calibrated in newtons, what do they

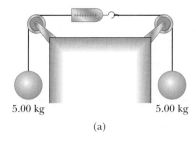

(a)

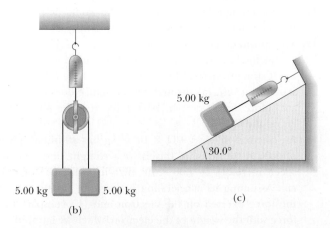

(b)   (c)

Figure P4.22

Figure P4.25

read? (Ignore the masses of the pulleys and strings, and assume the incline is frictionless.)

26. A 5.00-kg object placed on a frictionless, horizontal table is connected to a cable that passes over a pulley and then is fastened to a hanging 10.0-kg object, as in Figure P4.26. Draw free-body diagrams of both objects. Find the acceleration of the two objects and the tension in the cable.

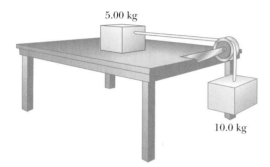

5.00 kg

10.0 kg

Figure P4.26

27. Two objects, of mass m_1 and m_2, situated on a frictionless, horizontal surface, are connected by a light string. A force **F** is exerted on one of the objects to the right (Fig. P4.27). Determine the acceleration of the system and the tension T in the string.

m_1 T m_2 **F**

Figure P4.27

28. Draw a free-body diagram of a block that slides down a frictionless plane having an inclination of $\theta = 15.0°$ (Fig. P4.28). If the block starts from rest at the top and the length of the incline is 2.00 m, find (a) the acceleration of the block and (b) its speed when it reaches the bottom of the incline.

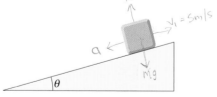

Figure P4.28   Problems 28 and 29

29. A block is given an initial speed of 5.00 m/s up a frictionless 20.0° incline (see Fig. P4.28). How far up the incline does the block slide before coming to rest?

30. One of the great dangers to mountain climbers is an avalanche, in which a large mass of snow and ice breaks loose and goes on an essentially frictionless ride down a mountainside on a cushion of compressed air. If you were on a 30.0° mountain slope and an avalanche started 400 m

up the slope, how much time would you have to get out of the way?

31. Two objects of mass 3.00 kg and 5.00 kg are connected by a light string that passes over a frictionless pulley, as in Figure 4.13a. Determine (a) the tension in the string, (b) the acceleration of each object, and (c) the distance each object will move in the first second of motion if they start from rest.

32. In the Atwood machine shown in Figure 4.13a, $m_1 = 2.00$ kg and $m_2 = 7.00$ kg. The masses of the pulley and string are negligible. The pulley turns without friction and the string does not stretch. The object of mass m_1 is released with a sharp push that sets it into motion at $v_i = 2.40$ m/s downward. (a) How far will m_1 descend below its initial level? (b) Find the velocity of m_1 after 1.80 s.

33. In the system shown in Figure P4.33, a horizontal force of magnitude F_x acts on the 8.00-kg object. The horizontal surface is frictionless. (a) For what values of F_x does the 2.00-kg object accelerate upward? (b) For what values of F_x is the tension in the cord zero? (c) Plot the acceleration of the 8.00-kg object versus F_x. Include values of F_x from -100 N to $+100$ N.

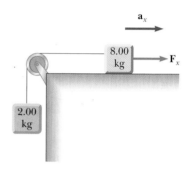

$\mathbf{a}_x$

8.00 kg $\mathbf{F}_x$

2.00 kg

Figure P4.33

34. Two objects are connected by a light string that passes over a frictionless pulley, as in Figure P4.34. If the incline

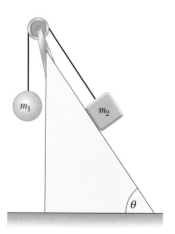

m_1 m_2

θ

Figure P4.34

is frictionless and if $m_1 = 2.00$ kg, $m_2 = 6.00$ kg, and $\theta = 55.0°$, find (a) the accelerations of the objects, (b) the tension in the string, and (c) the speed of each object 2.00 s after being released from rest. Draw free-body diagrams for both objects.

35. A 72.0-kg man stands on a spring scale in an elevator. Starting from rest, the elevator ascends, attaining its maximum speed of 1.20 m/s in 0.800 s. It travels with this constant speed for the next 5.00 s. The elevator then undergoes a uniform acceleration in the negative y direction for 1.50 s and comes to rest. What does the spring scale register (a) before the elevator starts to move? (b) during the first 0.800 s? (c) while the elevator is traveling at constant speed? (d) during the time it is slowing down?

36. A frictionless incline is 10.0 m long and inclined at 35.0°. A particle starts at the bottom with an initial speed of 5.00 m/s up the incline. When it reaches the point at which it momentarily stops, a second particle is released from the top of this incline with an initial speed v_i down the incline. Both particles reach the bottom of the incline at the same moment. (a) Determine the distance that the first particle traveled up the incline. (b) Determine the initial speed of the second particle.

Section 4.8 Context Connection — Controlling the Spacecraft in Empty Space

37. To model a spacecraft, a toy rocket engine is securely fastened to a large puck, which can glide with negligible friction over a horizontal surface, taken as the xy plane. The 4.00-kg puck has a velocity of $3.00\mathbf{i}$ m/s at one instant. Eight seconds later, its velocity is to be $(8.00\mathbf{i} + 10.0\mathbf{j})$ m/s. Assuming the rocket engine exerts a constant horizontal force, find (a) the components of the force and (b) its magnitude.

Additional Problems

38. An object of mass m_1 on a frictionless horizontal table is connected to an object of mass m_2 through a very light pulley P_1 and a light fixed pulley P_2 as shown in Figure P4.38. (a) If a_1 and a_2 are the accelerations of m_1 and m_2,

respectively, what is the relation between these accelerations? Express (b) the tensions in the strings and (c) the accelerations a_1 and a_2 in terms of the masses m_1 and m_2, and g.

39. An inventive child named Pat wants to reach an apple in a tree without climbing the tree. Sitting in a chair connected to a rope that passes over a frictionless pulley (Fig. P4.39), Pat pulls on the loose end of the rope with such a force that the spring scale reads 250 N. Pat's true weight is 320 N, and the chair weighs 160 N. (a) Draw free-body diagrams for Pat and the chair considered as separate systems, and another diagram for Pat and the chair considered as one system. (b) Show that the acceleration of the system is *upward* and find its magnitude. (c) Find the force Pat exerts on the chair.

Figure P4.39

40. Three blocks are in contact with one another on a frictionless, horizontal surface, as in Figure P4.40. A horizontal force $\mathbf{F}$ is applied to m_1. If $m_1 = 2.00$ kg, $m_2 = 3.00$ kg, $m_3 = 4.00$ kg, and $F = 18.0$ N, draw a separate free-body diagram for each block and find (a) the acceleration of the blocks, (b) the *net* force on each block, and (c) the magnitudes of the contact forces between the blocks. (d) You are working on a construction project. A coworker is nailing up plasterboard on one side of a light partition, and you are on the opposite side, providing "backing" by leaning against the wall with your back push-

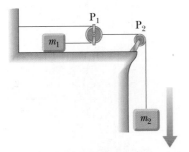

Figure P4.38

Figure P4.40

ing on it. Every blow makes your back sting. The supervisor helps you to place a heavy block of wood between the wall and your back. Using the situation analyzed in parts (a), (b), and (c) as a model, explain how this makes your job more comfortable.

41. A high diver of mass 70.0 kg jumps off a board 10.0 m above the water. If his downward motion is stopped 2.00 s after he enters the water, what average upward force did the water exert on him?

42. Any device that allows you to alter the force you exert is a kind of *machine*. Some machines, such as the prybar or the inclined plane, are very simple. Some machines do not even look like machines. An example is the following: Your car is stuck in the mud, and you can't pull hard enough to extract it. You have a long cable, however, which you connect between your front bumper and the trunk of a stout tree and pull taut as you tie it around the tree. You now pull sideways on the cable at its midpoint, exerting a force *f*. Each half of the cable is displaced through a small angle θ from the straight line between the ends of the cable. Deduce an expression for the magnitude of the force acting on the car. (b) Evaluate the cable tension for the case where $\theta = 7.00°$ and $f = 100$ N.

43. Two blocks of mass 3.50 kg and 8.00 kg are connected by a massless string passing over a frictionless pulley (Fig. P4.43). The inclines are frictionless. Find (a) the magnitude of acceleration of each block and (b) the tension in the string.

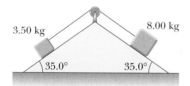

3.50 kg   8.00 kg

35.0°   35.0°

Figure P4.43

44. If you jump from a desk and land stiff-legged on a concrete floor, the chance is good that you will break a leg. To see how this happens, consider the average force stopping your body when you drop from rest from a height of 1.00 m and stop in a much shorter distance *d*. Your leg is likely to break where the cross-sectional area of the bone (the tibia) is smallest. This is at a point just above the ankle, where the cross-sectional area of one bone is about 1.60 cm². A bone fractures when the compressive stress on it exceeds about 1.60×10^8 N/m². If you land on both legs, the maximum force your ankles can safely exert on the rest of your body is then about

$$2 \,(1.60 \times 10^8 \text{ N/m}^2)\,(1.60 \times 10^{-4}\text{ m}^2) = 5.12 \times 10^4 \text{ N}.$$

Calculate the minimum stopping distance *d* that does not result in a broken leg if your mass is 60.0 kg. Don't try it! Bend your knees!

45. An inquisitive physics student, wishing to combine pleasure with scientific inquiry, rides on a roller coaster sitting on a bathroom scale. (Do not try this yourself on a roller coaster that forbids loose heavy packages.) The bottom of the seat in the roller coaster car is in a plane parallel to the track. The seat has a perpendicular back and a seat belt that fits around the student's chest in a plane parallel to the bottom of the seat. The student lifts his feet from the floor, so that the scale reads his weight, 200 lb, when the car is horizontal. At one point during the ride, the car zooms with negligible friction down a straight slope inclined at 30.0° below the horizontal. What does the scale read there?

46. An 8.40-kg object slides down a fixed, frictionless inclined plane. Use a computer to determine and tabulate the normal force exerted on the object and its acceleration for a series of incline angles (measured from the horizontal) ranging from 0° to 90° in 5° increments. Plot a graph of the normal force and the acceleration as functions of the incline angle. In the limiting cases of 0° and 90°, are your results consistent with the known behavior?

47. An object of mass *M* is held in place by an applied force **F** and a pulley system as shown in Figure P4.47. The pulleys are massless and frictionless. Find (a) the tension in each section of rope, T_1, T_2, T_3, T_4, and T_5, and (b) the magnitude of **F**. (*Hint:* Draw a free-body diagram for each pulley.)

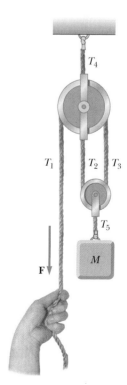

T_4

T_1 T_2 T_3

T_5

M

F

Figure P4.47

48. A mobile is formed by supporting four metal butterflies of equal mass m from a string of length L. The points of support are evenly spaced a distance ℓ apart as shown in Figure P4.48. The string forms an angle θ_1 with the ceiling at each end point. The center section of string is horizontal. (a) Find the tension in each section of string in terms of θ_1, m, and g. (b) Find the angle θ_2, in terms of θ_1, that the sections of string between the outside butterflies and the inside butterflies form with the horizontal. (c) Show that the distance D between the end points of the string is

$$D = \frac{L}{5} \left(2 \cos \theta_1 + 2 \cos \left[\tan^{-1} \left(\tfrac{1}{2} \tan \theta_1 \right) \right] + 1 \right)$$

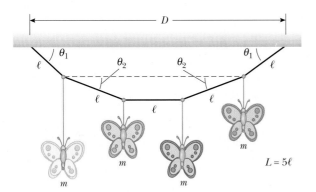

Figure P4.48

49. Review Problem. A block of mass $m = 2.00$ kg is released from rest at $h = 0.500$ m above the surface of a table, at the top of a $\theta = 30.0°$ incline as shown in Figure P4.49. The frictionless incline is fixed on a table of height $H = 2.00$ m. (a) Determine the acceleration of the block as it slides down the incline. (b) What is the speed of the block as it leaves the incline? (c) How far from the table will the block hit the floor? (d) How much time has elapsed between when the block is released and when it hits the

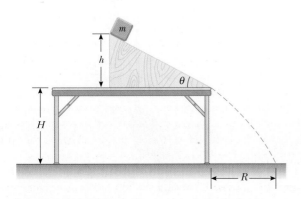

Figure P4.49   Problems 49 and 56.

floor? (e) Does the mass of the block affect any of the preceding calculations?

50. A student is asked to measure the acceleration of a cart on a "frictionless" inclined plane, using an air track, a stopwatch, and a meter stick. (Figure 4.11 shows a similar situation with a sled sliding on an incline on frictionless snow.) The height of the incline is measured to be 1.774 cm, and the total length of the incline is measured to be $d = 127.1$ cm. Hence, the angle of inclination θ is determined from the relation $\sin \theta = 1.774/127.1$. The cart is released from rest at the top of the incline, and its position x along the incline is measured as a function of time, where $x = 0$ refers to the initial position of the cart. For x values of 10.0 cm, 20.0 cm, 35.0 cm, 50.0 cm, 75.0 cm, and 100 cm, the measured times at these positions (averaged over five runs) are 1.02 s, 1.53 s, 2.01 s, 2.64 s, 3.30 s, and 3.75 s, respectively. Construct a graph of x versus t^2, and perform a linear least-squares fit to the data. Determine the acceleration of the cart from the slope of this graph, and compare it with the value you obtain using $a = g \sin \theta$, where $g = 9.80$ m/s^2.

51. What horizontal force must be applied to the cart shown in Figure P4.51 in order that the blocks remain stationary relative to the cart? Assume all surfaces, wheels, and pulley are frictionless. (*Hint:* Note that the force exerted by the string accelerates m_1.)

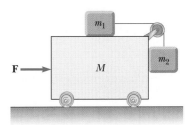

Figure P4.51   Problems 51 and 52.

52. Initially the system of objects shown in Figure P4.51 is held motionless. All surfaces, pulley, and wheels are frictionless. Let the force **F** be zero and assume that m_2 can move only vertically. At the instant after the system of objects is released, find (a) the tension T in the string, (b) the acceleration of m_2, (c) the acceleration of M, and (d) the acceleration of m_1. (*Note:* The pulley accelerates along with the cart.)

53. A van accelerates down a hill (Fig. P4.53), going from rest to 30.0 m/s in 6.00 s. During the acceleration, a toy ($m = 0.100$ kg) hangs by a string from the van's ceiling. The acceleration is such that the string remains perpendicular to the ceiling. Determine (a) the angle θ and (b) the tension in the string.

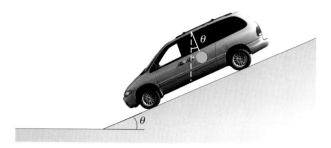

Figure P4.53

54. Cam mechanisms are used in many machines. For example, cams open and close the valves in your car engine to admit gasoline vapor to each cylinder and to allow the escape of exhaust. The principle of a cam is illustrated in Figure P4.54, showing a follower rod (also called a

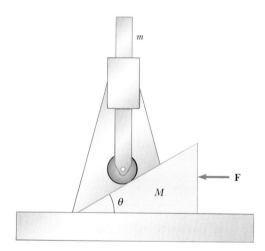

Figure P4.54

pushrod) of mass m resting on a wedge of mass M. The sliding wedge duplicates the function of a rotating eccentric disk on a camshaft in your car. Assume that no friction exists between the wedge and the base, between the pushrod and the wedge, or between the rod and the guide through which it slides. When the wedge is pushed to the left by the force **F**, the rod moves upward and does something such as opening a valve. By varying the shape of the wedge, the motion of the follower rod can be made quite complex, but assume that the wedge makes a constant angle of $\theta = 15.0°$. Suppose you want the wedge and the rod to start from rest and move with constant acceleration, with the rod moving upward 1.00 mm in 8.00 ms. Take $m = 0.250$ kg and $M = 0.500$ kg. What must be the magnitude F of the force applied to the wedge?

55. An orbiting shuttle spacecraft is in a nearly constant state of free-fall and hence produces a near zero-gravity environment inside it. The spacecraft experiences only small accelerations due to atmospheric drag and gaseous efflux from equipment. These accelerations produce a microgravity environment within the shuttle cabin, where the free-fall acceleration is measured in units of μg, where $1 \mu g = 1 \times 10^{-6} g$. On Earth, physics students sometimes drop their calculators. Find the time of fall and the impact speed of a calculator dropped through 1.30 m (a) at the Earth's surface and (b) in the cabin of a spacecraft where the free-fall acceleration is 8.00 μg.

56. In Figure P4.49 the incline has mass M and is fastened to the stationary horizontal tabletop. The block of mass m is placed near the bottom of the incline and is released with a quick push that sets it sliding upward. It stops near the top of the incline, as shown in the figure, and then slides down again, always without friction. Find the force the tabletop exerts on the incline throughout this motion.

ANSWERS TO QUICK QUIZZES

4.1 Motion requires no force. Newton's first law says an object in motion continues to move by itself in the absence of external forces. It is possible for forces to act on an object with no resulting motion if the forces are balanced.

4.2 (a) If a single force acts on an object, it must accelerate. (b) If an object accelerates, at least one force must act on it. (c) If an object has no acceleration, you cannot conclude that no forces act on it. In this case, you can only say that the *net* force on the object is zero.

4.3 If the object begins at rest or is moving with a velocity with only an x component, the net force in the x direction will cause the object to continue to move in the x direction. In any other case, however, the motion of the ob-

ject will involve velocity components in directions other than x. Thus, the direction of the velocity vector will not generally be along the x axis. What we can say with confidence is that a net force in the x direction causes the object to *accelerate* in the x direction.

4.4 Because the value of g is smaller on the Moon than on the Earth, more mass of gold would be required to represent 1 newton of weight on the Moon. Thus, your friend on the Moon is richer, by about a factor of 6!

4.5 The only force acting on the baseball during its motion is the downward gravitational force. The magnitude of this force is mg at all points in the motion, including the specific points mentioned in (a) and (b).

4.6 The car and truck experience forces that are equal in magnitude but in opposite directions, according to Newton's third law. A calibrated spring scale placed between the colliding vehicles reads the same whichever way it faces. Because the car has the smaller mass, it experiences the greater acceleration.

4.7 The force causing an automobile to move is the friction between the tires and the roadway as the automobile attempts to push the roadway backward. Notice that forces from the engine are not directly what accelerate the automobile. The force driving a propeller-driven airplane forward is the force of the air on the propeller as the rotating propeller pushes the air backward. The force propelling a rocket is the force of the exhausted gases from the back of the rocket, as the rocket pushes the gases backward. In a rowboat, the rower pushes the water backward with the oars. As a result, the water pushes forward on the oars, and the force is applied to the boat at the oarlocks. All of these situations involve the propulsion of an object due to Newton's third law.

4.8 According to Newton's third law, the force of the man on the boy and the force of the boy on the man are third law reaction forces, so they must be equal in magnitude. Both individuals experience the same magnitude of force, so the boy, with the smaller mass, experiences the larger acceleration. Both individuals experience an acceleration over the same time interval. The larger acceleration of the boy results in his moving away from the interaction with a larger velocity. Again, because of the larger acceleration, the boy moves through a larger distance during the interval while the hands are in contact.

4.9 (c). The scale is in equilibrium in both situations, so it experiences a net force of zero. Because each individual pulls with a force F and there is no acceleration, each individual is in equilibrium. Therefore, the tension in the ropes must be equal to F. In case (i), the individual on the right pulls with force F on a spring mounted rigidly to a brick wall. The resulting tension F in the rope causes the scale to read a force F. In case (ii), the individual on the left can be modeled as simply holding the rope tightly while the individual on the right pulls. Thus, the individual on the left is doing the same thing that the wall does in case (i). The resulting scale reading is the same whether a wall or a person is holding the left side of the scale.

4.10 (a). With the force applied toward the left on m_2, the contact force alone must accelerate m_1. (In the original example, the contact force alone accelerated m_2.) Because $m_1 > m_2$, this will require more force, so the magnitude of $\mathbf{P}_{12}$ is greater.

Calvin and Hobbes
by Bill Watterson

As these cyclists round a sharp turn in the Tour de France, they depend on the force of friction between the bicycle tires and the roadway to change the direction of their velocity.

(Alex Livesey/Allsport)

More Applications of Newton's Laws

n Chapter 4 we introduced Newton's laws of motion and applied them to situations in which we ignored friction. In this chapter, we shall expand our investigation to objects moving in the presence of friction, which will allow us to model situations more realistically. Such objects include those moving on rough surfaces and those moving through viscous media such as liquids and air. We also apply Newton's laws to the dynamics of circular motion so that we can understand more about various types of objects moving in circular paths.

5.1 • FORCES OF FRICTION

When an object moves either on a surface or through a viscous medium such as air or water, there is resistance to the motion because the object interacts with its surroundings. We call such resistance a **force of friction.** Forces of friction are very important in our everyday lives. They allow us to walk or run and are necessary for the motion of wheeled vehicles.

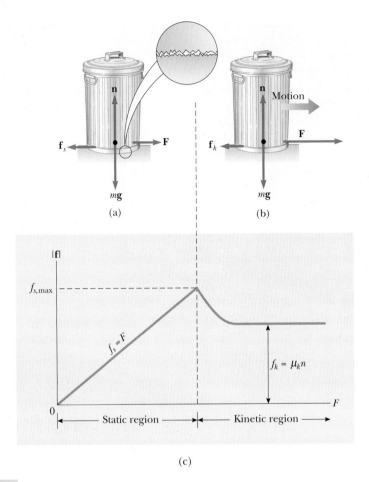

(a)   (b)

(c)

Figure 5.1

(a) The force of static friction $\mathbf{f}_s$ between a trash can and a concrete patio is opposite the applied force $\mathbf{F}$. The magnitude of the force of static friction equals that of the applied force. (b) When the magnitude of the applied force exceeds the magnitude of the force of kinetic friction $\mathbf{f}_k$, the trash can accelerates to the right. (c) A graph of the magnitude of the friction force versus that of the applied force. In our model, the force of kinetic friction is independent of the applied force and the relative speed of the surfaces. Note that $f_{s,\text{max}} > f_k$.

Imagine you are working in your garden and have filled a trash can with yard clippings. You then try to drag the trash can across the surface of your concrete patio as in Figure 5.1a. This is a *real* surface, not an idealized, frictionless surface in a simplification model. If we apply an external horizontal force $\mathbf{F}$ to the trash can, acting to the right, the trash can remains stationary if $\mathbf{F}$ is small. The force that counteracts $\mathbf{F}$ and keeps the trash can from moving acts to the left and is called the **force of static friction $\mathbf{f}_s$**. As long as the trash can is not moving, $f_s = F$. Thus, if $\mathbf{F}$ is increased, $\mathbf{f}_s$ also increases. Likewise, if $\mathbf{F}$ decreases, $\mathbf{f}_s$ also decreases. Experiments show that the friction force arises from the nature of the two surfaces; because of their roughness, contact is made only at a few points, as shown in the magnified view of the surface in Figure 5.1a.

If we increase the magnitude of $\mathbf{F}$, as in Figure 5.1b, the trash can eventually slips. When the trash can is on the verge of slipping, f_s is a maximum as shown in Figure 5.1c. When F exceeds $f_{s,\text{max}}$, the trash can moves and accelerates to the right. When the trash can is in motion, the friction force is less than $f_{s,\text{max}}$ (Fig. 5.1c). We call the friction force for an object in motion the **force of kinetic friction** $\mathbf{f}_k$. The net force $F - f_k$ in the x direction produces an acceleration to the right, according to Newton's second law. If $F = f_k$, the acceleration is zero, and the trash can moves to the right with constant speed. If the applied force is removed, the friction force acting to the left provides an acceleration of the trash can in the $- x$ direction and eventually brings it to rest.

Experimentally, one finds that, to a good approximation, both $f_{s,\text{max}}$ and f_k for an object on a surface are proportional to the normal force exerted by the surface on the object—thus we adopt a simplification model in which this approximation is assumed to be exact. The assumptions in this simplification model can be summarized as follows:

- The magnitude of the force of static friction between any two surfaces in contact can have the values

$$f_s \leq \mu_s n \qquad \textbf{[5.1]}$$

where the dimensionless constant μ_s is called the **coefficient of static friction** and n is the magnitude of the normal force. The equality in Equation 5.1 holds when the surfaces are on the verge of slipping, that is, when $f_s = f_{s,\text{max}} \equiv \mu_s n$. This situation is called *impending motion*. The inequality holds when the component of the applied force parallel to the surfaces is less than this value.

- The magnitude of the force of kinetic friction acting between two surfaces is

$$f_k = \mu_k n \qquad \textbf{[5.2]}$$

where μ_k is the **coefficient of kinetic friction.** In our simplification model, this coefficient is independent of the relative speed of the surfaces.

- The values of μ_k and μ_s depend on the nature of the surfaces, but μ_k is generally less than μ_s. Typical values of μ range from around 0.05 to 1.5. Table 5.1 lists some reported values.

- The direction of the friction force on an object is opposite to the actual motion (kinetic friction) or the impending motion (static friction) of the object relative to the surface with which it is in contact.

PITFALL PREVENTION 5.1

The equal sign is used in limited situations

In Equation 5.1, the equal sign is used *only* when the surfaces are just about to break free and begin sliding. Do not fall into the common trap of using $f_s = \mu_s n$ in *any* static situation.

- *Force of static friction*

- *Force of kinetic friction*

PITFALL PREVENTION 5.2

Friction equations

Keep in mind that Equations 5.1 and 5.2 are *not* vector equations. They are relationships between the *magnitudes* of the vectors representing the friction and normal forces. Because the friction and normal forces are perpendicular to each other, the vectors cannot be related by a multiplicative constant.

TABLE 5.1	**Coefficients of Friction**[a]	
	$\boldsymbol{\mu_s}$	$\boldsymbol{\mu_k}$
Steel on steel	0.74	0.57
Aluminum on steel	0.61	0.47
Copper on steel	0.53	0.36
Rubber on concrete	1.0	0.8
Wood on wood	0.25–0.5	0.2
Glass on glass	0.94	0.4
Waxed wood on wet snow	0.14	0.1
Waxed wood on dry snow	—	0.04
Metal on metal (lubricated)	0.15	0.06
Ice on ice	0.1	0.03
Teflon on Teflon	0.04	0.04
Synovial joints in humans	0.01	0.003

[a] All values are approximate.

Be careful with the direction of the friction force. Sometimes, an incorrect statement about the friction force between an object and a surface is made—"the friction force on an object is opposite to its motion or impending motion"—rather than the correct phrasing, "the friction force on an object is opposite to its motion or impending motion *relative to the surface.*" An example is a package placed on the bed of a pickup truck. As the truck accelerates forward, the natural inertia of the package tends to "leave it behind." Thus, the impending motion relative to the floor of the truck is toward the rear, and the friction force exerted on the package is forward, in the same direction in which it moves. If the truck accelerates rapidly enough, the package slides backward in the bed of the truck. In this case, we have real motion of the object rather than impending motion (relative to the surface), and the motion is backward. The friction force, then, is forward.

The approximate nature of Equations 5.1 and 5.2 is easily demonstrated by trying to arrange for an object to slide down an incline at constant speed. Especially at low speeds, the motion is likely to be characterized by alternate stick and slip episodes. These assumptions represent a simplification model, developed so that we can solve problems involving friction in a relatively straightforward way.

Now that we have identified the characteristics of the friction force, we can include the friction force in the net force on an object in the model of a particle under a net force.

Quick Quiz 5.1

You press your physics textbook flat against a vertical wall with your hand. What is the direction of the friction force on the book exerted by the wall? (a) Downward. (b) Upward. (c) Out from the wall. (d) Into the wall.

Quick Quiz 5.2

A crate is located at the center of a flatbed truck. The truck accelerates toward the east, and the crate moves with it, not sliding on the bed of the truck. What is the direction of the friction force exerted by the bed of the truck on the crate? (a) To the west. (b) To the east. (c) There is no friction force because the crate is not sliding.

Quick Quiz 5.3

You are playing with your daughter in the snow. She is sitting on a sled and asking you to slide her across a flat, horizontal field. You have a choice of pushing her from behind, by applying a force downward on her shoulders at 30° below the horizontal (Fig. 5.2a), or attaching a rope to the front of the sled and pulling with a force at 30° above the horizontal (Fig 5.2b). Which would be easier for you and why?

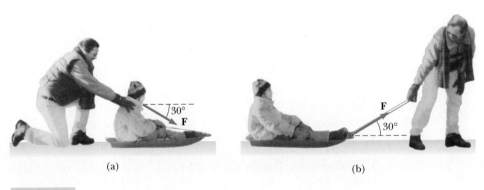

(a)　　　　　　　　　　　　　　　　(b)

Figure 5.2

(Quick Quiz 5.3) A father tries to push his daughter on a sled over snow by (a) pushing downward on her shoulders, or (b) pulling upward on a rope attached to the sled. Which is easier?

THINKING PHYSICS 5.1

In the motion picture *The Abyss* (Twentieth Century Fox, 1990), an underwater oil exploration rig is located at the ocean bottom in very deep water. It is connected to a ship on the ocean surface by an "umbilical cord" as suggested in Figure 5.3a. On the ship, the umbilical cord is attached to a gantry. During a hurricane, the gantry structure breaks loose from the ship, falls into the water, and sinks to the bottom, passing over the edge of an extremely deep abyss. As a result, the rig is dragged by the umbilical cord along the ocean bottom as described in Figure 5.3b. As the rig approaches the edge of the abyss, however, it is not pulled over the edge, but stops just short of the edge as shown in Figure 5.3c. Is this purely a cinematic edge-of-the-seat situation, or do the principles of physics suggest why the moving rig does not topple over the edge?

Reasoning Physics can explain this phenomenon. While the rig is being pulled across the ocean floor (Fig. 5.3b), it is pulled by the section of the umbilical cord that is almost horizontal—almost parallel to the ocean floor. Thus, the rig is subject to two horizontal forces: the tension in the umbilical cord pulling it forward and

WEB

For information about
The Abyss, visit
www.sciflicks.com/the__abyss

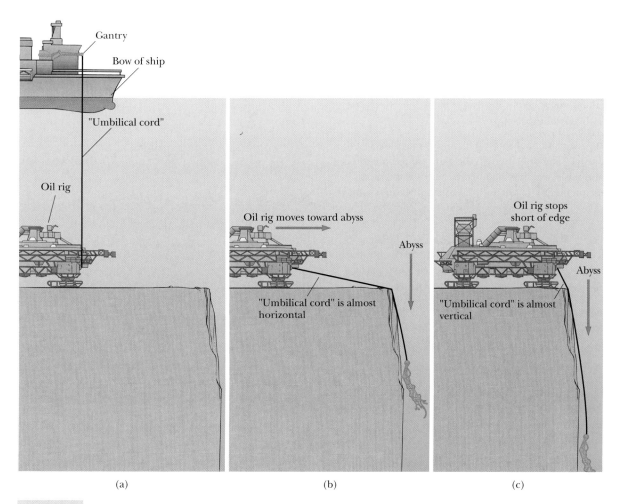

(a)   (b)   (c)

Figure 5.3

(Thinking Physics 5.1)

In Example 5.1, we find that the magnitudes of the normal force and the gravitational force are equal, $n = mg$. As pointed out in Pitfall Prevention 4.9 in Chapter 4, this is *not* a general result. This is such a common student pitfall that we point it out again here. *Always* use Newton's second law to find the relation between n and mg; it will give you the correct relationship. Consult Examples 5.2 and 5.4 for cases in which n is not equal to mg.

friction with the ocean floor pulling back. Let us assume that these forces are equal, so that the rig moves with constant speed. As the rig nears the edge of the abyss, the angle the umbilical cord makes with the horizontal increases. As a result, the component of the force from the cord parallel to the ocean floor decreases and the downward vertical component increases. As a result of the increased vertical force, the rig is pulled downward more strongly to the ocean floor, increasing the normal force on it and, in turn, increasing the friction force between the rig and the ocean floor. Thus, with less force pulling it forward (from the umbilical cord) and more force opposite to the motion (due to friction), the rig slows down. By the time the rig reaches the edge of the abyss, the force from the umbilical cord is almost *straight down* (Fig. 5.3c), resulting in little forward force. Furthermore, this large downward force pulls the rig into the ocean floor, resulting in a very large friction force that stops the rig.

If the coefficient of friction between the rig and ocean floor is relatively small, the forward velocity of the rig could be large enough so that the rig would be projected off the edge.

Example 5.1 The Skidding Truck

The driver of an empty speeding truck slams on the brakes and skids to a stop through a distance d. (a) If the truck carries a heavy load such that its mass is doubled, what would be its skidding distance if it starts from the same initial speed?

(b) If the initial speed of the empty truck is halved, what would be the skidding distance?

Solution (a) Figure 5.4 shows a free-body diagram for the skidding truck. The only force in the horizontal direction is the friction force, which is assumed to be constant in our simplification model for friction. Thus, from Newton's second law

$$\sum F_x = f_k = ma$$

where m is the mass of the truck. In the vertical direction, there is no acceleration, so we model the truck as a particle in equilibrium:

$$\sum F_y = n - mg = 0 \quad \longrightarrow \quad n = mg$$

Finally, from the relation between the friction force and the normal force, we combine these two equations:

$$f_k = \mu_k n \quad \longrightarrow \quad ma = \mu_k(mg) \quad \longrightarrow \quad a = \mu_k g$$

Because both μ_k and g are constant, the acceleration of the truck is constant. We therefore model the truck as a particle under constant acceleration. We use Equation 2.12 to find the position of the truck when the velocity is zero:

$$v_{xf}^2 = v_{xi}^2 + 2a_x(x_f - x_i) \quad \longrightarrow \quad 0 = v_{xi}^2 + 2(\mu_k g)(x_f - 0)$$

$$x_f = d = \frac{v_{xi}^2}{2\mu_k g}$$

We can argue from the mathematical representation as follows. The expression for the skidding distance d does not in-

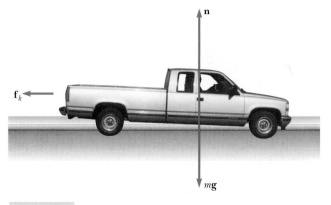

Figure 5.4

(Example 5.1) A truck skids to a stop.

clude the mass. Thus, the truck skids the same distance regardless of the mass. Conceptually, we can argue that the truck with twice the mass requires twice the friction force to exhibit the same acceleration and stop in the same distance. The normal force is equal to the doubled weight, and the friction force is proportional to the doubled normal force!

(b) This part of the problem is a comparison problem and can be solved by a ratio technique. We write the result from part (a) for the skidding distance d twice, once for the original situation, and once for the halved initial velocity:

$$d_1 = \frac{v_{1xi}^2}{2\mu_k g}$$

$$d_2 = \frac{v_{2xi}^2}{2\mu_k g} = \frac{(\frac{1}{2}v_{1xi})^2}{2\mu_k g} = \frac{1}{4}\frac{v_{1xi}^2}{2\mu_k g}$$

Dividing the first equation by the second, we have

$$\frac{d_1}{d_2} = 4 \quad \longrightarrow \quad d_2 = \tfrac{1}{4}d_1$$

Notice that a reduction of half the velocity reduces the skidding distance by 75%! This is an important safety consideration associated with the possibility of an accident when driving at high speed.

Example 5.2 Experimental Determination of μ_s and μ_k

The following is a simple method of measuring coefficients of friction. Suppose a block is placed on a rough surface inclined relative to the horizontal, as shown in Figure 5.5. The incline angle θ is increased until the block starts to move. (a) How is the coefficient of static friction related to the critical angle θ_c at which the block begins to move?

(b) How could we find the coefficient of kinetic friction?

Solution (a) The forces on the block, as shown in Figure 5.5, are the gravitational force $m\mathbf{g}$, the normal force $\mathbf{n}$, and the force of static friction $\mathbf{f}_s$. As long as the block is not moving, these forces are balanced and the block is in equilibrium. We choose a coordinate system with the positive x axis parallel to the incline and downhill and the positive y axis upward perpendicular to the incline. Applying Newton's second law in component form to the block gives

(1) $\quad \sum F_x = mg\sin\theta - f_s = 0$

(2) $\quad \sum F_y = n - mg\cos\theta = 0$

These equations are valid for any angle of inclination θ. At the critical angle θ_c at which the block is on the verge of slipping, the friction force has its maximum magnitude $\mu_s n$, so we rewrite (1) and (2) as

(3) $\quad mg\sin\theta_c = \mu_s n$

(4) $\quad mg\cos\theta_c = n$

Dividing (3) by (4), we have

$$\tan\theta_c = \mu_s$$

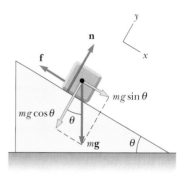

Figure 5.5

(Example 5.2) A block on an adjustable incline is used to determine the coefficients of friction.

Thus, the coefficient of static friction is equal to the tangent of the angle at which the block begins to slide.

(b) Once the block begins to move, the magnitude of the friction force is the kinetic value $\mu_k n$, which is smaller than that of the force of static friction. As a result, if the angle is maintained at the critical angle, the block accelerates down the incline. To restore the equilibrium situation in Equation (1), with f_s replaced by f_k, the angle must be reduced to a value θ_c' such that the block slides down the incline at *constant speed*. In this situation, Equations (3) and (4), with θ_c replaced by θ_c' and μ_s by μ_k, give us

$$\tan\theta_c' = \mu_k$$

Example 5.3 Connected Objects

A ball and a cube are connected by a light string that passes over a frictionless light pulley, as in Figure 5.6a. The coefficient of kinetic friction between the cube and the surface is 0.30. Find the acceleration of the two objects and the tension in the string.

Reasoning In the mental representation, imagine the ball moving downward and the cube sliding to the right, both accelerating from rest. We set up a simplified pictorial representation by drawing the free-body diagrams for the two objects as in Figures 5.6b and 5.6c. For the ball, no forces are exerted in the horizontal direction, and we apply Newton's second law in the vertical direction. For the cube, the acceleration is only horizontal, so we know the cube is in equilibrium in the vertical direction. We shall make use of the fact that the magnitude of the force of kinetic friction acting on the cube is proportional to the normal force according to $f_k = \mu_k n$. Because the pulley is light (massless) and frictionless,

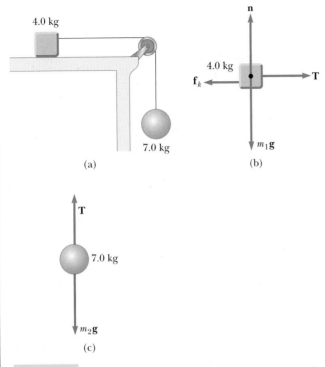

(a)

(b)

(c)

Figure 5.6

(Example 5.3) (a) Two objects connected by a light string that passes over a frictionless pulley. (b) Free-body diagram for the sliding cube. (c) Free-body diagram for the hanging ball.

the tension in the string is the same on both sides of the pulley. Because the tension acts on both objects, this is the common quantity that applies to both objects and allows us to combine separate equations for the two objects into one equation.

Solution   Let us address the cube of mass m_1 first. Newton's second law applied to the cube in component form, with the positive x direction to the right, gives

$$\sum F_x = m_1 a \quad \longrightarrow \quad T - f_k = m_1 a$$

$$\sum F_y = 0 \quad \longrightarrow \quad n - m_1 g = 0$$

where T is the tension in the string. Because $f_k = \mu_k n$ and $n = m_1 g$ from the equilibrium equation for the y direction, we have $f_k = \mu_k m_1 g$. Therefore, from the equation for the x direction,

$$(1) \quad T = \mu_k m_1 g + m_1 a$$

Now we apply Newton's second law to the ball moving in the vertical direction. Because the ball moves downward when the cube moves to the right, we choose the downward direction as positive for the ball:

$$(2) \quad \sum F_y = m_2 a \quad \longrightarrow \quad m_2 g - T = m_2 a$$

Substituting the expression for T from (1) into (2) gives us

$$m_2 g - (\mu_k m_1 g + m_1 a) = m_2 a$$

$$a = \frac{m_2 - \mu_k m_1}{m_1 + m_2} g$$

Now, substituting the known values,

$$a = \frac{7.0 \text{ kg} - 0.30(4.0 \text{ kg})}{7.0 \text{ kg} + 4.0 \text{ kg}} (9.80 \text{ m/s}^2) = \boxed{5.2 \text{ m/s}^2}$$

This is the *magnitude* of the acceleration of each of the two objects. For the ball, the acceleration *vector* is downward, and the vector is toward the right for the cube. When the magnitude of the acceleration is substituted into (1), we find the tension:

$$T = 0.30(4.0 \text{ kg})(9.80 \text{ m/s}^2) + (4.0 \text{ kg})(5.2 \text{ m/s}^2) = \boxed{33 \text{ N}}$$

EXERCISE   If the coefficient of kinetic friction were very large, the ball and cube could move with constant velocity. What coefficient is required for this situation to exist?

Answer   1.75

Example 5.4   The Sliding Crate

A warehouse worker places a crate on a sloped surface that is inclined at 30.0° with respect to the horizontal (Fig. 5.7a). If the crate slides down the incline with an acceleration of magnitude $g/3$, determine the coefficient of kinetic friction between the crate and the surface.

Solution   Figure 5.7b shows the forces acting on the crate. The x axis is chosen parallel to the incline and the y axis per-

pendicular. From Newton's second law,

$$(1) \quad \sum F_x = ma \quad \longrightarrow \quad mg \sin \theta - f_k = ma$$

$$(2) \quad \sum F_y = 0 \quad \longrightarrow \quad n - mg \cos \theta = 0$$

The friction force is $f_k = \mu_k n$ and, from (2), we find that $n = mg \cos \theta$. Thus, the friction force can be expressed as $f_k = \mu_k mg \cos \theta$. Substituting this into (1),

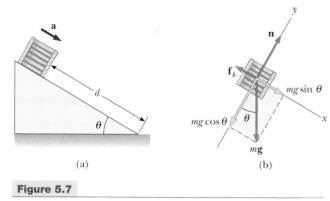

$$mg \sin \theta - \mu_k mg \cos \theta = ma \quad \longrightarrow \quad \mu_k = \frac{g \sin \theta - a}{g \cos \theta}$$

Substituting the known values, we have

$$\mu_k = \frac{\cancel{g} \sin 30.0° - \frac{1}{3}\cancel{g}}{\cancel{g} \cos 30.0°} = \frac{(0.500 - 0.333)}{0.867} = \boxed{0.192}$$

EXERCISE A block moves up a 45.0° incline with constant speed under the action of a force of 15.0 N applied *parallel* to the incline. If the coefficient of kinetic friction is 0.300, determine (a) the weight of the block and (b) the minimum force parallel to the incline required to allow it to move *down* the incline at constant speed.

Answer (a) 16.3 N (b) 8.07 N up the incline

Figure 5.7

(Example 5.4) (a) A crate of mass *m* slides down an incline.
(b) Free-body diagram for the sliding crate.

5.2 • NEWTON'S SECOND LAW APPLIED TO A PARTICLE IN UNIFORM CIRCULAR MOTION

Solving problems involving friction is just one of many applications of Newton's second law. Let us now consider another common situation, associated with a particle in uniform circular motion. In Chapter 3, we found that a particle moving in a circular path of radius *r* with uniform speed *v* experiences a centripetal acceleration of magnitude

$$a_c = \frac{v^2}{r}$$

The acceleration vector with this magnitude is directed toward the center of the circle and is *always* perpendicular to **v**.

According to Newton's second law, if an acceleration occurs, a net force must be causing it. Because the acceleration is toward the center of the circle, the net force must be toward the center of the circle. Thus, when a particle travels in a circular path, a force must be acting *inward* on the particle that causes the circular motion. We investigate the forces causing this type of acceleration in this section.

Consider an object of mass *m* tied to a string of length *r* and being whirled in a horizontal circular path on a frictionless table top as in Figure 5.8. Let us assume that the object moves with constant speed. The natural tendency of the object is to move in a straight-line path, according to Newton's first law; however, the string prevents this motion along a straight line by exerting a radial force **F**$_r$ on the object to make it follow a circular path. This force, which is the tension in the string, is directed along the length of the string toward the center of the circle as shown in Figure 5.8.

In this discussion, the tension in the string causes the circular motion. There are many other examples of forces causing objects to move in circular paths. These forces can be any of our familiar forces causing an object to follow a circular path and, therefore, to have a centripetal acceleration.

Regardless of the nature of the force acting on the particle in circular motion, we can apply Newton's second law to the particle along the radial direction:

$$\sum F = ma_c = m\frac{v^2}{r} \qquad [5.3]$$

See the *Core Concepts in Physics CD-ROM*, Screen 4.7

• *Centripetal acceleration*

PITFALL PREVENTION 5.5
Centrifugal force

The phrase "centrifugal force" is often heard, and described as a force pulling *outward* on an object moving in a circular path. It is likely that you have felt this effect, perhaps on an amusement park ride, or during a sharp turn in your car. This is not a real force, however. To convince yourself of this, remember that forces are always interactions between objects. If you are feeling a "centrifugal force," with what other object are you interacting? You cannot identify another object because this is not a real force. This is a "fictitious force" that occurs as a result of your being in a noninertial reference frame. (See "Inertial Frames" in Section 4.2—the acceleration of the puck on the table in the accelerating train is caused by another "fictitious force.")

The force causing centripetal acceleration is called *centripetal force* in some textbooks. This is a pitfall for many students. Giving the force causing circular motion a name—centripetal force—leads many students to consider this as a new *kind* of force, rather than a new *role* for force. A common mistake in force diagrams is to draw in all of the usual forces and then add another vector for the centripetal force. But it is not a separate force—it is simply one of our familiar forces *acting in the role of causing a circular motion.* For the motion of the Earth around the Sun, for example, the "centripetal force" is *gravity*. For a rock whirled on the end of a string, the "centripetal force" is the *tension* in the string. For an amusement park patron pressed against the inner wall of a rapidly rotating circular room, the "centripetal force" is the *normal force* from the wall. After this discussion, we shall no longer use the phrase *centripetal force*.

Study Figure 5.9 carefully. Many students have a misconception that the ball moves *radially* away from the center of the circle when the string is cut. The velocity of the ball is *tangent* to the circle. By Newton's first law, the ball simply continues to move in the direction that it is moving just as the force from the string disappears.

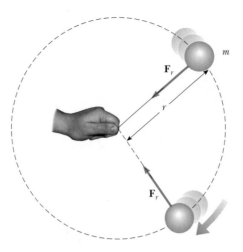

Figure 5.8

Overhead view of a ball moving in a circular path in a horizontal plane. A force $\mathbf{F}_r$ directed toward the center of the circle keeps the ball moving in its circular path.

In general, an object can move in a circular path under the influence of various types of forces, or a *combination* of forces, as we shall see in some of the examples that follow.

If the force acting on an object vanishes, the object no longer moves in its circular path; instead it moves along a straight-line path tangent to the circle. This idea is illustrated in Figure 5.9 for the case of the ball whirling in a circle at the end of a string. If the string breaks at some instant, the ball moves along the straight-line path tangent to the circle at the point where the string breaks.

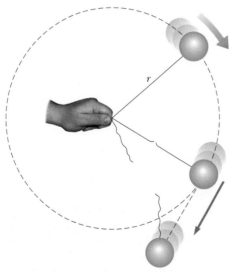

Figure 5.9

When the string breaks, the ball moves in the direction tangent to the circular path.

Quick Quiz 5.4

You are riding on a Ferris wheel that is rotating with constant speed. The car in which you ride always maintains its correct upward orientation—it does not invert. What is the direction of your centripetal acceleration when you are (a) at the top of the wheel? (b) At the bottom of the wheel? What is the direction of the normal

On a Ferris wheel, the force of the seat upward on a rider varies with the position of the rider. Where is this force the largest? *(Color Box/FPG)*

force on you from the seat when you are (c) at the top of the wheel? (d) At the bottom of the wheel? (e) At which point, top or bottom, is the normal force the largest in magnitude?

Quick Quiz 5.5

A car travels on a circular roadway of radius r. The roadway is flat, as in Figure 5.10a. The car travels at a high speed v, such that the friction force causing the centripetal acceleration is the maximum possible value. If this same car is now driven on another flat circular roadway of radius $2r$, and the coefficient of friction between the tires and the roadway is the same as on the first roadway, what is the maximum speed of the car such that it does not slide off the roadway?

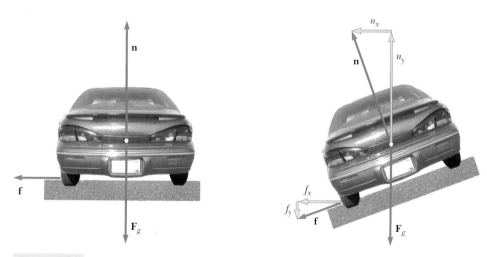

Figure 5.10

(a) (Quick Quiz 5.5) Free-body diagram for a car traveling on a curved roadway that is not banked. The center of curvature of the roadway is to the left. (b) (Thinking Physics 5.2) Free-body diagram for a car traveling on a curved roadway that is banked.

The cars of a corkscrew roller coaster must travel in tight loops. The normal force from the track contributes to the centripetal acceleration. The gravitational force, because it remains constant in direction, is sometimes in the same direction as the normal force, but is sometimes in the opposite direction. (*Robin Smith/Tony Stone Images*)

WEB

For discussions of the physics behind several amusement park rides, visit **www.learner.org/exhibits/ parkphysics/**

THINKING PHYSICS 5.2

High-speed curved roadways are *banked*—that is, the roadway is tilted toward the inside of tight curves. Why is this?

Reasoning   As an automobile rounds a curve, there must be a force on the car providing its centripetal acceleration. In the rolling motion of each tire, the bit of rubber meeting the road is instantaneously at rest relative to the road. It is prevented from skidding outward by a static friction force that acts radially inward, enabling the car to move in its circular path. This static friction force is parallel to the roadway surface and perpendicular to the centerline of the car, as shown in Figure 5.10a for a car on a flat roadway. This is the only horizontal force—the weight and normal force are both vertical. If the curve is tight or the speed is high, the required friction force may be larger than the maximum possible static friction force. In this case, Newton's first law takes over, and the car skids toward the side of the curved road. This situation is even more likely to occur if the road is icy, so that the friction coefficient is reduced. Banking the roadway "tilts" the normal force away from the vertical, so that it has a horizontal component toward the center of the circular roadway, as shown in Figure 5.10b. Thus, the combination of the normal force and the friction force works together to provide the needed centripetal acceleration. The larger the acceleration, the larger is the banking angle, to provide a larger component of the normal force in the horizontal direction.

THINKING PHYSICS 5.3

The Copernican theory of the solar system is a model in which the planets are assumed to travel around the Sun in circular orbits. Historically, this was a break from the Ptolemaic theory, a model in which the Earth was at the center. When the Copernican theory was proposed, a natural question arose—what keeps the Earth and other planets moving in their paths around the

Sun? An interesting response to this question comes from Richard Feynman*— "In those days, one of the theories proposed was that the planets went around because behind them there were invisible angels, beating their wings and driving the planets forward. . . . It turns out that in order to keep the planets going around, the invisible angels must fly in a different direction. . . ." What did Feynman mean by this?

Reasoning   The question asked by those at the time of Copernicus indicates that they did not have a proper understanding of inertia as described by Newton's first law. At that time in history, before Galileo and Newton, the interpretation was that *motion* was caused by force. This is different from our current understanding that *changes in motion* are caused by force. Thus, it was natural for Copernicus's contemporaries to ask what force propelled a planet in its orbit. According to our current understanding, it is equally natural for us to realize that no force tangent to the orbit is necessary—the motion simply continues due to inertia.

Thus, in Feynman's imagery, we do not need the angels pushing the planet *from behind*. We need the angels to push *inward*, to provide the centripetal acceleration associated with the orbital motion of the planet. Of course, the angels are not real, from a scientific point of view—they are a metaphor for the *gravitational force*.

Example 5.5   How Fast Can It Spin?

An object of mass 0.500 kg is attached to the end of a cord whose length is 1.50 m. The object is whirled in a horizontal circle as in Figure 5.8. If the cord can withstand a maximum tension of 50.0 N, what is the maximum speed the object can have before the cord breaks?

Solution   Because the magnitude of the force that provides the centripetal acceleration of the object in this case is the tension T exerted by the cord on the object, Newton's second law gives us for the inward radial direction,

$$\sum F_r = ma_c \quad \longrightarrow \quad T = m\frac{v^2}{r}$$

Solving for the speed v, we have

$$v = \sqrt{\frac{Tr}{m}}$$

The maximum speed that the object can have corresponds to the maximum value of the tension. Hence, we find

$$v_{max} = \sqrt{\frac{T_{max}r}{m}} = \sqrt{\frac{(50.0\ \text{N})(1.50\ \text{m})}{0.500\ \text{kg}}} = \boxed{12.2\ \text{m/s}}$$

EXERCISE   Calculate the tension in the cord if the speed of the object is 5.00 m/s.

Answer   8.33 N

Example 5.6   The Conical Pendulum

A small object of mass m is suspended from a string of length L. The object revolves in a horizontal circle of radius r with constant speed v, as in Figure 5.11a. (Because the string sweeps out the surface of a cone, the system is known as a *conical pendulum*.) Find (a) the speed of the object, and (b) the period of revolution, defined as the time needed to complete one revolution.

Solution   (a) The free-body diagram for the object of mass m is shown in Figure 5.11b, where the force **T** exerted by the string has been resolved into a vertical component $T\cos\theta$ and a horizontal component $T\sin\theta$ acting toward the center of rotation. Because the object does not accelerate in the ver-

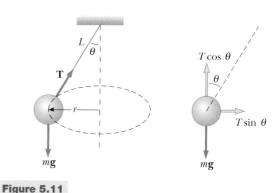

Figure 5.11

(Example 5.6) The conical pendulum and its free-body diagram.

* R. P. Feynman, R. B. Leighton, M. Sands, *The Feynman Lectures on Physics*, Vol. 1, Reading, Massachusetts, Addison-Wesley Publishing Co., 1963, p. 7-2.

tical direction, we model it as a particle in equilibrium in the vertical direction:

$$(1) \quad \sum F_y = 0 \quad \longrightarrow \quad T\cos\theta - mg = 0$$

$$\longrightarrow \quad T\cos\theta = mg$$

In the horizontal direction, we have a centripetal acceleration. Because the force that provides the centripetal acceleration in this example is the component $T\sin\theta$, from Newton's second law we have

$$(2) \quad \sum F_r = ma_c \quad \longrightarrow \quad T\sin\theta = m\frac{v^2}{r}$$

By dividing (2) by (1), we eliminate T and find that

$$\tan\theta = \frac{v^2}{rg} \quad \longrightarrow \quad v = \sqrt{rg\tan\theta}$$

From a triangle we can construct in the pictorial representation in Figure 5.11a, we note that $r = L\sin\theta$; therefore,

$$v = \sqrt{Lg\sin\theta\tan\theta}$$

(b) The object is traveling at constant speed around its circular path. Because the object travels a distance of $2\pi r$ (the circumference of the circular path) in a time equal to the period of revolution T (not to be confused with the magnitude T of the tension force), we find

$$(3) \quad T = \frac{2\pi r}{v} = \frac{2\pi r}{\sqrt{rg\tan\theta}} = 2\pi\sqrt{\frac{L\cos\theta}{g}}$$

The intermediate algebraic steps used in obtaining (3) are left to the reader. Note that the period is independent of the mass m!

Example 5.7   What Is the Maximum Speed of the Car?

A 1 500-kg car moving on a flat, horizontal road negotiates a curve whose radius is 35.0 m (Fig. 5.12a). If the coefficient of static friction between the tires and the dry pavement is 0.500, find the maximum speed the car can have in order to make the turn successfully.

Solution   The car is an extended object with four friction forces acting on it, one on each wheel, but we shall model it as a particle with only one net friction force. Figure 5.12b shows a free-body diagram for the car. From Newton's second law in the horizontal direction, we have

$$(1) \quad \sum F_x = ma \quad \longrightarrow \quad f_s = m\frac{v^2}{r}$$

The maximum speed that the car can have around the curve corresponds to the speed at which it is on the verge of skidding toward the side of the road. At this point, the friction force has its maximum value

$$f_{s,max} = \mu_s n$$

In the vertical direction, no acceleration occurs, so

$$\sum F_y = 0 \quad \longrightarrow \quad n - mg = 0$$

Thus, the magnitude of the normal force equals the weight in this case, and we find

$$f_{s,max} = \mu_s mg$$

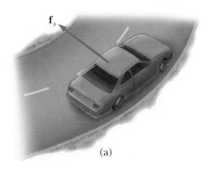

(a)

(b)

Figure 5.12

(Example 5.7) (a) The force of static friction directed toward the center of the curve keeps the car moving in a circular path. (b) Free-body diagram for the car.

Substituting this expression into (1), we find the maximum speed:

$$\mu_s mg = m\frac{v_{max}^2}{r} \quad \longrightarrow \quad v_{max} = \sqrt{\mu_s gr}$$

Substituting the numerical values,

$$v_{max} = \sqrt{(0.500)(9.80 \text{ m/s}^2)(35.0 \text{ m})} = \boxed{13.1 \text{ m/s}}$$

This is equivalent to 29.3 mi/h, which is less than a typical nonfreeway speed of 35 mi/h. Thus, this roadway could benefit greatly from some banking!

EXERCISE On a wet day, the car described in this example begins to skid on the curve when its speed reaches 8.00 m/s. What is the coefficient of static friction in this case?

Answer 0.187

Example 5.8 Let's Go Loop-the-Loop

A pilot of mass m in a jet aircraft executes a "loop-the-loop" maneuver as illustrated in Figure 5.13a. The aircraft moves in a vertical circle of radius 2.70 km at a *constant speed* of 225 m/s. Determine the force exerted by the seat on the pilot at (a) the bottom of the loop and (b) the top of the loop. Express the answers in terms of the weight mg of the pilot.

Reasoning This is the first numerical example in which the force causing the centripetal acceleration is a *combination* of forces rather than a single force. We shall model the pilot as a particle under a net force and analyze the situation at the bottom and top of the circular path.

Solution (a) The free-body diagram for the pilot at the bottom of the loop is shown in Figure 5.13b. The forces acting on the pilot are the downward gravitational force $m\mathbf{g}$ and the upward normal force $\mathbf{n}_{bot}$ exerted by the seat on the pilot. Because the net upward force at the bottom that provides the centripetal acceleration has a magnitude $n_{bot} - mg$,

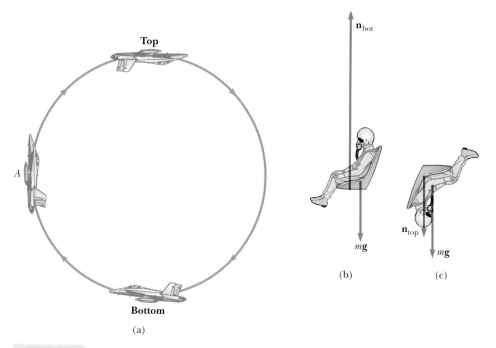

(a)

(b)   (c)

Figure 5.13

(Example 5.8) (a) An aircraft executes a loop-the-loop maneuver as it moves in a vertical circle at constant speed. (b) Free-body diagram for the pilot at the bottom of the loop. In this position, the pilot experiences a force from the seat that is larger than his weight. (c) Free-body diagram for the pilot at the top of the loop. Here the force from the seat is smaller than his weight.

Newton's second law for the radial (upward) direction gives

$$\sum F_y = ma \quad \longrightarrow \quad n_{bot} - mg = m\frac{v^2}{r}$$

$$n_{bot} = mg + m\frac{v^2}{r} = mg\left[1 + \frac{v^2}{rg}\right]$$

Substituting the values given for the speed and radius gives

$$n_{bot} = mg\left[1 + \frac{(225 \text{ m/s})^2}{(2.70 \times 10^3 \text{ m})(9.80 \text{ m/s}^2)}\right] = \boxed{2.91\, mg}$$

Hence, the force exerted by the seat on the pilot at the bottom of the loop is *greater* than his weight by a factor of 2.91.

(b) The free-body diagram for the pilot at the top of the loop is shown in Figure 5.13c. At this point, both the gravitational force and the force $\mathbf{n}_{top}$ exerted by the seat on the pilot act *downward*, so the net force downward that provides the centripetal acceleration has a magnitude $n_{top} + mg$. Applying Newton's second law gives

$$\sum F_y = ma \quad \longrightarrow \quad n_{top} + mg = m\frac{v^2}{r}$$

$$n_{top} = m\frac{v^2}{r} - mg = mg\left[\frac{v^2}{rg} - 1\right]$$

$$n_{top} = mg\left[\frac{(225 \text{ m/s})^2}{(2.70 \times 10^3 \text{ m})(9.80 \text{ m/s}^2)} - 1\right] = \boxed{0.911\, mg}$$

In this case, the force exerted by the seat on the pilot is *less* than the weight by a factor of 0.911. Hence, the pilot feels lighter at the top of the loop.

EXERCISE Calculate the force on the pilot causing the centripetal acceleration if the aircraft is at point *A* in Figure 5.13a, midway up the loop.

Answer $1.91\, mg$ directed toward the right.

EXERCISE An automobile moves at constant speed over the crest of a hill. The driver moves in a vertical circle of radius 18.0 m. At the top of the hill, she notices that she barely remains in contact with the seat. Find the speed of the vehicle.

Answer 13.3 m/s

5.3 • NONUNIFORM CIRCULAR MOTION

In Chapter 3 we found that if a particle moves with varying speed in a circular path, there is, in addition to the radial component of acceleration, a tangential component of magnitude dv/dt. Therefore, the net force acting on the particle must also have a radial and a tangential component as shown in Figure 5.14. That is, because the total acceleration is $\mathbf{a} = \mathbf{a}_r + \mathbf{a}_t$, the total force exerted on the particle is $\sum \mathbf{F} = \sum \mathbf{F}_r + \sum \mathbf{F}_t$. The component vector $\sum \mathbf{F}_r$ is directed toward the center of the circle and is responsible for the centripetal acceleration. The component vector $\sum \mathbf{F}_t$ tangent to the circle is responsible for the tangential acceleration, which causes the speed of the particle to change with time.

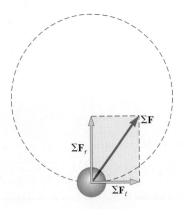

Figure 5.14

When the net force acting on a particle moving in a circular path has a tangential component vector $\sum \mathbf{F}_t$, its speed changes. The total force on the particle also has a component vector $\sum \mathbf{F}_r$ directed toward the center of the circular path. Thus, the total force is $\sum \mathbf{F} = \sum \mathbf{F}_t + \sum \mathbf{F}_r$.

Example 5.9 Follow the Rotating Ball

A small sphere of mass m is attached to the end of a cord of length R, which rotates under the influence of the gravitational force in a *vertical* circle about a fixed point O, as in Figure 5.15a. Let us determine the tension in the cord at any instant when the speed of the sphere is v and the cord makes an angle θ with the vertical.

Reasoning First note that the speed is *not* uniform because a tangential component of acceleration arises from the gravitational force on the sphere. Thus, although this example is similar to Example 5.8, it is not identical. From the free-body diagram in Figure 5.15a, we see that the only forces acting on the sphere are the gravitational force $m\mathbf{g}$ and the force $\mathbf{T}$ exerted by the cord.

Solution We resolve $m\mathbf{g}$ into a tangential component $mg \sin \theta$ and a radial component $mg \cos \theta$. Applying Newton's second law for the tangential direction gives

$$\sum F_t = ma_t \longrightarrow mg \sin \theta = ma_t$$

$$a_t = g \sin \theta$$

This component causes v to change in time, because $a_t = dv/dt$.

Applying Newton's second law to the forces in the radial direction (for which the outward direction is positive), we find

$$\sum F_r = ma_r \longrightarrow mg \cos \theta - T = -m \frac{v^2}{r}$$

$$T = m\left(\frac{v^2}{R} + g \cos \theta \right)$$

At the bottom of the path, where $\theta = 0$, we see that because $\cos 0 = 1$,

$$T_{\text{bot}} = m\left(\frac{v_{\text{bot}}^2}{r} + g \right)$$

This is the *maximum* value of T—as the sphere passes through the bottom point, the string is under the most tension. This is of interest to trapeze artists, because their support wires must withstand this largest tension at the bottom of the swing, as well as to Tarzan, in choosing a nice strong vine on which to swing to withstand this force.

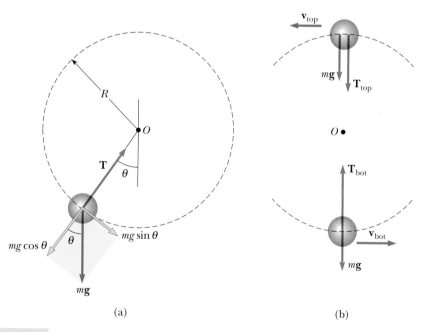

(a) (b)

Figure 5.15

(Example 5.9) (a) Forces acting on a sphere of mass m connected to a cord of length R and rotating in a vertical circle centered at O. (b) Forces acting on the sphere when it is at the top and bottom of the circle. The tension has its maximum value at the bottom and its minimum value at the top.

See Screen 4.9

5.4 • MOTION IN THE PRESENCE OF VELOCITY-DEPENDENT RESISTIVE FORCES

Earlier we described the friction force between a moving object and the surface along which it moves. So far, we have ignored any interaction between the object and the *medium* through which it moves. Now let us consider the effect of a medium such as a liquid or gas. The medium exerts a **resistive force R** on the object moving through it. You feel this force if you ride in a car at high speed with your hand out the window—the force you feel pushing your hand backward is the resistive force of the air rushing past the car. The magnitude of this force depends on the relative speed between the object and the medium, and the direction of **R** on the object is always opposite the direction of motion of the object relative to the medium. Some examples are the air resistance associated with moving vehicles (sometimes called air drag), the force of the wind on the sails of a sailboat, and the viscous forces that act on objects sinking through a liquid.

Generally, the magnitude of the resistive force increases with increasing speed. The resistive force can have a complicated speed dependence. In the following discussions, we consider two simplification models that allow us to analyze these situations. The first model assumes that the resistive force is proportional to the velocity; this is approximately the case for objects that fall through a liquid with low speed and for very small objects, such as dust particles, that move through air. The second model treats situations for which we assume that the magnitude of the resistive force is proportional to the square of the speed of the object. Large objects, such as a skydiver moving through air in free-fall, experience such a force.

Model 1: Resistive Force Proportional to Object Velocity

At low speeds, the resistive force acting on an object that is moving through a viscous medium is effectively modeled as being proportional to the object's velocity. The mathematical representation of the resistive force can be expressed as

$$\mathbf{R} = -b\mathbf{v} \qquad [5.4]$$

where **v** is the velocity of the object and b is a constant that depends on the properties of the medium and on the shape and dimensions of the object. If the object is a sphere of radius r, b is proportional to r. The negative sign represents the fact that the resistive force is opposite the velocity.

Consider a sphere of mass m released from rest in a liquid, as in Figure 5.16a. Assuming the only forces acting on the sphere are the resistive force **R** and the weight $m\mathbf{g}$ we can describe its motion using Newton's second law.* Considering the vertical motion and choosing the downward direction to be positive, we have

$$\sum F_y = ma_y \longrightarrow mg - bv = m\frac{dv}{dt}$$

Dividing this equation by the mass m gives

$$\frac{dv}{dt} = g - \frac{b}{m}v \qquad [5.5]$$

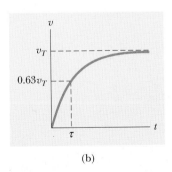

(a)

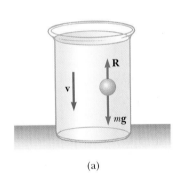

(b)

Figure 5.16

(a) A small sphere falling through a viscous fluid. (b) The speed–time graph for an object falling through a viscous medium. The object reaches a terminal speed v_T, and the time constant τ is the time it takes to reach $0.63v_T$.

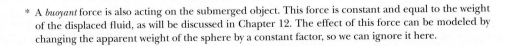

* A *buoyant* force is also acting on the submerged object. This force is constant and equal to the weight of the displaced fluid, as will be discussed in Chapter 12. The effect of this force can be modeled by changing the apparent weight of the sphere by a constant factor, so we can ignore it here.

Equation 5.5 is called a *differential equation*—it includes both the speed v and the derivative of the speed. The methods of solving such an equation may not be familiar to you as yet. However, note that if we define $t = 0$ when $v = 0$, the resistive force is zero at this time and the acceleration dv/dt is simply g. As t increases, the speed increases, the resistive force increases, and the acceleration decreases. Thus, this problem is one in which both the velocity and acceleration of the particle are not constant.

The acceleration becomes zero when the increasing resistive force eventually balances the weight. At this point, the object reaches its **terminal speed** v_T and from then on it continues to move with zero acceleration. After this point, the motion is that of a particle under constant velocity. The terminal speed can be obtained from Equation 5.5 by setting $a = dv/dt = 0$. This gives

• *Terminal speed*

$$mg - bv_T = 0 \quad \longrightarrow \quad v_T = \frac{mg}{b}$$

The expression for v that satisfies Equation 5.5 with $v = 0$ at $t = 0$ is

$$v = \frac{mg}{b}(1 - e^{-bt/m}) = v_T(1 - e^{-t/\tau}) \qquad \text{[5.6]}$$

where $e = 2.71828$ is the base of the natural logarithm. This expression for v can be verified by substituting this solution back into Equation 5.5. (Try it!) This function is plotted in Figure 5.16b.

The mathematical representation of the motion (Equation 5.6) indicates that the terminal speed is never reached, because the exponential function never becomes exactly equal to zero. For all practical purposes, however, when the exponential function is very small, the speed of the particle can be approximated as being constant.

We cannot compare different objects by means of the time to reach terminal speed, because, as we have just discussed, this time is infinite for all objects! We need some means to compare these exponential behaviors for different objects. We do this with a parameter called the **time constant.** The time constant $\tau = m/b$ that appears in Equation 5.6 is the time it takes for the factor in parentheses in Equation 5.6 to become equal to $1 - e^{-1} = 0.632$. Thus, the time constant represents the time required for the object to reach 63.2% of its terminal speed (Fig. 5.16b).

Example 5.10   A Sphere Falling in Oil

A small sphere of mass 2.00 g is released from rest in a large vessel filled with oil. The sphere approaches a terminal speed of 5.00 cm/s. Determine (a) the time constant τ and (b) the time it takes the sphere to reach 90.0% of its terminal speed.

Solution   (a) Because the terminal speed is given by $v_T = mg/b$, the coefficient b is

$$b = \frac{mg}{v_T} = \frac{(2.00 \times 10^{-3}\,\text{kg})(9.80\,\text{m/s}^2)}{5.00 \times 10^{-2}\,\text{m/s}} = 0.392\,\text{N·s/m}$$

Therefore, the time constant τ is

$$\tau = \frac{m}{b} = \frac{2.00 \times 10^{-3}\,\text{kg}}{0.392\,\text{N·s/m}} = 5.10 \times 10^{-3}\,\text{s}$$

(b) The speed of the sphere as a function of time is given by Equation 5.6. To find the time t at which the sphere is traveling at a speed of $0.900v_T$, we set $v = 0.900v_T$, substitute this into Equation 5.6, and solve for t:

$$0.900v_T = v_T(1 - e^{-t/\tau})$$

$$1 - e^{-t/\tau} = 0.900$$

$$e^{-t/\tau} = 0.100$$

$$-\frac{t}{\tau} = \ln 0.100 = -2.30$$

$$t = 2.30\tau = 2.30(5.10 \times 10^{-3}\,\text{s})$$

$$= 11.7 \times 10^{-3}\,\text{s} = \boxed{11.7\,\text{ms}}$$

EXERCISE   What is the sphere's speed through the oil at $t = 11.7$ ms? Compare this value with the speed the sphere would have if it were falling in a vacuum and so influenced only by gravity.

Answer   4.50 cm/s in oil versus 11.5 cm/s in free-fall

Model 2: Resistive Force Proportional to Object Speed Squared

For large objects moving at high speeds through air, such as airplanes, skydivers, and baseballs, the magnitude of the resistive force is modeled as being proportional to the square of the speed:

$$R = \tfrac{1}{2}D\rho Av^2 \qquad\qquad \text{[5.7]}$$

where ρ is the density of air, A is the cross-sectional area of the moving object measured in a plane perpendicular to its velocity, and D is a dimensionless empirical quantity called the *drag coefficient*. The drag coefficient has a value of about 0.5 for spherical objects moving through air but can be as high as 2 for irregularly shaped objects.

Consider an airplane in flight that experiences such a resistive force. Equation 5.7 shows that the force is proportional to the density of air and hence decreases with decreasing air density. Because air density decreases with increasing altitude, the resistive force on a jet airplane flying at a given speed will decrease with increasing altitude. Thus, airplanes tend to fly at very high altitudes to take advantage of this reduced resistive force. This allows them to fly faster for a given engine thrust. Of course, this higher speed *increases* the resistive force, in proportion to the square of the speed, so a balance is struck between flying at higher altitude and higher speed.

Now let us analyze the motion of a falling object subject to an upward air resistive force whose magnitude is given by Equation 5.7. Suppose an object of mass m is released from rest, as in Figure 5.17, from the position $y = 0$. The object experiences two external forces: the downward gravitational force $m\mathbf{g}$ and the upward resistive force $\mathbf{R}$. (There is also an upward buoyant force that we ignore, due to its small value.) Hence, using Newton's second law,

$$\sum F = ma \quad\longrightarrow\quad mg - \tfrac{1}{2}D\rho Av^2 = ma \qquad\qquad \text{[5.8]}$$

Solving for a, we find that the object has a downward acceleration of magnitude

$$a = g - \left(\frac{D\rho A}{2m}\right)v^2 \qquad\qquad \text{[5.9]}$$

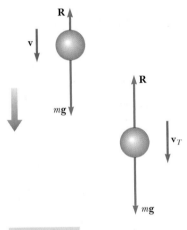

Figure 5.17

An object falling through air experiences a resistive drag force $\mathbf{R}$ and a gravitational force $\mathbf{F}_g = m\mathbf{g}$. The object reaches terminal speed (on the right) when the net force acting on it is zero, that is, when $\mathbf{R} = -\mathbf{F}_g$, or $R = mg$. Before this occurs, the acceleration varies with speed according to Equation 5.9.

Aerodynamic car. A streamlined body reduces air drag and increases fuel efficiency. *(Courtesy of Ford Motor Company)*

TABLE 5.2	Terminal Speeds for Various Objects Falling Through Air		
Object	**Mass (kg)**	**Cross-sectional Area (m²)**	v_T **(m/s)ᵃ**
Skydiver	75	0.70	60
Baseball (radius 3.7 cm)	0.145	4.2×10^{-3}	33
Golf ball (radius 2.1 cm)	0.046	1.4×10^{-3}	32
Hailstone (radius 0.50 cm)	4.8×10^{-4}	7.9×10^{-5}	14
Raindrop (radius 0.20 cm)	3.4×10^{-5}	1.3×10^{-5}	9.0

ᵃ The drag coefficient D is assumed to be 0.5 in each case.

Because $a = dv/dt$, this is another differential equation that provides us with the speed as a function of time.

Again, we can calculate the terminal speed v_T using the fact that when the gravitational force is balanced by the resistive force, the net force is zero and therefore the acceleration is zero. Setting $a = 0$ in Equation 5.9 gives

$$g - \left(\frac{D\rho A}{2m} \right) v_T^2 = 0$$

$$v_T = \sqrt{\frac{2mg}{D\rho A}} \qquad \textbf{[5.10]}$$

Table 5.2 lists the terminal speeds for several objects falling through air, all computed on the assumption that the drag coefficient is 0.5.

Quick Quiz 5.6

A baseball and a basketball, having the same mass, are dropped through air from rest from the same height above the ground. Which ball strikes the ground first?

Quick Quiz 5.7

Consider a sky surfer falling through air, as in Figure 5.18, before reaching her terminal speed. As the speed of the sky surfer increases, what happens to her acceleration? What is her acceleration once she reaches terminal speed?

Figure 5.18

(Quick Quiz 5.7) A sky surfer takes advantage of the upward force of the air on her board. (*Jump Run Productions/Image Bank*)

5.5 • NUMERICAL REPRESENTATIONS OF PARTICLE DYNAMICS*

So far, our mathematical representations for physical situations have involved analytical functions that could be represented as continuous curves on graph paper. This is due to the fact that we have adopted simplification models that have allowed the mathematical equations to be solved relatively simply. In the "real world," however, complications often arise that make analytical solutions for many practical situations difficult and perhaps beyond the mathematical abilities of most students taking introductory physics. For example, the force may depend on the

* The authors are most grateful to Colonel James Head of the U.S. Air Force Academy for preparing this section.

position of the particle, as in cases where the variation in the gravitational force with height must be taken into account. The force may also vary with speed, as we have seen in cases of resistive forces caused by motion through a liquid or gas. The force may depend on both position and speed, as in the case of an object falling through air where the resistive force depends on speed and on height (air density). In rocket motion, the mass changes with time, so even if the force is constant, the acceleration is not.

Another complication arises because the equations relating acceleration, velocity, position, and time are differential, not algebraic, equations. Differential equations are usually solved using integral calculus and other special techniques that you may not have mastered.

So how does one proceed to describe the motion of a particle under these complicated conditions? One answer is to solve such problems on personal computers, using elementary numerical methods. We describe one such technique in this section, which introduces us to the notion of a **numerical representation** of a problem. In a numerical representation, the calculations reduce to simple mathematical steps involving only addition, subtraction, multiplication, and division. Using the numerical representation, the position of a particle can be estimated step by step at many closely spaced instants. Of course, the cost of using this approach is that the calculations must be performed a very large number of times, which can be done with computers.

As we have seen in this and the preceding chapter, the study of dynamics of a particle focuses on describing the position, velocity, and acceleration due to a net force as functions of time. Cause-and-effect relationships exist between these quantities: A net force causes an acceleration, an acceleration causes velocity to change, a velocity causes position to change. Therefore, the study of motion usually begins with an evaluation of the net force on the particle.

In this section, we confine our discussion to one-dimensional motion, so boldface notation is not used for vector quantities. If a particle of mass m moves under the influence of a net force ΣF, Newton's second law tells us that the acceleration of the particle is given by $a = \Sigma F/m$. In general, we can then obtain the analytical solution to a dynamics problem (the position of the particle at any time) by using the following procedure:

- Sum all the forces on the particle to obtain the net force ΣF.
- Use this force to determine the acceleration, using $a = \Sigma F/m$.
- Use this acceleration to determine the velocity from $dv/dt = a$.
- Use this velocity to determine the position from $dx/dt = v$.

This procedure is straightforward for some physical situations, such as ones described by the model of a particle under constant acceleration. Let us now consider how we might solve a problem that is not so straightforward. Suppose, for example, that the acceleration is so complicated that we cannot easily integrate the acceleration function to find the velocity function. One approach to a numerical representation is the Euler method, named after the Swiss mathematician Leonhard Euler (1707–1783).

The Euler Method

In the **Euler method** of solving differential equations, derivatives are approximated as ratios of finite differences. Let us assume that we have determined the net force $\Sigma F(x, v, t)$ on the particle. The acceleration of the particle is determined from Newton's second law,

$$a(x, v, t) = \frac{\Sigma F(x, v, t)}{m} \qquad [5.11]$$

Notice that the net force, and therefore the acceleration, may depend explicitly on the position, velocity, or time. This is what makes the problem difficult to solve analytically.

Considering a small increment of time Δt, the relationship between velocity and acceleration may be approximated as

$$a(t) \approx \frac{\Delta v}{\Delta t} = \frac{v(t + \Delta t) - v(t)}{\Delta t}$$

Rearranging this equation, we find that the velocity of the particle at the end of the period Δt is approximately equal to the velocity at the beginning of the period plus the acceleration during the interval multiplied by Δt:

$$v(t + \Delta t) \approx v(t) + a(t)\,\Delta t \qquad [5.12]$$

Because the acceleration is a function of time, this estimate of $v(t + \Delta t)$ will be accurate only if the time interval Δt is short enough that the change in acceleration during this interval is very small. Thus, because the time interval we are considering is very short, we use the model of a particle under constant acceleration for this calculation. We assume that, despite the fact that the acceleration is not constant, it does not change by very much in a very short time interval.

The position can be found in the same manner:

$$v(t) \approx \frac{\Delta x}{\Delta t} = \frac{x(t + \Delta t) - x(t)}{\Delta t}$$

$$x(t + \Delta t) \approx x(t) + v(t)\,\Delta t \qquad [5.13]$$

Here, we assume the velocity to be constant, because it does not change by very much in a very short time interval. It may be tempting to add the term $\frac{1}{2}a(\Delta t)^2$ to this result to make it look like the familiar kinematics equation, but this term is not included in the Euler method of integration because Δt is assumed to be so small that $(\Delta t)^2$ is nearly zero.

If the acceleration at any instant t is known, the particle's velocity and position at $(t + \Delta t)$ can be calculated from Equations 5.12 and 5.13. The calculation can then proceed in a series of finite steps to determine the velocity and position at any later time. At the end of each time interval, the new value of the force is determined from the functional expression for the force, Equation 5.11 is used to find the new acceleration, Equation 5.12 is used to find the new velocity, and Equation 5.13 is used to find the new position.

It is convenient to set up the numerical solution to this kind of problem by numbering the steps and entering the calculations in a tabular representation. Table 5.3 illustrates how to do this in an orderly way. Many small increments can be taken, and accurate results can usually be obtained with the help of a computer. The equations provided in the table can be entered into a spreadsheet and the calculations performed row by row to determine the velocity, position, and acceleration as functions of time. The calculations can also be carried out using a programming language or with commercially available mathematics packages for personal computers. Graphical representations of velocity versus time or position versus time can be displayed to help you visualize the motion.

TABLE 5.3	The Euler Method for Solving Dynamics Problems			
Step	Time	Position	Velocity	Acceleration
0	t_0	x_0	v_0	$a_0 = F(x_0, v_0, t_0)/m$
1	$t_1 = t_0 + \Delta t$	$x_1 = x_0 + v_0\,\Delta t$	$v_1 = v_0 + a_0\,\Delta t$	$a_1 = F(x_1, v_1, t_1)/m$
2	$t_2 = t_1 + \Delta t$	$x_2 = x_1 + v_1\,\Delta t$	$v_2 = v_1 + a_1\,\Delta t$	$a_2 = F(x_2, v_2, t_2)/m$
3	$t_3 = t_2 + \Delta t$	$x_3 = x_2 + v_2\,\Delta t$	$v_3 = v_2 + a_2\,\Delta t$	$a_3 = F(x_3, v_3, t_3)/m$
$\vdots$	$\vdots$	$\vdots$	$\vdots$	$\vdots$
n	$t_n = t_{n-1} + \Delta t$	$x_n = x_{n-1} + v_{n-1}\,\Delta t$	$v_n = v_{n-1} + a_{n-1}\,\Delta t$	$a_n = F(x_n, v_n, t_n)/m$

The Euler method is completely reliable for infinitesimally small time increments, but for practical reasons a finite increment size must be chosen. For the finite difference approximation of Equation 5.11 to be valid, the time increment must be small enough that the acceleration can be approximated as being constant during the increment. We can determine an appropriate size for the time increment by examining the particular problem that is being investigated. The criterion for the size of the time increment may need to be changed during the course of the motion. In practice, however, we usually choose a time increment appropriate to the initial conditions and use the same value throughout the calculations.

The size of the time increment influences the accuracy of the result, but unfortunately it is not easy to determine the accuracy of a solution by the Euler method without a knowledge of the correct analytical solution. One method of determining the accuracy of the numerical solution is to repeat the calculations with a smaller time increment and compare results. If the two calculations agree to a certain number of significant figures, you can assume that the results are correct to that precision. The solution to Problem 27 at the end of the chapter, in the *Student Solutions Manual and Study Guide* or on the publisher's Web site, illustrates the use of the Euler method.

5.6 • THE FUNDAMENTAL FORCES OF NATURE

We have described a variety of forces experienced in our everyday activities, such as the gravitational force that acts on all objects at or near the Earth's surface and the force of friction as one surface slides over another. Newton's second law tells us how to relate the forces to the acceleration of the object or particle.

In addition to these familiar macroscopic forces in nature, forces also act in the atomic and subatomic world. For example, atomic forces within the atom are responsible for holding its constituents together, and nuclear forces act on different parts of the nucleus to keep its parts from separating.

Until recently, physicists believed that there were four fundamental forces in nature: the gravitational force, the electromagnetic force, the nuclear force, and the weak force. We shall discuss these individually and then consider the current view of fundamental forces.

The Gravitational Force

The **gravitational force** is the mutual force of attraction between any two objects in the Universe. It is interesting and rather curious that, although the gravitational force can be very strong between macroscopic objects, it is inherently the weakest of all the fundamental forces. For example, the gravitational force between the electron and proton in the hydrogen atom has a magnitude on the order of 10^{-47} N, whereas the electromagnetic force between these same two particles is on the order of 10^{-7} N.

In addition to his contributions to the understanding of motion, Newton studied gravity extensively. **Newton's law of universal gravitation** states that every particle in the Universe attracts every other particle with a force that is directly proportional to the product of the masses of the particles and inversely proportional to the square of the distance between them. If the particles have masses m_1 and m_2 and are separated by a distance r, as in Figure 5.19, the magnitude of the gravitational force is

$$F_g = G\frac{m_1 m_2}{r^2} \qquad \text{[5.14]}$$

where $G = 6.67 \times 10^{-11}$ N·m²/kg² is the **universal gravitational constant.** More detail on the gravitational force will be provided in Chapter 11.

The Electromagnetic Force

The **electromagnetic force** is the force that binds atoms and molecules in compounds to form ordinary matter. It is much stronger than the gravitational force. The force that causes a rubbed comb to attract bits of paper and the force that a magnet exerts on an iron nail are electromagnetic forces. Essentially all forces at work in our macroscopic world, apart from the gravitational force, are manifestations of the electromagnetic force. For example, friction forces, contact forces, tension forces, and forces in elongated springs are consequences of electromagnetic forces between charged particles in proximity.

The electromagnetic force involves two types of particles: those with positive charge and those with negative charge. (More information on these two types of charge is provided in Chapter 19.) Unlike the gravitational force, which is always an attractive interaction, the electromagnetic force can be either attractive or repulsive, depending on the charges on the particles.

Coulomb's law expresses the magnitude of the electrostatic force* F_e between two charged particles separated by a distance r:

$$F_e = k_e\frac{q_1 q_2}{r^2} \qquad \text{[5.15]}$$

where q_1 and q_2 are the charges on the two particles, measured in units called *coulombs* (C), and k_e ($= 8.99 \times 10^9$ N·m²/C²) is the Coulomb constant. Note that the electrostatic force has the same mathematical form as Newton's law of universal gravitation (see Eq. 5.14), with charge playing the mathematical role of mass, and the Coulomb constant being used in place of the universal gravitational constant. The electrostatic force is attractive if the two charges have opposite signs, and repulsive if the two charges have the same sign, as indicated in Figure 5.20.

The smallest amount of isolated charge found in nature (so far) is the charge on an electron or proton. This fundamental unit of charge is given the symbol e

---

* The electrostatic force is the electromagnetic force between two electric charges that are at rest. If the charges are moving, magnetic forces are also present; these will be studied in Chapter 22.

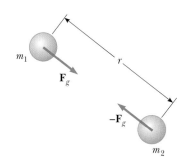

Figure 5.19

Two particles with masses m_1 and m_2 attract each other with a force of magnitude $Gm_1 m_2/r^2$.

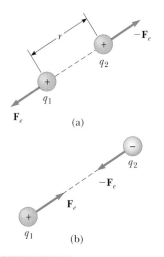

Figure 5.20

Two point charges separated by a distance r exert an electrostatic force on each other given by Coulomb's law. (a) When the charges are of the same sign, the charges repel each other. (b) When the charges are of opposite sign, the charges attract each other.

PITFALL PREVENTION 5.8

The symbol e

Take care not to confuse the symbol e for the fundamental electric charge with that for the base of the natural logarithm e. Both are represented by exactly the same symbol.

and has the magnitude $e = 1.60 \times 10^{-19}$ C. Theories developed in the latter half of the 20th century propose that protons and neutrons are made up of smaller particles called **quarks,** which have charges of either $\frac{2}{3}e$ or $-\frac{1}{3}e$ (discussed further in Chapter 31). Although experimental evidence has been found for such particles inside nuclear matter, free quarks have never been detected.

The Nuclear Force

An atom, as we currently model it, consists of an extremely dense positively charged nucleus surrounded by a cloud of negatively charged electrons, with the electrons attracted to the nucleus by the electric force. Because all nuclei except those of hydrogen are combinations of positively charged protons and neutral neutrons (collectively called nucleons), why does the repulsive electrostatic force between the protons not cause nuclei to break apart? Clearly, there must be an attractive force that counteracts the strong electrostatic repulsive force and is responsible for the stability of nuclei. This force that binds the nucleons to form a nucleus is called the **nuclear force.** Unlike the gravitational and electromagnetic forces, which depend on distance in an inverse-square fashion, the nuclear force is extremely short range; its strength decreases very rapidly outside the nucleus and is negligible for separations greater than approximately 10^{-14} m. For separations of approximately 10^{-15} m (a typical nuclear dimension), the nuclear force is about two orders of magnitude stronger than the electrostatic force.

- *The nuclear force*

The Weak Force

- *The weak force*

The **weak force** is a short-range force that tends to produce instability in certain nuclei. It was first observed in naturally occurring radioactive substances and was later found to play a key role in most radioactive decay reactions. The weak force is about 10^{25} times stronger than the gravitational force and about 10^{12} times weaker than the electromagnetic force.

The Current View of Fundamental Forces

For years physicists have searched for a simplification scheme that would reduce the number of fundamental forces in nature. In 1967 physicists predicted that the electromagnetic force and the weak force, originally thought to be independent of each other and both fundamental, are in fact manifestations of one force, now called the **electroweak** force. The prediction was confirmed experimentally in 1984. We shall discuss this more fully in Chapter 31.

We also now know that protons and neutrons are not fundamental particles; current models of protons and neutrons theorize that they are composed of simpler particles called quarks, as mentioned previously. The quark model has led to a modification of our understanding of the nuclear force. Scientists now define a **strong force** as that force which binds the quarks to one another in a nucleon (proton or neutron). This force is also referred to as a **color force,** in reference to a property of quarks called "color," which we shall investigate in Chapter 31. The previously defined nuclear force, that which acts between nucleons, is now interpreted as a secondary effect of the strong force between the quarks.

Scientists believe that the fundamental forces of nature are closely related to the origin of the Universe. The Big Bang theory states that the Universe began with a cataclysmic explosion 15 to 20 billion years ago. According to this theory,

PITFALL PREVENTION 5.9

The strong force

Historically, before the quark model was proposed, the force between nucleons was called the strong force. Now that the quark model is well accepted, the force between quarks is identified as the strong force. Thus, as you read other books, you may encounter the phrase "strong force" as meaning either the force between nucleons or the force between quarks. In this book, we use the phrase "nuclear force" to mean the force between nucleons and "strong force" to be the force between quarks.

the first moments after the Big Bang saw such extremes of energy that all the fundamental forces were unified into one force. Physicists are continuing their search for connections among the known fundamental forces—connections that could eventually prove that the forces are all merely different forms of a single superforce. This fascinating search continues to be at the forefront of physics.

5.7 • THE GRAVITATIONAL FIELD

When Newton first published his theory of gravitation, his contemporaries found it difficult to accept the concept of a force that one object could exert on another without anything happening in the space between them. How was it possible for two objects with mass to interact even though they were not in contact with each other? Although Newton himself could not answer this question, his theory was considered a success because it satisfactorily explained the motions of the planets.

An alternative mental representation of the gravitational force is to think of the gravitational interaction as a two-step process involving a *field* as discussed in Section 4.1. First, one object (a *source mass*) creates a **gravitational field g** throughout the space around it. Then a second object (a *test mass*) of mass m residing in this field experiences a force $\mathbf{F}_g = m\mathbf{g}$. In other words, we model the *field* as exerting a force on the test mass rather than the source mass exerting the force directly. The gravitational field is defined by

$$\mathbf{g} \equiv \frac{\mathbf{F}_g}{m}$$

[5.16] • *Definition of gravitational field*

That is, the gravitational field at a point in space equals the gravitational force that a test mass m experiences at that point divided by the mass. Consequently, if $\mathbf{g}$ is known at some point in space, a particle of mass m experiences a gravitational force $\mathbf{F}_g = m\mathbf{g}$ when placed at that point. We shall also use the model of a particle in a field for electricity and magnetism in later chapters, where it plays a much larger role than it does for gravity.

As an example, consider an object of mass m near the Earth's surface. The gravitational force on the object is directed toward the center of the Earth and has a magnitude mg. Thus, the gravitational field experienced by the object at some point has a magnitude equal to the free-fall acceleration at that point. Because the gravitational force on the object has a magnitude GM_Em/r^2 (where M_E is the mass of the Earth), the field $\mathbf{g}$ at a distance r from the center of the Earth is

$$\mathbf{g} = \frac{\mathbf{F}_g}{m} = -\frac{GM_E}{r^2}\hat{\mathbf{r}}$$

[5.17] • *Gravitational field of the Earth*

where $\hat{\mathbf{r}}$ is a unit vector pointing radially outward from the Earth, and the negative sign indicates that the field vector points toward the center of the Earth, as shown in Figure 5.21a. Note that the field vectors at different points surrounding the spherical mass vary in both direction and magnitude. In a small region near the Earth's surface, $\mathbf{g}$ is approximately constant and the downward field is uniform, as indicated in Figure 5.21b. Equation 5.17 is valid at all points outside the Earth's surface, assuming that the Earth is spherical and that rotation can be ignored. At the Earth's surface, where $r = R_E$, $\mathbf{g}$ has a magnitude of 9.80 m/s^2.

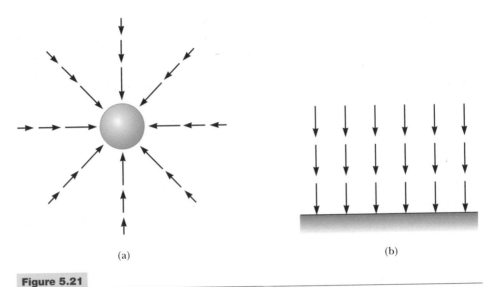

(a)   (b)

Figure 5.21

(a) The gravitational field vectors in the vicinity of a uniform spherical mass vary in both direction and magnitude. (b) The gravitational field vectors in a small region near the Earth's surface are uniform; that is, they all have the same direction and magnitude.

context connection

5.8 • THE EFFECT OF GRAVITY ON OUR SPACECRAFT

In the previous chapter, we considered the expenditure of fuel to cause a mid-course correction in the trajectory of our spacecraft on its way to Mars. Of course, we would like to use as little fuel as possible, because we don't want to have to depend on taking large amounts of heavy fuel with us. Most of our trip to Mars will be governed by the gravitational force between the spacecraft and the Sun. We will see the full details of this in the Context Conclusion after Chapter 11.

For now, let us just perform two calculations from Newton's law of universal gravitation that will be of interest to us. First, let's assume that we have fired ourselves out of the parking orbit around the Earth in Chapter 3 and are far enough away from the Earth so that the main gravitational interaction is with the Sun. How strong is this gravitational force? We can use data for the Sun and Newton's law of universal gravitation to find out. Let us assume that our spacecraft has a mass of 1.50×10^3 kg. Let us also assume that the position of the spacecraft is halfway between the orbits of Earth and Mars, which will put us at a distance of 1.89×10^{11} m from the Sun:

$$F_g = G \frac{M_{\text{Sun}} m_{\text{spacecraft}}}{r^2}$$

$$= (6.67 \times 10^{-11} \text{ N} \cdot \text{m}^2/\text{kg}^2) \frac{(1.99 \times 10^{30} \text{ kg})(1.50 \times 10^3 \text{ kg})}{(1.89 \times 10^{11} \text{ m})^2} = 5.56 \text{ N}$$

This is a very small force; on our spacecraft, this results in an acceleration of only 3.7×10^{-3} m/s². As the spacecraft moves farther from the Sun, this force becomes even smaller. But this force will act on the spacecraft *continuously* for a long time, so it can have a profound effect on the path of the spacecraft as we shall see later.

For our second calculation, let us imagine that we are on the Martian surface. Because Mars differs from Earth in radius and mass, the free-fall acceleration will be different. As discussed at the end of the previous section, the free-fall acceleration and the gravitational field are numerically equal, so let us calculate the magnitude of the gravitational field on the surface of Mars, using data for the planet:

$$g = \frac{GM_{Mars}}{r_{Mars}^2} = \frac{(6.67 \times 10^{-11}\ \text{N} \cdot \text{m}^2/\text{kg}^2)(6.42 \times 10^{23}\ \text{kg})}{(3.37 \times 10^6\ \text{m})^2} = 3.77\ \text{m/s}^2$$

Thus, objects on Mars will weigh significantly less than they do on the surface of the Earth, where $g = 9.80\ \text{m/s}^2$. We will need to plan for this in our activities after landing on Mars!

Earlier in this section, we mentioned the need for *fuel* for the rocket engines of the spacecraft. The fuel represents a source of *energy* for the spacecraft. In the next chapter, we will begin our study of the extremely important notions of energy and energy transfer.

SUMMARY

Forces of friction are complicated, but we design a simplification model for friction that allows us to analyze motion that includes the effects of friction. The **maximum force of static friction** $f_{s,max}$ between two surfaces is proportional to the normal force between the surfaces. This maximum force occurs when the surfaces are on the verge of slipping. In general, $f_s \leq \mu_s n$, where μ_s is the **coefficient of static friction** and n is the normal force. When an object slides over a rough surface, the **force of kinetic friction** f_k is opposite the direction of the velocity of the object relative to the surface and is also proportional to the normal force on the object. The magnitude of this force is given by $f_k = \mu_k n$, where μ_k is the **coefficient of kinetic friction.** Usually, $\mu_k < \mu_s$.

Newton's second law, applied to a particle moving in uniform circular motion, states that the net force in the inward radial direction must equal the product of the mass and the centripetal acceleration:

$$\sum F = ma_c = \frac{mv^2}{r} \qquad [5.3]$$

An object moving through a liquid or gas experiences a **resistive force** that is velocity dependent. This resistive force, which opposes the motion, generally increases with speed. The force depends on the shape of the object and on the properties of the medium through which the object is moving. In the limiting case for a falling object, when the resistive force balances the weight ($a = 0$), the object reaches its **terminal speed.**

The fundamental forces existing in nature can be expressed as the following four: the gravitational force, the electromagnetic force, the nuclear force, and the weak force.

Rather than considering the gravitational force as a direct interaction between two objects, we can imagine that one object sets up a **gravitational field g** in space. A second object in this field experiences a force $\mathbf{F}_g = m\mathbf{g}$ when placed in this field.

QUESTIONS

1. A book is pushed upward on a ramp (non-frictionless) so that it slides up to a point and then slides back down to the starting point. Does it take the same time to go up as it does to come down?

2. Pushing on a heavy box that is at rest requires some force **F** to start its motion. Once it is sliding, however, a smaller force is required to maintain that motion. Why?

3. Suppose you are driving a car along a highway at a high speed. Why should you avoid slamming on your brakes if you want to stop in the shortest distance? That is, why should you keep the wheels turning as you brake?

4. Twenty people are playing tug-of-war, with ten people on a team. The teams are so evenly matched that neither team wins. After they give up the game, they notice that one of the team members' cars is mired in mud. They attach the tug-of-war rope to the bumper of the car and the 20 people try to pull the car from the mud with the rope. The friction force between the car and the mud is huge, the car does not move, and the rope breaks. Why did the rope break in this situation when it did not break when 20 people pulled on it in the tug-of-war?

5. What causes a rotary lawn sprinkler to turn?

6. It has been suggested that rotating cylinders about 10 mi in length and 5 mi in diameter be placed in space and used as colonies. The purpose of the rotation is to simulate gravity for the inhabitants. Explain this concept for producing an effective gravity.

7. Consider a rotating space station, spinning with just the right speed such that the centripetal acceleration on the inner surface is *g*. Thus, astronauts standing on this inner surface would feel pressed to the surface as if they were pressed into the floor because of the Earth's gravitational force. Suppose an astronaut in this station holds a ball above her head and "drops" it to the floor. Will the ball fall just like it would on the Earth?

8. A pail of water can be whirled in a vertical path such that none is spilled. Why does the water stay in the pail, even when the pail is above your head?

9. How would you explain the force that pushes a rider toward the side of a car as the car rounds a corner?

10. Why does a pilot tend to black out when pulling out of a steep dive?

11. An object executes circular motion with a constant speed whenever a net force of constant magnitude acts perpendicular to the velocity. What happens to the speed if the force is not perpendicular to the velocity?

12. Suppose a race track is built on the Moon. For a given speed of a car around the track, would the roadway have to be banked at a steeper angle or a shallower angle than the same roadway on the Earth?

13. A skydiver in free-fall reaches terminal speed. After the parachute is opened, what parameters change to decrease this terminal speed?

14. Consider a small raindrop and a large raindrop falling through the atmosphere. Compare their terminal speeds. What are their accelerations when they reach terminal speed?

15. On long journeys, jet aircraft usually fly at high altitudes of about 30 000 ft. What is the main advantage of flying at these altitudes from an economic viewpoint?

16. If someone told you that astronauts are weightless in orbit because they are beyond the pull of gravity, would you accept the statement? Explain.

17. "If the current position and velocity of every particle in the Universe were known, together with the laws describing the forces that particles exert on one another, then the whole future of the Universe could be calculated. The future is determinate. Free will is an illusion." Do you agree with this thesis? Argue for or against it.

PROBLEMS

1, 2, 3 = straightforward, intermediate, challenging   □ = full solution available in the *Student Solutions Manual and Study Guide*

web = solution posted at **http://www.harcourtcollege.com/physics/**   🖥 = computer useful in solving problem

🖥 = Interactive Physics   ■ = paired numerical/symbolic problems   = life science application

Section 5.1 Forces of Friction

1. A 25.0-kg block is initially at rest on a horizontal surface. A horizontal force of 75.0 N is required to set the block in motion. After it is in motion, a horizontal force of 60.0 N is required to keep the block moving with constant speed. Find the coefficients of static and kinetic friction from this information.

2. A car is traveling at 50.0 mi/h on a horizontal highway. (a) If the coefficient of static friction between road and tires on a rainy day is 0.100, what is the minimum distance in which the car will stop? (b) What is the stopping distance when the surface is dry and $\mu_s = 0.600$?

3. To meet a U.S. Postal Service requirement, footwear must have a coefficient of static friction of 0.5 or more when in contact with a specified tile surface. A typical athletic shoe has a coefficient of 0.800. In an emergency, what is the minimum time for a person starting from rest to move 3.00 m on a tile surface if she is wearing (a) footwear meeting the Postal Service minimum? (b) A typical athletic shoe?

4. Before 1960 it was believed that the maximum attainable coefficient of static friction for an automobile tire was less than 1. Then about 1962, three companies independently developed racing tires with coefficients of 1.6. Since then, tires have improved, as illustrated in this problem. According to the 1990 *Guinness Book of Records,* the shortest time in which a piston-engine car initially at rest has covered a distance of one-quarter mile is 4.96 s. This record was set by

Figure P5.4 *(Mike Powell/AllSport USA)*

Shirley Muldowney in September 1989. (a) Assume that, as in Figure P5.4, the rear wheels lifted the front wheels off the pavement. What minimum value of μ_s is necessary to achieve the record time? (b) Suppose Muldowney were able to double her engine power, keeping other things equal. How would this change affect the elapsed time?

5. Consider a large truck carrying a heavy load, such as steel beams. A significant hazard for the driver is that the load may slide forward, crushing the cab, if the truck stops suddenly in an accident or even in braking. Assume, for example, that a 10 000-kg load is located on the flat bed of a 20 000-kg truck moving at 12.0 m/s. Assume the load is not tied down to the truck and assume that the coefficient of friction between the load and the truck bed is 0.500. (a) Calculate the minimum stopping distance for the truck for which the load will not slide forward relative to the truck. (b) Is any piece of data unnecessary for the solution?

6. A woman at an airport is towing her 20.0-kg suitcase at constant speed by pulling on a strap at an angle of θ above the horizontal (Fig. P5.6). She pulls on the strap with a 35.0-N force, and the frictional force on the suitcase is 20.0 N. Draw a free-body diagram of the suitcase. (a) What angle does the strap make with the horizontal? (b) What normal force does the ground exert on the suitcase?

Figure P5.6

7. A 3.00-kg block starts from rest at the top of a 30.0° incline and slides a distance of 2.00 m down the incline in 1.50 s. Find (a) the magnitude of the acceleration of the block, (b) the coefficient of kinetic friction between block and plane, (c) the friction force acting on the block, and (d) the speed of the block after it has slid 2.00 m.

8. A 9.00-kg hanging weight is connected by a string over a light pulley to a 5.00-kg block that is sliding on a flat table (Fig. P5.8). If the coefficient of kinetic friction is 0.200, find the tension in the string.

5.00 kg

9.00 kg

Figure P5.8

9. Two blocks connected by a rope of negligible mass are being dragged by a horizontal force **F** (see Fig. P4.27). Suppose that $F = 68.0$ N, $m_1 = 12.0$ kg, $m_2 = 18.0$ kg, and the coefficient of kinetic friction between each block and the surface is 0.100. (a) Draw a free-body diagram for each block. (b) Determine the tension T and the magnitude of the acceleration of the system.

10. To determine the coefficients of friction between rubber and various surfaces, a student uses a rubber eraser and an incline. In one experiment the eraser begins to slip down the incline when the angle of inclination is 36.0° and then moves down the incline with constant speed when the angle is reduced to 30.0°. From these data, determine the coefficients of static and kinetic friction for this experiment.

11. A block of mass 3.00 kg is pushed up against a wall by a force **P** that makes a 50.0° angle with the horizontal as shown in Figure P5.11. The coefficient of static friction between the block and the wall is 0.250. Determine the possible values for the magnitude of **P** that allow the block to remain stationary.

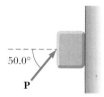

50.0°

P

Figure P5.11

Section 5.2 Newton's Second Law Applied to a Particle in Uniform Circular Motion

12. In the Bohr model of the hydrogen atom, the speed of the electron is approximately 2.20×10^6 m/s. Find (a) the force acting on the electron as it revolves in a circular orbit of radius 0.530×10^{-10} m and (b) the centripetal acceleration of the electron.

13. A light string can support a stationary hanging load of 25.0 kg before breaking. A 3.00-kg mass attached to the string rotates on a horizontal, frictionless table in a circle of radius 0.800 m. What range of speeds can the mass have before the string breaks?

14. A merry-go-round makes one complete revolution in 12.0 s. If a 45.0-kg child sits on the horizontal floor of the merry-go-round 3.00 m from the center, find (a) the child's acceleration and (b) the horizontal force of friction that acts on the child. (c) What minimum coefficient of static friction is necessary to keep the child from slipping?

15. A crate of eggs is located in the middle of the flat bed of a pickup truck as the truck negotiates an unbanked curve in the road. The curve may be regarded as an arc of a circle of radius 35.0 m. If the coefficient of static friction between crate and truck is 0.600, how fast can the truck be moving without the crate sliding?

16. Consider a conical pendulum with an 80.0-kg bob on a 10.0-m wire making an angle of 5.00° with the vertical (Fig. P5.16). Determine (a) the horizontal and vertical components of the force exerted by the wire on the pendulum and (b) the radial acceleration of the bob.

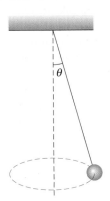

Figure P5.16

Section 5.3 Nonuniform Circular Motion

17. Tarzan (*m* = 85.0 kg) tries to cross a river by swinging from a vine. The vine is 10.0 m long, and his speed at the bottom of the swing (as he just clears the water) is 8.00 m/s. Tarzan doesn't know that the vine has a breaking strength of 1 000 N. Does he make it safely across the river?

18. A car of mass *m* passes over a bump in a road that follows the arc of a circle of radius *R* as in Figure P5.18. (a) What

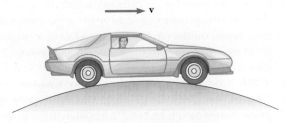

Figure P5.18

force does the road exert on the car as the car passes the highest point of the bump if the car travels at a speed *v*? (b) What is the maximum speed the car can have as it passes this highest point before losing contact with the road?

19. A pail of water is rotated in a vertical circle of radius 1.00 m. What is the minimum speed of the pail when it is upside down at the top of the circle if no water is to spill out?

20. A hawk flies in a horizontal arc of radius 12.0 m at a constant speed of 4.00 m/s. (a) Find its centripetal acceleration. (b) It continues to fly along the same horizontal arc but increases its speed at the rate of 1.20 m/s². Find the acceleration (magnitude and direction) under these conditions.

21. A roller coaster at the Six Flags Great America amusement park in Gurnee, Illinois, incorporates some of the latest design technology and some basic physics. Each vertical loop, instead of being circular, is shaped like a teardrop (Fig. P5.21). The cars ride on the inside of the loop at the top, and the speeds are high enough to ensure that the cars remain on the track. The biggest loop is 40.0 m high, with a maximum speed of 31.0 m/s (nearly 70 mph) at the bottom. Suppose the speed at the top is 13.0 m/s and the corresponding centripetal acceleration is 2*g*. (a) What is the radius of the arc of the teardrop at the top? (b) If the total mass of the cars plus people is *M*, what force does the rail exert on this total mass at the top? (c) Suppose the roller coaster had a circular loop of radius 20.0 m. If the

Figure P5.21 *(Frank Cezus/FPG International)*

cars have the same speed, 13.0 m/s at the top, what is the centripetal acceleration at the top? Comment on the normal force at the top in this situation.

Section 5.4 Motion in the Presence of Velocity-Dependent Resistive Forces

22. A small piece of Styrofoam packing material is dropped from a height of 2.00 m above the ground. Until it reaches terminal speed, the magnitude of its acceleration is given by $a = g - bv$. After falling 0.500 m, the Styrofoam effectively reaches terminal speed, and then takes 5.00 s more to reach the ground. (a) What is the value of the constant b? (b) What is the acceleration at $t = 0$? (c) What is the acceleration when the speed is 0.150 m/s?

23. A small, spherical bead of mass 3.00 g is released from rest at $t = 0$ in a bottle of liquid shampoo. The terminal speed is observed to be $v_T = 2.00$ cm/s. Find (a) the value of the constant b in Equation 5.5, (b) the time τ it takes to reach $0.632v_T$, and (c) the value of the resistive force when the bead reaches terminal speed.

24. (a) Estimate the terminal speed of a wooden sphere (density 0.830 g/cm³) falling through air if its radius is 8.00 cm. (b) From what height would a freely falling object reach this speed in the absence of air resistance?

25. A motorboat cuts its engine when its speed is 10.0 m/s and web coasts to rest. The equation describing the motion of the motorboat during this period is $v = v_i e^{-ct}$, where v is the speed at time t, v_i is the initial speed, and c is a constant. At $t = 20.0$ s, the speed is 5.00 m/s. (a) Find the constant c. (b) What is the speed at $t = 40.0$ s? (c) Differentiate the expression for $v(t)$ and thus show that the acceleration of the boat is proportional to the speed at any time.

26. Assume that the resistive force on a speed-skater is given by $f = -kmv^2$, where k is a constant and m is the skater's mass. The skater crosses the finish line of a straight-line race with speed v_f and then slows down by coasting on his skates. Show that the skater's speed at any time t after crossing the finish line is $v(t) = v_f/(1 + ktv_f)$.

Section 5.5 Numerical Representations of Particle Dynamics

27. A hailstone of mass 4.80×10^{-4} kg falls through the air and experiences a net force given by
web
$$F = -mg + Cv^2$$
where $C = 2.50 \times 10^{-5}$ kg/m. (a) Calculate the terminal speed of the hailstone. (b) Use Euler's method of numerical analysis to find the speed and position of the hailstone at 0.2-s intervals, taking the initial speed to be zero. Continue the calculation until the hailstone reaches 99% of terminal speed.

28. A 3.00-g leaf is dropped from a height of 2.00 m above the ground. Assume that the net downward force exerted on the leaf is given by $F = mg - bv$, where the drag factor is $b = 0.030\ 0$ N·s/m. (a) Calculate the terminal speed of the leaf. (b) Use Euler's method of numerical analysis to find the speed and position of the leaf, as functions of time, from the instant it is released until 99% of terminal speed is reached. (*Hint:* Try $\Delta t = 0.005$ s.)

29. A 0.142-kg baseball is dropped from rest and has a terminal speed of 42.5 m/s (95 mph). (a) If a baseball experiences a drag force of magnitude $R = Cv^2$, what is the value of the constant C? (b) What is the magnitude of the drag force when the speed of the baseball is 36.0 m/s? (c) Use a computer to determine the motion of a baseball thrown vertically upward at an initial speed of 36 m/s. What maximum height does the ball reach? How long is it in the air? What is its speed just before it hits the ground?

30. Consider a 10.0-kg projectile launched with an initial speed of 100 m/s, at an elevation angle of 35.0°. The resistive force is given by $\mathbf{R} = -b\mathbf{v}$, where $b = 10.0$ N·s/m. (a) Use a numerical method to determine the horizontal and vertical coordinates of the projectile as functions of time. (b) What is the range of this projectile? (c) Determine the elevation angle that gives the maximum range for the projectile. (*Hint:* Adjust the elevation angle by trial and error to find the greatest range.)

Section 5.6 The Fundamental Forces of Nature

31. Two identical isolated particles, each of mass 2.00 kg, are separated by a distance of 30.0 cm. What is the magnitude of the gravitational force exerted by one particle on the other?

32. In a thundercloud there may be electric charges of $+40.0$ C near the top of the cloud and -40.0 C near the bottom of the cloud. These charges are separated by 2.00 km. What is the electric force on the top charge?

33. Find the order of magnitude of the gravitational force you exert on another person 2 m away. In your solution state the quantities you measure or estimate and their values.

Section 5.7 The Gravitational Field

34. The Moon travels around the Earth in a nearly circular orbit of radius 3.84×10^8 m with a period of 27.3 days. (a) What is the speed of the Moon relative to the Earth? (b) Calculate the centripetal acceleration of the Moon from its orbital speed. (c) Use the law of universal gravitation to calculate the gravitational field created by the Earth at the location of the Moon. *Historical note:* Newton was the first person to solve this problem. He compared the results of parts (b) and (c) to check the validity of his law of gravitation, and found them to "answer pretty nearly."

35. When a falling meteor is at a distance above the Earth's surface of 3.00 times the Earth's radius, what is its free-fall acceleration due to the gravitational force exerted on it?

Section 5.8   Context Connection—The Effect of Gravity on Our Spacecraft

36. A satellite of mass 300 kg is in a circular orbit around the Earth at an altitude equal to the Earth's mean radius. Find (a) the satellite's orbital speed, (b) the period of its revolution, and (c) the gravitational force acting on it.

37. Whenever two Apollo astronauts were on the surface of the Moon, a third astronaut orbited the Moon. Assume the orbit to be circular and 100 km above the surface of the Moon. If the mass of the Moon is 7.40×10^{22} kg and its radius is 1.70×10^6 m, determine (a) the orbiting astronaut's acceleration, (b) his orbital speed, and (c) the period of the orbit.

Additional Problems

38. Consider the three connected objects shown in Figure P5.38. If the inclined plane is frictionless, and the system is in equilibrium, find (in terms of *m*, *g*, and *θ*) (a) the mass *M* and (b) the tensions T_1 and T_2. If the value of *M* is double the value found in part (a), find (c) the acceleration of each object, and (d) the tensions T_1 and T_2. If the coefficient of static friction between *m* and 2*m* and the inclined plane is μ_s, and the system is in equilibrium, find (e) the minimum value of *M* and (f) the maximum value of *M*. (g) Compare the values of T_2 when *M* has its minimum and maximum values.

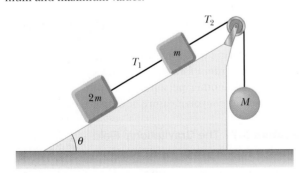

Figure P5.38

39. A crate of weight F_g is pushed by a force **P** on a horizontal floor. (a) If the coefficient of static friction is μ_s and **P** is directed at angle *θ* below the horizontal, show that the minimum value of *P* that will move the crate is given by

$$P = \frac{\mu_s F_g \sec \theta}{1 - \mu_s \tan \theta}$$

(b) Find the minimum value of *P* that can produce motion when $\mu_s = 0.400$, $F_g = 100$ N, and $\theta = 0°$, 15.0°, 30.0°, 45.0°, and 60.0°.

40. A 1.30-kg toaster is not plugged in. The coefficient of static friction between the toaster and a horizontal countertop is 0.350. To make the toaster start moving, you carelessly pull on its electric cord. (a) For the cord tension to be as small as possible, you should pull at what angle above the horizontal? (b) With this angle, how large must the tension be?

41. A 2.00-kg aluminum block and a 6.00-kg copper block are connected by a light string over a frictionless pulley. They sit on a steel surface, as shown in Figure P5.41, where $\theta = 30.0°$. When they are released from rest, will they start to move? If so, determine (a) their acceleration and (b) the tension in the string. If not, determine the sum of the magnitudes of the forces of friction acting on the blocks.

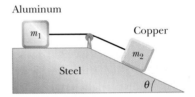

Figure P5.41

42. The system shown in Figure P4.43 (Chapter 4) has an acceleration of magnitude 1.50 m/s². Assume the coefficients of kinetic friction between block and incline are the same for both inclines. Find (a) the coefficient of kinetic friction and (b) the tension in the string.

43. A block of mass *m* = 2.00 kg rests on the left edge of a block of mass *M* = 8.00 kg. The coefficient of kinetic friction between the two blocks is 0.300, and the surface on which the 8.00-kg block rests is frictionless. A constant horizontal force of magnitude *F* = 10.0 N is applied to the 2.00-kg block, setting it in motion as shown in Figure P5.43a. If the distance *L* that the leading edge of the

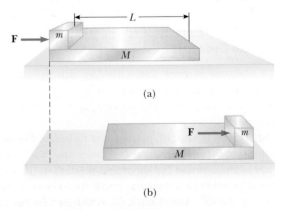

Figure P5.43

smaller block travels on the larger block is 3.00 m, (a) how long will it take before this block makes it to the right side of the 8.00-kg block, as shown in Figure P5.43b? (Note: Both blocks are set into motion when **F** is applied.) (b) How far does the 8.00-kg block move in the process?

44. A 5.00-kg block is placed on top of a 10.0-kg block (Fig. P5.44). A horizontal force of 45.0 N is applied to the 10-kg block, and the 5-kg block is tied to the wall. The coefficient of kinetic friction between all moving surfaces is 0.200. (a) Draw a free-body diagram for each block, and identify the action-reaction forces between the blocks. (b) Determine the tension in the string attached to the wall and the magnitude of the acceleration of the 10-kg block.

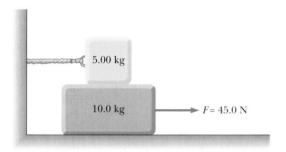

Figure P5.44

45. **Review Problem.** The roof of a building is sloped at an angle of 37.0° with the horizontal. A student throws a Frisbee onto the roof. It strikes with a speed of 15.0 m/s and does not bounce, but slides straight up the incline. The coefficient of kinetic friction between the Frisbee and the roof is 0.400. The Frisbee slides 10.0 m up the roof to its peak, where it goes into free fall, following a parabolic trajectory with negligible air resistance. Determine the maximum height the Frisbee reaches above the point where it struck the roof.

46. A student builds and calibrates an accelerometer, which she uses to determine the speed of her car around a certain unbanked highway curve. The accelerometer is a plumb bob with a protractor that she attaches to the roof of her car. A friend riding in the car with her observes that the plumb bob hangs at an angle of 15.0° from the vertical when the car has a speed of 23.0 m/s. (a) What is the centripetal acceleration of the car rounding the curve? (b) What is the radius of the curve? (c) What is the speed of the car if the plumb bob deflection is 9.00° while rounding the same curve?

47. An air puck of mass m_1 is tied to a string and allowed to revolve in a circle of radius R on a frictionless horizontal table. The other end of the string passes through a hole in the center of the table, and an object of mass m_2 is tied to it (Fig. P5.47). The suspended object remains in equilibrium while the puck on the tabletop revolves. What are (a)

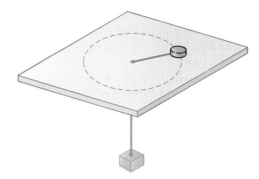

Figure P5.47

the tension in the string? (b) The net force acting on the puck? (c) The speed of the puck?

48. Materials such as rubber tires and shoe soles are tested for coefficients of static friction using an apparatus called a James tester. The pair of surfaces for which μ_s is to be measured are labeled B and C in Figure P5.48. Sample C is attached to a foot D at the lower end of a pivoting arm E, which makes an angle θ with the vertical. The upper end of the arm is hinged at F to a vertical rod G, which slides freely in a guide H fixed to the frame of the apparatus and supports a load I of 36.4 kg. The hinge pin at F is also the axle of a wheel that can roll vertically on the frame. All of the moving parts have weights negligible in comparison to the 36.4-kg load. The pivots are nearly frictionless. The test surface B is attached to a rolling platform A. An operator slowly moves the platform to the left in the picture until the sample C suddenly slips over surface B. At the critical point where sliding motion is ready to begin the operator notes the angle θ_s of the pivoting

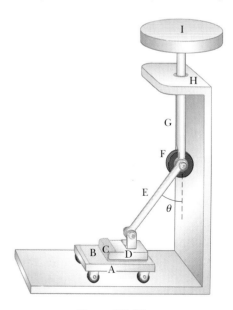

Figure P5.48

arm. (a) Make a free-body diagram of the pin at F, which is in equilibrium under three forces. These forces are the weight of the load I, a horizontal normal force exerted by the frame, and a force of compression directed upward along the arm E. (b) Draw a free-body diagram of the foot D and sample C, considered as one system. (c) Determine the normal force that the test surface B exerts on the sample for any angle θ. (d) Show that $\mu_s = \tan \theta_s$. (e) The protractor on the tester can record angles as large as 50.2°. What is the largest coefficient of friction it can measure?

49. Because the Earth rotates about its axis, a point on the **web** equator experiences a centripetal acceleration of 0.033 7 m/s², but a point at the poles experiences no centripetal acceleration. (a) Show that at the equator the gravitational force on an object must exceed the normal force required to support the object. That is, show that the object's true weight exceeds its apparent weight. (b) What is the apparent weight at the equator and at the poles of a person having a mass of 75.0 kg? (Assume the Earth is a uniform sphere and take $g = 9.800$ N/kg.)

50. An engineer wishes to design a curved exit ramp for an interstate highway in such a way that a car will not have to rely on friction to round the curve without skidding (Fig. P5.50). Suppose that a typical car rounds the curve with a speed of 30 mi/h (13.4 m/s) and that the radius of the curve is 50.0 m. At what angle should the curve be banked?

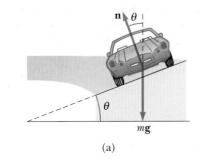

(a)

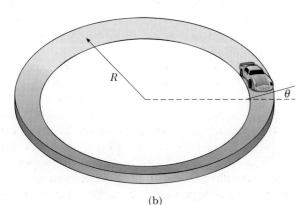

(b)

Figure P5.50

51. A space station, in the form of a wheel 120 m in diameter, rotates to provide an "artificial gravity" of 3.00 m/s² for persons walking around on the inner wall of the outer rim. Find the rate of rotation of the wheel (in revolutions per minute) that will produce this effect.

52. A 9.00-kg object starting from rest falls through a viscous medium and experiences a resistive force $\mathbf{R} = -b\mathbf{v}$, where $\mathbf{v}$ is the velocity of the object. If the object's speed reaches one-half its terminal speed in 5.54 s, (a) determine the terminal speed. (b) At what time is the speed of the object three-fourths the terminal speed? (c) How far has the object traveled in the first 5.54 s of motion?

53. An amusement park ride consists of a large vertical cylinder that spins about its axis fast enough that any person inside is held up against the wall when the floor drops away (Fig. P5.53). The coefficient of static friction between a person and the wall is μ_s, and the radius of the cylinder is R. (a) Show that the maximum period of revolution necessary to keep the person from falling is given by $T = (4\pi^2 R\mu_s/g)^{1/2}$. (b) Obtain a numerical value for T if $R = 4.00$ m and $\mu_s = 0.400$. How many revolutions per minute does the cylinder make?

Figure P5.53

54. The expression $F = arv + br^2v^2$ gives the magnitude of the resistive force (in newtons) exerted on a sphere of radius r (in meters) exerted by a stream of air moving at speed v (in meters per second), where a and b are constants with appropriate SI units. Their numerical values are $a = 3.10 \times 10^{-4}$ and $b = 0.870$. Using this expression, find the terminal speed for water droplets falling under

their own weight in air, taking the following values for the drop radii: (a) 10.0 μm, (b) 100 μm, (c) 1.00 mm. Note that for (a) and (c) you can obtain accurate answers without solving a quadratic equation, by considering which of the two contributions to the air resistance is dominant and ignoring the lesser contribution.

55. A model airplane of mass 0.750 kg flies in a horizontal circle at the end of a 60.0-m control wire, with a speed of 35.0 m/s. Compute the tension in the wire if it makes a constant angle of 20.0° with the horizontal. The forces exerted on the airplane are the pull of the control wire, its own weight, and aerodynamic lift, which acts at 20.0° inward from the vertical as shown in Figure P5.55.

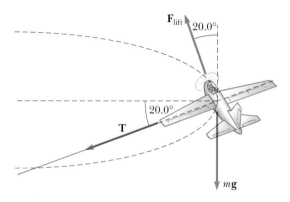

Figure P5.55

56. A single bead can slide with negligible friction on a wire that is bent into a circular loop of radius 15.0 cm, as in Figure P5.56. The circle is always in a vertical plane and rotates steadily about its vertical diameter with (a) a period of 0.450 s. The position of the bead is described by the angle θ that the radial line, from the center of the loop to the bead, makes with the vertical. At what angle up from the bottom of the circle can the bead stay motionless relative to the turning circle? (b) Repeat the problem if the period of the circle's rotation is 0.850 s.

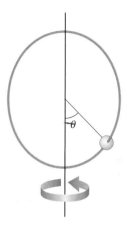

Figure P5.56

57. If a single constant force acts on an object of mass m that moves on a straight line, the object's velocity is a linear function of time. The equation $v = v_i + at$ gives its velocity v as a function of time, where a is its constant acceleration. Now assume that as the object moves through a resistive medium, its speed decreases as described by the equation $v = v_i - kx$, where x is the distance it has traveled since it had initial speed v_i and k is a constant coefficient. Find the law describing the total force acting on this object.

ANSWERS TO QUICK QUIZZES

5.1 (b). Friction forces are always parallel to the surfaces in contact, which, in this case, are the wall and the cover of the book. This tells us that the friction force is either upward or downward. Because the tendency of the book is to fall due to gravity, the friction force must be in the upward direction.

5.2 (b). The static friction force between the bottom surface of the crate and the surface of the truck bed is the net horizontal force on the crate that causes it to accelerate. It is in the same direction as the acceleration of the crate, to the east.

5.3 It is easier to attach the rope and pull. In this case, there is a component of your applied force that is upward. This reduces the normal force between the sled and the snow. In turn, this reduces the friction force between the sled and

the snow, making it easier to move. If you push from behind with a force having a downward component, the normal force is larger, the friction force is larger, and the sled is harder to move.

5.4 The centripetal acceleration is always directed toward the center of the circle. Thus, at the top (a), the centripetal acceleration is downward and at the bottom (b), it is upward. The normal force is always perpendicularly outward from the surface, so the normal force from the seat on you is upward at both the top (c) and the bottom (d). (e) The normal force is largest at the bottom because it is in the same direction as the centripetal acceleration.

5.5 On both roadways, the maximum friction force on the car is the same. From Newton's second law, this means that the centripetal acceleration is the same on both roadways.

Consequently, from the expression for centripetal acceleration, the maximum speed is proportional to the square root of the radius. Because we have doubled the radius, the maximum speed is $\sqrt{2}v$.

5.6 The baseball, because it has a smaller cross-sectional area, experiences a smaller resistive force as it falls through the air. Thus, the baseball strikes the ground first.

5.7 The forces exerted on the sky surfer are the downward gravitational force $m\mathbf{g}$ and an upward force of air resistance $\mathbf{R}$, the magnitude of which is less than her weight before she reaches terminal speed. As her downward speed increases, the force of air resistance increases. The vector sum of the gravitational force and the force of air resistance gives a total force that decreases with time, so her acceleration decreases. Once she reaches terminal speed, the two forces balance each other, the net force is zero, and her acceleration is zero.

A large earth-moving machine does work on a pile of dirt by moving it to a new location.

(© 1995 Paul Steel/The Stock Market)

Energy and Energy Transfer

I n the preceding chapters, we approached the motion of an object with quantities such as *position, velocity, acceleration,* and *force,* with which you are familiar from everyday life. We developed a number of models using these notions that allow us to solve a variety of problems. Some problems that, in theory, could be solved with Newton's laws, are very difficult to solve in practice, but can be made much simpler with a different approach. In this and the following two chapters, we shall investigate this new approach, which will introduce us to new analysis models for problem solving. This approach includes definitions of quantities that may not be familiar to you. You may be familiar with some quantities, but they may have more specific meanings in physics than in everyday life. We begin this discussion by exploring the notion of *energy.*

Energy is present in the Universe in various forms. *Every* physical process in the Universe involves energy and energy transfers or transformations. Thus, energy is an extremely important concept to understand. Unfortunately, despite its importance, it cannot be easily defined. The variables in previous chapters were relatively concrete—we have everyday experience with velocities and forces, for example. Although the *notion* of energy is more abstract, we do have *experiences* with energy, such as running out of gasoline or losing our electrical service if we forget to pay the bill.

The concept of energy can be applied to the dynamics of a mechanical system without resorting to Newton's laws. This "energy approach" to describing motion is especially useful when the force acting on a particle is not constant; in such a case, the acceleration is not constant, and we cannot apply the particle under constant acceleration model we developed in Chapter 2. Particles in nature are often subject to forces that vary with the particles' positions. These forces include gravitational forces and the force exerted on an object attached to a spring. We shall describe techniques for treating such situations with the help of an important concept called the *work–kinetic energy theorem*. The work–kinetic energy theorem is a special case of a global approach to problems involving energy and energy transfer. This approach extends well beyond physics and can be applied to biological organisms, technological systems, and engineering situations.

All of our analysis models in the earlier chapters were based on the motion of a particle or an object modeled as a particle. We begin our study of our new approach by identifying a new simplification model for a *system*.

6.1 • SYSTEMS AND ENVIRONMENTS

• *A system*

As mentioned in the introduction, an important new concept we will need in studying energy is that of a **system.** This is a simplification model, in that we focus our attention on a small region of the Universe and ignore details of the rest of the Universe outside the system. A critical skill in applying the energy approach to problems in the next three chapters is *correctly identifying the system.* A valid system may

- Be a single object or particle
- Be a collection of objects or particles
- Be a region of space (e.g., the interior of an automobile engine combustion cylinder)
- Vary in size and shape (e.g., a rubber ball that deforms upon striking a wall)

PITFALL PREVENTION 6.1

Identify the system

One of the most important steps to take in solving a problem using the energy approach is to correctly identify the system of interest. Be sure this is the *first* step you take in solving a problem.

A **system boundary,** which is an imaginary surface (often but not necessarily coinciding with a physical surface), divides the Universe between the system and the **environment** of the system.

• *The environment*

As an example, imagine a force applied to an object in empty space. We can define the object as the system, as in the first item in the previous list. The force applied to it influences the system from the environment, which acts across the system boundary. We will see how to analyze this situation from a system approach in a subsequent section of this chapter.

Another example occurs in Example 5.3. Here the system can be defined as the combination of the ball, the cube, and the string, consistent with the second item of the previous list. The influence from the environment includes the gravitational forces on the ball and the cube, the normal and friction forces on the cube, and the force of the pulley on the string. The forces exerted by the string on the ball and the cube are internal to the system and, therefore, are not included as an influence from the environment.

6.2 • WORK DONE BY A CONSTANT FORCE

See the *Core Concepts in Physics CD-ROM,* Screen 5.1

Let us begin our analysis of systems by introducing a term whose meaning in physics is distinctly different from its everyday meaning. This new term is **work.** Imagine that you are trying to push a heavy sofa across your living room floor. If you push on the sofa *and* it moves through a displacement, then you have done work on the sofa.

Consider a particle, which we identify as the system, that undergoes a displacement $\Delta \mathbf{r}$ along a straight line while acted on by a constant force $\mathbf{F}$ that makes an angle θ with $\Delta \mathbf{r}$, as in Figure 6.1. The force has accomplished something—it has moved the particle—so we say that work was done on the particle.

Notice that we only know the force and the displacement given in the description of the situation. We have no information about how long it took for this displacement to occur, nor any information about velocities or accelerations. This hints at the power of the energy approach, as well as how different it will be from our approach in previous chapters. We do not need these values to find the work done. Let us now define the work done on the system if the force is constant:

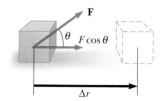

Figure 6.1

If an object undergoes a displacement **Δr**, the work done by the constant force **F** on the object is $(F \cos \theta)\Delta r$.

The work W done by an agent exerting a constant force on a system is the product of the component $F \cos \theta$ of the force along the direction of the displacement of the point of application of the force and the magnitude Δr of the displacement:

$$W \equiv F \Delta r \cos \theta \qquad [6.1]$$

• *Work done by a constant force*

Work is a scalar quantity—no direction is associated with it. Its units are those of force multiplied by length; therefore, the SI unit of work is the **newton·meter** (N·m). The newton·meter, when it refers to work or energy, is called the **joule** (J).

• *The joule*

From the definition in Equation 6.1, we see that a force does no work on a system if the point of application of the force does not move. In the mathematical representation, if $\Delta r = 0$, Equation 6.1 gives $W = 0$. In the mental representation, imagine pushing on the sofa mentioned earlier—if it doesn't move, no work has been done on the sofa. Of course, the work is also zero if the applied force is zero—if you don't push, you don't do any work!

Also note from Equation 6.1 that the work done by a force is zero when the force is perpendicular to the displacement. That is, if $\theta = 90°$, then $\cos 90° = 0$ and $W = 0$. For example, consider the free-body diagram for a block moving across a frictionless surface in Figure 6.2. The work done by the normal force and the gravitational force on the block during its horizontal displacement are both zero for the same reason—they are both perpendicular to the displacement.

In general, we will consider work done on a system. For example, in Chapter 17, we will consider work done in compressing a gas, which is modeled as a system of particles. For now, however, let us restrict our attention to a system consisting of a single particle. In the case of a single particle, the displacement of the point of application of the force is necessarily the same as the displacement of the particle.

PITFALL PREVENTION 6.2

What's the correct angle?

Keep in mind that the angle θ in Equation 6.1 is the angle between the force vector and the displacement vector. A given problem may provide a geometric construction that includes an angle, but this may not be the angle you want for Equation 6.1. Be sure you have correctly identified the proper angle θ before calculating the work.

The weight lifter does no work on the weights as he holds them at rest.
(Gerard Vandystadt/Photo Researchers, Inc.)

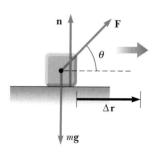

Figure 6.2

When an object is displaced horizontally on a flat table, the normal force **n** and the gravitational force $m\mathbf{g}$ do no work.

A force of **F** is applied to an object, which undergoes a displacement to the right. In each of the four cases, the magnitudes of the force and displacement are the same.

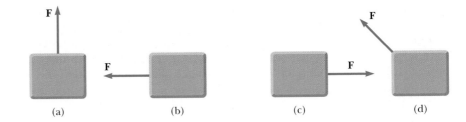

(a)   (b)   (c)   (d)

PITFALL PREVENTION 6.3

What is being displaced?

Note that the displacement in Equation 6.1 is that of the *point of application of the force*. For a force applied to a particle, it is the same as the displacement of the particle. For extended objects, however, this may differ from the displacement of the object. For example, imagine climbing a ladder. The force causing your body to be displaced upward is the force from the rungs upward on your feet. But this force undergoes no displacement, even though your body does. This force does no work! For now, we will restrict our attention to work done on particles, but keep this distinction in mind. We will see how to address problems like ladder climbing shortly.

PITFALL PREVENTION 6.4

Work is done by . . . on . . .

In Pitfall Prevention 6.1, we noted the importance of identifying the system. It is also important to identify the interaction of the system with the environment. When discussing work, try always to use the phrase, "the work done by . . . on. . . ." After "by," insert the part of the environment that is interacting directly with the system. After "on," insert the system. For example, "the work done by the hammer on the nail" identifies the nail as the system and the force from the hammer represents the interaction with the environment. This is similar to our discussion in Chapter 4 of taking care to say "the force exerted by . . . on. . . ."

In general, a particle may be moving under the influence of several forces. In that case, because work is a scalar quantity, the total work done as the particle undergoes some displacement is the algebraic sum of the work done by each of the forces.

The sign of the work depends on the direction of **F** relative to **Δr**. The work done by the applied force is positive when the vector component of magnitude $F \cos \theta$ is in the *same direction* as the displacement. For example, when an object is lifted, the work done by the applied force on the object is positive because the lifting force is upward, that is, in the same direction as the displacement. When the vector component of magnitude $F \cos \theta$ is in the direction *opposite* the displacement, W is *negative*. In the case of the object being lifted, for instance, the work done by the gravitational force on the object is negative.

If a constant applied force **F** acts parallel to the direction of the displacement, then $\theta = 0$ and $\cos 0 = 1$. In this case, Equation 6.1 gives

$$W = F \Delta r \qquad [6.2]$$

Both Equations 6.1 and 6.2 are special cases of a more generalized definition of work. Both equations assume a constant force, and Equation 6.2 assumes that the force is parallel to the displacement. In the next two sections, we shall consider the situation in which a force is not parallel to the displacement and the more general case of a varying force.

Quick Quiz 6.1

Consider a tug-of-war, in which the two teams pulling on the rope are evenly matched, so that no motion takes place. Is work done on the rope? On the pullers? On the ground? Is work done on any system?

Quick Quiz 6.2

Figure 6.3 shows four situations in which a force is applied to an object. In all four cases, the force has the same magnitude, and the displacement of the object is to the right and of the same magnitude. Rank the situations in order of the work done by the force on the object, from most positive to most negative.

THINKING PHYSICS 6.1

A person slowly lifts a heavy box of mass m a vertical height h, and then walks horizontally at constant velocity a distance d while holding the box, as in Figure 6.4. Determine the work done (a) by the person and (b) by the gravitational force on the box in this process.

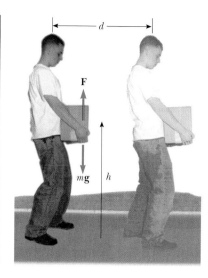

Figure 6.4

(Thinking Physics 6.1) A person lifts a heavy box of mass m a vertical distance h and then walks horizontally at constant velocity a distance d.

Reasoning (a) Assuming that the person lifts the box with a force of magnitude equal to the weight of the box mg, the work done by the person on the box during the vertical displacement is $W = F\,\Delta r = (mg)(h) = mgh$, because the force in this case is in the same direction as the displacement. For the horizontal displacement, we assume that the acceleration of the box is approximately zero. As a result, the work done by the person on the box during the horizontal displacement of the box is zero because the horizontal force is approximately zero, and the force to support the box's weight in this process is perpendicular to the displacement. Thus, the net work done by the person during the complete process is mgh.

(b) The work done by the gravitational force on the box during the vertical displacement of the box is $-mgh$, because this force is opposite the displacement. The work done by the gravitational force is zero during the horizontal displacement because this force is perpendicular to the displacement. Hence, the net work done by the gravitational force for the complete process is $-mgh$. The net work done by all forces on the box is zero, because $+mgh + (-mgh) = 0$.

THINKING PHYSICS 6.2

Roads going up mountains are formed into *switchbacks*, with the road weaving back and forth along the face of the slope, so that any portion of the roadway has only a gentle rise. Does this require that an automobile climbing the mountain do any less work than if it were driving on a roadway that runs straight up the slope? Why are the switchbacks used?

Reasoning If we ignore the effects of rolling friction on the tires of the car, the same amount of work would be done in driving up the switchbacks and driving straight up the mountain because the weight of the car is moved upward against the gravitational force by the same vertical distance in each case. So why do we use the switchbacks? The answer lies in the force required, not the work. The force needed from the engine to follow a gentle rise is much less than that required to drive straight up the hill. Roadways running straight uphill would require redesigning engines so as to enable them to apply much larger forces. This is similar to the ease with which a heavy object can be rolled up a ramp into a moving van truck, compared with lifting the object straight up from the ground.

PITFALL PREVENTION 6.5

Cause of the displacement

When we calculate the work done by a force on an object, we are *not* necessarily saying that the force is the cause of the displacement. In some cases, a causal relationship exists, as in our simple example of applying a horizontal force to an object sitting on a frictionless table. But in the case of the work done by the gravitational force in Thinking Physics 6.1, the gravitational force is clearly not causing upward motion of the box! Work done by a force can be calculated regardless of whether the force causes the displacement.

Example 6.1   Mr. Clean

A man cleaning his apartment pulls a vacuum cleaner with a force of magnitude $F = 50.0$ N. The force makes an angle of 30.0° with the horizontal as shown in Figure 6.5. The vacuum cleaner is displaced 3.00 m to the right. Calculate the work done by the 50.0-N force on the vacuum cleaner.

Solution Using the definition of work (Equation 6.1), we have

$$W = (F \cos \theta)\Delta r = (50.0 \text{ N})(\cos 30.0°)(3.00 \text{ m})$$
$$= 130 \text{ N} \cdot \text{m} = \boxed{130 \text{ J}}$$

Note that the normal force **n**, the gravitational force $m\mathbf{g}$, and the upward component of the applied force do *no* work because they are perpendicular to the displacement.

EXERCISE Find the work done by the man on the vacuum cleaner if he pulls it 3.00 m with a horizontal force of 32.0 N.

Answer 96.0 J

EXERCISE If a person lifts a 20.0-kg bucket very slowly from a well and does 6.00 kJ of work on the bucket, how deep is the well?

Answer 30.6 m

EXERCISE A 65.0-kg woman climbs a flight of 20 stairs, each 23.0 cm high. How much work is done by the gravitational force on the woman in the process?

Answer − 2.93 kJ

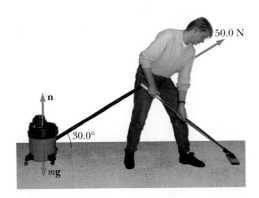

50.0 N

n

30.0°

$m\mathbf{g}$

Figure 6.5

(Example 6.5) A vacuum cleaner being pulled at an angle of 30.0° with the horizontal.

6.3 • THE SCALAR PRODUCT OF TWO VECTORS

See Screen 2.6

Based on Equation 6.1, it is convenient to express the definition of work in terms of a **scalar product** of the two vectors **F** and $\Delta \mathbf{r}$. The scalar product was introduced briefly in Section 1.9. We repeat its definition here:

> The scalar product of any two vectors **A** and **B** is a scalar quantity equal to the product of the magnitudes of the two vectors and the cosine of the angle θ between them:
>
> $$\mathbf{A} \cdot \mathbf{B} \equiv AB \cos \theta \qquad \text{[6.3]}$$
>
> where θ is the smaller of the two angles between **A** and **B**, as in Figure 6.6.

• *Scalar product of any two vectors* **A** *and* **B**

Note that **A** and **B** need not have the same units. The units of the scalar product are simply the product of the units of the two vectors. Because of the dot symbol, the scalar product is often called the *dot product*.

Notice that the right-hand side of Equation 6.3 has the same structure as the right-hand side of Equation 6.1. This allows us to write the definition of work as the scalar product **F** · $\Delta \mathbf{r}$. Thus, we can express Equation 6.1 as

• *Work expressed as a scalar product*

$$W = \mathbf{F} \cdot \Delta \mathbf{r} = F\Delta r \cos \theta \qquad \text{[6.4]}$$

Before continuing with our discussion of work, let us investigate some properties of the dot product, because we will need to use it later in the book as well. From Equation 6.3 we see that the scalar product is *commutative*. That is,

• *The order of the vectors in the scalar product can be reversed*

$$\mathbf{A} \cdot \mathbf{B} = \mathbf{B} \cdot \mathbf{A} \qquad \text{[6.5]}$$

In addition, the scalar product obeys the *distributive law of multiplication,* so that

$$\mathbf{A} \cdot (\mathbf{B} + \mathbf{C}) = \mathbf{A} \cdot \mathbf{B} + \mathbf{A} \cdot \mathbf{C} \qquad [6.6]$$

The dot product is simple to evaluate from Equation 6.3 when **A** is either perpendicular or parallel to **B**. If **A** is perpendicular to **B** ($\theta = 90°$), then $\mathbf{A} \cdot \mathbf{B} = 0$. (The equality $\mathbf{A} \cdot \mathbf{B} = 0$ also holds in the more trivial case when either **A** or **B** is zero.) If **A** and **B** point in the same direction ($\theta = 0°$), then $\mathbf{A} \cdot \mathbf{B} = AB$. If **A** and **B** point in opposite directions ($\theta = 180°$), then $\mathbf{A} \cdot \mathbf{B} = -AB$. The scalar product is negative when $90° < \theta < 180°$.

The unit vectors **i**, **j**, and **k**, which were defined in Chapter 1, lie in the positive *x*, *y*, and *z* directions, respectively, of a right-handed coordinate system. Therefore, it follows from the definition of $\mathbf{A} \cdot \mathbf{B}$ that the scalar products of these unit vectors are given by

$$\mathbf{i} \cdot \mathbf{i} = \mathbf{j} \cdot \mathbf{j} = \mathbf{k} \cdot \mathbf{k} = 1 \qquad [6.7]$$

$$\mathbf{i} \cdot \mathbf{j} = \mathbf{i} \cdot \mathbf{k} = \mathbf{j} \cdot \mathbf{k} = 0 \qquad [6.8]$$

Two vectors **A** and **B** can be expressed in component form as

$$\mathbf{A} = A_x\mathbf{i} + A_y\mathbf{j} + A_z\mathbf{k} \qquad \mathbf{B} = B_x\mathbf{i} + B_y\mathbf{j} + B_z\mathbf{k}$$

Therefore, using these expressions, Equations 6.7 and 6.8 reduce the scalar product of **A** and **B** to

$$\mathbf{A} \cdot \mathbf{B} = A_xB_x + A_yB_y + A_zB_z \qquad [6.9]$$

This equation and Equation 6.3 are alternative but equivalent expressions for the scalar product. Equation 6.3 is useful if you know the magnitudes and directions of the vectors, while Equation 6.9 is useful if you know the components of the vectors. In the special case where $\mathbf{A} = \mathbf{B}$, we see that

$$\mathbf{A} \cdot \mathbf{A} = A_x^2 + A_y^2 + A_z^2 = A^2$$

Quick Quiz 6.3

Which of the following statements is true about the relationship between the dot product of two vectors and the product of the magnitudes of the vectors? (a) $\mathbf{A} \cdot \mathbf{B}$ is larger than AB; (b) $\mathbf{A} \cdot \mathbf{B}$ is smaller than AB; (c) $\mathbf{A} \cdot \mathbf{B}$ could be larger or smaller than AB, depending on the angle between the vectors; (d) $\mathbf{A} \cdot \mathbf{B}$ could be equal to AB.

PITFALL PREVENTION 6.6

Work is a scalar

In Equation 6.4, we are defining the work in terms of two vectors, but keep in mind that *work is a scalar*—no direction is associated with it. In fact, all types of energy and energy transfer are scalars. This is one of the nice features of the energy approach—we don't need to concern ourselves so much about vectors!

• *Scalar products of unit vectors*

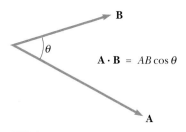

Figure 6.6

The scalar product $\mathbf{A} \cdot \mathbf{B}$ equals the magnitude of **A** multiplied by the magnitude of **B** and the cosine of the angle between **A** and **B**.

Example 6.2 The Scalar Product

The vectors **A** and **B** are given by $\mathbf{A} = 2\mathbf{i} + 3\mathbf{j}$ and $\mathbf{B} = -\mathbf{i} + 2\mathbf{j}$.

(a) Determine the scalar product $\mathbf{A} \cdot \mathbf{B}$.

Solution We will evaluate the scalar product directly using the unit vector notation:

$$\mathbf{A} \cdot \mathbf{B} = (2\mathbf{i} + 3\mathbf{j}) \cdot (-\mathbf{i} + 2\mathbf{j})$$
$$= -2\mathbf{i} \cdot \mathbf{i} + 2\mathbf{i} \cdot 2\mathbf{j} - 3\mathbf{j} \cdot \mathbf{i} + 3\mathbf{j} \cdot 2\mathbf{j}$$
$$= -2 + 6 = \boxed{4}$$

where we have used the facts that $\mathbf{i} \cdot \mathbf{i} = \mathbf{j} \cdot \mathbf{j} = 1$ and $\mathbf{i} \cdot \mathbf{j} =$

$\mathbf{j} \cdot \mathbf{i} = 0$. The same result is obtained using Equation 6.9 directly, where $A_x = 2$, $A_y = 3$, $B_x = -1$, and $B_y = 2$. Note that the result has no units because no units were specified on the original vectors **A** and **B**.

(b) Find the angle θ between **A** and **B**.

Solution The magnitudes of **A** and **B** are given by

$$A = \sqrt{A_x^2 + A_y^2} = \sqrt{(2)^2 + (3)^2} = \sqrt{13}$$
$$B = \sqrt{B_x^2 + B_y^2} = \sqrt{(1)^2 + (2)^2} = \sqrt{5}$$

Using Equation 6.3 and the result from (a) gives

$$\cos \theta = \frac{\mathbf{A} \cdot \mathbf{B}}{AB} = \frac{4}{\sqrt{13}\sqrt{5}} = \frac{4}{\sqrt{65}} = 0.496$$

$$\theta = \cos^{-1} 0.496 = \boxed{60.3°}$$

EXERCISE For $\mathbf{A} = 4\mathbf{i} + 3\mathbf{j}$ and $\mathbf{B} = -\mathbf{i} + 3\mathbf{j}$, find (a) $\mathbf{A} \cdot \mathbf{B}$ and (b) the angle between $\mathbf{A}$ and $\mathbf{B}$. Assume the vector components are known to three significant figures.

Answer (a) 5.00 (b) 71.6°

EXERCISE As a particle moves from the origin to $(3.0\mathbf{i} - 4.0\mathbf{j})$ m, it is acted upon by a force given by $(4.0\mathbf{i} - 5.0\mathbf{j})$ N. Calculate the work done by this force on the particle as it moves through the given displacement.

Answer 32 J

6.4 • WORK DONE BY A VARYING FORCE

See Screen 5.2

Consider a particle being displaced along the x axis under the action of a force of magnitude F_x in the x direction, which varies with position, as in the graphical representation in Figure 6.7. The particle is displaced in the direction of increasing x from $x = x_i$ to $x = x_f$. In such a situation, we cannot use Equation 6.1 to calculate the work done by the force because this relationship applies only when $\mathbf{F}$ is constant in magnitude and direction. As seen in Figure 6.7, we do not have a *single* value of the force to substitute into Equation 6.1. If, however, we imagine that the particle undergoes a small displacement $\Delta r = \Delta x$, shown in Figure 6.7a, then the x component F_x of the force is approximately constant over this interval, and we can approximate the work done by the force for this small displacement as

$$W_1 \approx F_x \Delta x \qquad [6.10]$$

This quantity is just the area of the shaded rectangle in Figure 6.7a. If we imagine that the curve described by F_x versus x is divided into a large number of such intervals, then the total work done for the displacement from x_i to x_f is approximately equal to the sum of a large number of such terms:

$$W \approx \sum_{x_i}^{x_f} F_x \Delta x$$

If the displacements Δx are allowed to approach zero, then the number of terms in the sum increases without limit, but the value of the sum approaches a definite value equal to the area under the curve bounded by F_x and the x axis in Figure 6.7b. As you probably have learned in calculus, this limit of the sum is called an *integral* and is represented by

$$\lim_{\Delta x \to 0} \sum_{x_i}^{x_f} F_x \Delta x = \int_{x_i}^{x_f} F_x \, dx$$

The limits on the integral $x = x_i$ to $x = x_f$ define what is called a **definite integral.** (An *indefinite integral* is the limit of a sum over an unspecified interval. Appendix B.7 gives a brief description of integration.) This definite integral is numerically equal to the area under the curve of F_x versus x between x_i and x_f. Therefore, we can express the work done by F_x for the displacement of the particle from x_i to x_f as

$$W = \int_{x_i}^{x_f} F_x \, dx \qquad [6.11]$$

This equation reduces to Equation 6.1 when $F_x = F \cos \theta$ is constant and $x_f - x_i = \Delta x$.

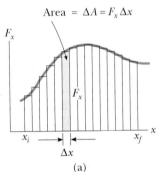

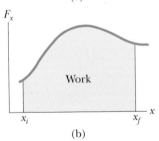

Figure 6.7

(a) The work done by a force of magnitude F_x for the small displacement Δx is $F_x \Delta x$, which equals the area of the shaded rectangle. The total work done for the displacement from x_i to x_f is approximately equal to the sum of the areas of all the rectangles. (b) The work done by the variable force F_x as the particle moves from x_i to x_f is *exactly* equal to the area under this curve.

If more than one force acts on a particle, the **total work done is just the scalar addition of the work done by each force.** Alternatively, the **total work is the work done by the net force.** If we express the net force in the x direction as $\Sigma\, F_x$, then the *net work* done on the particle as it moves from x_i to x_f is

$$W_{\text{net}} = \int_{x_i}^{x_f} \left(\sum F_x \right) dx \qquad \text{[6.12]}$$

The most general case for calculating work occurs when several forces are acting, the net force is not constant, and the net force is not parallel to the displacement. In this case, the expression for the work is

$$W_{\text{net}} = \int_{\mathbf{r}_1}^{\mathbf{r}_2} \left(\sum \mathbf{F} \right) \cdot d\mathbf{r}$$

• *Work done by a variable net force*

Example 6.3 Calculating Total Work Done from a Graph

A force acting on a particle varies with x as shown in Figure 6.8. Calculate the work done by the force on the particle as it moves from $x = 0$ to $x = 6.0$ m.

Solution The work done by the force is equal to the area under the curve from $x = 0$ to $x = 6.0$ m. We can model this geometrically as the area of the rectangular section from Ⓐ to Ⓑ combined with the area of the triangular section that extends from Ⓑ to Ⓒ. The area of the rectangle is $(5.0\ \text{N})(4.0\ \text{m}) = 20\ \text{J}$, and the area of the triangle is $\frac{1}{2}(5.0\ \text{N})(2.0\ \text{m}) = 5.0\ \text{J}$. Therefore, the total work done is $25\ \text{J}$.

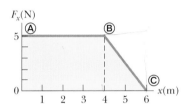

Figure 6.8

(Example 6.3) The force acting on a particle is constant for the first 4.0 m of motion and then decreases linearly with x from $x_B = 4.0$ m to $x_C = 6.0$ m. The net work done by this force is the area under this curve.

Work Done by a Spring

A common physical system for which the force varies with position is shown in Figure 6.9. A block on a horizontal, frictionless surface is connected to a spring. If the spring is stretched or compressed a small distance x from its equilibrium position $x = 0$, the spring exerts a force on the block given by

$$F_s = -kx \qquad \text{[6.13]}$$

• *Hooke's law*

See Screen 5.3

where k is a positive constant called the *force constant* (or *spring constant* or *stiffness constant*) of the spring. This force law for springs is known as **Hooke's law.** For many springs, Hooke's law can describe the behavior very accurately, provided that the displacement is not too large. The value of k is a measure of the stiffness of the spring. Stiff springs have larger k values, and weak springs have smaller k values. We shall employ a simplification model in which all springs, unless specified otherwise, obey Hooke's law.

The negative sign in Equation 6.13 signifies that the force exerted by the spring is always directed *opposite* the displacement from the equilibrium position $x = 0$. For example, when $x > 0$, such that the block is pulled to the right and the spring is stretched, as in Figure 6.9a, the spring force is to the left, or negative.

Figure 6.9

The force exerted by a spring on a block varies with the block's displacement from the equilibrium position $x = 0$. (a) When x is positive (stretched spring), the spring force is to the left. (b) When x is zero (natural length of the spring), the spring force is zero. (c) When x is negative (compressed spring), the spring force is to the right. (d) Graph of F_s versus x for the block–spring system. The work done by the spring force as the block moves from $-x_{max}$ to 0 is the area of the shaded triangle, $\frac{1}{2}kx_{max}^2$.

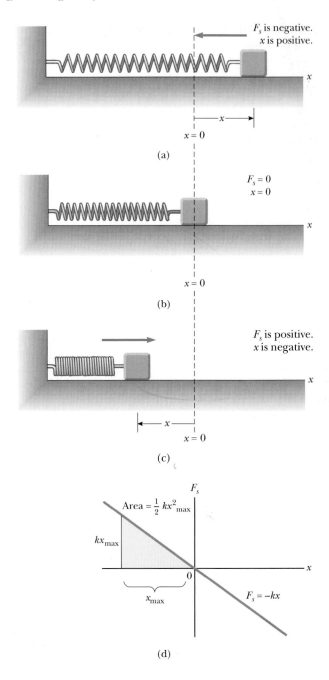

When $x < 0$, and the spring is compressed as in Figure 6.9c, the spring force is to the right, or positive. Of course, when $x = 0$, as in Figure 6.9b, the spring is unstretched and $F_s = 0$. Because the spring force always acts toward the equilibrium position, it is sometimes called a *restoring force*.

If the block is displaced to a position $-x_{max}$ and then released, it moves from $-x_{max}$ through zero to $+x_{max}$ (assuming a frictionless surface), and then turns around and returns to $-x_{max}$. The details of this oscillating motion will be given in Chapter 12. For our purposes here, let us calculate the work done by the spring force on the block as the block moves from $x_i = -x_{max}$ to $x_f = 0$. Applying the particle model to the block and using Equation 6.11, we have

$$W_s = \int_{x_i}^{x_f} F_s \, dx = \int_{-x_{max}}^{0} (-kx) \, dx = \tfrac{1}{2}kx_{max}^2 \qquad \text{[6.14]}$$

• *Work done by a spring*

The work done by the spring force on the block is positive because the spring force is in the same direction as the displacement (both are to the right). If we consider the work done by the spring force on the block as the block continues to move from $x_i = 0$ to $x_f = x_{max}$, we find that $W_s = -\tfrac{1}{2}kx_{max}^2$. This work is negative because for this part of the motion the displacement is to the right and the spring force is to the left. Therefore, the *net* work done by the spring force as the block moves from $x_i = -x_{max}$ to $x_f = x_{max}$ is *zero*.

If we plot F_s versus x, as in Figure 6.9d, we arrive at the same results. The work calculated in Equation 6.14 is equal to the area of the shaded triangle in Figure 6.9d, with base x_{max} and height kx_{max}. This area is $\tfrac{1}{2}kx_{max}^2$.

If the block undergoes an *arbitrary* displacement from $x = x_i$ to $x = x_f$, the work done by the spring force is

$$W_s = \int_{x_i}^{x_f} (-kx) \, dx = \tfrac{1}{2}kx_i^2 - \tfrac{1}{2}kx_f^2 \qquad \text{[6.15]}$$

From this equation we see that the work done by the spring force on the block is zero for any motion that ends where it began ($x_i = x_f$). We shall make use of this important result in Chapter 7, where we describe the motion of this system in more detail. Equation 6.15 also shows that the work done by the spring force is zero when the block moves between any two symmetric locations, $x_i = -x_f$. Consider the curve representing the spring force in Figure 6.9d; if the block moves from $x = -x_{max}$ to $x = +x_{max}$, the total work is zero because we are adding a positive area (for $-x_{max} < x < 0$) to a negative area (for $0 < x < +x_{max}$) of equal magnitude.

Equations 6.14 and 6.15 describe the work done by the spring force on the block. Now consider the work done by an *external agent* on the block as the spring is stretched *very slowly* from $x_i = 0$ to $x_f = x_{max}$, as in Figure 6.10. This work can be easily calculated by noting that the *applied force* $\mathbf{F}_{app}$ is of equal magnitude and opposite direction to the spring force $\mathbf{F}_s$ at any value of the displacement (because the block is not accelerating), so that $F_{app} = -(-kx) = +kx$. The work done by this applied force (the external agent) on the block is therefore

$$W_{F_{app}} = \int_{0}^{x_{max}} F_{app} \, dx = \int_{0}^{x_{max}} kx \, dx = \tfrac{1}{2}kx_{max}^2$$

Note that this work is equal to the negative of the work done by the spring force on the block for this displacement.

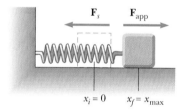

Figure 6.10

A block being pulled from $x_i = 0$ to $x_f = x_{max}$ on a frictionless surface by a force $\mathbf{F}_{app}$. If the process is carried out very slowly, the applied force is equal in magnitude and opposite in direction to the spring force at all times.

Example 6.4 Work Required to Stretch a Spring

One end of a horizontal spring ($k = 80$ N/m) is held fixed while an external force is applied to the free end, stretching it slowly from $x_A = 0$ to $x_B = 4.0$ cm. (a) Find the work done by the external force on the spring.

Solution Because we have not been told otherwise, we assume the spring obeys Hooke's law. We place the zero refer-

ence of the coordinate axis at the free end of the unstretched spring. The applied force is $F_{app} = kx = (80 \text{ N/m})(x)$. The work done by F_{app} is the area of the triangle from 0 to 4.0 cm in Figure 6.11:

$$W = \tfrac{1}{2}kx_B^2 = \tfrac{1}{2}(80 \text{ N/m})(0.040 \text{ m})^2 = \boxed{0.064 \text{ J}}$$

(b) Find the additional work done in stretching the spring from $x_B = 4.0$ cm to $x_C = 7.0$ cm.

Solution   The work done in stretching the spring the additional amount is the darker shaded area between these limits in Figure 6.11. Geometrically, it is the difference in area between the large and small triangles:

$$W = \tfrac{1}{2}kx_C{}^2 - \tfrac{1}{2}kx_B{}^2$$
$$= \tfrac{1}{2}(80 \text{ N/m})\,[(0.070 \text{ m})^2 - (0.040 \text{ m})^2] = \boxed{0.13 \text{ J}}$$

Using calculus, we find the same result:

$$W = \int_{x_B}^{x_C} F_{app}\, dx = \int_{0.04 \text{ m}}^{0.07 \text{ m}} (80 \text{ N/m})\,x\, dx$$

$$= \tfrac{1}{2}(80 \text{ N/m})\,(x^2)\Big|_{0.04 \text{ m}}^{0.07 \text{ m}}$$

$$W = \tfrac{1}{2}(80 \text{ N/m})\,[(0.070 \text{ m})^2 - (0.040 \text{ m})^2] = 0.13 \text{ J}$$

EXERCISE   If an applied force varies with position according to $F_x = 3.00x^2 - 5.00$, where x is in meters, how much work

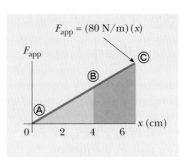

Figure 6.11

(Example 6.4) A graph of the applied force required to stretch a spring that obeys Hooke's law versus the elongation of the spring.

is done by this force on an object that moves from $x = 4.00$ m to $x = 7.00$ m?

Answer   264 J

6.5 • KINETIC ENERGY AND THE WORK–KINETIC ENERGY THEOREM

 See Screens 5.4 & 5.7

Now that we have explored various means of evaluating the work done by a force on a system, let us begin to explore the power of this approach. Solutions to problems using Newton's second law can be difficult if the forces in the problem are complicated. The alternative that enables us to understand and solve such problems relates the speed of the particle to its displacement under the influence of some net force from the environment. As we shall see in this section, if the work done by the net force on a particle can be calculated for a given displacement, the change in the particle's speed is easy to evaluate. Let's see how this is done.

Figure 6.12 shows an object modeled as a particle of mass m moving to the right along the x axis under the action of a net force $\Sigma\,\mathbf{F}$, also to the right. If the particle is moved through a displacement $\Delta x = x_f - x_i$, the work done by the force $\Sigma\,\mathbf{F}$ on the particle is

$$W_{net} = \int_{x_i}^{x_f} \Sigma\, F\, dx \qquad\qquad \textbf{[6.16]}$$

Using Newton's second law, we can substitute for the magnitude of the net force $\Sigma\, F = ma$, and then perform the following chain-rule manipulations on the integrand:

$$W_{net} = \int_{x_i}^{x_f} ma\, dx = \int_{x_i}^{x_f} m\frac{dv}{dt}\, dx = \int_{x_i}^{x_f} m\frac{dv}{dx}\frac{dx}{dt}\, dx = \int_{x_i}^{x_f} mv\, dv \qquad \textbf{[6.17]}$$

$$W_{net} = \tfrac{1}{2}mv_f{}^2 - \tfrac{1}{2}mv_i{}^2$$

This equation was generated for the specific situation of one-dimensional motion, but it is a general result. It tells us that the work done by the net force on a particle of mass m is equal to the difference between the initial and final values of

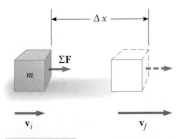

Figure 6.12

An object modeled as a particle undergoes a displacement of magnitude Δx and a change in speed under the action of a constant net force $\Sigma\,\mathbf{F}$.

a quantity $\frac{1}{2}mv^2$. This quantity $\frac{1}{2}mv^2$ is so important that we give it a special name—the **kinetic energy** K of a particle of mass m moving with a speed v is defined as

$$K \equiv \frac{1}{2}mv^2 \qquad \text{[6.18]}$$

- *Kinetic energy is energy associated with the motion of a particle*

Kinetic energy is a scalar quantity and has the same units as work. For example, an object of mass 2.0 kg moving with a speed of 4.0 m/s has a kinetic energy of 16 J. Table 6.1 shows some typical kinetic energies for various objects.

It is often convenient to write Equation 6.17 in the form

$$W_{\text{net}} = K_f - K_i = \Delta K \qquad \text{[6.19]}$$

- *The work–kinetic energy theorem*

Equation 6.19 is an important result known as the **work–kinetic energy theorem:**

> When work is done on a system and the only change in the system is in its speed, the work done by the net force equals the change in kinetic energy of the system.

The work–kinetic energy theorem indicates that the speed of a particle increases if the net work done on it is positive, because the final kinetic energy will be greater than the initial kinetic energy. The speed decreases if the net work is negative, because the final kinetic energy will be less than the initial kinetic energy.

The work–kinetic energy theorem will clarify some results we saw earlier in this chapter that may have seemed odd. In Thinking Physics 6.1, a person lifts a block and moves it horizontally. At the end of the Reasoning, it was mentioned that the net work done by all forces on the block is zero. This may seem strange, but it is correct. If we choose the block as the system, the net force on the system is zero because the upward lifting force is modeled as being equal to the gravitational force. Thus, the net force is zero, and zero net work is done, which is consistent with the fact that the kinetic energy of the block does not change. It may seem incorrect that no work was done, because something changed—the block was lifted—but this is correct *because we chose the block as the system.* If we had chosen the block and the Earth as the system, then we would have a different result—the work done on this system is not zero. We will explore this idea in the next chapter.

In Section 6.4, we also saw a result of zero work done, when we let a spring push a block from $x_i = -x_m$ to $x_f = x_m$. This is zero for a different reason from that for lifting the block. It is due to the combination of positive work and an equal amount of negative work done by the *same* force. It is also different from the

PITFALL PREVENTION 6.7

Conditions for the work–kinetic energy theorem

Always remember these special conditions for the work–kinetic energy theorem. We will see many situations in which other changes occur in the system besides its speed, and there are other interactions with the environment besides work. The work–kinetic energy theorem is important, but limited in its application—it is not a general principle. We shall present a general principle involving energy in Section 6.7.

PITFALL PREVENTION 6.8

The work–kinetic energy theorem: speed, not velocity

Note that the work–kinetic energy theorem states that work is related to a change in the *speed* of an object, not a change in velocity. For example, if an object is in uniform circular motion, the speed is constant so, even though the velocity is changing, no work is done by the force causing the circular motion. This is another reason why the energy approach is so attractive—we can handle many problems with only scalar calculations.

TABLE 6.1	Typical Kinetic Energies of Various Objects
Object	**Kinetic Energy (J)**
Earth	3×10^{33}
Moon	4×10^{28}
Space shuttle in orbit	3×10^{12}
Human sprinter	4×10^{3}
Flying bee	6×10^{-1}
Snail	6×10^{-8}
Electron in computer monitor	4×10^{-15}

block example in that the speed of the block is continually changing. The work–kinetic energy theorem only refers to the initial and final points for the speeds—it does not depend on details of the path followed between these points. We shall use this concept often in the remainder of this and the next chapter.

Example 6.5 A Block Pulled on a Frictionless Surface

A 6.00-kg block initially at rest is pulled to the right along a horizontal frictionless surface by a constant, horizontal force of 12.0 N, as in Figure 6.13. Find the speed of the block after it has moved 3.00 m.

Reasoning The block is the system, and three external forces interact with it. Neither the gravitational force nor

the normal force does work on the block because these forces are vertical and the displacement is horizontal. In the horizontal direction, there is no friction. Thus, the only external force that we must consider in the calculation is the 12.0-N force.

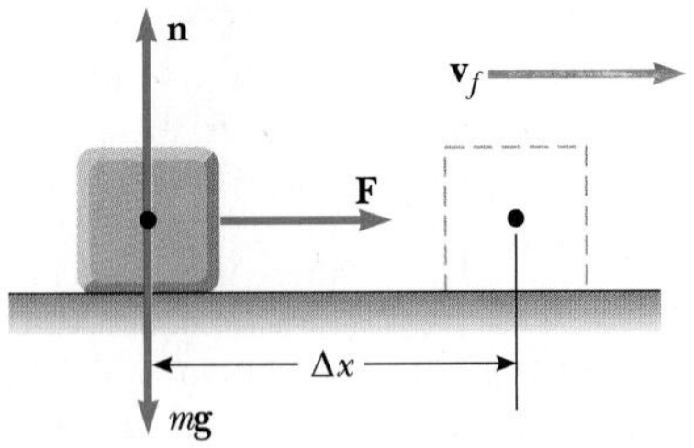

Solution The work done by the 12.0-N force is

$$W = F \Delta x = (12.0 \text{ N})(3.00 \text{ m}) = 36.0 \text{ N} \cdot \text{m} = 36.0 \text{ J}$$

Using the work–kinetic energy theorem and noting that the initial kinetic energy is zero, we find

$$W = K_f - K_i = \tfrac{1}{2}mv_f^2 - 0$$

$$v_f = \sqrt{\frac{2W}{m}} = \sqrt{\frac{2(36.0 \text{ J})}{6.00 \text{ kg}}} = \boxed{3.46 \text{ m/s}}$$

Notice that an energy calculation such as this will give you only the speed of the particle, not the velocity. In many cases, this is all you need. If you need the direction of the velocity vector, you may need to analyze the pictorial representation or perform other calculations.

Figure 6.13

(Example 6.5) A block is pulled to the right by a constant horizontal force on a frictionless surface.

Exa[mple]

A blo[ck]
const
comp
from

BORDERS.com

12,13,14,24 25,26

2A9

www.borders.com

through the equilibrium position $x = 0$ if the surface is frictionless.

Solution We identify the block as the system. Then, the force from the spring is the interaction with the environment that does work on the block. We use Equation 6.14 to find the work done by the spring on the block with $x_i = -x_{max} = -2.0 \text{ cm} = -2.0 \times 10^{-2}$ m:

$$W_s = \tfrac{1}{2}kx_{max}^2 = \tfrac{1}{2}(1.0 \times 10^3 \text{ N/m})(2.0 \times 10^{-2} \text{ m})^2 = 0.20 \text{ J}$$

Using the work–kinetic energy theorem with $v_i - 0$ gives

$$W_s = \tfrac{1}{2}mv_f^2 - \tfrac{1}{2}mv_i^2 = \tfrac{1}{2}mv_f^2 = 0$$

$$v_f = \sqrt{\frac{2W_s}{m}} = \sqrt{\frac{2(0.20 \text{ J})}{1.6 \text{ kg}}} = \boxed{0.50 \text{ m/s}}$$

Figu[re]

(Exam[ple]
2.0 cm

Example 6.7 Dropping a Block onto a Spring

A spring that has a force constant of 1.00×10^3 N/m is placed on a table in a vertical position as in Figure 6.15. A block of mass 1.60 kg is held 1.00 m above the free end of the spring. The block is dropped from rest so that it falls vertically onto the spring. By what distance does the spring compress?

Solution We identify the block as the system. We identify the initial condition as the release of the block from the height $y_i = h = 1.00$ m above the free end of the spring. The final condition occurs when the block is momentarily at rest with the spring compressed its maximum distance. For this condition, the block is located at $y_f = -d$, where d is the distance by which the spring is compressed. The net work on

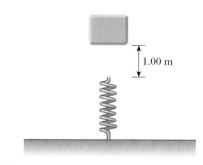

the block during this displacement is done by gravity (positive work) and the spring force (negative work):

$$
\begin{aligned}
W_{\text{net}} &= \mathbf{F}_g \cdot \Delta \mathbf{r} - \tfrac{1}{2}kd^2 = (-mg)\mathbf{j} \cdot (-d-h)\mathbf{j} - \tfrac{1}{2}kd^2 \\
&= mg(h+d) - \tfrac{1}{2}kd^2 \\
&= (1.60 \text{ kg})(9.80 \text{ m/s}^2)(1.00 \text{ m} + d) \\
&\quad - \tfrac{1}{2}(1.00 \times 10^3 \text{ N/m})d^2 \\
&= -500d^2 + 15.7d + 15.7
\end{aligned}
$$

The change in kinetic energy of the block is zero because it is at rest at both the initial and final conditions. Thus, from the work–kinetic energy theorem, the work done by the net force must be equal to zero:

$$-500d^2 + 15.7d + 15.7 = 0$$

This quadratic equation can be solved, and the solutions are $d = 0.19$ m and $d = -0.16$ m. Because we have chosen the value of d as a positive number, by claiming that $y = -d$ is *below* the initial position of the end of the spring, we need to choose the positive root, $d = 0.19$ m. (The negative root gives the position for the final condition as $y = -d = -(-0.16 \text{ m}) = +0.16$ m, which is the position above the *initial* $y = 0$ at which the spring again comes to rest in its oscillation, assuming that the block remains attached to the spring. These two positions are symmetric around $y = -0.016$ m, which is where the block would rest in equilibrium on the spring, according to Hooke's law.)

Figure 6.15

(Example 6.7) A mass is dropped onto a vertical spring, causing it to compress.

6.6 • THE NONISOLATED SYSTEM

We have seen a number of examples in which an object, modeled as a particle, is acted on by various forces, resulting in a change in its kinetic energy. This very simple situation is the first example of the **nonisolated system,** which is an important new analysis model for us. Physical problems for which this model is appropriate involve systems that interact with or are influenced by their environment, causing some kind of change in the system.

The work–kinetic energy theorem is our first introduction to the nonisolated system. The interaction is the work done by the external force, and the quantity related to the system that changes is its kinetic energy. Because the energy of the system changes, we conceptualize the work as a means of **energy transfer—work has the effect of transferring energy between the system and the environment.** If positive work is done on the system, energy is transferred to the system, whereas negative work indicates that energy is transferred from the system to the environment.

So far, we have discussed kinetic energy as the only type of energy in a system. We now argue the existence of a second type of energy. Consider a situation in

• *Work is energy transfer*

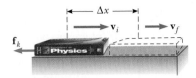

Figure 6.16

A book sliding to the right on a horizontal surface slows down in the presence of a force of kinetic friction acting to the left. The initial velocity of the book is $\mathbf{v}_i$, and its final velocity is $\mathbf{v}_f$. The normal force and gravitational force are not included in the diagram because they are perpendicular to the direction of motion and therefore do not influence the speed of the book.

• *Internal energy*

• *Methods of energy transfer*

which an object slides along a surface with friction. Clearly, work is done by the friction force because there is a force and a displacement. Keep in mind, however, that our equations for work involve the displacement of the *point of application of the force*. The friction force is spread out over the entire contact area of an object sliding on a surface, so the force is not localized at a point. In addition, the magnitudes of the friction forces at various points are constantly changing as spot welds occur, the surface and the object deform locally, and so on. The points of application of the friction force on the object are jumping all over the face of the object in contact with the surface. In general, this means that the displacement of the point of application of the friction force (assuming we could calculate it!) is not the same as the displacement of the object.*

The work–kinetic energy theorem is valid for a particle or an object that can be modeled as a particle. When an object is extended and cannot be treated as a particle, however, things can become more complicated.

Let us imagine the book in Figure 6.16 sliding to the right on the surface of a heavy table and slowing down due to the friction force. Suppose the *surface* is the system. Then the friction force from the sliding book does work on the surface. The force on the surface is to the right and the displacement of the point of application of the force is to the right—the work is positive. But the surface is not moving after the book has stopped. Positive work has been done on the surface, yet the kinetic energy of the surface does not increase. Is this a violation of the work–kinetic energy theorem?

It is indeed a violation, because this situation does not fit the description of the conditions given for the work–kinetic energy theorem. Work is done on the system of the surface, but the result of that work is *not* an increase in kinetic energy. From your everyday experience with sliding over surfaces with friction, you can probably guess that the surface will be *warmer* after the book slides over it (rub your hands together briskly to experience this!). Thus, the work done has gone into warming the surface rather than causing it to increase in speed. We use the phrase **internal energy** E_{int} for the energy associated with an object's temperature. (We will see a more general definition for internal energy in Chapter 17.) In this case, the work done on the surface does indeed represent energy transferred into the system, but it appears in the system as internal energy rather than kinetic energy.

We have now seen two methods of storing energy in a system: kinetic energy, related to motion of the system, and internal energy, related to its temperature. We have seen only one way to transfer energy into the system so far—work. Next, we introduce a few other ways to transfer energy into or out of a system, which will be studied in detail in other sections of the book. We will focus on the following six methods (Fig. 6.17) for transferring energy between the environment and the system.

Work (present chapter) is a method of transferring energy to a system by applying a force to the system and causing a displacement of the point of application of the force, as we have seen in the previous sections (Fig. 6.17a).

Mechanical waves (Chapter 13) are a means of transferring energy by allowing a disturbance to propagate through air or another medium. This is the method by which energy leaves a radio through the loudspeaker—sound—and by which energy enters your ears to stimulate the hearing process (Fig. 6.17b). Mechanical waves also include seismic waves and ocean waves.

---

* For more details on energy transfer situations involving forces of kinetic friction, see B. A. Sherwood and W. H. Bernard, *Am J Physics,* 52:1001, 1984; and R. P. Bauman, *Physics Teacher,* 30:264, 1992.

(a)

(b)

(c)

(d)

(e)

(f)

Figure 6.17

Energy transfer mechanisms. (a) Energy is transferred to the block by *work;* (b) energy leaves the radio by *mechanical waves;* (c) energy transfers up the handle of the spoon by *heat;* (d) energy enters the automobile gas tank by *matter transfer;* (e) energy enters the hair dryer by *electrical transmission;* and (f) energy leaves the light bulb by *electromagnetic radiation.* (*George Semple*)

Heat (Chapter 17) is a method of transferring energy by means of microscopic collisions; for example, the end of a metal spoon in a cup of coffee becomes hot because fast-moving electrons and atoms in the bowl of the spoon bump into slower ones in the nearby part of the handle (Fig. 6.17c). These particles move faster because of the collisions and bump into the next group of slow particles. Thus, the internal energy of the end of the spoon rises from energy transfer due to this bumping process. This process, also called *thermal conduction*, is caused by a temperature difference between two regions in space.*

***PITFALL PREVENTION 6.9**
Heat is not a form of energy

The word *heat* is one of the most misused words in our popular language. Note that heat is a method of *transferring* energy, it is *not* a form of storing energy. Thus, phrases such as "heat content," "the heat of the summer," and "the heat escaped" all represent *incorrect* uses of this word. We shall study heat in detail in Chapter 17.

---

* Many textbooks use the term *heat* to include *conduction, convection,* and *radiation.* Conduction is the only one of these three processes driven by a temperature difference alone, so we will restrict heat to this process in this book. Convection and radiation are included in other types of energy transfer in our list of six.

In **matter transfer** (Chapter 17), matter physically crosses the boundary of the system, carrying energy with it. Examples include filling the system of your automobile tank with gasoline (Fig. 6.17d) and carrying energy to the rooms of your home by means of circulating warm air from the furnace. Matter transfer occurs in several situations and is introduced in Chapter 17 by means of one example, *convection*.

Electrical transmission (Chapter 21) involves energy transfer by means of electric currents. This is how energy transfers into your stereo system or any other electrical device (Fig. 6.17e).

Electromagnetic radiation (Chapter 24) refers to electromagnetic waves such as light, microwaves, radio waves, and so on (Fig. 6.17f). Examples of this method of transfer include energy going into your baked potato in your microwave oven and light energy traveling from the Sun to the Earth through space.*

One of the central features of the energy approach is the notion that **we can neither create nor destroy energy—energy is *conserved*.** Thus, **if the amount of energy in a system changes, it can *only* be due to the fact that energy has crossed the boundary by a transfer mechanism such as one of the methods listed.** This is a general statement of the principle of **conservation of energy.** We can describe this idea mathematically as follows:

- *Conservation of energy—the continuity equation for energy*

$$\Delta E_{\text{system}} = \sum H \qquad \text{[6.20]}$$

where E_{system} is the total energy of the system, including all methods of energy storage (kinetic, internal, and another to be discussed in Chapter 7) and H is the amount of energy transferred across the system boundary by a transfer mechanism. Two of our transfer mechanisms have well-established symbolic notations. For work, $H_{\text{work}} = W$, as we have seen in the current chapter, and for heat, $H_{\text{heat}} = Q$, which we will see in detail in Chapter 17. The other four members of our list do not have established symbols, so we will call them H_{MW} (mechanical waves), H_{MT} (matter transfer), H_{ET} (electrical transmission), and H_{ER} (electromagnetic radiation).

We have seen how to calculate work in this chapter and the other types of transfers will be discussed in subsequent chapters. Equation 6.20 is called the **continuity equation for energy.** The full expansion of Equation 6.20, with kinetic and internal energy as the storage mechanisms, is

$$\Delta K + \Delta E_{\text{int}} = W + Q + H_{\text{MW}} + H_{\text{MT}} + H_{\text{ET}} + H_{\text{ER}}$$

This is the primary mathematical representation of the energy analysis of the non-isolated system. In most cases, this reduces to a much simpler equation because some of the terms are zero. A continuity equation arises in any situation where the change in a quantity in a system occurs solely because of transfers across the boundary (because the quantity is conserved), several examples of which occur in areas of physics, as we shall see.

This concept is no more complicated in theory than is that of balancing your checking account statement. If your account is the system, the change in the account balance for a given month is the sum of all the transfers—deposits, withdrawals, fees, interest, and checks written. It may be useful for you to think of energy as the *currency of nature!*

On a wind farm, the moving air does work on the blades of the windmills. The energy transferred into the system of the windmill appears as kinetic energy of the blades and the rotor of an electrical generator attached to the axis of the windmill. Energy is transferred out of the system of the windmill by electrical transmission. *(Billy Hustace/Stone)*

---

* Electromagnetic radiation and work done by field forces are the only energy transfer mechanisms that do not require molecules of the environment to be available at the system boundary. Thus, systems surrounded by a vacuum (such as planets) can only exchange energy with the environment by means of these two possibilities.

Suppose a force is applied to a nonisolated system and the point of application of the force moves through a displacement. Suppose further that the only effect on the system is to increase its speed. Then the only transfer mechanism is work (so that $\Sigma\, H$ in Equation 6.20 reduces to just W) and the only kind of energy in the system that changes is the kinetic energy (so that ΔE_{system} reduces to just ΔK). Equation 6.20 then becomes

$$\Delta K = W$$

which is the work–kinetic energy theorem, Equation 6.19. This theorem is a special case of the more general continuity equation for energy. We shall see several more examples of other special cases of the continuity equation for energy in future chapters.

Quick Quiz 6.4

By what transfer mechanisms does energy enter and leave (a) your television set? (b) Your gasoline-powered lawn mower? (c) Your hand-cranked pencil sharpener?

THINKING PHYSICS 6.3

An automobile strikes a tree and comes to rest. The total energy in the system (the car) is now less than before. How did the energy leave the system, and in what form is energy left in the system?

Reasoning   Energy transferred out of the system in a number of ways. The automobile did some *work* on the tree during the collision, causing the tree to deform. A large crash occurred during the collision, representing transfer of energy by *sound* (mechanical waves). If the gasoline tank leaks after the collision, the automobile is losing energy by means of *matter transfer*. After the car has come to rest, the system has no more kinetic energy. The automobile has more *internal energy* because its temperature is likely to be higher after the collision than before. We could express this process in terms of the continuity equation for energy as

$$\Delta E_{system} = \Sigma\, H \quad\longrightarrow\quad \Delta K + \Delta E_{int} = W + H_{MW} + H_{MT}$$

where ΔK is negative and ΔE_{int} is positive. All three of the energy transfer terms are negative because energy left the system by all three mechanisms.

THINKING PHYSICS 6.4

A toaster is turned on. Discuss the forms of energy and energy transfer occurring in the coils of the toaster.

Reasoning   We identify the coils as the system. The energy that changes in the system is *internal energy* because the temperature of the coils rises. The energy transfer mechanism for energy coming into the coils is *electrical transmission* through the wire plugged into the wall. Energy is transferring out of the coils by *electromagnetic radiation* because the coils are hot and glowing. Some transfer of energy also occurs by *heat* from the hot surfaces of the coils into the air. We could express this process in terms of the continuity equation for energy as

$$\Delta E_{system} = \Sigma\, H \quad\longrightarrow\quad \Delta E_{int} = Q + H_{ET} + H_{ER}$$

After a short warm-up period, the coils will stabilize in temperature, and the internal energy will no longer change. In this situation, the energy input and output will be balanced:

$$0 = Q + H_{ET} + H_{ER} \longrightarrow -H_{ET} = Q + H_{ER}$$

Note that Q and H_{ER} are both negative because they represent energy leaving the system, and H_{ET} is positive—energy continues to enter the system by electrical transmission.

THINKING PHYSICS 6.5

Soft steel can be made red hot by continued hammering on it. This doesn't work as well for hard steel. Why is there this difference?

Reasoning   The rise in temperature of the soft steel is an example of transferring energy into a system by work and having it appear as an increase in the internal energy of the system. This works well for the soft steel because it is *soft*. This softness results in a deformation of the steel under the blow of the hammer. Thus, the point of application of the force is displaced by the hammer, and positive work is done on the steel. With the hard steel, less deformation occurs; thus, there is less displacement of the point of application of the force, and less work done on the steel. The soft steel is therefore better at absorbing energy from the hammer by means of work, and its temperature rises more rapidly.

6.7 • SITUATIONS INVOLVING KINETIC FRICTION

In the preceding section, we discussed the complicated nature of the friction force as an object slides across a surface. For these kinds of situations, Newton's second law is still valid for the system, even though the work–kinetic energy theorem is not. In the case of a nondeformable object like our block in Figure 6.12 sliding on the surface,* we can handle this in a relatively straightforward way, as follows.

Consider a situation in which a constant force is applied to the block and follow a similar procedure to that in going from Equation 6.16 to Equation 6.17. We start by multiplying Newton's second law (*x* component only) by a displacement of the block:

$$\left(\sum F_x \right) \Delta x = (ma_x) \Delta x \qquad \text{[6.21]}$$

For an object under constant acceleration, we know that the following relationships (Eqs. 2.8 and 2.10) are valid:

$$a_x = \frac{v_f - v_i}{t} \qquad \Delta x = \tfrac{1}{2}(v_i + v_f) t$$

where v_i is the speed at $t = 0$ and v_f is the speed at time t. Substituting these expressions into Equation 6.21 gives

$$\left(\sum F_x \right) \Delta x = m \left(\frac{v_f - v_i}{t} \right) \tfrac{1}{2}(v_i + v_f) t$$

$$\left(\sum F_x \right) \Delta x = \tfrac{1}{2}mv_f^2 - \tfrac{1}{2}mv_i^2$$

---

* The overall shape of the block remains the same, which is why we say it is nondeformable. On a microscopic level, however, the face of the block is deformed as it slides over the surface.

This *looks* like the work–kinetic energy theorem, but *the left-hand side has not been called work*. The quantity Δx is the displacement of the block; it is *not* the displacement of the point of application of the friction force. The work–kinetic energy theorem can be applied only to a particle; the previous equation is true for an extended nondeformable object experiencing a constant force.

Let us now apply this equation to a block that has been projected across a surface. We imagine that the block has an initial velocity and slows down due to friction as the only force in the horizontal direction. The net force on the block is the kinetic friction force $\mathbf{f}_k$, which is directed opposite to the displacement Δx. Thus,

$$\left(\sum F_x\right)\Delta x = -f_k\,\Delta x = \tfrac{1}{2}mv_f^2 - \tfrac{1}{2}mv_i^2 = \Delta K \qquad \text{[6.22]}$$

which mathematically describes the decrease in kinetic energy due to the friction force.

If there are other forces besides friction acting on the object, the change in kinetic energy is the sum of that due to the other forces from the work–kinetic energy theorem and that due to friction:

$$\Delta K = -f_k\,\Delta x + \sum W_{\text{other forces}} \qquad \text{[6.23]}$$

Now, consider the larger system consisting of the block *and* the surface as the block slows down under the influence of a friction force alone. No work is done across the boundary of this system—the system does not interact with the environment. No other types of energy transfer occur across the boundary of the system, assuming we ignore the inevitable sound the sliding object makes—this is a simplification model! In this case, Equation 6.20 becomes

$$\Delta K + \Delta E_{\text{int}} = 0$$

The change in kinetic energy of this system is the same as the change in kinetic energy of the system of the block in Equation 6.22 because the block is the only part of the block–surface system that is moving. Thus,

$$\Delta K + \Delta E_{\text{int}} = 0 \quad\longrightarrow\quad -f_k\,\Delta x + \Delta E_{\text{int}} = 0$$

$$\Delta E_{\text{int}} = f_k\,\Delta x \qquad \text{[6.24]}$$

Therefore, the **increase in internal energy of the system is equal to the product of the friction force and the displacement of the block.**

The bottom line is that a friction force transforms kinetic energy to internal energy, and the increase in internal energy is equal to the decrease in kinetic energy.

Quick Quiz 6.5

You are traveling along a freeway at 65 mi/h. Your car has kinetic energy. You now brake to a stop because of congestion in traffic. Where is the kinetic energy that your car once had? That is, what form does it now take, and where is this form of energy located?

THINKING PHYSICS 6.6

In many situations, friction forces tend to reduce the kinetic energy of an object; however, friction forces can sometimes increase an object's kinetic energy. Describe a situation in which friction causes an increase in kinetic energy.

PITFALL PREVENTION 6.10

It sure looks like work

Although the left-hand side of Equation 6.22 does indeed look like work—force multiplied by displacement—we don't call it work, as discussed in the text. So despite the similarity, avoid saying that you are "calculating the work done by friction." The product $-f_k\,\Delta x$ represents "the change in kinetic energy of the system due to friction."

• *The change in kinetic energy of an object due to friction and other forces*

• *The increase in internal energy of a system due to friction*

Reasoning If a crate is located on the bed of a truck, and the truck accelerates to the east, the static friction force exerted on the crate by the truck acts to the east to give the crate the same acceleration as the truck (assuming the crate doesn't slip). Another example is a car that accelerates because of the friction forces exerted on the car's tires by the road. These forces act in the direction of the car's motion, and the sum of these forces causes an increase in the car's kinetic energy.

Example 6.8 A Block Pulled on a Rough Surface

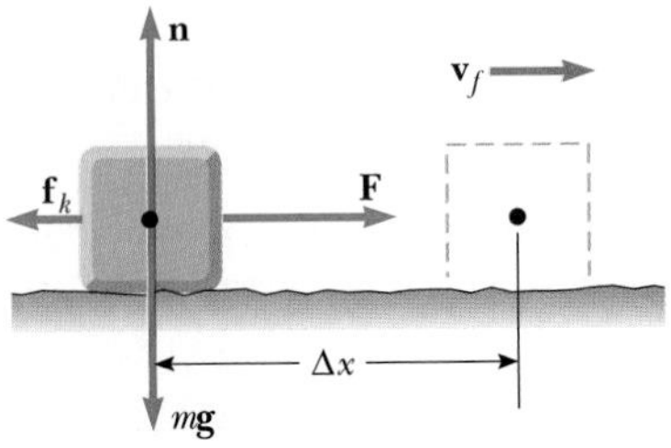

A block of mass 6.00 kg initially at rest is pulled to the right by a constant horizontal force with magnitude $F = 12.0$ N (Fig. 6.18). The coefficient of kinetic friction between the block and the surface is 0.150. Find the speed of the block after it has moved 3.00 m. (This is Example 6.5, modified so that the surface is no longer frictionless.)

Solution We define the system as the block and apply Equation 6.23:

$$\Delta K = -f_k \Delta x + \sum W_{\text{other forces}} = -\mu_k n \Delta x + F \Delta x$$

The block is modeled as a particle in equilibrium in the vertical direction, so that $n = mg$. Thus,

$$\Delta K = -\mu_k mg \Delta x + F \Delta x$$

Evaluating ΔK and solving for v_f, we have

$$\Delta K = \tfrac{1}{2}mv_f^2 - \tfrac{1}{2}mv_i^2 = -\mu_k mg \Delta x + F \Delta x$$

$$v_f = \sqrt{\left(\frac{2}{m}\right)\left(-\mu_k mg \Delta x + F \Delta x + \tfrac{1}{2}mv_i^2\right)}$$

Substituting the numerical values, we find,

$$v_f = \sqrt{\left(\frac{2}{6.00 \text{ kg}}\right)\left[-(0.150)(6.00 \text{ kg})(9.80 \text{ m/s}^2)(3.00 \text{ m}) + (12.0 \text{ N})(3.00 \text{ m}) + 0\right]}$$

$$= \boxed{1.78 \text{ m/s}}$$

Notice that this is less than the value calculated in Example 6.5, due to the effect of the friction force.

Figure 6.18

(Example 6.8) A block is pulled to the right by a constant horizontal force on a surface with friction.

6.8 • POWER

See Screen 5.8

We discussed transfers of energy across the boundary of a system by a number of methods. From a practical viewpoint, it is interesting to know not only the amount of energy transferred to a system but also the *rate* at which the energy is transferred. The time rate of energy transfer is called **power.**

We shall focus on work as our particular energy transfer method in this discussion, but keep in mind that the notion of power is valid for *any* means of energy transfer. If an external force is applied to an object (for which we will adopt the particle model), and if the work done by this force is W in the time interval Δt, then the **average power** during this interval is defined as

• *Average power when work is done*

$$\overline{\mathcal{P}} \equiv \frac{W}{\Delta t} \tag{6.25}$$

The **instantaneous power** $\mathscr{P}$ at a particular point in time is the limiting value of the average power as Δt approaches zero:

$$\mathscr{P} \equiv \lim_{\Delta t \to 0} \frac{W}{\Delta t} = \frac{dW}{dt} \qquad \text{[6.26]}$$

where we represent the infinitesimal value of the work done by dW. We know from Equation 6.4 that we can write the infinitesimal amount of work done over a displacement $d\mathbf{r}$ as $dW = \mathbf{F} \cdot d\mathbf{r}$. Therefore, the instantaneous power can be written

$$\mathscr{P} = \frac{dW}{dt} = \mathbf{F} \cdot \frac{d\mathbf{r}}{dt} = \mathbf{F} \cdot \mathbf{v} \qquad \text{[6.27]}$$

where we have used the fact that $\mathbf{v} = d\mathbf{r}/dt$.

In general, power is defined for any type of energy transfer. The most general expression for power is, therefore

$$\mathscr{P} = \frac{dE}{dt} \qquad \text{[6.28]}$$

• *General expression for power*

where dE/dt is the rate at which energy is crossing the boundary of the system by a given transfer mechanism.

The SI unit of power is joules per second (J/s), also called a **watt** (W) (after James Watt):

$$1 \text{ W} = 1 \text{ J/s} = 1 \text{ kg} \cdot \text{m}^2/\text{s}^3$$

• *The watt*

The unit of power in the British engineering system is the **horsepower** (hp):

$$1 \text{ hp} \equiv 550 \text{ ft} \cdot \text{lb/s} \equiv 746 \text{ W}$$

A new unit of energy can now be defined in terms of the unit of power. One **kilowatt hour** (kWh) is the energy transferred in 1 h at the constant rate of 1 kW. The numerical value of 1 kWh of energy is

$$1 \text{ kWh} = (10^3 \text{ W})(3600 \text{ s}) = 3.60 \times 10^6 \text{ J}$$

It is important to realize that a kilowatt hour is a unit of energy, not power. When you pay your electric bill, you are buying energy, and the amount of energy transferred by electrical transmission into a home during the period represented by the electric bill is usually expressed in kilowatt hours. For example, your bill may state that you used 900 kWh of energy during a month, and you are being charged at the rate of 10¢ per kWh. Your obligation is then $90 for this amount of energy. As another example, suppose an electric bulb is rated at 100 W. In 1.00 h of operation, it would have energy transferred to it by electrical transmission in the amount of $(0.100 \text{ kW})(1.00 \text{ h}) = 0.100 \text{ kWh} = 3.60 \times 10^5 \text{ J}$.

Quick Quiz 6.6

An older model car accelerates from 0 to speed v in 10 s. A newer, more powerful sports car accelerates from 0 to $2v$ in the same time. What is the ratio of powers for the two cars if they have the same mass?

PITFALL PREVENTION 6.11
Be careful with power

Students can be confused about power in two ways. First, do not confuse the symbol W for the watt with the italic symbol W for work. Second, students sometimes forget that the watt represents a rate of energy transfer and try to talk about "watts per second." We have seen rates expressed in earlier chapters as "something per second," as in the case of velocity and acceleration. Here, we have another rate, but the action of defining the joule per second as a watt has "hidden" the "per second" phrase in the definition. Always remember that the watt is *the same as* a joule per second.

Example 6.9 Power Delivered by an Elevator Motor

A 1 000-kg elevator carries a maximum load of 800 kg. A constant friction force of 4 000 N retards its motion upward, as in Figure 6.19. (a) What is the minimum power delivered by the motor to lift the elevator at a constant speed of 3.00 m/s?

Solution We use two analysis models for the elevator. First, we model it as a particle in equilibrium because it moves at constant speed. The motor must supply the force **T** that results in the tension in the cable that pulls the elevator upward. From Newton's second law and from the fact that $a = 0$ because v is constant, we have

$$T - f - Mg = 0$$

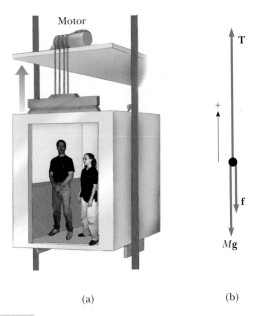

(a)

(b)

Figure 6.19

(Example 6.9) (a) A motor lifts an elevator car. (b) Free-body diagram for the elevator. The motor exerts an upward force **T** on the supporting cables. The magnitude of this force is T, the tension in the cables, which is applied in the upward direction on the elevator. The downward forces on the elevator are the friction force **f** and the gravitational force $\mathbf{F}_g = M\mathbf{g}$.

where M is the *total* mass (elevator plus load), equal to 1 800 kg. Therefore,

$$
\begin{aligned}
T &= f + Mg \\
&= 4.00 \times 10^3 \text{ N} + (1.80 \times 10^3 \text{ kg})(9.80 \text{ m/s}^2) \\
&= 2.16 \times 10^4 \text{ N}
\end{aligned}
$$

Now, we model the elevator as a nonisolated system. Work is being done on it by the tension force (as well as other forces). We can use Equation 6.27 to evaluate the power delivered by the motor, which is the rate at which work is done on the elevator by the tension force. Because **T** is in the same direction as **v**, we have

$$
\begin{aligned}
\mathcal{P} &= \mathbf{T} \cdot \mathbf{v} = Tv \\
&= (2.16 \times 10^4 \text{ N})(3.00 \text{ m/s}) = 6.48 \times 10^4 \text{ W} \\
&= \boxed{64.8 \text{ kW}}
\end{aligned}
$$

Because **T** is the force the motor applies to the cable, the preceding result represents the rate at which energy is being transferred out of the motor by doing work on the cable.

(b) What power must the motor deliver at any instant if it is designed to provide an upward acceleration of 1.00 m/s²?

Solution In this case, we expect the tension to be larger than in part (a) because the cable must now cause an upward acceleration of the elevator. Applying Newton's second law to the elevator gives

$$
\begin{aligned}
T - f - Mg &= Ma \\
T &= M(a + g) + f \\
&= (1.80 \times 10^3 \text{ kg})(1.00 \text{ m/s}^2 + 9.80 \text{ m/s}^2) \\
&\quad + 4.00 \times 10^3 \text{ N} \\
&= 2.34 \times 10^4 \text{ N}
\end{aligned}
$$

Therefore, using Equation 6.27, we have for the required power

$$\mathcal{P} = Tv = \boxed{(2.34 \times 10^4 \, v)}$$

where v is the instantaneous speed of the elevator in meters per second. Hence, the power required increases with increasing speed.

context connection

6.9 • A PROBE TO THE SUN

Suppose that a secondary purpose of our mission to Mars is to release a probe as we leave the Earth that will fall into the Sun, sending solar data back to Earth as it rushes toward its demise. The design plan for the probe is that we will release it from our spacecraft in such a way that it has zero velocity relative to the Sun. The probe has a mass of 200 kg and no rocket engine, so it simply falls into the Sun from rest due to the gravitational force from the Sun. We will form a simplification model by ignoring all other gravitational forces on the probe from other planets. Our question is this: With what speed will the probe arrive at the surface of the Sun?

If we imagine trying to model the probe as a particle under a net force, we run into trouble. Newton's second law allows us to find the acceleration of the probe. But because the gravitational force is not constant, the acceleration is not constant. We haven't discussed a kinematics model for a particle under varying acceleration. Thus, the problem facing us is complicated. In theory, we could solve this problem of Newton's second law using calculus, but the energy approach gives us a different path to follow, which will be very fruitful for us in solving future problems.

We will model the probe as a system consisting of a single particle. The gravitational force from the Sun acts as the interaction with the environment, doing work on the probe. The result of this work is the increasing kinetic energy of the probe as it falls toward the Sun. (Internal energy also increases as the probe warms up on approach to the Sun, but this is due to electromagnetic radiation—this is a separate process and will not be addressed here.) Thus, this situation is a candidate for use of the work–kinetic energy theorem.

Let us first calculate the amount of work done by the gravitational force on the probe as it falls to the surface of the Sun. The force on the probe is given by Newton's law of universal gravitation. The initial separation of the probe and the Sun is the radius of the Earth's orbit, 1.50×10^{11} m. The final separation is the radius of the Sun, 6.96×10^8 m. We use Equation 6.11 in the radial direction from the Sun to find the work done by the varying gravitational force:

$$W = \int_{r_i}^{r_f} F_r \, dr = \int_{r_i}^{r_f} - G \frac{M_{\text{Sun}} m_{\text{probe}}}{r^2} \, dr = - GM_{\text{Sun}} m_{\text{probe}} \int_{r_i}^{r_f} \frac{1}{r^2} \, dr$$

$$= - GM_{\text{Sun}} m_{\text{probe}} \left(-\frac{1}{r} \right)\Big|_{r_i}^{r_f} = GM_{\text{Sun}} m_{\text{probe}} \left(\frac{1}{r_f} - \frac{1}{r_i} \right)$$

In this calculation, the radial component of the gravitational force has been substituted as a negative expression because the radial component vector $\mathbf{F}_r$ is toward the Sun, but the radial unit vector $\hat{\mathbf{r}}$ points away from the Sun.

Now, let us substitute the numerical values for the masses and the initial and final separation distances:

$$W = (6.67 \times 10^{-11} \, \text{N} \cdot \text{m}^2/\text{kg}^2)(1.99 \times 10^{30} \, \text{kg})(200 \, \text{kg})$$

$$\times \left(\frac{1}{6.96 \times 10^8 \, \text{m}} - \frac{1}{1.50 \times 10^{11} \, \text{m}} \right)$$

$$= 3.80 \times 10^{13} \, \text{J}$$

Finally, we set this work done on the probe equal to the change in the kinetic energy of the probe, according to the work–kinetic energy theorem:

$$W = \Delta K = \tfrac{1}{2} m_{\text{probe}} v_f^2 - \tfrac{1}{2} m_{\text{probe}} v_i^2 = \tfrac{1}{2} m_{\text{probe}} v_f^2 - 0 \longrightarrow v_f = \sqrt{\frac{2W}{m_{\text{probe}}}}$$

Substituting the numerical values, we have,

$$v_f = \sqrt{\frac{2(3.80 \times 10^{13} \, \text{J})}{200 \, \text{kg}}} = 6.16 \times 10^5 \, \text{m/s}$$

This solution is about as much effort as that necessary in finding the acceleration from Newton's second law and using calculus to find the final velocity. In the next chapter, however, we will see some types of problems that are very tedious using calculus but become extremely simple when the energy approach is used. In particular, we shall use an energy approach to find an upper limit to the speed with which we need to leave the Earth to arrive at Mars.

SUMMARY

A **system** can be a single particle, a collection of particles, or a region of space. A **system boundary** separates the system from the **environment.** Many physics problems can be solved by considering the interaction of a system with its environment.

The **work** done by a *constant* force **F** on a particle is defined as the product of the component of the force in the direction of the particle's displacement and the magnitude of the displacement. If **F** makes an angle θ with the displacement $\Delta\mathbf{r}$, the work done by **F** is

$$W \equiv F\,\Delta r\cos\theta \qquad [6.1]$$

The **scalar,** or **dot, product** of any two vectors **A** and **B** is defined by the relationship

$$\mathbf{A}\cdot\mathbf{B} \equiv AB\cos\theta \qquad [6.3]$$

where the result is a scalar quantity and θ is the angle between the directions of the two vectors. The scalar product obeys the commutative and distributive laws.

The scalar product allows us to write the work done by a constant force **F** on a particle as

$$W = \mathbf{F}\cdot\Delta\mathbf{r} \qquad [6.4]$$

The work done by a *varying* force acting on a particle moving along the x axis from x_i to x_f is

$$W \equiv \int_{x_i}^{x_f} F_x\,dx \qquad [6.11]$$

where F_x is the component of force in the x direction. If several forces act on the particle, the net work done by all forces is the sum of the individual amounts of work done by each force.

The **kinetic energy** of a particle of mass m moving with a speed v is

$$K = \tfrac{1}{2}mv^2 \qquad [6.18]$$

The **work–kinetic energy theorem** states that when work is done on a system and the only change in the system is in its speed, the net work done on the system by external forces equals the change in kinetic energy of the system:

$$W_{\text{net}} = K_f - K_i = \tfrac{1}{2}mv_f^2 - \tfrac{1}{2}mv_i^2 \qquad [6.19]$$

For a nonisolated system, we can equate the change in the total energy stored in the system to the sum of all the transfers across the system boundary:

$$\Delta E_{\text{system}} = \sum H \qquad [6.20]$$

This is the **continuity equation for energy.** Methods of energy transfer (H) include **work** $(H = W)$, **mechanical waves** (H_{MW}), **heat** $(H = Q)$, **matter transfer** (H_{MT}), **electrical transmission** (H_{ET}), and **electromagnetic radiation** (H_{ER}). Storage mechanisms (E_{system}) seen in this chapter include **kinetic energy** K and **internal energy** E_{int}. The continuity equation arises because **energy is conserved**—we can neither create nor destroy energy. The work–kinetic energy theorem is a special case of the continuity equation for energy in situations in which work is the only transfer mechanism and kinetic energy is the only type of energy storage in the system.

In the case of an object sliding through a displacement Δx over a surface with friction, the change in kinetic energy of the system is

$$\Delta K = -f_k\,\Delta x + \sum W_{\text{other forces}} \qquad [6.23]$$

where f_k is the force of kinetic friction and $\sum W_{\text{other forces}}$ is the work done by all forces other than friction.

Average power is the time rate of energy transfer. If we use work as the energy transfer mechanism,

$$\overline{\mathcal{P}} \equiv \frac{W}{\Delta t} \qquad [6.25]$$

If an agent applies a force **F** to an object moving with a velocity **v**, the **instantaneous power** delivered by that agent is

$$\mathcal{P} \equiv \frac{dW}{dt} = \mathbf{F}\cdot\mathbf{v} \qquad [6.27]$$

Because power is defined for any type of energy transfer, the general expression for power is

$$\mathcal{P} = \frac{dE}{dt} \qquad [6.28]$$

QUESTIONS

1. Explain why the work done by the force of sliding friction is negative when an object undergoes a displacement on a rough surface.

2. When a particle rotates in a circle, a force acts on it directed toward the center of rotation. Why is it that this force does no work on the particle?

3. When a punter kicks a football, is he doing any work on the ball while his toe is in contact with it? Is he doing any work on the ball after it loses contact with his toe? Are any forces doing work on the ball while it is in flight?

4. A team of furniture movers wishes to load a truck using a ramp from the ground to the rear of the truck. One of the movers claims that less work would be required to load the truck if the length of the ramp were increased, reducing the angle of the ramp with respect to the horizontal. Is his claim valid? Explain.

5. Cite two examples in which a force is exerted on an object without doing any work on the object.

6. As a simple pendulum swings back and forth, the forces acting on the suspended mass are the gravitational force, the tension in the supporting cord, and air resistance. (a) Which of these forces, if any, does no work on the pendulum? (b) Which of these forces does negative work at all times during its motion? (c) Describe the work done by the gravitational force while the pendulum is swinging.

7. If the dot product of two vectors is positive, does this imply that the vectors must have positive rectangular components?

8. A hockey player pushes a puck with his hockey stick over frictionless ice. If the puck starts from rest and ends up moving at speed v, has the hockey player performed any work on the puck in the reference frame of the ice? An ant is on the puck and hangs on tightly while the puck is accelerated. In the reference frame of the ant, has any work been done by the hockey stick on the puck? Is the work–kinetic energy theorem satisfied for each observer?

9. Can kinetic energy be negative? Explain.

10. One bullet has twice the mass of a second bullet. If both are fired so that they have the same speed, which has more kinetic energy? What is the ratio of the kinetic energies of the two bullets?

11. If the speed of a particle is doubled, what happens to its kinetic energy?

12. A spring is stretched by a distance x_{max}, resulting in work W being done by the external agent. The spring is now cut in half and stretched by the same distance. How much work is done in this case?

13. You are working in a library, reshelving books. You lift a book from the floor to the top shelf. The kinetic energy of the book on the floor was zero, and the kinetic energy of the book on the top shelf is zero, so no change occurs in kinetic energy. Yet you did some work in lifting the book. Is the work–kinetic energy theorem violated?

14. What can be said about the speed of a particle if the net work done on it is zero?

15. Can the average power over a time interval ever equal the instantaneous power at an instant within the interval? Explain.

16. In Example 6.9, does the required power increase or decrease as the force of friction is reduced?

17. Sometimes, physics words are used in popular literature in interesting ways. For example, consider a description of a rock falling from the top of a cliff as "gathering force as it falls to the beach below." What does the phrase "gathering force" mean, and can you repair this phrase?

PROBLEMS

1, 2, 3 = straightforward, intermediate, challenging   □ = full solution available in the *Student Solutions Manual and Study Guide*

web = solution posted at **http://www.harcourtcollege.com/physics/**   💻 = computer useful in solving problem

🌐 = Interactive Physics   ▨ = paired numerical/symbolic problems   ▨ = life science application

Section 6.2   Work Done by a Constant Force

1. A block of mass 2.50 kg is pushed 2.20 m along a frictionless horizontal table by a constant 16.0-N force directed 25.0° below the horizontal. Determine the work done on the block by (a) the applied force, (b) the normal force exerted by the table, and (c) the gravitational force. (d) Determine the total work done on the block.

2. A shopper in a supermarket pushes a cart with a force of 35.0 N directed at an angle of 25.0° downward from the horizontal. Find the work done by the shopper on the cart as he moves down an aisle 50.0 m long.

3. Batman, whose mass is 80.0 kg, is dangling on the free **web** end of a 12.0-m rope, the other end of which is fixed to a tree limb above. He is able to get the rope in motion as only Batman knows how, eventually getting it to swing enough that he can reach a ledge when the rope makes a 60.0° angle with the vertical. How much work is done by the gravitational force in this maneuver?

4. A raindrop of mass 3.35×10^{-5} kg falls vertically at constant speed under the influence of gravity and air resistance. After the drop has fallen 100 m, what is the work done on the raindrop (a) by gravity and (b) by air resistance?

Section 6.3   The Scalar Product of Two Vectorsz

5. Vector **A** has a magnitude of 5.00 units, and **B** has a magnitude of 9.00 units. The two vectors make an angle of 50.0° with each other. Find **A** · **B**.
6. For any two vectors **A** and **B**, show that $\mathbf{A} \cdot \mathbf{B} = A_x B_x + A_y B_y + A_z B_z$. (*Hint:* Write **A** and **B** in unit vector form and use Equations 6.7 and 6.8.)

> In problems 7 through 10, calculate numerical answers to three significant figures.

7. A force $\mathbf{F} = (6\mathbf{i} - 2\mathbf{j})$ N acts on a particle that undergoes a displacement $\Delta \mathbf{r} = (3\mathbf{i} + \mathbf{j})$ m. Find (a) the work done by the force on the particle and (b) the angle between **F** and $\Delta \mathbf{r}$.

web

8. Find the scalar product of the vectors in Figure P6.8.

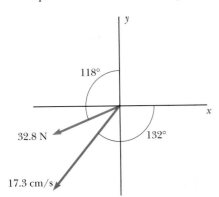

Figure P6.8

9. Using the definition of the scalar product, find the angles between (a) $\mathbf{A} = 3\mathbf{i} - 2\mathbf{j}$ and $\mathbf{B} = 4\mathbf{i} - 4\mathbf{j}$; (b) $\mathbf{A} = -2\mathbf{i} + 4\mathbf{j}$ and $\mathbf{B} = 3\mathbf{i} - 4\mathbf{j} + 2\mathbf{k}$; (c) $\mathbf{A} = \mathbf{i} - 2\mathbf{j} + 2\mathbf{k}$ and $\mathbf{B} = 3\mathbf{j} + 4\mathbf{k}$.
10. For $\mathbf{A} = 3\mathbf{i} + \mathbf{j} - \mathbf{k}$, $\mathbf{B} = -\mathbf{i} + 2\mathbf{j} + 5\mathbf{k}$, and $\mathbf{C} = 2\mathbf{j} - 3\mathbf{k}$, find $\mathbf{C} \cdot (\mathbf{A} - \mathbf{B})$.

Section 6.4   Work Done by a Varying Force

11. A particle is subject to a force F_x that varies with position as in Figure P6.11. Find the work done by the force on the object as it moves (a) from $x = 0$ to $x = 5.00$ m, (b) from $x = 5.00$ m to $x = 10.0$ m, and (c) from $x = 10.0$ m to $x = 15.0$ m. (d) What is the total work done by the force over the displacement $x = 0$ to $x = 15.0$ m?

web

12. The force (in newtons) acting on a particle is $F_x = (8.00x - 16.0)$, where x is in meters. (a) Make a plot of this force versus x from $x = 0$ to $x = 3.00$ m. (b) From your graph, find the net work done by this force on the particle as it moves from $x = 0$ to $x = 3.00$ m.

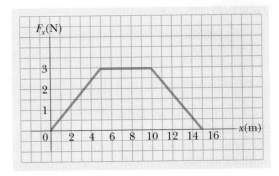

Figure P6.11   Problems 11 and 22.

13. When a 4.00-kg object is hung vertically on a certain light spring that obeys Hooke's law, the spring stretches 2.50 cm. If the 4.00-kg object is removed, (a) how far will the spring stretch if a 1.50-kg object is hung on it, and (b) how much work must an external agent do to stretch the same spring 4.00 cm from its unstretched position?
14. An archer pulls her bowstring back 0.400 m by exerting a force that increases uniformly from zero to 230 N. (a) What is the equivalent force constant of the bow? (b) How much work does the archer do in pulling the bow?
15. A 6 000-kg freight car rolls along rails with negligible friction. The car is brought to rest by a combination of two coiled springs, as illustrated in Figure P6.15. Both springs obey Hooke's law with $k_1 = 1\,600$ N/m and $k_2 =$

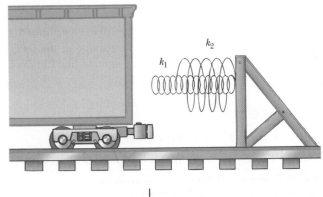

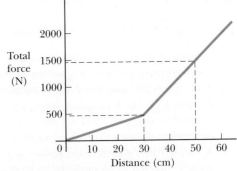

Figure P6.15

3 400 N/m. After the first spring compresses a distance of 30.0 cm, the second spring acts with the first to increase the force as additional compression occurs, as shown in the graph. If the car is brought to rest 50.0 cm after first contacting the two-spring system, find the car's initial speed.

16. A 100-g bullet is fired from a rifle having a barrel 0.600 m long. Assuming the origin is placed where the bullet begins to move, the force (in newtons) exerted on the bullet by the expanding gas is $15\,000 + 10\,000x - 25\,000x^2$, where x is in meters. (a) Determine the work done by the gas on the bullet as the bullet travels the length of the barrel. (b) If the barrel is 1.00 m long, how much work is done, and how does this value compare to the work calculated in part (a)?

17. A force (in newtons) $\mathbf{F} = (4x\mathbf{i} + 3y\mathbf{j})$ acts on an object as the object moves in the x direction from the origin to $x = 5.00$ m. Find the work done on the object by the force.

18. A small object of mass m is pulled to the top of a frictionless half-cylinder (of radius R) by a cord that passes over the top of the cylinder, as illustrated in Figure P6.18. (a) If the object moves at a constant speed, show that $F = mg\cos\theta$. (*Hint:* If the object moves at constant speed, the component of its acceleration tangent to the cylinder must be zero at all times.) (b) By directly integrating $W = \int \mathbf{F} \cdot d\mathbf{r}$, find the work done in moving the object at constant speed from the bottom to the top of the half-cylinder.

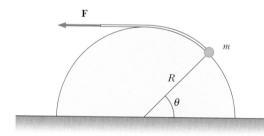

Figure P6.18

Section 6.5 Kinetic Energy and the Work–Kinetic Energy Theorem

19. A 0.600-kg particle has a speed of 2.00 m/s at point A and kinetic energy of 7.50 J at point B. What is (a) its kinetic energy at A? (b) Its speed at B? (c) The total work done on the particle as it moves from A to B?

20. A 0.300-kg ball has a speed of 15.0 m/s. (a) What is its kinetic energy? (b) If its speed were doubled, what would be its kinetic energy?

21. A 3.00-kg object has an initial velocity $\mathbf{v}_i = (6.00\mathbf{i} - 2.00\mathbf{j})$ m/s. (a) What is its kinetic energy at this time? (b) Find the total work done on the object if its velocity changes to $(8.00\mathbf{i} + 4.00\mathbf{j})$ m/s. (*Hint:* Remember that $v^2 = \mathbf{v} \cdot \mathbf{v}$.)

22. A 4.00-kg particle is subject to a total force that varies with position as shown in Figure P6.11. The particle starts from rest at $x = 0$. What is its speed at (a) $x = 5.00$ m, (b) $x = 10.0$ m, (c) $x = 15.0$ m?

23. A 2 100-kg pile driver is used to drive a steel I-beam into the ground. The pile driver falls 5.00 m before coming into contact with the top of the beam, and it drives the beam 12.0 cm farther into the ground before coming to rest. Using energy considerations, calculate the average force the beam exerts on the pile driver while the pile driver is brought to rest.

24. You can think of the work–kinetic energy theorem as a second theory of motion, parallel to Newton's laws in describing how outside influences affect the motion of an object. In this problem, work out parts (a) and (b) separately from parts (c) and (d) to compare the predictions of the two theories. In a rifle barrel, a 15.0-g bullet is accelerated from rest to a speed of 780 m/s. (a) Find the work done on the bullet. (b) If the rifle barrel is 72.0 cm long, find the magnitude of the average total force acted on it, as $F = W/(\Delta r \cos\theta)$. (c) Find the constant acceleration of a bullet that starts from rest and gains a speed of 780 m/s over a distance of 72.0 cm. (d) If the bullet has mass 15.0 g, find the total force that acted on it as $\Sigma F = ma$.

25. A block of mass 12.0 kg slides from rest down a frictionless $35.0°$ incline and is stopped by a strong spring with $k = 3.00 \times 10^4$ N/m. The block slides 3.00 m from the point of release to the point where it comes to rest against the spring. When the block comes to rest, how far has the spring been compressed?

26. In the neck of the picture tube of a certain black-and-white television set, an electron gun contains two charged metallic plates 2.80 cm apart. An electric force accelerates each electron in the beam from rest to 9.60% of the speed of light over this distance. (a) Determine the kinetic energy of the electron as it leaves the electron gun. Electrons carry this energy to a phosphorescent material on the inner surface of the television screen, making it glow. For an electron passing between the plates in the electron gun, determine (b) the magnitude of the constant force acting on the electron, (c) the acceleration, and (d) the time of flight.

27. An Atwood machine (see Fig. 4.13a) supports objects with masses of 0.200 kg and 0.300 kg. The objects are held at rest beside each other and then released. Ignoring friction, what is the speed of each object at the instant it has moved 0.400 m?

Section 6.7 Situations Involving Kinetic Friction

28. A sledge loaded with bricks has a total mass of 18.0 kg and is pulled at constant speed by a rope. The rope is inclined at $20.0°$ above the horizontal, and the cart moves a distance of 20.0 m on a horizontal surface. The coefficient of

kinetic friction between the sledge and surface is 0.500. (a) What is the tension in the rope? (b) How much work is done on the sledge by the rope? (c) What is the increase in internal energy in the sledge and the surface due to friction?

29. A 40.0-kg box initially at rest is pushed 5.00 m along a rough, horizontal floor with a constant applied horizontal force of 130 N. If the coefficient of friction between box and floor is 0.300, find (a) the work done by the applied force, (b) the increase in internal energy due to friction, (c) the work done by the normal force, (d) the work done by gravity, (e) the change in kinetic energy of the box, and (f) the final speed of the box.

30. A 15.0-kg block is dragged over a rough, horizontal surface by a 70.0-N force acting at 20.0° above the horizontal. The block is displaced 5.00 m, and the coefficient of kinetic friction is 0.300. Find the work done by (a) the 70-N force, (b) the normal force, and (c) the gravitational force. (d) What is the increase in internal energy due to friction? (e) Find the total change in the block's kinetic energy.

31. A sled of mass m is given a kick on a frozen pond. The kick imparts to it an initial speed of 2.00 m/s. The coefficient of kinetic friction between sled and ice is 0.100. Use energy considerations to find the distance the sled moves before it stops.

32. A 2.00-kg block is attached to a spring of force constant 500 N/m as in Figure 6.9. The block is pulled 5.00 cm to the right of equilibrium and released from rest. Find the speed of the block as it passes through equilibrium if (a) the horizontal surface is frictionless and (b) the coefficient of friction between block and surface is 0.350.

33. A crate of mass 10.0 kg is pulled up a rough incline with an initial speed of 1.50 m/s. The pulling force is 100 N parallel to the incline, which makes an angle of 20.0° with the horizontal. The coefficient of kinetic friction is 0.400, and the crate is pulled 5.00 m. (a) How much work is done by gravity? (b) Determine the increase in internal energy due to friction. (c) How much work is done by the 100-N force? (d) What is the change in kinetic energy of the crate? (e) What is the speed of the crate after being pulled 5.00 m?

Section 6.8   Power

34. Make an order-of-magnitude estimate of the power a car engine contributes to speeding the car up to highway speed. As a concrete example, consider your own car if you use one. In your solution state the physical quantities you take as data and the values you measure or estimate for them. The mass of the vehicle is given in the owner's manual. If you do not wish to estimate for a car, consider a bus or truck that you specify.

35. A 700-N Marine in basic training climbs a 10.0-m vertical rope at a constant speed in 8.00 s. What is his power output?

36. A certain automobile engine delivers 2.24×10^4 W (30.0 hp) to its wheels when moving at a constant speed of 27.0 m/s (≈ 60 mi/h). What is the resistive force acting on the automobile at that speed?

37. A skier of mass 70.0 kg is pulled up a slope by a motor-driven cable. (a) How much work is required to pull him a distance of 60.0 m up a 30.0° slope (assumed frictionless) at a constant speed of 2.00 m/s? (b) A motor of what power is required to perform this task?

38. A 650-kg elevator starts from rest. It moves upward for 3.00 s with constant acceleration until it reaches its cruising speed of 1.75 m/s. (a) What is the average power of the elevator motor during this period? (b) How does this power compare with the motor power when the elevator moves at its cruising speed?

39. Energy is conventionally measured in Calories as well as in joules. One Calorie in nutrition is equal to one kilocalorie, which we will define in Chapter 17 as 1 kcal = 4186 J. Metabolizing one gram of fat can release 9.00 kcal. A student decides to try to lose weight by exercising. She plans to run up and down the stairs in a football stadium as fast as she can and as many times as necessary. Is this in itself a practical way to lose weight? To evaluate the program, suppose she runs up a flight of 80 steps, each 0.150 m high, in 65.0 s. For simplicity, ignore the energy she uses in coming down (this is small). Assume that a typical efficiency for human muscles is 20.0%. This means that when your body converts 100 J from metabolizing fat, 20 J goes into doing mechanical work (climbing stairs). The remainder leaves the body by heat. Assume the student's mass is 50.0 kg. (a) How many times must she run the flight of stairs to lose 1 lb of fat? (b) What is her average power output, in watts and in horsepower, as she runs up the stairs?

40. For saving energy, bicycling and walking are far more efficient means of transportation than is travel by automobile. For example, when riding at 10.0 mi/h a cyclist uses food energy at a rate of about 400 kcal/h above what he would use if merely sitting still. (In exercise physiology, power is often measured in kilocalories per hour rather than in watts. Here 1 kcal = 1 dietary Calorie = 4186 J.) Walking at 3.00 mi/h requires about 220 kcal/h. It is interesting to compare these values with the energy consumption required for travel by car. Gasoline yields a value of about 1.30×10^8 J/gal. Find the fuel economy in equivalent miles per gallon for a person (a) walking and (b) bicycling.

Section 6.9   Context Connection — A Probe to the Sun

41. A payload is fired from the Earth's surface with an initial speed of 2.00×10^4 m/s. What will its speed be when it is very far from the Earth? (Ignore friction.)

42. (a) Determine the amount of work that must be done on a 100-kg object to elevate it to a height of 1 000 km above the Earth's surface. (b) Determine the amount of additional work required to put the object into a circular orbit at this elevation.

Additional Problems

43. A baseball outfielder throws a 0.150-kg baseball at a speed of 40.0 m/s and an initial angle of 30.0°. What is the kinetic energy of the baseball at the highest point of its trajectory?

44. The direction of an arbitrary vector **A** can be completely specified with the angles α, β, and γ that the vector makes with the x, y, and z axes, respectively. If $\mathbf{A} = A_x\mathbf{i} + A_y\mathbf{j} + A_z\mathbf{k}$, (a) find expressions for $\cos\alpha$, $\cos\beta$, and $\cos\gamma$ (these are known as *direction cosines*), and (b) show that these angles satisfy the relation $\cos^2\alpha + \cos^2\beta + \cos^2\gamma = 1$. (*Hint:* Take the scalar product of **A** with **i**, **j**, and **k** separately.)

45. A 4.00-kg particle moves along the x axis. Its position varies with time according to $x = t + 2.0t^3$, where x is in meters and t is in seconds. Find (a) the kinetic energy at any time t, (b) the acceleration of the particle and the force acting on it at time t, (c) the power being delivered to the particle at time t, and (d) the work done on the particle in the interval $t = 0$ to $t = 2.00$ s.

46. A traveler at an airport takes an escalator up one floor, as in Figure P6.46. The moving staircase would itself carry him upward with vertical velocity component v between entry and exit points separated by height h. However, while the escalator is moving, the hurried traveler climbs the steps of the escalator at a rate of n steps/s. Assume that the height of each step is h_s. (a) Determine the amount of work done by the traveler during his escalator ride, given that his mass is m. (b) Determine the work the escalator motor does on this person.

47. When a certain spring is stretched beyond its proportional limit, the restoring force of the spring satisfies the equation $F = -kx + \beta x^3$. If $k = 10.0$ N/m and $\beta = 100$ N/m^3, calculate the work done by this force when the spring is stretched 0.100 m.

48. In a control system, an accelerometer consists of a 4.70-g object sliding on a low-friction horizontal rail. A low-mass spring attaches the object to a flange at one end of the rail. When subject to a steady acceleration of $0.800g$, the mass is to assume a location 0.500 cm away from its equilibrium position. Find the force constant required for the spring.

49. A single constant force **F** acts on a particle of mass m. The particle starts at rest at $t = 0$. (a) Show that the instantaneous power delivered by the force at any time t is $(F^2/m)t$. (b) If $F = 20.0$ N and $m = 5.00$ kg, what is the power delivered at $t = 3.00$ s?

50. A particle is attached between two identical springs on a horizontal frictionless table. Both springs have force constant k and are initially unstressed. (a) If the particle is pulled a distance x along a direction perpendicular to the initial configuration of the springs, as in Figure P6.50, show that the force exerted on the particle by the springs is

$$\mathbf{F} = -2kx\left(1 - \frac{L}{\sqrt{x^2 + L^2}}\right)\mathbf{i}$$

(b) Determine the amount of work done by this force in moving the particle from $x = A$ to $x = 0$.

Figure P6.46 *(Ron Chapple/FPG)*

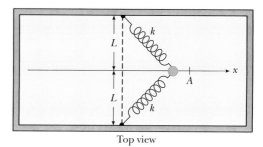

Top view

Figure P6.50

51. A 200-g block is pressed against a spring of force constant 1.40 kN/m until the block compresses the spring 10.0 cm. The spring rests at the bottom of a ramp inclined at 60.0° to the horizontal. Using energy considerations, determine

how far up the incline the block moves before it stops (a) if there is no friction between the block and the ramp and (b) if the coefficient of kinetic friction is 0.400.

52. When different weights are hung on a spring, the spring stretches to different lengths as shown in the following table. (a) Make a graph of the applied force versus the extension of the spring. By least-squares fitting, determine the straight line that best fits the data. (You may not want to use all the data points.) (b) From the slope of the best fit line, find the force constant k. (c) If the spring is extended to 105 mm, what force does it exert on the suspended weight?

F (N)	L (mm)
2.0	15
4.0	32
6.0	49
8.0	64
10	79
12	98
14	112
16	126
18	149
20	175
22	190

53. **Review Problem.** Two constant forces act on a 5.00-kg object moving in the *xy* plane, as shown in Figure P6.53. Force $\mathbf{F}_1$ is 25.0 N at 35.0°, while $\mathbf{F}_2$ = 42.0 N at 150°. At time $t = 0$, the object is at the origin and has velocity $4.00\mathbf{i} + 2.50\mathbf{j}$) m/s. (a) Express the two forces in unit-vector notation. Use unit-vector notation for your other answers. (b) Find the total force on the object. (c) Find the object's acceleration. Now, considering the instant $t = 3.00$ s, (d) find the object's velocity, (e) its position, (f) its kinetic energy from $\frac{1}{2}mv_f^2$, and (g) its kinetic energy from

$$\tfrac{1}{2}mv_i^2 + \sum \mathbf{F}\cdot\Delta\mathbf{r}$$

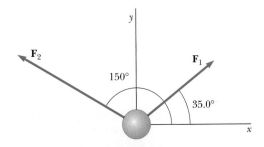

Figure P6.53

54. A 0.400-kg particle slides around a horizontal track. The track has a smooth vertical outer wall forming a circle with a radius of 1.50 m. The particle is given an initial speed of

8.00 m/s. After one revolution, its speed has dropped to 6.00 m/s because of friction with the rough floor of the track. (a) Find the energy transformed to internal energy due to friction in one revolution. (b) Calculate the coefficient of kinetic friction. (c) What is the total number of revolutions the particle makes before stopping?

55. The ball launcher in a pinball machine has a spring with a force constant of 1.20 N/cm (Fig. P6.55). The surface on which the ball moves is inclined 10.0° with respect to the horizontal. If the spring is initially compressed 5.00 cm, find the launching speed of a 100-g ball when the plunger is released. Friction and the mass of the plunger are negligible.

Figure P6.55

56. In diatomic molecules, the constituent atoms exert attractive forces on one another at large distances and repulsive forces at short distances. For many molecules, the Lennard–Jones law is a good approximation to the magnitude of these forces:

$$F = F_0\left[2\left(\frac{\sigma}{r}\right)^{13} - \left(\frac{\sigma}{r}\right)^{7}\right]$$

where r is the center-to-center distance between the atoms in the molecule, σ is a length parameter, and F_0 is the force when $r = \sigma$. For an oxygen molecule, the force is $F_0 = 9.60 \times 10^{-11}$ N and $\sigma = 3.50 \times 10^{-10}$ m. Determine the work done by this force if the atoms are pulled apart from $r = 4.00 \times 10^{-10}$ m to $r = 9.00 \times 10^{-10}$ m.

57. A horizontal string is attached to a 0.250-kg object lying on a rough, horizontal table. The string passes over a light frictionless pulley and a 0.400-kg object is then attached to its free end. The coefficient of sliding friction between the 0.250-kg object and table is 0.200. Using the work–kinetic energy theorem, determine (a) the speed of the objects after each has moved 20.0 m from rest and (b) the mass that must be added to the 0.250-kg object so that, given an initial velocity, the objects continue to move at a constant speed. (c) What mass must be removed from the 0.400-kg object so that the same constant-speed outcome as in part (b) is achieved?

58. As it plows a parking lot, a snowplow pushes an ever-growing pile of snow in front of it. Suppose a car moving through the air is similarly modeled as a cylinder pushing a growing plug of air in front of it. The originally stationary air is set into motion at the constant speed v of the cylinder, as in Figure P6.58. In a time Δt, a new disk of air

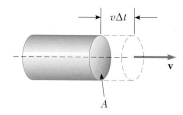

Figure P6.58

of mass Δm must be moved a distance $v\Delta t$ and hence must be given a kinetic energy $\frac{1}{2}(\Delta m)v^2$. Using this model, show that the automobile's power loss due to air resistance is $\frac{1}{2}\rho A v^3$ and that the resistive force acting on the car is

$\frac{1}{2}\rho A v^2$, where ρ is the density of air. Compare this result with the empirical expression $\frac{1}{2}D\rho A v^2$ for the resistive force.

59. A particle moves along the x axis from $x = 12.8$ m to $x = 23.7$ m under the influence of a force

$$F = \frac{375}{x^3 + 3.75x}$$

where F is in newtons and x is in meters. Using numerical integration, determine the total work done by this force during this displacement. Your result should be accurate to within 2%.

ANSWERS TO QUICK QUIZZES

6.1 Because no motion is taking place, the point of application of the forces from the pullers experiences no displacement. Thus, no work is done on the rope. For the same reason, no work is being done on the pullers or the ground. Work is being done only within the bodies of the pullers. For example, the heart of each puller is applying forces on the blood to move it through the body.

6.2 (c), (a), (d), (b). The work in (c) is positive and of the largest possible value because the angle between the force and the displacement is zero. The work done in (a) is zero because the force is perpendicular to the displacement. In (d) and (b), negative work is done by the applied force because in both cases there is a component of the force opposite the direction of the displacement. Situation (b) is the most negative value because the angle between the force and the displacement is 180°.

6.3 (d). Choice (a) represents something that cannot be true because the value of $\cos \theta$ cannot be larger than 1. For choice (b), although the dot product *can* be smaller than AB, it is not *necessarily* smaller. Part (c) cannot be true for the same reason as part (a). Choice (d) is true if the angle between **A** and **B** is zero.

6.4 (a) Enter: electrical transmission, electromagnetic radiation (for an antenna-based set); Leave: electromagnetic radiation (from the screen), mechanical waves (sound from the speaker); heat (from warm surfaces into the air). (b) Enter: matter transfer (gasoline); Leave: mechanical waves (sound), work (on grass), heat (from warm surfaces into the air). (c) Enter: work (when you crank it); Leave: mechanical waves (sound), work (on pencil).

6.5 The original kinetic energy of your car has been transformed to internal energy, located primarily in the brakes. Because your brakes are warmer than the surrounding air, the energy will leave the brakes by heat, going into the air and warming it. Eventually, this energy spreads throughout the atmosphere!

6.6 Because the time periods are the same for both cars, we need only compare the energy transferred from the engine. Because the sports car is moving twice as fast as the older car at the end of the time interval, it has four times the kinetic energy. Thus, four times as much energy was transferred from the engine, and the engine must have expended four times the power.

A strobe photograph of a pole
vaulter. During this process,
several types of energy
transformations occur.
*(©Harold E. Edgerton/Courtesy of Palm
Press, Inc.)*

Potential Energy

I n Chapter 6 we introduced the concepts of kinetic energy, which is associated
with the motion of an object or a particle, and internal energy, which is associated with the temperature of a system. In this chapter we introduce another
form of energy, called *potential energy,* which is associated with the configuration of a system of two or more interacting objects or particles. This new type of
energy will provide us with a powerful and universal fundamental principle.

7.1 • POTENTIAL ENERGY OF A SYSTEM

In Chapter 6, we defined a system in general but focused our attention on single
particles under the influence of an external force. In this chapter, we will consider
systems of two or more objects or particles interacting via a force that is *internal* to
the system. The kinetic energy of such a system is the algebraic sum of the kinetic
energies of all members of the system. In some systems, however, one object may be
so massive that it can be modeled as stationary and its kinetic energy can be ignored. For example, if we consider a ball–Earth system as the ball falls to the

ground, the kinetic energy of the system can be considered as only the kinetic energy of the ball. The Earth moves so slowly in this process that we can ignore its kinetic energy. (We will justify this in Chapter 8.) On the other hand, the kinetic energy of a system of two electrons must include the kinetic energies of both particles.

Let us imagine a system consisting of a book and the Earth, interacting via the gravitational force. We do some work on the system by lifting the book slowly through a height $\Delta y = y_b - y_a$, as in Figure 7.1. According to the continuity equation for energy (Eq. 6.20) introduced in Chapter 6, this work on the system must appear as an increase in energy of the system. The book is at rest before we perform the work and is at rest after we perform the work; thus, the kinetic energy of the system does not change. There is no reason the temperature of the book or the Earth should change, so the internal energy of the system experiences no change.

Because the energy change of the system is not in the form of kinetic energy or internal energy, it must appear as some other form of energy storage. After lifting the book, we could release it and let it fall to the ground. Notice that the book (and, therefore, the system) will now have kinetic energy, and its origin is in the work that was done in lifting the book. While the book was at the highest point, the energy of the system had the *potential* to become kinetic energy, but did not do so until the book was allowed to fall. Thus, we call the energy storage mechanism before we release the book **potential energy.** We will find that a potential energy can be associated with a number of types of forces. In this particular case, we are discussing **gravitational potential energy.**

Let us now derive an expression for the gravitational potential energy associated with an object at a given location above the surface of the Earth. To do this, consider an external agent lifting an object of mass m from an initial height y_a above the ground to a final height y_b as in Figure 7.1. We assume that the lifting is done slowly, with no acceleration, so that the lifting force can be modeled as equal to the weight of the object—the object is in equilibrium and moving at constant velocity. The work done by the external agent on the system (object and Earth) as the object undergoes this upward displacement is given by the product of the upward force $\mathbf{F} = m\mathbf{g}$ and the displacement $\Delta \mathbf{r} = \Delta y \mathbf{j}$:

$$W = (m\mathbf{g}) \cdot \Delta \mathbf{r} = (mg\mathbf{j}) \cdot [(y_b - y_a)\mathbf{j}] = mgy_b - mgy_a \qquad [7.1]$$

Notice how similar this equation is to Equation 6.17 in the preceding chapter. In each equation, the work done on a system equals a difference between the final and initial values of a quantity. Of course, both equations are nothing more than special cases of the continuity equation for energy. In Equation 6.17, the work represents a transfer of energy into the system, and the increase in energy of the system is kinetic in form. In Equation 7.1, the work represents a transfer of energy into the system, and the system energy appears in a different form, which we have called gravitational potential energy.

Thus, we can represent the quantity mgy to be the gravitational potential energy U_g:

$$U_g \equiv mgy \qquad [7.2]$$

The units of gravitational potential energy are joules, the same as those of work and kinetic energy. Potential energy, like work and kinetic energy, is a scalar quantity. Note that Equation 7.2 is valid only for objects near the surface of the Earth, where g is approximately constant.

See the *Core Concepts in Physics CD-ROM*, Screen 5.3

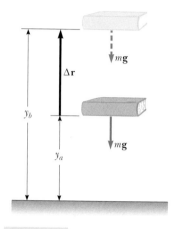

Figure 7.1

The work done by an external agent on the system of the book and the Earth as the book is lifted from y_a to y_b is equal to $mgy_b - mgy_a$.

PITFALL PREVENTION 7.1

Potential energy belongs to a system

Keep in mind that potential energy is always associated with a *system* of two or more interacting objects. In the gravitational case, in which a small object moves near the surface of the Earth, we may sometimes refer to the potential energy "associated with the object" rather than the more proper "associated with the system," because the Earth does not move significantly. We will not, however, refer to the potential energy "of the object" because this clearly ignores the role of the Earth in the potential energy.

• *Gravitational potential energy*

Using our definition of gravitational potential energy, we can now rewrite Equation 7.1 as

$$W = \Delta U_g$$

which mathematically describes that fact that the work done on the system in this situation appears as a change in the gravitational potential energy of the system.

The gravitational potential energy depends only on the vertical height of the object above the surface of the Earth. Thus, the same amount of work is done on an object–Earth system whether the object is lifted vertically from the Earth or it starts at the same point and is pushed up a frictionless incline, ending up at the same height. This can be shown in a mathematical representation by reperforming the work calculation in Equation 7.1 with a displacement having both vertical and horizontal components:

$$W = (m\mathbf{g}) \cdot \Delta \mathbf{r} = (mg\mathbf{j}) \cdot [(x_b - x_a)\mathbf{i} + (y_b - y_a)\mathbf{j}] = mgy_b - mgy_a$$

where no term involving x appears in the final result because $\mathbf{j} \cdot \mathbf{i} = 0$.

In solving problems, it is necessary to choose a reference location at which to set the gravitational potential energy equal to some reference value, which is normally zero. The choice of zero level is completely arbitrary because the important quantity is the *difference* in potential energy, and this difference is independent of the choice of zero level.

It is often convenient to choose the surface of the Earth as the reference position for zero potential energy, but this is not essential. Often, the statement of the problem suggests a convenient level to use.

Quick Quiz 7.1

Can the gravitational potential energy of a system be negative?

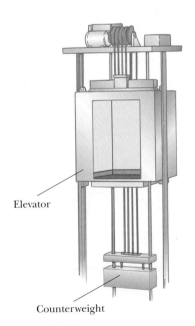

Elevator

Counterweight

Figure 7.2

(Thinking Physics 7.1)

THINKING PHYSICS 7.1

Elevators that are lifted by a cable have a counterweight, as shown in Figure 7.2. The elevator and the counterweight are connected by a cable that passes over a drive wheel. The counterweight moves in the direction opposite to that of the elevator. What is the purpose of the counterweight?

Reasoning Without the counterweight, the elevator motor would need to lift the weight of the elevator and the people riding it. This would represent a large increase in gravitational potential energy of the elevator–Earth system, which means a large transfer of energy to the system by work from the motor. With the counterweight, as the elevator moves upward, the counterweight moves down. Thus, the net change in separation distance between the total mass of the elevator–counterweight combination and the Earth is small. This results in a much smaller change in the system's potential energy than without the counterweight and less work from the motor.

EXERCISE What is the gravitational potential energy, relative to the ground, associated with a 0.15-kg baseball at the top of a 100-m-tall building?

Answer 147 J

7.2 • THE ISOLATED SYSTEM

See Screen 5.9

The introduction of potential energy allows us to generate a powerful and universally applicable principle for solving problems that are difficult to solve with Newton's laws. Let us develop this new principle by thinking about the book–Earth system in Figure 7.1 again. After we have lifted the book, there is gravitational potential energy stored in the system, which we can calculate from the work done by the external agent on the system, using $W = \Delta U_g$.

Let us now shift our focus to the work done on the book alone (Fig. 7.3) as it falls back to its original height under the influence of the gravitational force alone. As the book falls from y_b to y_a, the work done by the gravitational force on the book is

$$W_{\text{on book}} = (m\mathbf{g}) \cdot \Delta \mathbf{r} = (-mg\mathbf{j}) \cdot (y_a - y_b)\mathbf{j} = mgy_b - mgy_a \quad [7.3]$$

From the work–kinetic energy theorem of Chapter 6, the work done on the book is also

$$W_{\text{on book}} = \Delta K_{\text{book}}$$

Therefore, equating these two expressions for the work done on the book

$$\Delta K_{\text{book}} = mgy_b - mgy_a \quad [7.4]$$

Now, let us relate each side of this equation to the *system* of the book and the Earth. For the right-hand side,

$$mgy_b - mgy_a = -(mgy_a - mgy_b) = -(U_f - U_i) = -\Delta U_g$$

where U_g is the gravitational potential energy of the system. Because the book is the only part of the system that is moving, the left-hand side of Equation 7.4 becomes

$$\Delta K_{\text{book}} = \Delta K$$

where K is the kinetic energy of the system. Thus, replacing each side of Equation 7.4 with its system equivalent, the equation becomes

$$\Delta K = -\Delta U_g \quad [7.5]$$

This equation can be manipulated to provide a very important result for a new analysis model. First, we bring the change in potential energy to the left side of the equation:

$$\Delta K + \Delta U_g = 0$$

Notice that this is in the form of the continuity equation. On the left, we have a sum of changes of the energy stored in the system. The right-hand side in the continuity equation is the sum of the transfers across the boundary of the system. This is equal to zero in the present case because our book–Earth system is isolated from the environment.

Let us now write the changes in energy explicitly:

$$(K_f - K_i) + (U_f - U_i) = 0 \longrightarrow \boxed{K_f + U_f = K_i + U_i} \quad [7.6]$$

We define the sum of kinetic and potential energies as **mechanical energy:**

$$E_{\text{mech}} = K + U_g$$

Thus, Equation 7.6 is a statement of **conservation of mechanical energy** for an **isolated system.** An isolated system is one for which no energy transfers occur

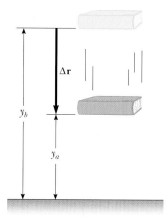

Figure 7.3

The work done by the gravitational force on the book as the book falls from y_b to y_a is equal to $mgy_b - mgy_a$.

PITFALL PREVENTION 7.2

Mechanical energy in an isolated system

Two comments are needed here. First, the isolated system model goes far beyond Equation 7.6. This is only the *mechanical energy* version of this model. We will see shortly how to include internal energy. In later chapters, we will see other isolated systems and generate new versions (and associated equations) related to such quantities as momentum, angular momentum, and electric charge. Second, Equation 7.6 is true for only one of two categories of forces, called *conservative forces,* which is discussed in the next section.

- *Conservation of mechanical energy for an isolated system*

Twin Falls on the Island of Kauai, Hawaii. The gravitational potential energy of the water–Earth system when the water is at the top of the falls is converted into kinetic energy as the water descends. *(Bruce Byers, FPG)*

across the boundary. Thus, the energy in the system is conserved—the sum of the kinetic and potential energies remains constant.

For the gravitational situation that we are describing, the conservation of mechanical energy equation can be written as

$$\tfrac{1}{2}mv_f^2 + mgy_f = \tfrac{1}{2}mv_i^2 + mgy_i$$

As the book falls to the Earth, the book–Earth system loses potential energy and gains kinetic energy, such that the total of the two types of energy always remains constant.

We will see other types of potential energy besides gravitational, so we can write the general form of the definition for mechanical energy without a subscript on *U*:

$$E_{mech} \equiv K + U \qquad\qquad [7.7]$$

Quick **Quiz** 7.2

Three identical balls are thrown from the top of a building, all with the same initial speed. The first ball is thrown horizontally, the second at some angle above the horizontal, and the third at some angle below the horizontal as in Figure 7.4. Ignoring air resistance, rank the speeds of the balls as they reach the ground.

Figure 7.4

(Quick Quiz 7.2) Three identical balls are thrown with the same initial speed from the top of a building.

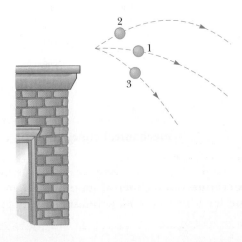

THINKING PHYSICS 7.2

You have graduated from college and are designing roller coasters for a living. You design a roller coaster in which a car is pulled to the top of a hill of height h and then, starting from a momentary rest, rolls freely down the hill and upward toward the peak of the next hill, which is at height $1.1h$. Will you have a long career in this business?

Reasoning   Your career will probably not be long, because this roller coaster will not work! At the top of the first hill, the roller coaster car has no kinetic energy, and the gravitational potential energy of the car–Earth system is that associated with a height h. If the car were to reach the top of the next hill, the system would have higher potential energy, that associated with height $1.1h$. This would violate the principle of conservation of mechanical energy. If this coaster were actually built, the car would move upward on the second hill to a height h (ignoring the effects of friction), stop short of the peak, and then start rolling backward, becoming trapped between the two hills.

Example 7.1   Ball in Free-Fall

A ball of mass m is dropped from rest at a height h above the ground, as in Figure 7.5. (a) Determine the speed of the ball when it is at a height y above the ground.

Reasoning   We will consider the ball and the Earth as the system. We adopt the simplification model in which the ball encounters no air resistance. Thus, the ball and the Earth do not experience any forces from the environment, and we can use the principle of conservation of mechanical energy. Initially, the system has potential energy and no kinetic energy.

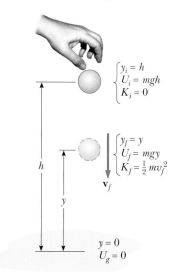

Figure 7.5

(Example 7.1) A ball is dropped from rest at a height h above the ground. Initially, the total energy of the ball–Earth system is gravitational potential energy, equal to mgh relative to the ground. At the elevation y, the total system energy is the sum of kinetic and potential energies.

As the ball falls, the total mechanical energy of the system (the sum of kinetic and potential energies) remains constant and equal to its initial potential energy. The potential energy of the system decreases, and the kinetic energy of the system (which is due only to the ball) increases.

Solution   Before the ball is released from rest at a height h above the ground, the kinetic energy of the system is $K_i = 0$ and the potential energy is $U_i = mgh$, where the y coordinate is measured from ground level. When the ball is at a distance y above the ground, its kinetic energy is $K_f = \frac{1}{2}mv_f^2$ and the potential energy of the system is $U_f = mgy$. Applying Equation 7.6, we have

$$K_i + U_i = K_f + U_f$$
$$0 + mgh = \tfrac{1}{2}mv_f^2 + mgy$$
$$v_f^2 = 2g(h - y)$$
$$v_f = \sqrt{2g(h - y)}$$

(b) Determine the speed of the ball at y if it is given an initial speed v_i at the initial altitude h.

Solution   In this case, the initial energy includes kinetic energy of the ball equal to $\frac{1}{2}mv_i^2$ and Equation 7.6 gives

$$\tfrac{1}{2}mv_i^2 + mgh = \tfrac{1}{2}mv_f^2 + mgy$$
$$v_f^2 = v_i^2 + 2g(h - y)$$
$$v_f = \sqrt{v_i^2 + 2g(h - y)}$$

This result is consistent with Equation 2.12 (Chapter 2), $v_{yf}^2 = v_{yi}^2 - 2g(y_f - y_i)$, for a particle under constant acceleration, where $y_i = h$. Furthermore, this result is valid even if the initial velocity is at an angle to the horizontal (the projectile situation), as discussed in Quick Quiz 7.2.

Example 7.2   One Way to Lift an Object

Two blocks are connected by a massless cord that passes over two frictionless pulleys, as in Figure 7.6. One end of the cord is attached to an object of mass $m_1 = 3.00$ kg that is a distance $R = 1.20$ m from the pulley on the left. The other end of the cord is connected to a block of mass $m_2 = 6.00$ kg resting on a table. From what angle θ (measured from the vertical) must the 3.00-kg mass be released in order to just lift the 6.00-kg block off the table?

Reasoning   Several models and concepts are needed to solve this problem. First, we consider the 3.00-kg object and the Earth as an isolated system. Even though an external force—that of the cord—is exerted on the 3.00-kg object, this force is always perpendicular to the displacement during the swing of the object, so no work is done across the boundary of the system by this force. Thus, we use conservation of mechanical energy to find the speed of the 3.00-kg object at the bottom of the circular path as a function of θ and the radius of the path. (In Example 5.9, we found that the tension in the cord is largest as the object passes through the lowest point.) Next, we use the particle under a net force model on the 3.00-kg object at the bottom of its path. At this position, Newton's second law will allow us to find the tension force exerted by the cord as a function of the given parameters. This force is also applied to the 6.00-kg block through the string. We apply the particle under a net force model again to the block. We note that the block lifts off the table when the force exerted on it by the cord exceeds the gravitational force acting on the block. This procedure enables us to find the required angle for the 3.00-kg object.

Solution   Applying conservation of energy to the system of the 3.00-kg object and the Earth gives

$$K_i + U_i = K_f + U_f$$

$$(1) \qquad 0 + m_1 g y_i = \tfrac{1}{2} m_1 v^2 + 0$$

where v is the speed of the object at the bottom of its path. (*Note:* $K_i = 0$ because the object starts from rest and $U_f = 0$ because we define the potential energy of the system as zero when the object is at the bottom of the circle.) From the geometry in Figure 7.6, we see that $y_i = R - R\cos\theta = R(1 - \cos\theta)$. Using this relation in (1) gives

$$(2) \qquad v^2 = 2gR(1 - \cos\theta)$$

Now we apply Newton's second law to the 3.00-kg object when it is at the bottom of the circular path:

$$T - m_1 g = m_1 \frac{v^2}{R}$$

$$(3) \qquad T = m_1 g + m_1 \frac{v^2}{R}$$

Notice that we have also used the centripetal acceleration equation for a particle in uniform circular motion here. The speed of the object is not uniform, but we can use the expression for the instantaneous acceleration at any speed.

This same tension force is transmitted to the 6.00-kg block, and if it is to be just lifted off the table, the normal force on it becomes zero. In this situation, Newton's second law gives us $T - m_2 g = 0$ or $T = m_2 g$. Using this condition, together with (2) and (3), gives

$$m_2 g = m_1 g + 2 m_1 g(1 - \cos\theta)$$

Solving for $\cos\theta$ and substituting in the given parameters, we find

$$\cos\theta = \frac{3m_1 - m_2}{2m_1} = \frac{3(3.00 \text{ kg}) - 6.00 \text{ kg}}{2(3.00 \text{ kg})} = 0.500$$

$$\theta = \boxed{60.0°}$$

EXERCISE   If the initial angle is $\theta = 40.0°$, find the speed of the 3.00-kg object and the tension in the cord when the 3.00-kg object is at the bottom of its circular path.

Answer   2.35 m/s; 43.2 N

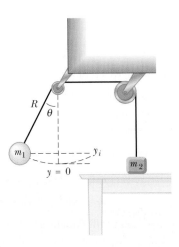

Figure 7.6

(Example 7.2)

7.3 • CONSERVATIVE AND NONCONSERVATIVE FORCES

In the preceding section, we showed that the mechanical energy of a system is conserved in a process in which the force between members of the system is the gravitational force. The gravitational force is one example of a category of forces for which the mechanical energy of a system is conserved. These are called **conservative forces.** The other possibility for energy storage in a system besides kinetic and potential is internal energy. Thus, a conservative force, for our purposes in mechanics, is a **force between members of a system that causes no transformation of mechanical energy to internal energy within the system.** If no energy is transformed to internal energy, the mechanical energy of the system is conserved, as described by Equation 7.6.

If a force is conservative, the work done by such a force has a special property as the members of the system move in response either to the force itself or to an external force: **The work done by a conservative force does not depend on the path followed by the members of the system, and depends only on the initial and final configurations of the system.**

• *Definition of a conservative force*

From this property, it follows that: **The work done by a conservative force when a member of the system is moved through a closed path is equal to zero.**

These statements can be mathematically demonstrated and serve as general mathematical definitions of conservative forces. Both of these statements can be seen for the gravitational force from Equation 7.3. The work done is expressed only in terms of the initial and final heights, with no indication of what path is followed. If the path is closed, the initial and final heights are the same in Equation 7.3, and the work is equal to zero.

Another example of a conservative force is the force of a spring on an object attached to the spring, where the spring force is given by Hooke's law, $F_s = -kx$. As we learned in Chapter 6 (Eq. 6.15), the work done by the spring force is

$$W_s = \tfrac{1}{2}kx_i^2 - \tfrac{1}{2}kx_f^2$$

where the initial and final positions of the object are measured from its equilibrium position $x = 0$, at which the spring is unstretched. Again we see that W_s depends only on the initial and final coordinates of the object and is zero for any closed path. Hence, the spring force is conservative.

In Section 7.1, we discussed the notion of an external agent lifting a book and storing energy as potential energy in the book–Earth system. In Section 6.4, we discussed an external agent pulling a block attached to a spring from $x = 0$ to $x = x_{max}$ and calculated the work that was done on the system as $\tfrac{1}{2}kx_{max}^2$. This is another situation, like that of the book in Section 7.1, in which work is done on a system but there is no change in kinetic energy of the system. Thus, the energy must be stored in the block–spring system as potential energy. The **elastic potential energy** associated with the spring force is defined by

$$U_s \equiv \tfrac{1}{2}kx^2 \qquad\qquad [7.8]$$

• *Elastic potential energy*

The elastic potential energy can be considered as the energy stored in the deformed spring (one that is either compressed or stretched from its equilibrium position). To visualize this, consider Figure 7.7a, which shows an undeformed spring on a frictionless, horizontal surface. When the block is pushed against the spring (Fig. 7.7b), compressing the spring a distance x, the elastic potential energy stored in the spring is $\tfrac{1}{2}kx^2$. When the block is released, the spring returns to its original

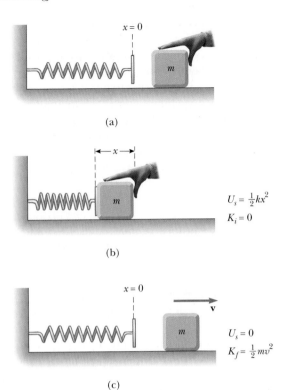

(a)

(b)

$U_s = \frac{1}{2}kx^2$
$K_i = 0$

$U_s = 0$
$K_f = \frac{1}{2}mv^2$

(c)

Figure 7.7

(a) An undeformed spring on a friction-less horizontal surface. (b) A block of mass m is pushed against a spring, compressing it through a distance x. (c) When the block is released from rest, the elastic potential energy stored in the system is transformed to kinetic energy of the block.

length, applying a force to the block. This force does work on the block, resulting in kinetic energy of the block (Fig. 7.7c). The elastic potential energy stored in the spring is zero whenever the spring is undeformed ($x = 0$). Finally, because the elastic potential energy is proportional to x^2, we see that U_s is always positive in a deformed spring.

In comparison to a conservative force, a **nonconservative force** in mechanics is **a force between members of a system that is not conservative,** that is, it causes transformation of mechanical energy to internal energy within the system. A common nonconservative force in mechanics is the friction force. If we consider a system consisting of a block and a surface and imagine an initially sliding block coming to a rest due to friction, we see the result of a nonconservative force. Initially, the system has kinetic energy (of the block). Afterward, nothing is moving. The friction force between the block and the surface transformed the mechanical energy into internal energy—the block and surface are both warmer than before.

Let us return to the notion of the work done over a path. The two statements claimed for conservative forces are not true for nonconservative forces. As an example, consider Figure 7.8. Suppose you displace a book between two points on a table. If the book is displaced in a straight line along the blue path between points Ⓐ and Ⓑ in Figure 7.8, you do a certain amount of work against the kinetic friction force to keep the book moving at a constant speed. Now, imagine that you push the book along the red semicircular path in Figure 7.8. You perform more work against friction along this longer path than along the straight path. The work done depends on the path, so the friction force cannot be conservative.

In Chapter 6, we discussed the fact that we cannot calculate the work done by friction because the actual friction force behaves in a very messy way. In this ex-

• *Definition of a nonconservative force*

Figure 7.8

The work done against the force of friction depends on the path taken as the book is moved from Ⓐ to Ⓑ and hence friction is a nonconservative force. The work required is greater along the red path than the blue path.

ample the particle model is not valid. Another example of a situation with non-particle objects is seen for deformable objects. Suppose, for example, that a rubber ball is flattened against a brick wall. The ball deforms during the pushing process. If the ball is suddenly released, it will jump away from the wall. The force causing the ball to accelerate is the normal force of the wall on the ball. But the point of application of this force does not move through space—it stays fixed at the point of contact between the ball and the wall. Thus, no work is done on the ball. Yet the ball has kinetic energy afterward. The work–kinetic energy theorem does not correctly describe this situation. It is more valuable to apply the isolated system model to this situation. With the ball as the system, there is no transfer of energy across the boundary as the ball springs off the wall. There is a *transformation* of energy, from elastic potential energy (stored in the ball when it was flattened) to kinetic energy. In the same way, if a skateboarder pushes off a wall to start rolling, no work is done by the force from the wall—the kinetic energy is transformed within the system from potential energy stored in the body from previous meals.

We also discussed in Chapter 6 the difficulties associated with the nonconservative force of friction in energy calculations. Recall that the result of a friction force is to transform kinetic energy in a system to internal energy, and the increase in internal energy is equal to the decrease in kinetic energy. If a potential energy is associated with the system, then the decrease in *mechanical* energy, kinetic plus potential, equals the increase in internal energy in the isolated system. Thus, we have

$$-f_k \Delta x = \Delta K + \Delta U = \Delta E_{mech} = -\Delta E_{int} \qquad [7.9]$$

We can recast this expression by putting all forms of energy storage on one side of the equation:

$$\Delta K + \Delta U + \Delta E_{int} = \Delta E_{system} = 0 \qquad [7.10]$$

This gives us the most general expression of the continuity equation for an isolated system. Note that ΔK may represent more than one term if two or more parts of the system are moving. Also, ΔU may represent more than one term if different types of potential energy (e.g., gravitational and elastic) are associated with the system. Equation 7.10 is equivalent to

$$K + U + E_{int} = \text{constant} \qquad [7.11]$$

• *General conservation of energy for an isolated system*

which claims that the total energy (kinetic, potential, and internal) of an isolated system is conserved. No violation of this critical conservation principle has ever been observed. If we consider the Universe as an isolated system, this statement claims that there is a fixed amount of energy in our Universe, and all processes within the Universe represent transformations of energy from one type to another.

Quick Quiz 7.3

A block of mass m is projected across a horizontal surface with a speed v. It slides until it stops due to the friction force between the block and the surface. The surface is now tilted at 30°, and the block is projected up the surface with the same initial speed v. When the block stops, how does the decrease in mechanical energy of the block–Earth system compare with that when the block slid over the horizontal surface? (a) It is the same. (b) It is larger. (c) It is smaller. (d) The relationship cannot be determined.

PROBLEM-SOLVING STRATEGY | **Isolated Systems**

Many problems in physics can be solved using the principle of conservation of energy for an isolated system. The following procedure should be used when you apply this principle:

1. Define your system, which may consist of more than one object and may or may not include springs or other possibilities for storage of potential energy. Choose situations to represent the initial and final conditions of the system.
2. Determine if any energy transfers occur across the boundary of your system. If so, use the nonisolated system model, $\Delta E_{\text{system}} = \Sigma H$. If not, use the isolated system model, $\Delta E_{\text{system}} = 0$, and continue with the following steps.
3. For each object that changes elevation, select a reference position for the zero point of gravitational potential energy. For a spring, the zero point of elastic potential energy is the position of the end of the spring when the spring force is zero. If there is more than one conservative force, write an expression for the potential energy associated with each force.
4. Determine whether any nonconservative forces are present. Remember that if friction or air resistance is present, mechanical energy *is not conserved.*
5. If mechanical energy is *conserved,* you can write the total initial system energy E_i at some point as the sum of the kinetic and potential energy at that point. Then write a similar expression for the total system energy at the final point that is of interest. Because mechanical energy is *conserved,* equate the two total energies and solve for the quantity that is unknown.
6. If nonconservative forces are present (and thus mechanical energy is not conserved), first write expressions for the total initial and total final mechanical energies. In this case, the difference between the total final mechanical energy and the total initial mechanical energy equals the energy transformed to or from internal energy by the nonconservative forces.

Example 7.3 Crate Sliding Down a Ramp

A 3.00-kg crate slides down a ramp at a loading dock. The ramp is 1.00 m in length, and inclined at an angle of 30.0°, as in Figure 7.9. The crate starts from rest at the top and experiences a constant friction force of magnitude 5.00 N. Use energy methods to determine the speed of the crate when it reaches the bottom of the ramp.

Reasoning We define the system as the crate, the Earth, and the ramp. This is an isolated system. If we had chosen the crate and the Earth as the system, we would need to use the nonisolated system model because the friction force between the crate and the ramp is an external influence. There would be work done across the boundary as well as flow of energy by heat between the crate and the ramp. This would

be a difficult problem to solve. In general, if a friction force acts, it is easiest to define the system so that the friction force is an internal force.

Solution Because $v_i = 0$ for the crate, the initial kinetic energy of the system is zero. If the y coordinate is measured from the bottom of the ramp, then $y_i = (1.00 \text{ m})\sin 30° = 0.500$ m for the crate. The total mechanical energy of the crate–Earth–ramp system when the crate is at the top is therefore the gravitational potential energy:

$$E_i = U_i = mgy_i$$

When the crate reaches the bottom, the gravitational potential energy of the system is *zero* because the elevation of

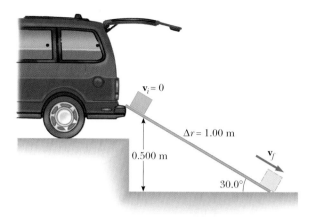

Figure 7.9

(Example 7.3) A crate slides down a ramp under the influence of gravity. The potential energy of the crate–Earth system decreases while the kinetic energy of the crate increases.

the crate is $y_f = 0$. The total mechanical energy when the crate is at the bottom is therefore kinetic energy:

$$E_f = K_f = \tfrac{1}{2}mv_f^2$$

We cannot say that $E_f = E_i$ in this case, however, because a nonconservative force—the force of friction—reduces the

mechanical energy of the system. In this case, the change in mechanical energy for the system is $\Delta E = -f_k\,\Delta r$, where Δr is the displacement of the crate along the ramp. With $f_k = 5.00$ N and $\Delta r = 1.00$ m, we have

$$\Delta E_{\text{mech}} = -f_k\,\Delta r = -(5.00\text{ N})(1.00\text{ m}) = -5.00\text{ J}$$

This is the change in the mechanical energy of the system. Because $\Delta E_{\text{mech}} = \Delta K + \Delta U = \tfrac{1}{2}mv_f^2 - mgy_i$ in this situation, Equation 7.9 gives

$$-f_k\Delta r = \tfrac{1}{2}mv_f^2 - mgy_i \longrightarrow \tfrac{1}{2}mv_f^2 = mgy_i - f_k\Delta r$$

$$\tfrac{1}{2}mv_f^2 = (3.00\text{ kg})(9.80\text{ m/s}^2)(0.500\text{ m}) - 5.00\text{ J} = 9.70\text{ J}$$

$$v_f = \sqrt{\frac{2(9.70\text{ J})}{3.00\text{ kg}}} = \boxed{2.54\text{ m/s}}$$

EXERCISE Use Newton's second law to find the acceleration of the crate along the ramp, and the equations of kinematics to determine the final speed of the crate.

Answer 3.23 m/s²; 2.54 m/s

EXERCISE If the ramp is assumed to be frictionless, find the final speed of the crate and its acceleration along the ramp.

Answer 3.13 m/s; 4.90 m/s²

Example 7.4 Motion on a Curved Track

A child of mass m takes a ride on an irregularly curved slide of height $h = 2.00$ m, as in Figure 7.10. The child starts from rest at the top. (a) Determine the speed of the child at the bottom, assuming no friction is present.

Reasoning We will define the system as the child and the Earth, and model the child as a particle. The normal force **n** does no work on the system because this force is always perpendicular to each element of the displacement. Furthermore, because no friction is present, no work is done by friction across the boundary of the system. Thus, we will use the isolated system model with no friction forces, for which mechanical energy is conserved; that is, $K + U = $ constant.

Solution If we measure the y coordinate for the child from the bottom of the slide, then $y_i = h$, $y_f = 0$, and we have for the system,

$$K_i + U_i = K_f + U_f$$

$$0 + mgh = \tfrac{1}{2}mv_f^2 + 0$$

$$v_f = \sqrt{2gh}$$

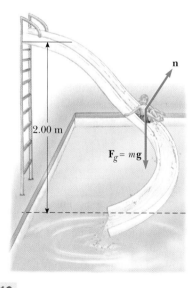

Figure 7.10

(Example 7.4) If the slide is frictionless, the speed of the child at the bottom depends only on the height of the slide.

The result is the same as if the child simply fell vertically through a distance h! In this example, $h = 2.00$ m, giving

$$v_f = \sqrt{2gh} = \sqrt{2(9.80 \text{ m/s}^2)(2.00 \text{ m})} = \boxed{6.26 \text{ m/s}}$$

(b) If a friction force acts on the 20.0-kg child, and he arrives at the bottom of the slide with a speed $v_f = 3.00$ m/s, by how much does the mechanical energy of the system decrease due to this force?

Solution   We define the system as the child, the Earth, and the slide. In this case, a nonconservative force acts within the system and mechanical energy is *not* conserved. We can find

the change in mechanical energy due to friction, given that the final speed at the bottom is known:

$$\Delta E_{mech} = K_f + U_f - K_i - U_i = \tfrac{1}{2}mv_f^2 + 0 - 0 - mgh$$

$$\Delta E_{mech} = \tfrac{1}{2}(20.0 \text{ kg})(3.00 \text{ m/s})^2$$
$$- (20.0 \text{ kg})(9.80 \text{ m/s}^2)(2.00 \text{ m})$$
$$= \boxed{-302 \text{ J}}$$

The change in mechanical energy ΔE_{mech} is negative because friction reduces the mechanical energy of the system. The change in internal energy in the system is $+302$ J.

Example 7.5   Block–Spring Collision

A block of mass 0.800 kg is given an initial velocity $v_A = 1.20$ m/s to the right and collides with a light spring of force constant $k = 50.0$ N/m, as in Figure 7.11. (a) If the surface is frictionless, calculate the maximum compression of the spring after the collision.

Reasoning   We define the system as the block and the spring. No transfers of energy occur across the boundary of this system, so we use the isolated system model. Before the collision, when the block is at Ⓐ, for example, the system has kinetic energy due to the moving block and the spring is uncompressed, so the potential energy stored in the system is zero. Thus, the total energy of the system before the collision is $\tfrac{1}{2}mv_A^2$. After the collision, and when the spring is fully compressed at point Ⓒ, the mass is momentarily at rest and has zero kinetic energy, whereas the potential energy stored in the spring has its maximum value $\tfrac{1}{2}kx_C^2$. The total mechanical energy of the system is conserved because no nonconservative forces act within the system.

Solution   Because mechanical energy is conserved, the mechanical energy of the system before the collision must equal the mechanical energy of the system when the spring is fully compressed, or

$$\tfrac{1}{2}mv_A^2 = \tfrac{1}{2}kx_C^2$$

$$x_C = \sqrt{\frac{m}{k}}\, v_A = \sqrt{\frac{0.800 \text{ kg}}{50.0 \text{ N/m}}}(1.20 \text{ m/s}) = \boxed{0.152 \text{ m}}$$

(b) If a constant force of kinetic friction acts between the block and the surface with $\mu_k = 0.500$, and if the speed of the block just as it collides with the spring is $v_A = 1.20$ m/s, what is the maximum compression in the spring?

Solution   We define the system as the block, the spring, and the surface. In this case, mechanical energy of the system is

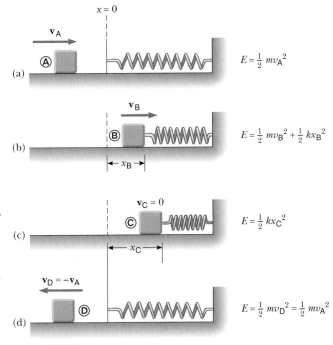

$$E = \tfrac{1}{2} mv_A^2$$

(a)

$$E = \tfrac{1}{2} mv_B^2 + \tfrac{1}{2} kx_B^2$$

(b)

$$E = \tfrac{1}{2} kx_C^2$$

(c)

$$E = \tfrac{1}{2} mv_D^2 = \tfrac{1}{2} mv_A^2$$

(d)

Figure 7.11

(Example 7.5) A block sliding on a smooth, horizontal surface collides with a light spring. (a) Initially the mechanical energy of the system is all kinetic energy. (b) The mechanical energy is the sum of the kinetic energy of the block and the elastic potential energy in the spring. (c) The mechanical energy is entirely potential energy. (d) The mechanical energy is transformed back to the kinetic energy of the block. The total energy of the block–spring system remains constant throughout the motion.

not conserved because of friction. The magnitude of the friction force is

$$f_k = \mu_k n = \mu_k mg = 0.500(0.800 \text{ kg})(9.80 \text{ m/s}^2) = 3.92 \text{ N}$$

where we have found $n = mg$ from Newton's second law in the vertical direction. Therefore, the decrease in mechanical energy due to friction as the block is displaced from $x_i = 0$ to the point $x_f = x_{max}$ at which the block stops is

$$\Delta E_{mech} = -f_k x_{max} = -3.92 x_{max}$$

Equation 7.9 gives

$$\Delta E_{mech} = E_f - E_i = (0 + \tfrac{1}{2} k x_{max}^2) - (\tfrac{1}{2} m v_A^2 + 0)$$

Substituting the numerical values and dropping the units, we have

$$-3.92 x_{max} = \frac{50.0}{2} x_{max}^2 - \tfrac{1}{2}(0.800)(1.20)^2$$

$$25.0 x_{max}^2 + 3.92 x_{max} - 0.576 = 0$$

Solving the quadratic equation for x_{max} gives $x_{max} = 0.0924$ m and $x_{max} = -0.249$ m. We choose the positive root $x_{max} = 0.0924$ m because the block must be to the right of the origin when it comes to rest. Note that 0.0924 m is less than the

distance obtained in the frictionless case (a). This result is what we expect because friction retards the motion of the system.

EXERCISE A 3.00-kg block starts at a height $h = 60.0$ cm on a plane that has an inclination angle of 30.0°. Upon reaching the bottom, the block slides along a horizontal surface. If the coefficient of friction for both surfaces is $\mu_k = 0.200$, how far does the block slide on the horizontal surface before coming to rest?

Answer 1.96 m

EXERCISE A child starts from rest at the top of a slide of height 4.00 m. (a) What is her speed at the bottom if the incline is frictionless? (b) If she reaches the bottom with a speed of 6.00 m/s, what percentage of the total mechanical energy when she was at the top of the slide was transformed to internal energy as a result of friction?

Answer (a) 8.85 m/s (b) 54.1%

7.4 • CONSERVATIVE FORCES AND POTENTIAL ENERGY

Let us return to the falling book discussed in Section 7.2. We found that the work done on the book by the gravitational force can be expressed as the negative of the difference between two quantities that we called the initial and final potential energies of the system of the book and the Earth:

$$W_{on\ book} = mgy_b - mgy_a = -\Delta U \qquad \text{[7.12]}$$

This is the hallmark of a conservative force—we can identify a potential energy function, such that the work done by the force on a member of the system in which the force acts depends only on the difference in the initial and final values of the function. This cannot be done for a nonconservative force because the work done depends on the particular path followed between the initial and final points.

For a conservative force, this notion allows us to generate a mathematical relationship between a force and its potential energy function. From the definition of work, we can write Equation 7.12 for a general force in the x direction as

$$W = \int_{x_i}^{x_f} F_x \, dx = -\Delta U = -(U_f - U_i) = -U_f + U_i \qquad \text{[7.13]}$$

• *Work done by a conservative force*

Thus, the potential energy function can be written as

$$U_f = -\int_{x_i}^{x_f} F_x \, dx + U_i \qquad \text{[7.14]}$$

• *Finding the potential energy from the force*

This expression allows us to calculate the potential energy function associated with a conservative force if we know the force function. The value of U_i is often taken to be zero at some arbitrary reference point. It really doesn't matter what value we assign to U_i because any value simply shifts U_f by a constant, and it is the *change* in potential energy that is physically meaningful.

As an example, let us calculate the potential energy function for the spring force. We model the spring as obeying Hooke's law, so the force the spring exerts is $F_s = -kx$. The potential energy stored in a block–spring system is

$$U_f = -\int_{x_i}^{x_f} (-kx)\, dx + U_i = \tfrac{1}{2}kx_f^2 - \tfrac{1}{2}kx_i^2 + U_i$$

As mentioned earlier, we can choose the value of the zero of potential energy arbitrarily. Let us choose $U_i = 0$ when the block is at the position $x_i = 0$. Then,

$$U_f = \tfrac{1}{2}kx_f^2 - \tfrac{1}{2}kx_i^2 + U_i = \tfrac{1}{2}kx_f^2 - 0 + 0 \quad \longrightarrow \quad U_f = U_s = \tfrac{1}{2}kx^2$$

This is the potential energy function we have already recognized (see Eq. 7.8) for a spring that obeys Hooke's law.

In the previous discussion, we have seen how to find a potential energy function if we know the force function. Let us now turn this around. Suppose we know the potential energy function. Can we find the force function? We start from the basic definition of work done by a conservative force for an infinitesimal displacement $d\mathbf{r} = dx\mathbf{i}$ in the x direction:

$$dW = \mathbf{F} \cdot d\mathbf{r} = \mathbf{F} \cdot dx\mathbf{i} = F_x\, dx = -dU$$

This can be rewritten as

- *Finding the force from the potential energy*

$$F_x = -\frac{dU}{dx} \qquad\qquad \text{[7.15]}$$

Thus, the **conservative force acting between parts of a system equals the negative derivative of the potential energy associated with that system.**[*]

In the case of an object located a distance y above some reference point, the gravitational potential energy function is given by $U_g = mgy$, and it follows from Equation 7.15 that (considering the y direction rather than x)

$$F_y = -\frac{dU_g}{dy} = -\frac{d}{dy}(mgy) = -mg$$

which is the correct expression for the gravitational force.

7.5 • THE NONISOLATED SYSTEM IN STEADY STATE

We have seen two approaches related to systems so far. In a nonisolated system, the energy stored in the system changes due to transfers across the boundaries of the system. Thus, nonzero terms occur on both sides of the continuity equation for energy, $\Delta E_{\text{system}} = \Sigma\, H$. For an isolated system, no energy transfer takes place across the boundary, so the right-hand side of the continuity equation is zero; that is, $\Delta E_{\text{system}} = 0$.

---

[*] In three dimensions, the appropriate expression is

$$\mathbf{F} = -\mathbf{i}\frac{\partial U}{\partial x} - \mathbf{j}\frac{\partial U}{\partial y} - \mathbf{k}\frac{\partial U}{\partial z}$$

where $\partial U/\partial x$, and so on are *partial derivatives*. In the language of vector calculus, $\mathbf{F}$ equals the negative of the *gradient* of the scalar potential energy function $U(x, y, z)$.

Another possibility exists that we have not yet addressed. It is possible for no change to occur in the energy of the system even though nonzero terms are present on the right-hand side of the continuity equation, $0 = \Sigma\ H$. This can only occur if the rate at which energy is entering the system is equal to the rate at which it is leaving. In this case, the system is in steady state under the effects of two or more competing transfers, and we describe this as a **nonisolated system in steady state.** The system is nonisolated because it is interacting with the environment, but it is in steady state because the system energy remains constant.

We could identify a number of examples of this type of situation. First, consider your home as a nonisolated system. Ideally, you would like to keep the temperature of your home constant for the comfort of the occupants. Thus, your goal is to keep the internal energy in the home fixed.

The energy transfer mechanisms for the home are numerous, as we can see in Figure 7.12. Solar electromagnetic radiation is absorbed by the roof and walls of the home and enters the home through the windows. Energy enters by electrical transmission to operate electrical devices. Leaks in the walls, windows, and doors allow warm or cold air to enter and leave, carrying energy across the boundary of the system by matter transfer. Matter transfer also occurs if any devices in the home operate from natural gas—energy is carried in with the gas. Energy transfers through the walls, windows, floor, and roof by heat due to temperature differences between the inside and outside of the home. Thus, we have a variety of transfers, but the energy in the home remains constant in the idealized case. In reality, this is a system in *quasi-steady state* because some small temperature variations actually occur over a 24-h period, but we can imagine an idealized situation that conforms to the nonisolated system in steady-state model.

• *The nonisolated system in steady state*

WEB

For general information on energy and tips on conserving energy from the U.S. Department of Energy, visit **www.energy.gov**

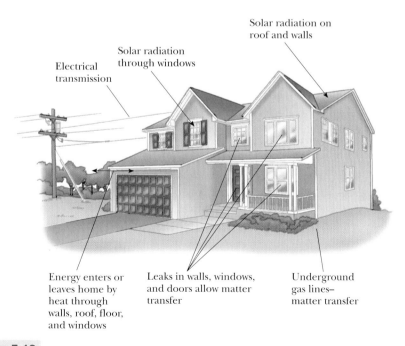

Figure 7.12

Energy enters and leaves a home by several mechanisms. The home can be modeled as a nonisolated system in steady state.

As a second example, consider the Earth as a system. Because the Earth is located in the vacuum of space, the only possible types of energy transfers involve no contact between the system and external molecules in the environment. As mentioned in Chapter 6, only two types of transfer do not depend on contact with molecules: work done by field forces and electromagnetic radiation. The Earth as a system exchanges energy with the rest of the Universe only by means of electromagnetic radiation (ignoring work done by field forces, and some small matter transfer due to cosmic ray particles and meteoroids entering the system and spacecraft leaving the system!). The primary input radiation is that from the Sun, and the output radiation is primarily infrared radiation emitted from the ground. These transfers are balanced so that the Earth maintains a constant temperature. The transfers are not *exactly* balanced, however, so the Earth is in quasi-steady state—the temperature does appear to be changing. The change in temperature is very gradual and currently appears to be in the positive direction. This is the essence of the social issue of global warming.

In this chapter and Chapter 6, we explored three analysis models related to systems. We described these models only in terms of energy and energy transfers. Other physical quantities can transfer across the boundary of a system, and we will return to these system models again throughout the book. Additional quantities for which we will discuss versions of system models include momentum (Chapter 8), angular momentum (Chapter 10), volume of fluid (Chapter 15), and electric charge (Chapter 19).

Quick Quiz 7.4

Model your body as a nonisolated system in steady state. What types of energy transfer occur?

7.6 • POTENTIAL ENERGY FOR GRAVITATIONAL AND ELECTRIC FORCES

Earlier in this chapter we introduced the concept of gravitational potential energy, that is, the energy associated with a system of objects interacting via the gravitational force. We emphasized the fact that the gravitational potential energy function, Equation 7.2, is valid only when the object of mass m is near the Earth's surface. We would like to find a more general expression for the gravitational potential energy that is valid for all separation distances. Because the free-fall acceleration g varies as $1/r^2$, it follows that the general dependence of the potential energy function of the system on separation distance is more complicated than our simple expression, Equation 7.2.

Consider a particle of mass m moving between two points Ⓐ and Ⓑ above the Earth's surface, as in Figure 7.13. The gravitational force on the particle due to the Earth, first introduced in Section 5.6, can be written in vector form as

$$\mathbf{F}_g = -\frac{GM_E m}{r^2}\hat{\mathbf{r}} \qquad [7.16]$$

where $\hat{\mathbf{r}}$ is a unit vector directed from the Earth toward the particle, and the negative sign indicates that the force is downward toward the Earth. This expression shows that the gravitational force depends on the polar coordinate r. Further-

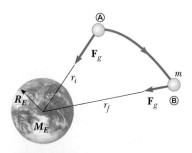

Figure 7.13

As a particle of mass m moves from Ⓐ to Ⓑ above the Earth's surface, the potential energy of the particle–Earth system changes according to Equation 7.16.

more, the gravitational force is conservative. Equation 7.14 gives

$$U_f = -\int_{r_i}^{r_f} F(r)\,dr + U_i = GM_E m \int_{r_i}^{r_f} \frac{dr}{r^2} + U_i = GM_E m\left(-\frac{1}{r}\right)_{r_i}^{r_f} + U_i$$

or

$$U_f = -GM_E m\left(\frac{1}{r_f} - \frac{1}{r_i}\right) + U_i \qquad [7.17]$$

As always, the choice of a reference point for the potential energy is completely arbitrary. It is customary to locate the reference point where the force is zero. Letting $U_i \rightarrow 0$ as $r_i \rightarrow \infty$, we obtain the important result

$$U_g = -\frac{GM_E m}{r} \qquad [7.18]$$

Because of our choice of the reference point for zero potential energy, the function U_g is always negative (Fig. 7.14).

Although Equation 7.18 was derived for the particle–Earth system, it can be applied to *any* two particles. For *any pair* of particles of masses m_1 and m_2 separated by a distance r, the gravitational force of attraction is given by Equation 5.14, and the gravitational potential energy of the system of two particles is

$$U_g = -\frac{Gm_1 m_2}{r} \qquad [7.19]$$

This expression also applies to larger objects *if they are spherically symmetric*, as first shown by Newton.

Equation 7.19 shows that the gravitational potential energy for any pair of particles varies as $1/r$ (whereas the force between them varies as $1/r^2$). Furthermore, the potential energy is *negative* because the force is attractive and we have chosen the potential energy to be zero when the particle separation is infinity. Because the force between the particles is attractive, we know that an external agent must do positive work to increase the separation between the two particles. The work done by the external agent produces an increase in the potential energy as the two particles are separated. That is, U_g becomes less negative as r increases.

When the two particles are separated by a distance r, an external agent would have to supply an energy *at least* equal to $+Gm_1 m_2/r$ in order to separate the particles by an infinite distance. (In practice, any distance larger than r by at least three orders of magnitude can be modeled as an infinite distance.) It is convenient to think of the absolute value of the gravitational potential energy defined this way as the *binding energy* of the system. An external agent needs to provide this energy to separate the particles by an infinite distance. If the external agent supplies an energy *greater than* the binding energy $Gm_1 m_2/r$, the additional energy of the system is in the form of kinetic energy when the particles are at an infinite separation.

We can extend this concept to three or more particles. In this case, the total potential energy of the system is the sum over all *pairs* of particles. Each pair contributes a term of the form given by Equation 7.19. For example, if the system contains three particles, as in Figure 7.15, we find that

$$U_{\text{total}} = U_{12} + U_{13} + U_{23} = -G\left(\frac{m_1 m_2}{r_{12}} + \frac{m_1 m_3}{r_{13}} + \frac{m_2 m_3}{r_{23}}\right) \qquad [7.20]$$

The absolute value of U_{total} represents the work needed to separate the particles by an infinite distance.

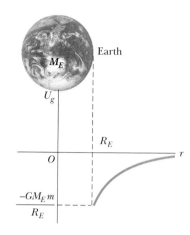

Figure 7.14

Graph of the gravitational potential energy U_g versus r for a particle above the Earth's surface. The potential energy of the system goes to zero as r approaches infinity.

• *Gravitational potential energy*

PITFALL PREVENTION 7.4

Gravitational potential energy

Be careful! Equation 7.19 looks similar to Equation 7.15 for the gravitational force, but there are two major differences. The gravitational force is a vector, whereas the gravitational potential energy is a scalar. The gravitational force varies as the *inverse square* of the separation distance, whereas the gravitational potential energy varies as the simple *inverse* of the separation distance.

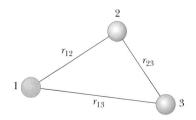

Figure 7.15

Three interacting particles.

THINKING PHYSICS 7.3

Why is the Sun hot?

Reasoning The Sun was formed when a cloud of gas and dust coalesced, due to gravitational attraction, into a massive astronomical object. Let us define this cloud as our system and model the gas and dust as particles. Initially, the particles of the system were widely scattered, representing a large amount of gravitational potential energy. In principle, we could calculate this potential energy by summing the gravitational potential energy for all pairs of particles in the system:

$$U = \sum_{\text{all pairs } i, j} - G\frac{m_i m_j}{r}$$

As the particles moved together to form the Sun, the gravitational potential energy of the system decreased. According to the isolated system model, this potential energy was transformed to kinetic energy as the particles fell toward the center. As the speeds of the particles increased, many collisions occurred between particles, randomizing their motion and transforming the kinetic energy to internal energy, representing an increase in temperature. If enough particles come together, the temperature can rise to a point at which nuclear reactions occur, and the object becomes a star. If not enough particles are present, the temperature rises, but not to a point at which nuclear reactions occur—the object is a planet if it is in orbit around a star. Jupiter is an example of a large planet that might have been a star if more particles had been available.

Example 7.6 The Change in Potential Energy

A particle of mass m is displaced through a small vertical distance Δy near the Earth's surface. Show that the general expression for the change in gravitational potential energy reduces to the familiar relationship $\Delta U_g = mg\,\Delta y$.

Solution We can express Equation 7.17 in the form

$$\Delta U_g = - GM_E m \left(\frac{1}{r_f} - \frac{1}{r_i} \right) = GM_E m \left(\frac{r_f - r_i}{r_i r_f} \right)$$

If both the initial and final positions of the particle are close to the Earth's surface, then $r_f - r_i = \Delta y$ and $r_i r_f \approx R_E^2$. (Recall that r is measured from the center of the Earth.) The change in potential energy therefore, becomes

$$\Delta U_g \approx \frac{GM_E m}{R_E^2}\Delta y = mg\,\Delta y$$

where we have used the fact that $g = GM_E/R_E^2$.

EXERCISE A satellite of the Earth has a mass of 100 kg and its altitude is 2.0×10^6 m. (a) What is the gravitational potential energy of the satellite–Earth system? (b) What is the magnitude of the force on the satellite?

Answer (a) -4.8×10^9 J   (b) 5.7×10^2 N

In Chapter 5, we discussed the electrostatic force between two point particles, which is Coulomb's law,

$$F_e = k_e \frac{q_1 q_2}{r^2} \qquad [7.21]$$

Because this looks so similar to Newton's law of universal gravitation, we would expect that the generation of a potential energy function for this force would proceed in a similar way. This is indeed the case, and this procedure results in the **electric potential energy** function,

$$U_e = k_e \frac{q_1 q_2}{r} \qquad \text{[7.22]}$$

As with the gravitational potential energy, the electric potential energy is defined as zero when the charges are infinitely far apart. Comparing this expression with that for the gravitational potential energy, we see the obvious differences in the constants and the use of charges instead of masses, but there is one more difference. The gravitational expression has a negative sign, but the electrical expression doesn't. For systems of objects that experience an attractive force, the potential energy decreases as the objects are brought closer together. Because we have defined zero potential energy at infinite separation, all real separations are finite and the energy must decrease from a value of zero. Thus, all potential energies for systems of objects that attract must be negative. In the gravitational case, attraction is the only possibility. The constant, the masses, and the separation distance are all positive, so the negative sign must be included explicitly, as it is in Equation 7.19.

The electric force can be either attractive or repulsive. Attraction occurs between *unlike* charges. Thus, for the two charges in Equation 7.22, one is positive and one is negative if the force is attractive. The product of the charges provides the negative sign for the potential energy mathematically, and we do not need an explicit negative sign in the potential energy expression. In the case of like charges, either a product of two negative charges or two positive charges will be positive, leading to a positive potential energy. This is reasonable, because to cause repelling particles to move together from infinite separation requires work to be done on the system, so the potential energy increases.

7.7 • ENERGY DIAGRAMS AND STABILITY OF EQUILIBRIUM

The motion of a system can often be understood qualitatively by analyzing a graphical representation of the system's potential energy curve. An **energy diagram** shows the potential energy of the system as a function of the position of one of the members of the system (or as a function of the separation distance between two members of the system). Consider the potential energy function for the block–spring system, given by $U_s = \frac{1}{2}kx^2$. This function is plotted versus x in Figure 7.16a.

The spring force is related to U_s through Equation 7.15:

$$F_s = -\frac{dU_s}{dx} = -kx$$

That is, the force is equal to the negative of the *slope* of the U_s versus x curve. When the block is placed at rest at the equilibrium position ($x = 0$), where $F_s = 0$, it will remain there unless some external force acts on it. If the spring in Figure 7.16b is stretched to the right from equilibrium, x is positive and the slope dU_s/dx is positive; therefore, F_s is negative and the block accelerates back toward $x = 0$. If the spring is compressed, x is negative and the slope is negative; therefore, F_s is positive and again the block accelerates toward $x = 0$.

From this analysis, we conclude that the $x = 0$ position is one of **stable equilibrium.** That is, any movement away from this position results in a force directed

• *Stable equilibrium*

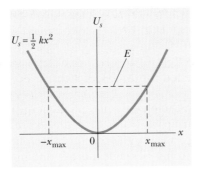

$U_s = \frac{1}{2} kx^2$

$-x_{max}$　　0　　x_{max}

(a)

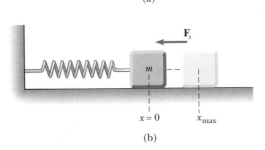

$x = 0$　　x_{max}

(b)

Figure 7.16

(a) Potential energy as a function of x for the block–spring system shown in part (b). The block oscillates between the turning points, which have the coordinates $x = \pm x_{max}$. The restoring force exerted by the spring always acts toward $x = 0$, the position of stable equilibrium.

back toward $x = 0$. (We described this type of force in Chapter 6 as a *restoring force*.) In general, **positions of stable equilibrium correspond to those values of x for which $U(x)$ has a relative minimum value on an energy diagram.**

From Figure 7.16 we see that if the block is given an initial displacement x_{max} and released from rest, the total initial energy of the system is the potential energy stored in the spring, given by $\frac{1}{2}kx_{max}^2$. As motion commences, the system acquires kinetic energy and loses an equal amount of potential energy. From an energy viewpoint, the energy of the system cannot exceed $\frac{1}{2}kx_{max}^2$; therefore, the block must stop at the points $x = \pm x_{max}$ and, because of the spring force, accelerate toward $x = 0$. The block oscillates between the two points $x = \pm x_{max}$, called the *turning points*. The block cannot be farther from equilibrium than $\pm x_{max}$ because the potential energy of the system beyond these points would be larger than the total energy, an impossible situation in classical physics. Because there is no transformation of mechanical energy to internal energy (no friction), the block oscillates between $- x_{max}$ and $+ x_{max}$ forever. (We shall discuss these oscillations further in Chapter 12).

Now consider an example in which the curve of U versus x is as shown in Figure 7.17. In this case, $F_x = 0$ at $x = 0$, and so the particle is in equilibrium at this point. However, this is a position of **unstable equilibrium** for the following reason. Suppose the particle is displaced to the *right* from the origin. Because the slope is negative for $x > 0$, $F_x = - dU/dx$ is positive and the particle accelerates away from $x = 0$. Now suppose the particle is displaced to the *left* from the origin. In this case the force is negative because the slope is positive for $x < 0$, and the particle again accelerates away from the equilibrium position. The $x = 0$ position in this situation is called a position of unstable equilibrium because, for any displacement from this point, the force pushes the particle farther away from equilibrium. In fact, the force pushes the particle toward a position of lower potential energy of the system. A ball placed on the top of an inverted spherical bowl is in a position of unstable equilibrium. If the ball is displaced slightly from the top and

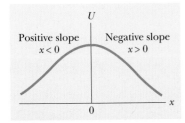

U

Positive slope　　Negative slope
$x < 0$　　　　$x > 0$

0

Figure 7.17

A plot of U versus x for a particle that has a position of unstable equilibrium, located at $x = 0$. For any finite displacement of the particle, the force on the particle is directed away from $x = 0$.

released, it will roll off the bowl. In general, **positions of unstable equilibrium correspond to those values of *x* for which *U*(*x*) has a relative maximum value on an energy diagram.***

• *Unstable equilibrium*

Finally, a situation may arise in which *U* is constant over some region, and hence *F* = 0. A point in this region is called a position of **neutral equilibrium.** Small displacements from this position produce neither restoring nor disrupting forces. A ball lying on a flat horizontal surface is an example of an object in neutral equilibrium.

• *Neutral equilibrium*

context connection
7.8 • ESCAPE SPEED FROM THE SUN

Our central question is based on the requirements for sending a spacecraft from Earth to Mars. We have discussed some initial steps toward this goal: launching the spacecraft from the surface of the Earth and placing it in a circular parking orbit around the Earth. The next step is to fire our rocket engines and begin our journey to Mars. But with what speed should we leave the parking orbit?

We cannot answer this question yet because it requires information not covered until later chapters. But we can establish an *upper limit* to the required speed. Let us ask this question: With what speed would we need to leave the orbit to be able to completely escape from the gravitational force of the Sun? Surely, the speed necessary to travel to Mars must be less than this speed.

Using our notions of energy, we can find the speed necessary to escape from the Sun. We will model the spacecraft and the Sun as an isolated system. As a simplification model, we will ignore the interaction between the spacecraft and all planets and stars in the universe, including Earth. In this isolated system, the total amount of energy must remain constant. Before the rocket engines are fired, the system has kinetic energy due to the motion of the spacecraft in its orbit around the Sun, gravitational potential energy related to the force between the spacecraft and the Sun, and energy stored in the fuel in the rocket. Let us simplify things further by imagining that the spacecraft does not move very far relative to the Sun while the rocket engine is burning. Thus, we will imagine the rocket engine fires for a very short time to give the rocket some speed and then is shut off, and the rocket drifts through space.

After the firing, the spacecraft–Sun system has kinetic energy and gravitational potential energy, and no nonconservative forces act. The energy in the fuel that was burned has been transformed into kinetic energy of the spacecraft. To completely escape from the Sun, the spacecraft must keep drifting away, approaching an infinite separation distance from the Sun. The minimum speed to produce this is found by imagining that the spacecraft keeps drifting away forever, with its speed asymptotically approaching zero. The mechanical energy of the system after the rocket engine is fired must equal the mechanical energy at infinite separation. For a spacecraft of mass *m*, this gives us

$$\tfrac{1}{2}mv^2_{\text{at Earth}} - G\frac{mm_{\text{Sun}}}{r_{\text{Earth–Sun}}} = \tfrac{1}{2}mv^2_{\text{at infinity}} - G\frac{mm_{\text{Sun}}}{r_{\text{infinity}}}$$

As the separation distance approaches infinity, the last term on the right approaches zero. If the velocity asymptotically approaches zero as the separation

---

* You can mathematically test whether an extreme of *U* is stable or unstable by examining the sign of d^2U/dx^2.

distance approaches infinity, the kinetic energy term on the right also becomes zero. With this condition, the speed at the position of the Earth becomes the escape speed v_{esc}. Thus, we have,

$$\tfrac{1}{2}mv_{esc}^2 - G\frac{mm_{Sun}}{r_{Earth-Sun}} = 0 \quad \longrightarrow \quad v_{esc} = \sqrt{2G\frac{m_{Sun}}{r_{Earth-Sun}}} \qquad [7.23]$$

When we substitute values for the mass of the Sun and the initial distance of the spacecraft from the Sun, we have

$$v_{esc} = \sqrt{2(6.67 \times 10^{-11}\ \text{N}\cdot\text{m}^2/\text{kg}^2)\,\frac{1.99 \times 10^{30}\ \text{kg}}{1.50 \times 10^{11}\ \text{m}}} = 4.21 \times 10^4\ \text{m/s}$$

Thus, we must move with at least this speed from Earth orbit to escape from the Sun. As mentioned previously, the speed necessary to send the spacecraft to Mars is less than this value. In fact, it is about one-tenth this value.

We have calculated an upper limit to a speed with which we need to leave the Earth and have indicated that we can determine the actual speed with material from upcoming chapters. Assuming that we know this required speed, how do we determine how much fuel to burn to attain this speed? We will do this in the next chapter by considering isolated systems again, but in terms of a new quantity called *momentum*. This quantity is critical to determining our fuel requirements for our trip to Mars.

SUMMARY

If a particle of mass m is elevated a distance y from a reference point $y = 0$ near the Earth's surface, the **gravitational potential energy** of the particle–Earth system can be defined as

$$U_g \equiv mgy \qquad [7.2]$$

The **total mechanical energy of a system** is defined as the sum of the kinetic energy and potential energy:

$$E_{mech} \equiv K + U \qquad [7.7]$$

If no energy transfers occur across the boundary of the system, the system is modeled as an **isolated system.** In this model, the principle of **conservation of mechanical energy** states that the total mechanical energy of the system is constant if all of the forces in the system are conservative:

$$K_i + U_i = K_f + U_f \qquad [7.6]$$

A force is **conservative** if the work it does on a particle is independent of the path the particle takes between two given points. A conservative force in mechanics does not cause a transformation of mechanical energy to internal energy. A force that does not meet these criteria is said to be **nonconservative.**

The **elastic potential energy** stored in a spring of force constant k is

$$U_s \equiv \tfrac{1}{2}kx^2 \qquad [7.8]$$

If some of the forces acting within a system are not conservative, the mechanical energy of the system does not remain constant. In the case of a common nonconservative force, friction, the change in mechanical energy is equal to the product of the kinetic friction force and the displacement:

$$-f_k\,\Delta x = \Delta K + \Delta U \qquad [7.9]$$

This decrease in mechanical energy in the system is equal to the increase in internal energy,

$$\Delta E_{int} = f_k\,\Delta x \qquad [7.9]$$

A potential energy function U can be associated only with a conservative force. If a conservative force **F** acts on a particle that moves along the x axis from x_i to x_f, the **change in the potential energy equals the negative of the work done by that force:**

$$U_f - U_i = -\int_{x_i}^{x_f} F_x\,dx \qquad [7.14]$$

If we know the potential energy function, a conservative force is given by the negative of the derivative of the potential energy function:

$$F_x = -\frac{dU}{dx} \qquad [7.15]$$

In some situations, a system may have energy crossing the boundary with no change in the energy stored in the system. In this case, the energy input in any time interval equals the energy output, and we describe this system as a **nonisolated system in steady state.**

The **gravitational potential energy** associated with two particles or uniform spherical distributions of mass separated by a distance *r* is

$$U_g = -\frac{Gm_1m_2}{r} \qquad [7.19]$$

where U_g is taken to approach zero as $r \rightarrow \infty$.

The **electric potential energy** associated with two charged particles separated by a distance *r* is

$$U_e = k_e\frac{q_1q_2}{r} \qquad [7.22]$$

where U_e is taken to approach zero as $r \rightarrow \infty$.

In an energy diagram, a point of **stable equilibrium** is one at which the potential energy is a minimum. A point of **unstable equilibrium** is one at which the potential energy is a maximum. **Neutral equilibrium** exists if the potential energy function is constant.

QUESTIONS

1. One person drops a ball from the top of a building while another person at the bottom observes its motion. Will these two people agree on the value of the gravitational potential energy of the ball–Earth system? On the change in potential energy? On the kinetic energy?

2. Discuss the changes in mechanical energy of an object–Earth system in (a) lifting the object, (b) holding the object at a fixed position, and (c) lowering the object slowly. Include the muscles in your discussion.

3. In Chapter 6, the work–kinetic energy theorem, $W = \Delta K$, was introduced. This equation states that work done on a system appears as a change in kinetic energy. This is a special case equation—it is valid if no changes occur in any other type of energy, such as potential or internal. Give some examples in which work is done but the change in energy is not that of kinetic energy.

4. A skier prepares to take off down a snow-covered hill. Is it correct to say that gravity provides the energy for the trip down the hill?

5. You walk leisurely up a flight of stairs. You then return to the bottom and run up the same set of stairs as fast as you can. Have you performed a different amount of work in these two cases?

6. A rock is thrown vertically in the air. Ignoring air friction, it takes the same time to go up as it does to come down. Is this still true in the presence of air friction?

7. A driver brings an automobile to a stop. If the brakes lock, so that the car skids, where is the original kinetic energy of the car, and in what form is it, after the car stops? Answer the same question for the case in which the brakes do not lock but the wheels continue to turn.

8. In an earthquake, a large amount of energy is "released" and spreads outward, potentially causing severe damage.

In what form does this energy exist before the earthquake, and by what energy transfer mechanism does it travel?

9. You ride a bicycle. In what sense is your bicycle solar powered?

10. A bowling ball is suspended from the ceiling of a lecture hall by a strong cord. The bowling ball is drawn away from its equilibrium position and released from rest at the tip of the demonstrator's nose. If the demonstrator remains stationary, explain why she will not be struck by the ball on its return swing. Would the demonstrator be safe if she pushed the ball as she released it?

11. A pile driver is a device used to drive objects into the Earth by repeatedly dropping a heavy weight on them. By how much does the energy of the pile driver–Earth system increase when the weight it drops is doubled? (Assume the weight is dropped from the same height each time.)

12. Our body muscles exert forces when we lift, push, run, jump, and so forth. Are these forces conservative?

13. A block is connected to a spring suspended from the ceiling. If the block is set in motion and air resistance is ignored, describe the energy transformations that occur within the system consisting of the block and spring.

14. What would the curve of *U* versus *x* look like if a particle were in a region of neutral equilibrium?

15. A ball rolls on a horizontal surface. Is the ball in stable, unstable, or neutral equilibrium?

16. Discuss the energy transformations that occur during the operation of an automobile.

17. A ball is thrown straight up into the air. At what position is its kinetic energy a maximum? At what position is the gravitational potential energy of the ball–Earth system a maximum?

PROBLEMS

1, 2, 3 = straightforward, intermediate, challenging   □ = full solution available in the *Student Solutions Manual and Study Guide*

web = solution posted at **http://www.harcourtcollege.com/physics/**   ▨ = computer useful in solving problem

▨ = Interactive Physics   ▨ = paired numerical/symbolic problems   ▨ = life science application

Section 7.1   Potential Energy of a System

1. A 1 000-kg roller coaster is initially at the top of a rise, at point Ⓐ. It then moves 135 ft, at an angle of 40.0° below the horizontal, to a lower point Ⓑ. (a) Choose point Ⓑ to be the zero level for gravitational potential energy. Find the potential energy of the roller coaster–Earth system at points Ⓐ and Ⓑ and the change in potential energy as the coaster moves. (b) Repeat part (a), setting the zero reference level at point Ⓐ.

2. A 400-N child is in a swing that is attached to ropes 2.00 m long. Find the gravitational potential energy of the child–Earth system relative to the child's lowest position when (a) the ropes are horizontal, (b) the ropes make a 30.0° angle with the vertical, and (c) the child is at the bottom of the circular arc.

3. A person with a remote mountain cabin plans to install her own hydroelectric plant. A nearby stream is 3.00 m wide and 0.500 m deep. Water flows at 1.20 m/s over the brink of a waterfall 5.00 m high. The manufacturer promises only 25.0% efficiency in converting the potential energy of the water–Earth system into electric energy. Find the power she can generate. (Large-scale hydroelectric plants, with a much larger drop, are more efficient.)

Section 7.2   The Isolated System

4. A particle of mass 0.500 kg is shot from *P* as shown in Figure P7.4. The particle has an initial velocity $\mathbf{v}_i$ with a horizontal component of 30.0 m/s. The particle rises to a maximum height of 20.0 m above *P*. Using the law of conservation of energy, determine (a) the vertical component of $\mathbf{v}_i$, (b) the work done by the gravitational force on the particle during its motion from *P* to *B*, and (c) the horizontal and the vertical components of the velocity vector when the particle reaches *B*.

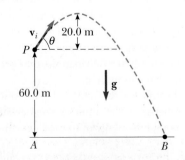

Figure P7.4

5. A bead slides without friction around a loop-the-loop (Fig. P7. 5). The bead is released from a height $h = 3.50R$. (a) What is its speed at point *A*? (b) How large is the normal force on it at point *A* if its mass is 5.00 g?

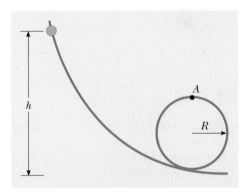

Figure P7.5

6. Dave Johnson, the bronze medalist at the 1992 Olympic decathlon in Barcelona, leaves the ground at the high jump with vertical velocity component 6.00 m/s. How far does his center of mass move up as he makes the jump?

7. A simple pendulum consists of a small object suspended by a string. The object is modeled as a particle. The string, with its top end fixed, has negligible mass and does not stretch. In the absence of air friction, the system oscillates by swinging back and forth in a vertical plane. If the string is 2.00 m long and makes an initial angle of 30.0° with the vertical, calculate the speed of the particle (a) at the lowest point in its trajectory and (b) when the angle is 15.0°.

8. A circus trapeze consists of a bar suspended by two parallel ropes, each of length ℓ, allowing performers to swing in a vertical circular arc (Figure P7.8). Suppose a performer with mass *m* holds the bar and steps off an elevated platform, starting from rest with the ropes at an angle of θ_i with respect to the vertical. Suppose the size of the performer's body is small compared with the length ℓ, that she does not pump the trapeze to swing higher, and that air resistance is negligible. (a) Show that when the ropes make an angle θ with the vertical, the performer must exert a force of magnitude $mg(3 \cos \theta - 2 \cos \theta_i)$ in order to hang on. (b) Determine the angle θ_i for which the force needed to hang on at the bottom of the swing is twice the performer's weight.

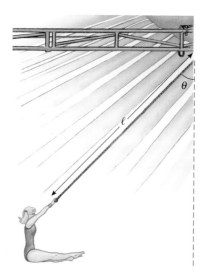

Figure P7.8

9. Two objects are connected by a light string passing over a light, frictionless pulley as shown in Figure P7.9. The 5.00-kg object is released from rest. Using the law of conservation of energy, (a) determine the speed of the 3.00-kg object just as the 5.00-kg object hits the ground. (b) Find the maximum height to which the 3.00-kg object rises.

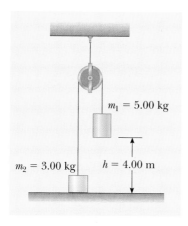

$m_1 = 5.00$ kg

$m_2 = 3.00$ kg $h = 4.00$ m

Figure P7.9

10. A light rigid rod is 77.0 cm long. Its top end is pivoted on a low-friction horizontal axle. The rod hangs straight down at rest, with a small massive ball attached to its bottom end. You strike the ball, suddenly giving it a horizontal velocity so that it swings around in a full circle. What minimum speed at the bottom is required to make the ball go over the top of the circle?

Section 7.3 Conservative and Nonconservative Forces

11. A system contains one particle that is free to move. At time t_i, the kinetic energy of the particle is 30.0 J and the po-

tential energy of the system is 10.0 J. At some later time t_f, the kinetic energy of the particle is 18.0 J. (a) If only conservative forces act on the particle, what are the potential energy and the total energy of the system at time t_f? (b) If the potential energy of the system at time t_f is 5.00 J, are any nonconservative forces acting on the particle? Explain.

12. (a) Suppose that a constant force acts on an object. The force does not vary with time, nor with the position or the velocity of the object. Start with the general definition for work done by a force

$$W = \int_i^f \mathbf{F} \cdot d\mathbf{r}$$

and show that the force is conservative. (b) As a special case, suppose that the force $\mathbf{F} = (3\mathbf{i} + 4\mathbf{j})$ N acts on a particle that moves from O to C in Figure P7.12. Calculate the work done by $\mathbf{F}$ if the particle moves along each one of the three paths OAC, OBC, and OC. (Your three answers should be identical.)

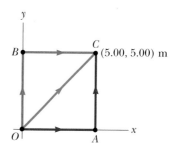

Figure P7.12   Problems 12 and 13.

13. A force acting on a particle moving in the xy plane is given by $\mathbf{F} = (2y\mathbf{i} + x^2\mathbf{j})$ N, where x and y are in meters. The particle moves from the origin to a final position having coordinates $x = 5.00$ m and $y = 5.00$ m, as in Figure P7.12. Calculate the work done by $\mathbf{F}$ along (a) OAC, (b) OBC, (c) OC. (d) Is $\mathbf{F}$ conservative or nonconservative? Explain.

14. In her hand a softball pitcher swings a ball of mass 0.250 kg around a vertical circular path of radius 60.0 cm before releasing it. The pitcher maintains a component of force on the ball of constant magnitude 30.0 N in the direction of motion around the complete path. The speed of the ball at the top of the circle is 15.0 m/s. If she releases the ball at the bottom of the circle, what is its speed upon release?

15. The world's biggest locomotive is the MK5000C, a 160-metric-ton behemoth driven by the most powerful engine ever used for rail transportation, a Caterpillar diesel capable of 5 000 hp. Such a huge machine provides certain gains in efficiency, but its large inertia presents challenges as well. The engine driver finds that the locomotive handles

differently from conventional ones, notably in braking and climbing hills. To investigate the loss of speed that can occur on an uphill stretch of track, consider the locomotive pulling no train, but traveling at 27.0 m/s on a level track while operating with an output power 1 000 hp. It comes to a 5.00% grade (a slope that rises 5.00 m for every 100 m along the track). If the throttle is not advanced, so that the power level is held steady, to what value will the speed drop? Assume that friction forces do not depend on the speed.

16. A glider of mass 0.150 kg moves on a horizontal frictionless air track. It is permanently attached to one end of a massless horizontal spring, which has a force constant of 10.0 N/m. The other end of the spring is fixed. The glider is moved so as to compress the spring by 0.180 m and then released from rest. Calculate the speed of the glider at the point where it has moved (a) 0.180 m from its starting point, so that the spring is momentarily exerting no force, and (b) 0.250 m from its starting point.

17. An object of mass m starts from rest and slides down a frictionless incline of angle θ. After sliding a distance d, it contacts the end of an unstressed spring of negligible mass as shown in Figure P7.17. The object slides an additional distance x as it is brought momentarily to rest by compression of the spring (of force constant k). Find the initial separation d between the object and the end of the spring.

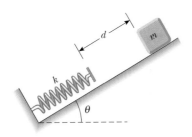

Figure P7.17

18. A 120-g object is attached to the bottom end of an unstressed spring. The spring is hanging vertically and has a force constant 40.0 N/m. The object is dropped. (a) What is its maximum speed? (b) How far does it drop before coming to rest momentarily?

19. A block of mass 0.250 kg is placed on top of a light vertical spring of force constant 5 000 N/m and pushed downward, so that the spring is compressed by 0.100 m. After the block is released from rest, it travels upward and then leaves the spring. To what maximum height above the point of release does it rise?

20. A boy in a wheelchair (total mass 47.0 kg) wins a race with a skateboarder. He has speed 1.40 m/s at the crest of a slope 2.60 m high and 12.4 m long. At the bottom of the slope his speed is 6.20 m/s. If air resistance and rolling resistance can be modeled as a constant friction force of 41.0 N, find the work he did in pushing forward on his wheels during the downhill ride.

21. The coefficient of friction between the 3.00-kg block and the surface in Figure P7.21 is 0.400. The system starts from rest. What is the speed of the 5.00-kg ball when it has fallen 1.50 m?

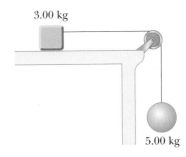

Figure P7.21

22. An 80.0-kg skydiver jumps out of a balloon at an altitude of 1 000 m and opens the parachute at an altitude of 200 m. (a) Assuming that the total retarding force on the diver is constant at 50.0 N with the parachute closed and constant at 3 600 N with the parachute open, what is the speed of the diver when he lands on the ground? (b) Do you think the skydiver will be injured? Explain. (c) At what height should the parachute be opened so that the final speed of the skydiver when he hits the ground is 5.00 m/s? (d) How realistic is the assumption that the total retarding force is constant? Explain.

23. A 5.00-kg block is set into motion up an inclined plane with an initial speed of 8.00 m/s (Fig. P7.23). The block comes to rest after traveling 3.00 m along the plane, which is inclined at an angle of 30.0° to the horizontal. For this motion determine (a) the change in the block's kinetic energy, (b) the change in the potential energy of the block–Earth system, and (c) the friction force exerted on the block (assumed to be constant). (d) What is the coefficient of kinetic friction?

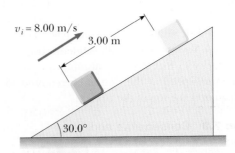

Figure P7.23

24. A toy cannon uses a spring to project a 5.30-g soft rubber ball. The spring is originally compressed by 5.00 cm and has a force constant of 8.00 N/m. When it is fired, the ball moves 15.0 cm through the horizontal barrel of the cannon, and a constant frictional force of 0.032 0 N exists between the barrel and the ball. (a) With what speed does the projectile leave the barrel of the cannon? (b) At what point does the ball have maximum speed? (c) What is this maximum speed?

25. A 1.50-kg object is held 1.20 m above a relaxed massless vertical spring with a force constant of 320 N/m. The object is dropped onto the spring. (a) How far does it compress the spring? (b) How far does it compress the spring if the same experiment is performed on the Moon, where $g = 1.63$ m/s^2? (c) Repeat part (a), but this time assume a constant air resistance force of 0.700 N acts on the mass during its motion.

26. A 75.0-kg skysurfer is falling straight down with terminal speed 60.0 m/s. Determine the rate at which the skysurfer–Earth system is losing mechanical energy.

Section 7.4 Conservative Forces and Potential Energy

27. A single conservative force acts on a 5.00-kg particle. The equation $F_x = (2x + 4)$ N describes the force, where x is in meters. As the particle moves along the x axis from $x = 1.00$ m to $x = 5.00$ m, calculate (a) the work done by this force, (b) the change in the potential energy of the system, and (c) the kinetic energy of the particle at $x = 5.00$ m if its speed is 3.00 m/s at $x = 1.00$ m.

28. A single conservative force acting on a particle varies as $\mathbf{F} = (-Ax + Bx^2)\mathbf{i}$ N, where A and B are constants and x is in meters. (a) Calculate the potential energy function $U(x)$ associated with this force, taking $U(x) = 0$ at $x = 0$. (b) Find the change in potential energy of the system and the change in kinetic energy of the particle as it moves from $x = 2.00$ m to $x = 3.00$ m.

29. The potential energy of a system of two particles separated by a distance r is given by $U(r) = A/r$, where A is a constant. Find the radial force $\mathbf{F}_r$ that each particle exerts on the other.

30. A potential energy function for a two-dimensional force on a particle is of the form $U = 3x^3y - 7x$. Find the force that acts on the particle at the point (x, y).

Section 7.6 Potential Energy for Gravitational and Electric Forces

31. A satellite of the Earth has a mass of 100 kg and is at an altitude of 2.00×10^6 m. (a) What is the potential energy of the satellite–Earth system? (b) What is the magnitude of the gravitational force exerted by the Earth on the satellite? (c) What force does the satellite exert on the Earth?

32. How much energy is required to move a 1 000-kg mass from the Earth's surface to an altitude twice the Earth's radius?

33. At the Earth's surface a projectile is launched straight up at a speed of 10.0 km/s. To what height will it rise? Ignore air resistance.

34. A system consists of three particles, each of mass 5.00 g, located at the corners of an equilateral triangle with sides of 30.0 cm. (a) Calculate the gravitational potential energy of the system. (b) If the particles are released simultaneously, where will they collide?

Section 7.7 Energy Diagrams and Stability of Equilibrium

35. For the potential energy curve shown in Figure P7.35, (a) determine whether the force F_x is positive, negative, or zero at the five points indicated. (b) Indicate points of stable, unstable, and neutral equilibrium. (c) Sketch the curve for F_x versus x from $x = 0$ to $x = 9.5$ m.

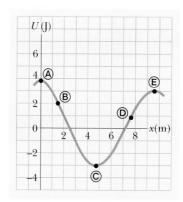

Figure P7.35

36. A particle moves along a line where the potential energy of the system depends on the position r of the particle as graphed in Figure P7.36. In the limit as r increases without bound, $U(r)$ approaches $+1$ J. (a) Identify each equilibrium position for this particle. Indicate whether each is a point of stable, unstable, or neutral equilibrium. (b) The particle will be bound if the total energy of the system is in what range? Now suppose that the system has energy -3 J. Determine (c) the range of positions where it can be found, (d) its maximum kinetic energy, (e) the location where it has maximum kinetic energy, and (f) its *binding energy*—that is, the additional energy that it would have to be given in order for it to move out to $r \rightarrow \infty$.

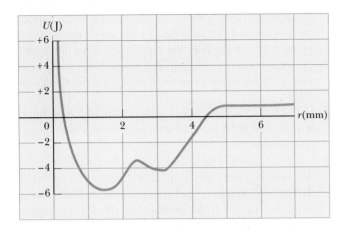

Figure P7.36

37. A particle of mass 1.18 kg is attached between two identical springs on a horizontal frictionless tabletop. The springs have force constant k and each is initially unstressed. (a) If the particle is pulled a distance x along a direction perpendicular to the initial configuration of the springs, as in Figure P7.37, show that the potential energy of the system is

$$U(x) = kx^2 + 2kL(L - \sqrt{x^2 + L^2})$$

(*Hint:* See Problem 50 in Chapter 6.) (b) Make a plot of $U(x)$ versus x, and identify all equilibrium points. Assume that $L = 1.20$ m and $k = 40.0$ N/m. (c) If the particle is pulled 0.500 m to the right and then released, what is its speed when it reaches the equilibrium point $x = 0$?

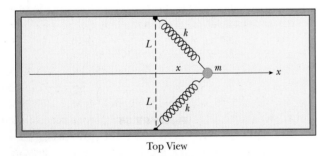

Top View

Figure P7.37

Section 7.8   Context Connection — Escape Speed from the Sun

38. *Voyager 1* achieved a maximum speed of 125 000 km/h on its way to photograph Jupiter. At what distance from the Sun is this speed sufficient for a spacecraft to escape the solar system?

39. An asteroid is on a collision course with Earth. An astronaut lands on the rock to bury explosive charges that will blow the asteroid apart. Most of the small fragments will miss the Earth, and those that fall into the atmosphere will produce only a beautiful meteor shower. The astronaut finds that the density of the spherical asteroid is equal to the average density of the Earth. To ensure its pulverization, she incorporates into the explosives the rocket fuel and oxidizer intended for her return journey. What maximum radius must the asteroid have for her to be able to leave it entirely simply by jumping straight up? On Earth she can jump to a height of 0.500 m.

Additional Problems

40. In the dangerous "sport" of bungee-jumping, a thrill-seeker jumps from a great height with an elastic cord attached to his ankles. Consider a 70.0-kg person leaping from the top of the world's tallest hotel, the 226-m Westin Stamford in Raffles City, Singapore. Instead of a specially designed elastic cord, suppose that he plans to use 9.00-mm-diameter nylon mountain climber's rope, with effective force constant 4 900 N/m. (a) What length of rope does he need to stop his plunge 10.0 m above the sidewalk? His height is negligible compared with the distance he falls. (b) What maximum force will the rope exert on the daredevil? (c) Express the maximum force as a multiple of the jumper's weight. Has he made a good plan?

41. Make an order-of-magnitude estimate of your power output as you climb stairs. In your solution state the physical quantities you take as data and the values you measure or estimate for them. Do you consider your peak power or your sustainable power?

42. A 200-g particle is released from rest at point Ⓐ along the horizontal diameter on the inside of a frictionless, hemispherical bowl of radius $R = 30.0$ cm (Fig. P7.42). Calculate (a) the gravitational potential energy of the particle–Earth system when the particle is at point Ⓐ relative to point Ⓑ, (b) the kinetic energy of the particle at point Ⓑ, (c) its speed at point Ⓑ, and (d) its kinetic energy and the potential energy of the system when the particle is at point Ⓒ.

Figure P7.42   Problems 42 and 43.

43. The particle described in Problem 42 (see Fig. P7.42) is
web released from rest at Ⓐ, and the surface of the bowl is
rough. The speed of the particle at Ⓑ is 1.50 m/s. (a)
What is its kinetic energy at Ⓑ? (b) How much mechanical energy is transformed due to friction as the particle
moves from Ⓐ to Ⓑ? (c) Is it possible to determine the coefficient of friction from these results in any simple manner? Explain.

44. A 50.0-kg block and a 100-kg block are connected by a
string as in Figure P7.44. The pulley is frictionless and of
negligible mass. The coefficient of kinetic friction between
the 50-kg block and incline is 0.250. Determine the
change in the kinetic energy of the 50-kg block as it moves
from Ⓐ to Ⓑ, a distance of 20.0 m.

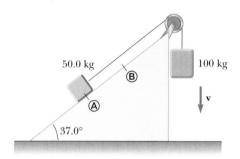

Figure P7.44

45. Assume that you attend a state university that started out
as an agricultural college. Close to the center of the campus is a tall silo topped with a hemispherical cap. The cap
is frictionless when wet. Someone has somehow balanced
a pumpkin at the highest point. The line from the center
of curvature of the cap to the pumpkin makes an angle
$\theta_i = 0°$ with the vertical. While you happen to be standing
nearby in the middle of a rainy night, a breath of wind
makes the pumpkin start sliding downward from rest. It
loses contact with the cap when the line from the center
of the hemisphere to the pumpkin makes a certain angle
with the vertical. What is this angle?

46. A child's pogo stick (Fig. P7.46) stores energy in a spring
with a force constant of 2.50×10^4 N/m. At position Ⓐ,
($x_A = -0.100$ m), the spring compression is a maximum

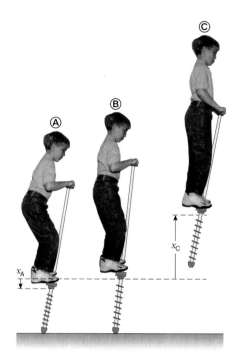

Figure P7.46

and the child is momentarily at rest. At position Ⓑ ($x_B = 0$), the spring is relaxed and the child is moving upward.
At position Ⓒ, the child is again momentarily at rest at the
top of the jump. The combined mass of child and pogo
stick is 25.0 kg. (a) Calculate the total energy of the system
if both the gravitational and elastic potential energies are
zero at $x = 0$. (b) Determine x_C. (c) Calculate the speed
of the child at Ⓑ. (d) Determine the value of x for which
the kinetic energy of the system is a maximum. (e) Calculate the child's maximum upward speed.

47. A 10.0-kg block is released from point Ⓐ in Figure P7.47.
The track is frictionless except for the portion between
points Ⓑ and Ⓒ, which has a length of 6.00 m. The block
travels down the track, hits a spring of force constant
2 250 N/m, and compresses the spring 0.300 m from its
equilibrium position before coming to rest momentarily.
Determine the coefficient of kinetic friction between the
block and the rough surface between Ⓑ and Ⓒ.

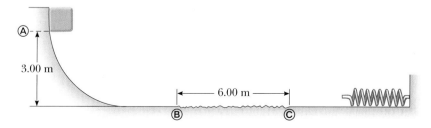

Figure P7.47

48. The potential energy function for a system is given by $U(x) = -x^3 + 2x^2 + 3x$. (a) Determine the force F_x as a function of x. (b) For what values of x is the force equal to zero? (c) Plot $U(x)$ versus x and F_x versus x, and indicate points of stable and unstable equilibrium.

49. A 20.0-kg block is connected to a 30.0-kg block by a string that passes over a light frictionless pulley. The 30.0-kg block is connected to a spring that has negligible mass and a force constant of 250 N/m, as shown in Figure P7.49. The spring is unstretched when the system is as shown in the figure, and the incline is frictionless. The 20.0-kg block is pulled 20.0 cm down the incline (so that the 30.0-kg block is 40.0 cm above the floor) and released from rest. Find the speed of each block when the 30.0-kg block is 20.0 cm above the floor (i.e., when the spring is unstretched).

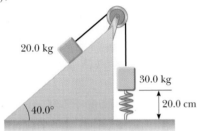

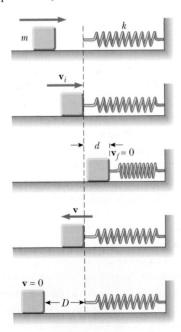

Figure P7.49

50. A 1.00-kg object slides to the right on a surface having a coefficient of kinetic friction 0.250 (Fig. P7.50). The object has a speed of $v_i = 3.00$ m/s when it makes contact

Figure P7.50

with a light spring that has a force constant of 50.0 N/m. The object comes to rest after the spring has been compressed a distance d. The object is then forced toward the left by the spring and continues to move in that direction beyond the spring's unstretched position. Finally the object comes to rest a distance D to the left of the unstretched spring. Find (a) the distance of compression d, (b) the speed v at the unstretched position when the object is moving to the left, and (c) the distance D.

51. A block of mass 0.500 kg is pushed against a horizontal spring of negligible mass until the spring is compressed a distance Δx (Fig. P7.51). The force constant of the spring is 450 N/m. When it is released, the block travels along a frictionless, horizontal surface to point B, the bottom of a vertical circular track of radius $R = 1.00$ m, and continues to move up the track. The speed of the block at the bottom of the track is $v_B = 12.0$ m/s, and the block experiences an average frictional force of 7.00 N while sliding up the track. (a) What is Δx? (b) What speed do you predict for the block at the top of the track? (c) Does the block actually reach the top of the track, or does it fall off before reaching the top?

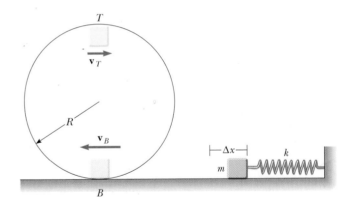

Figure P7.51

52. A uniform chain of length 8.00 m initially lies stretched out on a horizontal table. (a) If the coefficient of static friction between chain and table is 0.600, show that the chain will begin to slide off the table if at least 3.00 m of it hangs over the edge of the table. (b) Determine the speed of the chain as all of it leaves the table, given that the coefficient of kinetic friction between the chain and the table is 0.400.

53. A child slides without friction from a height h along a curved water slide (Fig. P7.53). She is launched from a height $h/5$ into the pool. Determine her maximum airborne height y in terms of h and θ.

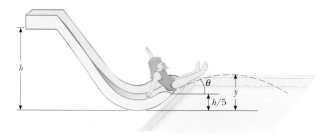

Figure P7.53

54. Jane, whose mass is 50.0 kg, needs to swing across a river (having width D) filled with crocodiles to save Tarzan from danger. She must swing into a wind exerting constant horizontal force **F**, on a vine having length L, and initially making an angle θ with the vertical (Fig. P7.54). Taking $D = 50.0$ m, $F = 110$ N, $L = 40.0$ m, and $\theta = 50.0°$, (a) with what minimum speed must Jane begin her swing to just make it to the other side? (*Hint:* First determine the potential energy associated with the wind force.) (b) Once the rescue is complete, Tarzan and Jane must swing back across the river. With what minimum speed must they begin their swing? Assume that Tarzan has a mass of 80.0 kg.

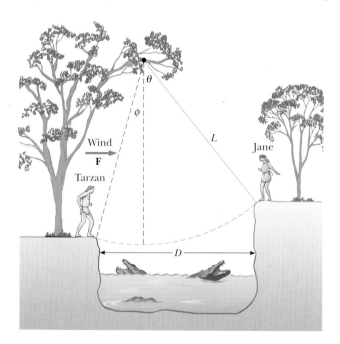

Figure P7.54

55. A 5.00-kg block free to move on a horizontal, frictionless surface is attached to one end of a light horizontal spring. The other end of the spring is held fixed. The spring is compressed 0.100 m from equilibrium and released. The speed of the block is 1.20 m/s when it passes the equilib-

rium position of the spring. The same experiment is now repeated with the frictionless surface replaced by a surface for which the coefficient of kinetic friction is 0.300. Determine the speed of the block at the equilibrium position of the spring.

56. A block of mass M rests on a table. It is fastened to the lower end of a light vertical spring. The upper end of the spring is fastened to a block of mass m. The upper block is pushed down by an additional force $3mg$, so that the spring compression is $4mg/k$. In this configuration the upper block is released from rest. The spring lifts the lower block off the table. In terms of m, what is the greatest possible value for M?

57. A roller coaster car is released from rest at the top of the first rise and then moves freely with negligible friction. The roller coaster shown in Figure P7.57 has a circular

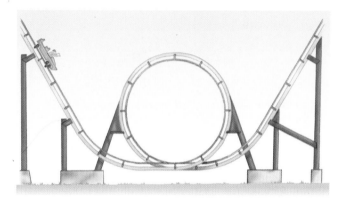

Figure P7.57

loop of radius R in a vertical plane. (a) Suppose first that the car barely makes it around the loop: At the top of the loop the riders are upside down and feel weightless. Find the required height of the release point above the bottom of the loop, in terms of R. (b) Now assume that the release point is at or above the minimum required height. Show that the normal force on the car at the bottom of the loop exceeds the normal force at the top of the loop by six times the weight of the car. The normal force on each rider follows the same rule. Such a large normal force is dangerous and very uncomfortable for the riders. Roller coasters are therefore not built with circular loops in vertical planes. Figure P5.21 and the photograph on page 150 show two actual designs.

58. **Review Problem.** In 1887 in Bridgeport, Connecticut, C. J. Belknap built the water slide shown in Figure P7.58. A rider on a small sled, of total mass 80.0 kg, pushed off to start at the top of the slide (point A) with a speed of 2.50 m/s. The chute was 9.76 m high at the top, 54.3 m long,

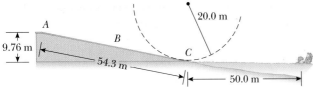

and 0.51 m wide. Along its length, 725 wheels made friction negligible. Upon leaving the chute horizontally at its bottom end (point *C*), the rider skimmed across the water of Long Island Sound for as much as 50 m, "skipping along like a flat pebble," before at last coming to rest and swimming ashore, pulling his sled after him. According to *Scientific American,* "The facial expression of novices taking their first adventurous slide is quite remarkable, and the sensations felt are correspondingly novel and peculiar." (a) Find the speed of the sled and rider at point *C*. (b) Model the force of water friction as a constant retarding force acting on a particle. Find the work done by water friction in stopping the sled and rider. (c) Find the magnitude of the force the water exerts on the sled. (d) Find the magnitude of the force the chute exerts on the sled at point *B*. (e) At point *C* the chute is horizontal but curving in the vertical plane. Assume its radius of curvature is 20.0 m. Find the force the chute exerts on the sled at point *C*.

Figure P7.58 (*Photo from* Scientific American, *July 1988*)

ANSWERS TO QUICK QUIZZES

7.1 Yes. Unlike kinetic energy, which is always positive, the gravitational potential energy of an object–Earth system can be negative if the object is located below the level at which *U* is defined to be zero.

7.2 All three balls have the same speed as they reach the ground. Each system of a ball and the Earth has the same potential energy at the top of the building, and all three balls start off from that point with the same kinetic energy. Because energy is conserved, all three balls will have the same kinetic energy and, therefore, the same speed when they strike the ground.

7.3 (c). The decrease in mechanical energy of the system is $f_k \Delta x$. This is smaller than the value on the horizontal surface for two reasons: (1) the force of kinetic friction f_k is smaller because the normal force is smaller, and (2) the displacement Δx is smaller because a component of the gravitational force is pulling on the block in the direction opposite to its velocity.

7.4 Energy transfer methods include work (you apply forces on objects that move), heat (your body is warmer than the surrounding air), matter transfer (breathing, eating), mechanical waves (you speak and hear), and electromagnetic radiation (you see, as well as absorb and emit radiation from your skin).

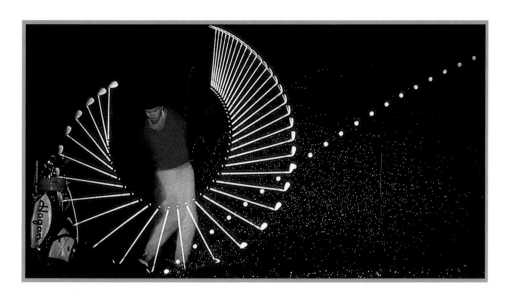

(Courtesy of Michael Hans/Photo
Researchers, Inc.)

chapter
8

During the brief time that a golf
club is in contact with the ball, the
speed of the ball is changed by a
large amount. During that time,
Newton's third law tells us that the
force on the club has the same
magnitude as the force on the
ball. But the speed of the club appears to change very little upon
hitting the ball in this photograph.
In this chapter, we will learn to understand such events.

Momentum and Collisions

Consider what happens when a golf ball is struck by a club as in the
opening photograph for this chapter. The ball is given a very large
initial velocity as a result of the collision; consequently, it is able to
travel a large distance through the air. Because the ball experiences
this change in velocity over a very short time interval, the average force on it during the collision is very large. By Newton's third law, the club experiences a reaction force equal in magnitude to and opposite the force on the ball. This reaction
force produces a change in the velocity of the club. Because the club is much
more massive than the ball, however, the change in velocity of the club is much
less than the change in velocity of the ball.

One of the main objectives of this chapter is to enable you to understand and
analyze such events. As a first step, we shall introduce the concept of *momentum,* a
term used to describe objects in motion. The concept of momentum leads us to a
second conservation law, conservation of momentum, and momentum approaches
for treating isolated and nonisolated systems. This conservation law is especially
useful for treating problems that involve collisions between objects.

8.1 • LINEAR MOMENTUM AND ITS CONSERVATION

In previous chapters, we discussed the velocity of an object as a measure of its motion. We will find in this chapter that this definition does not completely specify
the nature of motion. As an illustration, imagine that you have two objects moving
toward you, each traveling at 2 m/s. One is a Ping-Pong ball and the other is a
large truck. You most likely have an intuitive notion about which one you would

See the *Core Concepts in
Physics CD-ROM,* Screen 6.2

rather have strike your body. They both travel with the same speed, but something about the truck represents a larger measure of motion. We can use the notion of momentum to describe the difference between these two moving objects. The **linear momentum** (which we often call simply **momentum**) of an object of mass m moving with a velocity $\mathbf{v}$ is defined to be the product of the mass and velocity:*

- *Definition of linear momentum*

$$\mathbf{p} = m\mathbf{v} \qquad [8.1]$$

Because momentum equals the product of a scalar m and a vector $\mathbf{v}$, it is a vector quantity. Its direction is the same as that for $\mathbf{v}$, and it has dimensions ML/T. In the SI system, momentum has the units kg · m /s.

If an object is moving in an arbitrary direction in three-dimensional space, $\mathbf{p}$ has three components, and Equation 8.1 is equivalent to the component equations

$$p_x = mv_x \qquad p_y = mv_y \qquad p_z = mv_z \qquad [8.2]$$

As you can see from its definition, the concept of momentum provides a quantitative distinction between objects of different masses moving at the same velocity. For example, the momentum of the truck moving at 2 m/s is much greater in magnitude than that of the Ping-Pong ball moving at the same speed. Newton called the product $m\mathbf{v}$ the *quantity of motion*, perhaps a more graphic description than *momentum*, which comes from the Latin word for movement.

Quick Quiz 8.1

Two objects have equal kinetic energies. How do the magnitudes of their momenta compare? (a) $p_1 < p_2$, (b) $p_1 = p_2$, (c) $p_1 > p_2$, (d) not enough information to determine the answer.

Let us use the particle model for an object in motion. By using Newton's second law of motion, we can relate the linear momentum of a particle to the net force acting on the particle. In Chapter 4 we learned that Newton's second law can be written as $\Sigma\,\mathbf{F} = m\mathbf{a}$. This form applies only when the mass of the particle remains constant, however. In situations where the mass is changing with time, one must use an alternative statement of Newton's second law: **The time rate of change of momentum of a particle is equal to the net force acting on the particle.**

- *Newton's second law for a particle*

$$\Sigma\,\mathbf{F} = \frac{d\mathbf{p}}{dt} \qquad [8.3]$$

If the mass of the particle is constant, the preceding equation reduces to our previous expression for Newton's second law:

$$\Sigma\,\mathbf{F} = \frac{d\mathbf{p}}{dt} = \frac{d(m\mathbf{v})}{dt} = m\frac{d\mathbf{v}}{dt} = m\mathbf{a}$$

It is difficult to imagine a particle whose mass is changing, but if we consider objects, a number of examples emerge. These include a rocket that is ejecting its fuel as it operates, a snowball rolling down a hill and picking up additional snow, and a watertight pickup truck whose bed is collecting water as it moves in the rain.

---

* This expression is nonrelativistic and is valid only when $v \ll c$, where c is the speed of light. In the next chapter, we discuss momentum for high-speed particles.

From Equation 8.3 we see that if the net force on an object is zero, the time derivative of the momentum is zero, and therefore the momentum of the object must be constant. This should sound familiar because it is the case of a particle in equilibrium, expressed in terms of momentum. Of course, if the particle is *isolated* (that is, if it does not interact with its environment), then no forces act on it and **p** remains unchanged—this is Newton's first law.

Momentum and Isolated Systems

To see how to apply this new quantity, momentum, consider a system of two particles that can interact with each other but are isolated from their surroundings (Fig. 8.1). That is, the particles exert forces on each other, but no *external* forces are exerted on the system.

Suppose that at some instant, the momentum of particle 1 is **p**$_1$ and the momentum of particle 2 is **p**$_2$. Applying Newton's second law to each particle gives

$$\mathbf{F}_{21} = \frac{d\mathbf{p}_1}{dt} \quad \text{and} \quad \mathbf{F}_{12} = \frac{d\mathbf{p}_2}{dt}$$

where $\mathbf{F}_{21}$ is the force exerted by particle 2 on particle 1, and $\mathbf{F}_{12}$ is the force exerted by particle 1 on particle 2. (These forces could be gravitational forces, or they could have some other origin. The source of the forces isn't important for the present discussion.) Newton's third law tells us that $\mathbf{F}_{12}$ and $\mathbf{F}_{21}$ are equal in magnitude and opposite in direction. That is, they form an action–reaction pair, and $\mathbf{F}_{12} = -\mathbf{F}_{21}$. We can also express this condition as

$$\mathbf{F}_{21} + \mathbf{F}_{12} = 0$$

or, using Equation 8.3,

$$\frac{d\mathbf{p}_1}{dt} + \frac{d\mathbf{p}_2}{dt} = \frac{d}{dt}(\mathbf{p}_1 + \mathbf{p}_2) = 0$$

Because the time derivative of the total system momentum $\mathbf{p}_{\text{tot}} = \mathbf{p}_1 + \mathbf{p}_2$ is *zero*, we conclude that the *total* momentum $\mathbf{p}_{\text{tot}}$ must remain constant:

$$\mathbf{p}_{\text{tot}} = \text{constant} \qquad [8.4]$$

or, equivalently,

$$\mathbf{p}_{1i} + \mathbf{p}_{2i} = \mathbf{p}_{1f} + \mathbf{p}_{2f} \qquad [8.5]$$

where $\mathbf{p}_{1i}$ and $\mathbf{p}_{2i}$ are initial values and $\mathbf{p}_{1f}$ and $\mathbf{p}_{2f}$ are final values of the momentum during a period over which the particles interact. Equation 8.5 in component form states that the momentum components of the isolated system in the *x*, *y*, and *z* directions are all *independently constant;* that is,

$$\sum_{\text{system}} p_{ix} = \sum_{\text{system}} p_{fx} \quad \sum_{\text{system}} p_{iy} = \sum_{\text{system}} p_{fy} \quad \sum_{\text{system}} p_{iz} = \sum_{\text{system}} p_{fz} \qquad [8.6]$$

This result is known as the law of **conservation of linear momentum.** It is the mathematical representation of the momentum version of the isolated system model. It is considered one of the most important laws of mechanics. We have generated this law for a system of two interacting particles, but it can be shown to be true for a system of any number of particles. We can state it as follows: **The total momentum of an isolated system remains constant.**

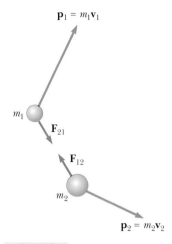

Figure 8.1

A system of two interacting particles. At some instant, the momentum of the particle with mass m_1 is $\mathbf{p}_1 = m_1\mathbf{v}_1$ and the momentum of the particle with mass m_2 is $\mathbf{p}_2 = m_2\mathbf{v}_2$. Note that $\mathbf{F}_{12} = -\mathbf{F}_{21}$. The total momentum $\mathbf{p}_{\text{tot}}$ of the system of two particles is $\mathbf{p}_1 + \mathbf{p}_2$.

• *Principle of conservation of momentum for an isolated system*

Notice that we have made no statement concerning the nature of the forces acting. The only requirement is that the forces must be *internal* to the system. Thus, momentum is conserved for an isolated system *regardless* of the nature of the internal forces, *even if the forces are nonconservative.*

PITFALL PREVENTION 8.1

Momentum of a system is conserved

Remember that the momentum of an isolated *system* is conserved. The momentum of one particle within an isolated system is not necessarily conserved, because other particles in the system may be interacting with it. Always apply conservation of momentum to an isolated *system.*

THINKING PHYSICS 8.1

A baseball is projected into the air at an upward angle to the ground. As it moves through its trajectory, its velocity and therefore its momentum constantly change. Is this a violation of conservation of momentum?

Reasoning The principle of conservation of momentum states that the momentum of an isolated system of particles is conserved. The statement of the question suggests that the ball is to be considered the system. A ball projected through the air is subject to the gravitational force as an external force, so it is *not* an isolated system, and we would not expect its momentum to be conserved. If we consider this in a little more detail, the gravitational force is in the vertical direction, so it is actually only the vertical component of the momentum that changes due to this force. In the horizontal direction, no force is exerted (ignoring air friction), so the horizontal component of the momentum is conserved. In Chapter 3, we used the same idea in stating that the horizontal component of the *velocity* of a projectile remains constant.

If we consider the baseball and the Earth as a system of two particles, the gravitational force is an internal force to this system. The momentum of the ball–Earth system is conserved. The downward force from the thrower's feet as the ball is thrown gives the Earth an initial motion. As the ball rises and falls, the Earth initially sinks and then rises (although imperceptibly!), due to the upward gravitational force from the ball, so that the total momentum of the system remains unchanged.

Example 8.1   Can We Really Ignore the Kinetic Energy of the Earth?

In Section 7.1, we claimed that we can ignore the kinetic energy of the Earth when considering the energy of a system consisting of the Earth and a dropped ball. Verify this claim.

Solution We will verify this claim by setting up a ratio of the kinetic energy of the Earth to that of the ball:

$$(1) \quad \frac{K_E}{K_b} = \frac{\frac{1}{2} m_E v_E^2}{\frac{1}{2} m_b v_b^2} = \left(\frac{m_E}{m_b} \right)\left(\frac{v_E}{v_b} \right)^2$$

where v_E and v_b are the speeds of the Earth and the ball, respectively, after the ball has fallen through some distance. Now we find a relationship between these two speeds by considering conservation of momentum in the vertical direction for the system of the ball and the Earth. The initial momentum of the system is zero, so the final momentum must also be zero:

$$p_i = p_f \quad \longrightarrow \quad 0 = m_b v_b + m_E v_E$$
$$\longrightarrow \quad \frac{v_E}{v_b} = -\left(\frac{m_b}{m_E} \right)$$

Substituting for v_E/v_b in (1), we have

$$\frac{K_E}{K_b} = \left(\frac{m_E}{m_b} \right)\left(-\frac{m_b}{m_E} \right)^2 = \frac{m_b}{m_E}$$

Substituting order-of-magnitude numbers for the masses, this ratio becomes

$$\frac{K_E}{K_b} = \frac{m_b}{m_E} \sim \frac{1 \text{ kg}}{10^{24} \text{ kg}} \sim 10^{-24}$$

The kinetic energy of the Earth is a very small fraction of the kinetic energy of the ball, so we are justified in ignoring it in the kinetic energy of the system. A similar calculation will verify the statement in Thinking Physics 8.1 that the Earth moves imperceptibly when a ball is thrown.

Example 8.2 The Recoiling Pitching Machine

A baseball player wishes to maintain his physical condition during the winter. He uses a 50.0-kg pitching machine to help him, placing the machine on the pitcher's mound as in Figure 8.2. The ground is covered with a thin layer of ice, so that friction between the ground and the machine is negligible. The machine fires a 0.15-kg baseball horizontally with a speed of 36 m/s. What is the recoil speed of the machine?

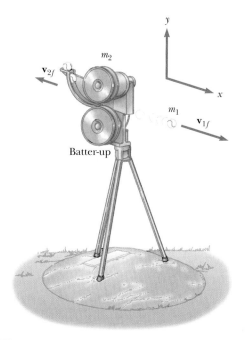

Figure 8.2

(Example 8.2) When the baseball is projected to the right, the pitching machine recoils to the left. The momentum of the baseball–machine system is conserved.

Reasoning We identify the isolated system as the baseball and the pitching machine. Because of the gravitational force and the normal force from the ground, the system is not truly isolated. Both of these forces are in the *y* direction, however, and therefore are directed perpendicularly to the motion of the members of the system. In addition, there is no friction between the machine and the ground. Momentum is therefore conserved in the *x* direction because there are no external forces in this direction.

We will restrict our attention to the *x* direction and use the symbol *v* for *components* of the velocity in the *x* direction. (We will not use a subscript *x* because we will already have two subscripts, one to identify the particle and one representing initial or final values!) Thus, for example, a negative value for *v* means that the velocity component is in the negative direction.

Solution The total momentum of the system before firing is zero. The total momentum after firing must therefore be zero; that is,

$$m_1 v_{1f} + m_2 v_{2f} = 0$$

With $m_1 = 0.15$ kg, $v_{1i} = 36$ m/s, and $m_2 = 50$ kg, solving for v_{2f}, we find the recoil speed of the pitching machine to be

$$v_{2f} = -\frac{m_1}{m_2} v_{1f} = -\left(\frac{0.15 \text{ kg}}{50 \text{ kg}}\right)(36 \text{ m/s}) = \boxed{-0.11 \text{ m/s}}$$

As alluded to in the Reasoning section, the negative sign for v_{2f} indicates that the pitching machine is moving to the left after firing, in the direction opposite the direction of motion of the baseball. Because the pitching machine is much more massive than the ball, the acceleration and consequent speed of the pitching machine are much smaller than the acceleration and speed of the ball.

Example 8.3 Decay of the Kaon at Rest

One type of nuclear particle, called the *neutral kaon* (K^0), decays into a pair of other particles called *pions* (π^+ and π^-), which are oppositely charged but equal in mass, as in Figure 8.3. Assuming the kaon is initially at rest, prove that the two pions must have momenta that are equal in magnitude and opposite in direction.

Solution The isolated system is the kaon before the decay and the two pions afterward. The decay of the kaon, represented in Figure 8.3, can be written

$$K^0 \longrightarrow \pi^+ + \pi^-$$

If we let $\mathbf{p}^+$ be the momentum of the positive pion and $\mathbf{p}^-$ be the momentum of the negative pion after the decay, the

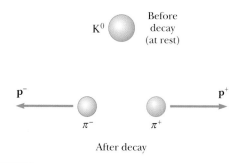

Figure 8.3

(Example 8.3) A kaon at rest decays into a pair of oppositely charged pions. The pions move apart with momenta of equal magnitudes but opposite directions.

final momentum of the isolated system of two pions can be written

$$\sum_{\text{system}} \mathbf{p}_f = \mathbf{p}^+ + \mathbf{p}^-$$

Because the kaon is at rest before the decay, we know that $\Sigma_{\text{system}} \mathbf{p}_i = 0$. Furthermore, because momentum is conserved, $\Sigma_{\text{system}} \mathbf{p}_i = \Sigma_{\text{system}} \mathbf{p}_f = 0$, so that $\mathbf{p}^+ + \mathbf{p}^- = 0$, or

$$\mathbf{p}^+ = -\mathbf{p}^-$$

Thus, we see that the two momentum vectors of the pions are equal in magnitude and opposite in direction.

EXERCISE A 60-kg boy and a 40-kg girl, both wearing skates, face each other at rest. The girl pushes the boy, sending him eastward with a speed of 4.0 m/s. Describe the subsequent motion of the girl. (Ignore friction.)

Answer She moves westward with a speed of 6.0 m/s.

8.2 • IMPULSE AND MOMENTUM

See Screens 6.3 & 6.4

As described by Equation 8.3, the momentum of a particle changes if a net force acts on the particle. Let us assume that a net force $\Sigma \mathbf{F}$ acts on a particle and that this force may vary with time. According to Equation 8.3,

$$d\mathbf{p} = \sum \mathbf{F} \, dt \qquad [8.7]$$

We can integrate this expression to find the change in the momentum of a particle during the time interval $\Delta t = t_f - t_i$. Integrating Equation 8.7 gives

$$\Delta \mathbf{p} = \mathbf{p}_f - \mathbf{p}_i = \int_{t_i}^{t_f} \sum \mathbf{F} \, dt \qquad [8.8]$$

The integral of a force over the time during which it acts is called the **impulse** of the force. The impulse of the net force $\Sigma \mathbf{F}$ is a vector defined by

• *Impulse of a net force*

$$\mathbf{I} \equiv \int_{t_i}^{t_f} \sum \mathbf{F} \, dt = \Delta \mathbf{p} \qquad [8.9]$$

Thus, the total **impulse** of the net force $\Sigma \mathbf{F}$ on a particle equals the change in the momentum of the particle. This statement, known as the **impulse–momentum theorem,** is equivalent to Newton's second law. It also applies to a system of particles, in which we consider that the net force external to the system causes a change in the total momentum of the system:

• *Impulse–momentum theorem*

$$\mathbf{I} \equiv \int_{t_i}^{t_f} \sum \mathbf{F}_{\text{ext}} \, dt = \Delta \mathbf{p}_{\text{tot}} \qquad [8.10]$$

Impulse is an interaction between the system and its environment. As a result of this interaction, the momentum of the system changes. This is an analog to the continuity equation for energy, in which an interaction with the environment causes the energy of the system to change. Therefore, when we say that an impulse is given to a system, we imply that momentum is transferred from an external agent to that system. In many situations, the system can be modeled as a particle, so Equation 8.9 can be used rather than the more general Equation 8.10.

From the definition, we see that impulse is a vector quantity having a magnitude equal to the area under the curve of the magnitude of the net force versus time, as described in Figure 8.4. In this figure it is assumed that the net force

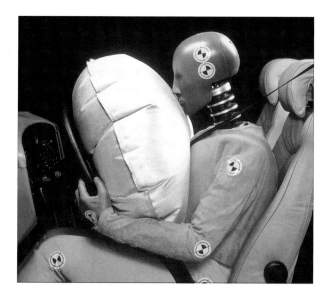

You experience the same change in momentum in a collision whether your car has air bags or not. The air bag allows you to experience that change in momentum over a longer time interval, however, reducing the peak force on you and increasing your chances of escaping without injury. *(Courtesy of Saab)*

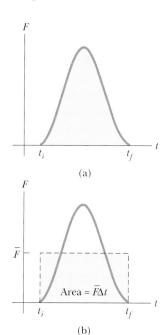

(a)

(b)

Figure 8.4

(a) A net force acting on a particle may vary in time. The impulse is the area under the curve of the magnitude of the net force versus time. (b) The average force (*horizontal dashed line*) gives the same impulse to the particle in the time Δt as the time-varying force described in part (a). The area of the rectangle is the same as the area under the curve.

varies in time in the general manner shown and is nonzero in the time interval $\Delta t = t_f - t_i$. The direction of the impulse vector is the same as the direction of the change in momentum. Impulse has the dimensions of momentum, ML/T.

Because the force can generally vary in time as in Figure 8.4a, it is convenient to define a time-averaged net force $\sum \overline{\mathbf{F}}$ given by

$$\sum \overline{\mathbf{F}} \equiv \frac{1}{\Delta t} \int_{t_i}^{t_f} \sum \mathbf{F} \, dt \qquad \text{[8.11]}$$

where $\Delta t = t_f - t_i$. Therefore, we can express Equation 8.9 as

$$\mathbf{I} = \Delta \mathbf{p} = \sum \overline{\mathbf{F}} \, \Delta t \qquad \text{[8.12]}$$

The magnitude of this average net force, described in Figure 8.4b, can be thought of as the magnitude of the constant net force that would give the same impulse to the particle in the time interval Δt as the actual time-varying net force gives over this same interval.

In principle, if $\sum \mathbf{F}$ is known as a function of time, the impulse can be calculated from Equation 8.9. The calculation becomes especially simple if the net force acting on the particle is constant. In this case, $\sum \overline{\mathbf{F}}$ over a time interval is the same as the constant $\sum \mathbf{F}$ at any instant, and Equation 8.12 becomes

$$\mathbf{I} = \Delta \mathbf{p} = \sum \mathbf{F} \, \Delta t \qquad \text{[8.13]}$$

In many physical situations, we shall use what is called the **impulse approximation: We assume that one of the forces exerted on a particle acts for a short time but is much greater than any other force present.** This simplification model allows us to ignore the effects of other forces, because these effects are be small during the short time during which the large force acts. This approximation is especially useful in treating collisions, in which the duration of the collision is very short. When this approximation is made, we refer to the force as an *impulsive force*. For example, when a baseball is struck with a bat, the duration of the

• *Impulse approximation*

WEB

For information on the use of air bags and seatbelts in automobiles, visit the National Safety Council at **www.nsc.org/airbag.htm**

collision is about 0.01 s, and the average force the bat exerts on the ball during this time interval is typically several thousand newtons. This is much greater than the gravitational force, so we ignore any change in velocity related to the gravitational force during the collision. It is important to remember that $\mathbf{p}_i$ and $\mathbf{p}_f$ represent the momenta *immediately* before and after the collision, respectively. In the impulse approximation, therefore, very little motion of the particle takes place during the collision.

Quick Quiz 8.2

Two objects are at rest on a frictionless surface. Object 1 has a larger mass than object 2. When a force is applied to object 1, it accelerates through a displacement Δx. The force is removed from object 1 and is applied to object 2. At the moment when object 2 accelerates through the same displacement Δx, which statements are true? (a) $p_1 < p_2$, (b) $p_1 = p_2$, (c) $p_1 > p_2$, (d) $K_1 < K_2$, (e) $K_1 = K_2$, (f) $K_1 > K_2$.

APPLICATION

Boxing and brain injury

Quick Quiz 8.3

In boxing matches of the 19th century, bare fists were used. In modern boxing, fighters wear padded gloves. How does this better protect the brain of the boxer from injury?

THINKING PHYSICS 8.2

A race car travels rapidly around a circular race track at constant speed. For a given portion of the track, what is the direction of the impulse vector? Once the car returns to the starting point, there is no net change in momentum for the entire trip around the track. Thus, according to the impulse—momentum theorem, there must be no impulse. Yet a nonzero force acted over the time interval for one rotation around the track. How can the impulse be zero?

Reasoning For the movement around a short portion of the track, the direction of the impulse vector is the same as that of the change in momentum. The direction of the vector representing the change in momentum is the same as the direction of the vector representing the change in velocity. For a particle in uniform circular motion, we know that the direction of the vector representing the change in velocity is that of the acceleration vector, which is toward the center of the circle. Thus, the impulse vector is directed toward the center of the circle.

If the car makes one rotation around the track, there is indeed no net change in momentum, so there must be zero net impulse. The force on the car causing the circular motion is the friction between the tires and the roadway, and this force is always directed toward the center of the circle. For every location of the car on the track, another point of its motion is diametrically opposed across the circle; at this point the force vector is directed in the opposite direction. Thus, as we add up the vector impulses around the circle to find the total impulse, we discover that they cancel in pairs, for a total impulse of zero.

Example 8.4   How Good Are the Bumpers?

In a crash test, an automobile of mass 1 500 kg collides with a wall, as in Figure 8.5. The initial and final velocities of the automobile are $\mathbf{v}_i = -15.0\mathbf{i}$ m/s and $\mathbf{v}_f = 2.60\mathbf{i}$ m/s. If the collision lasts for 0.150 s, find the impulse due to the collision and the average force exerted on the automobile.

Solution   We identify the automobile as the system. The initial and final momenta of the automobile are

$$\mathbf{p}_i = m\mathbf{v}_i = (1\,500 \text{ kg})\,(-15.0\mathbf{i} \text{ m/s})$$
$$= -2.25 \times 10^4\mathbf{i} \text{ kg} \cdot \text{m/s}$$

$$\mathbf{p}_f = m\mathbf{v}_f = (1\,500 \text{ kg})\,(2.60\mathbf{i} \text{ m/s})$$
$$= 0.390 \times 10^4\mathbf{i} \text{ kg} \cdot \text{m/s}$$

Hence, the impulse is

$$\mathbf{I} = \Delta\mathbf{p} = \mathbf{p}_f - \mathbf{p}_i$$
$$= 0.390 \times 10^4\mathbf{i} \text{ kg} \cdot \text{m/s} - (-2.25 \times 10^4\mathbf{i} \text{ kg} \cdot \text{m/s})$$

$$\mathbf{I} = \boxed{2.64 \times 10^4\mathbf{i} \text{ kg} \cdot \text{m/s}}$$

The average force exerted on the automobile is

$$\overline{\mathbf{F}} = \frac{\Delta\mathbf{p}}{\Delta t} = \frac{2.64 \times 10^4\mathbf{i} \text{ kg} \cdot \text{m/s}}{0.150 \text{ s}} = \boxed{1.76 \times 10^5\mathbf{i} \text{ N}}$$

EXERCISE   A 0.150-kg baseball is thrown with a speed of 40.0 m/s. It is hit straight back at the pitcher with a speed of 50.0 m/s. (a) What impulse is delivered to the baseball? (b) Find the average force exerted by the bat on the ball if the two are in contact for 2.00×10^{-3} s.

Answer   (a) 13.5 kg·m/s   (b) 6.75 kN

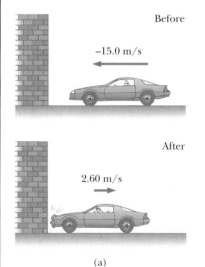

Before

–15.0 m/s

After

2.60 m/s

(a)

(b)

Figure 8.5

(Example 8.4) (a) The car's momentum changes as a result of its collision with the wall. (b) In a crash test, the large force exerted by the wall on the car produces extensive damage to the front end of the car. *(Courtesy of General Motors)*

8.3 • COLLISIONS

In this section we use the law of conservation of momentum to describe what happens when two objects collide. The forces due to the collision are assumed to be much larger than any external forces present, so we use the simplification model we call the impulse approximation. The general goal in collision problems is to relate the final conditions of the system to the initial conditions.

A collision may be the result of physical contact between two objects, as described in Figure 8.6a. This is a common observation when two macroscopic objects, such as two billiard balls or a baseball and a bat, collide.

The notion of what we mean by *collision* must be generalized because "contact" on a microscopic scale is ill defined. To understand the distinction between

See Screens 6.5 & 6.6

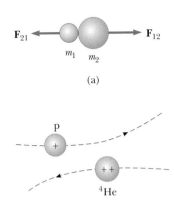

(a)

(b)

Figure 8.6

(a) A collision between two objects as the result of direct contact. (b) A "collision" between two charged particles that do not make contact.

PITFALL PREVENTION 1.1

Momentum and kinetic energy in collisions

 Momentum of an isolated system is conserved in *all* collisions. Kinetic energy of an isolated system is conserved *only* in elastic collisions. This is true because kinetic energy can be transformed into several types of energy or be transferred out of the system (so that the system may *not* be isolated in terms of energy during the collision), but momentum comes in only one type.

---

macroscopic and microscopic collisions, consider the collision of a proton with an alpha particle (the nucleus of the helium atom), illustrated in Figure 8.6b. Because the two particles are positively charged, they repel each other. A collision has occurred, but the colliding particles were never in "contact."

When two particles of masses m_1 and m_2 collide, the collision forces may vary in time in a complicated way, as seen in Figure 8.4. As a result, an analysis of the situation with Newton's second law could be very complicated. We find, however, that the momentum concept is similar to the energy concept in Chapters 6 and 7, in that it provides us with a much easier method to solve problems involving isolated systems.

According to Equation 8.5, the momentum of an isolated system is conserved during some interaction event, such as a collision. The kinetic energy of the system, however, is generally *not* conserved in a collision. We define an **inelastic collision** as one in which the kinetic energy of the system is not conserved (even though momentum is conserved). The collision of a rubber ball with a hard surface is inelastic because some of the kinetic energy of the ball is transformed to internal energy when the ball is deformed while in contact with the surface.

When two objects collide and stick together after a collision, the maximum possible fraction of the initial kinetic energy is transformed, and the collision is called **perfectly inelastic.** For example, if two vehicles collide and become entangled, they move with some common velocity after the perfectly inelastic collision. If a meteorite collides with the Earth, it becomes buried in the ground, and the collision is perfectly inelastic.

An **elastic collision** is defined as one in which the kinetic energy of the system is conserved (as well as momentum). Real collisions in the macroscopic world, such as those between billiard balls, are only approximately elastic because some transformation of kinetic energy takes place, and some energy leaves the system by mechanical waves—sound. Imagine a billiard game with truly elastic collisions. The opening break would be completely silent! Truly elastic collisions do occur between atomic and subatomic particles. Elastic and perfectly inelastic collisions are *limiting* cases; a large number of collisions fall in the range between them.

In the remainder of this section we treat collisions in one dimension and consider the two extreme cases: perfectly inelastic and elastic collisions. The important distinction between these two types of collisions is that **momentum is conserved in all cases, but kinetic energy is conserved only in elastic collisions.**

Quick Quiz 8.4

A skater is using very low friction rollerblades. A friend throws a Frisbee at her, along the straight line along which she is skating. Describe each of the following events as an elastic, an inelastic, or a perfectly inelastic collision between the skater and the Frisbee: (a) she catches the Frisbee and holds it, (b) she tries to catch the Frisbee but it bounces off her hands and falls to the ground at her feet, (c) she catches the Frisbee and immediately throws it back with the same speed (relative to the ground) to her friend.

One-Dimensional Perfectly Inelastic Collisions

Consider two objects of masses m_1 and m_2 moving with initial velocity components v_{1i} and v_{2i} along a straight line, as in Figure 8.7. If the two objects collide head-on, stick together, and move with some common velocity component v_f after the collision, the collision is perfectly inelastic. Because the total momentum of the two-object isolated system before the collision equals the total momentum of the combined-object system after the collision, we have

$$m_1 v_{1i} + m_2 v_{2i} = (m_1 + m_2) v_f \qquad \text{[8.14]}$$

$$v_f = \frac{m_1 v_{1i} + m_2 v_{2i}}{m_1 + m_2} \qquad \text{[8.15]}$$

Thus, knowing the initial velocity components of the two objects, we can use this single equation to determine the final common velocity component.

One-Dimensional Elastic Collisions

Now consider two objects that undergo an elastic head-on collision (Fig. 8.8) in one dimension. In this collision, both momentum and kinetic energy are conserved; therefore, we can write*

$$m_1 v_{1i} + m_2 v_{2i} = m_1 v_{1f} + m_2 v_{2f} \qquad \text{[8.16]}$$

$$\tfrac{1}{2} m_1 v_{1i}^2 + \tfrac{1}{2} m_2 v_{2i}^2 = \tfrac{1}{2} m_1 v_{1f}^2 + \tfrac{1}{2} m_2 v_{2f}^2 \qquad \text{[8.17]}$$

In a typical problem involving elastic collisions, two unknown quantities occur, and Equations 8.16 and 8.17 can be solved simultaneously to find them. An alternative approach, employing a little mathematical manipulation of Equation 8.17, often simplifies this process. To see this, let us cancel the factor of $\tfrac{1}{2}$ in Equation 8.17 and rewrite the equation as

$$m_1(v_{1i}^2 - v_{1f}^2) = m_2(v_{2f}^2 - v_{2i}^2)$$

Here we have moved the terms containing m_1 to one side of the equation and those containing m_2 to the other. Next, let us factor both sides:

$$m_1(v_{1i} - v_{1f})(v_{1i} + v_{1f}) = m_2(v_{2f} - v_{2i})(v_{2f} + v_{2i}) \qquad \text{[8.18]}$$

We now separate the terms containing m_1 and m_2 in the equation for the conservation of momentum (Eq. 8.16) to obtain

$$m_1(v_{1i} - v_{1f}) = m_2(v_{2f} - v_{2i}) \qquad \text{[8.19]}$$

To obtain our final result, we divide Equation 8.18 by Equation 8.19 and obtain

$$v_{1i} + v_{1f} = v_{2f} + v_{2i}$$

or, gathering initial and final values on opposite sides of the equation,

$$v_{1i} - v_{2i} = -(v_{1f} - v_{2f}) \qquad \text{[8.20]}$$

This equation, in combination with the condition for conservation of momentum, Equation 8.16, can be used to solve problems dealing with one-dimensional

---

* Notice that the kinetic energy of the system is the sum of the kinetic energies of the two particles. In our energy conservation examples in Chapter 7 involving a falling object and the Earth, we ignored the kinetic energy of the Earth because it was so small. Thus, the kinetic energy of the *system* was just the kinetic energy of the falling *object*. This was a special case, in which the mass of one of the objects (the Earth) was so immense that ignoring its kinetic energy introduced no measurable error. In problems such as those described here, however, and for the particle decay problems we will see in Chapters 30 and 31, we need to include the kinetic energies of *all* particles in the system.

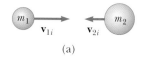

Before collision

(a)

After collision

$m_1 + m_2$

(b)

Figure 8.7

A perfectly inelastic head-on collision between two particles: (a) before the collision and (b) after the collision.

PITFALL PREVENTION 8.3
Perfectly inelastic collisions

Keep in mind the distinction between inelastic and perfectly inelastic collisions. If the colliding particles stick together, the collision is perfectly inelastic. If they bounce off each other (and kinetic energy is not conserved), the collision is inelastic. Generally, inelastic collisions are hard to analyze unless additional information is provided. This appears in the mathematical representation as having more unknowns than equations.

Before collision

(a)

After collision

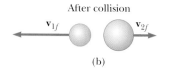

(b)

Figure 8.8

An elastic head-on collision between two particles: (a) before the collision and (b) after the collision.

elastic collisions between two objects. According to Equation 8.20, the relative speed* $v_{1i} - v_{2i}$ of the two objects before the collision equals the negative of their relative speed after the collision, $-(v_{1f} - v_{2f})$.

Suppose that the masses and the initial velocity components of both objects are known. Equations 8.16 and 8.20 can be solved for the final velocity components in terms of the initial values, because we have two equations and two unknowns:

$$v_{1f} = \left(\frac{m_1 - m_2}{m_1 + m_2}\right) v_{1i} + \left(\frac{2m_2}{m_1 + m_2}\right) v_{2i} \qquad [8.21]$$

$$v_{2f} = \left(\frac{2m_1}{m_1 + m_2}\right) v_{1i} + \left(\frac{m_2 - m_1}{m_1 + m_2}\right) v_{2i} \qquad [8.22]$$

It is important to remember that the appropriate signs for the velocity components v_{1i} and v_{2i} must be included in Equations 8.21 and 8.22. For example, if m_2 is moving to the left initially, as in Figure 8.8, then v_{2i} is negative.

Let us consider some special cases. If $m_1 = m_2$, then Equations 8.21 and 8.22 show us that $v_{1f} = v_{2i}$ and $v_{2f} = v_{1i}$. That is, the objects exchange speeds if they have equal masses. This is what one observes in head-on billiard ball collisions: The initially moving ball stops and the initially stationary ball moves away with approximately the same speed.

If m_2 is initially at rest, $v_{2i} = 0$, and Equations 8.21 and 8.22 become

$$v_{1f} = \left(\frac{m_1 - m_2}{m_1 + m_2}\right) v_{1i} \qquad [8.23]$$

$$v_{2f} = \left(\frac{2m_1}{m_1 + m_2}\right) v_{1i} \qquad [8.24]$$

If m_1 is very large compared with m_2, we see from Equations 8.23 and 8.24 that $v_{1f} \approx v_{1i}$ and $v_{2f} \approx 2v_{1i}$. That is, when a very heavy object collides head-on with a very light one initially at rest, the heavy object continues its motion unaltered after the collision, but the light object rebounds with a speed equal to about twice the initial speed of the heavy object. An example of such a collision is that of a moving heavy atom, such as uranium, with a light atom, such as hydrogen.

If m_2 is much larger than m_1, and m_2 is initially at rest, then we find from Equations 8.23 and 8.24 that $v_{1f} \approx -v_{1i}$ and $v_{2f} \approx 0$. That is, when a very light object collides head-on with a very heavy object initially at rest, the velocity of the light object is reversed, and the heavy object remains approximately at rest. For example, imagine what happens when a marble hits a stationary bowling ball.

Quick Quiz 8.5

In a perfectly inelastic collision between two identical objects, under what conditions is all of the initial kinetic energy of the system of two objects transformed to other forms of energy?

* See Section 3.6 for a review of relative speed.

Example 8.5   Kinetic Energy in a Perfectly Inelastic Collision

We claimed that the maximum amount of kinetic energy was transformed to other forms in a perfectly inelastic collision. Prove this statement mathematically for a one-dimensional two-particle collision.

Solution   We will assume that the maximum kinetic energy is transformed and prove that the collision must be perfectly inelastic. We set up the fraction f of the final kinetic energy after the collision to the initial kinetic energy:

$$f = \frac{K_f}{K_i} = \frac{\frac{1}{2}m_1 v_{1f}^2 + \frac{1}{2}m_2 v_{2f}^2}{\frac{1}{2}m_1 v_{1i}^2 + \frac{1}{2}m_2 v_{2i}^2} = \frac{m_1 v_{1f}^2 + m_2 v_{2f}^2}{m_1 v_{1i}^2 + m_2 v_{2i}^2}$$

The *maximum* amount of energy transformed to other forms corresponds to the *minimum* value of f. For fixed initial conditions, we imagine that the final velocity components v_{1f} and v_{2f} are variables. We minimize the fraction f by taking the derivative of f with respect to v_{1f} and setting the result equal to zero:

$$(1) \quad \frac{df}{dv_{1f}} = \frac{d}{dv_{1f}}\left(\frac{m_1 v_{1f}^2 + m_2 v_{2f}^2}{m_1 v_{1i}^2 + m_2 v_{2i}^2} \right)$$

$$= \frac{m_1 2v_{1f} + m_2 2v_{2f}\, \dfrac{dv_{2f}}{dv_{1f}}}{m_1 v_{1i}^2 + m_2 v_{2i}^2} = 0$$

$$\longrightarrow \quad m_1 v_{1f} + m_2 v_{2f}\, \frac{dv_{2f}}{dv_{1f}} = 0$$

From the conservation of momentum condition, we can evaluate the derivative in (1). We differentiate Equation 8.16 with respect to v_{1f}:

$$\frac{d}{dv_{1f}}(m_1 v_{1i} + m_2 v_{2i}) = \frac{d}{dv_{1f}}(m_1 v_{1f} + m_2 v_{2f})$$

$$\longrightarrow \quad 0 = m_1 + m_2\, \frac{dv_{2f}}{dv_{1f}} \quad \longrightarrow \quad \frac{dv_{2f}}{dv_{1f}} = -\frac{m_1}{m_2}$$

Substituting this expression for the derivative into (1), we find,

$$m_1 v_{1f} - m_2 v_{2f}\, \frac{m_1}{m_2} = 0 \quad \longrightarrow \quad v_{1f} = v_{2f}$$

If the particles come out of the collision with the same velocity components, they are joined together and this is a perfectly inelastic collision, which is what we set out to prove.

Example 8.6   Carry Collision Insurance

An 1 800-kg car stopped at a traffic light is struck from the rear by a 900-kg car and the two become entangled. If the smaller car was moving at 20.0 m/s before the collision, what is the speed of the entangled cars after the collision?

Reasoning   The total momentum of the system (the two cars) before the collision equals the total momentum of the system after the collision because the system is isolated (in the impulse approximation). Notice that we ignore friction with the road in the impulse approximation. Thus, the result we obtain for the final speed will only be approximately true just after the collision. For longer time intervals after the collision, we would use Newton's second law to describe the slowing down of the system due to friction. Because the cars "become entangled," this is a perfectly inelastic collision.

Solution   The magnitude of the total momentum of the system before the collision is equal to that of only the smaller car because the larger car is initially at rest:

$$p_i = m_1 v_i = (900 \text{ kg})(20.0 \text{ m/s}) = 1.80 \times 10^4 \text{ kg} \cdot \text{m/s}$$

After the collision, the mass that moves is the sum of the masses of the cars. The magnitude of the momentum of the combination is

$$p_f = (m_1 + m_2)v_f = (2\,700 \text{ kg})v_f$$

Equating the initial momentum to the final momentum and solving for v_f, the speed of the entangled cars, we have

$$v_f = \frac{p_i}{m_1 + m_2} = \frac{1.80 \times 10^4 \text{ kg} \cdot \text{m/s}}{2\,700 \text{ kg}} = \boxed{6.67 \text{ m/s}}$$

Example 8.7   Slowing Down Neutrons by Collisions

In a nuclear reactor, neutrons are produced when $^{235}_{92}U$ atoms split in a process called *fission*. These neutrons are moving at about 10^7 m/s and must be slowed down to about 10^3 m/s before they take part in another fission event. They are slowed down by being passed through a solid or liquid material called a *moderator*. The slowing-down process involves elastic collisions. Let us show that a neutron can lose most of its kinetic energy if it collides elastically with a moderator containing light nuclei, such as deuterium (in "heavy water," D_2O).

Reasoning   We identify the system as the neutron and a moderator nucleus. Because the momentum and kinetic energy of this system are conserved in an elastic collision, Equations 8.23 and 8.24 can be applied to a one-dimensional collision of these two particles.

Solution   Let us assume that the moderator nucleus of mass m_m is at rest initially and that the neutron of mass m_n and initial speed v_{ni} collides head-on with it. The initial kinetic energy of the neutron is

$$K_{ni} = \tfrac{1}{2}m_n v_{ni}^2$$

After the collision, the neutron has kinetic energy $\tfrac{1}{2}m_n v_{nf}^2$, where v_{nf} is given by Equation 8.23.

$$K_{nf} = \tfrac{1}{2}m_n v_{nf}^2 = \tfrac{1}{2}m_n \left(\frac{m_n - m_m}{m_n + m_m}\right)^2 v_{ni}^2$$

Therefore, the fraction of the total kinetic energy possessed by the neutron after the collision is

$$(1) \quad f_n = \frac{K_{nf}}{K_{ni}} = \frac{\tfrac{1}{2}m_n \left(\dfrac{m_n - m_m}{m_n + m_m}\right)^2 v_{ni}^2}{\tfrac{1}{2}m_n v_{ni}^2} = \left(\frac{m_n - m_m}{m_n + m_m}\right)^2$$

From this result, we see that the final kinetic energy of the neutron is small when m_m is close to m_n and is zero when $m_m = m_n$.

We can calculate the kinetic energy of the moderator nucleus after the collision using Equation 8.24:

$$K_{mf} = \tfrac{1}{2}m_m v_{mf}^2 = \frac{2m_n^2 m_m}{(m_n + m_m)^2} v_{ni}^2$$

Hence, the fraction of the total kinetic energy transferred to the moderator nucleus is

$$(2) \quad f_{trans} = \frac{K_{mf}}{K_{ni}} = \frac{\dfrac{2m_n^2 m_m}{(m_n + m_m)^2} v_{ni}^2}{\tfrac{1}{2}m_n v_{ni}^2} = \frac{4m_n m_m}{(m_n + m_m)^2}$$

If $m_m \approx m_n$, we see that $f_{trans} \approx 1 = 100\%$. Because the system's kinetic energy is conserved, (2) can also be obtained from (1) with the condition that $f_n + f_m = 1$, so that $f_m = 1 - f_n$.

For collisions of the neutrons with deuterium nuclei in D_2O ($m_m = 2m_n$), $f_n = 1/9$ and $f_{trans} = 8/9$. That is, 89% of the neutron's kinetic energy is transferred to the deuterium nucleus. In practice, the moderator efficiency is reduced because head-on collisions are very unlikely to occur.

Example 8.8   Two Blocks and a Spring

A block of mass $m_1 = 1.60$ kg, initially moving to the right with a speed of 4.00 m/s on a frictionless horizontal track, collides with a massless spring attached to a second block of mass $m_2 = 2.10$ kg, moving to the left with a speed of 2.50 m/s, as in Figure. 8.9a. The spring has a spring constant of 600 N/m. (a) At the instant when m_1 is moving to the right with a speed of 3.00 m/s, as in Figure 8.9b, determine the speed of m_2.

Solution   We identify the system as the two blocks and the spring. The system satisfies the conditions of the isolated system model, so the total momentum of the system is conserved. We have

$$m_1 v_{1i} + m_2 v_{2i} = m_1 v_{1f} + m_2 v_{2f}$$

$$(1.60 \text{ kg})(4.00 \text{ m/s}) + (2.10 \text{ kg})(-2.50 \text{ m/s})$$

$$= (1.60 \text{ kg})(3.00 \text{ m/s}) + (2.10 \text{ kg})v_{2f}$$

$$v_{2f} = -1.74 \text{ m/s}$$

Note that the initial velocity component of m_2 is -2.50 m/s because its direction is to the left. The negative value for v_{2f} means that m_2 is still moving to the left at the instant we are considering.

(b) Determine the distance the spring is compressed at that instant.

Solution   Because this is an isolated system, we can also approach this problem from the energy version of the isolated system model to determine the compression x in the spring shown in Figure 8.9b. No nonconservative forces are acting within the system, so mechanical energy is conserved:

$$E_i = E_f$$

$$\tfrac{1}{2}m_1 v_{1i}^2 + \tfrac{1}{2}m_2 v_{2i}^2 = \tfrac{1}{2}m_1 v_{1f}^2 + \tfrac{1}{2}m_2 v_{2f}^2 + \tfrac{1}{2}kx^2$$

Substituting the given values and the result to part (a) into this expression gives

$$x = \boxed{0.173 \text{ m}}$$

(c) Determine the maximum distance by which the spring is compressed during the collision.

Solution The maximum compression of the spring occurs when the two blocks are not moving relative to each other. For their relative velocity to be zero, they must be moving with the same velocity in our reference frame as we watch the collision. Thus, we can model the collision *up to this point* as a perfectly inelastic collision:

$$m_1 v_{1i} + m_2 v_{2i} = (m_1 + m_2) v_f$$
$$(1.60 \text{ kg})(4.00 \text{ m/s}) + (2.10 \text{ kg})(-2.50 \text{ m/s})$$
$$= (1.60 \text{ kg} + 2.10 \text{ kg}) v_f$$
$$v_f = 0.311 \text{ m/s}$$

As in part (b), mechanical energy is conserved, so we set up a conservation of mechanical energy expression:

$$E_i = E_f$$
$$\tfrac{1}{2} m_1 v_{1i}^2 + \tfrac{1}{2} m_2 v_{2i}^2 = \tfrac{1}{2}(m_1 + m_2) v_f^2 + \tfrac{1}{2} k x^2$$

Substituting the values into this expression gives

$$x = \boxed{0.358 \text{ m}}$$

EXERCISE Find the velocity of m_1 and the compression in the spring at the instant that m_2 is at rest.

Answer 0.719 m/s to the right; 0.251 m

EXERCISE A 2.50-kg object, moving initially with a speed of 10.0 m/s, makes a perfectly inelastic head-on collision with a 5.00-kg object that is initially at rest. (a) Find the final speed of the composite object. (b) How much of the system kinetic energy is transformed to other forms in the collision?

Answer (a) 3.33 m/s (b) 83.4 J

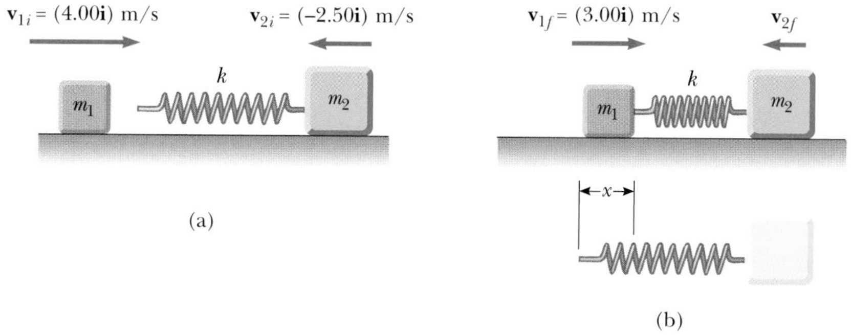

$\mathbf{v}_{1i} = (4.00\mathbf{i})$ m/s $\mathbf{v}_{2i} = (-2.50\mathbf{i})$ m/s $\mathbf{v}_{1f} = (3.00\mathbf{i})$ m/s $\mathbf{v}_{2f}$

(a)

(b)

Figure 8.9

(Example 8.8) A moving block collides with another moving block with a spring attached: (a) before the collision and (b) at one instant during the collision.

8.4 • TWO-DIMENSIONAL COLLISIONS

In Section 8.1 we showed that the total momentum of a system is conserved when the system is isolated (i.e., when no external forces act on the system). For a general collision of two objects in three-dimensional space, the principle of conservation of momentum implies that the total momentum in each direction is conserved. An important subset of collisions takes place in a plane. The game of billiards is a familiar example involving multiple collisions of objects moving on a two-dimensional surface. Let us restrict our attention to a single two-dimensional collision between two objects that takes place in a plane. For such collisions, we obtain two component equations for the conservation of momentum:

$$m_1 v_{1ix} + m_2 v_{2ix} = m_1 v_{1fx} + m_2 v_{2fx}$$
$$m_1 v_{1iy} + m_2 v_{2iy} = m_1 v_{1fy} + m_2 v_{2fy}$$

where we use three subscripts in this general equation to represent, respectively, (1) the identification of the object, (2) initial and final values, and (3) the velocity component.

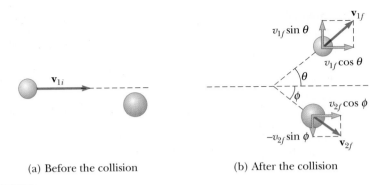

(a) Before the collision   (b) After the collision

Figure 8.10

An elastic glancing collision between two particles.

Consider a two-dimensional problem in which an object of mass m_1 collides with an object of mass m_2 that is initially at rest, as in Figure 8.10. After the collision, m_1 moves at an angle θ with respect to the horizontal, and m_2 moves at an angle ϕ with respect to the horizontal. This is called a *glancing* collision. Applying the law of conservation of momentum in component form and noting that the initial y component of momentum is zero, we have

- *Component equations for conservation of momentum in a two-dimensional, two-particle collision*

$$x \text{ component:} \quad m_1 v_{1i} + 0 = m_1 v_{1f} \cos \theta + m_2 v_{2f} \cos \phi \qquad [8.25]$$

$$y \text{ component:} \quad 0 + 0 = m_1 v_{1f} \sin \theta - m_2 v_{2f} \sin \phi \qquad [8.26]$$

If the collision is elastic, we can write a third equation for conservation of kinetic energy, in the form

- *Conservation of kinetic energy in a two-dimensional, two-particle collision*

$$\tfrac{1}{2} m_1 v_{1i}{}^2 = \tfrac{1}{2} m_1 v_{1f}{}^2 + \tfrac{1}{2} m_2 v_{2f}{}^2 \qquad [8.27]$$

If we know the initial velocity v_{1i} and the masses, we are left with four unknowns $(v_{1f}, v_{2f}, \theta, \text{and } \phi)$. Because we have only three equations, one of the four remaining quantities must be given to determine the motion after the collision from conservation principles alone.

If the collision is inelastic, kinetic energy is *not* conserved, and Equation 8.27 does *not* apply.

PROBLEM-SOLVING STRATEGY   **Two-Dimensional Collisions**

The following procedure is recommended when dealing with problems involving collisions between two objects.

1. Set up a coordinate system and define your velocities with respect to that system. It is convenient to have the *x* axis coincide with one of the initial velocities.
2. In your sketch of the coordinate system, draw and label all velocity vectors and include all the given information. Write expressions for the *x* and *y* components of the momentum of each object before and after the colli-

sion. Remember to include the appropriate signs for the components of the velocity vectors. It is essential that you pay careful attention to signs.

3. Write expressions for the *total* momentum in the *x* direction *before* and *after* the collision, and equate the two. Repeat this procedure for the total momentum in the *y* direction.

4. Proceed to solve the momentum equations for the unknown quantities. If the collision is inelastic, kinetic energy is *not* conserved, and additional information is probably required. If the collision is perfectly inelastic, the final velocities of the two objects are equal.

5. If the collision is elastic, kinetic energy is conserved, and you can equate the total kinetic energy before the collision to the total kinetic energy after the collision. This provides an additional relationship between the velocity magnitudes.

Example 8.9 Proton–Proton Collision

A proton collides elastically with another proton that is initially at rest. The incoming proton has an initial speed of 3.5×10^5 m/s and makes a glancing collision with the second proton, as in Figure 8.10. (At close separations, the protons exert a repulsive electrostatic force on each other.) After the collision, one proton moves off at an angle of 37° to the original direction of motion, and the second deflects at an angle of ϕ to the same axis. Find the final speeds of the two protons and the angle ϕ.

Solution The isolated system is the pair of protons. Both momentum and kinetic energy of the system are conserved in this glancing elastic collision. Because $m_1 = m_2$, $\theta = 37°$, and we are given that $v_{1i} = 3.5 \times 10^5$ m/s, Equations 8.25, 8.26, and 8.27 become

(1) $v_{1f} \cos 37° + v_{2f} \cos \phi = 3.5 \times 10^5$ m/s

(2) $v_{1f} \sin 37° - v_{2f} \sin \phi = 0$

(3) $v_{1f}^2 + v_{2f}^2 = (3.5 \times 10^5 \text{ m/s})^2 = 1.2 \times 10^{11} \text{ m}^2/\text{s}^2$

We rewrite (1) and (2) as follows:

$$v_{2f} \cos \phi = 3.5 \times 10^5 \text{ m/s} - v_{1f} \cos 37°$$

$$v_{2f} \sin \phi = v_{1f} \sin 37°$$

Now we square these two equations and add them:

$$v_{2f}^2 \cos^2 \phi + v_{2f}^2 \sin^2 \phi = 1.2 \times 10^{11} \text{ m}^2/\text{s}^2$$
$$- (7.0 \times 10^5 \text{ m/s}) v_{1f} \cos 37°$$
$$+ v_{1f}^2 \cos^2 37° + v_{1f}^2 \sin^2 37°$$
$$\longrightarrow v_{2f}^2 = 1.2 \times 10^{11} - (5.6 \times 10^5) v_{1f} + v_{1f}^2$$

Substituting this expression into (3),

$$v_{1f}^2 + [1.2 \times 10^{11} - (5.6 \times 10^5) v_{1f} + v_{1f}^2] = 1.2 \times 10^{11}$$
$$\longrightarrow 2v_{1f}^2 - (5.6 \times 10^5) v_{1f} = (2v_{1f} - 5.6 \times 10^5) v_{1f} = 0$$

One possibility for the solution of this equation is $v_{1f} = 0$, which corresponds to a head-on collision—the first proton stops and the second continues with the same speed in the same direction. This is not what we want. The other possibility is

$$2v_{1f} - 5.6 \times 10^5 = 0 \longrightarrow v_{1f} = \boxed{2.8 \times 10^5 \text{ m/s}}$$

From (3),

$$v_{2f} = \sqrt{1.2 \times 10^{11} - v_{1f}^2} = \sqrt{1.2 \times 10^{11} - (2.8 \times 10^5)^2}$$
$$= \boxed{2.1 \times 10^5 \text{ m/s}}$$

and from (2),

$$\phi = \sin^{-1}\left(\frac{v_{1f} \sin 37°}{v_{2f}}\right) = \sin^{-1}\left(\frac{(2.8 \times 10^5) \sin 37°}{2.1 \times 10^5}\right)$$
$$= \boxed{53°}$$

It is interesting that $\theta + \phi = 90°$. This result is *not* accidental. Whenever two objects of equal mass collide elastically in a glancing collision and one of them is initially at rest, their final velocities are at right angles to each other.

Example 8.10 Collision at an Intersection

A 1 500-kg car traveling east with a speed of 25.0 m/s collides at an intersection with a 2 500-kg van traveling north at a speed of 20.0 m/s, as shown in Figure 8.11. Find the direction and magnitude of the velocity of the wreckage after the collision, assuming that the vehicles undergo a perfectly inelastic collision (i.e., they stick together).

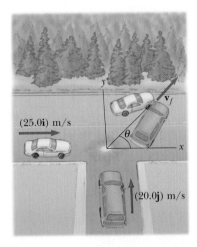

Figure 8.11

(Example 8.10) An eastbound car colliding with a northbound van.

Solution Let us choose east to be along the positive *x* direction and north to be along the positive *y* direction, as in Figure 8.11. Before the collision, the only object having momentum in the *x* direction is the car. Thus, the magnitude of the total initial momentum of the system (car plus van) in the *x* direction is

$$\Sigma \, p_{xi} = (1\,500 \text{ kg})(25.0 \text{ m/s}) = 3.75 \times 10^4 \text{ kg} \cdot \text{m/s}$$

The wreckage moves at an angle θ and speed v after the collision. The magnitude of the total momentum in the *x* direction after the collision is

$$\Sigma \, p_{xf} = (4\,000 \text{ kg}) \, v \cos \theta$$

Because the total momentum in the *x* direction is conserved, we can equate these two equations to obtain

$$(1) \quad (3.75 \times 10^4) \text{ kg} \cdot \text{m/s} = (4000 \text{ kg}) \, v \cos \theta$$

Similarly, the total initial momentum of the system in the *y* direction is that of the van, whose magnitude is equal to $(2\,500 \text{ kg})(20.0 \text{ m/s})$. Applying conservation of momentum to the *y* direction, we have

$$\Sigma \, p_{yi} = \Sigma \, p_{yf}$$

$$(2\,500 \text{ kg})(20.0 \text{ m/s}) = (4\,000 \text{ kg}) v \sin \theta$$

$$(2) \quad 5.00 \times 10^4 \text{ kg} \cdot \text{m/s} = (4\,000 \text{ kg}) v \sin \theta$$

If we divide (2) by (1), we find

$$\tan \theta = \frac{5.00 \times 10^4}{3.75 \times 10^4} = 1.33$$

$$\theta = \boxed{53.1^\circ}$$

When this angle is substituted into (2), the value of v is

$$v = \frac{5.00 \times 10^4 \text{ kg} \cdot \text{m/s}}{(4\,000 \text{ kg}) \sin 53.1^\circ} = \boxed{15.6 \text{ m/s}}$$

EXERCISE A 3.00-kg object with an initial velocity of $5.00\mathbf{i}$ m/s collides with and sticks to a 2.00-kg object with an initial velocity of $-3.00\mathbf{j}$ m/s. Find the final velocity of the composite object.

Answer $(3.00\mathbf{i} - 1.20\mathbf{j})$ m/s

8.5 • THE CENTER OF MASS

See Screen 6.7

In this section we describe the overall motion of a system of particles in terms of a very special point called the **center of mass** of the system. The notion of the center of mass gives us confidence in the particle model, because we will see that the center of mass accelerates as if all of the system's mass were concentrated at that point and all external forces act there.

Consider a system consisting of a pair of particles connected by a light, rigid rod (Fig. 8.12). The center of mass as indicated in the figure is located on the rod and closer to the larger mass in the figure—we will see why soon. If a single force is applied at some point on the rod that is above the center of mass, the system rotates clockwise (Fig. 8.12a) as it translates through space. If the force is applied at a point on the rod below the center of mass, the system rotates counterclockwise (Fig. 8.12b). If the force is applied exactly at the center of mass, the system moves in the direction of **F** without rotating (Fig. 8.12c), as if the system is behaving as a particle. Thus, in theory, the center of mass can be located with this experiment.

If we were to analyze the motion in Figure 8.12c, we would find that the system moves as if all its mass were concentrated at the center of mass. Furthermore, if the external net force on the system is Σ **F** and the total mass of the system is M, the center of mass moves with an acceleration given by **a** $= \Sigma$ **F**$/M$. That is, the system moves as if the resultant external force were applied to a single particle of mass M located at the center of mass. This justifies our particle model for extended objects. We have ignored all rotational effects for extended objects so far, implicitly assuming that forces were provided at just the right position so as to cause no rotation. We will study rotational motion in Chapter 10, where we will apply forces that do not pass through the center of mass.

The position of the center of mass of a system can be described as being the *average position* of the system's mass. For example, the center of mass of the pair of particles described in Figure 8.13 is located on the x axis, somewhere between the particles. The x coordinate of the center of mass in this case is

$$x_{CM} = \frac{m_1 x_1 + m_2 x_2}{m_1 + m_2} \qquad \text{[8.28]}$$

For example, if $x_1 = 0$, $x_2 = d$, and $m_2 = 2m_1$, we find $x_{CM} = \frac{2}{3}d$. That is, the center of mass lies closer to the more massive particle. If the two masses are equal, the center of mass lies midway between the particles.

We can extend the concept of center of mass to a system of many particles in three dimensions. The x coordinate of the center of mass of n particles is defined to be

$$x_{CM} \equiv \frac{m_1 x_1 + m_2 x_2 + m_3 x_3 + \cdots + m_n x_n}{m_1 + m_2 + m_3 + \cdots + m_n} = \frac{\sum\limits_i m_i x_i}{\sum\limits_i m_i} = \frac{\sum\limits_i m_i x_i}{M} \qquad \text{[8.29]}$$

where x_i is the x coordinate of the ith particle and $\Sigma_i m_i$ is the *total mass M* of the system. The y and z coordinates of the center of mass are similarly defined by the equations

$$y_{CM} \equiv \frac{\sum\limits_i m_i y_i}{M} \qquad \text{and} \qquad z_{CM} \equiv \frac{\sum\limits_i m_i z_i}{M} \qquad \text{[8.30]}$$

The center of mass can also be located by its position vector, $\mathbf{r}_{CM}$. The rectangular coordinates of this vector are x_{CM}, y_{CM}, and z_{CM}, defined in Equations 8.29 and 8.30. Therefore,

$$\mathbf{r}_{CM} = x_{CM}\mathbf{i} + y_{CM}\mathbf{j} + z_{CM}\mathbf{k} = \frac{\sum\limits_i m_i x_i \mathbf{i} + \sum\limits_i m_i y_i \mathbf{j} + \sum\limits_i m_i z_i \mathbf{k}}{M}$$

$$\mathbf{r}_{CM} = \frac{\sum\limits_i m_i \mathbf{r}_i}{M} \qquad \text{[8.31]}$$

where $\mathbf{r}_i$ is the position vector of the ith particle, defined by

$$\mathbf{r}_i \equiv x_i\mathbf{i} + y_i\mathbf{j} + z_i\mathbf{k}$$

Although locating the center of mass for an extended object is somewhat more cumbersome than locating the center of mass of a system of particles, this location

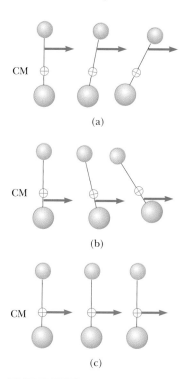

(a)

(b)

(c)

Figure 8.12

Two particles of unequal mass are connected by a light, rigid rod. (a) The system rotates clockwise when a force is applied between the less massive particle and the center of mass. (b) The system rotates counterclockwise when a force is applied between the more massive particle and the center of mass. (c) The system moves in the direction of the force without rotating when a force is applied at the center of mass.

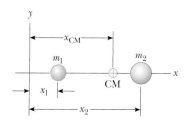

Figure 8.13

The center of mass of two particles having unequal mass is located on the x axis at x_{CM}, a point between the particles, closer to the one having the larger mass.

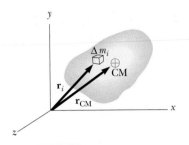

Figure 8.14

An extended object can be modeled as a distribution of small elements of mass Δm_i. The center of mass of the object is located at the vector position $\mathbf{r}_{CM}$, which has coordinates x_{CM}, y_{CM}, and z_{CM}.

is based on the same fundamental ideas. We can model the extended object as a system containing a large number of elements (Fig. 8.14). Each element is modeled as a particle of mass Δm_i, with coordinates x_i, y_i, z_i. The particle separation is very small, so this model is a good representation of the continuous mass distribution of the object. The x coordinate of the center of mass of the particles representing the object, and therefore of the approximate center of mass of the object, is

$$x_{CM} \approx \frac{\sum_i x_i \Delta m_i}{M}$$

with similar expressions for y_{CM} and z_{CM}. If we let the number of elements approach infinity (and, as a consequence, the size and mass of each element approach zero), the model becomes indistinguishable from the continuous mass distribution, and x_{CM} is given precisely. In this limit, we replace the sum by an integral and Δm_i by the differential element dm:

$$x_{CM} = \lim_{\Delta m_i \to 0} \frac{\sum_i x_i \Delta m_i}{M} = \frac{1}{M} \int x \, dm \qquad \text{[8.32]}$$

where the integration is over the length of the object in the x direction. Likewise, for y_{CM} and z_{CM} we obtain

$$y_{CM} = \frac{1}{M} \int y \, dm \qquad \text{and} \qquad z_{CM} = \frac{1}{M} \int z \, dm \qquad \text{[8.33]}$$

We can express the vector position of the center of mass of an extended object as

$$\mathbf{r}_{CM} = \frac{1}{M} \int \mathbf{r} \, dm \qquad \text{[8.34]}$$

which is equivalent to the three expressions in Equations 8.32 and 8.33.

The center of mass of a homogeneous, symmetric body must lie on an axis of symmetry. For example, the center of mass of a homogeneous rod must lie midway between the ends of the rod. The center of mass of a homogeneous sphere or a homogeneous cube must lie at the geometric center of the object.

The center of mass of a system is often confused with the **center of gravity** of a system. Each portion of a system is acted on by the gravitational force. The net effect of all of these forces is equivalent to the effect of a single force $M\mathbf{g}$ acting at a special point called the center of gravity. The center of gravity is the average position of the gravitational forces on all parts of the object. If $\mathbf{g}$ is uniform over the system, the center of gravity coincides with the center of mass. If the gravitational field over the system is not uniform, the center of gravity and the center of mass are different. In most cases, for objects or systems of reasonable size, the two points can be considered to be coincident.

One can experimentally determine the center of gravity of an irregularly shaped object, such as a wrench, by suspending the wrench from two different points (Fig. 8.15). An object of this size has virtually no variation in the gravitational field over its dimensions, so this method also locates the center of mass. The wrench is first hung from point A, and a vertical line AB is drawn (which can be established with a plumb bob) when the wrench is in equilibrium. The wrench

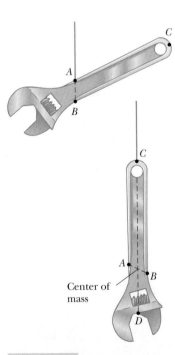

Figure 8.15

An experimental technique for determining the center of gravity of a wrench. The wrench is hung freely from two different pivots, A and C. The intersection of the two vertical lines AB and CD locates the center of gravity.

is then hung from point *C*, and a second vertical line *CD* is drawn. The center of mass coincides with the intersection of these two lines. In fact, if the wrench is hung freely from any point, the vertical line through that point will pass through the center of mass.

Quick **Quiz** 8.6

Figure 8.16 shows a one-bottle wine holder, which remains stationary in the position shown in the photograph. Describe where the center of mass of the system consisting of the bottle and the wooden holder is located.

Figure 8.16

(Quick Quiz 8.6) *(Charles D. Winters)*

Example **8.11** The Center of Mass of Three Particles

A system consists of three particles located at the corners of a right triangle as in Figure 8.17. Find the center of mass of the system.

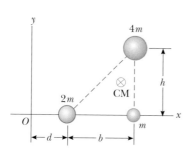

Figure 8.17

(Example 8.11) Locating the center of mass for a system of three particles.

Solution Using the basic defining equations for the coordinates of the center of mass, and noting that $z_{CM} = 0$, we have

$$x_{CM} = \frac{\Sigma_i m_i x_i}{M} = \frac{2md + m(d + b) + 4m(d + b)}{7m}$$

$$= d + \frac{5}{7}\,b$$

$$y_{CM} = \frac{\Sigma_i m_i y_i}{M} = \frac{2m(0) + m(0) + 4mh}{7m} = \frac{4}{7}\,h$$

Therefore, we can express the position vector to the center of mass measured from the origin as

$$\mathbf{r}_{CM} = x_{CM}\mathbf{i} + y_{CM}\mathbf{j} + z_{CM}\mathbf{k} = \left(d + \frac{5}{7}\,b\right)\mathbf{i} + \frac{4}{7}\,h\mathbf{j}$$

Example **8.12** The Center of Mass of a Right Triangle

You have been asked to hang a metal sign from a single vertical wire. The sign is of the triangular shape shown in Figure 8.18a. The bottom of the sign is to be parallel to the ground. At what distance from the left end of the sign should you attach the wire?

Reasoning We will need to attach the wire at a point directly above the center of gravity of the sign, which is the same as its center of mass because it is in a uniform gravitational field. We model the sign as a perfect triangle. We assume that the sign has a uniform density and total mass *M*. Because the sign is a continuous distribution of mass, we will need to use the integral expression in Equation 8.32 to find the *x* coordinate of the center of mass.

Solution We divide the triangle into narrow strips of width *dx* and height *y* as shown in Figure 8.18b, where *y* is the

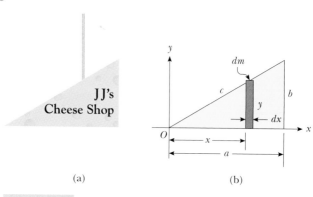

(a)

(b)

Figure 8.18

(Example 8.12) (a) A triangular sign to be hung from a single wire. (b) Geometric construction for locating the center of mass.

height to the hypotenuse of the triangle above the x axis for a given value of x. The mass of each strip is the product of the volume of the strip and the density ρ of the material from which the sign is made: $dm = \rho y t\, dx$, where t is the thickness of the metal sign. The density of the material is the total mass of the sign divided by its total volume (area of the triangle times thickness), so

$$dm = \rho y t\, dx = \left(\frac{M}{\frac{1}{2}abt}\right) y t\, dx = \frac{2My}{ab}\, dx$$

Using Equation 8.32 to find the x coordinate of the center of mass,

$$x_{CM} = \frac{1}{M}\int x\, dm = \frac{1}{M}\int_0^a x\,\frac{2My}{ab}\, dx = \frac{2}{ab}\int_0^a xy\, dx$$

To proceed further and evaluate the integral, we must express y in terms of x. The line representing the hypotenuse of

the triangle in Figure 8.18b has a slope of b/a and passes through the origin, so the equation of this line is $y = (b/a)x$. With this substitution for y in the integral, we have

$$x_{CM} = \frac{2}{ab}\int_0^a x\left(\frac{b}{a}x\right) dx = \frac{2}{a^2}\int_0^a x^2\, dx = \frac{2}{a^2}\left[\frac{x^3}{3}\right]_0^a$$

$$= \tfrac{2}{3}a$$

Thus, the wire must be attached to the sign at a distance two thirds of the length of the bottom edge from the left end. We could also find the y coordinate of the center of mass of the sign, but this is not needed to determine where the wire should be attached.

EXERCISE Find the y coordinate of the center of mass of the sign.

Answer $\tfrac{1}{3}b$

8.6 • MOTION OF A SYSTEM OF PARTICLES

See Screen 6.8

We can begin to understand the physical significance and utility of the center of mass concept by taking the time derivative of the position vector $\mathbf{r}_{CM}$ of the center of mass, given by Equation 8.31. Assuming that M remains constant—that is, no particles enter or leave the system—we find the following expression for the **velocity of the center of mass:**

• *Velocity of the center of mass*

$$\mathbf{v}_{CM} = \frac{d\mathbf{r}_{CM}}{dt} = \frac{1}{M}\sum_i m_i\frac{d\mathbf{r}_i}{dt} = \frac{1}{M}\sum_i m_i\mathbf{v}_i \qquad [8.35]$$

where $\mathbf{v}_i$ is the velocity of the ith particle. Rearranging Equation 8.35 gives

• *Total momentum of a system of particles*

$$M\mathbf{v}_{CM} = \sum_i m_i\mathbf{v}_i = \sum_i \mathbf{p}_i = \mathbf{p}_{tot} \qquad [8.36]$$

This result tells us that **the total momentum of the system equals its total mass multiplied by the velocity of its center of mass**—in other words, the total momentum of a single particle of mass M moving with a velocity $\mathbf{v}_{CM}$—this is the particle model.

If we now differentiate Equation 8.35 with respect to time, we find the **acceleration of the center of mass:**

• *Acceleration of the center of mass for a system of particles*

$$\mathbf{a}_{CM} = \frac{d\mathbf{v}_{CM}}{dt} = \frac{1}{M}\sum_i m_i\frac{d\mathbf{v}_i}{dt} = \frac{1}{M}\sum_i m_i\mathbf{a}_i \qquad [8.37]$$

Rearranging this expression and using Newton's second law, we have

$$M\mathbf{a}_{CM} = \sum_i m_i\mathbf{a}_i = \sum_i \mathbf{F}_i \qquad [8.38]$$

where $\mathbf{F}_i$ is the force on particle i.

The forces on any particle in the system may include both external and internal forces. By Newton's third law, however, the force exerted by particle 1 on particle 2, for example, is equal in magnitude and opposite the force exerted by parti- cle 2 on particle 1. When we sum over all internal forces in Equation 8.38, they cancel in pairs. The net force on the system is therefore due *only* to external forces. Thus, we can write Equation 8.38 in the form

$$\sum \mathbf{F}_{ext} = M\mathbf{a}_{CM} = \frac{d\mathbf{p}_{tot}}{dt} \qquad [8.39]$$

•   *Newton's second law for a system of particles*

That is, the external net force on the system of particles equals the total mass of the system multiplied by the acceleration of the center of mass, or the time rate of change of the momentum of the system. If we compare this to Newton's second law for a single particle, we see that the center of mass moves like an imag- inary particle of mass M under the influence of the external net force on the sys- tem. In the absence of external forces, the center of mass moves with uniform ve- locity, as in the case of the translating and rotating wrench in Figure 8.19. If the net force acts along a line through the center of mass of an extended body such as the wrench, the body is accelerated without rotation. If the net force does not act through the center of mass, the body will undergo rotation in addition to translation. The linear acceleration of the center of mass is the same in either case, as given by Equation 8.39.

Finally, we see that if the external net force is zero, then from Equation 8.39 it follows that

$$\frac{d\mathbf{p}_{tot}}{dt} = M\mathbf{a}_{CM} = 0$$

so that

$$\mathbf{p}_{tot} = M\mathbf{v}_{CM} = \text{constant} \quad \left(\text{when } \sum \mathbf{F}_{ext} = 0\right) \qquad [8.40]$$

That is, the total linear momentum of a system of particles is constant if no ex- ternal forces act on the system. It follows that, for an *isolated* system of particles, the total momentum is conserved. The law of conservation of momentum that was de- rived in Section 8.1 for a two-particle system is thus generalized to a many-particle system.

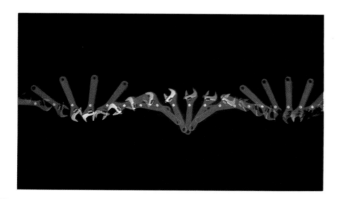

Figure 8.19

Strobe photograph showing an overhead view of a wrench moving on a horizontal surface. The center of mass of the wrench (marked with a white dot) moves in a straight line as the wrench rotates about this point. *(Richard Megna, Fundamental Photographs)*

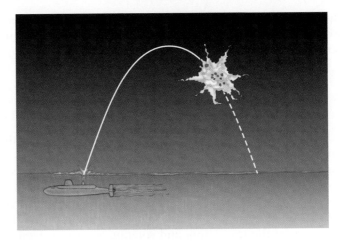

Figure 8.20

(Quick Quiz 8.7)

Quick Quiz 8.7

A projectile is fired into the air and suddenly explodes into several fragments (Fig. 8.20). What can be said about the motion of the center of mass of the fragments after the explosion?

THINKING PHYSICS 8.3

A boy stands at one end of a canoe that is stationary relative to the shore (Fig. 8.21). He then walks to the opposite end of the canoe, away from the shore. Does the canoe move?

Reasoning Yes, the canoe moves toward the shore. Ignoring friction between the canoe and water, no horizontal force acts on the system consisting of the boy and canoe. The center of mass of the system therefore remains fixed relative to the shore (or any stationary point). As the boy moves away from the shore, the canoe must move toward the shore such that the center of mass of the system remains fixed in position.

Figure 8.21

(Thinking Physics 8.3)

EXERCISE A 5.0-kg particle moves along the x axis with a velocity of 3.0 m/s. A 2.0-kg particle moves along the x axis with a velocity of -2.5 m/s. Find (a) the velocity of the center of mass and (b) the total momentum of the system.

Answer   (a) 1.4 m/s to the right   (b) 10 kg·m/s to the right

EXERCISE A 2.0-kg particle has a velocity $(2.0\mathbf{i} - 3.0\mathbf{j})$ m/s, and a 3.0-kg particle has a velocity $(1.0\mathbf{i} + 6.0\mathbf{j})$ m/s. Find (a) the velocity of the center of mass and (b) the total momentum of the system.

Answer   (a) $(1.4\mathbf{i} + 2.4\mathbf{j})$ m/s   (b) $(7.0\mathbf{i} + 12\mathbf{j})$ kg·m/s

context connection
8.7 • ROCKET PROPULSION

On our trip to Mars, we will need to control our spacecraft by firing the rocket engines. When ordinary vehicles, such as automobiles and locomotives, are propelled, the driving force for the motion is one of friction. In the case of the automobile, the driving force is the force exerted by the road on the car. A locomotive "pushes" against the tracks; hence, the driving force is the reaction force exerted by the tracks on the locomotive. A rocket moving in space, however, has no road or tracks to "push" against. The source of the propulsion of a rocket must therefore be different. **The operation of a rocket depends on the law of conservation of momentum as applied to a system, where the system is the rocket plus its ejected fuel.**

The propulsion of a rocket can be understood by first considering the pitching machine on ice in Example 8.2. As a baseball is fired from the machine, the baseball receives momentum $m\mathbf{v}$ in one direction, and the machine receives a momentum of equal magnitude in the opposite direction. As each baseball is fired, the machine moves faster, so that a large velocity can be established by firing many baseballs.

In a similar manner, as a rocket moves in free space (a vacuum), its momentum changes when some of its mass is released in the form of ejected gases. Because the ejected gases acquire some momentum, the rocket receives a compensating momentum in the opposite direction. The rocket therefore is accelerated as a result of the "push," or thrust, from the exhaust gases. Note that the rocket represents the *inverse* of an inelastic collision; that is, momentum is conserved, but the kinetic energy of the system is *increased* (at the expense of energy stored in the fuel of the rocket).

Suppose that at some time t, the magnitude of the momentum of the rocket plus the fuel is $(M + \Delta m)v$ (Fig. 8.22a). During a short time interval Δt, the rocket ejects fuel of mass Δm and the rocket's speed therefore increases to $v + \Delta v$ (Fig. 8.22b). If the fuel is ejected with velocity v_e *relative to the rocket*, the speed of the fuel relative to a stationary frame of reference is $v - v_e$, according to our discussion of relative velocity in Section 3.6. Thus, if we equate the total initial momentum of the system with the total final momentum, we have

$$(M + \Delta m)v = M(v + \Delta v) + \Delta m(v - v_e)$$

Simplifying this expression gives

$$M\,\Delta v = \Delta m(v_e)$$

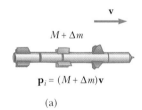

(a)

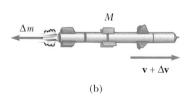

(b)

Figure 8.22

Rocket propulsion. (a) The initial mass of the rocket and fuel is $M + \Delta m$ at a time t, and its speed is v. (b) At a time $t + \Delta t$, the rocket's mass has been reduced to M, and an amount of fuel Δm has been ejected. The rocket's speed increases by an amount Δv.

If we now take the limit as Δt goes to zero, $\Delta v \rightarrow dv$ and $\Delta m \rightarrow dm$. Furthermore, the increase dm in the exhaust mass corresponds to an equal decrease in the rocket mass, so that $dm = -dM$. Note that the negative sign is introduced into the equation because dM represents a decrease in mass. Using this fact, we have

$$M \, dv = -v_e \, dM \qquad \qquad [8.41]$$

Integrating this equation, and taking the initial mass of the rocket plus fuel to be M_i and the final mass of the rocket plus its remaining fuel to be M_f, we have

$$\int_{v_i}^{v_f} dv = -v_e \int_{M_i}^{M_f} \frac{dM}{M}$$

- *Basic expression for velocity change in rocket propulsion*

$$v_f - v_i = v_e \ln\left(\frac{M_i}{M_f}\right) \qquad \qquad [8.42]$$

This is the basic expression for rocket propulsion. It tells us that the increase in speed is proportional to the exhaust speed v_e. The exhaust speed should therefore be very high.

The **thrust** on the rocket is the force exerted on the rocket by the ejected exhaust gases. We can obtain an expression for the instantaneous thrust from Equation 8.41:

- *Rocket thrust*

$$\text{Instantaneous thrust} = Ma = M \frac{dv}{dt} = \left| v_e \frac{dM}{dt} \right| \qquad \qquad [8.43]$$

Here we see that the thrust increases as the exhaust speed increases and as the rate of change of mass (burn rate) increases.

We now can determine the amount of fuel needed to set us on our journey to Mars. The fuel requirements are well within the capabilities of current technology, as evidenced by the fact that several missions to Mars have already been accomplished. What if we wanted to visit another *star*, however, rather than another *planet*? This raises many new technological challenges, including the requirement to consider the effects of relativity, which we investigate in the next chapter.

The thrust from a device that expels nitrogen gas allows an astronaut to maneuver in space without using a restrictive tether. *(Courtesy of NASA)*

WEB

For an essay on the history of rockets and rocket propulsion by a NASA scientist, visit **www.science.ksc.nasa.gov/ history/rocket-history.txt**

THINKING PHYSICS 8.4

When Robert Goddard proposed the possibility of rocket propelled vehicles, the *New York Times* agreed that such vehicles would be useful and successful within the Earth's atmosphere ("Topics of the Times," *New York Times*, January 13, 1920, p. 12). But the *Times* balked at the idea of using such a rocket in the vacuum of space, noting that, ". . . its flight would be neither accelerated nor maintained by the explosion of the charges it then might have left. To claim that it would be is to deny a fundamental law of dynamics, and only Dr. Einstein and his chosen dozen, so few and fit, are licensed to do that . . . That Professor Goddard, with his 'chair' in Clark College and the countenancing of the Smithsonian Institution, does not know the relation of action to reaction, and of the need to have something better than a vacuum against which to react—to say that would be absurd. Of course, he only seems to lack the knowledge ladled out daily in high schools." What did the writer of this passage overlook?

Reasoning The writer of this passage was making a common mistake in believing that a rocket works by expelling gases which push on something, propelling the rocket forward. With this belief, it is impossible to see how a rocket fired in empty space would work.

Gases do not need to push on anything—it is the act itself of expelling the gases that pushes the rocket forward. This can be argued from Newton's third law—the rocket pushes the gases backward, resulting in the gases pushing the rocket forward. It can also be argued from conservation of momentum—as the gases gain momentum in one direction, the rocket must gain momentum in the opposite direction to conserve the original momentum of the rocket–gas system.

The *New York Times* did publish a retraction 49 years later ("A Correction," *New York Times,* July 17, 1969, p. 43) while the *Apollo 11* astronauts were on their way to the Moon. It appeared on a page with two other articles entitled, "Fundamentals of Space Travel," and "Spacecraft, Like Squid, Maneuver by 'Squirts'," and contained the following passages: ". . . an editorial feature of the *New York Times* dismissed the notion that a rocket could function in a vacuum and commented on the ideas of Robert H. Goddard . . . Further investigation and experimentation have confirmed the findings of Isaac Newton in the 17th century, and it is now definitely established that a rocket can function in a vacuum as well as in an atmosphere. The *Times* regrets the error."

Example 8.13 A Rocket in Space

A rocket moving in free space has a speed of 3.0×10^3 m/s relative to Earth. Its engines are turned on, and fuel is ejected in a direction opposite the rocket's motion at a speed of 5.0×10^3 m/s relative to the rocket. (a) What is the speed of the rocket relative to Earth once its mass is reduced to one half its mass before ignition?

Solution Applying Equation 8.42, we have

$$v_f = v_i + v_e \ln\left(\frac{M_i}{M_f}\right)$$

$$= 3.0 \times 10^3 \text{ m/s} + (5.0 \times 10^3 \text{ m/s})\ln\left(\frac{M_i}{0.5M_i}\right)$$

$$= 6.5 \times 10^3 \text{ m/s}$$

(b) What is the thrust on the rocket if it burns fuel at the rate of 50 kg/s?

Solution Using Equation 8.43, we have

$$\text{Thrust} = \left|v_e \frac{dM}{DT}\right| = (5.0 \times 10^3 \text{ m/s})(50\text{kg/s})$$

$$= 2.5 \times 10^5 \text{ N}$$

SUMMARY

The linear momentum of any object of mass m moving with a velocity **v** is

$$\mathbf{p} = m\mathbf{v} \qquad [8.1]$$

Conservation of momentum applied to two interacting objects states that if the two objects form an isolated system, the total momentum of the system at all times equals its initial total momentum:

$$\mathbf{p}_{1i} + \mathbf{p}_{2i} = \mathbf{p}_{1f} + \mathbf{p}_{2f} \qquad [8.5]$$

The **impulse** of a net force Σ **F** is defined as the integral of the force over the time interval during which it acts. The total impulse on any object is equal to the change in the momentum of the object and is given by

$$\mathbf{I} \equiv \int_{t_i}^{t_f} \Sigma \mathbf{F} \, dt = \Delta \mathbf{p} = \mathbf{p}_f - \mathbf{p}_i \qquad [8.9]$$

This is known as the **impulse–momentum theorem.**

When two objects collide, the total momentum of the isolated system before the collision always equals the total momentum after the collision, regardless of the nature of the collision. An **inelastic collision** is one for which kinetic energy is not conserved. A **perfectly inelastic collision** is one in which the colliding objects stick together after the collision. An **elastic collision** is one in which both momentum and kinetic energy are conserved.

In a two- or three-dimensional collision, the components of momentum in each of the directions are conserved independently.

The **vector position of the center of mass of a system of particles** is defined as

$$\mathbf{r}_{CM} \equiv \frac{\Sigma_i m_i \mathbf{r}_i}{M} \qquad [8.31]$$

where $M = \Sigma_i m_i$ is the total mass of the system and $\mathbf{r}_i$ is the vector position of the ith particle.

The **velocity of the center of mass for a system of parti-** cles is

$$\mathbf{v}_{CM} = \frac{1}{M} \sum_i m_i \mathbf{v}_i \qquad [8.35]$$

The total momentum of a system of particles equals the total mass multiplied by the velocity of the center of mass, that is, $\mathbf{p}_{tot} = M\mathbf{v}_{CM}$.

Newton's second law applied to a system of particles is

$$\sum \mathbf{F}_{ext} = M\mathbf{a}_{CM} = \frac{d\mathbf{p}_{tot}}{dt} \qquad [8.39]$$

where $\mathbf{a}_{CM}$ is the acceleration of the center of mass and the sum is over all external forces. The center of mass therefore moves like an imaginary particle of mass M under the influence of the resultant external force on the system.

QUESTIONS

1. Does a large force always produce a larger impulse on an object than a smaller force does? Explain.

2. If two objects collide and one is initially at rest, is it possible for both to be at rest after the collision? Is it possible for one to be at rest after the collision? Explain.

3. Explain how linear momentum is conserved when a ball bounces from a floor.

4. Consider a perfectly inelastic collision between a car and a large truck. Which vehicle experiences a larger change in kinetic energy as a result of the collision?

5. Your physical education teacher, who knows something about physics, throws you a tennis ball at a certain velocity, and you catch it. You are now given the following choice: The teacher can throw you a bowling ball, which is much more massive than the tennis ball, with the same velocity as the tennis ball, the same momentum, or the same kinetic energy. Which choice would you make in order to make the easiest catch, and why?

6. You are watching a movie about a superhero and notice that the superhero hovers in the air and throws a piano at some bad guys while remaining stationary in the air. What's wrong with this scenario?

7. Can the center of mass of an object be located at a position at which there is no mass? If so, give examples.

8. Is the center of mass of a mountain higher or lower than the center of gravity?

9. In golf, novice players are often advised to be sure to "follow through" with their swing. Why does this make the ball travel a longer distance?

10. A sharpshooter fires a rifle while standing with the butt of the gun against his shoulder. If the forward momentum of a bullet is the same as the backward momentum of the gun, why isn't it as dangerous to be hit by the gun as by the bullet?

11. A pole-vaulter falls from a height of 6.0 m onto a foam rubber pad. Can you calculate his speed just before he reaches the pad? Could you calculate the force exerted on him due to the collision? Explain.

12. Firefighters must apply large forces to hold a fire hose steady (Fig. Q8.12). What factors related to the projection of the water determine the magnitude of the force needed to keep the end of the fire hose stationary?

Figure Q8.12 Firefighters attack a burning house with a hose line. (© *Bill Stormont/The Stock Market*)

13. Does the center of mass of a rocket in free space accelerate? Explain. Can the speed of a rocket exceed the exhaust speed of the fuel? Explain.

14. A large bed sheet is held vertically by two students. A third student, who happens to be the star pitcher on the baseball team, throws a raw egg at the sheet. Explain why the egg does not break when it hits the sheet, regardless of its initial speed. (If you try this one, make sure the pitcher hits the sheet near its center, and do not allow the egg to fall on the floor after being caught.)

15. On the subject of the following positions, state your own view and argue to support it: (a) The best theory of motion is that force causes acceleration. (b) The true measure of a force's effectiveness is the work it does, and the best theory of motion is that work on an object changes its energy. (c) The true measure of a force's effect is impulse, and the best theory of motion is that impulse injected into an object changes its momentum.

PROBLEMS

1, 2, 3 = straightforward, intermediate, challenging   □ = full solution available in the *Student Solutions Manual and Study Guide*

web = solution posted at **http://www.harcourtcollege.com/physics/**    = computer useful in solving problem

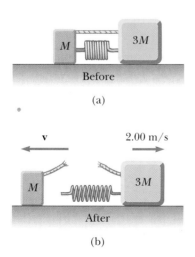 = Interactive Physics   ▓ = paired numerical/symbolic problems   ◪ = life science application

Section 8.1 Linear Momentum and Its Conservation

1. A 3.00-kg particle has a velocity of $(3.00\mathbf{i} - 4.00\mathbf{j})$ m/s. (a) Find its x and y components of momentum. (b) Find the magnitude and direction of its momentum.

2. A 40.0-kg child standing on a frozen pond throws a 0.500-kg stone to the east with a speed of 5.00 m/s. Ignoring friction between the child and the ice, find the recoil velocity of the child.

3. How fast can you set the Earth moving? In particular, when you jump straight up as high as you can, what is the order of magnitude of the maximum recoil speed that you give to the Earth? Model the Earth as a perfectly solid object. In your solution, state the physical quantities you take as data, and the values you measure or estimate for them.

4. Two blocks of masses M and $3M$ are placed on a horizontal, frictionless surface. A light spring is attached to one of them, and the blocks are pushed together with the spring

between them (Fig. P8.4). A cord initially holding the blocks together is burned; after this, the block of mass $3M$ moves to the right with a speed of 2.00 m/s. (a) What is the speed of the block of mass M? (b) Find the original elastic potential energy in the spring if $M = 0.350$ kg.

5. (a) A particle of mass m moves with momentum p. Show that the kinetic energy of the particle is given by $K = p^2/2m$. (b) Express the magnitude of the particle's momentum in terms of its kinetic energy and mass.

6. In research in cardiology and exercise physiology it is often important to know the mass of blood pumped by a person's heart in one stroke. These data can be obtained by means of a *ballistocardiograph*. The instrument works as follows. The subject lies on a horizontal pallet floating on a film of air. Friction on the pallet is negligible. Initially the momentum of the system is zero. When the heart beats, it expels a mass m of blood into the aorta with velocity v, and the body and platform move in the opposite direction with velocity V. The blood velocity can be determined independently (e.g., by observing the Doppler shift of ultrasound). Assume it is 50.0 cm/s in one typical trial. The mass of the subject plus the pallet is 54.0 kg. The pallet moves 6.00×10^{-5} m in 0.160 s after one heartbeat. Calculate the mass of blood that leaves the heart. Assume the mass of blood is negligible compared with the total mass of the person. This simplified example illustrates the principle of ballistocardiography, but in practice a more sophisticated model of heart function is used.

Section 8.2 Impulse and Momentum

7. An estimated force–time curve for a baseball struck by a bat is shown in Figure P8.7. From this curve, determine (a) the impulse delivered to the ball, (b) the average force exerted on the ball, and (c) the peak force exerted on the ball.

Before

(a)

v   2.00 m/s

After

(b)

Figure P8.4

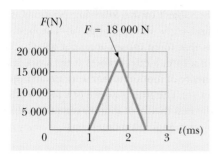

Figure P8.7

Figure P8.12

8. A car is stopped for a traffic signal. When the light turns green, the car accelerates, increasing its speed from 0 to 5.20 m/s in 0.832 s. What linear impulse and average force does a 70.0-kg passenger in the car experience?

9. A 3.00-kg steel ball strikes a wall with a speed of 10.0 m/s
web   at an angle of 60.0° with the surface. It bounces off with the same speed and angle (Fig. P8.9). If the ball is in contact with the wall for 0.200 s, what is the average force exerted on the ball by the wall?

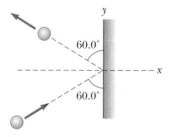

Figure P8.9

10. A tennis player receives a volley with the ball (0.060 0 kg) traveling horizontally at 50.0 m/s and returns the ball so that it travels horizontally at 40.0 m/s in the opposite direction. (a) What is the impulse delivered to the ball by the racquet? (b) What work does the racquet do on the ball?

11. In a slow-pitch softball game, a 0.200-kg softball crosses the plate at 15.0 m/s at an angle of 45.0° below the horizontal. The batter hits the ball toward center field, giving it a velocity of 40.0 m/s at 30.0° above the horizontal. (a) Determine the impulse delivered to the ball. (b) If the force on the ball increases linearly for 4.00 ms, holds constant for 20.0 ms, and then decreases to zero linearly in another 4.00 ms, what is the maximum force on the ball?

12. A garden hose is held as shown in Figure P8.12. The hose is originally full of motionless water. What additional force is necessary to hold the nozzle stationary, after the water is turned on, if the discharge rate is 0.600 kg/s with a speed of 25.0 m/s?

13. A glider of mass m is free to slide along a horizontal air track. It is pushed against a launcher at one end of the track. Model the launcher as a light spring of force constant k, compressed by a distance x. The glider is released from rest. (a) Show that the glider attains a speed given by $v = x(k/m)^{1/2}$. (b) Is a greater speed attained by a large or a small mass? (c) Show that the impulse imparted to the glider is given by $I = x(km)^{1/2}$. (d) Is a greater impulse injected into a large or a small mass? (e) Is more work done on a large or a small mass?

Section 8.3   Collisions

14. A 7.00-kg bowling ball collides head-on with a single 2.00-kg bowling pin. The pin flies forward with a speed of 3.00 m/s. If the ball continues forward with a speed of 1.80 m/s, what was the initial speed of the ball? Ignore rotation of the ball.

15. A 45.0-kg girl is standing on a plank that has a mass of 150 kg. The plank, originally at rest, is free to slide on a frozen lake, which is a flat, frictionless supporting surface. The girl begins to walk along the plank at a constant speed of 1.50 m/s relative to the plank. (a) What is her speed relative to the ice surface? (b) What is the speed of the plank relative to the ice surface?

16. Gayle runs at a speed of 4.00 m/s and dives onto a sled, which is initially at rest on the top of a frictionless snow-covered hill. After she has descended a vertical distance of 5.00 m, her brother, who is initially at rest, hops onto her back and together they continue down the hill. What is their speed at the bottom of the hill if the total vertical drop is 15.0 m? Gayle's mass is 50.0 kg, the sled has a mass of 5.00 kg, and her brother has a mass of 30.0 kg.

17. A railroad car of mass 2.50×10^4 kg is moving with a speed of 4.00 m/s. It collides and couples with three other coupled railroad cars, each of the same mass as the single car and moving in the same direction with an initial speed of 2.00 m/s. (a) What is the speed of the four cars after

the collision? (b) How much mechanical energy is lost in the collision?

18. Four railroad cars, each of mass 2.50×10^4 kg, are coupled together and coasting along horizontal tracks at speed v_i toward the south. A very strong but foolish movie actor, riding on the second car, uncouples the front car and gives it a big push, increasing its speed to 4.00 m/s southward. The remaining three cars continue moving south, now at 2.00 m/s. (a) Find the initial speed of the cars. (b) How much work did the actor do? (c) State the relationship between the process described here and the process in Problem 17.

19. A neutron in a nuclear reactor makes an elastic head-on collision with the nucleus of a carbon atom initially at rest. (a) What fraction of the neutron's kinetic energy is transferred to the carbon nucleus? (b) If the initial kinetic energy of the neutron is 1.60×10^{-13} J, find its final kinetic energy and the kinetic energy of the carbon nucleus after the collision. (The mass of the carbon nucleus is nearly 12.0 times the mass of the neutron.)

web

20. Consider a frictionless track ABC as shown in Figure P8.20. A block of mass $m_1 = 5.00$ kg is released from A. It makes a head-on elastic collision at B with a block having a mass of $m_2 = 10.0$ kg that is initially at rest. Calculate the maximum height to which m_1 rises after the collision.

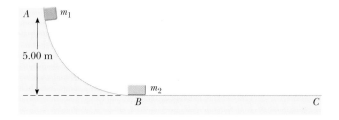

Figure P8.20

21. A 12.0-g wad of sticky clay is hurled horizontally at a 100-g wooden block initially at rest on a horizontal surface. The clay sticks to the block. After impact, the block slides 7.50 m before coming to rest. If the coefficient of friction between the block and the surface is 0.650, what was the speed of the clay immediately before impact?

22. Most of us know intuitively that in a head-on collision between a large dump truck and a subcompact car, you are better off being in the truck than in the car. Why is this? Many people imagine that the collision force exerted on the car is much greater than that experienced by the truck. To substantiate this view, they point out that the car is crushed, whereas the truck is only dented. This idea of unequal forces, of course, is false. Newton's third law tells us that both objects experience forces of the same magnitude. The truck suffers less damage because it is made of stronger metal. But what about the two drivers? Do they experience the same forces? To answer this question, suppose that each vehicle is initially moving at 8.00 m/s and that they undergo a perfectly inelastic head-on collision. Each driver has a mass of 80.0 kg. Including the drivers, the total vehicle masses are 800 kg for the car and 4 000 kg for the truck. If the collision time is 0.120 s, what average force does the seatbelt exert on each driver?

23. As shown in Figure P8.23, a bullet of mass m and speed v passes completely through a pendulum bob of mass M. The bullet emerges with a speed of $v/2$. The pendulum bob is suspended by a stiff rod of length ℓ and negligible mass. What is the minimum value of v such that the pendulum bob will barely swing through a complete vertical circle?

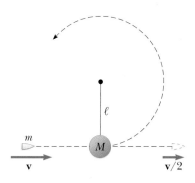

Figure P8.23

24. A 7.00-g bullet, when fired from a gun into a 1.00-kg block of wood held in a vise, will penetrate the block to a depth of 8.00 cm. This block of wood is placed on a frictionless horizontal surface, and a 7.00-g bullet is fired from the gun into the block. To what depth will the bullet penetrate the block in this case?

Section 8.4 Two-Dimensional Collisions

25. A 90.0-kg fullback running east with a speed of 5.00 m/s is tackled by a 95.0-kg opponent running north with a speed of 3.00 m/s. If the collision is perfectly inelastic, (a) calculate the speed and direction of the players just after the tackle and (b) determine the mechanical energy lost as a result of the collision. Account for the missing energy.

26. Two automobiles of equal mass approach an intersection. One vehicle is traveling with velocity 13.0 m/s toward the east, and the other is traveling north with speed v_{2i}. Neither driver sees the other. The vehicles collide in the intersection and stick together, leaving parallel skid marks at an angle of 55.0° north of east. The speed limit for both roads is 35 mi/h, and the driver of the northward-moving vehicle claims he was within the speed limit when the collision occurred. Is he telling the truth?

27. A billiard ball moving at 5.00 m/s strikes a stationary ball of the same mass. After the collision, the first ball moves at 4.33 m/s, at an angle of 30.0° with respect to the original line of motion. Assuming an elastic collision (and ignoring friction and rotational motion), find the struck ball's velocity.

28. A proton, moving with a velocity of $v_i\mathbf{i}$, collides elastically with another proton that is initially at rest. If the two protons have equal speeds after the collision, find (a) the speed of each proton after the collision in terms of v_i and (b) the direction of the velocity vectors after the collision.

29. A 3.00-kg object with an initial velocity of $5.00\mathbf{i}$ m/s collides with and sticks to a 2.00-kg object with an initial velocity of $-3.00\mathbf{j}$ m/s. Find the final velocity of the composite object.

30. A 0.300-kg puck, initially at rest on a horizontal, frictionless surface, is struck by a 0.200-kg puck moving initially along the x axis with a speed of 2.00 m/s. After the collision, the 0.200-kg puck has a speed of 1.00 m/s at an angle of $\theta = 53.0°$ to the positive x axis (see Figure 8.10). (a) Determine the velocity of the 0.300-kg puck after the collision. (b) Find the fraction of kinetic energy lost in the collision.

31. An unstable atomic nucleus of mass 17.0×10^{-27} kg initially at rest disintegrates into three particles. One of the particles, of mass 5.00×10^{-27} kg, moves along the y axis with a velocity of 6.00×10^{6} m/s. Another particle, of mass 8.40×10^{-27} kg, moves along the x axis with a speed of 4.00×10^{6} m/s. Find (a) the velocity of the third particle and (b) the total kinetic energy increase in the process.

32. Two shuffleboard disks of equal mass, one orange and the other yellow, are involved in an elastic, glancing collision. The yellow disk is initially at rest and is struck by the orange disk moving with a speed v_i. After the collision, the orange disk moves along a direction that makes an angle θ with its initial direction of motion. The velocities of the two disks are perpendicular after the collision. Determine the final speed of each disk.

Section 8.5   The Center of Mass

33. Four objects are situated along the y axis as follows: a 2.00-kg object is at $+3.00$ m, a 3.00-kg object is at $+2.50$ m, a 2.50-kg object is at the origin, and a 4.00-kg object is at -0.500 m. Where is the center of mass of this system of objects?

34. A water molecule consists of an oxygen atom with two hydrogen atoms bound to it (Fig. P8.34). The angle between the two bonds is 106°. If the bonds are 0.100 nm long, where is the center of mass of the molecule?

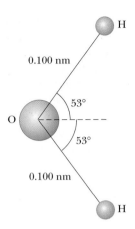

Figure P8.34

35. A uniform piece of sheet steel is shaped as in Figure P8.35. Compute the x and y coordinates of the center of mass of the piece.

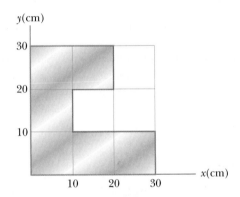

Figure P8.35

36. A rod of length 30.0 cm has linear density (mass per unit length) given by

$$\lambda = 50.0 \text{ g/m} + (20.0 \text{ g/m}^2)x$$

where x is the distance from one end, measured in meters. (a) What is the mass of the rod? (b) How far from the $x = 0$ end is its center of mass?

Section 8.6   Motion of a System of Particles

37. A 2.00-kg particle has a velocity $(2.00\mathbf{i} - 3.00\mathbf{j})$ m/s, and a 3.00-kg particle has a velocity $(1.00\mathbf{i} + 6.00\mathbf{j})$ m/s. Find (a) the velocity of the center of mass and (b) the total momentum of the system.

38. Consider a system of two particles in the xy plane: $m_1 = 2.00$ kg is at the location $\mathbf{r}_1 = (1.00\mathbf{i} + 2.00\mathbf{j})$ m and has a velocity of $(3.00\mathbf{i} + 0.500\mathbf{j})$ m/s; $m_2 = 3.00$ kg is at $\mathbf{r}_2 = (-4.00\mathbf{i} - 3.00\mathbf{j})$ m and has velocity $(3.00\mathbf{i} - 2.00\mathbf{j})$ m/s.

(a) Plot these particles on a grid or graph paper. Draw their position vectors and show their velocities. (b) Find the position of the center of mass of the system and mark it on the grid. (c) Determine the velocity of the center of mass and also show it on the diagram. (d) What is the total linear momentum of the system?

39. Romeo (77.0 kg) entertains Juliet (55.0 kg) by playing his guitar from the rear of their boat at rest in still water, 2.70 m away from Juliet who is in the front of the boat. After the serenade, Juliet carefully moves to the rear of the boat (away from shore) to plant a kiss on Romeo's cheek. How far does the 80.0-kg boat move toward the shore it is facing?

40. A ball of mass 0.200 kg has a velocity of $1.50\mathbf{i}$ m/s; a ball of mass 0.300 kg has a velocity of $-0.400\mathbf{i}$ m/s. They meet in a head-on elastic collision. (a) Find their velocities after the collision. (b) Find the velocity of their center of mass before and after the collision.

Section 8.7   Context Connection—Rocket Propulsion

41. The first stage of a Saturn V space vehicle consumes fuel and oxidizer at the rate of 1.50×10^4 kg/s, with an exhaust speed of 2.60×10^3 m/s. (a) Calculate the thrust produced by these engines. (b) Find the acceleration of the vehicle just as it leaves the launch pad if its initial mass is 3.00×10^6 kg. (*Hint:* You must include the gravitational force to solve part (b).)

42. Model rocket engines are sized by thrust, thrust duration, and total impulse, among other characteristics. A size C5 model rocket engine has an average thrust of 5.26 N, a fuel mass of 12.7 g, and an initial mass of 25.5 g. The duration of its burn is 1.90 s. (a) What is the average exhaust speed of the engine? (b) If this engine is placed in a rocket body of mass 53.5 g, what is the final velocity of the rocket if it is fired in outer space? Assume the fuel burns at a constant rate.

43. A rocket for use in deep space is to have the capability of boosting a total load (payload plus rocket frame and engine) of 3.00 metric tons to a speed of 10 000 m/s. (a) It has an engine and fuel designed to produce an exhaust speed of 2 000 m/s. How much fuel plus oxidizer is required? (b) If a different fuel and engine design could give an exhaust speed of 5 000 m/s, what amount of fuel and oxidizer would be required for the same task?

44. A 3 500-kg spacecraft experiences an acceleration of 2.50 μg = 2.45×10^{-5} m/s² due to a leak from one of its hydraulic control systems. The fluid is known to escape with a speed of 70.0 m/s into the vacuum of space. How much fluid will be lost in 1.00 h if the leak is not stopped?

45. A rocket has total mass M_i = 360 kg, including 330 kg of fuel and oxidizer. In interstellar space it starts from rest.

Its engine is turned on at time t = 0, and it puts out exhaust with relative speed v_e = 1 500 m/s at the constant rate 2.50 kg/s. The burn lasts until the fuel runs out, at time 330 kg/(2.50 kg/s) = 132 s. Set up and carry out a computer analysis of the motion according to Euler's method. Find (a) the final velocity of the rocket and (b) the distance it travels during the burn.

Additional Problems

46. **Review Problem.** A 60.0-kg person running at an initial speed of 4.00 m/s jumps onto a 120-kg cart initially at rest (Figure P8.46). The person slides on the cart's top surface and finally comes to rest relative to the cart. The coefficient of kinetic friction between the person and the cart is 0.400. Friction between the cart and ground can be ignored. (a) Find the final velocity of the person and cart relative to the ground. (b) Find the friction force acting on the person while he is sliding across the top surface of the cart. (c) How long does the friction force act on the person? (d) Find the change in momentum of the person and the change in momentum of the cart. (e) Determine the displacement of the person relative to the ground while he is sliding on the cart. (f) Determine the displacement of the cart relative to the ground while the person is sliding. (g) Find the change in kinetic energy of the person. (h) Find the change in kinetic energy of the cart. (i) Explain why the answers to parts (g) and (h) differ. (What kind of collision is this, and what accounts for the loss of mechanical energy?)

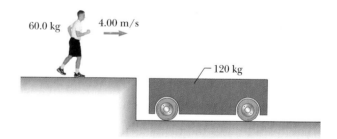

60.0 kg   **4.00 m/s**   **120 kg**

Figure P8.46

47. Tarzan, whose mass is 80.0 kg, swings from a 3.00-m vine that is horizontal when he starts. At the bottom of his arc, he picks up 60.0-kg Jane in a perfectly inelastic collision. What is the height of the highest tree limb they can reach on their upward swing?

48. A small block of mass m_1 = 0.500 kg is released from rest at the top of a curve-shaped frictionless wedge of mass m_2 = 3.00 kg, which sits on a frictionless horizontal surface as in Figure P8.48a. When the block leaves the wedge, its velocity is measured to be 4.00 m/s to the right, as in

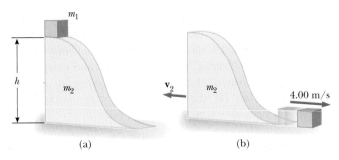

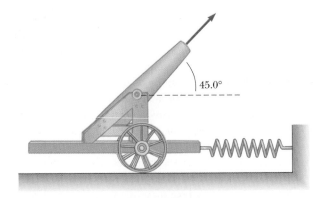

Figure P8.48

Figure P8.48b. (a) What is the velocity of the wedge after the block reaches the horizontal surface? (b) What is the height *h* of the wedge?

49. A jet aircraft is traveling at 500 mi/h (223 m/s) in horizontal flight. The engine takes in air at a rate of 80.0 kg/s and burns fuel at a rate of 3.00 kg/s. If the exhaust gases are ejected at 600 m/s relative to the aircraft, find the thrust of the jet engine and the delivered power.

50. An 8.00-g bullet is fired into a 2.50-kg block initially at rest at the edge of a frictionless table of height 1.00 m (Fig. P8.50). The bullet remains in the block, and after impact the block lands 2.00 m from the bottom of the table. Determine the initial speed of the bullet.

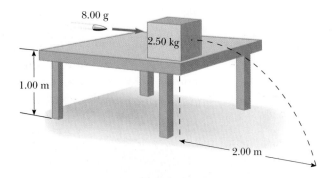

Figure P8.50

51. A cannon is rigidly attached to a carriage, which can move along horizontal rails but is connected to a post by a large spring, initially unstretched and with force constant $k = 2.00 \times 10^4$ N/m, as in Figure P8.51. The cannon fires a 200-kg projectile at a velocity of 125 m/s directed 45.0° above the horizontal. (a) If the mass of the cannon and its carriage is 5 000 kg, find the recoil speed of the cannon. (b) Determine the maximum extension of the spring. (c) Find the maximum force the spring exerts on the carriage. (d) Consider the system consisting of the cannon, carriage, and shell. Is the momentum of this system conserved during the firing? Why or why not?

Figure P8.51

52. Two objects with masses m and $3m$ are moving toward each other along the x axis with the same initial speeds v_i. The object with mass m is traveling to the left, and the object with mass $3m$ is traveling to the right. They undergo a head-on elastic collision and each rebounds along the same line as it approached. Find the final speeds of the objects.

53. Two objects with masses m and $3m$ are moving toward each other along the x axis with the same initial speeds v_i. The object with mass m is traveling to the left, and the object with mass $3m$ is traveling to the right. They undergo an elastic glancing collision such that the object with mass m is moving downward after the collision at a right angle from its initial direction. (a) Find the final speeds of the two objects. (b) What is the angle θ at which the object with mass $3m$ is scattered?

54. A 40.0-kg child stands at one end of a 70.0-kg boat that is 4.00 m in length (Fig. P8.54). The boat is initially 3.00 m from the pier. The child notices a turtle on a rock at the far end of the boat and proceeds to walk to that end to catch the turtle. Ignoring friction between boat and water, (a) describe the subsequent motion of the system (child

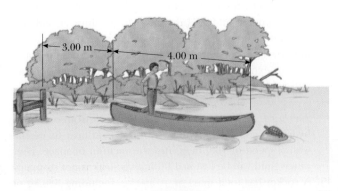

Figure P8.54

plus boat). (b) Where is the child *relative to the pier* when he reaches the far end of the boat? (c) Will he catch the turtle? (Assume he can reach out 1.00 m from the end of the boat.)

55. **Review Problem.** One can say that there are three coequal theories of motion. They are: Newton's second law, stating that the net force on an object causes its acceleration; the work–kinetic energy theorem, stating that the total work on an object causes its change in kinetic energy; and the impulse–momentum theorem, stating that the total impulse on an object causes its change in momentum. In this problem, compare predictions of the three theories in one particular case. A 3.00-kg object has velocity $7.00\mathbf{j}$ m/s. Then, a total force $12.0\mathbf{i}$ N acts on the object for 5.00 s. (a) Calculate the object's final velocity, using the impulse–momentum theorem. (b) Calculate its acceleration from $\mathbf{a} = (\mathbf{v}_f - \mathbf{v}_i)/t$. (c) Calculate its acceleration from $\mathbf{a} = \Sigma \mathbf{F}/m$. (d) Find the object's vector displacement from $\Delta\mathbf{r} = \mathbf{v}_i t + \frac{1}{2}\mathbf{a}t^2$. (e) Find the work done on the object from $W = \mathbf{F} \cdot \Delta\mathbf{r}$. (f) Find the final kinetic energy from $\frac{1}{2}mv_f^2 = \frac{1}{2}m\mathbf{v}_f \cdot \mathbf{v}_f$. (g) Find the final kinetic energy from $\frac{1}{2}mv_i^2 + W$.

56. Two gliders are set in motion on an air track. A spring of force constant k is attached to the near side of one glider. The first glider of mass m_1 has velocity $\mathbf{v}_1$ and the second glider of mass m_2 moves more slowly, with velocity $\mathbf{v}_2$, as in Figure P8.56. When m_1 collides with the spring attached to m_2 and compresses the spring to its maximum compres-

sion x_{max}, the velocity of the gliders is $\mathbf{v}$. In terms of $\mathbf{v}_1$, $\mathbf{v}_2$, m_1, m_2, and k, find (a) the velocity $\mathbf{v}$ at maximum compression, (b) the maximum compression x_{max}, and (c) the velocity of each glider after m_1 has lost contact with the spring.

57. A chain of length L and total mass M is released from rest with its lower end just touching the top of a table, as in Figure P8.57a. Find the force exerted by the table on the chain after the chain has fallen through a distance x, as in Figure P8.57b. (Assume each link comes to rest the instant it reaches the table.)

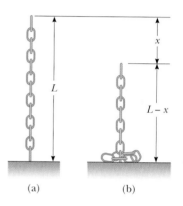

(a)   (b)

Figure P8.57

58. Sand from a stationary hopper falls onto a moving conveyor belt at the rate of 5.00 kg/s as in Figure P8.58. The conveyor belt is supported by frictionless rollers and moves at a constant speed of 0.750 m/s under the action of a constant horizontal external force $\mathbf{F}_{ext}$ supplied by the motor that drives the belt. Find (a) the sand's rate of change of momentum in the horizontal direction, (b) the force of friction exerted by the belt on the sand, (c) the external force $\mathbf{F}_{ext}$, (d) the work done by $\mathbf{F}_{ext}$ in 1.00 s, and (e) the kinetic energy acquired by the falling sand each second due to the change in its horizontal motion. (f) Why are the answers to parts (d) and (e) different?

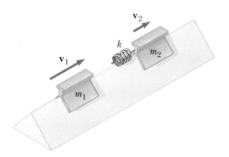

Figure P8.56

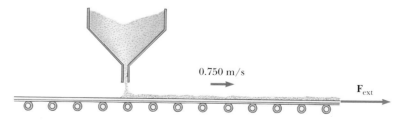

0.750 m/s

$\mathbf{F}_{ext}$

Figure P8.58

ANSWERS TO QUICK QUIZZES

8.1 (d). We are given no information about the masses of the objects. If the masses are the same, the speeds must be the same (so that they have equal kinetic energies), and then $p_1 = p_2$. If the masses are not the same, the speeds will be different, as will the momenta, and either $p_1 < p_2$, or $p_1 > p_2$, depending on which particle has more mass. Without information about the masses, we cannot choose among these possibilities.

8.2 (c) and (e). Because object 1 has larger mass, its acceleration due to the applied force is smaller, and it takes a longer time interval Δt to experience the displacement Δx than does object 2. Thus, the magnitude of the impulse $F\Delta t$ on object 1 is larger than that on object 2. Consequently, object 1 will experience a larger change in momentum than object 2, which tells us that (c) is true. The same force acts on both objects through the same displacement. Thus, the same work is done on each object, so that each must experience the same change in kinetic energy, which tells us that (e) is true.

8.3 The brain is immersed in a cushioning fluid inside the skull. If the head is struck suddenly by a bare fist, the skull accelerates rapidly. The brain matches this acceleration only because of the large impulsive force by the inner surface of the skull on the brain. This large, sudden force can severely injure the brain. Padded gloves extend the time over which the force is applied to the head. Thus, for a given impulse, the gloves provide a longer time interval than the bare fist, decreasing the average force. Because the average force is decreased, the acceleration of the skull is decreased, reducing (but not eliminating) the chance for brain injury.

8.4 (a) Perfectly inelastic—the two "particles," skater and Frisbee, are combined after the collision; (b) Inelastic— because the Frisbee bounced back with almost no speed, kinetic energy has been transformed to other forms of energy; (c) Inelastic—the kinetic energy of the Frisbee is the same before and after the collision. Because momentum of the skater–Frisbee system is conserved, however, the skater must be moving after the catch and the throw, so that the final kinetic energy of the system is larger than the initial kinetic energy. This extra energy comes from the muscles of the skater.

8.5 If all of the initial kinetic energy is transformed, then nothing is moving after the collision. Consequently, the final momentum of the system is necessarily zero. Because momentum of the system is conserved, the initial momentum of the system must be zero. Finally, because the objects are identical, they must have been initially moving toward each other along the same line with the same speed.

8.6 The center of mass of the system must be vertically above the support point on the table. If this were not so, because the center of gravity and the center of mass are the same for small objects, the gravitational force acting at the center of gravity would tip the system over.

8.7 Ignoring air resistance, the only external force on the projectile is the gravitational force. If the projectile did not explode, it would follow the complete parabolic path (solid and dashed) shown in Figure 8.21. In the case of the exploding projectile, all forces involved in the explosion are internal to the system. (The system before the explosion is the projectile; the system after the explosion is all of the fragments.) Thus, the center of mass of the system will follow the same path as if the explosion had not occurred. The center of mass of the system of fragments will land at the same place as the unexploded projectile would.

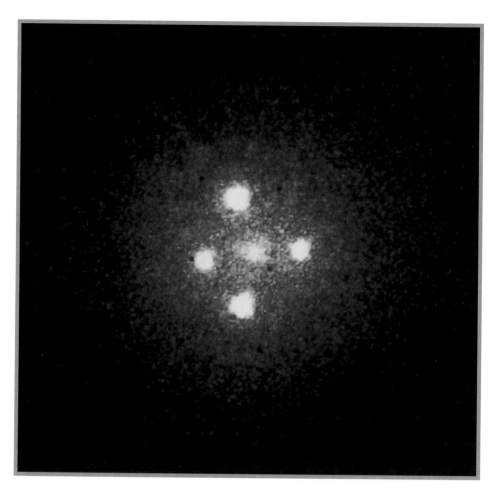

Einstein's cross. The four bright spots are images of the same galaxy formed by light that is bent around a massive object located between the galaxy and the Earth. The massive object acts as a lens, causing the rays of light that were diverging from the distant galaxy to converge at the Earth. If the intervening massive object were perfectly uniform and symmetric, we would see a bright ring of light instead of discrete spots.

(Courtesy of NASA)

Relativity

Most of our everyday experiences and observations are with objects that move at speeds much less than that of light in a vacuum, $c = 3.00 \times 10^8$ m/s. Models based on Newtonian mechanics and early ideas on space and time were formulated to describe the motion of such objects. This formalism is very successful in describing a wide range of phenomena that occur at low speeds, as we have seen in previous chapters. It fails, however, when applied to particles whose speeds approach that of light. Experimentally, the predictions of Newtonian theory can be tested by accelerating electrons or other particles to very high speeds. For example, it is possible to accelerate an electron to a speed of $0.99c$. According to the Newtonian definition of kinetic energy, if the energy transferred to such an electron were increased by a factor of 4, the electron speed should double to $1.98c$. Relativistic calculations, however, show that the speed of the electron—as well as the speeds of all other particles in the Universe—remains less than the speed of light, regardless of the particle energy. Because it places no upper limit on speed, Newtonian mechanics is contrary to modern theoretical predictions and experimental results, and the Newtonian models that we have developed are limited to objects moving much slower than the speed of light. Because Newtonian mechanics does not correctly predict the results of experiments carried out on particles moving at high speeds, we need a new formalism that is valid for these particles.

In 1905, at the age of only 26, Einstein published his *special theory of relativity*, which is the subject of most of this chapter. Regarding the theory, Einstein wrote:

> The relativity theory arose from necessity, from serious and deep contradictions in the old theory from which there seemed no escape. The strength of the new theory lies in the consistency and simplicity with which it solves all these difficulties, using only a few very convincing assumptions. . . .*

Although Einstein made many important contributions to science, special relativity alone represents one of the greatest intellectual achievements of the 20th century. With special relativity, experimental observations can be correctly predicted over the range of speeds from rest to speeds approaching the speed of light. This chapter gives an introduction to special relativity, with emphasis on some of its consequences.

9.1 • THE PRINCIPLE OF NEWTONIAN RELATIVITY

We will begin by considering the notion of relativity at low speeds. This discussion was actually begun in Section 3.6, in which we discussed relative velocity. At that time, we discussed the importance of the observer. In a similar way here, we will generate equations that allow us to express one observer's measurements in terms of the other's. This will lead to some rather unexpected and startling results about our understanding of space and time.

As we have mentioned previously, to describe a physical event, it is necessary to establish a frame of reference. You should recall from earlier chapters that Newton's laws are valid in all inertial frames of reference. Because an inertial frame is defined as one in which Newton's first law is valid, it can be said that an **inertial frame is one in which an object experiences no acceleration if no forces act on it.** Furthermore, any frame moving with constant velocity with respect to an inertial frame must also be an inertial frame. The laws predicting the results of an experiment performed in a vehicle moving with uniform velocity will be identical for the driver of the vehicle and a hitchhiker on the side of the road. The formal statement of this result is called the principle of **Newtonian relativity:**

* *Inertial frame of reference*

* *Principle of Newtonian relativity*

> The laws of mechanics must be the same in all inertial frames of reference.

The following observation illustrates the equivalence of the laws of mechanics in different inertial frames. Consider a pickup truck moving with a constant velocity, as in Figure 9.1a. If a passenger in the truck throws a ball straight up in the air, the passenger observes that the ball moves in a vertical path (ignoring air resistance). The motion of the ball appears to be precisely the same as if the ball were thrown by a person at rest on Earth and observed by that person. The kinematic equations of Chapter 3 describe the results correctly whether the truck is at rest or in uniform motion. Now consider the ball thrown in the truck as viewed by an observer at rest on Earth. This observer sees the path of the ball as a parabola, as in Figure 9.1b. Furthermore, according to this observer, the ball has a horizontal component of velocity equal to the speed of the truck. Although the two observers measure different velocities and see different paths of the ball, they see the same forces

---

* A. Einstein and L. Infeld, *The Evolution of Physics*, New York, Simon & Schuster, 1961.

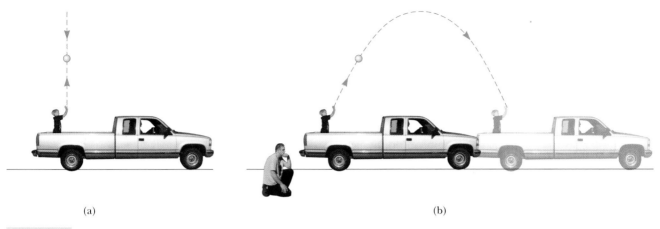

Figure 9.1

(a) The observer in the truck sees the ball move in a vertical path when thrown upward. (b) The Earth observer sees the path of the ball to be a parabola.

on the ball and agree on the validity of Newton's laws as well as classical principles such as conservation of energy and conservation of momentum. Their measurements differ, but the measurements satisfy the same laws. All differences between the two views stem from the relative motion of one frame with respect to the other.

Suppose that some physical phenomenon, which we call an **event,** occurs. The event's location in space and time of occurrence can be specified by an observer with the coordinates (x, y, z, t). We would like to be able to transform these coordinates from one inertial frame to another moving with uniform relative velocity. This will allow us to express one observer's measurements in terms of the other's.

Consider two inertial frames S and S' (Fig. 9.2). The frame S' moves with a constant velocity **v** along the common x and x' axes, where **v** is measured relative to S. We assume that the origins of S and S' coincide at $t = 0$. Thus, at time t, the origin of frame S' is to the right of the origin of S by a distance vt. An event occurs at point P. An observer in S describes the event with space–time coordinates (x, y, z, t), and an observer in S' uses (x', y', z', t') to describe the same event. As we can see from Figure 9.2, a simple geometric model shows that the space coordinates are related by the equations

$$x' = x - vt \qquad y' = y \qquad z' = z \qquad \text{[9.1]}$$

Time is assumed to be the same in both inertial frames. That is, within the framework of classical mechanics, all clocks run at the same rate, regardless of their velocity, so that the time at which an event occurs for an observer in S is the same as the time for the same event in S':

$$t' = t \qquad \text{[9.2]}$$

Equations 9.1 and 9.2 constitute what is known as the **Galilean transformation of coordinates.**

Now suppose that a particle moves a distance dx in a time interval dt as measured by an observer in S. It follows from the first of Equations 9.1 that the corresponding distance dx' measured by an observer in S' is $dx' = dx - v\,dt$. Because $dt = dt'$ (Eq. 9.2), we find that

$$\frac{dx'}{dt'} = \frac{dx}{dt} - v$$

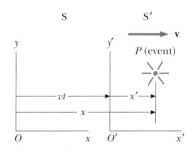

Figure 9.2

An event occurs at a point P. The event is seen by two observers O and O' in inertial frames S and S', where S' moves with a velocity **v** relative to S.

- *Galilean transformation of coordinates*

PITFALL PREVENTION 9.1
The relationship between the S and S′ frames

Keep in mind the relationship between the S and S′ frames. Otherwise, many of the mathematical representations in this chapter are not true. The *x* and *x*′ axes coincide, except that their origins are different. The *y* and *y*′ axes (and the *z* and *z*′ axes), are parallel, but do not coincide due to the displacement of the origin of S′ with respect to that of S. We choose the time $t = 0$ to be the instant at which the origins of the two coordinate systems coincide. If the S′ frame is moving in the positive *x* direction relative to S, *v* is positive; otherwise it is negative.

Albert A. Michelson (1852–1931)

A German–American physicist, Michelson spent much of his life making accurate measurements of the speed of light, In 1907 he was the first American to be awarded the Nobel prize in physics, which he received for his work in optics. His most famous experiment, conducted with Edward Morley in 1887, indicated that is was impossible to measure the absolute velocity of the Earth with respect to the ether. *(AIP Emilio Segrè Visual Archives, Michelson Collection)*

or

$$u'_x = u_x - v \qquad [9.3]$$

where u_x and u'_x are the instantaneous *x* components of velocity of the particle* relative to S and S′, respectively. This result, which is called the **Galilean addition law for velocities** (or Galilean velocity transformation), is used in everyday observations and is consistent with our intuitive notion of time and space. It is the same equation we generated in Section 3.6 (Eq. 3.22) when we first discussed relative velocity in one dimension. We find, however, that it leads to serious contradictions when applied to electromagnetic waves.

9.2 • THE MICHELSON–MORLEY EXPERIMENT

Many experiments similar to throwing the ball in the pickup truck, described in the preceding section, show us that the laws of classical mechanics are the same in all inertial frames of reference. When similar inquiries are made into the laws of other branches of physics, however, the results are contradictory. In particular, the laws of electricity and magnetism are found to depend on the frame of reference used. It might be argued that these laws are wrong, but that is difficult to accept because the laws are in total agreement with known experimental results. The Michelson–Morley experiment was one of many attempts to investigate this dilemma.

The experiment stemmed from a misconception early physicists had concerning the manner in which light propagates. The properties of mechanical waves, such as water and sound waves, were well known, and all of these waves require a *medium* to support the propagation of the disturbance, as we shall discuss in Chapter 13. For sound from your stereo system, the medium is the air, and for ocean waves, the medium is the water surface. In the 19th century, physicists subscribed to a model for light in which electromagnetic waves also require a medium through which to propagate. They proposed that such a medium exists, and they named it the **luminiferous ether.** The ether would define an **absolute frame of reference** in which the speed of light is *c*.

The most famous experiment designed to show the presence of the ether was performed in 1887 by A. A. Michelson (1852–1931) and E. W. Morley (1838–1923). The objective was to determine the speed of the Earth with respect to the ether, and the experimental tool used was a device called the *interferometer,* shown schematically in Figure 9.3.

Light from the source at the left encounters a beam splitter M_0, which is a partially silvered mirror. Part of the light passes through toward mirror M_2, and the other part is reflected upward toward mirror M_1. Both mirrors are the same distance from the beam splitter. After reflecting from these mirrors, the light returns to the beam splitter, and part of each light beam propagates toward the observer at the bottom.

Suppose one arm of the interferometer (M_0M_2, Arm 2, in Fig. 9.3) is aligned along the direction of the velocity **v** of the Earth through space and, therefore, through the ether. The "ether wind" blowing in the direction opposite the Earth's motion should cause the speed of light, as measured in the Earth's frame of refer-

---

* We have used *v* for the speed of the S′ frame relative to the S frame. To avoid confusion, we will use *u* for the speed of an object or particle.

ence, to be $c - v$ as the light approaches mirror M_2 in Figure 9.3, and $c + v$ after reflection.

The other arm (M_0M_1, Arm 1) is perpendicular to the ether wind. For light to travel in this direction, the vector **c** must be aimed "upstream" so that the vector addition of **c** and **v** gives the speed of the light perpendicular to the ether wind as $\sqrt{c^2 - v^2}$. This is similar to Example 3.6, in which a boat crossed a river with a current. The boat is a model for the light beam in the Michelson–Morley experiment, and the river current is a model for the ether wind.

Because they travel in perpendicular directions with different speeds, light beams leaving the beam splitter will arrive back at the beam splitter at different times. The interferometer is designed to detect this time difference. Measurements failed, however, to show any time difference! The Michelson–Morley experiment was repeated by other researchers under varying conditions and at different locations, but the results were always the same: *No time difference of the magnitude required was ever observed.**

The negative result of the Michelson–Morley experiment not only contradicted the ether hypothesis; it also meant that it was impossible to measure the absolute speed of the Earth with respect to the ether frame. From a theoretical viewpoint, this meant that it was impossible to find the absolute frame. As we shall see in the next section, however, Einstein offered a postulate that places a different interpretation on the negative result. In later years, when more was known about the nature of light, the idea of an ether that permeates all space was abandoned. **Light is now understood to be an electromagnetic wave that requires no medium for its propagation.** As a result, an ether through which light travels is an unnecessary construct.

Modern versions of the Michelson–Morley experiment have placed an upper limit of about 5 cm/s on ether wind velocity. These results have shown quite conclusively that the motion of the Earth has no effect on the measured speed of light!

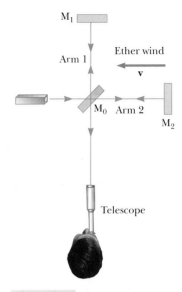

Figure 9.3

In the Michelson interferometer, the ether theory claims that the time of travel for a light beam traveling from the beam splitter to mirror M_1 will be different from that for a light beam traveling from the beam splitter to mirror M_2. The interferometer is sufficiently sensitive to detect this time difference.

9.3 • EINSTEIN'S PRINCIPLE OF RELATIVITY

In the preceding section we noted the failure of experiments to measure the speed of the ether with respect to the Earth. Albert Einstein proposed a theory that boldly removed these difficulties and at the same time completely altered our notion of space and time.[†] Einstein based his relativity theory on two postulates:

1. **The principle of relativity:** All the laws of physics are the same in all inertial reference frames.
2. **The constancy of the speed of light:** The speed of light in vacuum has the same value in all inertial frames, regardless of the velocity of the observer or the velocity of the source emitting the light.

WEB

For information from the American Institute of Physics on Albert Einstein, visit **www.aip.org/history/einstein/**

• *The postulates of special relativity*

---

* From an Earth observer's point of view, changes in the Earth's speed and direction of motion in the course of a year are viewed as ether wind shifts. Even if the speed of the Earth with respect to the ether were zero at some time, six months later the speed of the Earth would be 60 km/s with respect to the ether and a clear time difference should be detected. None has ever been observed, however.

[†] A. Einstein, "On the Electrodynamics of Moving Bodies," *Ann. Physik* **17**:891, 1905. For an English translation of this article and other publications by Einstein, see the book by H. Lorentz, A. Einstein, H. Minkowski, and H. Weyl, *The Principle of Relativity,* Dover, 1958.

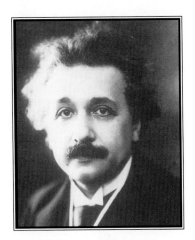

Albert Einstein (1879–1955)

Einstein, one of the greatest physicists of all times, was born in Ulm, Germany. In 1905, at the age of 26, he published four scientific papers that revolutionized physics. Two of these papers were concerned with what is now considered his most important contribution of all, the special theory of relativity.

In 1916, Einstein published his work on the general theory of relativity. The most dramatic prediction of this theory is the degree to which light is deflected by a gravitational field. Measurements made by astronomers on bright stars in the vicinity of the eclipsed sun in 1919 confirmed Einstein's prediction, and Einstein suddenly became a world celebrity.

Einstein was deeply disturbed by the development of quantum mechanics in the 1920s despite his own role as a scientific revolutionary. In particular, he could never accept the probabilistic view of events in nature that is a central feature of quantum theory. The last few decades of his life were devoted to an unsuccessful search for a unified theory that would combine gravitation and electromagnetism into one picture. *(AIP Niels Bohr Library)*

These postulates form the basis of **special relativity**, which is the relativity theory applied to observers moving with constant velocity. The first postulate asserts that *all* the laws of physics, those dealing with mechanics, electricity and magnetism, optics, thermodynamics, and so on, are the same in all reference frames moving with constant velocity relative to each other. This postulate is a sweeping generalization of the principle of Newtonian relativity that only refers to the laws of mechanics. From an experimental point of view, Einstein's principle of relativity means that any kind of experiment performed in a laboratory at rest must agree with the same laws of physics as when performed in a laboratory moving at constant velocity relative to the first one. Hence no preferred inertial reference frame exists and it is impossible to detect absolute motion.

Note that postulate 2, the principle of the constancy of the speed of light, is required by postulate 1: If the speed of light were not the same in all inertial frames, it would be possible to experimentally distinguish between inertial frames, and a preferred, absolute frame in which the speed of light is *c* could be identified, in contradiction to postulate 1. Postulate 2 also eliminates the problem of measuring the speed of the ether by denying the existence of the ether and boldly asserting that light always moves with speed *c* relative to all inertial observers.

Quick Quiz 9.1

You are in a speedboat on a lake. You see ahead of you the crest of a wave, caused by the previous passage of another boat, moving away from you. You accelerate, catch up with, and pass the crest. Is this scenario possible if you are in a spacecraft and you detect the crest of a wave of *light* ahead of you?

9.4 • CONSEQUENCES OF SPECIAL RELATIVITY

If we accept the postulates of special relativity, we must conclude that relative motion is unimportant when measuring the speed of light—this is the lesson of the Michelson–Morley experiment. At the same time, we must alter our commonsense notion of space and time and be prepared for some very unexpected consequences, as we shall see now.

Simultaneity and the Relativity of Time

A basic premise of Newtonian mechanics is that a universal time scale exists that is the same for all observers. In fact, Newton wrote that "Absolute, true, and mathematical time, of itself, and from its own nature, flows equably without relation to anything external." Thus, Newton and his followers simply took simultaneity for granted. In his development of special relativity, Einstein abandoned the notion that two events that appear simultaneous to one observer appear simultaneous to all observers. According to Einstein, a **time measurement depends on the reference frame in which the measurement is made.**

Einstein devised the following thought experiment to illustrate this point. A boxcar moves with uniform velocity, and two lightning bolts strike its ends, as in Figure 9.4a, leaving marks on the boxcar and on the ground. The marks on the boxcar are labeled A' and B', and those on the ground are labeled A and B. An observer at O' moving with the boxcar is midway between A' and B', and a ground observer at O is midway between A and B. The events recorded by the observers are the light signals from the lightning bolts.

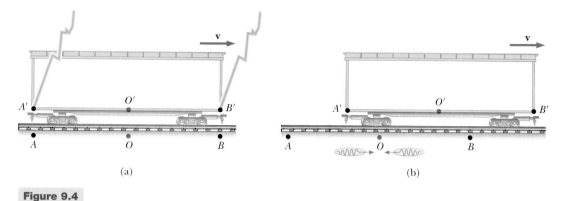

Figure 9.4

(a) Two lightning bolts strike the ends of a moving boxcar. (b) The events appear to be simultaneous to the observer at *O*, who is standing on the ground midway between *A* and *B*. The events do not appear to be simultaneous to the observer *O′* riding on the boxcar, who claims that the front of the car is struck before the rear. Note that the leftward-traveling light signal from *B′* has already passed observer *O′*, but the rightward-traveling light signal from *A′* has not yet reached *O′*.

The two light signals reach observer *O* at the same time, as indicated in Figure 9.4b. Because of this, *O* concludes that the events at *A* and *B* occurred simultaneously. Now consider the same events as viewed by the observer on the boxcar at *O′*. From our frame of reference, at rest with respect to the tracks in Figure 9.4, we see the lightning strikes occur as *A′* passes *A*, *O′* passes *O*, and *B′* passes *B*. By the time the light has reached observer *O*, observer *O′* has moved as indicated in Figure 9.4b. Thus, the light signal from *B′* has already swept past *O′* because it had less distance to travel, but the light from *A′* has not yet reached *O′*. According to Einstein, **observer *O′* must find that light travels at the same speed as that measured by observer *O*.** The observer *O′* therefore concludes that the lightning struck the front of the boxcar before it struck the back. This thought experiment clearly demonstrates that the two events, which appear to be simultaneous to observer *O*, do not appear to be simultaneous to observer *O′*. In general, two events separated in space and observed to be simultaneous in one reference frame are not observed to be simultaneous in a second frame moving relative to the first. That is, simultaneity is not an absolute concept but one that depends upon the state of motion of the observer.

PITFALL PREVENTION 9.2
Who's right?

At this point, you might wonder which observer is correct concerning the two events. *Both are correct*, because the principle of relativity states that *no inertial frame of reference is preferred*. Although the two observers reach different conclusions, both are correct in their own reference frame because the concept of simultaneity is not absolute. This, in fact, is the central point of relativity—any uniformly moving frame of reference can be used to describe events and do physics.

Time Dilation

According to our thought experiment, observers in different inertial frames measure different time intervals between a pair of events. This can be illustrated by considering a vehicle moving to the right with a speed *v* as in the pictorial representation in Figure 9.5a. A mirror is fixed to the ceiling of the vehicle and observer *O′*, at rest in a frame attached to the vehicle, holds a flashlight a distance *d* below the mirror. At some instant, the flashlight is turned on momentarily and emits a pulse of light (event 1) directed toward the mirror. At some later time after reflecting from the mirror, the pulse arrives back at the flashlight (event 2). Observer *O′* carries a clock that she uses to measure the time interval Δt_p between these two events. (The subscript "*p*" stands for "proper," as will be discussed shortly.) Because the light pulse has a constant speed *c*, the time it takes the pulse to travel

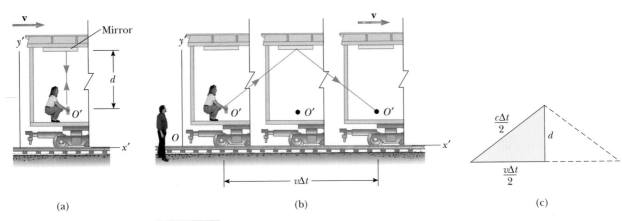

(a)   (b)   (c)

Figure 9.5

A mirror is fixed to a moving vehicle, and two observers measure the time interval between two events: the leaving of a light pulse from a flashlight and the arrival of the reflected light pulse back at the flashlight. (a) Observer O', riding on the vehicle, sees the light pulse travel a total distance of $2d$ and measures a time interval between the events of Δt_p. (b) Observer O is standing on the Earth and sees the mirror and O' move to the right with speed v. Observer O measures the distance that the light pulse travels to be greater than $2d$ and measures a time interval between the events of Δt. (c) The right triangle for calculating the relationship between Δt and Δt_p.

from O' to the mirror and back to O' can be found by modeling the light pulse as a particle under constant speed from Chapter 2:

$$\Delta t_p = \frac{s}{v} = \frac{2d}{c}$$
[9.4]

This time interval Δt_p is measured by O', for whom the two events occur at the same spatial position.

    Now consider the same pair of events as viewed by observer O in a second frame attached to the ground as in Figure 9.5b. According to this observer, the mirror and flashlight are moving to the right with a speed v. The geometry appears to be entirely different as viewed by this observer. By the time the light from the flashlight reaches the mirror, the mirror has moved horizontally a distance $v\,\Delta t/2$, where Δt is the time it takes the light to travel from the flashlight to the mirror and back to the flashlight as measured by observer O. In other words, the second observer concludes that because of the motion of the vehicle, if the light is to hit the mirror, it must leave the flashlight at an angle with respect to the vertical direction. Comparing Figures 9.5a and 9.5b, we see that the light must travel farther when observed in the second frame than in the first frame.

    According to the second postulate of special relativity, both observers must measure c for the speed of light. Because the light travels farther in the second frame, but at the same speed, it follows that the time interval Δt measured by the observer in the second frame is longer than the time interval Δt_p measured by the observer in the first frame. To obtain a relationship between these two time intervals, it is convenient to use the right triangle geometric model shown in Figure 9.5c. The Pythagorean theorem applied to the triangle gives

$$\left(\frac{c\,\Delta t}{2}\right)^2 = \left(\frac{v\,\Delta t}{2}\right)^2 + d^2$$

Solving for Δt gives

$$\Delta t = \frac{2d}{\sqrt{c^2 - v^2}} = \frac{2d}{c\sqrt{1 - \dfrac{v^2}{c^2}}}$$ [9.5]

Because $\Delta t_p = 2d/c$, we can express Equation 9.5 as

$$\Delta t = \frac{\Delta t_p}{\sqrt{1 - \dfrac{v^2}{c^2}}} = \gamma \Delta t_p$$ [9.6]

where $\gamma = (1 - v^2/c^2)^{-1/2}$. This result says that the **time interval Δt measured by O is longer than the time interval Δt_p measured by O'** because γ is always greater than unity. That is, $\Delta t > \Delta t_p$. This effect is known as **time dilation.**

We can see that time dilation is not observed in our everyday lives by considering the factor γ. This factor deviates significantly from a value of 1 only for very high speeds, as shown in Table 9.1. For example, for a speed of $0.1c$, the value of γ is 1.005. Thus, a time dilation of only 0.5% occurs at one-tenth the speed of light. Speeds we encounter on an everyday basis are far slower than this, so we do not see time dilation in normal situations.

The time interval Δt_p in Equation 9.6 is called the **proper time interval.** In general, the proper time interval is defined as the **time interval between two events as measured by an observer for whom the events occur at the same point in space.** In our case, observer O' measures the proper time interval. For us to be able to use Equation 9.6, the events must occur at the same spatial position in *some* inertial frame. Thus, for example, this equation cannot be used for the lightning example described at the beginning of this section.

Time dilation is a verifiable phenomenon—let us look at an example of a situation in which the effects of time dilation can be observed and that served as an important historical confirmation of the predictions of relativity. Muons are unstable elementary particles that have a charge equal to that of the electron and a mass 207 times that of the electron. Muons can be produced as a result of collisions of cosmic radiation with atoms high in the atmosphere. Slow-moving muons in the laboratory have a lifetime measured to be the proper time interval $\Delta t_p = 2.2$ μs. If we assume that the speed of atmospheric muons is close to the speed of light, we find that these particles can travel a distance of approximately $(3.0 \times 10^8 \text{ m/s})(2.2 \times 10^{-6} \text{ s}) \approx 6 \times 10^2$ m before they decay (Fig. 9.6a). Hence, they are unlikely to reach the surface of the Earth from high in the atmosphere where they are produced; however, experiments show that a large number of muons *do* reach the surface. The phenomenon of time dilation explains this effect. As measured by an observer on Earth, the muons have a dilated lifetime equal to $\gamma \Delta t_p$. For example, for $v = 0.99c$, $\gamma \approx 7.1$ and $\gamma \Delta t_p \approx 16$ μs. Hence, the average distance traveled by the muons in this time as measured by an observer on Earth is approximately $(3.0 \times 10^8 \text{ m/s})(16 \times 10^{-6} \text{ s}) \approx 5 \times 10^3$ m, as shown in Figure 9.6b.

The results of an experiment reported by J. C. Hafele and R. E. Keating provided direct evidence of time dilation.* The experiment involved the use of very stable atomic clocks. Time intervals measured with four such clocks in jet flight

TABLE 9.1

Value for γ at Various Speeds

v/c	γ
0.0010	1.000
0.010	1.000
0.10	1.005
0.20	1.021
0.30	1.048
0.40	1.091
0.50	1.155
0.60	1.250
0.70	1.400
0.80	1.667
0.90	2.294
0.92	2.552
0.94	2.931
0.96	3.571
0.98	5.025
0.99	7.089
0.995	10.01
0.999	22.37

PITFALL PREVENTION 9.3
The proper time interval

Read this definition carefully. It is *very* important in relativistic calculations to correctly identify the observer who measures the proper time interval. **The proper time interval between two events is always the time interval measured by an observer for whom the two events take place at the same position.**

* J. C. Hafele and R. E. Keating, "Around the World Atomic Clocks: Relativistic Time Gains Observed," *Science*, July 14, 1972, p. 168.

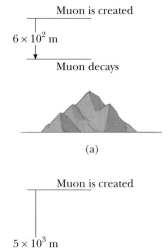

Figure 9.6

(a) Without relativistic considerations, muons created in the atmosphere and traveling downward with a speed of $0.99c$ would travel only about 6×10^2 m before decaying with an average lifetime of 2.2 μs. Thus, very few muons would reach the surface of the Earth. (b) With relativistic considerations, the muon's lifetime is dilated according to an observer on Earth. As a result, according to this observer, the muon can travel about 5×10^3 m before decaying. This results in many of them arriving at the surface.

were compared with time intervals measured by reference clocks located at the U.S. Naval Observatory. Their results were in good agreement with the predictions of special relativity and can be explained in terms of the relative motion between the Earth's rotation and the jet aircraft. In their paper, Hafele and Keating report the following: "Relative to the atomic time scale of the U.S. Naval Observatory, the flying clocks lost 59 ± 10 ns during the eastward trip and gained 273 ± 7 ns during the westward trip. . . ."

Quick Quiz 9.2

A crew watches a movie that is 2 h long in a spacecraft that is moving at high speed through space. Will an Earth-bound observer measure the duration of the movie to be (a) longer than, (b) shorter than, or (c) equal to 2 h?

The Twins Paradox

Because of time dilation, the time interval between ticks of a clock moving with respect to an observer is observed to be longer than the time interval between ticks of an identical clock at rest with respect to the observer—it is often said, *"A moving clock runs slower than a clock at rest by a factor γ."* This is true for ordinary mechanical clocks as well as for the light clock on the vehicle in Figure 9.5. In fact, we can generalize these results by stating that **all physical processes, including chemical and biological ones, are measured by an observer to slow down when those processes occur in a frame moving with respect to the observer.** For example, the heartbeat of an astronaut moving through space will be measured by the astronaut to be the same as when he was at rest on Earth. The astronaut will not have any sensation of life slowing down in the spacecraft. The astronaut's heartbeat, however, will appear to slow down when measured by a clock on Earth.

An intriguing biological consequence of time dilation is the so-called *twins paradox*. Consider an experiment involving a set of twins named Speedo and Goslo (Fig. 9.7) who are 20 years old. Speedo, the more adventuresome of the two, sets out on an epic journey to Planet X, located 20 lightyears (ly) from Earth. Furthermore, his spacecraft is capable of reaching a speed of $0.95c$ relative to the inertial

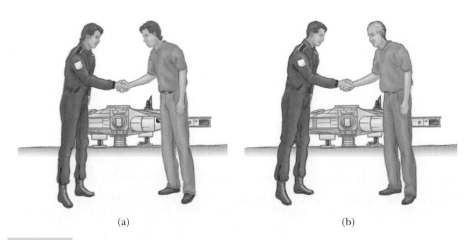

(a)   (b)

Figure 9.7

(a) As Speedo leaves his twin brother Goslo on the Earth, both are the same age. (b) When Speedo returns from his journey to Planet X, he is younger than Goslo.

frame of his twin brother on Earth. After reaching Planet X, Speedo becomes homesick and immediately returns to Earth at the same high speed as on the outbound journey. On his return, Speedo is shocked to discover that many things have changed on Earth in his absence. To Speedo, the most significant change to have occurred is that his twin brother Goslo has aged 42 years and is now 62 years old. Speedo, on the other hand, has aged by only 13 years.

At this point, it is fair to raise the following questions: Which twin is the traveler, and which twin is really younger as a result of this experiment? From Goslo's frame of reference, he was at rest while his brother traveled at a high speed out and back. From Speedo's perspective, it is he who was at rest while Goslo and the Earth raced away on a 6.5-year journey and then headed back for another 6.5 years. This leads to an apparent contradiction. Each twin observed the other in motion and might claim that the other's clock runs slow. Which twin has *actually* aged more?

In reality, Speedo does return younger than Goslo. How can we resolve this apparent paradox? Despite our discussion in the previous paragraph, the situation is *not* symmetric. Speedo, the space traveler, must experience a series of accelerations during his journey in order to turn around and come back. As a result, his speed is not always uniform and, as a consequence, Speedo is not in a single inertial frame during the entire trip. He cannot be regarded as always being at rest while Goslo is in uniform motion, because he changes his state of motion at least once. Therefore no paradox exists—only Goslo remains in a single inertial frame and can correctly predict results based on special relativity.

The conclusion that Speedo does not remain in a single inertial frame is inescapable. The time required to accelerate and decelerate Speedo's spacecraft may be made very small using large rocket engines, so that Speedo can claim that he spends most of his time traveling to and from Planet X at $0.95c$ in an inertial frame. However, Speedo must slow down, reverse his motion, and return to Earth in an altogether different inertial frame. Only Goslo, who is in a single inertial frame, can apply the simple time dilation equation to Speedo's trip. Thus, Goslo finds that instead of aging 42 years, Speedo ages only $(1 - v^2/c^2)^{1/2}(42 \text{ years}) = 13 \text{ years}$.

Quick Quiz 9.3

Suppose astronauts were paid according to the time spent traveling in space. After a long voyage traveling at a speed near that of light, a crew of astronauts return to Earth and open their pay envelopes. What will their reaction be?

THINKING PHYSICS 9.1

Suppose a student explains time dilation with the following argument: If you start running away from a clock at 12:00 at a speed very close to the speed of light, you would not see the time change, because the light from the clock representing 12:01 would never reach you. What is the flaw in this argument?

Reasoning The implication in this argument is that the velocity of light relative to the runner is approximately *zero*—"the light . . . would never reach you." This is a Galilean point of view, in which the relative velocity is a simple subtraction of running velocity from the light velocity. From the point of view of special relativity, one of the fundamental postulates is that the speed of light is the same for all observers, *including one running away from the light source at the speed of light*. Thus, the light from 12:01 will move toward the runner at the speed of light, as measured by all observers, including the runner.

Example 9.1   What is the Period of the Pendulum?

The period T of a pendulum on the Earth is measured to be 3.0 s in the rest frame of the Earth. What is the period when measured by an observer in a spacecraft moving at a speed of $0.95c$ relative to the pendulum?

Reasoning   We identify two events: two sequential arrivals of the pendulum at the same maximum deflection point. The proper time interval between these events is the period $T_p = 3.0$ s. The observer in the spacecraft will measure a dilated time interval.

Solution   For the period measured by the observer in the spacecraft, Equation 9.6 gives

$$T = \gamma T_p = \frac{1}{\sqrt{1 - \frac{(0.95c)^2}{c^2}}} \, T_p = (3.2)(3.0\text{ s}) = \boxed{9.6\text{ s}}$$

That is, measurements by this observer show that it takes longer to complete an oscillation compared with an observer on Earth.

EXERCISE   Muons move in a circular orbit of radius 500 m at a speed of $0.9994c$. If a muon at rest decays into other particles after 2.2 μs (the proper time interval), how many trips around the circular path do we expect the muons to make before they decay?

Answer   6.1 revolutions

Length Contraction

PITFALL PREVENTION 9.4

The proper length

As with the proper time interval, it is *very* important in relativistic calculations to correctly identify the observer who measures the proper length. **The proper length between two points in space is always the length measured by an observer at rest with respect to the length.** It is often the case that the proper time interval and the proper length are *not* measured by the same observer.

The measured distance between two points also depends on the frame of reference. The **proper length** of an object is defined as the **distance in space between the endpoints of the object measured by someone who is at rest relative to the object.** An observer in a reference frame that is moving with respect to the object will measure a length along the direction of the velocity that is always less than the proper length. This effect is known as **length contraction.** Although we have introduced this through the mental representation of an object, the object is not necessary. **The distance between *any* two points in space is measured by an observer to be contracted along the direction of the velocity of the observer relative to the points.**

Consider a spacecraft traveling with a speed v from one star to another. We will consider the time interval between two events: (1) the leaving of the spacecraft from the first star and (2) the arrival of the spacecraft at the second star. There are two observers: one on Earth and the other in the spacecraft. The observer at rest on Earth (and also at rest with respect to the two stars) measures the distance between the stars to be L_p, the proper length. According to this observer, the time it takes the spacecraft to complete the voyage is $\Delta t = L_p / v$. What does an observer in the moving spacecraft measure for the distance between the stars? This observer measures the proper time interval, because the passage of each of the two stars by his spacecraft occurs at the same position in his reference frame—at his spacecraft. Therefore, because of time dilation, the time of travel measured by the space traveler will be smaller than that for the Earth-bound observer. Using the time dilation expression, the proper time interval between events is $\Delta t_p = \Delta t / \gamma$. The space traveler claims to be at rest and sees the destination star moving toward the spacecraft with speed v. Because the space traveler reaches the star in the time $\Delta t_p < \Delta t$, he concludes that the distance L between the stars is shorter than L_p. This distance measured by the space traveler is

$$L = v\,\Delta t_p = v\,\frac{\Delta t}{\gamma}$$

Because $L_p = v \, \Delta t$, we see that

$$L = \frac{L_p}{\gamma} = L_p \left(1 - \frac{v^2}{c^2}\right)^{1/2}$$ [9.7] • *Length contraction*

Because $(1 - v^2/c^2)^{1/2}$ is less than 1, the space traveler measures a length that is shorter than the proper length. Thus, an observer in motion with respect to two points in space measures the length L between the points (along the direction of motion) to be shorter than the length L_p measured by an observer at rest with respect to the points (the proper length).

Note that **length contraction takes place only along the direction of motion.** For example, suppose a stick moves past an Earth observer with speed v, as in Figure 9.8. The length of the stick as measured by an observer in a frame attached to the stick is the proper length L_p, as in Figure 9.8a. The length L of the stick measured by the Earth observer is shorter than L_p by the factor $(1 - v^2/c^2)^{1/2}$, but the width is the same. Furthermore, length contraction is a symmetric effect: If the stick is at rest on Earth, an observer in the moving frame would also measure its length to be shorter by the same factor $(1 - v^2/c^2)^{1/2}$.

It is important to emphasize that the proper length and proper time interval are defined differently. The proper length is measured by an observer at rest with respect to the endpoints of the length. The proper time interval is measured by someone for whom the events occur at the same position. As an example of this point, let us return to the decaying muons moving at speeds close to the speed of light. An observer in the muon's reference frame would measure the proper lifetime, and an Earth-based observer would measure the proper length (the distance from creation to decay in Figure 9.6). In the muon's reference frame, no time dilation occurs, but the distance of travel is observed to be shorter when measured in this frame. Likewise, in the Earth observer's reference frame, a time dilation does occur, but the distance of travel is measured to be the proper length. Thus, when calculations on the muon are performed in both frames, the outcome of the experiment in one frame is the same as the outcome in the other frame: More muons reach the surface than would be predicted without relativistic calculations.

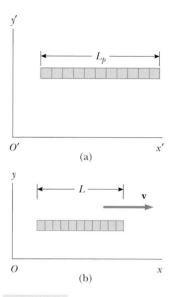

Figure 9.8

(a) According to an observer in a frame attached to the stick (that is, both the stick and the frame have the same velocity), the stick is measured to have its proper length L_p. (b) According to an observer in a frame in which the stick has a velocity **v** relative to the frame, the stick is measured to be *shorter* than the proper length L_p by a factor $(1 - v^2/c^2)^{1/2}$.

Quick Quiz 9.4

If you were somehow able to travel at the speed of light, how big would the Universe be according to your measurements?

Quick Quiz 9.5

You are packing for a trip to another star, to which you will be traveling at $0.99c$. Should you buy smaller sizes of your clothing because you will be skinnier on your trip due to length contraction? Can you sleep in a smaller cabin than usual because you will be shorter when you lie down?

Quick Quiz 9.6

You are observing a spacecraft moving away from you. You notice that it is measured to be shorter than when it was at rest on the ground next to you, and, through the spacecraft window, you can see a clock. You observe that the passage of time on the clock is measured to be slower than that of the watch on your wrist. What if the spacecraft turns around and comes toward you? Will it appear to be *longer* and will the clock in the spacecraft run *faster*?

Example 9.2   A Voyage to Sirius

An astronaut takes a trip to Sirius, located 8.00 lightyears from Earth. A lightyear is the distance that light travels in one year: 1 ly = (3.00 × 10⁸ m/s)(3.156 × 10⁷ s) = 9.46 × 10¹⁵ m. The astronaut measures the time of the one-way journey to be 6.00 yr. If the spacecraft moves at a constant speed of 0.800*c*, how can the 8.00-lightyear distance be reconciled with the 6.00-yr duration measured by the astronaut?

Reasoning   The 8.00 lightyears represents the proper length (the distance from Earth to Sirius) measured by an observer seeing both Earth and Sirius at rest. The astronaut sees Sirius approaching her at 0.800*c* but also sees the distance contracted to

$$L = \frac{L_p}{\gamma} = \frac{8.00 \text{ ly}}{\gamma} = (8.00 \text{ ly})\sqrt{1 - \frac{v^2}{c^2}}$$

$$= (8.00 \text{ ly})\sqrt{1 - \frac{(0.800c)^2}{c^2}} = 4.80 \text{ ly}$$

So the travel time measured on her clock is

$$\Delta t = \frac{L}{v} = \frac{4.80 \text{ ly}}{0.800c} = \frac{4.80 \text{ ly}}{0.800(1.00 \text{ ly/yr})} = 6.00 \text{ yr}$$

Notice that we have used the speed of light as *c* = 1.00 ly/yr to determine this last result.

Example 9.3   The Triangular Spacecraft

A spacecraft in the form of a triangle flies by an observer on Earth with a speed of 0.95*c* along the *x* direction. According to an observer on the spacecraft (Fig. 9.9a), the distances L_p and *y* are measured to be 52 m and 25 m, respectively. What are the dimensions of the spacecraft as measured by the Earth observer when the spacecraft is in motion along the direction shown in Figure 9.9b?

Solution   In Figure 9.9a, we show the shape of the spacecraft as measured* by the observer on the spacecraft. The proper length along the direction of motion is L_p = 52 m. The Earth observer watching the moving spacecraft measures the horizontal length of the spacecraft to be contracted to

$$L = L_p\sqrt{1 - \frac{v^2}{c^2}} = (52 \text{ m})\sqrt{1 - \frac{(0.95c)^2}{c^2}} = \boxed{16 \text{ m}}$$

The 25-m vertical height is unchanged because it is perpendicular to the direction of relative motion between observer and spacecraft. Figure 9.9b represents the shape of the spacecraft as measured by the Earth observer.

EXERCISE   A spacecraft moves at 0.900*c*. If its length is L_0 when measured from inside the spacecraft, what is its length as measured by a ground observer?

Answer   0.436 L_0

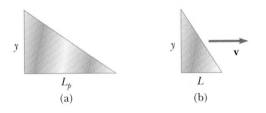

Figure 9.9

(Example 9.3) (a) When the spacecraft is at rest, its shape is measured as shown. (b) The spacecraft is measured to have this shape when it moves to the right with a speed *v*. Note that only its *x* dimension is contracted in this case.

---

* Notice that we are careful here to say the shape "as measured" by an observer, rather than "as seen" by an observer. What an observer *sees* when he looks at an object is the set of light rays entering the eye at a given instant. These rays left different parts of the object at different times because different parts of the object are at different distances from the eye. Fig. 9.9b is a representation of the light rays leaving different parts of the object *simultaneously*. Viewing an object moving at high speed introduces additional changes to the object besides length contraction, including apparent rotations.

9.5 • THE LORENTZ TRANSFORMATION EQUATIONS

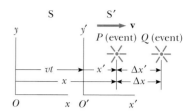

Figure 9.10

Events occur at points P and Q and are observed by an observer at rest in the S frame and another in the S′ frame, which is moving to the right with a speed v.

Suppose an event that occurs at some point P is reported by two observers: one at rest in a frame S and another in a frame S′ that is moving to the right with speed v, as in Figure 9.10. The observer in S reports the event with space–time coordinates (x, y, z, t), and the observer in S′ reports the same event using the coordinates (x', y', z', t'). If two events occur at P and Q, Equation 9.1 predicts that $\Delta x = \Delta x'$; that is, the distance between the two points in space at which the events occur does not depend on the motion of the observer. Because this is contradictory to the notion of length contraction, the Galilean transformation is not valid when v approaches the speed of light. In this section, we state the correct transformation equations that apply for all speeds in the range $0 \leq v < c$.

The equations that relate these measurements and enable us to transform coordinates from S to S′ are the **Lorentz transformation equations:**

$$x' = \gamma(x - vt)$$
$$y' = y$$
$$z' = z \qquad \text{[9.8]}$$
$$t' = \gamma\left(t - \frac{v}{c^2}x\right)$$

• *Lorentz transformation for* $S \rightarrow S'$

These transformation equations were developed by Hendrik A. Lorentz (1853–1928) in 1890 in connection with electromagnetism. Einstein, however, recognized their physical significance and took the bold step of interpreting them within the framework of special relativity.

We see that the value for t' assigned to an event by observer O' depends both on the time t and on the coordinate x as measured by observer O. Thus, in relativity, space and time are not separate concepts but rather are closely interwoven with each other in what we call **space–time.** This is unlike the case of the Galilean transformation in which $t = t'$.

If we wish to transform coordinates in the S′ frame to coordinates in the S frame, we simply replace v by $-v$ and interchange the primed and unprimed coordinates in Equation 9.8:

$$x = \gamma(x' + vt')$$
$$y = y'$$
$$z = z' \qquad \text{[9.9]}$$
$$t = \gamma\left(t' + \frac{v}{c^2}x'\right)$$

• *Inverse Lorentz transformation for* $S' \rightarrow S$

When $v \ll c$, the Lorentz transformation reduces to the Galilean transformation. To check this, note that if $v \ll c$, $v^2/c^2 \ll 1$, so that γ approaches 1 and Equation 9.8 reduces in this limit to Equations 9.1 and 9.2:

$$x' = x - vt \qquad y' = y \qquad z' = z \qquad t' = t$$

Lorentz Velocity Transformation

Let us now derive the Lorentz velocity transformation, which is the relativistic counterpart of the Galilean velocity transformation, Equation 9.3. Once again S is our first frame of reference and S′ is our second frame of reference that moves at a speed v relative to S along the common x and x' axes. Suppose that an object is measured in S' to have an instantaneous velocity component u'_x given by

$$u'_x = \frac{dx'}{dt'} \qquad \text{[9.10]}$$

Using Equations 9.8, we have

$$dx' = \gamma(dx - v\,dt) \qquad \text{and} \qquad dt' = \gamma\left(dt - \frac{v}{c^2}\,dx\right)$$

Substituting these values into Equation 9.10 gives

$$u'_x = \frac{dx'}{dt'} = \frac{dx - v\,dt}{dt - \frac{v}{c^2}\,dx} = \frac{\frac{dx}{dt} - v}{1 - \frac{v}{c^2}\frac{dx}{dt}}$$

But dx/dt is the velocity component u_x of the object measured in S, and so this expression becomes

$$u'_x = \frac{u_x - v}{1 - \frac{u_x v}{c^2}} \qquad \text{[9.11]}$$

- *Lorentz velocity transformation for S → S'*

Similarly, if the object has velocity components along y and z, the components in S' are

$$u'_y = \frac{u_y}{\gamma\left(1 - \frac{u_x v}{c^2}\right)} \qquad \text{and} \qquad u'_z = \frac{u_z}{\gamma\left(1 - \frac{u_x v}{c^2}\right)} \qquad \text{[9.12]}$$

When u_x or v is much smaller than c (the nonrelativistic case), the denominator of Equation 9.11 approaches unity and so $u'_x \approx u_x - v$. This corresponds to the Galilean velocity transformations. In the other extreme, when $u_x = c$, Equation 9.11 becomes

$$u'_x = \frac{c - v}{1 - \frac{cv}{c^2}} = \frac{c\left(1 - \frac{v}{c}\right)}{1 - \frac{v}{c}} = c$$

From this result, we see that an object whose speed approaches c relative to an observer in S also has a speed approaching c relative to an observer in S'—independent of the relative motion of S and S'. Note that this conclusion is consistent with Einstein's second postulate, namely, that the speed of light must be c relative to all inertial frames of reference.

To obtain u_x in terms of u'_x, we replace v by $-v$ in Equation 9.11 and interchange the roles of primed and unprimed variables:

$$u_x = \frac{u'_x + v}{1 + \frac{u'_x v}{c^2}} \qquad \text{[9.13]}$$

- *Inverse Lorentz velocity transformation for S' → S*

Example 9.4   Relative Velocity of Spacecraft

Two spacecraft A and B are moving directly toward each other, as in Figure 9.11. An observer on Earth measures the speed of A to be $0.750c$ and the speed of B to be $0.850c$. Find the speed of B with respect to A.

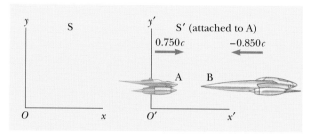

Figure 9.11

(Example 9.4) Two spacecraft A and B move in opposite directions.

Solution   This problem can be solved by taking the S′ frame as being attached to A, so that $v = 0.750c$ relative to the Earth observer (in the S frame). Spacecraft B can be considered as an object moving with a velocity component $u_x = -0.850c$ relative to the Earth observer. Hence, the speed of B with respect to A is the speed of B as measured by the observer $O′$ on A, and can be obtained using Equation 9.11:

$$u'_x = \frac{u_x - v}{1 - \dfrac{u_x v}{c^2}} = \frac{-0.850c - 0.750c}{1 - \dfrac{(-0.850c)(0.750c)}{c^2}} = \boxed{-0.980c}$$

The negative sign in the result indicates that spacecraft B is moving in the negative x direction as observed by A.

Example 9.5   Relativistic Leaders of the Pack

Two motorcycle pack leaders named David and Emily are racing at relativistic speeds along perpendicular paths as in Figure 9.12. How fast does Emily recede over David's right shoulder as seen by David?

Solution   Figure 9.12 represents the situation as seen by a police officer at rest in frame S, who observes the following:

$$\text{David:} \quad u_x = 0.75c \qquad u_y = 0$$
$$\text{Emily:} \quad u_x = 0 \qquad u_y = -0.90c$$

To determine Emily's speed of recession as seen by David, we take S′ to move along with David and we calculate u'_x and u'_y for Emily using Equations 9.11 and 9.12:

$$u'_x = \frac{u_x - v}{1 - \dfrac{u_x v}{c^2}} = \frac{0 - 0.75c}{1 - \dfrac{(0)(0.75c)}{c^2}} = -0.75c$$

$$u'_y = \frac{u_y}{\gamma\left(1 - \dfrac{u_x v}{c^2}\right)} = \frac{\sqrt{1 - \dfrac{(0.75c)^2}{c^2}}\,(-0.90c)}{\left(1 - \dfrac{(0)(0.75c)}{c^2}\right)} = -0.60c$$

Thus, the speed of Emily as observed by David is

$$u' = \sqrt{(u'_x)^2 + (u'_y)^2} = \sqrt{(-0.75c)^2 + (-0.60c)^2} = \boxed{0.96c}$$

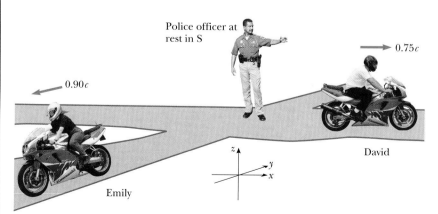

Police officer at rest in S

0.90c

0.75c

Emily

David

Figure 9.12

(Example 9.5) David moves to the east with a speed $0.75c$ relative to the police officer, and Emily travels south at a speed $0.90c$ relative to the officer.

9.6 • RELATIVISTIC MOMENTUM AND THE RELATIVISTIC FORM OF NEWTON'S LAWS

We have seen that to properly describe the motion of particles within the framework of special relativity, the Galilean transformation must be replaced by the Lorentz transformation. Because the laws of physics must remain unchanged under the Lorentz transformation, we must generalize Newton's laws and the definitions of momentum and energy to conform to the Lorentz transformation and the principle of relativity. These generalized definitions should reduce to the classical (nonrelativistic) definitions for $v \ll c$ or $u \ll c$. (As we have done previously, we will use v for the speed of one reference frame relative to another, and u for the speed of a particle.)

First, recall that the total momentum of an isolated system of particles remains constant. Suppose a collision between two particles is described in a reference frame S in which the momentum of the system is observed to be conserved. If the velocities in a second reference frame S′ are calculated using the Lorentz transformation and the Newtonian definition of momentum, $\mathbf{p} = m\mathbf{u}$, it is found that the momentum of the system is *not* observed to be conserved in the second reference frame. This violates one of Einstein's postulates: The laws of physics are the same in all inertial frames. In view of this, and assuming the Lorentz transformation to be correct, we must modify the definition of momentum.

The relativistic equation for the momentum of a particle of mass m that maintains the principle of conservation of momentum is

• *Definition of relativistic momentum*

$$\mathbf{p} \equiv \frac{m\mathbf{u}}{\sqrt{1 - \dfrac{u^2}{c^2}}} \qquad [9.14]$$

where $\mathbf{u}$ is the velocity of the particle. When u is much less than c, the denominator of Equation 9.14 approaches unity, so that $\mathbf{p}$ approaches $m\mathbf{u}$. Therefore, the relativistic equation for $\mathbf{p}$ reduces to the classical expression when u is small compared with c. Equation 9.14 is often written in simpler form as

$$\mathbf{p} = \gamma m\mathbf{u} \qquad [9.15]$$

using our previously defined expression for γ.*

The relativistic force $\mathbf{F}$ on a particle whose momentum is $\mathbf{p}$ is defined as

$$\mathbf{F} \equiv \frac{d\mathbf{p}}{dt} \qquad [9.16]$$

where $\mathbf{p}$ is given by Equation 9.14. This expression preserves both classical mechanics in the limit of low velocities and conservation of momentum for an isolated system ($\Sigma\mathbf{F}_{\text{ext}} = 0$) both relativistically and classically.

We leave it to Problem 53 at the end of the chapter to show that the acceleration $\mathbf{a}$ of a particle decreases under the action of a constant force, in which case $a \propto (1 - u^2/c^2)^{3/2}$. From this proportionality, note that as the particle's speed approaches c, the acceleration caused by any finite force approaches zero. It is therefore impossible to accelerate a particle from rest to a speed $v \geq c$.

PITFALL PREVENTION 9.6

Watch out for "relativistic mass"

Some older treatments of relativity maintained the conservation of momentum principle at high speeds by using a model in which the mass of a particle increases with speed. You might still encounter this notion of "relativistic mass" in your outside reading, especially in older books. Be aware that this notion is no longer widely accepted, and mass is considered as *invariant*, independent of speed. The mass of an object in all frames is considered to be the mass as measured by an observer at rest with respect to the object.

---

* We defined γ previously in terms of the speed v of one frame relative to another frame. The same symbol is also used for $(1 - u^2/c^2)^{-1/2}$ where u is the speed of a particle.

This means that c is an upper limit for the speed of any particle. In fact, it is possible to show that **no matter, energy, or information can travel through space faster than c.** Note that the relative speeds of the two spacecraft in Example 9.4 and the two motorcyclists in Example 9.5 were both less than c. If we had attempted to solve these examples with Galilean transformations, we would have obtained relative speeds larger than c in both cases.

Example 9.6   Momentum of an Electron

An electron, which has a mass of 9.11×10^{-31} kg, moves with a speed of $0.750c$. Find its relativistic momentum and compare this with the momentum calculated from the classical expression.

Solution   Using Equation 9.14 with $u = 0.750c$, we have

$$p = \frac{m_e u}{\sqrt{1 - \dfrac{u^2}{c^2}}}$$

$$p = \frac{(9.11 \times 10^{-31} \text{ kg})(0.750 \times 3.00 \times 10^8 \text{ m/s})}{\sqrt{1 - \dfrac{(0.750c)^2}{c^2}}}$$

$$= 3.10 \times 10^{-22} \text{ kg} \cdot \text{m/s}$$

The classical expression, if used (inappropriately) for this high-speed particle, gives

$$p = m_e u = 2.05 \times 10^{-22} \text{ kg} \cdot \text{m/s}$$

Hence, the correct relativistic result is more than 50% larger than the classical result!

9.7 • RELATIVISTIC ENERGY

We have seen that the definition of momentum requires generalization to make it compatible with the principle of relativity. We find that the definition of kinetic energy must also be modified.

To derive the relativistic form of the work–kinetic energy theorem, let us start with the definition of the work done by a force of magnitude F on a particle initially at rest. Recall from Chapter 6 that the work–kinetic energy theorem states that the work done by a net force acting on a particle equals the change in kinetic energy of the particle. Because the initial kinetic energy is zero, we conclude that the work W done in accelerating a particle from rest is equivalent to the relativistic kinetic energy K of the particle:

$$W = \Delta K = K - 0 = K = \int_{x_1}^{x_2} F \, dx = \int_{x_1}^{x_2} \frac{dp}{dt} \, dx \qquad \textbf{[9.17]}$$

where we are considering the special case of force and displacement vectors along the x axis for simplicity. To perform this integration and find the relativistic kinetic energy as a function of u, we first evaluate dp/dt, using Equation 9.14:

$$\frac{dp}{dt} = \frac{d}{dt} \frac{mu}{\sqrt{1 - \dfrac{u^2}{c^2}}} = \frac{m(du/dt)}{\left(1 - \dfrac{u^2}{c^2}\right)^{3/2}}$$

Substituting this expression for dp/dt and $dx = u\,dt$ into Equation 9.17 gives

$$K = \int_0^t \frac{m(du/dt)\,u\,dt}{\left(1 - \dfrac{u^2}{c^2}\right)^{3/2}} = m\int_0^u \frac{u}{\left(1 - \dfrac{u^2}{c^2}\right)^{3/2}}\,du$$

Evaluating the integral, we find that

- *Relativistic kinetic energy*

$$K = \frac{mc^2}{\sqrt{1 - \dfrac{u^2}{c^2}}} - mc^2 = \gamma mc^2 - mc^2 = (\gamma - 1)\,mc^2 \qquad \textbf{[9.18]}$$

At low speeds, where $u/c \ll 1$, Equation 9.18 should reduce to the classical expression $K = \frac{1}{2}mu^2$. We can check this by using the binomial expansion $(1 - x^2)^{-1/2} \approx 1 + \frac{1}{2}x^2 + \cdots$ for $x \ll 1$, where the higher order powers of x are ignored in the expansion because they are so small. In our case, $x = u/c$, so that

$$\gamma = \frac{1}{\sqrt{1 - \dfrac{u^2}{c^2}}} = \left(1 - \frac{u^2}{c^2}\right)^{-1/2} \approx 1 + \frac{1}{2}\frac{u^2}{c^2} + \cdots$$

Substituting this into Equation 9.19 gives

$$K \approx \left(1 + \frac{1}{2}\frac{u^2}{c^2} + \cdots\right) mc^2 - mc^2 = \frac{1}{2}mu^2$$

which agrees with the classical result. Figure 9.13 shows a comparison of the speed–kinetic energy relationships for a particle using the nonrelativistic expression for K (the blue curve) and the relativistic expression for K (the brown curve). The curves are in good agreement at low speeds, but deviate at higher speeds. The nonrelativistic expression indicates a violation of special relativity because it suggests that sufficient energy can be added to the particle to accelerate it to a speed larger than c. In the relativistic case, the particle speed never exceeds c, regardless of the kinetic energy, which is consistent with experimental results. When an object's speed is less than one-tenth the speed of light, the classical kinetic energy equation differs by less than 1% from the relativistic equation (which is experi-

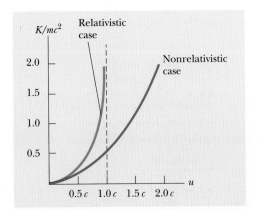

Figure 9.13

A graph comparing relativistic and nonrelativistic kinetic energy. The energies are plotted as a function of speed. In the relativistic case, u is always less than c.

mentally verified at all speeds). Thus, for practical calculations it is valid to use the classical equation when the object's speed is less than $0.1c$.

The constant term mc^2 in Equation 9.18, which is independent of the speed, is called the **rest energy** E_R of the particle:

$$E_R = mc^2 \qquad \text{[9.19]}$$

• *Rest energy*

The term γmc^2 in Equation 9.18 depends on the particle speed and is the sum of the kinetic and rest energies. We define γmc^2 to be the **total energy** E, that is, total energy = kinetic energy + rest energy

$$E = \gamma mc^2 = K + mc^2 = K + E_R \qquad \text{[9.20]}$$

or, when γ is replaced by its equivalent,

$$E = \frac{mc^2}{\sqrt{1 - \dfrac{u^2}{c^2}}} \qquad \text{[9.21]}$$

• *Definition of total energy*

The relation $E_R = mc^2$ shows that **mass is a manifestation of energy.** The relation shows that a small mass corresponds to an enormous amount of energy. This concept is fundamental to much of the field of nuclear physics.

In many situations, the momentum or energy of a particle is measured rather than its speed. It is therefore useful to have an expression relating the total energy E to the relativistic momentum p. This is accomplished by using the expressions $E = \gamma mc^2$ and $p = \gamma mu$. By squaring these equations and subtracting, we can eliminate u (see Problem 35). The result, after some algebra, is

$$E^2 = p^2c^2 + (mc^2)^2 \qquad \text{[9.22]}$$

• *Energy–momentum relationship*

When the particle is at rest, $p = 0$, and so $E = E_R = mc^2$. That is, the total energy equals the rest energy.

For the case of particles that have zero mass, such as photons (massless, chargeless particles of light to be discussed further in Chapter 28), we set $m = 0$ in Equation 9.22, and we see that

$$E = pc \qquad \text{[9.23]}$$

• *Energy–momentum relationship for a photon*

This equation is an exact expression relating energy and momentum for photons, which always travel at the speed of light.

When dealing with subatomic particles, it is convenient to express their energy in a unit called an electron volt (eV). The equality between electron volts and our standard energy unit is

$$1 \text{ eV} = 1.60 \times 10^{-19} \text{ J}$$

For example, the mass of an electron is 9.11×10^{-31} kg. Hence, the rest energy of the electron is

$$E_R = m_e c^2 = (9.11 \times 10^{-31} \text{ kg})(3.00 \times 10^8 \text{ m/s})^2 = 8.20 \times 10^{-14} \text{ J}$$

Converting this to eV, we have

$$E_R = m_e c^2 = (8.20 \times 10^{-14} \text{ J}) \left(\frac{1 \text{ eV}}{1.60 \times 10^{-19} \text{ J}} \right) = 0.511 \text{ MeV}$$

PITFALL PREVENTION 9.7

The electron volt

The electron volt is one of a small number of units in physics whose names are misleading. We will introduce the volt as a unit of *electric potential* in Chapter 20, but the electron volt is *not* a unit of electric potential. It is a unit of *energy*.

The following *pairs* of energies represent the rest energy and total energy of three different particles: particle 1: E, $2E$; particle 2: E, $3E$; particle 3: $2E$, $4E$. Rank the particles, from greatest to least, according to their (a) mass; (b) kinetic energy; (c) speed.

Example 9.7 The Energy of a Speedy Proton

Let us consider the relativistic motion of a proton. (a) Find the proton's rest energy in electron volts. (b) The total energy of a proton is three times its rest energy. With what speed is the proton moving? (c) Determine the kinetic energy of the proton in part (b) in electron volts. (d) What is the magnitude of the proton's momentum in part (b)?

Solution (a) To find the rest energy, we use Equation 9.19,

$$E_R = m_p c^2 = (1.67 \times 10^{-27} \text{ kg})(3.00 \times 10^8 \text{ m/s})^2$$

$$= (1.50 \times 10^{-10} \text{ J}) \left(\frac{1.00 \text{ eV}}{1.60 \times 10^{-19} \text{ J}} \right)$$

$$= \boxed{938 \text{ MeV}}$$

(b) Because the total energy E is three times the rest energy, $E = \gamma m c^2$ (Eq. 9.20) gives

$$E = 3m_p c^2 = \frac{m_p c^2}{\sqrt{1 - \dfrac{u^2}{c^2}}}$$

$$3 = \frac{1}{\sqrt{1 - \dfrac{u^2}{c^2}}}$$

Solving for u gives

$$\left(1 - \frac{u^2}{c^2} \right) = \frac{1}{9} \quad \text{or} \quad \frac{u^2}{c^2} = \frac{8}{9}$$

$$u = \frac{\sqrt{8}}{3} c = \boxed{2.83 \times 10^8 \text{ m/s}}$$

(c) We use Equation 9.20,

$$K = E - m_p c^2 = 3m_p c^2 - m_p c^2 = 2m_p c^2$$

Because $m_p c^2 = 938$ MeV, $K = \boxed{1.88 \text{ GeV}}$

(d) We can use Equation 9.22 to calculate the momentum with $E = 3m_p c^2$:

$$E^2 = p^2 c^2 + (m_p c^2)^2 = (3m_p c^2)^2$$

$$p^2 c^2 = 9(m_p c^2)^2 - (m_p c^2)^2 = 8(m_p c^2)^2$$

$$p = \sqrt{8} \, \frac{m_p c^2}{c} = \sqrt{8} \, \frac{(938 \text{ MeV})}{c} = \boxed{2.65 \, \frac{\text{GeV}}{c}}$$

The unit of momentum is written MeV/c, which is a momentum unit often used in particle studies.

EXERCISE Find the speed of a particle whose total energy is twice its rest energy.

Answer $0.866c$

EXERCISE An electron moves with a speed $u = 0.850c$. Find its (a) total energy and (b) kinetic energy in electron volts.

Answer (a) 0.970 MeV (b) 0.459 MeV

9.8 • MASS AND ENERGY

Equation 9.20, $E = \gamma m c^2$, which represents the total energy of a particle, suggests that even when a particle is at rest ($\gamma = 1$), it still possesses enormous energy through its mass. The clearest experimental proof of the equivalence of mass and energy occurs in nuclear and elementary particle interactions in which the conversion of mass into kinetic energy takes place. Because of this, in relativistic situations, we cannot use the principle of conservation of energy as outlined in Chapter 7. We must include rest energy as another form of energy storage.

This concept is important in atomic and nuclear processes, in which the change in mass during the process is on the order of the initial mass. For example, in a conventional nuclear reactor, the uranium nucleus undergoes *fission*, a reaction that results in several lighter fragments having considerable kinetic energy. In the case of a ^{235}U atom, which is used as fuel in nuclear power plants, the fragments are two lighter nuclei and a few neutrons. The total mass of the fragments is less than that of the ^{235}U by an amount Δm. The corresponding energy Δmc^2 associated with this mass difference is exactly equal to the total kinetic energy of the fragments. The kinetic energy is absorbed as the fragments move through water, raising the internal energy of the water. This internal energy is used to produce steam for the generation of electric power.

Next, consider a basic *fusion* reaction in which two deuterium atoms combine to form one helium atom. The decrease in mass that results from the creation of one helium atom from two deuterium atoms is $\Delta m = 4.25 \times 10^{-29}$ kg. Hence, the corresponding energy that results from one fusion reaction is $\Delta mc^2 = 3.83 \times 10^{-12}$ J $= 23.9$ MeV. To appreciate the magnitude of this result, if 1 g of deuterium is converted to helium, the energy released is on the order of 10^{12} J! At the year 2001 cost of electric energy, this would be worth about \$30 000. We will see more details of these nuclear processes in Chapter 30.

Example 9.8   Mass Change in a Radioactive Decay

The ^{216}Po nucleus is unstable and exhibits radioactivity, which we will study in detail in Chapter 30. It decays to ^{212}Pb by emitting an alpha particle, which is a helium nucleus, ^{4}He. Find (a) the mass change in this decay and (b) the energy that this represents.

Solution   Using values in Table A.3, we see that the initial and final masses are

$$m_i = m(^{216}\text{Po}) = 216.001\ 889 \text{ u}$$

$$m_f = m(^{212}\text{Pb}) + m(^4\text{He}) = 211.991\ 872 \text{ u} + 4.002\ 602 \text{ u}$$

$$= 215.994\ 474 \text{ u}$$

Thus, the mass change is,

$$\Delta m = 216.001\ 889 \text{ u} - 215.994\ 474 \text{ u} = 0.007\ 415 \text{ u}$$

$$= 1.23 \times 10^{-29} \text{ kg}$$

(b) The energy associated with this mass change is,

$$E = \Delta mc^2 = (1.23 \times 10^{-29} \text{ kg})(3.00 \times 10^8 \text{ m/s})^2$$

$$= 1.11 \times 10^{-12} \text{ J} = 6.92 \text{ MeV}$$

This energy appears as the kinetic energy of the alpha particle and the ^{212}Pb nucleus after the decay.

9.9 • GENERAL RELATIVITY

Up to this point, we have sidestepped a curious puzzle. Mass has two seemingly different properties: It determines a force of mutual gravitational attraction between two objects (Newton's law of universal gravitation) and also represents the resistance of a single object to being accelerated (Newton's second law), regardless of the type of force producing the acceleration. How can one quantity have two such different properties? An answer to this question, which puzzled Newton and many other physicists over the years, was provided when Einstein published his theory of gravitation, known as *general relativity*, in 1916. Because it is a mathematically complex theory, we merely offer a hint of its elegance and insight.

In Einstein's view, the dual behavior of mass was evidence for a very intimate and basic connection between the two behaviors. He pointed out that no mechanical experiment (e.g., dropping a mass) could distinguish between the two situations illustrated in Figures 9.14a and 9.14b. In Figure 9.14a, a person is standing in an elevator on the surface of a planet, and feels pressed into the floor, due to the

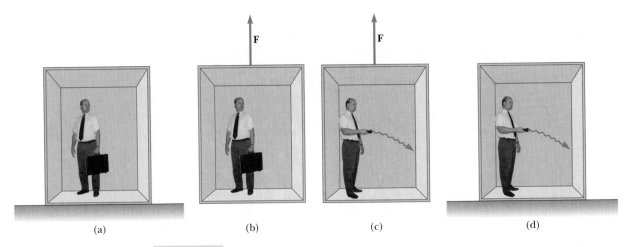

Figure 9.14

(a) The observer is at rest in a uniform gravitational field **g**, directed downward. (b) The observer is in a region where gravity is negligible, but the frame is accelerated by an external force **F** that produces an acceleration **g** directed upward. According to Einstein, the frames of reference in parts (a) and (b) are equivalent in every way. No local experiment can distinguish any difference between the two frames. (c) In the accelerating frame, a ray of light would appear to bend downward due to the acceleration. (d) If parts (a) and (b) are truly equivalent, as Einstein proposed, then part (c) suggests that a ray of light would bend downward in a gravitational field.

gravitational force. In Figure 9.14b, the person is in an elevator in empty space accelerating upward with $a = g$. The person feels pressed into the floor with the same force as in Figure 9.14a. In each case, a mass released by the observer undergoes a downward acceleration of magnitude g relative to the floor. In Figure 9.14a, the person is in an inertial frame in a gravitational field. In Figure 9.14b, the person is in a noninertial frame accelerating in gravity-free space. Einstein's claim is that these two situations are completely equivalent.

Einstein carried this idea further and proposed that *no* experiment, mechanical or otherwise, could distinguish between the two cases. This extension to include all phenomena (not just mechanical ones) has interesting consequences. For example, suppose that a light pulse is sent horizontally across the box, as in Figure 9.14c, in which the elevator is accelerating upward in empty space. From the point of view of an observer in an inertial frame outside the elevator, the light travels in a straight line while the floor of the elevator accelerates upward. According to the observer on the elevator, however, the trajectory of the light pulse bends downward as the floor of the elevator (and the observer) accelerates upward. Therefore, based on the equality of parts (a) and (b) of the figure for all phenomena, Einstein proposed that a **beam of light should also be bent downward by a gravitational field,** as in Figure 9.14d.

The two postulates of Einstein's **general relativity** are as follows:

• *Postulates of general relativity*

- All the laws of nature have the same form for observers in any frame of reference, whether accelerated or not.
- In the vicinity of any given point, a gravitational field is equivalent to an accelerated frame of reference in the absence of gravitational effects. (This is the **principle of equivalence.**)

One interesting effect predicted by general relativity is that the passage of time is altered by gravity. A clock in the presence of gravity runs more slowly than one in which gravity is negligible. Consequently, the frequencies of radiation emitted by atoms in the presence of a strong gravitational field are shifted to lower frequencies compared with the same emissions in a weak field. This gravitational shift has been detected in light emitted by atoms in massive stars. It has also been verified on the Earth by comparing the frequencies of gamma rays (a high-energy form of electromagnetic radiation) emitted from nuclei separated vertically by about 20 m.

The second postulate suggests that a gravitational field may be "transformed away" at any point if we choose an appropriate accelerated frame of reference—a freely falling one. Einstein developed an ingenious method of describing the acceleration necessary to make the gravitational field "disappear." He specified a certain quantity, the *curvature of space–time,* that describes the gravitational effect of a mass. In fact, the curvature of space–time completely replaces Newton's gravitational theory. According to Einstein, there is no such thing as a gravitational force. Rather, the presence of a mass causes a curvature of space–time in the vicinity of the mass, and this curvature dictates the space–time path that all freely moving objects must follow.

One important test of general relativity is the prediction that a light ray passing near the Sun should be deflected by some angle. This prediction was confirmed by astronomers as bending of starlight during a total solar eclipse shortly following World War I (Fig. 9.15).

As an example of the effects of curved space–time, imagine two travelers moving on parallel paths a few meters apart on the surface of the Earth and maintaining an exact northward heading along two longitude lines. As they observe each other near the equator, they will claim that their paths are exactly parallel. As they approach the North Pole, however, they may notice that they are moving closer together, and they will actually meet at the North Pole. Thus, they will claim that they moved along parallel paths, but moved toward each other, *as if there were an attractive force between them.* They will make this conclusion based on their everyday experience of moving on flat surfaces. From our mental representation, however, we realize that they are walking on a curved surface, and the geometry of the curved surface causes them to converge, rather than an attractive force. In a similar way, general relativity replaces the notion of forces with the movement of objects through curved space–time.

This Global Positioning System (GPS) unit incorporates relativistically corrected time calculations in its analysis of signals it receives from orbiting satellites. These corrections allow the unit to determine its position on the Earth's surface to within a few meters. If the corrections were not made, the location error would be about 1 km. *(Courtesy of Trimble Navigation Limited)*

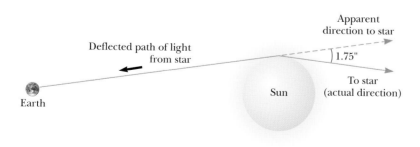

Figure 9.15

Deflection of starlight passing near the Sun. Because of this effect, the Sun and other remote objects can act as a *gravitational lens.* In his general theory of relativity, Einstein calculated that starlight just grazing the Sun's surface should be deflected by an angle of 1.75″.

If the concentration of mass becomes very great, as is believed to occur when a large star exhausts its nuclear fuel and collapses to a very small volume, a **black hole** may form. Here the curvature of space–time is so extreme that, within a certain distance from the center of the black hole, all matter and light become trapped. We will say more about black holes in Chapter 11.

THINKING PHYSICS 9.2

Atomic clocks are extremely accurate; in fact, an error of 1 s in 3 million years is typical. This error can be described as about 1 part in 10^{14}. On the other hand, the atomic clock in Boulder, Colorado, near Denver, is often 15 ns faster than the one in Washington after only one day. This is an error of about 1 part in 6×10^{12}, which is about 17 times larger than the previously expressed error. If atomic clocks are so accurate, why does a clock in Boulder not remain in synchronization with one in Washington? (*Hint:* Denver is known as the Mile High City.)

Reasoning According to the general theory of relativity, the passage of time depends on gravity—time runs more slowly in strong gravitational fields. Washington is at an elevation very close to sea level, but Boulder is about a mile higher in altitude. This difference results in a weaker gravitational field at Boulder than at Washington. As a result, time runs more rapidly in Boulder than in Washington.

context connection
9.10 • FROM MARS TO THE STARS

In this chapter, we have discussed the strange effects of traveling at high speeds. Do we need to consider these effects in our planned mission to Mars?

To answer this question, let us look back at our Context Connection for Chapter 7. In that discussion, we calculated an upper limit for the speed of our spacecraft to Mars, based on a calculation of the escape speed from the Sun. The result was $u = 4.21 \times 10^4$ m/s. Let us evaluate γ for this speed:

$$\gamma = \frac{1}{\sqrt{1 - \dfrac{u^2}{c^2}}} = \frac{1}{\sqrt{1 - \dfrac{(4.21 \times 10^4 \text{ m/s})^2}{(3.00 \times 10^8 \text{ m/s})^2}}} = 1.00000001$$

where we have completely ignored the rules of significant figures, so that we could find the first nonzero digit to the right of the decimal place!

It is clear from this result that relativistic considerations are not important for our trip to Mars. But what about deeper travels into space—suppose we wish to travel to another star? This distance is several orders of magnitude larger. The nearest star is about 4.2 lightyears from Earth. In comparison, Mars is 4.0×10^{-5} lightyears at its farthest from Earth. Thus, we are talking about a distance to the nearest star that is five orders of magnitude larger than the distance to Mars. Very long travel times will be needed to reach even the nearest star. At the escape speed from the Sun, for example, assuming that this speed is maintained during the entire trip, the travel time is 30 000 years to the nearest star. This is clearly prohibitive, especially if we would like the people who leave Earth to be the same people who arrive at the star!

We can use the principles of relativity to reduce this travel time by traveling at very high speeds. Suppose our spacecraft travels at a constant speed of $0.99c$. Then, we can use length contraction to calculate the distance from the Earth to the nearest star as measured by the spacecraft occupants:

$$L = \frac{L_p}{\gamma} = L_p \sqrt{1 - \frac{v^2}{c^2}} = (4.2 \text{ ly}) \sqrt{1 - \frac{(0.99c)^2}{c^2}} = 0.59 \text{ ly}$$

The time required to reach the star can now be calculated:

$$\Delta t = \frac{\Delta x}{v} = \frac{0.59 \text{ ly}}{0.99(1.0 \text{ ly/yr})} = 0.60 \text{ yr}$$

This is clearly a reduction in time from the low-speed trip!

There are two major problems with this scenario, however. The first is the technological challenge of designing and building a spacecraft and rocket engine assembly that can attain a speed of $0.99c$. Beyond that is the design of a safety system that will provide early warnings about running into asteroids, meteoroids, or other bits of matter while traveling at almost light speed through space. Even a small piece of rock could be disastrous if struck at $0.99c$. The second problem is related to the twins paradox discussed earlier in this chapter. During the trip to the star, 4.3 years will pass on Earth. If the travelers return to the Earth, another 4.3 years will pass. Thus, the travelers will have aged by only $2(0.6 \text{ y}) = 1.2$ years, but 8.6 years will have passed on Earth. For stars farther away than the nearest star, these effects could result in the personnel assisting with the liftoff from Earth no longer being alive when the travelers return. In conclusion, we see that travel to the stars will be an enormous challenge!

WEB

For a discussion of "warp drive" and other inventions of science fiction by a NASA scientist, visit **ssdoo.gsfc.nasa.gov/education/ just_for_fun/startrek.html**

SUMMARY

The two basic postulates of **special relativity** are

- All the laws of physics are the same in all inertial reference frames.
- The speed of light in vacuum has the same value, in all inertial frames, regardless of the velocity of the observer or the velocity of the source emitting the light.

Three consequences of special relativity are

- Events that are simultaneous for one observer may not be simultaneous for another observer who is in motion relative to the first.
- Clocks in motion relative to an observer are measured to be slowed down by a factor γ. This is known as **time dilation.**
- Lengths of objects in motion are measured to be shorter in the direction of motion. This is known as **length contraction.**

To satisfy the postulates of special relativity, the Galilean transformations must be replaced by the **Lorentz transformations:**

$$x' = \gamma(x - vt)$$

$$y' = y \qquad \text{[9.8]}$$

$$z' = z$$

$$t' = \gamma\left(t - \frac{v}{c^2} x\right)$$

where $\gamma = (1 - v^2/c^2)^{-1/2}$.

The relativistic form of the **velocity transformation** is

$$u_x' = \frac{u_x - v}{1 - \frac{u_x v}{c^2}} \qquad \text{[9.11]}$$

where u_x is the speed of an object as measured in the S frame and u_x' is its speed measured in the S′ frame.

The relativistic expression for the **momentum** of a particle moving with a velocity **u** is

$$\mathbf{p} \equiv \frac{m\mathbf{u}}{\sqrt{1 - \frac{u^2}{c^2}}} = \gamma m\mathbf{u} \qquad \text{[9.14]}$$

The relativistic expression for the **kinetic energy** of a particle is

$$K = \gamma mc^2 - mc^2 = (\gamma - 1)mc^2 \qquad \text{[9.18]}$$

where $E_R = mc^2$ is the **rest energy** of the particle.

The **total energy** E of a particle is related to its mass through the expression:

$$E = \gamma mc^2 = \frac{mc^2}{\sqrt{1 - \dfrac{u^2}{c^2}}} \qquad \text{[9.21]}$$

The relativistic momentum is related to the total energy through the equation

$$E^2 = p^2 c^2 + (mc^2)^2 \qquad \text{[9.22]}$$

The **general theory of relativity** claims that no experiment can distinguish between a gravitational field and an accelerating reference frame. It correctly predicts that the path of light is affected by a gravitational field.

QUESTIONS

1. What two speed measurements do two observers in relative motion *always* agree on?
2. The speed of light in water is 2.3×10^8 m/s. Suppose an electron is moving through water at 2.5×10^8 m/s. Does this violate the principles of relativity?
3. Two identical clocks are synchronized. One is put in orbit directed eastward around the Earth, and the other remains on Earth. Which clock runs slower? When the moving clock returns to Earth, do the two clocks still read the same time?
4. A train is approaching you as you stand next to the tracks. Just as an observer on the train passes you, you both begin to play the same Beethoven symphony on portable compact disc players. (a) According to you, whose CD player finishes the symphony first? (b) According to the observer on the train, whose CD player finishes the symphony first? (c) Whose CD player *really* finishes the symphony first?
5. A spacecraft in the shape of a sphere moves past an observer on Earth with a speed $0.5c$. What shape does the observer measure for the spacecraft as it moves past?
6. Explain why it is necessary, when defining length, to specify that the positions of the ends of a rod are to be measured simultaneously.
7. Does saying that a moving clock runs slower than a stationary one imply that something is physically unusual about the moving clock?
8. Give a physical argument that shows that it is impossible to accelerate an object of mass m to the speed of light, even with a continuous force acting on it.
9. List some ways our everyday lives would change if the speed of light were only 50 m/s.
10. It is said that Einstein, in his teenage years, asked the question, "What would I see in a mirror if I carried it in my hands and ran at the speed of light?" How would you answer this question?
11. Some distant astronomical objects, called *quasars*, are receding from us at half the speed of light (or greater). What is the speed of the light we receive from these quasars?
12. How is it possible that photons of light, which have zero mass, have momentum?
13. Relativistic quantities must make a smooth transition to their Newtonian counterparts as the speed of a system becomes small relative to the speed of light. Explain.
14. If the speed of a particle is doubled, what effect does it have on its momentum? (Assume that its initial speed is less than $c/2$.)
15. The upper limit on the speed of an electron is the speed of light c. Does that mean that the momentum of the electron has an upper limit?

PROBLEMS

1, 2, 3 = straightforward, intermediate, challenging   □ = full solution available in the *Student Solutions Manual and Study Guide*

web = solution posted at **http://www.harcourtcollege.com/physics/**   🖥 = computer useful in solving problem

🖳 = Interactive Physics   ▦ = paired numerical/symbolic problems   ▨ = life science application

Section 9.1   The Principle of Newtonian Relativity

1. A 2 000-kg car moving at 20.0 m/s collides and locks together with a 1 500-kg car at rest at a stop sign. Show that momentum is conserved in a reference frame moving at 10.0 m/s in the direction of the moving car.
2. A ball is thrown at 20.0 m/s inside a boxcar moving along the tracks at 40.0 m/s. What is the speed of the ball relative to the ground if the ball is thrown (a) forward? (b) backward? (c) out the side door?
3. In a laboratory frame of reference, an observer notes that Newton's second law is valid. Show that it is also valid for an observer moving at a constant speed, small compared with the speed of light, relative to the laboratory frame.

4. Show that Newton's second law is *not* valid in a reference frame moving past the laboratory frame of Problem 3 with a constant acceleration.

Section 9.4 Consequences of Special Relativity

5. How fast must a meter stick be moving relative to an observer if the observer measures its length as 0.500 m?

6. At what speed does a clock move if it is measured to run at a rate that is one-half the rate of a clock at rest with respect to an observer?

7. An astronaut is traveling in a space vehicle that has a speed of 0.500c relative to the Earth. The astronaut measures his pulse rate at 75.0 beats per minute. Signals generated by the astronaut's pulse are radioed to Earth when the vehicle is moving in a direction perpendicular to the line that connects the vehicle with an observer on the Earth. (a) What pulse rate does the Earth observer measure? (b) What would be the pulse rate if the speed of the space vehicle were increased to 0.990c?

8. An astronomer on Earth observes a meteoroid in the southern sky approaching the Earth at a speed of 0.800c. At the time of its discovery the meteoroid is 20.0 lightyears from the Earth. Calculate (a) the time required for the meteoroid to reach the Earth as measured by the Earthbound astronomer, (b) this time as measured by a tourist on the meteoroid, and (c) the distance to the Earth as measured by the tourist.

9. An atomic clock moves at 1 000 km/h for 1.00 h as measured by an identical clock fixed to the Earth. How many nanoseconds slow will the moving atomic clock be compared with the Earth clock at the end of the 1-h interval?

10. A muon formed high in the Earth's atmosphere travels at speed $v = 0.990c$ for a distance of 4.60 km before it decays. (a) How long does the muon live, as measured in its reference frame? (b) How far does the muon travel, as measured in its frame?

11. A spacecraft with a proper length of 300 m takes 0.750 μs to pass an Earth observer. Determine the speed of the spacecraft as measured by the Earth observer.
web

12. (a) An object of proper length L_p takes time Δt to pass an Earth observer. Determine the speed of the object as measured by the Earth observer. (b) A column of tanks, 300 m long, takes 75.0 s to pass a child waiting at a street corner on her way to school. Determine the speed of the armored vehicles. (c) Show that the answer to part (a) includes the answer to Problem 11 as a special case and includes the answer to part (b) of this problem as another special case.

13. For what value of v does $\gamma = 1.010\ 0$? Observe that for speeds less than this value, time dilation and length contraction are effects amounting to less than 1.00%.

14. A friend passes by you in a spacecraft traveling at a high speed. He tells you that his spacecraft is 20.0 m long and that the identically constructed spacecraft you are sitting in is 19.0 m long. According to your observations, (a) how long is your spacecraft, (b) how long is your friend's spacecraft, and (c) what is the speed of your friend's spacecraft?

15. An interstellar space probe is launched from Earth. After a brief period of acceleration, it moves with a constant velocity, with a magnitude of 70.0% of the speed of light. Its nuclear-powered batteries supply the energy to keep its data transmitter active continuously. The batteries have a lifetime of 15.0 yr as measured in a rest frame. (a) How long do the batteries on the space probe last as measured by Mission Control on Earth? (b) How far is the probe from Earth when its batteries fail, as measured by Mission Control? (c) How far is the probe from Earth when its batteries fail, as measured by its built-in trip odometer? (d) For what total time after launch is data received from the probe by Mission Control? Note that radio waves travel at the speed of light and fill the space between the probe and Earth at the time of battery failure.

Section 9.5 The Lorentz Transformation Equations

16. A moving rod is observed to have a length of 2.00 m and to be oriented at an angle of 30.0° with respect to the direction of motion, as in Figure P9.16. The rod has a speed of 0.995c. (a) What is the proper length of the rod? (b) What is the orientation angle in the proper frame?

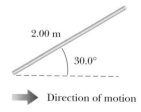

2.00 m

30.0°

Direction of motion

Figure P9.16

17. A spacecraft travels at 0.750c relative to Earth. If the spacecraft fires a small rocket in the forward direction, how fast (relative to the spacecraft) must it be fired for it to travel at 0.950c relative to Earth?

18. A Klingon space ship moves away from the Earth at a speed of 0.800c (Fig. P9.18). The starship *Enterprise* pursues at a speed of 0.900c relative to the Earth. Observers on Earth see the *Enterprise* overtaking the Klingon ship at a relative speed of 0.100c. With what speed is the *Enterprise* overtaking the Klingon ship as seen by the crew of the *Enterprise*?

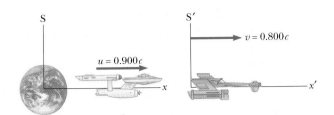

Figure P9.18

19. Two jets of material from the center of a radio galaxy are ejected in opposite directions. Both jets move at $0.750c$ relative to the galaxy. Determine the speed of one jet relative to the other.

20. An observer in reference frame S sees two events as simultaneous. Event *A* occurs at the point (50.0 m, 0, 0) at the instant 9:00:00 Universal time, 15 January 2001. Event *B* occurs at the point (150 m, 0, 0) at the same moment. A second observer, moving past with a velocity of $0.800\,c\,\mathbf{i}$, also observes the two events. In her reference frame S′, which event occurred first and what time elapsed between the events?

21. A red light flashes at position $x_R = 3.00$ m and time $t_R = 1.00 \times 10^{-9}$ s, and a blue light flashes at $x_B = 5.00$ m and $t_B = 9.00 \times 10^{-9}$ s (all values measured in the S reference frame.) Reference frame S′ has its origin at the same point as S at $t = t' = 0$; frame S′ moves uniformly to the right. Both flashes are observed to occur at the same place in S′. (a) Find the relative speed between S and S′. (b) Find the location of the two flashes in frame S′. (c) At what time does the red flash occur in the S′ frame?

22. A spacecraft is launched from the surface of the Earth with a velocity of magnitude $0.600c$ at an angle of $50.0°$ above the horizontal positive x-axis. Another spacecraft is moving past, with a velocity of magnitude $0.700c$ in the negative x direction. Determine the magnitude and direction of the velocity of the first spacecraft as measured by the pilot of the second spacecraft.

Section 9.6   Relativistic Momentum and the Relativistic Form of Newton's Laws

23. Calculate the momentum of an electron moving with a speed of (a) $0.010\,0c$, (b) $0.500c$, (c) $0.900c$.

24. The nonrelativistic expression for the magnitude of the momentum of a particle, $p = mu$, can be used if $u \ll c$. For what speed does the use of this equation give an error in the momentum of (a) 1.00% and (b) 10.0%?

25. A golf ball travels with a speed of 90.0 m/s. By what fraction does its relativistic momentum magnitude p differ from its classical value mu? That is, find the ratio $(p - mu)/mu$.

26. Show that the speed of an object having momentum of magnitude p and mass m is

$$u = \frac{c}{\sqrt{1 + (mc/p)^2}}$$

27. An unstable particle at rest breaks into two fragments of *unequal* mass. The mass of one fragment is 2.50×10^{-28} kg, and that of the other is 1.67×10^{-27} kg. If the lighter fragment has a speed of $0.893c$ after the breakup, what is the speed of the heavier fragment?

Section 9.7   Relativistic Energy

28. Determine the energy required to accelerate an electron from (a) $0.500c$ to $0.900c$ and (b) $0.900c$ to $0.990c$.

29. Find the momentum of a proton in MeV/c units if its total energy is twice its rest energy.

30. Show that, for any object moving at less than one-tenth the speed of light, the relativistic kinetic energy agrees with the result of the classical expression $K = \frac{1}{2}mu^2$ to within less than 1%. Thus, for most purposes, the classical equation is good enough to describe these objects, whose motion we call *nonrelativistic.*

31. A proton moves at $0.950c$. Calculate its (a) rest energy, (b) total energy, and (c) kinetic energy.

32. An electron has a kinetic energy five times greater than its rest energy. Find (a) its total energy and (b) its speed.

33. A cube of steel has a volume of 1.00 cm³ and a mass of 8.00 g when at rest on the Earth. If this cube is now given a speed $u = 0.900c$, what is its density as measured by a stationary observer? Note that relativistic density is defined as E_R/c^2V.

34. An unstable particle with a mass of 3.34×10^{-27} kg is initially at rest. The particle decays into two fragments that fly off along the x axis with velocity components of $0.987c$ and $-0.868c$. Find the masses of the fragments. (*Hint:* Conserve both mass–energy and momentum.)

35. Show that the energy–momentum relationship $E^2 = p^2c^2 + (mc^2)^2$ follows from the expressions $E = \gamma mc^2$ and $p = \gamma mu$.

36. Find the kinetic energy of a 78.0-kg spacecraft launched out of the Solar System with speed 106 km/s by using (a) the classical equation $K = \frac{1}{2}mu^2$ and (b) the relativistic equation.

37. A pion at rest ($m_\pi = 270\ m_e$) decays to a muon ($m_\mu = 207\ m_e$) and an antineutrino ($m_{\overline{v}} \approx 0$). The reaction is written $\pi^- \rightarrow \mu^- + \overline{v}$. Find the kinetic energy of the muon and the energy of the antineutrino in electron volts. (*Hint:* Relativistic momentum is conserved.)

Section 9.8   Mass and Energy

38. When 1.00 g of hydrogen combines with 8.00 g of oxygen, 9.00 g of water is formed. During this chemical reaction,

2.86×10^5 J of energy is released. How much mass do the constituents of this reaction lose? Is the loss of mass likely to be detectable?

39. The power output of the Sun is 3.77×10^{26} W. How much mass is converted to energy in the Sun each second?

40. In a nuclear power plant, the fuel rods last 3.00 yr before they are replaced. If a plant with rated thermal power 1.00 GW operates at 80.0% capacity for the 3.00 yr, what is the loss of mass of the fuel?

41. A gamma ray (a high-energy photon) can produce an electron (e^-) and a positron (e^+) when it enters the electric field of a heavy nucleus: $\gamma \rightarrow e^+ + e^-$. What minimum gamma-ray energy is required to accomplish this task? (*Hint:* The masses of the electron and the positron are equal.)

Section 9.10 Context Connection—From Mars to the Stars

42. Review Problem. In 1962, when Mercury astronaut Scott Carpenter orbited the Earth 22 times, the press stated that for each orbit he aged 2 millionths of a second less than he would have had he remained on Earth. (a) Assuming that he was 160 km above the Earth in a circular orbit, determine the time difference between someone on Earth and the orbiting astronaut for the 22 orbits. You will need to use the approximation $\sqrt{1-x} \approx 1 - x/2$, for small x. (b) Did the press report accurate information? Explain.

43. An astronaut wishes to visit the Andromeda galaxy, making a one-way trip that will take 30.0 yr in the spacecraft's frame of reference. Assume that the galaxy is 2.00 million lightyears away and that his speed is constant. (a) How fast must he travel relative to the Earth? (b) What will be the kinetic energy of his 1 000-metric-ton spacecraft? (c) What is the cost of this energy if it is purchased at a typical consumer price for electric energy: 13.0¢/kWh?

Additional Problems

44. An electron has a speed of $0.750c$. Find the speed of a proton that has (a) the same kinetic energy as the electron; (b) the same momentum as the electron.

45. The cosmic rays of highest energy are protons that have kinetic energy on the order of 10^{13} MeV. (a) How long would it take a proton of this energy to travel across the Milky Way galaxy, having a diameter $\sim 10^5$ lightyears, as measured in the proton's frame? (b) From the point of view of the proton, how many kilometers across is the galaxy?

46. The net nuclear fusion reaction inside the Sun can be written as $4\,^1\text{H} \rightarrow\, ^4\text{He} + \Delta E$. The rest energy of each hydrogen atom is 938.78 MeV and the rest energy of the helium-4 atom is 3 728.4 MeV. What is the percentage of the starting mass that is transformed to other forms of energy?

47. A rechargeable AA battery with a mass of 25.0 g can supply a power of 1.20 W for 50.0 min. (a) What is the difference in mass between a charged and an uncharged battery? (b) What fraction of the total mass is this mass difference?

48. An alien spacecraft traveling at $0.600c$ toward the Earth launches a landing craft with an advance guard of purchasing agents. The lander travels in the same direction with a speed of $0.800c$ relative to the spacecraft. As observed on the Earth, the spacecraft is 0.200 ly from Earth when the lander is launched. (a) With what speed is the lander observed to be approaching by observers on the Earth? (b) What is the distance to the Earth at the time of lander launch, as observed by the aliens? (c) How long does it take the lander to reach the Earth as observed by the aliens on the mother ship? (d) If the lander has a mass of 4.00×10^5 kg, what is its kinetic energy as observed in the Earth reference frame?

49. A physics professor on Earth gives an exam to her students, who are on a spacecraft traveling at speed v relative to Earth. The moment the spacecraft passes the professor, she signals the start of the exam. She wishes her students to have time T_0 (spacecraft time) to complete the exam. Show that she should wait a time (Earth time) of

$$T = T_0 \sqrt{\frac{1 - v/c}{1 + v/c}}$$

before sending a light signal telling them to stop. (*Hint:* Remember that it takes some time for the second light signal to travel from the professor to the students.)

50. Energy reaches the upper atmosphere of the Earth from the Sun at the rate of 1.79×10^{17} W. If all of this energy were absorbed by the Earth and not re-emitted, how much would the mass of the Earth increase in 1.00 yr?

51. A supertrain (proper length 100 m) travels at a speed of $0.950c$ as it passes through a tunnel (proper length 50.0 m). As seen by a trackside observer, is the train ever completely within the tunnel? If so, with how much space to spare?

52. Imagine that the entire Sun collapses to a sphere of radius R_g such that the work required to remove a small mass m from the surface would be equal to its rest energy mc^2. This radius is called the *gravitational radius* for the Sun. Find R_g. (It is believed that the ultimate fate of very massive stars is to collapse beyond their gravitational radii into black holes.)

53. A particle with electric charge q moves along a straight line in a uniform electric field **E** with a speed of u. The electric force exerted on the charge is q**E**. If the motion and the electric field are both in the x direction, (a) show that the acceleration of the charge q in the x direction is given by

$$a = \frac{du}{dt} = \frac{qE}{m}\left(1 - \frac{u^2}{c^2}\right)^{3/2}$$

(b) Discuss the significance of the dependence of the acceleration on the speed. (c) If the particle starts from rest at $x = 0$ at $t = 0$, how would you proceed to find the speed of the particle and its position at time t?

54. An observer in a spacecraft moves toward a mirror at speed v relative to the reference frame labeled by S in Figure P9.54. The mirror is stationary with respect to S. A light pulse emitted by the spacecraft travels toward the mirror and is reflected back to the spacecraft. The front of the spacecraft is a distance d from the mirror (as measured by observers in S) at the moment the light pulse leaves the spacecraft. What is the total travel time of the pulse as measured by observers in (a) the S frame and (b) the front of the spacecraft?

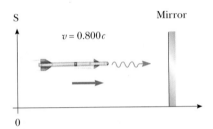

Figure P9.54

55. The creation and study of new elementary particles is an important part of contemporary physics. Especially interesting is the discovery of a very massive particle. To create a particle of mass M requires an energy Mc^2. With enough energy, an exotic particle can be created by allowing a fast-moving particle of ordinary matter, such as a proton, to collide with a similar target particle. Let us consider a perfectly inelastic collision between two protons: an incident proton with mass m, kinetic energy K, and momentum magnitude p joins with an originally stationary target proton to form a single product particle of mass M. You might think that the creation of a new product particle, nine times more massive than in a previous experiment, would require just nine times more energy for the incident proton. Unfortunately not all of the kinetic energy of the incoming proton is available to create the product particle, because conservation of momentum requires that after the collision the system as a whole still must have some kinetic energy. Only a fraction of the energy of the incident particle is thus available to create a new particle. You will determine how the energy available for particle creation

depends on the energy of the moving proton. Show that the energy available to create a product particle is given by

$$Mc^2 = 2mc^2\sqrt{1 + \frac{K}{2mc^2}}$$

From this result, when the kinetic energy K of the incident proton is large relative to its rest energy mc^2, we see that M approaches $(2mK)^{1/2}/c$. Thus if the energy of the incoming proton is increased by a factor of 9, the mass you can create increases only by a factor of 3. This disappointing result is the main reason that most modern accelerators, such as those at CERN (in Europe), at Fermilab (near Chicago), at SLAC (in California), and at DESY (in Germany), use *colliding beams*. Here the total momentum of a pair of interacting particles can be zero. The center of mass can be at rest after the collision, so in principle all of the initial kinetic energy can be used for particle creation, according to

$$Mc^2 = 2mc^2 + K = 2mc^2\left(1 + \frac{K}{2mc^2}\right)$$

where K is the total kinetic energy of two identical colliding particles. Here if $K \gg mc^2$, we have M directly proportional to K, as we would desire. These machines are difficult to build and to operate, but they open new vistas in physics.

56. Ted and Mary are playing a game of catch in frame S′, which is moving at $0.600c$, while Jim in frame S watches the action (Fig. P9.56). Ted throws the ball to Mary at $0.800c$ (according to Ted) and their separation (measured in S′) is 1.80×10^{12} m. (a) According to Mary, how fast is the ball moving? (b) According to Mary, how long does it take the ball to reach her? (c) According to Jim, how far apart are Ted and Mary, and how fast is the ball moving? (d) According to Jim, how long does it take the ball to reach Mary?

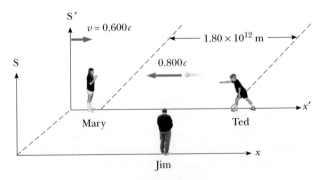

Figure P9.56

57. Suppose our Sun is about to explode. In an effort to escape, we depart in a spacecraft at $v = 0.800c$ and head to-

ward the star Tau Ceti, 12.0 ly away. When we reach the midpoint of our journey from the Earth, we see our Sun explode and, unfortunately, at the same instant we see Tau Ceti explode as well. (a) In the spacecraft's frame of reference, should we conclude that the two explosions occurred simultaneously? If not, which occurred first? (b) In a frame of reference in which the Sun and Tau Ceti are at rest, did they explode simultaneously? If not, which exploded first?

58. Prepare a graph of the relativistic kinetic energy and the classical kinetic energy, both as a function of speed, for an object with a mass of your choice. At what speed does the classical kinetic energy underestimate the experimental value by 1%? By 5%? By 50%?

ANSWERS TO QUICK QUIZZES

9.1. This scenario is not possible with light. Water waves move through a medium—the surface of the water—and therefore have an absolute speed relative to the reference frame in which this medium is at rest. Thus, the relative speed of the waves and your boat is determined by Galilean relativity. This makes it possible for you to move faster than the waves. Light waves, which do not require a medium, are described by the principles of special relativity. As you detect the light wave ahead of you and moving away from you (which would be a pretty good trick—think about it!), its speed relative to you is c. No matter how fast you accelerate in that direction, trying to catch up to it, its speed relative to you is still c. Thus, you will not be able to catch up to the light wave.

9.2 (a). The two events are the beginning and the end of the movie, both of which take place at rest with respect to the spacecraft crew. Thus, the crew measures the proper time interval of 2 h. Any observer in motion with respect to the spacecraft, which includes the observer on Earth, will measure a longer time interval due to time dilation.

9.3 Assuming that their on-duty time was kept on Earth, they will be pleasantly surprised with a large paycheck. Less time will have passed for the astronauts in their frame of reference than for their employer back on Earth.

9.4 According to Equation 9.7, the length of the Universe would contract to zero size. It would maintain the original size in the directions perpendicular to your velocity.

9.5 The answers to both of these questions is *no*. Both your body and your sleeping cabin are at rest in your reference frame; thus, they will have their proper length according to you. There will be no change in measured lengths of objects within your spacecraft. Another observer, on a spacecraft traveling at a high speed relative to yours, will measure you as thinner (if your body is oriented in a direction *perpendicular* to your velocity vector relative to the other observer) or will claim that you are able to fit into a shorter sleeping cabin (if your body is oriented in a direction *parallel* to your velocity vector relative to the other observer).

9.6 No. Time dilation and length contraction depend only on the relative speed of one observer relative to another, not on whether the observers are receding or approaching each other.

9.7 (a) $m_3 > m_2 = m_1$. The rest energy of particle 3 is $2E$, and it is E for particles 1 and 2. (b) $K_3 = K_2 > K_1$; The kinetic energy is the difference between the total energy and the rest energy. The kinetic energy is $4E - 2E = 2E$ for particle 3, $3E - E = 2E$ for particle 2, and $2E - E = E$ for particle 1. (c) $u_2 > u_3 = u_1$; From Equation 9.21, $E = \gamma E_R$. Solving this for the square of the particle speed u, we find, $u^2 = c^2(1 - (E_R/E)^2)$. Thus, the particle with the smallest ratio of rest energy to total energy will have the largest speed. Particles 1 and 3 have the same ratio as each other, and the ratio of particle 2 is smaller.

chapter 10

Mark Ruiz undergoes a rotation during a dive at the U. S. Olympic trials in June, 2000. He spins at a higher rate when he curls up and grabs his ankles.

(Otto Greule/Allsport)

Rotational Motion

In previous chapters, we have discussed the motion of particles and systems and have developed a number of analysis models to help us solve problems involving various types of situations. So far, however, we have only considered *translational* motion—motion through space in which the position of the center of mass of the object changes. We do not yet know how to describe the motion of a *rotating* object, such as a compact disc in a stereo system. We will address this type of motion in this chapter.

We will investigate the dynamics of two types of rotational situations. The first type is that of a particle, or an object that can be modeled as a particle, moving along a circular path. We have seen this situation before, beginning in Chapter 3, with the particle in uniform circular motion. The second type of situation involves an extended object (*not* modeled as a particle) rotating about a fixed axis, which we will see for the first time in this chapter. The Earth exhibits both of these types of motion. In its orbit around the Sun, it can be modeled as a particle in a circular path, as we discussed in Chapter 1. In addition, it behaves as an extended object in rotation as it spins about its own axis. Notice that we clearly cannot use the particle model in this second situation. We will develop a new simplification model for objects that we will call the *rigid body model*. The results we derive here will enable us to understand the rotational motions of a diverse range of objects in our environment, from an electron orbiting a nucleus to clusters of galaxies orbiting a common center.

10.1 • ANGULAR SPEED AND ANGULAR ACCELERATION

We began our study of translational motion in Chapter 2 by defining the terms *position, velocity,* and *acceleration*. For example, we locate a particle in one-dimensional space with the position variable *x*. In this chapter, we will insert the word *translational* before our previously studied kinematic variables to distinguish them from the analogous rotational variables that we will develop.

Let us think about a rotating object—how would we describe its position in its rotational motion? We do that by describing its *orientation* relative to some fixed reference direction. For example, imagine two soldiers performing a military about-face maneuver. They both begin by facing due north. One soldier, who has been practicing diligently, ends up after the maneuver with his body facing due south. The second, who has not been practicing, ends up facing in a southeast direction. We could describe their respective rotational positions after the maneuver by reporting the angle through which each turned from the original direction. The first soldier turned through 180°, but the second through only 135°. Thus, we can use an angle measured from a reference direction as a measure of **rotational position,** or **angular position,** which is our starting point for our description of rotational motion.

We will now follow up on this mental representation of using an angle to describe rotational position with a more mathematical representation. Let us begin by considering a planar object rotating about a fixed axis that is perpendicular to the object and goes through the point *O* (Fig. 10.1). A particle on the object, indicated by the black dot, is at a fixed distance *r* from the origin and rotates about *O* in a circle of radius *r*. In fact, **every particle on the object undergoes circular motion about O.** As a result, there is a strong link between rotational motion of an object and the motion of a particle in a circular path. A rotating object is made up of particles, all undergoing circular motion. The circular motion may not be uniform, however, because the rate of spin of the object may change.

It is convenient to represent the position of the particle with its polar coordinates: (*r*, *θ*). The origin of the polar coordinates is chosen to coincide with the center of the circular paths of the particles. In this representation, the only coordinate for a given particle that changes in time is the angular position *θ*; *r* remains constant. As a particle on the object moves along the circle of radius *r* from the positive *x* axis (*θ* = 0) to the point *P*, it moves through an arc of length *s*, which is related to *θ* through the relation

$$s = r\theta \qquad \text{[10.1a]}$$

$$\theta = \frac{s}{r} \qquad \text{[10.1b]}$$

It is important to note the units of *θ* as expressed by Equation 10.1b. The angle *θ* is the ratio of an arc length and the radius of the circle and, hence, is a pure (dimensionless) number. We commonly refer to the unit of *θ* as a **radian** (rad), however. One radian is the angle subtended by an arc length equal to the radius of the arc. Because the circumference of a circle is $2\pi r$, it follows that 360° corresponds to an angle of $2\pi r/r$ rad or 2π rad. Hence, 1 rad = $360°/2\pi \approx 57.3°$. To convert an angle in degrees to an angle in radians, we can use the fact that 2π radians = 360°, or π radians = 180°; hence,

$$\theta(\text{rad}) = \left(\frac{2\pi}{360°}\right)\theta(\text{deg}) = \left(\frac{\pi}{180°}\right)\theta(\text{deg})$$

• *Angular position*

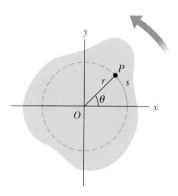

Figure 10.1

An object rotating about a fixed axis (the *z* axis) through *O* perpendicular to the plane of the figure. A particle on the object at *P* rotates in a circle of radius *r* centered at *O*.

• *The radian*

PITFALL PREVENTION 10.1
Remember the radian

Keep in mind that Equation 10.1b defines an angle expressed in *radians*. Don't fall into the trap of using this equation for angles measured in degrees. Also, be sure to set your calculator in radian mode when doing problems in rotation.

Michael Johnson starts in the 400-m final at the 1996 Olympics (he would go on to win the gold medal). In a short event such as this, in which runners must stay in their lanes, the runners start from staggered positions. If they all started from the same position, Equation 10.1a tells us that the runners farther from the center of a circular portion of the track would have to run farther than those on an inside lane. *(Stephen Dunn/Allsport)*

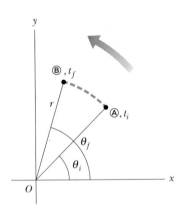

Figure 10.2

A particle on a rotating rigid body moves from Ⓐ to Ⓑ along the arc of a circle. In the time interval $\Delta t = t_f - t_i$, the radius vector sweeps out an angle $\Delta \theta = \theta_f - \theta_i$.

For example, 60° equals $\pi/3$ rad, and 45° equals $\pi/4$ rad.

In Figure 10.2, as the particle travels from Ⓐ to Ⓑ in a time Δt, the radius vector sweeps out an angle of $\Delta \theta = \theta_f - \theta_i$, which equals the **angular displacement** during the time interval Δt. Note that neither the angular position nor the angular displacement is limited to the range $0 < \theta < 2\pi$. For example, if the particle in Figure 10.2 starts at polar coordinates $(r, 0)$, makes two trips around a circular path, and ends up where it started, its angular displacement is described as $\theta = 4\pi$. Thus, we do not "reset" the angular position at zero each time the particle crosses the x axis. The **number of revolutions** the particles makes in a time interval is the angular displacement during the interval divided by 2π.

We define the **average angular speed** $\overline{\omega}$ (Greek omega) as the ratio of the angular displacement to the time interval Δt:

- *Average angular speed*

$$\overline{\omega} \equiv \frac{\theta_f - \theta_i}{t_f - t_i} = \frac{\Delta \theta}{\Delta t} \qquad \text{[10.2]}$$

By analogy with translational speed, the **instantaneous angular speed** ω is defined as the limit of the ratio in Equation 10.2 as Δt approaches zero:

- *Instantaneous angular speed*

$$\omega \equiv \lim_{\Delta t \to 0} \frac{\Delta \theta}{\Delta t} = \frac{d\theta}{dt} \qquad \text{[10.3]}$$

Angular speed has units of rad/s (or s^{-1} because radians are not dimensional). Let us adopt the convention that the fixed axis of rotation for an object is the z axis, as in Figure 10.1. We shall take ω to be positive when θ is increasing (counterclockwise motion in Figure 10.1) and negative when θ is decreasing (clockwise motion).

If the instantaneous angular speed of a particle changes from ω_i to ω_f in the time interval Δt, the particle has an angular acceleration. The **average angular acceleration** $\overline{\alpha}$ (Greek alpha) of a particle moving in a circular path is defined as the ratio of the change in the angular speed to the time interval Δt:

$$\overline{\alpha} \equiv \frac{\omega_f - \omega_i}{t_f - t_i} = \frac{\Delta \omega}{\Delta t}$$

[10.4]　　• *Average angular acceleration*

By analogy with translational acceleration, the **instantaneous angular acceleration** is defined as the limit of the ratio $\Delta\omega/\Delta t$ as Δt approaches zero:

$$\alpha \equiv \lim_{\Delta t \to 0} \frac{\Delta \omega}{\Delta t} = \frac{d\omega}{dt}$$

[10.5]　　• *Instantaneous angular acceleration*

Angular acceleration has units of rad/s^2, or s^{-2}.

Let us now generalize our argument from the rotating object in Figure 10.1 to any rigid body. A **rigid body** is any system of particles in which the particles remain fixed in position with respect to one another. We will call this simplification model, similar to the particle model, the **rigid body model.** If we were to imagine a rotating block of gelatin, the motion is very complicated because of the combination of rotation of the particles and the movement of particles within the deformable block relative to one another. Another example is our own Sun—the region of the Sun near the solar equator is moving with a higher angular speed than the region near the poles—it does not rotate as a rigid body. We shall not analyze such messy problems, however. Instead, our analyses will address only rigid bodies.

• *The rigid body model*

A color-enhanced, infrared image of Hurricane Mitch, which devastated large areas of Honduras and Nicaragua in October 1998. The hurricane does not rotate as a rigid body—regions near the interior rotate with a different angular speed than regions near the edge. *(Courtesy of NOAA)*

WEB

For more storm images and information on storms, visit the National Oceanic & Atmospheric Administration (NOAA) at **www.noaa.gov**

In solving rotation problems, you will need to specify an axis of rotation—we did not see this new feature in our study of translational motion. The choice is arbitrary, but once you make it, you need to maintain that choice consistently throughout the problem. In some problems, a natural axis is suggested by the physical situation, such as the center of an automobile wheel. In other problems, the choice may not be obvious, and you will need to choose an axis.

As we investigate rotational motion, we shall develop a number of analysis models for rigid bodies that have analogs in our analysis models for particles.

With our simplification model of a rigid body, we can make a statement about the various particles in the rigid body—**for rotation about a fixed axis, every particle on a rigid body has the same angular speed and the same angular acceleration.** That is, the quantities ω and α that we have discussed for particles characterize the rotational motion of the *entire* rigid body. Using these quantities, we can greatly simplify the analysis of rigid body rotation. Such a statement cannot be made for a nonrigid body, which is why studying the rotating block of gelatin or the Sun would be so difficult.

The angular position θ, angular speed ω, and angular acceleration α are analogous to translational position x, translational speed v, and translational acceleration a, respectively, for the corresponding one-dimensional motion discussed in Chapter 2. The variables θ, ω, and α differ dimensionally from the variables x, v, and a only by a length factor, as we shall see shortly.

We have not associated any direction with the angular speed and angular acceleration.* Strictly speaking, these variables are the magnitudes of the angular velocity and angular acceleration vectors $\boldsymbol{\omega}$ and $\boldsymbol{\alpha}$. Because we are considering rotation about a fixed axis, we can indicate the directions of these vectors by assigning a positive or negative sign to ω and α, as discussed for ω after Equation 10.3. For rotation about a fixed axis, the only direction in space that uniquely specifies the rotational motion is the direction along the axis; however, we still must specify one of the two directions along this axis as positive.

The direction of $\boldsymbol{\omega}$ is along the axis of rotation, which is the z axis in Figure 10.1. By convention, we take the direction of $\boldsymbol{\omega}$ to be *out of* the plane of the diagram when the rotation is counterclockwise and *into* the plane of the diagram when the rotation is clockwise. To further illustrate this convention, it is convenient to use the **right-hand rule** illustrated by Figure 10.3. The four fingers of the right hand are wrapped in the direction of the rotation. The extended right thumb points in the direction of $\boldsymbol{\omega}$.

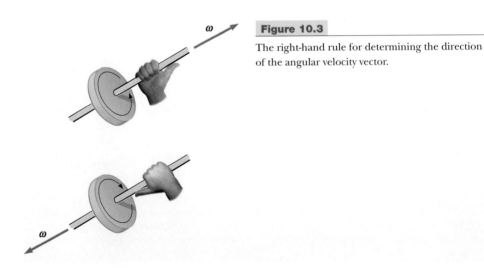

Figure 10.3

The right-hand rule for determining the direction of the angular velocity vector.

---

* Although we do not verify it here, the instantaneous angular velocity and instantaneous angular acceleration are vector quantities, but the corresponding average values are not. This is because angular displacements do not add as vector quantities for finite rotations.

The direction of $\boldsymbol{\alpha}$ follows from its vector definition as $d\boldsymbol{\omega}/dt$. For rotation about a fixed axis, the direction of $\boldsymbol{\alpha}$ is the same as $\boldsymbol{\omega}$ if the angular speed (the magnitude of $\boldsymbol{\omega}$) is increasing in time, and antiparallel to $\boldsymbol{\omega}$ if the angular speed is decreasing in time.

The full vector treatment of rotational motion is beyond the scope of this book and not necessary for our level of understanding, so we will use nonvector notation for most of this chapter.

Quick Quiz 10.1

A rigid body is rotating in a counterclockwise sense around a fixed axis. Each of the following pairs of quantities represents an initial angular position and a final angular position of the rigid body. Which of the sets can *only* occur if the rigid body rotates through more than 180°?   (a) 3 rad, 6 rad;   (b) −1 rad, 1 rad; (c) 1 rad, 5 rad.

Quick Quiz 10.2

Suppose that the change in angular position for each of the pairs of values in Quick Quiz 10.1 occurred in 1 s. Which choice represents the lowest average angular speed?

10.2 • ROTATIONAL KINEMATICS: THE RIGID BODY UNDER CONSTANT ANGULAR ACCELERATION

In our study of one-dimensional motion, we found that the simplest accelerated motion to analyze is motion under constant translational acceleration (Chapter 2). Likewise, for rotational motion about a fixed axis, the simplest accelerated motion to analyze is motion under constant angular acceleration.

If we write Equation 10.5 in the form $d\omega = \alpha\,dt$ and let $\omega = \omega_i$ at $t_i = 0$, we can integrate this expression directly to find the final angular speed ω_f:

See the Core Concepts in Physics CD-ROM, Screen 7.2

$$\int_{\omega_i}^{\omega_f} d\omega = \int_0^t \alpha\,dt \quad\longrightarrow\quad \omega_f = \omega_i + \alpha t \qquad [10.6]$$

• *Rotational kinematic equations* (α = *constant*)

Likewise, rewriting Equation 10.3 and substituting Equation 10.6, we can integrate once more (with $\theta = \theta_i$ at $t_i = 0$):

$$\int_{\theta_i}^{\theta_f} d\theta = \int_0^t \omega\,dt = \int_0^t (\omega_i + \alpha t)\,dt \quad\longrightarrow\quad \theta_f = \theta_i + \omega_i t + \tfrac{1}{2}\alpha t^2 \qquad [10.7]$$

If we eliminate t from Equations 10.6 and 10.7, we obtain

$$\omega_f^2 = \omega_i^2 + 2\alpha(\theta_f - \theta_i) \qquad [10.8]$$

If we eliminate α, we find

$$\theta_f = \theta_i + \tfrac{1}{2}(\omega_i + \omega_f)t \qquad [10.9]$$

Table 10.1 suggests that rotational kinematics is just like translational kinematics. That is almost true, but keep in mind two differences that you must address. (1) In rotational kinematics, as suggested in Pitfall Prevention 10.2, you need to specify a rotation axis. (2) In rotational motion, the object keeps returning to its original orientation—thus, you may be asked for the number of revolutions made by a rigid body, a concept that has no meaning in translational motion.

TABLE 10.1	A Comparison of Equations for Rotational and Translational Motion: Kinematic Equations
Rotational Motion About a Fixed Axis with α = Constant (Variables: θ_f and ω_f)	**Translational Motion with a = Constant (Variables: x_f and v_f)**
$\omega_f = \omega_i + \alpha t$	$v_f = v_i + at$
$\theta_f = \theta_i + \omega_i t + \frac{1}{2}\alpha t^2$	$x_f = x_i + v_i t + \frac{1}{2}at^2$
$\theta_f = \theta_i + \frac{1}{2}(\omega_i + \omega_f)t$	$x_f = x_i + \frac{1}{2}(v_i + v_f)t$
$\omega_f^2 = \omega_i^2 + 2\alpha(\theta_f - \theta_i)$	$v_f^2 = v_i^2 + 2a(x_f - x_i)$

Notice that these kinematic expressions for rotational motion under constant angular acceleration are of the *same mathematical form* as those for translational motion under constant translational acceleration, with the substitutions $x \rightarrow \theta$, $v \rightarrow \omega$, and $a \rightarrow \alpha$. The similarities between rotational and translational kinematic equations are shown in Table 10.1.

Quick Quiz 10.3

Consider again the pairs of angular positions for the rigid body in Quick Quiz 10.1. If the object starts from rest at the initial angular position, moves counterclockwise with constant angular acceleration, and arrives at the final angular position with the same angular speed in all three cases, for which choice is the angular acceleration the highest?

Example 10.1 Rotating Wheel

A wheel rotates with a constant angular acceleration of 3.50 rad/s². If the angular speed of the wheel is 2.00 rad/s at $t = 0$, (a) through what angle does the wheel rotate between $t = 0$ and $t = 2.00$ s?

Solution We assume that the wheel is perfectly rigid, so that we can use the rigid body model. Because the angular acceleration in the problem is given as constant, we can use the rotational kinematic equations. We use Equation 10.7, setting $\theta_i = 0$,

$$\theta_f = \theta_i + \omega_i t + \frac{1}{2}\alpha t^2$$
$$= 0 + (2.00 \text{ rad/s})(2.00 \text{ s}) + \frac{1}{2}(3.50 \text{ rad/s}^2)(2.00 \text{ s})^2$$
$$= \boxed{11.0 \text{ rad}}$$

This is equivalent to 11.0 rad/$(2\pi$ rad/rev$) = 1.75$ rev.

(b) What is the angular speed of the wheel at $t = 2.00$ s?

Solution We use Equation 10.6:

$$\omega_f = \omega_i + \alpha t = 2.00 \text{ rad/s} + (3.50 \text{ rad/s}^2)(2.00 \text{ s})$$
$$= \boxed{9.00 \text{ rad/s}}$$

We could also obtain this result using Equation 10.8 and the results of part (a). Try it!

EXERCISE Find the angle through which the wheel rotates between $t = 2.00$ s and $t = 3.00$ s.

Answer 10.8 rad

EXERCISE A wheel starts from rest and rotates with constant angular acceleration to an angular speed of 12.0 rad/s in 3.00 s. Find (a) the angular acceleration of the wheel and (b) the angle in radians through which it rotates in this time.

Answer (a) 4.00 rad/s² (b) 18.0 rad

10.3 • RELATIONS BETWEEN ROTATIONAL AND TRANSLATIONAL QUANTITIES

In this section we shall derive some useful relations between the angular speed and angular acceleration of a particle on a rotating rigid body and its translational speed and translational acceleration. Keep in mind that when a rigid body rotates about a fixed axis, *every* particle of the body moves in a circle whose center is the axis of rotation.

Consider a particle on a rotating rigid body, moving in a circle of radius r about the z axis, as in Figure 10.4. Because the particle moves along a circular path, its translational velocity vector **v** is always tangent to the path; hence, we often call this quantity **tangential velocity.** The magnitude of the tangential velocity of the particle is, by definition, the **tangential speed,** given by $v = ds/dt$, where s is the distance traveled by the particle along the circular path. Recalling from Equation 10.1a that $s = r\theta$, and noting that r is a constant, we have

$$v = \frac{ds}{dt} = r\frac{d\theta}{dt}$$

$$v = r\omega \qquad [10.10]$$

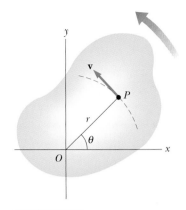

Figure 10.4

As a rigid body rotates about the fixed axis through O, the point P has a translational velocity **v** that is always tangent to the circular path of radius r.

That is, the tangential speed of the particle equals the distance of the particle from the axis of rotation multiplied by the particle's angular speed.

We can relate the angular acceleration of the particle to its tangential acceleration a_t — which is the component of its acceleration tangent to the path of motion — by taking the time derivative of v:

$$a_t = \frac{dv}{dt} = r\frac{d\omega}{dt}$$

$$a_t = r\alpha \qquad [10.11]$$

That is, the tangential component of the translational acceleration of a particle undergoing circular motion equals the distance of the particle from the axis of rotation multiplied by the angular acceleration.

In Chapter 3 we found that a particle rotating in a circular path undergoes a centripetal, or radial, acceleration of magnitude v^2/r directed toward the center of rotation (Fig. 10.5). Because $v = r\omega$, we can express the centripetal acceleration of the particle in terms of the angular speed as

$$a_c = \frac{v^2}{r} = r\omega^2 \qquad [10.12]$$

The *total translational acceleration* of the particle is $\mathbf{a} = \mathbf{a}_t + \mathbf{a}_r$. The magnitude of the total translational acceleration of the particle is therefore

$$a = \sqrt{a_t^2 + a_r^2} = \sqrt{r^2\alpha^2 + r^2\omega^4} = r\sqrt{\alpha^2 + \omega^4} \qquad [10.13]$$

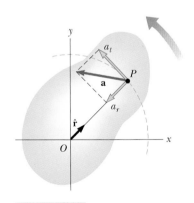

Figure 10.5

As a rigid body rotates about a fixed axis through O, a particle at point P experiences a tangential component of translational acceleration a_t and a radial component of translational acceleration a_r. The total translational acceleration of this particle is $\mathbf{a} = \mathbf{a}_t + \mathbf{a}_r$, where $\mathbf{a}_r = -a_c\,\hat{\mathbf{r}}$.

Quick Quiz 10.4

When a wheel of radius R rotates about a fixed axis as in Figure 10.6, (a) do all points on the wheel (including the spokes) have the same angular speed? (b) Do they all have the same tangential speed? (c) If the angular speed is constant and

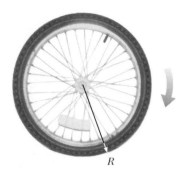

Figure 10.6

(Quick Quiz 10.4) A wheel of radius R rotating about its axis.

equal to ω, describe the tangential speeds and total translational accelerations of the points located at $r = 0$, $r = R/2$, and $r = R$, where the points are measured from the center of the wheel.

THINKING PHYSICS 10.1

A phonograph record (LP, for *long-playing*) is rotated at a constant *angular* speed. A compact disc (CD) is rotated so that the surface sweeps past the laser at a constant *tangential* speed. Consider two circular grooves of information on an LP—one near the outer edge and one near the inner edge. Suppose the outer groove "contains" 1.8 s of music. Does the inner groove also contain 1.8 s of music? How about the CD—do the inner and outer "grooves" contain the same time interval of music?

Reasoning On the LP, the inner and outer grooves must both rotate once in the same time period. Thus, each groove, regardless of where it is on the record, contains the same time interval of information. Of course, on the inner grooves, this same information must be compressed into a smaller circumference. On a CD, the constant tangential speed requires that no such compression occur—the digital pits representing the information are spaced uniformly everywhere on the surface. Thus, there is more information in an outer "groove," due to its larger circumference, and, as a result, a longer time interval of music than in the inner "groove."

THINKING PHYSICS 10.2

The launch area for the European Space Agency is not in Europe—it is in South America. Why?

Reasoning Placing a satellite in Earth orbit requires providing a large tangential speed to the satellite, which is the task of the rocket propulsion system. Anything that reduces the requirements on the propulsion system is a welcome contribution. The surface of the Earth is already traveling toward the east at a high speed, due to the rotation of the Earth. Thus, if rockets are launched toward the east, the rotation of the Earth provides some initial tangential speed, reducing somewhat the requirements on the propulsion system. If rockets were launched from Europe, which is at a relatively large latitude, the contribution of the Earth's rotation is relatively small because the distance between Europe and the rotation axis of the Earth is relatively small. The ideal place for launching is at the equator, which is as far as one can be from the rotation axis of the Earth and still be on the surface of the Earth. This results in the largest possible tangential speed due to the Earth's rotation. The European Space Agency exploits this advantage by launching from French Guiana, which is only a few degrees north of the equator.

A second advantage of this launch location is that launching toward the east takes the spacecraft over water. In the event of an accident or a failure, the wreckage will fall into the ocean rather than into populated areas as it would if launched to the east from Europe. This is a primary reason why the United States launches spacecraft from Florida rather than California, despite the more favorable weather conditions in California.

WEB

Visit the European Space agency at
www.esrin.esa.it

10.4 • ROTATIONAL KINETIC ENERGY

Imagine that you begin a workout session on a stationary exercise bicycle. You apply a force with your feet on the pedals, moving them through a displacement—you have done work. The result of this work is the spinning of the wheel. This rotational motion represents kinetic energy because mass is in motion. In this section, we will investigate this kinetic energy. In a later section, we will consider the work done in rotational motion and develop a rotational version of the work–kinetic energy theorem.

Let us model a collection of particles, such as the particles making up the wheel on our exercise bicycle, with the rigid body model, and let us assume that the body rotates about the fixed z axis with an angular speed of ω (Fig. 10.7). Each particle of the rigid body is in motion and thus has some kinetic energy, determined by its mass and tangential speed. If the mass of the ith particle is m_i and its tangential speed is v_i, the kinetic energy of this particle is

$$K_i = \tfrac{1}{2}m_iv_i^2$$

We can express the *total* kinetic energy K_R of the rotating rigid body as the sum of the kinetic energies of the individual particles. Thus, incorporating Equation 10.10,

$$K_R = \sum_i K_i = \sum_i \tfrac{1}{2}m_iv_i^2 = \tfrac{1}{2}\sum_i m_ir_i^2\omega^2$$

$$= \tfrac{1}{2}\left(\sum_i m_ir_i^2\right)\omega^2$$

where we have factored ω^2 from the sum because it is common to every particle. The quantity in parentheses is called the **moment of inertia** I of the rigid body:

$$I = \sum_i m_ir_i^2 \qquad\qquad \textbf{[10.14]}$$

Therefore, we can express the kinetic energy of the rotating rigid body as

$$K_R = \tfrac{1}{2}I\omega^2 \qquad\qquad \textbf{[10.15]}$$

From the definition of moment of inertia, we see that it has dimensions of ML^2 ($\text{kg}\cdot\text{m}^2$ in SI units). The moment of inertia is a measure of a system's *resistance to change in its angular speed.* Thus, it plays a role in rotational motion identical to that which mass plays in translational motion. Notice that moment of inertia depends not only on the mass of the rigid body but also on *how the mass is distributed around the rotation axis.*

Although we shall commonly refer to the quantity $\tfrac{1}{2}I\omega^2$ as the **rotational kinetic energy,** it is not a new form of energy. It is ordinary kinetic energy because it was derived from a sum over individual kinetic energies of the particles contained in the rigid body. It is a new role for kinetic energy for us, however, because we have only considered kinetic energy associated with translation through space so far. **On the storage side of the continuity equation for energy (see Eq. 6.20), we should now consider that the kinetic energy term should be the sum of the changes in both translational and rotational kinetic energy.** Thus, in energy versions of system models, we should keep in mind the possibility of rotational kinetic energy.

See Screen 7.3

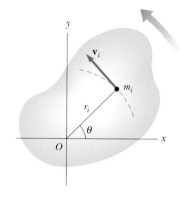

Figure 10.7

A rigid body rotating about the z axis with angular speed ω. The kinetic energy of the particle of mass m_i is $\tfrac{1}{2}m_iv_i^2$. The kinetic energy of the rigid body is called its rotational kinetic energy.

• *Moment of inertia for a system of particles*

• *Kinetic energy of a rotating rigid body*

PITFALL PREVENTION 10.4

No single moment of inertia

We have pointed out that moment of inertia is analogous to mass, but there is one major difference. Mass is an inherent property of an object. The moment of inertia of an object depends on your choice of rotation axis; therefore an object has no single value of the moment of inertia. An object does have a minimum value of the moment of inertia, which is that calculated around an axis passing through the center of mass of the object.

Equation 10.14 gives the moment of inertia of a collection of particles. For an extended, continuous object, we can calculate the moment of inertia by dividing the object into many small elements with mass Δm_i. Then, the moment of inertia is approximately $I \approx \Sigma_i r_i^2 \Delta m_i$, where r_i is the perpendicular distance of the element of mass Δm_i from the rotation axis. Now we take the limit as $\Delta m_i \to 0$, in which case the sum becomes an integral:

$$I = \lim_{\Delta m_i \to 0} \sum_i r_i^2 \, \Delta m_i = \int r^2 dm \qquad [10.16]$$

It is usually easier to calculate moments of inertia in terms of the volume of the elements rather than their mass, and we can easily make this change by using Equation 1.1, $\rho = m/V$, where ρ is the density of the object and V is its volume. We can express the mass of an element by writing Equation 1.1 in differential form, $dm = \rho \, dV$. Using this, Equation 10.16 becomes

$$I = \int \rho r^2 dV \qquad [10.17]$$

• *Moment of inertia for an extended, continuous object*

If the object is homogenous, the density ρ is constant and the integral can be evaluated for a given geometry. If ρ is not uniform over the volume of the object, its variation with position must be known in order to perform the integration.

For symmetric objects, the moment of inertia can be expressed in terms of the total mass of the object and one or more dimensions of the object. Table 10.2 shows the moments of inertia of various common symmetric objects.

Quick Quiz 10.5

Two spheres, one hollow and one solid, are rotating with the same angular speed about their centers. Both spheres have the same mass and radius. Which one, if either, has the higher rotational kinetic energy?

Example 10.2 The Oxygen Molecule

Consider the diatomic oxygen molecule O_2, which is rotating in the xy plane about the z axis passing through its center, perpendicular to its length. The mass of each oxygen atom is 2.66×10^{-26} kg, and at room temperature, the average separation between the two oxygen atoms is $d = 1.21 \times 10^{-10}$ m. (a) Calculate the moment of inertia of the molecule about the z axis.

Solution We model the molecule as a rigid body, consisting of two particles (the two oxygen atoms), in rotation. Because the distance of each particle from the z axis is $d/2$, the moment of inertia about the z axis is

$$I = \sum_i m_i r_i^2 = m\left(\frac{d}{2}\right)^2 + m\left(\frac{d}{2}\right)^2 = \frac{md^2}{2}$$

$$= \frac{(2.66 \times 10^{-26} \text{ kg})(1.21 \times 10^{-10} \text{ m})^2}{2}$$

$$= \boxed{1.95 \times 10^{-46} \text{ kg} \cdot \text{m}^2}$$

(b) A typical angular speed of a molecule is 4.60×10^{12} rad/s. If the oxygen molecule is rotating with this angular speed about the z axis, what is its rotational kinetic energy?

Solution We use Equation 10.15:

$$K_R = \tfrac{1}{2}I\omega^2$$

$$= \tfrac{1}{2}(1.95 \times 10^{-46} \text{ kg} \cdot \text{m}^2)(4.60 \times 10^{12} \text{ rad/s})^2$$

$$= \boxed{2.06 \times 10^{-21} \text{ J}}$$

TABLE 10.2	Moments of Inertia of Homogeneous Rigid Bodies with Different Geometries

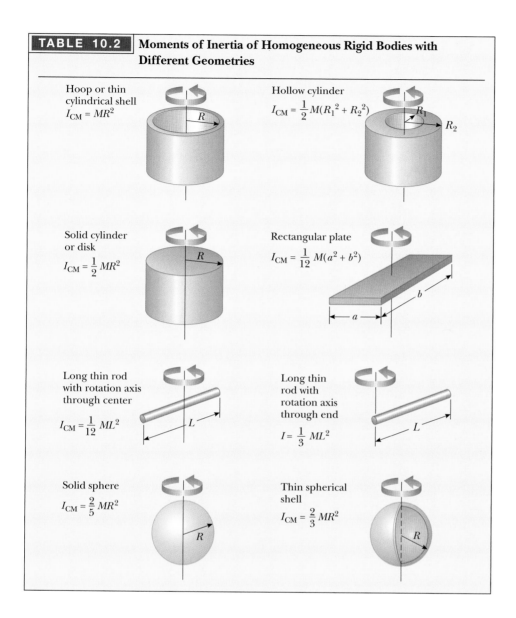

Hoop or thin cylindrical shell
$I_{CM} = MR^2$

Hollow cylinder
$I_{CM} = \frac{1}{2} M(R_1^2 + R_2^2)$

Solid cylinder or disk
$I_{CM} = \frac{1}{2} MR^2$

Rectangular plate
$I_{CM} = \frac{1}{12} M(a^2 + b^2)$

Long thin rod with rotation axis through center
$I_{CM} = \frac{1}{12} ML^2$

Long thin rod with rotation axis through end
$I = \frac{1}{3} ML^2$

Solid sphere
$I_{CM} = \frac{2}{5} MR^2$

Thin spherical shell
$I_{CM} = \frac{2}{3} MR^2$

Example 10.3 Four Rotating Masses

Four small spheres are fastened to the corners of a frame of negligible mass lying in the *xy* plane (Fig. 10.8). (a) If the rotation of the system occurs about the *y* axis with an angular speed ω, find the moment of inertia I_y about the *y* axis and the rotational kinetic energy about this axis.

Solution First, note that the two spheres of mass *m* that lie on the *y* axis do not contribute to I_y. Because they are modeled as particles, $r_i = 0$ for these spheres about this axis. Applying Equation 10.14, we have for the two spheres on the *x* axis,

$$I_y = \sum_i m_i r_i^2 = Ma^2 + Ma^2 = 2Ma^2$$

Therefore, the rotational kinetic energy about the *y* axis is

$$K_R = \tfrac{1}{2} I_y \omega^2 = \tfrac{1}{2}(2Ma^2)\omega^2 = \boxed{Ma^2\omega^2}$$

The fact that the spheres of mass *m* do not enter into this result makes sense, because they have no motion about the chosen axis of rotation; hence, they have no kinetic energy.

(b) Suppose the system rotates in the *xy* plane about an axis through *O* (the *z* axis). Calculate the moment of inertia about the *z* axis and the rotational energy about this axis.

Solution Because r_i in Equation 10.14 is the perpendicular distance to the axis of rotation, we have

$$I_z = \sum_i m_i r_i^2 = Ma^2 + Ma^2 + mb^2 + mb^2 = \boxed{2Ma^2 + 2mb^2}$$

$$K_R = \tfrac{1}{2}I_z\omega^2 = \tfrac{1}{2}(2Ma^2 + 2mb^2)\omega^2 = \boxed{(Ma^2 + mb^2)\omega^2}$$

Comparing the results for (a) and (b), we see explicitly that the moment of inertia and therefore the rotational energy associated with a given angular speed depend on the axis of rotation. In (b), we expect the result to include all masses and distances because all four spheres are in motion for rotation in the *xy* plane. Furthermore, the fact that the rotational energy in (a) is smaller than in (b) indicates that there is less resistance to changes in rotational motion about the *y* axis than about the *z* axis.

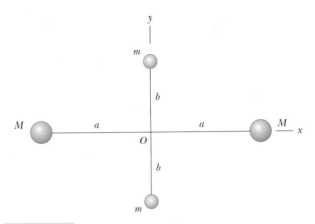

Figure 10.8

(Example 10.3) All four spheres are at fixed positions as shown. The moment of inertia of the system depends on the axis about which it is evaluated.

Example 10.4 Moment of Inertia of a Uniform Solid Cylinder

A uniform solid cylinder has a radius *R*, mass *M*, and length *L*. Calculate its moment of inertia about its central axis (the *z* axis shown in Figure 10.9).

Solution The integral can be evaluated relatively simply by dividing the cylinder into many cylindrical shells of radius *r*, thickness *dr*, and length *L*, as shown in Figure 10.9. The volume *dV* of a shell is its cross-sectional area multiplied by its length: $dV = (dA)L = (2\pi r \, dr)L$. Equation 10.17 gives the moment of inertia,

$$I = \int \rho r^2 dV = \int_0^R \rho r^2 (2\pi rL)\, dr = 2\pi\rho L \int_0^R r^3 dr = \tfrac{1}{2}\pi\rho LR^4$$

The volume of the entire cylinder is $\pi R^2 L$, so the density is $\rho = M/V = M/\pi R^2 L$. Substituting this value of ρ in the above result gives

$$I = \tfrac{1}{2}\pi\left(\frac{M}{\pi R^2 L}\right)Lr^4 = \boxed{\tfrac{1}{2}MR^2}$$

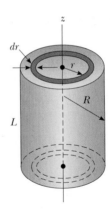

Figure 10.9

(Example 10.4) The geometry for calculating the moment of inertia about the central axis of a uniform solid cylinder.

Note that this result does not depend on *L*. Thus, it applies equally well to a long cylinder and a flat disk.

Example 10.5 Rotating Rod

A uniform rod of length *L* and mass *M* is free to rotate on a frictionless pin through one end (Fig. 10.10). The rod is released from rest in the horizontal position. **(a)** What is the angular speed of the rod at its lowest position?

Reasoning We consider the rod and the Earth as an isolated system and use the energy version of the isolated system model. Consider the mechanical energy of the system. When the rod is horizontal, as in Figure 10.10, it has no rotational

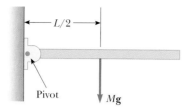

Figure 10.10

(Example 10.5) A uniform rod rotates freely under the influence of gravity around a pivot at the left end.

kinetic energy. Let us also define this position of the rod as representing the zero of gravitational potential energy of the system. When the center of mass of the rod is at the lowest position, the potential energy of the system is $-MgL/2$ and the rod has rotational kinetic energy $\frac{1}{2}I\omega^2$, where I is the moment of inertia about the pivot.

Solution Because $I = \frac{1}{3}ML^2$ (see Table 10.2) for a geometric model of a long, thin rod, and because mechanical en-

ergy of the isolated system is conserved, we have

$$K_i + U_i = K_f + U_f$$

$$0 + 0 = -\tfrac{1}{2}MgL + \tfrac{1}{2}I\omega^2 = -\tfrac{1}{2}MgL + \tfrac{1}{2}(\tfrac{1}{3}ML^2)\omega^2$$

$$\omega = \sqrt{\frac{3g}{L}}$$

(b) Determine the tangential speed of the center of mass and the tangential speed of the lowest point on the rod in the vertical position.

Solution Using Equation 10.10, we have

$$v_{CM} = r\omega = \frac{L}{2}\omega = \tfrac{1}{2}\sqrt{3gL}$$

The lowest point on the rod, because it is twice as far from the pivot as the center of mass, has a tangential speed equal to $2v_{CM} = \sqrt{3gL}$

10.5 • TORQUE AND THE VECTOR PRODUCT

Recall our stationary exercise bicycle from the preceding section. We caused the rotational motion of the wheel by applying forces to the pedals. When a force is exerted on a rigid body pivoted about some axis, and the line of action* of the force does not pass through the pivot, the body tends to rotate about that axis. For example, when you push on a door, the door rotates about an axis through the hinges. The tendency of a force to rotate a body about some axis is measured by a vector quantity called **torque.** Torque is the cause of changes in rotational motion, and is analogous to force, which causes changes in translational motion. Consider the wrench pivoted about the axis through O in Figure 10.11. The applied force $\mathbf{F}$ generally can act at an angle ϕ with respect to the position vector $\mathbf{r}$ locating the point of application of the force. We define the torque τ (Greek letter tau) resulting from the force $\mathbf{F}$ with the expression[†]

$$\tau \equiv rF \sin \phi \qquad [10.18]$$

It is very important to recognize that **torque is defined only when a reference axis is specified,** from which the distance r is determined. We can interpret Equation 10.18 in two different ways. Looking at the force components in Figure 10.11, we see that the component $F\cos\phi$ parallel to $\mathbf{r}$ will not cause a rotation around the pivot point, because its line of action passes right through the pivot

See Screen 7.6

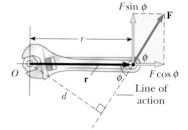

Figure 10.11

The force $\mathbf{F}$ has a greater rotating tendency about O as F increases and as the moment arm d increases. The component $F\sin\phi$ tends to rotate the system about O.

---

* The line of action of a force is an imaginary line colinear with the force vector and extending to infinity in both directions.

[†] In general, torque is a vector. For rotation about a fixed axis, however, we will use italic, nonbold notation and specify the direction with a positive or a negative sign, as we did for angular speed and acceleration in Section 10.1. We will treat the vector nature of torque briefly in a short while.

point. You cannot open a door by pushing on the hinges! Thus, only the perpendicular component causes a rotation about the pivot. In this case, we can write Equation 10.18 as

$$\tau = r(F \sin \phi)$$

so that the torque is the product of the distance to the point of application of the force and the perpendicular component of the force. In some problems, this is the easiest way to interpret the calculation of the torque.

The second way to interpret Equation 10.18 is to associate the sine function with the distance *r*, so that we can write

$$\tau = F(r \sin \phi) = Fd$$

- *Moment arm*

The quantity $d = r \sin \phi$, called the **moment arm** (or *lever arm*) of the force **F**, represents the perpendicular distance from the rotation axis to the line of action of **F**. In some problems, this approach to the calculation of the torque is easier than that of resolving the force into components.

If two or more forces are acting on a rigid body, as in Figure 10.12, each has a tendency to produce a rotation about the pivot at *O*. For example, if the body is initially at rest, **F**$_2$ tends to rotate the body clockwise and **F**$_1$ tends to rotate the body counterclockwise. We shall use the convention that the sign of the torque resulting from a force is positive if its turning tendency is counterclockwise around the rotation axis and negative if its turning tendency is clockwise. For example, in Figure 10.12, the torque resulting from **F**$_1$, which has a moment arm of d_1, is *positive* and equal to $+F_1 d_1$; the torque from **F**$_2$ is *negative* and equal to $-F_2 d_2$. Hence, the *net* torque acting on the rigid body about an axis through *O* is

$$\tau_{net} = \tau_1 + \tau_2 = F_1 d_1 - F_2 d_2$$

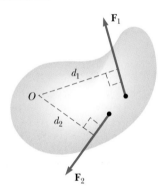

Figure 10.12

The force **F**$_1$ tends to rotate the object counterclockwise about an axis through *O*, and **F**$_2$ tends to rotate the object clockwise.

From the definition of torque, we see that the rotating tendency increases as *F* increases and as *d* increases. For example, we cause more rotation of a door if (a) we push harder or (b) we push at the doorknob rather than at a point close to the hinges. **Torque should not be confused with force.** Torque *depends* on force, but it also depends on *where the force is applied*. Torque has units of force times length— N · m in SI units.*

So far, we have not discussed the vector nature of torque, aside from assigning a positive or negative value to τ. Consider a force **F** acting on a particle located at the vector position **r** (Fig. 10.13). The *magnitude* of the torque due to this force relative to an axis through the origin is $|rF \sin \phi|$, where ϕ is the angle between **r** and **F**. The axis about which **F** would tend to produce rotation is perpendicular to the plane formed by **r** and **F**. If the force lies in the *xy* plane, as in Figure 10.13, the torque is represented by a vector parallel to the *z* axis. The force in Figure 10.13 creates a torque that tends to rotate the body counterclockwise when we are looking down the *z* axis—we define this as the vector τ being in the positive *z* direction (i.e., coming toward your eyes). If we reverse the direction of **F** in Figure 10.13, τ is in the negative *z* direction. With this choice, the torque vector can be defined to be equal to the **vector product,** or **cross product,** of **r** and **F**:

- *Definition of torque using the cross product*

$$\tau \equiv \mathbf{r} \times \mathbf{F}$$ [10.19]

---

* In Chapter 6, we saw the product of newtons and meters when we defined work, and we called this product a *joule*. We do not use this term here because the joule is only to be used when discussing energy. For torque, the unit is simply the N · m.

We now give a formal definition of the vector product, first introduced in Section 1.9. Given any two vectors **A** and **B**, the vector product **A** × **B** is defined as a third vector **C**, the *magnitude* of which is $AB \sin \theta$, where θ is the angle between **A** and **B**:

$$\mathbf{C} = \mathbf{A} \times \mathbf{B} \qquad \text{[10.20]}$$

$$C = |\mathbf{C}| \equiv AB \sin \theta \qquad \text{[10.21]}$$

Note that the quantity $AB \sin \theta$ is equal to the area of the parallelogram formed by **A** and **B**, as shown in Figure 10.14. The *direction* of **A** × **B** is perpendicular to the plane formed by **A** and **B** and is determined by the right-hand rule illustrated in Figure 10.14. The four fingers of the right hand are pointed along **A** and then "wrapped" into **B** through the angle θ. The direction of the upright thumb is the direction of **A** × **B**. Because of the notation, **A** × **B** is often read "**A** cross **B**"; hence the term *cross product*.

Some properties of the vector product follow from its definition:

- Unlike the case of the scalar product, the vector product is not commutative; in fact,

$$\mathbf{A} \times \mathbf{B} = -(\mathbf{B} \times \mathbf{A}) \qquad \text{[10.22]}$$

Therefore, if you change the order of the vector product, you must change the sign. One could easily verify this relation with the right-hand rule (see Fig. 10.14).

- If **A** is parallel to **B** ($\theta = 0°$ or $180°$), then **A** × **B** = 0; therefore, it follows that **A** × **A** = 0.

- If **A** is perpendicular to **B**, then $|\mathbf{A} \times \mathbf{B}| = AB$. It is left to Problem 22 to show, from Equations 10.20 and 10.21 and the definition of unit vectors, that the cross products of the unit vectors **i**, **j**, and **k** obey the following expressions:

$$\mathbf{i} \times \mathbf{i} = \mathbf{j} \times \mathbf{j} = \mathbf{k} \times \mathbf{k} = 0$$

$$\mathbf{i} \times \mathbf{j} = -\mathbf{j} \times \mathbf{i} = \mathbf{k}$$

$$\mathbf{j} \times \mathbf{k} = -\mathbf{k} \times \mathbf{j} = \mathbf{i} \qquad \text{[10.23]}$$

$$\mathbf{k} \times \mathbf{i} = -\mathbf{i} \times \mathbf{k} = \mathbf{j}$$

Signs are interchangeable. For example, $\mathbf{i} \times (-\mathbf{j}) = -\mathbf{i} \times \mathbf{j} = -\mathbf{k}.$

Quick Quiz 10.6

Both torque and work are products of force and distance. How are they different?

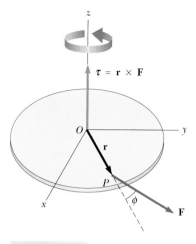

Figure 10.13

The torque vector $\boldsymbol{\tau}$ lies in a direction perpendicular to the plane formed by the position vector **r** and the applied force vector **F**.

PITFALL PREVENTION 10.6

The cross product is a vector

Remember that the result of taking a cross product between two vectors is a *third vector*. Equation 10.21 gives only the magnitude of this vector.

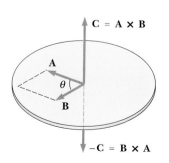

Right-hand rule

Figure 10.14

The vector product **A** × **B** is a third vector **C** having a magnitude $AB \sin \theta$ equal to the area of the parallelogram shown. The vector **C** is perpendicular to the plane formed by **A** and **B**, and its direction is determined by the right-hand rule.

Quick Quiz 10.7

If you are trying to loosen a stubborn screw from a piece of wood with a screwdriver and fail, should you find a screwdriver for which the handle is (a) longer or (b) fatter? Why?

Example 10.6   The Net Torque on a Cylinder

A one-piece cylinder is shaped as in Figure 10.15, with a core section protruding from the larger drum. The cylinder is free to rotate around the central axis shown in the drawing. A rope wrapped around the drum, of radius R_1, exerts a force $\mathbf{T}_1$ to the right on the cylinder. A rope wrapped around

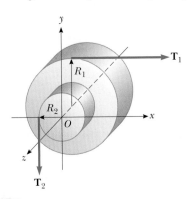

Figure 10.15

(Example 10.6) A solid cylinder pivoted about the z axis through O. The moment arm of $\mathbf{T}_1$ is R_1, and the moment arm of $\mathbf{T}_2$ is R_2.

the core, of radius R_2, exerts a force $\mathbf{T}_2$ downward on the cylinder. (a) What is the net torque acting on the cylinder about the rotation axis (which is the z axis in Fig. 10.15)?

Solution   The torque due to $\mathbf{T}_1$ is $-R_1 T_1$. It is negative because it tends to produce a clockwise rotation from the point of view in Figure 10.15. The torque due to $\mathbf{T}_2$ is $+R_2 T_2$ and is positive because it tends to produce a counterclockwise rotation. Therefore, the net torque about the rotation axis is

$$\tau_{net} = \tau_1 + \tau_2 = \boxed{R_2 T_2 - R_1 T_1}$$

(b) Suppose $T_1 = 5.0$ N, $R_1 = 1.0$ m, $T_2 = 6.0$ N, and $R_2 = 0.50$ m. What is the net torque about the rotation axis and which way does the cylinder rotate if it starts from rest?

Solution   We substitute numerical values in the result from part (a),

$$\tau_{net} = (6.0\ \text{N})(0.50\ \text{m}) - (5.0\ \text{N})(1.0\ \text{m}) = \boxed{-2.0\ \text{N} \cdot \text{m}}$$

Because the net torque is negative, the cylinder rotates clockwise from rest.

Example 10.7   The Vector Product

Two vectors lying in the xy plane are given by the equations $\mathbf{A} = 2\mathbf{i} + 3\mathbf{j}$ and $\mathbf{B} = -\mathbf{i} + 2\mathbf{j}$. Find $\mathbf{A} \times \mathbf{B}$, and verify explicitly that $\mathbf{A} \times \mathbf{B} = -\mathbf{B} \times \mathbf{A}$.

Solution   Using Equation 10.23 for the vector product of unit vectors gives

$$\mathbf{A} \times \mathbf{B} = (2\mathbf{i} + 3\mathbf{j}) \times (-\mathbf{i} + 2\mathbf{j})$$
$$= 2\mathbf{i} \times (-\mathbf{i}) + 2\mathbf{i} \times 2\mathbf{j} + 3\mathbf{j} \times (-\mathbf{i}) + 3\mathbf{j} \times 2\mathbf{j}$$
$$= 0 + 4\mathbf{k} + 3\mathbf{k} + 0 = \boxed{7\ \mathbf{k}}$$

$$\mathbf{B} \times \mathbf{A} = (-\mathbf{i} + 2\mathbf{j}) \times (2\mathbf{i} + 3\mathbf{j})$$
$$= -\mathbf{i} \times 2\mathbf{i} + (-\mathbf{i}) \times 3\mathbf{j} + 2\mathbf{j} \times 2\mathbf{i} + 2\mathbf{j} \times 3\mathbf{j}$$
$$= 0 - 3\mathbf{k} - 4\mathbf{k} + 0 = -7\mathbf{k}$$

Therefore, $\mathbf{A} \times \mathbf{B} = -\mathbf{B} \times \mathbf{A}$.

EXERCISE   Use the results of this example and Equation 10.21 to find the angle between $\mathbf{A}$ and $\mathbf{B}$.

Answer   60.3°

EXERCISE   A particle is located at the vector position $\mathbf{r} = (\mathbf{i} + 3\mathbf{j})$ m, and the force acting on it is $\mathbf{F} = (3\mathbf{i} + 2\mathbf{j})$ N. What is the torque on the particle about an axis through (a) the origin and (b) the point having coordinates (0, 6) m?]

Answer   (a) $-7\mathbf{k}$ N·m   (b) $11\mathbf{k}$ N·m

EXERCISE   If $|\mathbf{A} \times \mathbf{B}| = \mathbf{A} \cdot \mathbf{B}$, what is the angle between $\mathbf{A}$ and $\mathbf{B}$?

Answer   45.0°

10.6 • THE RIGID BODY IN EQUILIBRIUM

We have defined a rigid body and have discussed torque as the cause of changes in rotational motion of a rigid body. We can now establish models for a rigid body subject to torques that are analogous to those for a particle subject to forces. We begin by imagining a rigid body with balanced torques.

Consider two forces of equal magnitude and opposite direction applied to an object as shown in Figure 10.16a. The force directed to the right tends to rotate the object clockwise about an axis perpendicular to the diagram through *O*, whereas the force directed to the left tends to rotate it counterclockwise about that axis. Because the forces are of equal magnitude and act at the same perpendicular distance from *O*, their torques are equal in magnitude. Thus, the net torque on the rigid body is zero. The situation shown in Figure 10.16b is another case in which the net torque about *O* is zero (although the net *force* on the object is not zero), and we can devise many more cases.

With no net torque, no change occurs in rotational motion, and the rotational motion of the rigid body remains in its original state. This is an equilibrium situation, analogous to translational equilibrium, discussed in Chapter 4.

We now have **two conditions for complete equilibrium of an object,** which can be stated as follows:

- The net external force must equal zero

$$\sum \mathbf{F} = 0 \qquad\qquad [10.24]$$

- The net external torque must be zero about *any* axis

$$\sum \boldsymbol{\tau} = 0 \qquad\qquad [10.25]$$

The first condition is a statement of translational equilibrium. The second condition is a statement of rotational equilibrium. In the special case of **static equilibrium,** the object is at rest, so that it has no translational or angular speed (i.e., $v_{CM} = 0$ and $\omega = 0$).

The two vector expressions given by Equations 10.24 and 10.25 are equivalent, in general, to six scalar equations: three from the first condition of equilibrium, and three from the second (corresponding to *x*, *y*, and *z* components). Hence, in a complex system involving several forces acting in various directions, you would be faced with solving a set of equations with many unknowns. Here, we restrict our discussion to situations in which all the forces lie in the *xy* plane. (Forces whose vector representations are in the same plane are said to be *coplanar.*) With this restriction, we need to deal with only three scalar equations. Two of these come from balancing the forces in the *x* and *y* directions. The third comes from the torque equation, namely, that the net torque about an axis through *any* point in the *xy* plane must be zero. Hence, the two conditions of equilibrium provide the equations

$$\sum F_x = 0 \qquad \sum F_y = 0 \qquad \sum \tau_z = 0 \qquad [10.26]$$

where the axis of the torque equation is arbitrary.

Quick Quiz 10.8

(a) Is it possible for a situation to exist in which Equation 10.24 is satisfied, but Equation 10.25 is not? (b) Can Equation 10.25 be satisfied, but Equation 10.24 is not?

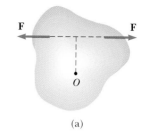

(a)

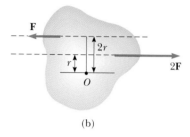

(b)

Figure 10.16

(a) The two forces acting on the object are equal in magnitude and opposite in direction. Because they also act along the same line of action, the net torque is zero and the object is in equilibrium. (b) Another situation in which two forces act on an object to produce zero net torque (but *not* zero net force).

PITFALL PREVENTION 10.7
Zero torque

Keep in mind that zero net torque does not mean an absence of rotational motion. An object rotating at a constant angular speed can be under the influence of a net torque of zero. This is analogous to the translational situation—zero net force does not mean an absence of translational motion.

In working static equilibrium problems, it is important to recognize all external forces acting on the object. Failure to do so will result in an incorrect analysis. The following procedure is recommended when analyzing an object in equilibrium under the action of several external forces:

PROBLEM-SOLVING STRATEGY | **Rigid Body in Equilibrium**

1. Make a sketch of the rigid body under consideration.
2. Draw a free-body diagram, and label all external forces acting on the object. Try to guess the correct direction for each force. If you select a direction that leads to a negative sign in your solution for a force, do not be alarmed; this merely means that the direction of the force is the opposite of what you guessed.
3. Resolve all forces into rectangular components, choosing a convenient coordinate system. Then apply the first condition for equilibrium, Equation 10.24. Remember to keep track of the signs of the various force components.
4. Choose a convenient axis for calculating the net torque on the rigid body. Remember that the choice of the axis for the torque equation is arbitrary; therefore, choose an axis that will simplify your calculation as much as possible. Usually, the most convenient axis for calculating torques is one through a point at which several forces act, so that their torques around this axis are zero. If you don't know a force or don't need to know a force, it is often beneficial to choose an axis through the point at which this force acts. Apply the second condition for equilibrium, Equation 10.25.
5. Solve the simultaneous equations for the unknowns in terms of the known quantities.

Example 10.8   Standing on a Horizontal Beam

A uniform horizontal beam of length 8.00 m and weight 200 N is attached to a wall by a pin connection. Its far end is supported by a cable that makes an angle of 53.0° with the horizontal (Fig. 10.17a). If a 600-N man stands 2.00 m from the wall, find the tension in the cable and the force exerted by the wall on the beam.

Solution   The beam–man system is at rest and remains at rest, so it is clearly in static equilibrium. First we must identify all the external forces acting on the system, which we do in the free-body diagram in Figure 10.17b. These are the weights of the beam and the man, the force **T** exerted by the cable, the force **R** exerted by the wall at the pivot (the direction of this force is unknown). (The force between the man and the beam is internal to the system, so it is not included in the free-body diagram.) Notice that we have imagined the

gravitational force on the beam as acting at its center of gravity. Because the beam is uniform, the center of gravity is at the geometric center. If we resolve **T** and **R** into horizontal and vertical components (Figure 10.17c) and apply the first condition for equilibrium for the system, we have

(1) $\sum F_x = R \cos\theta - T \cos 53.0° = 0$

(2) $\sum F_y = R \sin\theta + T \sin 53.0° - 600 \text{ N} - 200 \text{ N} = 0$

Because we have three unknowns—R, T, and θ—we cannot obtain a solution from these two expressions alone.

To generate a third expression, let us invoke the condition for rotational equilibrium because the system can be modeled as a rigid body in equilibrium. A convenient axis to choose for our torque equation is the one that passes

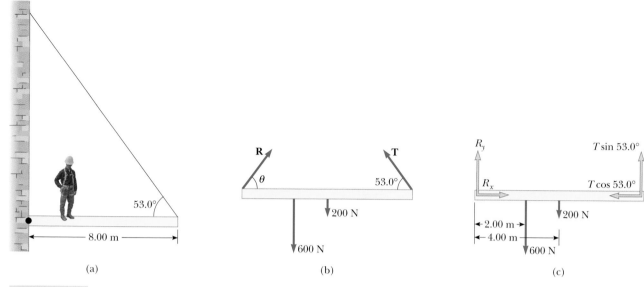

Figure 10.17

(Example 10.8) (a) A uniform beam supported by a cable. A man walks out on the beam. (b) The free-body diagram for the beam–man system. (c) The free-body diagram with forces resolved into horizontal and vertical components.

through the pivot at the wall. The feature that makes this point so convenient is that the force **R** and the horizontal component of **T** both have a lever arm of zero, and hence zero torque, about this pivot. Recalling our convention for the sign of the torque about an axis and noting that the lever arms of the 600-N, 200-N, and $T \sin 53°$ forces are 2.00 m, 4.00 m, and 8.00 m, respectively, we have

$$\sum \tau = (T \sin 53.0°)(8.00 \text{ m}) - (600 \text{ N})(2.00 \text{m})$$
$$- (200 \text{ N})(4.00 \text{ m}) = 0$$

$$T = \boxed{313 \text{ N}}$$

The torque equation gives us one of the unknowns directly! This is due to our judicious choice of the axis. This value is substituted into (1) and (2) to give

$$R \cos \theta = 188 \text{ N}$$

$$R \sin \theta = 550 \text{ N}$$

We divide these two equations to find

$$\tan \theta = \frac{550 \text{ N}}{188 \text{ N}} = 2.93$$

$$\theta = 71.1°$$

Finally,

$$R = \frac{188 \text{ N}}{\cos \theta} = \frac{188 \text{ N}}{\cos 71.1°} = \boxed{581 \text{ N}}$$

If we had selected some other axis for the torque equation, the results would have been the same, although the details of the solution would be somewhat different. For example, if we had chosen to have the axis pass through the center of gravity of the beam, the torque equation would involve both T and R. However, this equation, coupled with (1) and (2), could still be solved for the unknowns T, R, and θ, yielding the same results. Try it!

Example 10.9   The Leaning Ladder

A uniform ladder of length ℓ and mass m rests against a smooth, vertical wall (Fig. 10.18a). If the coefficient of static friction between ladder and ground is $\mu_s = 0.40$, find the minimum angle θ_{min} such that the ladder does not slip.

Solution   The ladder is at rest and remains at rest, so we model it as a rigid body in equilibrium. The free-body diagram showing all the external forces acting on the ladder is illustrated in Figure 10.18b. The reaction **R** exerted by the ground on the ladder is the vector sum of a normal force **n** and the force of static friction **f**. The reaction force **P**

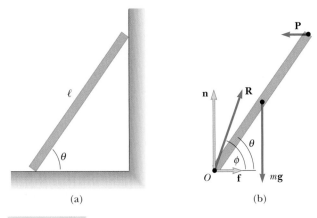

Figure 10.18

(Example 10.9) (a) A uniform ladder at rest, leaning against a frictionless wall. (b) The free-body diagram for the ladder.

exerted by the wall on the ladder is horizontal, because the wall is smooth, meaning that it is frictionless. Thus, **P** is simply the normal force on the ladder from the wall. From the first condition of equilibrium applied to the ladder, we have

$$\sum F_x = f - P = 0$$

$$\sum F_y = n - mg = 0$$

We see from the second equation that $n = mg$. Furthermore, when the ladder is on the verge of slipping, the force of static friction must be a maximum, given by $f_{s,\,max} = \mu_s n$.

To find θ, we use the second condition of equilibrium. When the torques are taken about the origin O at the bottom of the ladder, we have

$$\sum \tau_O = P\ell \sin \theta - mg\frac{\ell}{2} \cos \theta = 0$$

This expression gives

$$\tan \theta_{min} = \frac{mg}{2P} = \frac{n}{2f_{s,\,max}} = \frac{n}{2(\mu_s n)} = \frac{1}{2(0.40)} = 1.25$$

$$\theta_{min} = \boxed{51°}$$

It is interesting that the result does not depend on ℓ or m. The answer depends only on μ_s.

EXERCISE Two children weighing 500 N and 350 N are on a uniform board weighing 40.0 N supported at its center (a "see-saw"). If the 500-N child is 1.50 m from the center, determine (a) the upward force exerted on the board by the support and (b) where the 350-N child must sit to balance the system.

Answer   (a) 890 N   (b) 2.14 m from the center

10.7 • THE RIGID BODY UNDER A NET TORQUE

See Screen 7.6

In the preceding section, we investigated the equilibrium situation in which the net torque on a rigid body is zero. What if the net torque on a rigid body is not zero? In analogy with Newton's second law for translational motion, we should expect the angular speed of the rigid body to change. The net torque will cause angular acceleration of the rigid body. We investigate this notion in this section.

Let us imagine a rotating rigid body again as a collection of particles. The rigid body will be subject to a number of forces applied at various locations on the rigid body, at which individual particles will be located. Thus, we can imagine that the forces on the rigid body are exerted on individual particles of the rigid body. We will calculate the net torque on the object due to the torques resulting from these forces around the rotation axis of the rotating body. Any applied force can be represented by its radial component and its tangential component. The radial component of an applied force provides no torque because its line of action goes through the rotation axis. Thus, only the tangential component of an applied force contributes to the torque.

On any given particle, described by index variable i, within the rigid body, we can use Newton's second law to describe the tangential acceleration of the particle:

$$F_{ti} = m_i a_{ti}$$

where the t subscript refers to tangential components. Let us multiply both sides of this expression by r_i, the distance of the particle from the rotation axis:

$$r_i F_{ti} = r_i m_i a_{ti}$$

Using Equation 10.11, and recognizing the definition of torque ($\tau = rF \sin \phi = rF_t$ in this case), we can rewrite this as

$$\tau_i = m_i r_i^2 \alpha_i$$

Now, let us add up the torques on all particles of the rigid body:

$$\sum_i \tau_i = \sum_i m_i r_i^2 \alpha_i$$

The left side is the net torque on all particles of the rigid body. The net torque associated with *internal* forces is zero, however. To understand this, recall that Newton's third law tells us that the internal forces occur in equal and opposite pairs that lie along the line of separation of each pair of particles. The torque due to each action–reaction force pair is therefore zero. On summation of all torques, we see that the *net internal torque vanishes*. The term on the left, then, reduces to the net *external* torque.

On the right, we impose the rigid body model by demanding that all particles have the same angular acceleration. Thus, this equation becomes

$$\sum \tau = \left(\sum_i m_i r_i^2 \right) \alpha$$

where the torque and angular acceleration no longer have subscripts because they refer to quantities associated with the rigid body as a whole rather than to individual particles. We recognize the quantity in parentheses as the moment of inertia I. Therefore,

$$\sum \tau = I\alpha \qquad \text{[10.27]}$$

• *Relationship between net torque and angular acceleration*

That is, **the net torque acting on the rigid body is proportional to its angular acceleration,** and the proportionality constant is the moment of inertia. It is important to note that $\sum \tau = I\alpha$ is the rotational analog of Newton's second law of motion, $\sum F = ma$.

Quick Quiz 10.9

If you turn off your workshop grinding wheel at the same time as your electric drill, the grinding wheel takes much longer to stop rotating. Why?

THINKING PHYSICS 10.3

When an automobile driver steps on the accelerator, the nose of the car moves upward. When the driver brakes, the nose moves downward. Why do these effects occur?

Reasoning When the driver steps on the accelerator, there is an increased force on the tires from the roadway. This force is parallel to the roadway and directed toward the front of the automobile, as suggested in Figure 10.19a. This force provides a torque that tends to cause the car to rotate in the clockwise direction

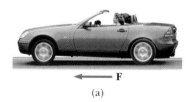
(a)

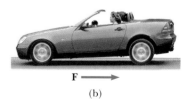
(b)

Figure 10.19

(Thinking Physics 10.3) (a) When you step on the gas, the car "noses up." (b) When you step on the brake, the car "noses down."

around the center of mass of the car. The result of this rotation is a "nosing up" of the car.

When the driver steps on the brake, there is an increased force on the tires from the roadway, directed toward the rear of the automobile, as suggested in Figure 10.19b. This force results in a torque that causes a counterclockwise rotation and the subsequent "nosing down" of the automobile.

Example 10.10 An Atwood Machine with a Massive Pulley

In Example 4.4, we analyzed an Atwood machine in which two objects with unequal masses hang from a string that passes over a light, frictionless pulley. Suppose the pulley, which is modeled as a disk, has mass M and radius R, and the pulley surface is not frictionless, so that the string does not slide on the pulley (Fig. 10.20a). We will assume that the frictional torque acting at the bearing of the pulley is negligible. Calculate the magnitude of the acceleration of the two objects.

Reasoning Because counterclockwise angular acceleration of the pulley is defined as positive, we define the positive directions for m_1 and m_2 as shown in Figure 10.20 so that all accelerations, translational and rotational, are positive if m_1 accelerates downward. If the pulley has mass and friction, the tensions T_1 and T_2 in the string on either side of the pulley are not equal in magnitude, as they are in Example 4.4. Indeed, it is the difference in torque due to these different tensions that provides the net torque to cause the angular acceleration of the pulley (Fig. 10.20b). Consequently, some of what we do here will look similar to Example 4.4, with the exception of incorporating T_1 and T_2 in our mathematical representation instead of just T. The forces $M\mathbf{g}$ and $\mathbf{F}$ (the force supporting the pulley) both act through the pulley axle, so these forces do not contribute to the torque on the pulley.

Solution With the help of the free-body diagrams in Figure 10.20b, we apply Newton's second law to m_1, so that

$$(1) \quad \sum F_y = m_1 g - T_1 = m_1 a$$

and for m_2,

$$(2) \quad \sum F_y = T_2 - m_2 g = m_2 a$$

We cannot solve these two equations for a, as is done in Example 4.4, because we have three unknowns: a, T_1, and T_2. We can find a third equation by applying Equation 10.27 to the pulley (Fig. 10.20b):

$$(3) \quad \sum \tau = T_1 R - T_2 R = I\alpha = (\tfrac{1}{2}MR^2)\left(\frac{a}{R}\right)$$

$$= \tfrac{1}{2}MRa \quad \longrightarrow \quad T_1 - T_2 = \tfrac{1}{2}Ma$$

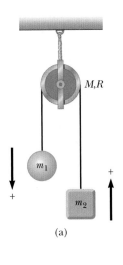

(a)

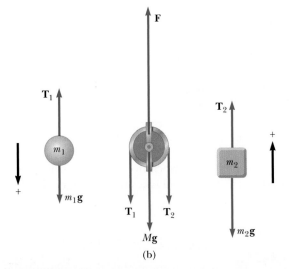

(b)

Figure 10.20

(Example 10.10) (a) An Atwood machine with a massive pulley. The pulley is modeled as a disk. (b) Free-body diagrams for the two hanging objects and the pulley.

We substitute into (3) expressions for T_1 and T_2 from (1) and (2):

$$(m_1g - m_1a) - (m_2a + m_2g) = \tfrac{1}{2}Ma$$

$$a = \left(\frac{m_1 - m_2}{m_1 + m_2 + \tfrac{1}{2}M}\right)g$$

Notice that this differs from the result for Example 4.4 only in the extra term $\tfrac{1}{2}M$ in the denominator. If the pulley mass $M \to 0$, this expression reduces to that in Example 4.4.

Work and Energy in Rotational Motion

In translational motion, we found energy concepts, and in particular the reduction of the continuity equation for energy called the work–kinetic energy theorem, to be extremely useful in describing the motion of a system. Energy concepts can be equally useful in simplifying the analysis of rotational motion, as we saw in the isolated system analysis in Example 10.5. From the continuity equation for energy, we expect that for rotation of an object about a fixed axis, the work done by external forces on the object will equal the change in the rotational kinetic energy, as long as energy is not stored by any other means. To show that this is in fact the case, we begin by finding an expression for the work done by a torque.

Consider a rigid body pivoted at the point O in Figure 10.21. Suppose a single external force $\mathbf{F}$ is applied at the point P and $d\mathbf{s}$ is the displacement of the point of application of the force. The small amount of work dW done by $\mathbf{F}$ as the point of application rotates through an infinitesimal distance $ds = r\,d\theta$ in a time dt is

$$dW = \mathbf{F} \cdot d\mathbf{s} = (F\sin\phi)\,r\,d\theta$$

where $F\sin\phi$ is the tangential component of $\mathbf{F}$, or the component of the force along the displacement. Note from Figure 10.21 that the **radial component of F does no work because it is perpendicular to the displacement.**

Because the magnitude of the torque due to $\mathbf{F}$ about the origin is defined as $rF\sin\phi$, we can write the work done for the infinitesimal rotation in the form

$$dW = \tau\,d\theta \qquad \textbf{[10.28]}$$

Notice that this is the product of torque and angular displacement, making it analogous to the work done in translational motion, which is the product of force and translational displacement (Eq. 6.2).

Now, we will combine this result with the rotational form of Newton's second law, $\tau = I\alpha$. Using the chain rule from the calculus, we can express the torque as

$$\tau = I\alpha = I\frac{d\omega}{dt} = I\frac{d\omega}{d\theta}\frac{d\theta}{dt} = I\frac{d\omega}{d\theta}\omega$$

Rearranging this expression and noting that $\tau\,d\theta = dW$ from Equation 10.28, we have

$$\tau\,d\theta = dW = I\omega\,d\omega$$

Integrating this expression, we find the total work done by the torque:

$$W = \int_{\theta_i}^{\theta_f} \tau\,d\theta = \int_{\omega_i}^{\omega_f} I\omega\,d\omega = \tfrac{1}{2}I\omega_f^2 - \tfrac{1}{2}I\omega_i^2 \qquad \textbf{[10.29]}$$

• *Work–kinetic energy theorem for pure rotation*

Notice that this equation, which can be written as $W = \Delta K_R$, has exactly the same mathematical form as the work–kinetic energy theorem for translation.

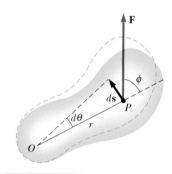

Figure 10.21

A rigid body rotates about an axis through O under the action of an external force $\mathbf{F}$ applied at P.

We finish this discussion of energy concepts for rotation by investigating the *rate* at which work is being done by **F** on an object rotating about a fixed axis. This rate is obtained by dividing the left and right sides of Equation 10.28 by dt:

$$\frac{dW}{dt} = \tau \frac{d\theta}{dt} \qquad \text{[10.30]}$$

The quantity dW/dt is, by definition, the instantaneous power $\mathcal{P}$ delivered by the force. Furthermore, because $d\theta/dt = \omega$, Equation 10.30 reduces to

$$\mathcal{P} = \frac{dW}{dt} = \tau\omega \qquad \text{[10.31]}$$

This expression is analogous to $\mathcal{P} = Fv$ in the case of translational motion.

Example 10.11   Work on an Atwood Machine

Consider the Atwood machine of Example 10.10. Calculate the rate at which work is done on the system of the pulley and the hanging objects by the gravitational force.

Reasoning   The only means by which energy enters the system is by work done by the gravitational force. The only types of energy in the system are translational and rotational kinetic energy. Thus, we can apply the reduction of the continuity equation called the work–kinetic energy theorem, $W = \Delta K$. The rate at which work is being done on the system is equal to the rate at which the kinetic energy of the system is changing, so we will evaluate the power as $\mathcal{P} = dK/dt$.

Solution   Because the acceleration of the hanging objects in Example 10.10 is constant, we can find the speed of either object at time t by using Equation 2.8. If the objects start from rest, the speed of either at time t is

$$v_f = v_i + at = \left(\frac{m_1 - m_2}{m_1 + m_2 + \frac{1}{2}M}\right)gt$$

Using Equation 10.10, the angular speed of the pulley at time t is

$$\omega_f = \frac{v_f}{R} = \frac{1}{R}\left(\frac{m_1 - m_2}{m_1 + m_2 + \frac{1}{2}M}\right)gt$$

Thus, the kinetic energy of the system is

$$K_{\text{system}} = \tfrac{1}{2}m_1 v_{1f}^2 + \tfrac{1}{2}m_2 v_{2f}^2 + \tfrac{1}{2}I\omega_f^2$$

$$= \tfrac{1}{2}(m_1 + m_2)\left(\frac{m_1 - m_2}{m_1 + m_2 + \frac{1}{2}M}\right)^2 g^2 t^2$$

$$+ \tfrac{1}{2}(\tfrac{1}{2}MR^2)\frac{1}{R^2}\left(\frac{m_1 - m_2}{m_1 + m_2 + \frac{1}{2}M}\right)^2 g^2 t^2$$

$$= \tfrac{1}{2}(m_1 + m_2 + \tfrac{1}{2}M)\left(\frac{m_1 - m_2}{m_1 + m_2 + \frac{1}{2}M}\right)^2 g^2 t^2$$

Figure 10.22

(Example 10.11) The tension in the cord produces a torque about the axle passing through O.

$$= \tfrac{1}{2}\frac{(m_1 - m_2)^2}{m_1 + m_2 + \frac{1}{2}M}g^2 t^2$$

Finally, we differentiate this expression with respect to time:

$$\mathcal{P} = \frac{dK_{\text{system}}}{dt} = \frac{d}{dt}\left[\tfrac{1}{2}\frac{(m_1 - m_2)^2}{m_1 + m_2 + \frac{1}{2}M}g^2 t^2\right]$$

$$= \frac{(m_1 - m_2)^2}{m_1 + m_2 + \frac{1}{2}M}g^2 t$$

Thus, the power increases with time, because the angular speed increases in time.

EXERCISE   The wheel in Figure 10.22 is a solid disk with $M = 2.00$ kg, $R = 30.0$ cm, and $I = 0.0900$ kg·m². The suspended object has a mass $m = 0.500$ kg. If the suspended object starts from rest and descends to a point 1.00 m lower, what is its speed at this point?

Answer   2.56 m/s

10.8 • ANGULAR MOMENTUM

Imagine an object rotating in space with no motion of its center of mass. Each particle in the object is moving in a circular path, so momentum is associated with the motion of each particle. Although the object has no linear momentum (its center of mass is not moving through space), a "quantity of motion" is associated with its rotation. We will investigate **angular momentum** in this section.

Let us first consider a particle of mass m, situated at the vector position **r** and moving with a momentum **p**, as shown in Figure 10.23. For now, this is not a particle on a rigid body; it is any particle moving with momentum **p**. We will apply the result to a rotating rigid body shortly. The **instantaneous angular momentum L** of the particle relative to the origin O is defined by the vector product of its instantaneous vector position **r** and the instantaneous linear momentum **p**:

$$\mathbf{L} = \mathbf{r} \times \mathbf{p} \qquad [10.32]$$

The SI units of angular momentum are $\text{kg} \cdot \text{m}^2/\text{s}$. It is important to note that both the magnitude and the direction of **L** depend on the choice of origin. The direction of **L** is perpendicular to the plane formed by **r** and **p**, and the sense of **L** is governed by the right-hand rule. For example, in Figure 10.23, **r** and **p** are assumed to be in the xy plane, and **L** points in the z direction. Because $\mathbf{p} = m\mathbf{v}$, the magnitude of **L** is

$$L = mvr \sin \phi \qquad [10.33]$$

where ϕ is the angle between **r** and **p**. It follows that **L** is zero when **r** is parallel to **p** ($\phi = 0°$ or $180°$). In other words, when the particle moves along a line that passes through the origin, it has zero angular momentum with respect to the origin. This is equivalent to stating that the momentum vector is not tangent to *any* circle drawn about the origin. On the other hand, if **r** is perpendicular to **p** ($\phi = 90°$), L is a maximum and equal to mvr. In fact, at that instant the particle moves exactly as though it were on the rim of a wheel rotating about the origin in a plane defined by **r** and **p**. A particle has nonzero angular momentum about some point if the position vector of the particle measured from that point rotates about the point as the particle moves.

For translational motion, we found that the net force on a particle equals the time rate of change of the particle's linear momentum (Eq. 8.3). We shall now show that Newton's second law implies an analogous situation for rotation—that the net torque acting on a particle equals the time rate of change of the particle's angular momentum. Let us start by writing the torque on the particle in the form

$$\boldsymbol{\tau} = \mathbf{r} \times \mathbf{F} = \mathbf{r} \times \frac{d\mathbf{p}}{dt} \qquad [10.34]$$

where we have used the fact that $\mathbf{F} = d\mathbf{p}/dt$ (Eq. 8.3). Now let us differentiate Equation 10.32 with respect to time, using the product rule for differentiation:

$$\frac{d\mathbf{L}}{dt} = \frac{d}{dt}(\mathbf{r} \times \mathbf{p}) = \mathbf{r} \times \frac{d\mathbf{p}}{dt} + \frac{d\mathbf{r}}{dt} \times \mathbf{p}$$

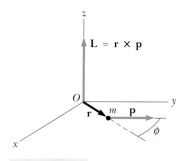

Figure 10.23

The angular momentum **L** of a particle of mass m and linear momentum **p** located at the position **r** is a vector given by $\mathbf{L} = \mathbf{r} \times \mathbf{p}$. The value of **L** depends on the origin about which it is measured and is a vector perpendicular to both **r** and **p**.

PITFALL PREVENTION 10.8

Is rotation necessary for angular momentum?

Notice that we can define angular momentum even though there is no motion of the particle in a circular path. Even a particle moving in a straight line has angular momentum about any axis displaced from the path of the particle.

It is important to adhere to the order of factors in the vector product because the vector product is not commutative.

The last term on the right in the preceding equation is zero because $\mathbf{v} = d\mathbf{r}/dt$ is parallel to $\mathbf{p}$. Therefore,

$$\frac{d\mathbf{L}}{dt} = \mathbf{r} \times \frac{d\mathbf{p}}{dt} \qquad \text{[10.35]}$$

Comparing Equations 10.34 and 10.35, we see that

- *Torque equals time rate of change of angular momentum*

$$\boxed{\tau = \frac{d\mathbf{L}}{dt}} \qquad \text{[10.36]}$$

This result is the rotational analog of Newton's second law, $\mathbf{F} = d\mathbf{p}/dt$. Equation 10.36 says that the torque acting on a particle is equal to the time rate of change of the particle's angular momentum. It is important to note that Equation 10.36 is valid only if the origins of τ and $\mathbf{L}$ are the *same*. Equation 10.36 is also valid when several forces are acting on the particle, in which case τ is the *net* torque on the particle. Of course, the same origin must be used in calculating all torques as well as the angular momentum.

Now, let us apply these ideas to a system of particles. The total angular momentum $\mathbf{L}$ of the system of particles about some point is defined as the vector sum of the angular momenta of the individual particles:

$$\mathbf{L} = \mathbf{L}_1 + \mathbf{L}_2 + \cdots + \mathbf{L}_n = \sum_i \mathbf{L}_i$$

where the vector sum is over all of the n particles in the system.

Because the individual angular momenta of the particles may change in time, the total angular momentum may also vary in time. In fact, from Equations 10.34 and 10.35, we find that the time rate of change of the total angular momentum of the system equals the vector sum of *all* torques, including those associated with internal forces between particles and those associated with external forces.

As we found in our discussion of the rigid body under a net torque, however, the sum of the internal torques is zero. Thus, we conclude that the total angular momentum can vary with time *only* if there is a net *external* torque on the system, so that we have

- *Net external torque on a system equals time rate of change of angular momentum*

$$\sum \tau_{\text{ext}} = \sum_i \frac{d\mathbf{L}_i}{dt} = \frac{d}{dt} \sum_i \mathbf{L}_i = \frac{d\mathbf{L}_{\text{tot}}}{dt} \qquad \text{[10.37]}$$

That is, the time rate of change of the total angular momentum of the system about some origin in an inertial frame equals the net external torque acting on the system about that origin. Note that Equation 10.37 is the rotational analog of $\sum \mathbf{F}_{\text{ext}} = d\mathbf{p}_{\text{tot}}/dt$ (Eq. 8.39) for a system of particles.

This result is valid for a system of particles that change their positions with respect to one another, that is, a nonrigid body. In this discussion of angular momentum of a system of particles, notice that we never imposed the rigid body condition.

TABLE 10.3	A Comparison of Equations for Rotational and Translational Motion: Dynamic Equations[a]	
	Rotational Motion About a Fixed Axis	**Translational Motion**
Kinetic energy	$K_R = \frac{1}{2}I\omega^2$	$K = \frac{1}{2}mv^2$
Equilibrium	$\sum \tau = 0$	$\sum \mathbf{F} = 0$
Newton's second law	$\sum \tau = I\alpha$	$\sum \mathbf{F} = m\mathbf{a}$
Newton's second law	$\sum \tau = \dfrac{d\mathbf{L}}{dt}$	$\sum \mathbf{F} = \dfrac{d\mathbf{p}}{dt}$
Momentum	$L = I\omega$	$\mathbf{p} = m\mathbf{v}$
Conservation principle	$L_i = L_f$	$\mathbf{p}_i = \mathbf{p}_f$
Power	$\mathcal{P} = \tau\omega$	$\mathcal{P} = Fv$

[a] Equations in translation motion expressed in terms of vectors have rotational analogs in terms of vectors. Because the full vector treatment of rotation is beyond the scope of this book, however, some rotational equations are given in nonvector form.

Equation 10.37 is the primary equation in the angular momentum version of the nonisolated system model. The system's angular momentum changes in response to an interaction with the environment, described by means of the net torque on the system.

One final result can be obtained for angular momentum, which will serve as an analog to the definition of linear momentum. Let us imagine a rigid body rotating about an axis. Each particle of mass m_i in the rigid body moves in a circular path of radius r_i, with a tangential speed v_i. Thus, the total angular momentum of the rigid body is

$$L = \sum_i m_i v_i r_i$$

Let us now replace the tangential speed with the product of the radial distance and the angular speed (Eq. 10.10):

$$L = \sum_i m_i v_i r_i = \sum_i m_i (r_i \omega) r_i = \left(\sum_i m_i r_i^2 \right) \omega$$

We recognize the combination in the parentheses as the moment of inertia, so we can write the angular momentum of the rigid body as

$$L = I\omega$$

• *Angular momentum of an object with moment of inertia I*

This is the rotational analog to $p = mv$. Table 10.3 is a continuation of Table 10.1, with additional translational and rotational analogs that we have developed in the past few sections.

Example 10.12   The Atwood Machine Once Again

Consider again the Atwood machine with the massive pulley in Examples 10.10 and 10.11. Determine the acceleration of the two objects using an angular momentum approach.

Reasoning   This is an example of a nonrigid body experiencing a net torque, so we use the nonisolated system model. We will evaluate the angular momentum of the system at any time and then differentiate the angular momentum, setting it equal to the net external torque. We will solve the resulting expression for the acceleration of the objects.

Solution   Let us first calculate the angular momentum of the system, which consists of the two objects plus the pulley. At the instant m_1 and m_2 have a speed v, the angular momentum of m_1 around the axle of the pulley is $m_1 vR$, and that of m_2 is $m_2 vR$. At the same instant, the angular momentum of the pulley around its center is $L = I\omega = Iv/R$. Therefore, the total angular momentum of the system is

$$(1) \quad L = m_1 vR + m_2 vR + I\frac{v}{R} = m_1 vR + m_2 vR + (\tfrac{1}{2}MR^2)\frac{v}{R}$$

$$= (m_1 + m_2 + \tfrac{1}{2}M)vR$$

Now let us evaluate the total external torque on the system about the axle. The weight of the pulley and the force of the axle upward on the pulley have zero moment arm around the center of the pulley, so they do not contribute to the torque. The external forces on the system that produce torques about the axle are $m_1\mathbf{g}$, with a torque of $m_1 gR$, and $m_2\mathbf{g}$, with a torque of $-m_2 gR$. Combining the net external torque with (1) and Equation 10.37 gives us

$$\tau_{\text{ext}} = \frac{dL}{dt}$$

$$m_1 gR - m_2 gR = \frac{d}{dt}\left[(m_1 + m_2 + \tfrac{1}{2}M)vR\right]$$

$$(2) \quad m_1 gR - m_2 gR = (m_1 + m_2 + \tfrac{1}{2}M)R\frac{dv}{dt}$$

Because $dv/dt = a$, we can solve Equation (2) for a to find

$$a = \left(\frac{m_1 - m_2}{m_1 + m_2 + \tfrac{1}{2}M}\right)g$$

which is the same result as that obtained in Example 10.10. You may wonder why we did not include the tension forces that the cord exerts on the objects in evaluating the net torque about the axle. The reason is that these forces are *internal* to the system under consideration. Only the *external* torques contribute to the change in angular momentum.

EXERCISE   A car of mass 1 500 kg moves on a circular race track of radius 50 m with a speed of 40 m/s. What is the magnitude of its angular momentum relative to the center of the track?

Answer 3.0×10^6 kg·m²/s

EXERCISE   What is the angular momentum of the Earth due to its (a) revolution about the Sun, relative to the Sun, and (b) rotation about its axis, relative to the axis? Assume the Earth is a sphere of uniform density.

Answer   (a) 2.7×10^{40} kg·m²/s   (b) 7.1×10^{33} kg·m²/s

10.9 • CONSERVATION OF ANGULAR MOMENTUM

In Chapter 8 we found that the total linear momentum of a system of particles remains constant when the net external force acting on the system is zero. In rotational motion, we have an analogous conservation law that states that the **total angular momentum of a system remains constant if the net external torque acting on the system is zero.**

Because the net torque acting on the system equals the time rate of change of the system's angular momentum, we see that if

$$\sum \boldsymbol{\tau}_{\text{ext}} = \frac{d\mathbf{L}_{\text{tot}}}{dt} = 0 \qquad [10.38]$$

then

• *Conservation of angular momentum for an isolated system*

$$\mathbf{L}_{\text{tot}} = \text{constant} \quad \longrightarrow \quad \mathbf{L}_{\text{tot},\,i} = \mathbf{L}_{\text{tot},\,f} \qquad [10.39]$$

Equation 10.39 represents a third conservation law to add to our list of fundamental conservation principles. We can now state that the **total energy, linear momentum, and angular momentum of an isolated system are all conserved.** We have focused our attention in this chapter on rigid bodies; the conservation of angular momentum principle, however, is a general result of the isolated system model. Thus, **angular momentum of an isolated system is conserved whether the system is a rigid body or not.**

The angular momentum of a rigid body about a fixed axis has a magnitude given by $L = I\omega$, where I is the moment of inertia of the rigid body about the axis. In this case, if the net external torque on the body is zero, we can express the conservation of angular momentum principle as $I\omega =$ constant.

Many examples can be used to demonstrate conservation of angular momentum; some of them should be familiar to you. You may have observed a figure skater spinning (Fig. 10.24). The angular speed of the skater increases as he pulls his hands and feet close to the trunk of his body. Ignoring friction between skater and ice, we see that there are no external torques on the skater. The moment of inertia of his body decreases as his hands and feet are brought in closer to his body.* The resulting change in angular speed is accounted for as follows. Because angular momentum must be conserved, the product $I\omega$ remains constant, and a decrease in I causes a corresponding increase in ω.

An interesting astrophysical example of conservation of angular momentum occurs when, at the end of its lifetime, a massive star uses up all its fuel and collapses under the influence of gravitational forces, causing a gigantic outburst of energy called a supernova explosion. The best studied example of a remnant of a supernova explosion is the Crab Nebula, a chaotic, expanding mass of gas (Fig. 10.25). In a supernova, part of the star's mass is released into space, where it

Figure 10.24

Angular momentum is conserved as figure skater Todd Eldridge pulls his arms toward his body. *(© 1998 David Madison)*

Figure 10.25

The Crab Nebula, in the constellation Taurus. This nebula is the remnant of a supernova explosion, which was seen on Earth in the year A.D. 1054. It is located some 6 300 lightyears away and is approximately 6 lightyears in diameter, still expanding outward. *(Max Planck Institute for Astronomy and Calar Alto Observatory; K. Meisenheimer and A. Quetz)*

WEB

For detailed Hubble Space Telescope images of the Crab Nebula, visit **www.seds.org/messier/more/moo1_hst.html**

---

* We derived $L = I\omega$ for a rigid body. The skater is deformable, so is therefore not a rigid body. On the other hand, we apply $L = I\omega$ only at the instant before he pulls his arms in and the instant after he finishes pulling his arms in. At these two instants, no deformation occurs and he can be modeled as a rigid body.

eventually condenses into new stars and planets. Most of what is left behind typically collapses into a **neutron star,** an extremely dense sphere of matter with a diameter of about 10 km in comparison with the 10^6-km diameter of the original star, and containing a large fraction of the original mass of the star. As the moment of inertia of the system decreases during the collapse, the star's rotational speed increases, similar to the increase in speed of the skater in Figure 10.24. More than 700 rapidly rotating neutron stars have been identified since the first discovery of such astronomical bodies in 1967, with periods of rotation ranging from a millisecond to several seconds. The neutron star is a most dramatic system—an object with a mass greater than the Sun, rotating about its axis many times each second!

Quick Quiz 10.10

A skater moves along a straight line on frictionless ice and then grabs hold of a pole in the middle of the ice, so that she rotates around the pole. This represents a change from translational motion to rotational motion. Is the angular momentum of the skater conserved in this process?

Example 10.13 A Revolving Ball on a Horizontal, Frictionless Surface

A ball of mass m on a horizontal, frictionless table is connected to a string that passes through a small hole in the table. The ball is set into circular motion of radius R, at which time its speed is v_i (Fig. 10.26). (a) If the string is pulled from the bottom so that the radius of the circular path is decreased to r, what is the final speed v_f of the ball?

Solution We identify the system as the ball. We will calculate torque about the center of rotation O. Note that the gravitational force acting on the ball is balanced by the up-

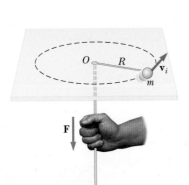

Figure 10.26

(Example 10.13) When the string is pulled downward, the speed of the ball changes.

ward normal force, and so these forces cancel, resulting in zero net torque from these forces. The force **F** of the string on the ball acts toward the center of rotation, and the vector position **r** is directed away from O. Thus, we see that $\boldsymbol{\tau} = \mathbf{r} \times \mathbf{F} = 0$, so no torque is applied on the ball due to this force. Three forces are acting on the ball, but zero net torque occurs. Thus, **L** is a constant of the motion. The ball can be modeled as a particle moving in a circular path, so $L = mv_iR = mv_fr$, or

$$v_f = \frac{v_iR}{r}$$

From this result, we see that as r decreases, the speed v increases.

(b) Is the kinetic energy of the ball conserved in this process?

Solution We set up the ratio of the final kinetic energy to the initial kinetic energy:

$$\frac{K_f}{K_i} = \frac{\frac{1}{2}mv_f^2}{\frac{1}{2}mv_i^2} = \frac{1}{v_i^2}\left(\frac{v_iR}{r}\right)^2 = \frac{R^2}{r^2}$$

Because this ratio is not equal to 1, kinetic energy is not conserved. Furthermore, because $R > r$, the kinetic energy of the ball has increased. This increase entered the system of the ball by the work done by the person pulling the string.

Example 10.14   Rotation Period of a Neutron Star

A star undergoes a supernova explosion. The material left behind forms a sphere of radius 8.0×10^6 m just after the explosion with a rotation period of 15 h. This remaining material collapses into a neutron star of radius 8.0 km. What is the rotation period T of the neutron star?

Solution   We model the star as an isolated system, so that angular momentum is conserved. As the moment of inertia of the stellar core decreases during its collapse, the angular speed increases. Thus,

$$I_i \omega_i = I_f \omega_f$$

We have no information about the variation in the density of material with radius in either the initial star or the neutron star, so we choose a simplification model in which the density

is uniform. Our result will most likely not be entirely accurate, but we can consider it an estimate. Using the moment of inertia of a sphere of uniform density (see Table 10.2),

$$(\tfrac{2}{5} M R_i{}^2)\omega_i = (\tfrac{2}{5} M R_f{}^2)\omega_f \quad \longrightarrow \quad \omega_f = \left(\frac{R_i}{R_f}\right)^2 \omega_i$$

Now, because $\omega = 2\pi/T$, we have

$$\frac{2\pi}{T_f} = \left(\frac{R_i}{R_f}\right)^2 \frac{2\pi}{T_i} \quad \longrightarrow \quad T_f = \left(\frac{R_f}{R_i}\right)^2 T_i$$

Substituting numerical values, we have

$$T_f = \left(\frac{8.0 \times 10^3 \text{ m}}{8.0 \times 10^6 \text{ m}}\right)^2 (15 \text{ h}) = 1.5 \times 10^{-5} \text{ h} = \boxed{0.054 \text{ s}}$$

10.10 • PRECESSIONAL MOTION OF GYROSCOPES

Angular momentum is the basis of the operation of a **gyroscope,** which is a spinning object used to control or maintain the orientation in space of the object or a system containing the object. As an example, consider a quarterback passing a football. If he imparts no spin to the ball, there is no angular momentum to be conserved, and forces from the air might cause the ball to tumble as it moves through its trajectory. If a spin is imparted to the ball along the long axis of the football, however, the angular momentum vector stays fixed in direction and the football maintains its orientation throughout the trajectory, resulting in much less air resistance and a longer pass. In this application, the football is acting as a gyroscope to maintain its own orientation in space.

An unusual and fascinating type of motion you probably have observed is that of a top spinning rapidly about its axis of symmetry, as shown in Figure 10.27a. The top is acting as a gyroscope and one might expect the orientation to remain fixed in space. If the top is leaning over, however, it is observed that the symmetry axis rotates about the z axis, sweeping out a cone (see Fig. 10.27b). This phenomenon is called **precessional motion.** The angular speed of the symmetry axis about the vertical is usually slow relative to the angular speed of the top about the symmetry axis.

It is quite natural to wonder why the top does not maintain its direction of spin. Because the center of mass of the top is not directly above the pivot point O, a net torque is acting on the top about O—a torque resulting from the gravitational force $M\mathbf{g}$. The top would certainly fall over if it were not spinning. Because it is spinning, however, it has an angular momentum $\mathbf{L}$ directed along its symmetry axis. As we shall show, the motion of this symmetry axis about the z axis (the precessional motion) occurs because the torque produces a change in the *direction* of the symmetry axis. This is an excellent example of the importance of the directional nature of angular momentum.

The two forces acting on the top are the downward gravitational force $M\mathbf{g}$ and the normal force $\mathbf{n}$ acting upward at the pivot point O. The normal force produces no torque about the pivot because its moment arm through that point is zero. However, the gravitational force produces a torque $\boldsymbol{\tau} = \mathbf{r} \times M\mathbf{g}$ about O, where

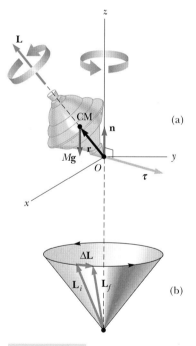

Figure 10.27

Precessional motion of a top spinning about its symmetry axis. (a) The only external forces acting on the top are the normal force $\mathbf{n}$ and the gravitational force $M\mathbf{g}$. The direction of the angular momentum $\mathbf{L}$ is along the axis of symmetry. The right-hand rule indicates that $\boldsymbol{\tau} = \mathbf{r} \times \mathbf{F} = \mathbf{r} \times M\mathbf{g}$ is in the xy plane. (b). The direction of $\Delta\mathbf{L}$ is parallel to that of $\boldsymbol{\tau}$ in part (a). The fact that $\mathbf{L}_f = \mathbf{L}_i + \Delta\mathbf{L}$ indicates that the top precesses about the z axis.

the direction of $\boldsymbol{\tau}$ is perpendicular to the plane formed by **r** and $M\mathbf{g}$. By necessity, the vector $\boldsymbol{\tau}$ lies in a horizontal plane perpendicular to the angular momentum vector. The net torque and angular momentum of the top are related through Equation 10.37:

$$\sum \boldsymbol{\tau} = \frac{d\mathbf{L}}{dt}$$

From this expression, we see that the nonzero torque produces a change in angular momentum $d\mathbf{L}$—a change that is in the same direction as $\sum \boldsymbol{\tau}$. Therefore, like the torque vector, $d\mathbf{L}$ must also be perpendicular to **L**. Figure 10.27b illustrates the resulting precessional motion of the symmetry axis of the top. In a time Δt, the change in angular momentum is $\Delta \mathbf{L} = \mathbf{L}_f - \mathbf{L}_i = \sum \boldsymbol{\tau} \, \Delta t$. Because $\Delta \mathbf{L}$ is perpendicular to **L**, the magnitude of **L** does not change ($|\mathbf{L}_i| = |\mathbf{L}_f|$). Rather, what is changing is the *direction* of **L**. Because the change in angular momentum $\Delta \mathbf{L}$ is in the direction of $\sum \boldsymbol{\tau}$, which lies in the xy plane, the top undergoes precessional motion.

With careful manufacturing tolerances, precession due to gravitational torque can be made very small and gyroscopes can be used for guidance systems in vehicles—a change in the direction of the velocity of a vehicle is detected as a change between the direction of the angular momentum of the gyroscope and a reference direction attached to the vehicle. With proper electronic feedback, the deviation from the desired direction of motion can be removed, bringing the angular momentum back in line with the reference direction. Precession rates for highly specialized military gyroscopes are as low as 0.02° per day.

10.11 • ROLLING MOTION OF RIGID BODIES

In this section we shall investigate the special case of rotational motion in which a round object rolls on a surface. Many everyday examples exist for such motion, including automobile tires rolling on roads and bowling balls rolling toward the pins.

Suppose a cylinder is rolling on a straight path, as in Figure 10.28. The center of mass moves in a straight line, but a point on the rim moves in a more complex path called a *cycloid*. Let us further assume that the cylinder of radius R is uniform

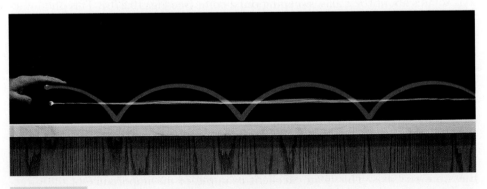

Figure 10.28

Light sources at the center and rim of a rolling cylinder illustrate the different paths these points take. The center moves in a straight line *(green line),* while a point on the rim moves in the path of a cycloid *(red curve). (Henry Leap and Jim Lehman)*

and rolls on a surface with friction. We make a rather odd, but valid, simplification model here for rolling objects. The surfaces must exert friction forces on each other; otherwise the cylinder would simply slide rather than roll. If the friction force on the cylinder is large enough, the cylinder rolls without slipping. In this situation, the friction force is static rather than kinetic because the contact point of the cylinder with the surface is at rest relative to the surface at any instant. The static friction force acts through no displacement, so that it does no work on the cylinder and causes no decrease in mechanical energy of the cylinder. In real rolling objects, deformations of the surfaces result in some rolling resistance. If both surfaces are hard, however, they will deform very little, and rolling resistance can be negligibly small. Thus, we can model the rolling motion as maintaining constant mechanical energy. The wheel was a great invention!

As the cylinder rotates through an angle θ, its center of mass moves a distance of $s = r\theta$. Therefore, the speed and acceleration of the center of mass for **pure rolling motion** are

$$v_{CM} = \frac{ds}{dt} = R\frac{d\theta}{dt} = R\omega \qquad \text{[10.40]}$$

$$a_{CM} = \frac{dv_{CM}}{dt} = R\frac{d\omega}{dt} = R\alpha \qquad \text{[10.41]}$$

• *Relations between translational and rotational variables for a rolling object*

The translational velocities of various points on the rolling cylinder are illustrated in Figure 10.29. Note that the translational velocity of any point is in a direction perpendicular to the line from that point to the contact point. At any instant, the point P is at rest relative to the surface because sliding does not occur.

We can express the total kinetic energy of a rolling object of mass M and moment of inertia I as the combination of the rotational kinetic energy around the center of mass plus the translational kinetic energy of the center of mass:

$$K = \tfrac{1}{2}I_{CM}\omega^2 + \tfrac{1}{2}Mv_{CM}^2 \qquad \text{[10.42]}$$

• *Total kinetic energy of a rolling object*

A useful theorem called the **parallel axis theorem** enables us to express this energy in terms of the moment of inertia I_p through any axis parallel to the axis through the center of mass of an object. This theorem states that

$$I_p = I_{CM} + MD^2 \qquad \text{[10.43]}$$

where D is the distance from the center-of-mass axis to the parallel axis, and M is the total mass of the object. Let us use this theorem to express the moment of inertia around the contact point P between the rolling object and the surface. The distance from this point to the center of mass of the symmetric object is its radius, so

$$I_p = I_{CM} + MR^2$$

If we write the translational speed of the center of mass of the object in Equation 10.42 in terms of the angular speed, we have

$$K = \tfrac{1}{2}I_{CM}\omega^2 + \tfrac{1}{2}MR^2\omega^2 = \tfrac{1}{2}(I_{CM} + MR^2)\omega^2 = \tfrac{1}{2}I_p\omega^2 \qquad \text{[10.44]}$$

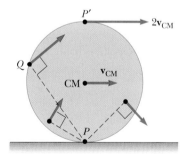

Figure 10.29

All points on a rolling object move in a direction perpendicular to an axis through the instantaneous point of contact P. The center of the object moves with a velocity $\mathbf{v}_{CM}$, whereas the point P' moves with a velocity $2\mathbf{v}_{CM}$.

Thus, the kinetic energy of the rolling object can be considered as equivalent to a purely rotational kinetic energy of the object rotating around its contact point.

We can use the energy version of the isolated system model to treat a class of problems concerning the rolling motion of a rigid body down a rough incline. In these types of problems, gravitational potential energy of the body–Earth system decreases as the rotational and translational kinetic energies of the body increase.

As this old-style bicycle rolls along a roadway, the centers of mass of both wheels travel with the same translational speed. Which wheel has the higher angular speed? *(© Steve Lovegrove/Tasmanian Photo Library)*

Quick Quiz 10.11

Which arrives at the bottom first: a ball rolling without sliding down incline A or a box of the same mass as the ball sliding down a frictionless incline B having the same dimensions as incline A?

THINKING PHYSICS 10.4

Consider the wheel in Figure 10.29 to be the tire of a truck, moving to the right, with pieces of mud stuck between the treads. As the mud becomes dislodged from the treads and free of the tire, it is thrown in the direction of the tangential velocity vectors shown in Figure 10.29. All of these vectors have a component in the forward direction, that is, in the same direction as the truck is moving. So why do trucks have mud flaps *behind* the tires?

Reasoning   Consider a piece of mud released from the tire just to the left of point *P* in Figure 10.29. Although this piece of mud does have a component of velocity in the forward direction, it is smaller than v_{CM}. Thus, the tire will move horizontally more rapidly than the mud. Relative to the truck, then, the piece of mud moves *backward*. Without a mud flap, this piece of mud could strike a car driving behind the truck.

Example 10.15   Sphere Rolling Down an Incline

If the object in Figure 10.30 is a solid sphere, (a) calculate the speed of its center of mass at the bottom and (b) determine the magnitude of the translational acceleration of the center of mass.

Solution   (a) We shall consider the sphere and the Earth as an isolated system and use the energy version of the isolated system model. The energy of the system when the sphere is at the top of the incline is gravitational potential energy only. We choose the zero of gravitational potential energy to be when the sphere is at the bottom of the incline. Thus, conservation of mechanical energy gives us

$$K_i + U_i = K_f + U_f$$
$$0 + Mgh = (\tfrac{1}{2}Mv_{CM,f}^2 + \tfrac{1}{2}I_{CM}\omega_f^2) + 0$$

Using Equation 10.40 to relate the translational and angular speeds, and substituting the moment of inertia for a sphere, we have

$$Mgh = \tfrac{1}{2}Mv_{CM,f}^2 + \tfrac{1}{2}\left(\tfrac{2}{5}MR^2\right)\frac{v_{CM,f}^2}{R^2}$$
$$= \tfrac{1}{2}Mv_{CM,f}^2 + \tfrac{1}{5}Mv_{CM,f}^2$$
$$= \tfrac{7}{10}Mv_{CM,f}^2 \quad\longrightarrow\quad v_{CM,f} = \sqrt{\tfrac{10}{7}gh}$$

(b) To find the acceleration, let us recognize that the constant gravitational force should cause a constant acceleration of the center of mass of the sphere. From Equation 2.12,

$$v_{CM,f}^2 = v_{CM,i}^2 + 2a_{CM}(x_{CM,f} - x_{CM,i})$$

we can solve for the acceleration

$$a_{CM} = \frac{v_{CM,f}^2 - v_{CM,i}^2}{2(x_{CM,f} - x_{CM,i})} = \frac{\tfrac{10}{7}gh - 0}{2\left(\dfrac{h}{\sin\theta}\right)} = \tfrac{5}{7}g\sin\theta$$

These results are quite interesting in that both the speed and the acceleration of the center of mass are independent of the mass and radius of the sphere. That is, *all* homogeneous solid spheres experience the same speed and acceleration on a given incline!

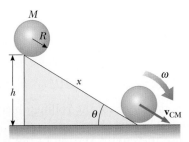

Figure 10.30

A round object rolling down an incline. Mechanical energy of the object–Earth system is conserved if no slipping occurs and there is no rolling resistance.

If we repeated the calculations for a hollow sphere, a solid cylinder, or a hoop, we would obtain similar results with different numerical factors appearing in the expressions for $v_{CM,f}$ and a_{CM}. These factors depend only on the moment of inertia about the center of mass for the specific object. In all cases, the acceleration of the center of mass is less than $g \sin \theta$, the value it would have if the plane were frictionless and no rolling occurred.

context connection
10.12 • GYROSCOPES IN SPACE

Let us follow up on the gyroscope discussion of Section 10.10. Our technological society offers several examples of gyroscopes. We shall focus on just three such examples that relate to our Context. First, let us imagine that an additional assignment before we set out for Mars is to place a satellite in orbit around the Earth. If you simply hold the stationary satellite outside the orbiting spacecraft and let it go, you will inevitably impart some rotation to it when you let it go, and it will spin according to this influence. If it happens to catch on your spacesuit sleeve just as you release it, it could go into an uncontrolled tumble. This is undesirable. To stabilize the satellite as you let it go, it is more desirable to set it into rotation, so that the satellite has an angular momentum (Fig. 10.31). Then, the direction of the angular momentum vector can be determined before you release the satellite, and this direction will remain fixed according to the principle of conservation of angular momentum. The satellite acts as a gyroscope, and its spin remains stable. The space shuttle has a turntable on which satellites are placed and set into rotation before being released. We would want such a turntable on our spacecraft also.

Our second example is related to the Context Connection discussed in Chapter 4, in which we determined the direction in which to fire our rocket engines so as to return to our desired trajectory. We needed to turn the spacecraft around in

Figure 10.31

A spinning satellite is released from the space shuttle. The gyroscopic action due to its spin keeps its rotation axis fixed in direction. *(NASA)*

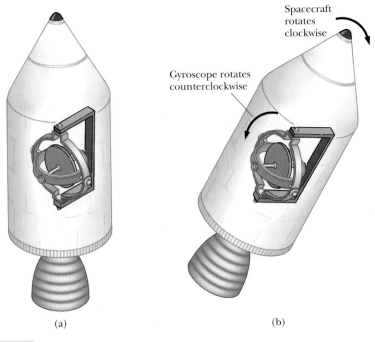

Spacecraft rotates clockwise

Gyroscope rotates counterclockwise

(a)

(b)

Figure 10.32

(a) A spacecraft carries a gyroscope that is not spinning. (b) When the gyroscope is set into rotation, the spacecraft turns the other way so that the angular momentum of the system is conserved.

order to fire the engines in the correct direction. But how do you turn a spacecraft around in empty space? One way is to have small rocket engines that fire perpendicularly out the side of the spacecraft, providing a torque around its center of mass. This is desirable, and many spacecraft have such rockets.

Let us consider, however, another possibility related to angular momentum. Suppose the spacecraft carries a gyroscope that is not rotating, as in Figure 10.32a. In this case, the angular momentum of the spacecraft about its center of mass is zero. Suppose the gyroscope is set into rotation. Now, it would appear that the spacecraft system has a nonzero angular momentum, due to the rotation of the gyroscope. But there is no external torque on the system, so the angular momentum of the isolated system must remain zero according to the principle of conservation of angular momentum. This principle can be satisfied by realizing that the spacecraft will turn in the direction opposite to that of the gyroscope, so that the angular momentum vectors of the gyroscope and the spacecraft cancel, resulting in no angular momentum of the system. The result of rotating the gyroscope, as in Figure 10.32b, is that the spacecraft turns around! By including three gyroscopes with mutually perpendicular axles, any desired rotation in space can be achieved. Once the desired gravitation is achieved, the rotation of the gyroscope is halted.

This effect occurred in an undesirable situation with the *Voyager* 2 spacecraft during its flight. The spacecraft carried a tape recorder whose reels rotated at high speeds. Each time the tape recorder was turned on, the reels acted as gyroscopes, and the spacecraft started an undesirable rotation in the opposite direction. This had to be counteracted by Mission Control by using the sideward-firing jets to stop the rotation!

For our third example of gyroscopic motion, let us not look at our spacecraft, but at the Earth. The spinning Earth acts as a gyroscope, and we normally consider

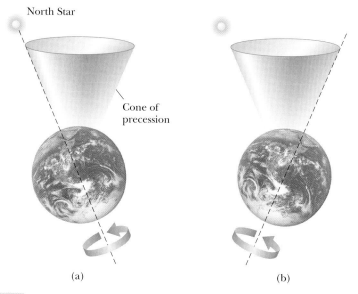

North Star

Cone of precession

(a)   (b)

Figure 10.33

(a) At present, the spin axis of the Earth points toward the North Star. (b) Torque on the spinning Earth will cause it to precess, so that the spin axis will no longer be pointing in this direction in the future.

the rotation axis to be fixed in space, pointing at a location near the North Star. In our first two examples, we applied the principle of conservation of angular momentum because the satellite and the gyroscope–spacecraft system had no external torque on them. We might guess that the Earth rotates torque-free also. Here is a case, however, in which a simplification model of the Earth as a perfect sphere works well over relatively short time periods, but not over a long one. In reality, the Earth is not a perfect sphere. The radius of the Earth is larger at the equator than at the poles, due to the bulging of the Earth at the equator, which is a consequence of its rotation. We could geometrically model the Earth as a perfect sphere with an extra belt of mass around the equator. The gravitational force of the Sun and Moon on this extra mass results in a torque on the Earth.

As discussed in Section 10.10, the torque on a spinning object will cause precession—the angular momentum vector moves along the surface of a cone (Figure 10.33). The time required for the axis to spin around once and trace out a complete cone depends on several factors, including the amount of torque. The torque that the Sun and Moon exert on the Earth is relatively small for an object as massive as the Earth, so that the period for the precession is 2.6×10^4 years. Thus, 13 000 years from now, the axis of the Earth will not be pointing at the North Star, and 13 000 years later, it will return to the direction it has at present.

SUMMARY

The **instantaneous angular speed** of a particle rotating in a circle or of a rigid body rotating about a fixed axis is

$$\omega = \frac{d\theta}{dt} \qquad \text{[10.3]}$$

where ω is in rad/s or in s^{-1}.

The **instantaneous angular acceleration** of a particle rotating in a circle or of a rigid body rotating about a fixed axis is

$$\alpha = \frac{d\omega}{dt} \qquad \text{[10.5]}$$

and has units of rad/s^2 or s^{-2}.

When a rigid body rotates about a fixed axis, every part of the body has the same angular speed and the same angular acceleration. However, different parts of the body, in general, have different translational speeds and different translational accelerations.

If a particle (or object) undergoes rotational motion about a fixed axis under constant angular acceleration α, one can apply equations of kinematics by analogy with kinematic equations for translational motion with constant translational acceleration:

$$\omega_f = \omega_i + \alpha t \qquad [10.6]$$

$$\theta_f = \theta_i + \omega_i t + \tfrac{1}{2}\alpha t^2 \qquad [10.7]$$

$$\omega_f^2 = \omega_i^2 + 2\alpha(\theta_f - \theta_i) \qquad [10.8]$$

$$\theta_f = \theta_i + \tfrac{1}{2}(\omega_i + \omega_f)t \qquad [10.9]$$

When a particle rotates about a fixed axis, the angular speed and angular acceleration are related to the tangential speed and tangential acceleration through the relationships

$$v = r\omega \qquad [10.10]$$

$$a_t = r\alpha \qquad [10.11]$$

The **moment of inertia** of a system of particles is

$$I = \sum_i m_i r_i^2 \qquad [10.14]$$

If a rigid body rotates about a fixed axis with angular speed ω, its **rotational kinetic energy** can be written

$$K_R = \tfrac{1}{2}I\omega^2 \qquad [10.15]$$

where I is the moment of inertia about the axis of rotation.

The moment of inertia of a continuous object of density ρ is

$$I = \int \rho r^2 dV \qquad [10.17]$$

The **torque** τ due to a force $\mathbf{F}$ about an origin in an inertial frame is defined to be

$$\boldsymbol{\tau} \equiv \mathbf{r} \times \mathbf{F} \qquad [10.19]$$

where $\mathbf{r}$ is the position vector of the point of application of the force.

Given two vectors $\mathbf{A}$ and $\mathbf{B}$, their **vector product** or **cross product** $\mathbf{A} \times \mathbf{B}$ is a vector $\mathbf{C}$ having the magnitude

$$C \equiv AB \sin \theta \qquad [10.21]$$

where θ is the angle between $\mathbf{A}$ and $\mathbf{B}$. The direction of $\mathbf{C}$ is perpendicular to the plane formed by $\mathbf{A}$ and $\mathbf{B}$, and is determined by the right-hand rule.

The net torque acting on an object is proportional to the angular acceleration of the object, and the proportionality constant is the moment of inertia I:

$$\sum \tau = I\alpha \qquad [10.27]$$

The **angular momentum L** of a particle with linear momentum $\mathbf{p} = m\mathbf{v}$ is

$$\mathbf{L} \equiv \mathbf{r} \times \mathbf{p} \qquad [10.32]$$

where $\mathbf{r}$ is the vector position of the particle relative to the origin. If ϕ is the angle between $\mathbf{r}$ and $\mathbf{p}$, the magnitude of $\mathbf{L}$ is

$$L = mvr \sin \phi \qquad [10.33]$$

The net external torque acting on a system is equal to the time rate of change of its angular momentum:

$$\sum \boldsymbol{\tau}_{\text{ext}} = \frac{d\mathbf{L}_{\text{tot}}}{dt} \qquad [10.37]$$

The law of conservation of angular momentum states that the total angular momentum of a system remains constant if the net external torque acting on the system is zero:

$$\mathbf{L}_{\text{tot}, i} = \mathbf{L}_{\text{tot}, f} \qquad [10.39]$$

The **total kinetic energy** of a rigid body, such as a cylinder, that is rolling on a rough surface without slipping equals the rotational kinetic energy $\tfrac{1}{2}I_{\text{CM}}\omega^2$ about the body's center of mass plus the translational kinetic energy $\tfrac{1}{2}Mv_{\text{CM}}^2$ of the center of mass:

$$K = \tfrac{1}{2}I_{\text{CM}}\omega^2 + \tfrac{1}{2}Mv_{\text{CM}}^2 \qquad [10.42]$$

In this expression, v_{CM} is the speed of the center of mass and $v_{\text{CM}} = R\omega$ for pure rolling motion.

QUESTIONS

1. In a tape recorder, the tape is pulled past the read and write heads at a constant speed by the drive mechanism. Consider the reel from which the tape is pulled—as the tape is pulled off it, the radius of the roll of remaining tape decreases. How does the torque on the reel change with time? How does the angular speed of the reel change with time? If the tape mechanism is suddenly turned on so that the tape is quickly pulled with a large force, is the tape more likely to break when pulled from a nearly full reel or a nearly empty reel?

2. If a car's wheels are replaced with wheels of a larger diameter, will the reading of the speedometer change? Explain.

3. What is the magnitude of the angular velocity $\boldsymbol{\omega}$ of the second hand of a clock? What is the direction of $\boldsymbol{\omega}$ as you

view a clock hanging vertically? What is the magnitude of the angular acceleration α of the second hand?

4. If you see an object rotating, does a net torque necessarily act on it?

5. Suppose a pencil is balanced on a perfectly frictionless table. If it falls over, what is the path followed by the center of mass of the pencil?

6. For a helicopter to be stable as it flies, it must have two propellers. Why?

7. In some motorcycle races, the riders drive over small hills, and the motorcycle becomes airborne for a short time. If the motorcycle racer keeps the throttle open while leaving the hill and going into the air, the motorcycle tends to nose upward. Why does this happen?

8. The moment of inertia of an object depends on the choice of rotation axis, as suggested by the parallel axis theorem. Argue that an axis passing through the center of mass of an object must be the axis with the smallest moment of inertia.

9. Suppose you remove two eggs from the refrigerator, one hard-boiled and the other uncooked. You wish to determine which is the hard-boiled egg without breaking the eggs. This can be done by spinning the two eggs on the floor and comparing the rotational motions. Which egg spins faster? Which rotates more uniformly? Explain.

10. A ladder rests inclined against a wall. Would you feel safer climbing up the ladder if you were told that the floor is frictionless but the wall is rough, or that the wall is frictionless but the floor is rough? Justify your answer.

11. Often when a high diver wants to flip in midair, she will draw her legs up against her chest. Why does this make her rotate faster? What should she do when she wants to stop rotating?

12. If the net force acting on a system is zero, then is it necessarily true that the net torque on it is also zero?

13. Why do tightrope walkers carry a long pole to help balance themselves?

14. Two uniform solid spheres are rolled down a hill: a large, massive sphere and a small sphere with low mass. Which one reaches the bottom of the hill first? Next, we roll a large, low-density sphere, and a small high-density sphere, and both spheres have the same mass. Which one wins in this case?

15. Suppose you are designing a car for a coasting race. The cars in this race have no engines; they simply coast down a hill. Do you want large wheels or small wheels? Do you want solid, disk-like wheels or hoop-like wheels? Should the wheels be heavy or light?

16. Consider an object in the shape of a hoop lying in the *xy* plane with all of its mass concentrated on its rim. In two separate experiments, the hoop is rotated by an external agent from rest to an angular speed ω. In one experiment, the rotation occurs about the *z* axis through the center of the hoop. In the other experiment, the rotation occurs about an axis parallel to *z* passing through a point *P* on the rim of the hoop. Which rotation requires more work?

17. Vector **A** is in the negative *y* direction, and vector **B** is in the negative *x* direction. What are the directions of (a) **A** × **B** and (b) **B** × **A**?

18. If global warming occurs over the next century, it is likely that some polar ice will melt and the water will be distributed closer to the equator. How would this change the moment of inertia of the Earth? Would the length of the day (one revolution) increase or decrease?

PROBLEMS

1, 2, 3 = straightforward, intermediate, challenging   □ = full solution available in the *Student Solutions Manual and Study Guide*

web = solution posted at **http://www.harcourtcollege.com/physics/**   🖥 = computer useful in solving problem

📲 = Interactive Physics   ▓ = paired numerical/symbolic problems   📐 = life science application

Section 10.1 Angular Speed and Angular Acceleration

Section 10.2 Rotational Kinematics: The Rigid Body Under Constant Angular Acceleration

1. A motor rotating a grinding wheel at 100 rev/min is web switched off. Assuming constant negative angular acceleration of magnitude 2.00 rad/s^2, (a) how long does it take the wheel to stop? (b) Through how many radians does it turn while it is slowing down?

2. An airliner arrives at the terminal, and its engines are shut off. The rotor of one of the engines has an initial clockwise angular speed of 2 000 rad/s. The engine's rotation slows with an angular acceleration of magnitude 80.0 rad/s^2. (a) Determine the angular speed after 10.0 s. (b) How long does it take the rotor to come to rest?

3. A dentist's drill starts from rest. After 3.20 s of constant angular acceleration, the drill turns at a rate of 2.51 × 10^4 rev/min. (a) Find the drill's angular acceleration. (b) Determine the angle (in radians) through which the drill rotates during this period.

4. The angular position of a swinging door is described by $\theta = 5.00 + 10.0t + 2.00t^2$ rad. Determine the angular position, angular speed, and angular acceleration of the door (a) at $t = 0$ and (b) at $t = 3.00$ s.

5. The tub of a washer goes into its spin cycle, starting from rest and gaining angular speed uniformly for 8.00 s, at which time it is turning at 5.00 rev/s. At this point the person doing the laundry opens the lid, and a safety switch turns off the washer. The tub smoothly slows to rest in 12.0 s. Through how many revolutions does the tub turn while it is in motion?

6. A rotating wheel requires 3.00 s to rotate 37.0 rev. Its angular speed at the end of the 3.00-s interval is 98.0 rad/s. What is the constant angular acceleration of the wheel?

7. (a) Find the angular speed of the Earth's rotation on its axis. As the Earth turns toward the east, we see the sky turning toward the west at this same rate.
   (b) *The rainy Pleiads wester*
       And seek beyond the sea
       The head that I shall dream of
         That shall not dream of me.
         —A. E. Housman (© Robert E. Symons)
   Cambridge, England, is at longitude 0° and Saskatoon, Saskatchewan, is at longitude 107° west. How much time elapses after the Pleiades set in Cambridge until these stars fall below the western horizon in Saskatoon?

Section 10.3   Relations Between Rotational and Translational Quantities

8. Make an order-of-magnitude estimate of the number of revolutions through which a typical automobile tire turns in 1 yr. State the quantities you measure or estimate and their values.

9. A disk 8.00 cm in radius rotates about its central axis at a constant rate of 1 200 rev/min. Determine (a) its angular speed, (b) the tangential speed at a point 3.00 cm from its center, (c) the radial acceleration of a point on the rim, and (d) the total distance a point on the rim moves in 2.00 s.
web

10. A digital audio compact disc carries data, each bit of which occupies 0.6 μm along a continuous spiral track from the inner circumference of the disc to the outside edge. A CD player turns the disc to carry the track counterclockwise above a lens at a constant speed of 1.30 m/s. Find the required angular speed (a) at the beginning of the recording, where the spiral has a radius of 2.30 cm, and (b) at the end of the recording, where the spiral has a radius of 5.80 cm. (c) A full-length recording lasts for 74 min 33 s. Find the average angular acceleration of the disc. (d) Assuming that the acceleration is constant, find the total angular displacement of the disc as it plays. (e) Find the total length of the track.

11. A car accelerates uniformly from rest and reaches a speed of 22.0 m/s in 9.00 s. If the diameter of a tire is 58.0 cm, find (a) the number of revolutions the tire makes during this motion, assuming that no slipping occurs. (b) What is the final rotational speed of a tire in revolutions per second?

12. A wheel 2.00 m in diameter lies in a vertical plane and rotates with a constant angular acceleration of 4.00 rad/s². The wheel starts at rest at $t = 0$, and the radius vector of a certain point P on the rim makes an angle of 57.3° with the horizontal at this time. At $t = 2.00$ s, find (a) the angular speed of the wheel, (b) the tangential speed and the total acceleration of the point P, and (c) the angular position of the point P.

Section 10.4   Rotational Kinetic Energy

13. Rigid rods of negligible mass lying along the y axis connect three small particles (Fig. P10.13). If the system rotates about the x axis with an angular speed of 2.00 rad/s, find (a) the moment of inertia about the x axis and the total rotational kinetic energy evaluated from $\frac{1}{2}I\omega^2$ and (b) the tangential speed of each particle and the total kinetic energy evaluated from $\sum_i \frac{1}{2}m_i v_i^2$.

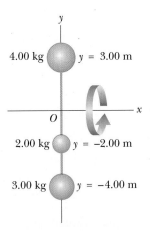

Figure P10.13

14. Big Ben, the Parliament tower clock in London, shown in Figure P10.14, has an hour hand 2.70 m long with a mass of 60.0 kg, and a minute hand 4.50 m long with a mass of 100 kg. Calculate the total rotational kinetic energy of the two hands about the axis of rotation. (You may model the hands as long thin uniform rods.)

Figure P10.14   Problems 14, 40, and 71. (*John Lawrence/Stone*)

15. This problem describes one experimental method of determining the moment of inertia of an irregularly shaped object such as the payload for a satellite. Figure P10.15 shows a cylinder of mass m suspended by a cord that is wound around a spool of radius r, forming part of a turntable supporting the object. When the cylinder is released from rest, it descends through a distance h, acquiring a speed v. Show that the moment of inertia I of the equipment (including the turntable) is $mr^2(2gh/v^2 - 1)$.

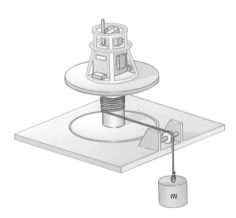

Figure P10.15

16. Consider two objects with $m_1 > m_2$ connected by a light string that passes over a pulley having a moment of inertia of I about its axis of rotation, as in Figure P10.16. The string does not slip on the pulley. The pulley turns without friction. The objects are released from rest separated by a vertical distance $2h$. Use the principle of conservation of energy to find the translational speeds of the objects as they pass each other. Find the angular speed of the pulley at this time.

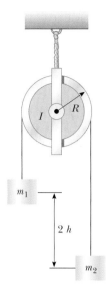

Figure P10.16

Section 10.6   Torque and the Vector Product

17. Find the net torque on the wheel in Figure P10.17 about the axle through O if $a = 10.0$ cm and $b = 25.0$ cm.

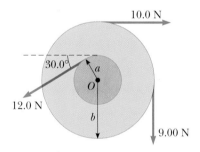

Figure P10.17

18. The fishing pole in Figure P10.18 makes an angle of $20.0°$ with the horizontal. What is the torque exerted by the fish

Figure P10.18

about an axis perpendicular to the page and passing through the fisher's hand?

19. Given $\mathbf{M} = 6\mathbf{i} + 2\mathbf{j} - \mathbf{k}$ and $\mathbf{N} = 2\mathbf{i} - \mathbf{j} - 3\mathbf{k}$, calculate the vector product $\mathbf{M} \times \mathbf{N}$.

20. For the vectors $\mathbf{A} = -3\mathbf{i} + 7\mathbf{j} - 4\mathbf{k}$ and $\mathbf{B} = 6\mathbf{i} - 10\mathbf{j} + 9\mathbf{k}$, evaluate the quantities (a) $\cos^{-1}[\mathbf{A}\cdot\mathbf{B}/AB]$ and (b) $\sin^{-1}[|\mathbf{A} \times \mathbf{B}|/AB]$. (c) Which give(s) the angle between the vectors?

21. A force of $\mathbf{F} = (2.00\mathbf{i} + 3.00\mathbf{j})$ N is applied to an object that is pivoted about a fixed axle aligned along the z coordinate axis. If the force is applied at the point $\mathbf{r} = (4.00\mathbf{i} + 5.00\mathbf{j} + 0\mathbf{k})$ m, find (a) the magnitude of the net torque about the z axis and (b) the direction of the torque vector $\boldsymbol{\tau}$.

22. Use the definition of the vector product and the definitions of the unit vectors $\mathbf{i}$, $\mathbf{j}$, and $\mathbf{k}$ to prove Equations 10.23. You may assume that the x axis points to the right, the y axis up, and the z axis toward you (not away from you). This choice is said to make the coordinate system *right handed*.

Section 10.6   The Rigid Body in Equilibrium

23. A uniform beam of mass m_b and length ℓ supports blocks with masses m_1 and m_2 at two positions, as in Figure P10.23. The beam rests on two knife edges. For what value of x will the beam be balanced at P such that the normal force at O is zero?

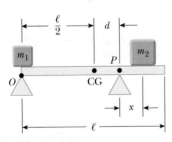

Figure P10.23

24. A 1 500-kg automobile has a wheel base (the distance between the axles) of 3.00 m. The center of mass of the automobile is on the center line at a point 1.20 m behind the front axle. Find the force exerted by the ground on each wheel.

25. A uniform ladder of length L and mass m_1 rests against a frictionless wall. The ladder makes an angle θ with the horizontal. (a) Find the horizontal and vertical forces the ground exerts on the base of the ladder when a firefighter of mass m_2 is a distance x from the bottom. (b) If the ladder is just on the verge of slipping when the firefighter is a distance d from the bottom, what is the coefficient of static friction between ladder and ground?

26. Figure P10.26 shows a claw hammer as it is being used to pull a nail out of a horizontal board. If a force of magnitude 150 N is exerted horizontally as shown, find (a) the force exerted by the hammer claws on the nail and (b) the force exerted by the surface on the point of contact with the hammer head. Assume that the force the hammer exerts on the nail is parallel to the nail.

Figure P10.26

27. A uniform sign of weight F_g and width $2L$ hangs from a light, horizontal beam, hinged at the wall and supported by a cable (Fig. P10.27). Determine (a) the tension in the cable and (b) the components of the reaction force exerted by the wall on the beam, in terms of F_g, d, L, and θ.

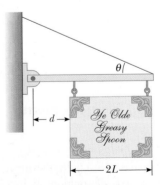

Figure P10.27

28. A crane of mass 3 000 kg supports a load of 10 000 kg, as in Figure P10.28. The crane is pivoted with a frictionless pin at A and rests against a smooth support at B. Find the reaction forces at A and B.

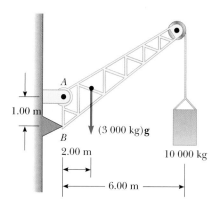

Figure P10.28

Section 10.7   The Rigid Body Under a Net Torque

29. A model airplane with mass 0.750 kg is tethered by a wire so that it flies in a circle 30.0 m in radius. The airplane engine provides a net thrust of 0.800 N perpendicular to the tethering wire. (a) Find the torque the net thrust produces about the center of the circle. (b) Find the angular acceleration of the airplane when it is in level flight. (c) Find the tangential acceleration of the airplane.

30. The combination of an applied force and a friction force produces a constant total torque of 36.0 N · m on a wheel rotating about a fixed axis. The applied force acts for 6.00 s. During this time the angular speed of the wheel increases from 0 to 10.0 rad/s. The applied force is then removed, and the wheel comes to rest in 60.0 s. Find (a) the moment of inertia of the wheel, (b) the magnitude of the frictional torque, and (c) the total number of revolutions of the wheel.

31. An electric motor turns a flywheel through a drive belt that joins a pulley on the motor and a pulley that is rigidly attached to the flywheel, as shown in Figure P10.31. The flywheel is a solid disk with a mass of 80.0 kg and a diameter of 1.25 m. It turns on a frictionless axle. Its pulley has much smaller mass and a radius of 0.230 m. If the tension in the upper (taut) segment of the belt is 135 N and the

flywheel has a clockwise angular acceleration of 1.67 rad/s^2, find the tension in the lower (slack) segment of the belt.

32. A potter's wheel—a thick stone disk with a radius of 0.500 m and a mass of 100 kg—is freely rotating at 50.0 rev/min. The potter can stop the wheel in 6.00 s by pressing a wet rag against the rim and exerting a radially inward force of 70.0 N. Find the effective coefficient of kinetic friction between wheel and rag.

33. Two blocks, as shown in Figure P10.33, are connected by a string of negligible mass passing over a pulley of radius 0.250 m and moment of inertia I. The block on the frictionless incline is moving up with a constant acceleration of 2.00 m/s^2. (a) Determine T_1 and T_2, the tensions in the two parts of the string. (b) Find the moment of inertia of the pulley.

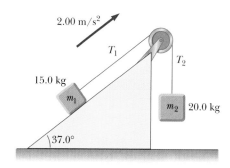

Figure P10.33

34. In Figure P10.34 the sliding block has a mass of 0.850 kg, the counterweight has a mass of 0.420 kg, and the pulley is a hollow cylinder with a mass of 0.350 kg, an inner radius of 0.020 0 m, and an outer radius of 0.030 0 m. The coefficient of kinetic friction between the block and the horizontal surface is 0.250. The pulley turns without friction on its axle. The light cord does not stretch and does not slip on the pulley. The block has a velocity of 0.820 m/s toward the pulley when it passes through a photogate. (a) Use energy methods to predict its speed after it has moved

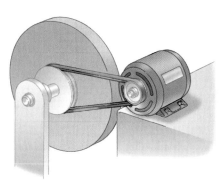

Figure P10.31

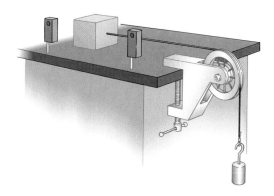

Figure P10.34

to a second photogate, 0.700 m away. (b) Find the angular speed of the pulley at the same moment.

35. An object with weight 50.0 N is attached to the free end of a light string wrapped around a reel with a radius of 0.250 m and a mass of 3.00 kg. The reel is a solid disk, free to rotate in a vertical plane about the horizontal axis passing through its center. The object is released 6.00 m above the floor. (a) Determine the tension in the string, the acceleration of the object, and the speed with which the object hits the floor. (b) Using the principle of conservation of energy, find the speed with which the object hits the floor.

36. A constant torque of 25.0 N·m is applied to a grindstone with a moment of inertia of 0.130 kg·m², which starts from rest and turns without friction. Using energy principles, find the angular speed after the grindstone has made 15.0 rev.

37. A uniform rod of length L and mass M is free to rotate about a frictionless pivot at one end, as in Figure P10.37. The rod is released from rest in the horizontal position. What are the *initial* angular acceleration of the rod and the *initial* translational acceleration of the right end of the rod?

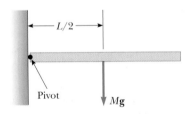

Figure P10.37

Section 10.8 Angular Momentum

38. Heading straight toward the summit of Pike's Peak, an airplane of mass 12 000 kg flies over the plains of Kansas at nearly constant altitude 4.30 km, with constant velocity 175 m/s west. (a) What is the airplane's vector angular momentum relative to a wheat farmer on the ground directly below the airplane? (b) Does this value change as the airplane continues its motion along a straight line? (c) What is its angular momentum relative to the summit of Pike's Peak?

39. The position vector of a particle of mass 2.00 kg is given as a function of time by $\mathbf{r} = (6.00\mathbf{i} + 5.00t\mathbf{j})$ m. Determine the angular momentum of the particle about the origin, as a function of time.

40. The hour and minute hands of Big Ben, the Parliament Building tower clock in London, are 2.70 m and 4.50 m long and have masses of 60.0 kg and 100 kg, respectively (Fig. P10.14). Calculate the total angular momentum of

these hands about the center point. Model the hands as long thin uniform rods.

41. A particle of mass 0.400 kg is attached to the 100-cm mark of a meter stick of mass 0.100 kg. The meter stick rotates on a horizontal, frictionless table with an angular speed of 4.00 rad/s. Calculate the angular momentum of the system when the stick is pivoted about an axis (a) perpendicular to the table through the 50.0-cm mark and (b) perpendicular to the table through the 0-cm mark.

42. A 4.00-kg object is attached to a light cord, which is wound around a spool (see Figure 10.22). The spool is a uniform solid cylinder of radius 8.00 cm and mass 2.00 kg. (a) What is the net torque on the system about the point O? (b) When the object has a speed v, the spool has an angular speed $\omega = v/R$. Determine the total angular momentum of the system about O. (c) Using the fact that $\tau = d\mathbf{L}/dt$ and your result from (b), calculate the acceleration of the object.

Section 10.9 Conservation of Angular Momentum

43. A cylinder with a moment of inertia of I_1 rotates about a vertical, frictionless axle with angular speed ω_i. A second cylinder that has a moment of inertia of I_2 and initially is not rotating drops onto the first cylinder (Fig. P10.43). Because of friction between the surfaces, the two eventually reach the same angular speed ω_f. (a) Calculate ω_f. (b) Show that the kinetic energy of the system decreases in this interaction and calculate the ratio of the final rotational energy to the initial rotational energy.

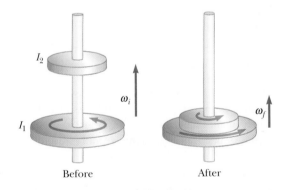

Figure P10.43

44. A student sits on a freely rotating stool holding two weights, each of which has a mass of 3.00 kg. When his arms are extended horizontally, the weights are 1.00 m from the axis of rotation and he rotates with an angular speed of 0.750 rad/s. The moment of inertia of the student plus stool is 3.00 kg·m² and is assumed to be constant. The student pulls the weights inward horizontally to a position 0.300 m from the rotation axis. (a) Find the new angular speed of the student. (b) Find the kinetic en-

ergy of the rotating system before and after he pulls the weights inward.

45. A playground merry-go-round of radius $R = 2.00$ m has a moment of inertia $I = 250$ kg·m² and is rotating at 10.0 rev/min about a frictionless vertical axle. Facing the axle, a 25.0-kg child hops onto the merry-go-round from the ground and manages to sit down on its edge. What is the new angular speed of the merry-go-round?

46. A puck with a mass of 80.0 g and a radius of 4.00 cm slides along an air table at a speed of 1.50 m/s, as shown in Figure P10.46a. It makes a glancing collision with a second puck having a radius of 6.00 cm and a mass of 120 g (initially at rest) such that their rims just touch. Because their rims are coated with instant-acting glue, the pucks stick together and spin after the collision (Fig. P10.46b). (a) What is the angular momentum of the system relative to the center of mass? (b) What is the angular speed about the center of mass?

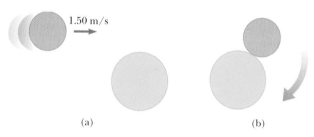

1.50 m/s

(a)   (b)

Figure P10.46

47. A space station shaped like a giant wheel has a radius of 100 m and a moment of inertia of 5.00×10^8 kg·m². A crew of 150 are living on the rim, and the station's rotation causes the crew to experience an acceleration of $1g$ (Fig. P10.47). When 100 people move to the center of the station for a union meeting, the angular speed changes.

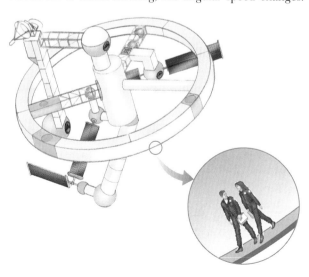

Figure P10.47

What acceleration is experienced by the managers remaining at the rim? Assume that the average mass of each inhabitant is 65.0 kg.

48. The ball in Figure 10.26 has a mass of 0.120 kg. The distance of the ball from the center of rotation is originally 40.0 cm, and the ball is moving with a speed of 80.0 cm/s. The string is pulled downward 15.0 cm through the hole in the frictionless table. Determine the work done on the ball. (*Hint:* Consider the change of kinetic energy.)

Section 10.11 Rolling of Rigid Bodies

49. A cylinder of mass 10.0 kg rolls without slipping on a horizontal surface. At the instant its center of mass has a speed of 10.0 m/s, determine (a) the translational kinetic energy of its center of mass, (b) the rotational energy about its center of mass, and (c) its total energy.

50. A uniform solid disk and a uniform hoop are placed side by side at the top of an incline of height *h*. If they are released from rest and roll without slipping, determine their speeds when they reach the bottom. Which object reaches the bottom first?

51. A metal can containing condensed mushroom soup has a mass of 215 g, a height of 10.8 cm, and a diameter of 6.38 cm. It is placed at rest on its side at the top of a 3.00-m-long incline that is at 25.0° to the horizontal, and is then released to roll straight down. Assuming energy conservation, calculate the moment of inertia of the can if it takes 1.50 s to reach the bottom of the incline. Which pieces of data, if any, are unnecessary for calculating the solution?

52. A tennis ball is a hollow sphere with a thin wall. It is set rolling without slipping at 4.03 m/s on the horizontal section of a track, as shown in Figure P10.52. It rolls around the inside of a vertical circular loop 90.0 cm in diameter and finally leaves the track at a point 20.0 cm below the horizontal section. (a) Find the speed of the ball at the top of the loop. Demonstrate that it will not fall from the track.

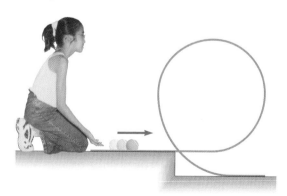

Figure P10.52

(b) Find its speed as it leaves the track. (c) Suppose that static friction between ball and track were negligible, so that the ball slid instead of rolling. Would its speed then be higher, lower, or the same at the top of the loop? Explain.

Section 10.12 Context Connection — Gyroscopes in Space

53. A spacecraft is in empty space. It carries on board a gyroscope with a moment of inertia $I_g = 20.0 \text{ kg} \cdot \text{m}^2$ around the axle of the gyroscope. The moment of inertia of the spacecraft around the axle of the gyroscope is $I_s = 5.00 \times 10^5 \text{ kg} \cdot \text{m}^2$. Neither the spacecraft nor the gyroscope is rotating. The gyroscope can be powered up in a negligible period of time to an angular speed of 100 s^{-1}. If the orientation of the spacecraft is to be changed by $30.0°$, for how long should the gyroscope be operated?

54. In a precessing gyroscope, the angular momentum vector sweeps out a cone, as in Figure 10.27, with a *precessional frequency*, given by

$$\omega_p = \frac{\tau}{L}$$

where τ is the magnitude of the torque on the gyroscope and L is the magnitude of its angular momentum. Given that the precession period of the Earth is 2.58×10^4 years, calculate the torque on the Earth that is causing this precession. (Model the Earth as a uniform sphere.)

Additional Problems

55. A 60.0-kg woman stands at the rim of a horizontal turntable having a moment of inertia of $500 \text{ kg} \cdot \text{m}^2$ and a radius of 2.00 m. The turntable is initially at rest and is free to rotate about a frictionless, vertical axle through its center. The woman then starts walking around the rim clockwise (as viewed from above the system) at a constant speed of 1.50 m/s relative to the Earth. (a) In what direction and with what angular speed does the turntable rotate? (b) How much work does the woman do to set herself and the turntable into motion?

56. Two astronauts (Fig. P10.56), each having a mass M, are connected by a rope of length d having negligible mass. They are isolated in space, orbiting their center of mass at speeds v. Treating the astronauts as particles, calculate (a) the magnitude of the angular momentum of the system and (b) the rotational energy of the system. By pulling on the rope, one of the astronauts shortens the distance between them to $d/2$. (c) What is the new angular momentum of the system? (d) What are the astronauts' new speeds? (e) What is the new rotational energy of the system? (f) How much work does the astronaut do in shortening the rope?

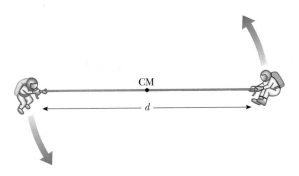

Figure P10.56

57. A long, uniform rod of length L and mass M is pivoted about a horizontal, frictionless pin passing through one end. The rod is released from rest in a vertical position, as shown in Figure P10.57. At the instant the rod is horizontal, find (a) its angular speed, (b) the magnitude of its angular acceleration, (c) the x and y components of the acceleration of its center of mass, and (d) the components of the reaction force at the pivot.

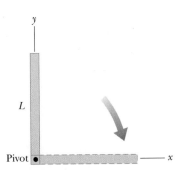

Figure P10.57

58. **Review Problem.** A mixing beater consists of three thin rods. Each is 10.0 cm long, diverges from a central hub, and is separated from the others by $120°$. All turn in the same plane. A ball is attached to the end of each rod. Each ball has cross-sectional area 4.00 cm^2 and is so shaped that it has a drag coefficient of 0.600. Calculate the power input required to spin the beater at 1 000 rev/min (a) in air and (b) in water.

59. A uniform, hollow, cylindrical spool has inside radius $R/2$, outside radius R, and mass M (Fig. P10.59). It is mounted so that it rotates on a stationary horizontal axle. An object of mass m is connected to the end of a string wound around the spool. The object falls from rest through a distance y in time t. Show that the torque due to the frictional forces between spool and axle is

$$\tau_f = R \left[m \left(g - \frac{2y}{t^2} \right) - M \frac{5y}{4t^2} \right]$$

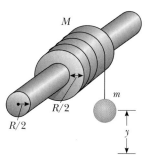

Figure P10.59

60. The reel shown in Figure P10.60 has radius R and moment of inertia I. The object of mass m is connected between a spring of force constant k and a cord wrapped around the reel. The reel axle and the incline are frictionless. The reel is wound counterclockwise so that the spring stretches a distance d from its unstretched position and is then released from rest. Find (a) the angular speed of the reel when the spring is again unstretched and (b) a numerical value for the angular speed at this point if $I = 1.00$ kg·m^2, $R = 0.300$ m, $k = 50.0$ N/m, $m = 0.500$ kg, $d = 0.200$ m, and $\theta = 37.0°$.

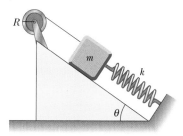

Figure P10.60

61. A block of mass $m_1 = 2.00$ kg and a block of mass $m_2 = 6.00$ kg are connected by a massless string over a pulley in the shape of a solid disk having radius $R = 0.250$ m and mass $M = 10.0$ kg. These blocks are allowed to move on a fixed block wedge of angle $\theta = 30.0°$, as in Figure P10.61. The coefficient of kinetic friction for both blocks is 0.360. Draw free-body diagrams of both blocks and of the pulley. Determine (a) the acceleration of the two blocks, and (b) the tensions in the string on both sides of the pulley.

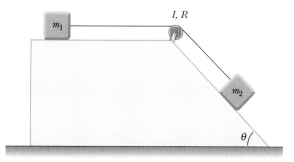

Figure P10.61

62. (a) Without the wheels, a bicycle frame has a mass of 8.44 kg. Each of the wheels can be roughly modeled as a uniform solid disk with a mass of 0.820 kg and a radius of 0.343 m. Find the kinetic energy of the whole bicycle when it is moving forward at 3.35 m/s. (b) Before the invention of a wheel turning on an axle, ancient people moved heavy loads by placing rollers under them. (Modern people use rollers too. Any hardware store will sell you a roller bearing for a lazy susan.) A stone block of mass 844 kg moves forward at 0.335 m/s, supported by two uniform cylindrical tree trunks, each of mass 82.0 kg and radius 0.343 m. No slipping occurs between the block and the rollers or between the rollers and the ground. Find the total kinetic energy of the moving objects.

63. A force acts on a uniform rectangular cabinet weighing 400 N, as in Figure P10.63. (a) If the cabinet slides with constant speed when $F = 200$ N and $h = 0.400$ m, find the coefficient of kinetic friction and the position of the resultant normal force. (b) If $F = 300$ N, find the value of h for which the cabinet just begins to tip.

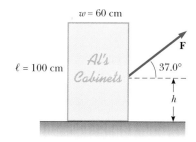

Figure P10.63

64. A stepladder of negligible weight is constructed as shown in Figure P10.64. A painter of mass 70.0 kg stands on the

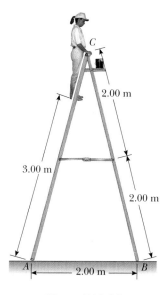

Figure P10.64

ladder 3.00 m from the bottom. Assuming the floor is frictionless, find (a) the tension in the horizontal bar connecting the two halves of the ladder, (b) the normal forces at *A* and *B*, and (c) the components of the reaction force at the single hinge *C* that the left half of the ladder exerts on the right half. (*Hint:* Treat each half of the ladder separately.)

65. Figure P10.65 shows a vertical force applied tangentially to a uniform cylinder of weight F_g. The coefficient of static friction between the cylinder and both surfaces is 0.500. In terms of F_g, find the maximum force **P** that can be applied that does not cause the cylinder to rotate. (*Hint:* When the cylinder is on the verge of slipping, both friction forces are at their maximum values. Why?)

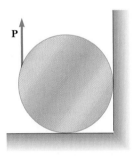

Figure P10.65

66. In exercise physiology studies it is sometimes important to determine the location of the center of mass of a person. This can be done with the arrangement shown in Figure P10.66. A light plank rests on two scales, which read $F_{g1} = 380$ N and $F_{g2} = 320$ N. The scales are separated by a distance of 2.00 m. How far from the woman's feet is her center of mass?

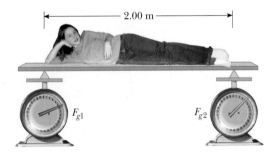

Figure P10.66

67. A person bending forward to lift a load "with his back," rather than "with his knees," can be injured by large forces exerted on the muscles and vertebrae. The spine pivots mainly at the fifth lumbar vertebra, with the principal supporting force provided by the erector spinalis muscle in the back. To see the magnitude of the forces involved and

to understand why back problems are so common among humans, consider the model for a person bending forward to lift a 200-N weight shown in Figure P10.67. The spine and upper body are represented as a uniform horizontal rod of weight 350 N, pivoted at the base of the spine. The erector spinalis muscle, attached at a point two thirds of the way up the spine, maintains the position of the back. The angle between the spine and this muscle is 12.0°. Find the tension in the back muscle and the compressional force in the spine.

(a)   (b)

Figure P10.67

68. A wad of sticky clay of mass *m* and velocity $\mathbf{v}_i$ is fired at a solid cylinder of mass *M* and radius *R* (Fig. P10.68). The cylinder is initially at rest and is mounted on a fixed horizontal axle that runs through the center of mass. The line of motion of the projectile is perpendicular to the axle and at a distance *d*, less than *R*, from the center. (a) Find the angular speed of the system just after the clay strikes and sticks to the surface of the cylinder. (b) Is mechanical energy conserved in this process? Explain your answer.

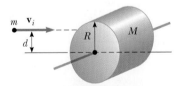

Figure P10.68

69. A string is wound around a uniform disk of radius *R* and mass *M*. The disk is released from rest with the string vertical and its top end tied to a fixed bar (Fig. P10.69). Show

that (a) the tension in the string is one-third the weight of the disk, (b) the magnitude of the acceleration of the center of mass is $2g/3$, and (c) the speed of the center of mass is $(4gh/3)^{1/2}$ as the disk descends. Verify your answer to part (c) using the energy approach.

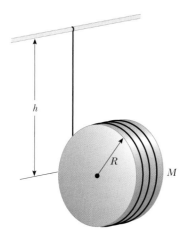

Figure P10.69

70. A common demonstration, illustrated in Figure P10.70, consists of a ball resting at one end of a uniform board of length ℓ, hinged at the other end and elevated at an angle θ. A light cup is attached to the board at r_c so that it will catch the ball when the support stick is suddenly removed. (a) Show that the ball will lag behind the end of the falling board when θ is less than 35.3°. (b) If the board is 1.00 m long and is supported at this limiting angle, show that the cup must be 18.4 cm from the moving end.

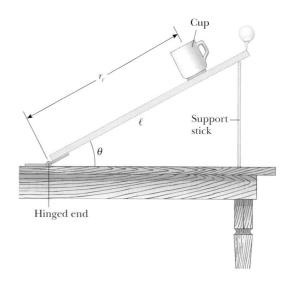

Figure P10.70

71. The hour hand and the minute hand of Big Ben, the famous Parliament tower clock in London, are 2.70 m long and 4.50 m long and have masses of 60.0 kg and 100 kg, respectively (see Figure P10.14). (a) Determine the total torque due to the weight of these hands about the axis of rotation when the time reads (i) 3:00, (ii) 5:15, (iii) 6:00, (iv) 8:20, (v) 9:45. (You may model the hands as long thin uniform rods.) (b) Determine all times when the total torque about the axis of rotation is zero. Determine the times to the nearest second, solving a transcendental equation numerically.

ANSWERS TO QUICK QUIZZES

10.1 (c). For a rotation of more than 180°, the angular displacement must be larger than $\pi = 3.14$ rad. The angular displacements in the three choices are (a) 6 rad − 3 rad = 3 rad, (b) 1 rad − (− 1) rad = 2 rad, (c) 5 rad − 1 rad = 4 rad.

10.2 (b). Because all angular displacements occurred in the same time interval, the displacement with the lowest value will be associated with the lowest average angular speed.

10.3 (b). In Equation 10.8, both the initial and final angular speeds are the same in all three cases. As a result, the angular acceleration is inversely proportional to the angular displacement. Thus, the highest angular acceleration is associated with the lowest angular displacement.

10.4 (a) Yes. All points on the wheel make one full rotation in the same time interval. (b) No. Points farther from the rotation axis have higher tangential speeds. (c) The point at $r = 0$ has zero tangential speed and zero acceleration; a point at $r = R/2$ has a tangential speed $v = R\omega/2$ and a total translational acceleration equal to the centripetal acceleration $v^2/(R/2) = R\omega^2/2$ (the tangential acceleration is zero at all points because ω is constant). A point on the rim at $r = R$ has a tangential speed $v = R\omega$ and a total translational acceleration $R\omega^2$.

10.5 The hollow sphere has the higher rotational kinetic energy. Its mass is located near the spherical surface, and the mass of the solid sphere is located throughout the volume. This results in a higher moment of inertia for the hollow sphere, as can be seen in Table 10.2.

10.6 Torque and work have two major differences. The primary difference is that the vector displacement in the expression for work is directed *parallel* to the force, but the position vector in the torque expression is *perpendicular* to the force. The second difference depends on whether there is motion or not—in the case of work, work is done only if the force succeeds in causing a displacement of the point of application of the force. On the other hand, a force applied at a perpendicular distance from a rotation axis results in a torque whether there is motion or not.

10.7   (b). You are trying to change the rotational motion of the screw, which requires torque. You have failed with the original screwdriver, so you have not been able to supply enough torque to exceed the frictional torque holding the screw in place in the wood. To apply more torque with the same force, you need to apply the force farther from the rotation axis, which you can do by applying the force to a screwdriver with a fatter handle.

10.8   (a) Yes. If two forces of equal magnitude but opposite direction are applied to an object along different lines of action, the net force on the object is zero but the net torque is nonzero. (b) Yes. If a number of forces are applied to an object and the lines of action of the forces intersect at a common point, the net torque is zero, but the forces may result in a net force on the object.

10.9   Both the grinding wheel and the drill slow down because of frictional torque in the bearings of the motor. Let us estimate that this frictional torque is approximately the same in both tools. The grinding wheel is a large, massive disk, and the spinning bit in an electric drill is a small cylinder of relatively low mass. Thus, the grinding wheel has a larger moment of inertia than the drill bit, because of both its larger mass and its shape. According to Equation 10.27, the larger moment of inertia will result in a smaller angular acceleration for the same torque and a correspondingly longer time to come to rest.

10.10   Yes. Even though the skater is not moving in a circular path before grabbing the pole, she has angular momentum about the pole, as pointed out in Pitfall Prevention 10.8. After she grabs the pole, the force on the skater is along her arm, which is parallel to the position vector from the pole to her body. Thus, this force exerts no torque on the skater, so she can be modeled as an isolated system with no external torques. As a result, her angular momentum is conserved.

10.11   The box. Both the ball–Earth system and the box–Earth system have the same potential energy before the ball and box are released. As the items move down the inclines, this potential energy is transformed to kinetic energy. For the ball, the transformation is into both rotational and translational kinetic energy. The box has only translational kinetic energy. Because the kinetic energy of the ball is split into two types, its translational kinetic energy is necessarily less than that of the box. Consequently, its translational speed is less than that of the box, so the ball will lag behind.

Astronauts F. Story Musgrave and Jeffrey A. Hoffman, along with the Hubble Space Telescope and the space shuttle *Endeavor*, are all falling around the Earth.

(Courtesy NASA)

WEB

For information on the Hubble Space Telescope and links to images, visit **www.stsci.edu**

Gravity, Planetary Orbits, and the Hydrogen Atom

At the beginning of our discussion of mechanics in Chapter 1, we introduced the notion of modeling and defined four categories of models: geometric, simplification, analysis, and structural. In this chapter, we apply our analysis models to two very common *structural models.* We shall discuss a structural model for a large system—the Solar System—and a structural model for a small system—the hydrogen atom.

In this chapter we return to Newton's law of universal gravitation—one of the fundamental force laws in nature—and show how it, together with our analysis models, enables us to understand the motions of planets and Earth satellites.

We conclude this chapter with a discussion of Niels Bohr's model of the hydrogen atom, which represents an interesting mixture of classical and nonclassical physics. Despite the hybrid nature of the model, some of its predictions agree with experimental measurements made on hydrogen atoms. This will be our first major venture into the area of *quantum physics*, which we will continue in Chapter 28.

11.1 • NEWTON'S LAW OF UNIVERSAL GRAVITATION REVISITED

Prior to 1686, many data had been collected on the motions of the Moon and the planets, but a clear understanding of the forces involved with the motions was not yet attainable. In that year, Isaac Newton provided the key that unlocked the secrets of the heavens. He knew, from the first law of motion, that a net force had to be acting on the Moon. If not, the Moon would move in a straight-line path rather than in its almost circular orbit. Newton reasoned that this force between the Moon and the Earth was an attractive force. He also concluded that there could be nothing special about the Earth–Moon system or the Sun and its planets that would cause gravitational forces to act on them alone.

As you should recall from Chapter 5, every particle in the Universe attracts every other particle with a force that is directly proportional to the product of their masses and inversely proportional to the square of the distance between them. If two particles have masses m_1 and m_2 and are separated by a distance r, the magnitude of the gravitational force between them is

- *Law of universal gravitation*

$$F_g = G \frac{m_1 m_2}{r^2} \quad [11.1]$$

where G is the *gravitational constant*, whose value in SI units is

$$G = 6.673 \times 10^{-11} \ \text{N} \cdot \text{m}^2/\text{kg}^2 \quad [11.2]$$

The force law given by Equation 11.1 is often referred to as an **inverse-square law** because the magnitude of the force varies as the inverse square of the separation of the particles. We can express this attractive force in vector form by defining a unit vector $\hat{\mathbf{r}}_{12}$ directed from m_1 toward m_2, as shown in Figure 11.1. The force exerted by m_1 on m_2 is

$$\mathbf{F}_{12} = -G \frac{m_1 m_2}{r^2} \hat{\mathbf{r}}_{12} \quad [11.3]$$

where the negative sign indicates that particle 1 is attracted toward particle 2. Likewise, by Newton's third law, the force exerted by m_2 on m_1, designated $\mathbf{F}_{21}$, is equal in magnitude to $\mathbf{F}_{12}$ and in the opposite direction. That is, these forces form an action–reaction pair, and

$$\mathbf{F}_{21} = -\mathbf{F}_{12}$$

As Newton demonstrated, the **gravitational force exerted by a finite-sized, spherically symmetric mass distribution on a particle outside the distribution is the same as if the entire mass of the distribution were concentrated at its center.** For example, the force on a particle of mass m at the Earth's surface has the magnitude

$$F_g = G \frac{M_E m}{R_E^2}$$

where M_E is the Earth's mass and R_E is the Earth's radius. This force is directed toward the center of the Earth.

In formulating his law of universal gravitation, Newton built on an observation that suggests the inverse square nature of the gravitational force between two bodies.

PITFALL PREVENTION 11.1

Be clear on g and G

Be sure you understand the difference between g and G. The symbol g represents the magnitude of the free-fall acceleration near a planet. At the surface of the Earth, g has the value $9.80 \ \text{m/s}^2$. On the other hand, G is a universal constant that has the same value everywhere in the Universe.

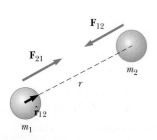

Figure 11.1

The gravitational force between two particles is attractive. The unit vector $\hat{\mathbf{r}}_{12}$ is directed from particle 1 toward particle 2. Note that $\mathbf{F}_{21} = -\mathbf{F}_{12}$.

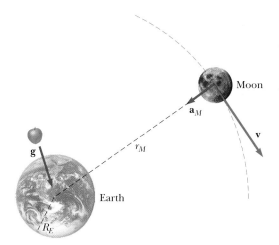

Figure 11.2

As it revolves about the Earth, the Moon experiences a centripetal acceleration $\mathbf{a}_M$ directed toward the Earth. An object near the Earth's surface, such as the apple shown here, experiences an acceleration $\mathbf{g}$. (Dimensions are not to scale.)

He wrote in his notes, "I deduced that the forces which keep the planets in their orbs must be reciprocally as the squares of their distances from the centers about which they revolve; and thereby compared the force requisite to keep the Moon in her orb with force of gravity at the surface of the Earth; and found them answer pretty nearly."

We will perform some calculations that resemble the steps Newton took in deducing his law of gravitation and that will allow us to understand the phrase "and found them answer pretty nearly." Let us compare the acceleration of the Moon in its orbit with the acceleration of an object falling near the Earth's surface, such as the apple in Figure 11.2. We assume that the two accelerations have the same cause, namely, the gravitational attraction of the Earth. From the inverse-square law, Newton found that the acceleration of the Moon toward the Earth should be proportional to $1/r_M{}^2$, where r_M is the separation between the centers of the Earth and Moon. Furthermore, the acceleration of the apple toward the Earth should be proportional to $1/R_E{}^2$, where R_E is the radius of the Earth. When the values $r_M = 3.84 \times 10^8$ m and $R_E = 6.37 \times 10^6$ m are used, the ratio of the Moon's acceleration a_M to the apple's acceleration g is predicted to be

$$\frac{a_M}{g} = \frac{\left(G\dfrac{M_E}{r_M{}^2} \right)}{\left(G\dfrac{M_E}{R_E{}^2} \right)} = \left(\frac{R_E}{r_M} \right)^2 = \left(\frac{6.37 \times 10^6 \text{ m}}{3.84 \times 10^8 \text{ m}} \right)^2 = 2.75 \times 10^{-4}$$

Therefore,

$$a_M = (2.75 \times 10^{-4})(9.80 \text{ m/s}^2) = 2.70 \times 10^{-3} \text{ m/s}^2$$

The centripetal acceleration of the Moon can also be calculated. We know the Moon's orbital period $T = 27.32$ days $= 2.36 \times 10^6$ s, and its mean distance r_M from the Earth. Using Equation 10.12,

$$a_M = r_M \omega^2 = r_M \left(\frac{2\pi}{T} \right)^2 = \frac{4\pi^2 r_M}{T^2} = \frac{4\pi^2 (3.84 \times 10^8 \text{ m})}{(2.36 \times 10^6 \text{ s})^2} = 2.72 \times 10^{-3} \text{ m/s}^2$$

This agreement between the results of these two *independent* calculations of the acceleration of the Moon— Newton's "answering pretty nearly"—provides strong evidence that the inverse-square law of force is correct.

Artificial Earth satellites behave in the same way that the Moon does. A satellite in orbit around the Earth has a centripetal acceleration equal to the free-fall acceleration at the particular position of the satellite.

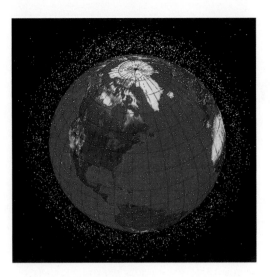

Humans have put many artificial satellites in orbit around the Earth. This diagram shows a plot of all known unclassified satellites and satellite debris larger in size than a baseball as of 1995. *(U.S. Space Command, NORAD)*

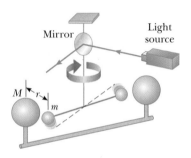

Figure 11.3

Schematic diagram of the Cavendish apparatus for measuring G. As the small spheres of mass *m* are attracted to the large spheres of mass *M*, the rod rotates through a small angle. A light beam reflected from a mirror on the rotating apparatus measures the angle of rotation. The dashed line represents the original position of the rod. (In reality, the length of wire above the mirror is much larger than that below it.)

Measurement of the Gravitational Constant

The gravitational constant *G* was first measured in an important experiment by Sir Henry Cavendish in 1798. The apparatus he used consists of two small spheres, each of mass *m*, fixed to the ends of a light horizontal rod suspended by a thin wire, as in Figure 11.3. Two large spheres, each of mass *M*, are then placed near the smaller spheres. The attractive force between the smaller and larger spheres causes the rod to rotate and twist the wire. If the system is oriented as shown in Figure 11.3, the rod rotates clockwise when viewed from the top. The angle through which it rotates is measured by the deflection of a light beam that is reflected from a mirror attached to the wire. The experiment is carefully repeated with different masses at various separations. In addition to providing a value for *G*, the results confirm that the force is attractive, proportional to the product *mM*, and inversely proportional to the square of the distance *r*.

It is interesting that *G* is the least well known of the fundamental constants, with a percentage uncertainty thousands of times larger than those for other constants, such as the speed of light *c* and the fundamental electric charge *e*. Several measurements of *G* made in the 1990s have varied significantly from the previous value and from one another! The search for a more precise value of *G* continues to be an area of active research.

THINKING PHYSICS 11.1

The novel *Icebound*, by Dean Koontz (Ballantine Books, New York, 1995), is a story of a group of scientists trapped on a floating iceberg near the North Pole. One of the devices that the scientists have with them is a transmitter with which they can fix their position with "the aid of a geosynchronous polar satellite." Can a satellite in a *polar* orbit be *geosynchronous*?

Reasoning A geosynchronous satellite is one that stays over one location on the Earth's surface at all times. Thus, an antenna on the surface that receives signals from the satellite, such as a television dish, can stay pointed in a fixed direction toward the sky. The satellite must be in an orbit with the correct radius such that its orbital period is the same as that of the Earth's rotation. This results in the satellite appearing to have no east–west motion relative to the observer at the chosen location. Another requirement is that a geosynchronous satellite *must be in orbit over the* *equator.* Otherwise it would appear to undergo a north–south oscillation during one orbit. Thus, it would be impossible to have a geosynchronous satellite in a *polar* orbit. Even if such a satellite were at the proper distance from the Earth, it would be moving rapidly in the north–south direction, resulting in the necessity of accurate tracking equipment. What's more, it would be below the horizon for long periods of time, making it useless for determining one's position.

THINKING PHYSICS 11.2

A satellite is in a circular orbit in space. Is it necessary that a massive object be at the center of the orbit?

Reasoning   This is the usual situation—satellites are normally in orbit around the Earth, a massive object. Imagine, however, an analogy—tie a rock somewhere along a length of string, and twirl the rock in a circle, as shown in Figure 11.4a. In this case, there is no entity at the center of the circular path of the rock, nor along its radius, that is exerting a force on the rock. The net force holding the rock in the circular orbit is a combination of the radial components of two forces: one from the string on each side of the rock.

The question posed is not about rocks on strings; it is about satellites. Does this situation have a gravitational counterpart? Yes—imagine moving a satellite from the Earth toward the Sun along the center line between them. At some point along that line, the gravitational force of attraction to the Earth will cancel just enough of the force from the Sun that the net force on the satellite can produce circular motion around the Sun with a period of one year, equal to the period of the Earth. As seen from the Earth, the satellite appears to be stationary at this so-called *Lagrange point*

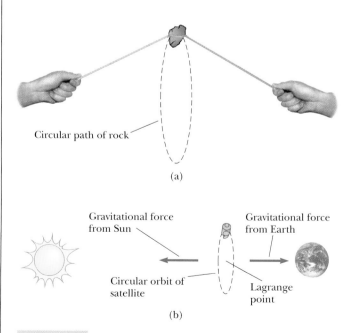

Circular path of rock

(a)

Gravitational force from Sun

Gravitational force from Earth

Circular orbit of satellite

Lagrange point

(b)

Figure 11.4

(Thinking Physics 11.2)

between the Earth and the Sun. This would look similar to the balance along the string of the tension on either side of the rock. Now, imagine pulling the satellite perpendicularly off the Earth–Sun line a small amount. As long as the distance is relatively small, the components of the forces from the Earth and Sun along the Sun–Earth centerline can add as before to cause the satellite to orbit the Sun with a period of one year. The components of the two forces perpendicular to the centerline will add, to provide a net force on the satellite back toward the centerline. Now, suppose we set the satellite into circular motion with just the right velocity. The satellite will be in orbit around the Lagrange point with no mass located there as shown in Figure 11.4b.

The International Sun–Earth Explorer 3 satellite (ISEE-3) was launched in 1978 and placed in just such an orbit in order to study the solar wind. The advantage of this orbit is that the satellite always resides on the side of the Sun toward the Earth, so that it is constantly in communication with Earth. In the absence of a specific mission to Comet Halley from Earth, the ISEE-3 satellite was removed from this orbit in 1982 and placed in orbit around the Moon, in anticipation of a flyby of Comet Halley. On the way to this flyby, it passed through the tail of Comet Giacobini–Zinner and successfully gathered important data on both comets.

Example 11.1   The Mass of the Earth

Use the gravitational force law to find an approximate value for the mass of the Earth.

Solution   Figure 11.5, obviously not to scale, shows a baseball falling toward the Earth at a location where the free-fall acceleration is g. We know from Chapter 5 that the magnitude of the gravitational force exerted on the baseball by the Earth is the same as the weight $m_b g$ of the ball. That is,

$$m_b g = G \frac{M_E m_b}{R_E{}^2}$$

We divide each side of this equation by m_b and solve for the mass of the Earth M_E:

$$(1) \quad M_E = \frac{g R_E{}^2}{G}$$

The falling baseball is close enough to the Earth that the distance of separation between the center of the ball and the center of the Earth can be taken as the radius of the Earth, 6.37×10^6 m. Thus, the mass of the Earth is

$$M_E = \frac{(9.80 \text{ m/s}^2)(6.37 \times 10^6 \text{ m})^2}{6.67 \times 10^{-11} \text{ N·m}^2/\text{kg}^2} = \boxed{5.98 \times 10^{24} \text{ kg}}$$

Notice also that (1) can be written as

$$g = G \frac{M_E}{r^2}$$

which we have seen before as Equation 5.17. This indicates that the free-fall acceleration at a point above the Earth's sur-

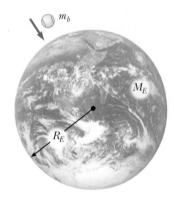

Figure 11.5

(Example 11.1) A baseball falling toward the Earth (not drawn to scale).

face decreases as the inverse square of the distance between the point and the center of the Earth.

EXERCISE   If an object weighs 270 N at the Earth's surface, what will it weigh at an altitude equal to twice the radius of the Earth?

Answer   30 N

EXERCISE   Determine the magnitude of the free-fall acceleration at an altitude of 500 km. By what percentage is the weight of a body reduced at this altitude?

Answer   8.43 m/s²; 14%

Example 11.2 An Earth Satellite

A satellite of mass m moves in a circular orbit about the Earth with a constant speed v, at a height of $h = 1\ 000$ km above the Earth's surface, as in Figure 11.6. (For clarity, this figure is not drawn to scale.) Find the orbital speed of the satellite.

Solution The only external force on the satellite is the gravitational force exerted by the Earth. This force is directed to-

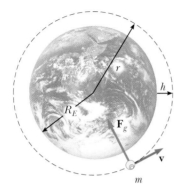

Figure 11.6

(Example 11.2) A satellite of mass m moving around the Earth in a circular orbit of radius r with constant speed v. The only force acting on the satellite is the gravitational force $\mathbf{F}_g$. (Not drawn to scale.)

ward the center of the satellite's circular path. We apply Newton's second law to the satellite modeled as a particle in uniform circular motion. Because the magnitude of the gravitational force between the Earth and the satellite is GM_Em/r^2 we find that

$$F_g = G\frac{M_E m}{r^2} = m\frac{v^2}{r}$$

$$v = \sqrt{\frac{GM_E}{r}}$$

In this expression, the distance r is the Earth's radius plus the height of the satellite; that is, $r = R_E + h = 7.37 \times 10^6$ m, so that

$$v = \sqrt{\frac{(6.67 \times 10^{-11}\ \text{N}\cdot\text{m}^2/\text{kg}^2)(5.98 \times 10^{24}\ \text{kg})}{7.37 \times 10^6\ \text{m}}}$$

$$= 7.36 \times 10^3\ \text{m/s} \approx \boxed{16\ 400\ \text{mi/h}}$$

Note that v is independent of the mass of the satellite!

EXERCISE Calculate the period of revolution T of the satellite.

Answer 105 min

11.2 • STRUCTURAL MODELS

In Chapter 1, we mentioned that we would discuss four categories of models. The fourth category is **structural models.** In these models, we propose theoretical structures in an attempt to understand the behavior of a system with which we cannot interact directly because it is far different in scale from our macroscopic world—either much smaller or much larger.

One of the earliest structural models to be explored was that of the place of the Earth in the Universe. The movements of the planets, stars, and other celestial bodies have been observed by people for thousands of years. Early in history, scientists regarded the Earth as the center of the Universe. This is a structural model for the Universe called the *geocentric model.* It was elaborated and formalized by the Greek astronomer Claudius Ptolemy in the second century A.D. and was accepted for the next 1400 years. In 1543, the Polish astronomer Nicolaus Copernicus (1473–1543) offered a different structural model in which the Earth is part of a local Solar System, suggesting that the Earth and the other planets revolve in perfectly circular orbits about the Sun (the *heliocentric model*).

In general, a structural model contains the following features:

1. A description of the physical components of the system; in the heliocentric model, the components are the planets and the Sun.
2. A description of where the components are located relative to one another and how they interact; in the heliocentric model, the planets and Sun interact via the gravitational force and the planets are in orbit around the Sun.
3. A description of the time evolution of the system; the heliocentric model assumes a steady-state Solar System, with planets revolving in orbits around the Sun with fixed periods.

• *Features of structural models*

4. A description of the agreement between predictions of the model and actual observations and, possibly, predictions of new effects that have not yet been observed; the heliocentric model predicts Earth-based observations of Mars that are in agreement with historical and present measurements. The geocentric model also was able to find agreement between predictions and observations, but only at the expense of a very complicated structural model in which the planets moved in circles built on other circles. The heliocentric model, along with Newton's law of universal gravitation, predicted that a spacecraft could be sent from the Earth to Mars long before it was actually done in the 1970s.

In Sections 11.3 and 11.4, we explore some of the details of the structural model of the Solar System. In Section 11.5, we investigate a structural model of the hydrogen atom.

11.3 • KEPLER'S LAWS

The Danish astronomer Tycho Brahe (1546–1601) made accurate astronomical measurements over a period of 20 years and provided the basis for the currently accepted structural model of the Solar System. It is interesting to note that these precise observations, made on the planets and 777 stars, were carried out with nothing more elaborate than a large sextant and compass; the telescope had not yet been invented.

The German astronomer Johannes Kepler, who was Brahe's assistant, acquired Brahe's astronomical data and spent about 16 years trying to deduce a mathematical model for the motions of the planets. After many laborious calculations, he found that Brahe's precise data on the revolution of Mars about the Sun provided the answer. Kepler's analysis first showed that the concept of circular orbits about the Sun in the heliocentric model had to be abandoned. He discovered that the orbit of Mars could be accurately described by a curve called an *ellipse*. He then generalized this analysis to include the motions of all planets. The complete analysis is summarized in three statements, known as **Kepler's laws,** each of which are discussed in the following sections.

Newton demonstrated that these laws are consequences of the gravitational force that exists between any two masses. Newton's law of universal gravitation, together with his laws of motion, provides the basis for a full mathematical representation of the motion of planets and satellites.

Kepler's First Law

We are familiar with circular orbits of objects around gravitational force centers from our discussions in this chapter. Kepler's first law indicates that the circular orbit is a very special case, and that elliptical orbits are the general situation:*

Each planet in the Solar System moves in an elliptical orbit with the Sun at one focus.

Johannes Kepler (1571–1630)

The German astronomer Kepler is best known for developing the laws of planetary motion based on the careful observations of Tycho Brahe. *(Art Resource)*

• *Kepler's first law*

---

* We choose a simplification model in which a body of mass m is in orbit around a body of mass M, with $M \gg m$. In this way, we model the body of mass M to be stationary. In reality, this is not true—both M and m move around the center of mass of the system of two objects. This is how we indirectly detect planets around other stars—we see the "wobbling" motion of the star as the planet and the star rotate about the center of mass.

Figure 11.7 shows the geometry of an ellipse, which serves as our geometric model for the elliptical orbit of a planet.* An ellipse is mathematically defined by choosing two points, F_1 and F_2, each of which is a called a **focus,** and then drawing a curve through points for which the sum of the distances r_1 and r_2 from F_1 and F_2 is a constant. The longest distance through the center between points on the ellipse (and passing through both foci) is called the **major axis,** and this distance is $2a$. In Figure 11.7, the major axis is drawn along the x direction. The distance a is called the **semimajor axis.** Similarly, the shortest distance through the center between points on the ellipse is called the **minor axis** of length $2b$, where the distance b is the **semiminor axis.** Either focus of the ellipse is located at a distance c from the center of the ellipse, where $a^2 = b^2 + c^2$. In the elliptical orbit of a planet around the Sun, the Sun is at one focus of the ellipse. Nothing is at the other focus.

The **eccentricity** of an ellipse is defined as $e = c/a$ and describes the general shape of the ellipse. For a circle, $c = 0$, and the eccentricity is therefore zero. The smaller b is than a, the shorter the ellipse is along the y direction compared with its extent in the x direction in Figure 11.7. As b decreases, c increases, and the eccentricity e increases. Thus, higher values of eccentricity correspond to longer and thinner ellipses. The range of values of the eccentricity for an ellipse is $e \leq 1$. Eccentricities higher than 1 correspond to hyperbolas.

Eccentricities for planetary orbits vary widely in the Solar System. The eccentricity of the Earth's orbit is 0.017, which makes it nearly circular. On the other hand, the eccentricity of Pluto's orbit is 0.25, the highest of all the nine planets. Figure 11.8a shows an ellipse with the eccentricity of that of Pluto's orbit. Notice

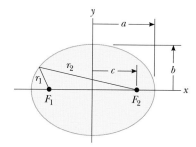

Figure 11.7

Plot of an ellipse. The semimajor axis has length a, and the semiminor axis has length b. A focus is located at a distance c from the center on each side of the center.

PITFALL PREVENTION 11.2

Where is the Sun?

 The Sun is located at one focus of the elliptical orbit of a planet. It is *not* located at the center of the ellipse.

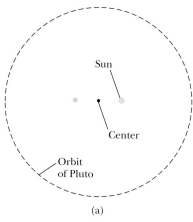

(a)

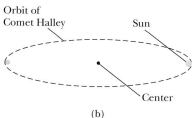

(b)

Figure 11.8

(a) The shape of the orbit of Pluto, which has the highest eccentricity ($e = 0.25$) among the planets in the Solar System. The Sun is located at the large yellow dot, which is a focus of the ellipse. Nothing physical is located at the center of the orbit *(the small dot)* or the other focus *(the blue dot).* (b) The shape of the orbit of Comet Halley.

---

* Actual orbits show perturbations due to moons in orbit around the planet and passages of the planet near other planets. We will ignore these perturbations and adopt a simplification model in which the planet follows a perfectly elliptical orbit.

that even this highest eccentricity orbit is difficult to distinguish from a circle. This is why Kepler's first law is an admirable accomplishment.

The eccentricity of the orbit of Comet Halley is 0.97, describing an orbit whose major axis is much longer than its minor axis, as shown in Figure 11.8b. As a result, Comet Halley spends much of its 76-year period far from the Sun and invisible from the Earth. It is only visible to the naked eye during a small part of its orbit when it is near the Sun.

Let us imagine now a planet in an elliptical orbit such as that shown in Figure 11.7, with the Sun at focus F_2. When the planet is at the far left in the diagram, the distance between the planet and the Sun is $a + c$. This point is called the *aphelion*, where the planet is the farthest away from the Sun that it can be in the orbit (for an object in orbit around the Earth, this point is called the *apogee*). Conversely, when the planet is at the right end of the ellipse, the point is called the *perihelion* (for an Earth orbit, the *perigee*), and the distance between the planet and the Sun is $a - c$.

Kepler's first law is a direct result of the inverse-square nature of the gravitational force. We have discussed circular and elliptical orbits. These are the allowed shapes of orbits for objects that are *bound* to the gravitational force center. These objects include planets, asteroids, and comets that move repeatedly around the Sun, as well as moons orbiting a planet. *Unbound* objects also occur such as a meteoroid from deep space that might pass by the Sun once and then never return. The gravitational force between the Sun and these objects also varies as the inverse square of the separation distance, and the allowed paths for these objects are parabolas and hyperbolas.

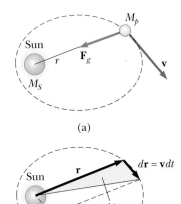

(a)

(b)

Figure 11.9

(a) The gravitational force acting on a planet acts toward the Sun, along the radius vector. (b) As a planet orbits the Sun, the area swept out by the radius vector in a time interval dt is equal to one-half the area of the parallelogram formed by the vectors **r** and $d\mathbf{r} = \mathbf{v}\,dt$.

Kepler's Second Law

The second of Kepler's laws is

The radius vector drawn from the Sun to any planet sweeps out equal areas in equal time intervals.

This can be shown to be a consequence of angular momentum conservation as follows. Consider a planet of mass M_p moving about the Sun in an elliptical orbit (Fig. 11.9a). Let us consider the planet as a system. We shall assume that the Sun is much more massive than the planet, so that the Sun does not move. The gravitational force acting on the planet is a central force, always along the radius vector, directed toward the Sun. The torque on the planet due to this central force is zero because **F** is parallel to **r**. That is,

$$\boldsymbol{\tau} = \mathbf{r} \times \mathbf{F} = \mathbf{r} \times F(r)\hat{\mathbf{r}} = 0$$

Recall that the external net torque on a system equals the time rate of change of angular momentum of the system; that is, $\boldsymbol{\tau} = d\mathbf{L}/dt$. Therefore, because $\boldsymbol{\tau} = 0$, the angular momentum **L** of the planet is a constant of the motion:

$$\mathbf{L} = \mathbf{r} \times \mathbf{p} = M_p \mathbf{r} \times \mathbf{v} = \text{constant}$$

We can relate this result to the following geometric consideration. In a time interval dt, the radius vector **r** in Figure 11.9b sweeps out the area dA, which equals one-half the area $|\mathbf{r} \times d\mathbf{r}|$ of the parallelogram formed by the vectors **r** and $d\mathbf{r}$. Because the displacement of the planet in the time interval dt is given by $d\mathbf{r} = \mathbf{v}dt$, we have

$$dA = \tfrac{1}{2}\left|\mathbf{r} \times d\mathbf{r}\right| = \tfrac{1}{2}\left|\mathbf{r} \times \mathbf{v}dt\right| = \frac{L}{2M_p}\,dt$$

$$\frac{dA}{dt} = \frac{L}{2M_p} = \text{constant} \qquad \qquad \textbf{[11.4]}$$

where L and M_p are both constants. Thus, we conclude that the radius vector from the Sun to any planet sweeps out equal areas in equal times.

It is important to recognize that this result is a consequence of the fact that the gravitational force is a central force, which in turn implies that angular momentum of the planet is constant. Therefore, the law applies to *any* situation that involves a central force, whether inverse-square or not.

EXERCISE   At its aphelion the planet Mercury is 6.99×10^{10} m from the Sun, and at its perihelion it is 4.60×10^{10} m from the Sun. If its orbital speed is 3.88×10^{4} m/s at the aphelion, what is its orbital speed at the perihelion?

Answer 5.90×10^{4} m/s

WEB

For information on Kepler's life as well as an animated demonstration of Kepler's second law, visit **www.kepler.arc.nasa.gov/ johannes.html**

THINKING PHYSICS 11.3

The Earth is closer to the Sun when it is winter in the Northern Hemisphere than in summer. July and January both have 31 days. In which month, if either, does the Earth move through a longer distance in its orbit?

Reasoning   The Earth is in a slightly elliptical orbit around the Sun. Because of angular momentum conservation, the Earth moves more rapidly when it is close to the Sun and more slowly when it is farther away. Thus, because it is closer to the Sun in January, it is moving faster and will cover more distance in its orbit than it will in July.

Kepler's Third Law

Kepler's third law is

The square of the orbital period of any planet is proportional to the cube of the semimajor axis of the elliptical orbit.

This can be shown easily for circular orbits. Consider a planet of mass M_p that is assumed to be moving about the Sun (mass M_S) in a circular orbit, as in Figure 11.10. Because the gravitational force provides the centripetal acceleration of the planet as it moves in a circle, we use Newton's second law for a particle in uniform circular motion,

$$\frac{GM_SM_p}{r^2} = \frac{M_pv^2}{r}$$

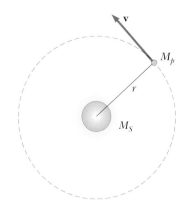

Figure 11.10

A planet of mass M_p moving in a circular orbit about the Sun. Kepler's third law relates the period of the orbit to the radius.

The orbital speed of the planet is $2\pi r/T$, where T is the period; therefore, the preceding expression becomes

$$\frac{GM_S}{r^2} = \frac{(2\pi r/T)^2}{r}$$

$$T^2 = \left(\frac{4\pi^2}{GM_S}\right)r^3 = K_S r^3$$

where K_S is a constant given by

$$K_S = \frac{4\pi^2}{GM_S} = 2.97 \times 10^{-19}\ \text{s}^2/\text{m}^3$$

For elliptical orbits Kepler's third law is expressed by starting with $T^2 = K_S r^3$ and replacing r with the length a of the semimajor axis (see Fig. 11.7):

$$T^2 = \left(\frac{4\pi^2}{GM_S}\right)a^3 = K_S a^3 \qquad\qquad \text{[11.5]}$$

Equation 11.5 is Kepler's third law. Because the semimajor axis of a circular orbit is its radius, Equation 11.5 is valid for both circular and elliptical orbits. Note that the constant of proportionality K_S is independent of the mass of the planet. Equation 11.5 is therefore valid for *any* planet. If we were to consider the orbit of a satellite about the Earth, such as the Moon, then the constant would have a different value, with the Sun's mass replaced by the Earth's mass, that is, $K_E = 4\pi^2/GM_E$.

Table 11.1 is a collection of useful planetary data. The last column verifies that the ratio T^2/r^3 is constant. The small variations in the values in this column are due to uncertainties in the data measured for the periods and semimajor axes of the planets.

Quick Quiz 11.1

Kepler's first law indicates that planets travel in elliptical orbits. This is intimately related to the fact that the gravitational force varies as the inverse square of the

TABLE 11.1	Useful Planetary Data[a]				
Body	**Mass (kg)**	**Mean Radius (m)**	**Period (s)**	**Mean Distance from Sun (m)**	$\dfrac{T^2}{r^3}\left(\dfrac{\text{s}^2}{\text{m}^3}\right)$
Mercury	3.18×10^{23}	2.43×10^6	7.60×10^6	5.79×10^{10}	2.97×10^{-19}
Venus	4.88×10^{24}	6.06×10^6	1.94×10^7	1.08×10^{11}	2.99×10^{-19}
Earth	5.98×10^{24}	6.37×10^6	3.156×10^7	1.496×10^{11}	2.97×10^{-19}
Mars	6.42×10^{23}	3.37×10^6	5.94×10^7	2.28×10^{11}	2.98×10^{-19}
Jupiter	1.90×10^{27}	6.99×10^7	3.74×10^8	7.78×10^{11}	2.97×10^{-19}
Saturn	5.68×10^{26}	5.85×10^7	9.35×10^8	1.43×10^{12}	2.99×10^{-19}
Uranus	8.68×10^{25}	2.33×10^7	2.64×10^9	2.87×10^{12}	2.95×10^{-19}
Neptune	1.03×10^{26}	2.21×10^7	5.22×10^9	4.50×10^{12}	2.99×10^{-19}
Pluto	$\approx 1.4 \times 10^{22}$	$\approx 1.5 \times 10^6$	7.82×10^9	5.91×10^{12}	2.96×10^{-19}
Moon	7.36×10^{22}	1.74×10^6	—	—	—
Sun	1.991×10^{30}	6.96×10^8	—	—	—

[a] For a more complete set of data, see, for example, the *Handbook of Chemistry and Physics*, Boca Raton, FL, The Chemical Rubber Publishing Co.

separation between the Sun and the planet. Suppose the gravitational force varied as the inverse *cube*, rather than the inverse square. This allows for some new types of orbits. What about the second and third laws? Will they change if the force depends on the inverse cube of the separation?

Quick **Quiz** 11.2

A comet is in a highly elliptical orbit around the Sun. The period of the comet's orbit is 90 days. Which of the following statements is true about the possibility of a collision between this comet and the Earth? (a) Collision is not possible. (b) Collision is possible. (c) Not enough information is available to determine whether a collision is possible.

Quick **Quiz** 11.3

A satellite moves in an elliptical orbit about the Earth such that, at perigee and apogee positions, the distances from the Earth's center are, respectively, D and $4D$. The relationship between the speeds at these two positions is (a) $v_p = v_a$; (b) $v_p = 4v_a$; (c) $v_a = 4v_p$; (d) $v_p = 2v_a$; (e) $v_a = 2v_p$.

11.4 • ENERGY CONSIDERATIONS IN PLANETARY AND SATELLITE MOTION

So far we have approached orbital mechanics from the point of view of forces and angular momentum. Let us now investigate the motion of planets in orbit from the *energy* point of view.

Consider an object of mass m moving with a speed v in the vicinity of a massive object of mass $M \gg m$. This two-object system might be a planet moving around the Sun, a satellite orbiting the Earth, or a comet making a one-time flyby past the Sun. We will treat the two objects of mass m and M as an isolated system. If we assume that M is at rest in an inertial reference frame (because $M \gg m$), then the total mechanical energy E of the two-object system is the sum of the kinetic energy of the object of mass m and the gravitational potential energy of the system:

$$E = K + U_g$$

Recall from Chapter 7 that the gravitational potential energy U_g associated with *any pair* of particles of masses m_1 and m_2 separated by a distance r is given by

$$U_g = -\frac{Gm_1 m_2}{r}$$

where we have defined $U_g \to 0$ as $r \to \infty$; therefore, in our case, the mechanical energy of the system of m and M is

$$E = \tfrac{1}{2} mv^2 - \frac{GMm}{r} \qquad \text{[11.6]}$$

Equation 11.6 shows that E may be positive, negative, or zero, depending on the value of v at a particular separation distance r. If we consider the energy diagram method of Section 7.7, we can show the potential and total energies of the system as a function of r as in Figure 11.11. A planet moving around the Sun and a satellite in orbit around the Earth are *bound* systems, such as those we discussed in Section 11.3—the Earth will always stay near the Sun and the satellite near the

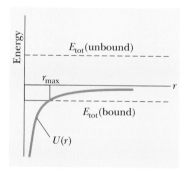

Figure 11.11

The lower total energy line represents a bound system. The separation distance r between the two gravitationally bound objects never exceeds r_{max}. The upper total energy line represents an unbound system of two objects interacting gravitationally. The separation distance r between the two objects can have any value.

Earth. In Figure 11.11, these are represented by a total energy that is negative. The point at which the total energy line intersects the potential energy curve is a turning point—the maximum separation distance r_{max} between the two bound objects.

A one-time meteoroid flyby represents an unbound system—the meteoroid interacts with the Sun but is not bound to it. Thus, the meteoroid can in theory move infinitely far away from the Sun. This is represented in Figure 11.11 by a total energy line in the positive region of the graph. This line never intersects the potential energy curve, so all values of r are possible.

For a bound system, such as the Earth and Sun, E is necessarily less than zero because we have chosen the convention that $U_g \rightarrow 0$ as $r \rightarrow \infty$. We can easily establish that $E < 0$ for the system consisting of an object of mass m moving in a circular orbit about an object of mass $M \gg m$. Applying Newton's second law to the object of mass m in uniform circular motion,

$$\sum F = ma \quad \longrightarrow \quad \frac{GMm}{r^2} = \frac{mv^2}{r}$$

Multiplying both sides by r and dividing by 2,

$$\tfrac{1}{2} mv^2 = \frac{GMm}{2r} \tag{11.7}$$

Substituting this into Equation 11.6, we obtain

$$E = \frac{GMm}{2r} - \frac{GMm}{r}$$

$$E = -\frac{GMm}{2r} \tag{11.8}$$

This result clearly shows that the **total mechanical energy must be negative in the case of circular orbits.** Furthermore, Equation 11.7 shows that the **kinetic energy of an object in a circular orbit is equal to one-half the magnitude of the potential energy of the system** (when the potential energy is chosen to be zero at infinite separation).

The total mechanical energy is also negative in the case of elliptical orbits. The expression for E for elliptical orbits is the same as Equation 11.8, with r replaced by the semimajor axis length, a:

• *Total energy of a planet–star system*

$$E = -\frac{GMm}{2a} \tag{11.9}$$

The total energy, the total angular momentum, and the total linear momentum of a planet–star system are constants of the motion, according to the isolated system model.

Quick Quiz 11.4

A comet moves in an elliptical orbit around the Sun. Which point in its orbit represents the highest value of (a) the speed of the comet, (b) the potential energy of the comet–Sun system, (c) the kinetic energy of the comet, and (d) the total energy of the comet–Sun system?

Example 11.3 A Satellite in an Elliptical Orbit

A satellite moves in an elliptical orbit about the Earth, as in Figure 11.12. The minimum and maximum distances from the surface of the Earth are 400 km and 3 000 km, respectively. Find the speeds of the satellite at apogee and perigee.

Solution Because the mass of the satellite is negligible compared with the Earth's mass, we take the center of mass of the Earth to be at rest. Gravity is a central force, and so the angular momentum of the satellite about the Earth's center of mass remains constant with time. With subscripts a and p

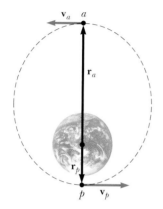

Figure 11.12

(Example 11.3) A satellite in an elliptical orbit about the Earth.

for the apogee and perigee positions, conservation of angular momentum for the satellite gives $L_p = L_a$, or

$$mv_p r_p = mv_a r_a$$

$$v_p r_p = v_a r_a$$

Using the Earth's radius of 6.37×10^6 m and the given data, we find that $r_a = 9.37 \times 10^6$ m and $r_p = 6.77 \times 10^6$ m. Thus,

$$(1) \quad \frac{v_p}{v_a} = \frac{r_a}{r_p} = \frac{9.37 \times 10^6 \text{ m}}{6.77 \times 10^6 \text{ m}} = 1.38$$

Because the satellite and the Earth form an isolated system, we can apply conservation of energy and obtain $E_p = E_a$, or

$$U_p + K_p = U_a + K_a$$

$$-G\frac{M_E m}{r_p} + \tfrac{1}{2}mv_p^2 = -G\frac{M_E m}{r_a} + \tfrac{1}{2}mv_a^2$$

$$(2) \quad 2GM_E\left(\frac{1}{r_a} - \frac{1}{r_p}\right) = (v_a^2 - v_p^2)$$

Because we know the numerical values of G, M_E, r_p, and r_a, we can use Equations (1) and (2) to determine the two unknowns v_p and v_a. Solving the equations simultaneously, we obtain

$$v_a = \boxed{8.27 \text{ km/s}} \qquad v_a = \boxed{5.98 \text{ km/s}}$$

Escape Speed

Suppose an object of mass m is projected vertically from the Earth's surface with an initial speed v_i, as in Figure 11.13. We can use energy considerations to find the minimum value of the initial speed such that the object will continue to move away from the Earth forever. Equation 11.6 gives the total energy of the object–Earth system at any point when the speed of the object and its distance from the center of the Earth are known. At the surface of the Earth, $r_i = R_E$. When the object reaches its maximum altitude, $v_f = 0$ and $r_f = r_{max}$. Because the total energy of the system is conserved, substitution of these conditions into Equation 11.6 gives

$$\tfrac{1}{2}mv_i^2 - \frac{GM_E m}{R_E} = -\frac{GM_E m}{r_{max}}$$

Solving for v_i^2 gives

$$v_i^2 = 2GM_E\left(\frac{1}{R_E} - \frac{1}{r_{max}}\right) \qquad \textbf{[11.10]}$$

If the initial speed is known, this expression can therefore be used to calculate the maximum altitude h because we know that $h = r_{max} - R_E$.

We are now in a position to calculate the minimum speed the object must have at the Earth's surface in order to continue to move away forever. This is

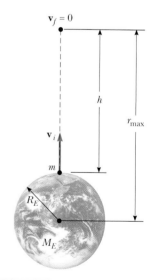

Figure 11.13

An object of mass m projected upward from the Earth's surface with an initial speed v_i reaches a maximum altitude $h = r_{max} - R_E$.

• *Escape speed*

PITFALL PREVENTION 11.3

You can't really escape

Although Equation 11.11 provides the escape speed, remember that this conventional name is misleading. It is impossible to *completely* escape from the Earth's gravitational influence, because the gravitational force is of infinite range. No matter how far away you are, you will always feel some gravitational force due to the Earth. In practice, however, this force will be much smaller than forces due to other astronomical objects closer to you, so the gravitational force from the Earth can be ignored.

called the **escape speed** v_{esc} and results in the speed asymptotically approaching *zero*. Letting $r_{max} \rightarrow \infty$ in Equation 11.10 and setting $v_f = 0$ and $v_i = v_{esc}$, we have

$$v_{esc} = \sqrt{\frac{2GM_E}{R_E}}$$ [11.11]

Note that this expression for v_{esc} is independent of the mass of the object projected from the Earth. For example, a spacecraft has the same escape speed as a molecule. Furthermore, the result is independent of the *direction* of the velocity.

Note also that Equations 11.10 and 11.11 can be applied to objects projected from *any* planet. That is, in general, the escape speed from any planet of mass M and radius R is

$$v_{esc} = \sqrt{\frac{2GM}{R}}$$ [11.12]

A list of escape speeds for the planets, the Moon, and the Sun is given in Table 11.2. Note that the values vary from 1.1 km/s for Pluto to about 618 km/s for the Sun. These results, together with some ideas from the kinetic theory of gases (Chapter 16), explain why our atmosphere does not contain significant amounts of hydrogen, which is the most abundant element in the Universe. As we shall see later, gas molecules have an average kinetic energy that depends on the temperature of the gas. Lighter molecules in an atmosphere have translational speeds that are closer to the escape speed than more massive molecules, so they have a higher probability of escaping from the planet, and the lighter molecules diffuse into space. This mechanism explains why the Earth does not retain hydrogen molecules and helium atoms in its atmosphere but does retain much heavier molecules, such as oxygen and nitrogen. On the other hand, Jupiter has a very large escape speed (60 km/s), which enables it to retain hydrogen, the primary constituent of its atmosphere.

TABLE 11.2

Escape Speeds from the Surfaces of the Planets, the Moon, and the Sun

Planet	v_{esc} (km/s)
Mercury	4.3
Venus	10.3
Earth	11.2
Mars	5.0
Jupiter	60
Saturn	36
Uranus	22
Neptune	24
Pluto	1.1
Moon	2.3
Sun	618

Quick Quiz 11.5

If you were a planetary prospector and discovered gold on an asteroid, it probably would not be a good idea to jump up and down in excitement over your find. Why?

Black Holes

In Chapter 10 we briefly described a rare event called a supernova—the catastrophic explosion of a very massive star. The material that remains in the central core of such an object continues to collapse, and the core's ultimate fate depends on its mass. If the core has a mass less than 1.4 times the mass of our Sun, it gradually cools down and ends its life as a white dwarf star. However, if the core's mass is greater than this, it may collapse further due to gravitational forces. What remains is a neutron star, discussed in Chapter 10, in which the mass of a star is compressed to a radius of about 10 km. (On Earth, a teaspoon of this material would weigh about 5 billion tons!)

An even more unusual star death may occur when the core has a mass greater than about three solar masses. The collapse may continue until the star becomes a very small object in space, commonly referred to as a **black hole.** In effect, black holes are the remains of stars that have collapsed under their own gravitational force. If an object such as a spacecraft comes close to a black hole, it experiences an extremely strong gravitational force and is trapped forever.

The escape speed from any spherical body depends on the mass and radius of the body. The escape speed for a black hole is very high, due to the concentration of the mass of the star into a sphere of very small radius. If the escape speed exceeds the speed of light c, radiation from the body (e.g., visible light) cannot escape, and the body appears to be black; hence the origin of the term *black hole*. The critical radius R_S at which the escape speed is c is called the **Schwarzschild radius** (Fig. 11.14). The imaginary surface of a sphere of this radius surrounding the black hole is called the **event horizon.** This is the limit of how close you can approach the black hole and hope to be able to escape.

Although light from a black hole cannot escape, light from events taking place near the black hole should be visible. For example, it is possible for a binary star system to consist of one normal star and one black hole. Material surrounding the ordinary star can be pulled into the black hole, forming an **accretion disk** around the black hole, as suggested in Figure 11.15. Friction among particles in the accretion disk results in transformation of mechanical energy into internal energy. As a result, the orbital height of the material above the event horizon decreases and the temperature rises. This high-temperature material emits a large amount of radiation, extending well into the x-ray region of the electromagnetic spectrum. These x-rays are characteristic of a black hole. Several possible candidates for black holes have been identified by observation of these x-rays.

Evidence also supports the existence of supermassive black holes at the centers of galaxies, with masses very much larger than the Sun. (The evidence is strong for a supermassive black hole of mass 2 to 3 million solar masses at the center of our galaxy.) Theoretical models for these bizarre objects predict that jets of material should be evident along the rotation axis of the black hole. Figure 11.16 shows a Hubble Space Telescope photograph of galaxy M87. The jet of material

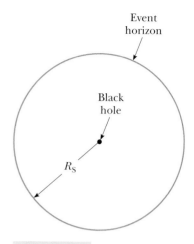

Figure 11.14

A black hole. The distance R_S equals the Schwarzschild radius. Any event occurring within the boundary of radius R_S, called the event horizon, is invisible to an outside observer.

Figure 11.15

A binary star system consisting of an ordinary star on the left and a black hole on the right. Matter pulled from the ordinary star forms an accretion disk around the black hole, in which matter is raised to very high temperatures, resulting in the emission of x-rays.

WEB

Stephen Hawking is one of the world's foremost experts on black holes. For information on black holes, visit Stephen Hawking's Web site at
www.hawking.org.uk/home/hindex.html

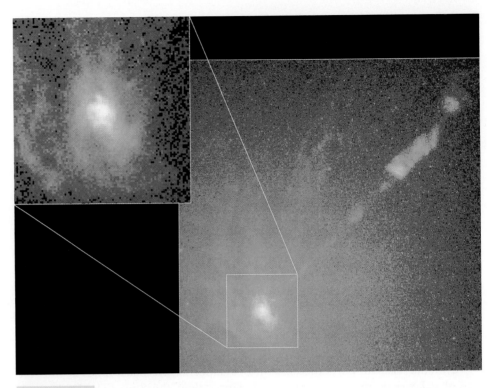

Figure 11.16

Hubble Space Telescope images of the galaxy M87. The inset shows the center of the galaxy. The wider view shows a jet of material moving away from the center of the galaxy toward the upper right of the figure at about one-tenth the speed of light. Such jets are believed to be evidence of a supermassive black hole at the galaxy's center. *(H. Ford et al. & NASA)*

WEB

For information on the current status of LIGO, visit **www.ligo.caltech.edu/**

coming from this galaxy is believed to be evidence for a supermassive black hole at the center of the galaxy.

Black holes are of considerable interest to those searching for **gravity waves,** which are ripples in space–time caused by changes in a gravitational system. These ripples can be caused by a star collapsing into a black hole, a binary star consisting of a black hole and a visible companion, and supermassive black holes at a galaxy center. A gravity wave detector, the Laser Interferometer Gravitational Wave Observatory (LIGO), is currently being built in the United States and hopes are high for detecting gravitational waves with this instrument.

EXERCISE The escape speed from the surface of the Earth is 11.2 km/s. Estimate the escape speed for a spacecraft from the surface of the Moon. The Moon has a mass 1/81 that of Earth and a radius $\frac{1}{4}$ that of Earth.

Answer 2.49 km/s

EXERCISE (a) Calculate the minimum energy required to send a 3 000-kg spacecraft from the Earth to a distant point in space where Earth's gravity is negligible. (b) If the journey is to take three weeks, what *average* power will the engines have to supply over the three-week period?

Answer (a) 1.88×10^{11} J (b) 103 kW

11.5 • ATOMIC SPECTRA AND THE BOHR THEORY OF HYDROGEN

In the preceding sections, we described a structural model for a large scale system—the Solar System. Let us now do the same for a very small scale system—the hydrogen atom. We shall find that a Solar System model of the atom, with a few extra features, provides explanations for some of the experimental observations made on the hydrogen atom.

As you may have already learned in a chemistry course, the hydrogen atom is the simplest known atomic system and an especially important one to understand. Much of what is learned about the hydrogen atom (which consists of one proton and one electron) can be extended to single-electron ions such as He^+ and Li^{2+}. Furthermore, a thorough understanding of the physics underlying the hydrogen atom can then be used to describe more complex atoms and the periodic table of the elements.

Atomic systems can be investigated by observing *electromagnetic waves* emitted from the atom. Our eyes are sensitive to visible light, one type of electromagnetic wave. As you shall learn in more detail in Chapter 13, a mechanical wave is a disturbance that transports energy as it moves through a medium without transporting matter. The wave will be one of our four simplification models around which we will identify analysis models, as we have done for a particle, a system, and a rigid body. A common form of periodic wave is the sinusoidal wave, whose shape is depicted in Figure 11.17. The distance between two consecutive crests of the wave is called the **wavelength** λ. As the wave travels to the right with a speed v, any point on the wave travels a distance of one wavelength in a time interval of one period T (the time for one cycle), so the wave speed is given by $v = \lambda / T$. The inverse of the period, $1/T$, is called the **frequency** f of the wave; it represents the number of cycles per second. Thus, the speed of the wave is often written as $v = \lambda f$. In this section, because we shall deal with electromagnetic waves—which travel at the speed of light c—the appropriate relation is

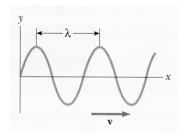

Figure 11.17

A sinusoidal wave traveling to the right with wave speed v. Any point on the wave moves a distance of one wavelength λ in a time interval equal to the period T of the wave.

$$c = \lambda f$$ [11.13]

• *Relation between wavelength, frequency, and wave speed*

Suppose an evacuated glass tube is filled with hydrogen (or some other gas). If a voltage applied between metal electrodes in the tube is great enough to produce an electric current in the gas, the tube emits light with colors that are characteristic of the gas. (This is how a neon sign works.) When the emitted light is analyzed with a device called a spectroscope, a series of discrete **spectral lines** is observed, each line corresponding to a different wavelength, or color, of light. Such a series of spectral lines is commonly referred to as an **emission spectrum.** The wavelengths contained in a given spectrum are characteristic of the element emitting the light. Figure 11.18 is a semigraphical representation of the spectra of various elements. It is semigraphical because the horizontal axis is linear in wavelength, but the vertical axis has no significance. Because no two elements emit the same line spectrum, this phenomenon represents a marvelous and reliable technique for identifying elements in a substance.

The emission spectrum of hydrogen shown in Figure 11.19 includes four prominent lines that occur at wavelengths of 656.3 nm, 486.1 nm, 434.1 nm, and 410.2 nm. In 1885, Johann Balmer (1825–1898) found that the wavelengths of

PITFALL PREVENTION 11.4

Why lines?

The phrase "spectral lines" is used very often when discussing the radiation from atoms. The shape of a *line* is due to the fact that the light is passed through a long and very narrow slit before being separated by wavelength. You will see many references to "lines" in both physics and chemistry.

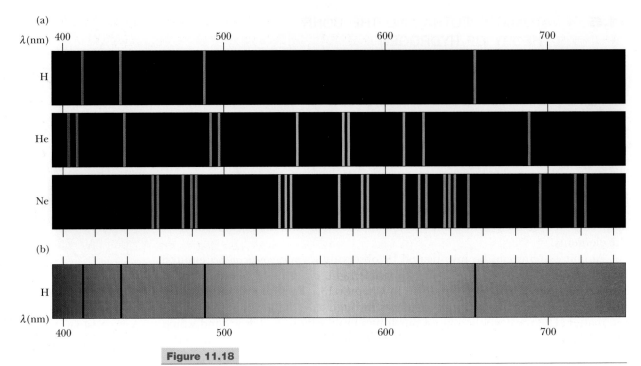

(a)

λ(nm)

(b)

Figure 11.18

Visible spectra. (a) Line spectra produced by emission in the visible range for the elements hydrogen, helium, and neon. (b) The absorption spectrum for hydrogen. The dark absorption lines occur at the same wavelengths as the emission lines for hydrogen shown in (a).

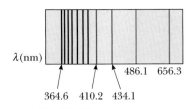

λ(nm)

486.1 656.3

364.6 410.2 434.1

Figure 11.19

A series of spectral lines for atomic hydrogen. The prominent lines labeled are part of the Balmer series.

• *Rydberg equation*

these and less prominent lines can be described by the following simple empirical equation:

$$\lambda = 364.56\,\frac{n^2}{n^2 - 4} \qquad n = 3, 4, 5, \ldots$$

in which n is an integer starting at 3, and the wavelengths given by this expression are in nanometers. These spectral lines are called the **Balmer series.** The first line in the Balmer series, at 656.3 nm, corresponds to $n = 3$; the line at 486.1 nm corresponds to $n = 4$; and so on. At the time this equation was formulated, it had no valid theoretical basis; it simply predicted the wavelengths correctly. Thus, this equation is not based on a model—it is simply a trial-and-error equation that happens to work. A few years later, Johannes Rydberg (1854–1919) recast the equation in the following form:

$$\frac{1}{\lambda} = R_H\left(\frac{1}{2^2} - \frac{1}{n^2}\right) \qquad n = 3, 4, 5, \ldots \qquad \text{[11.14]}$$

where n may have integral values of 3, 4, 5, . . . , and R_H is a constant, now called the **Rydberg constant.** If the wavelength is in meters, R_H has the value

$$R_H = 1.0973732 \times 10^7\,\text{m}^{-1}$$

Equation 11.14 is no more based on a model than is Balmer's equation. In this form, however, we can compare it with the predictions of a structural model of the hydrogen atom, which is described as follows.

In addition to emitting light at specific wavelengths, an element can also absorb light at specific wavelengths. The spectral lines corresponding to this process form what is known as an **absorption spectrum.** An absorption spectrum can be obtained by passing a continuous radiation spectrum (one containing all wavelengths) through a vapor of the element being analyzed. The absorption spectrum consists of a series of dark lines superimposed on the otherwise continuous spectrum.

At the beginning of the 20th century, scientists were perplexed by the failure of classical physics to explain the characteristics of atomic spectra. Why did atoms of a given element emit only certain wavelengths of radiation, so that the emission spectrum displayed discrete lines? Furthermore, why did the atoms absorb many of the same wavelengths which they emitted? In 1913, Niels Bohr provided an explanation of atomic spectra that includes some features of the currently accepted theory. Using the simplest atom, hydrogen, Bohr described a structural model for the atom. His model of the hydrogen atom contains some classical features that can be related to our analysis models, as well as some revolutionary postulates that could not be justified within the framework of classical physics. The basic assumptions of the Bohr model as it applies to the hydrogen atom are as follows:

1. The electron moves in a circular orbit about the proton under the influence of the electric force of attraction, as in Figure 11.20. This is a purely classical notion and is very similar to our previous discussion of planets in orbit around the Sun in our structural model of the Solar System.

2. Only certain electron orbits are stable, and these are the only orbits in which we find the electron. In these orbits the hydrogen atom does not emit energy in the form of radiation. Hence, the total energy of the atom remains constant, and classical mechanics can be used to describe the electron's motion. This restriction to certain orbits is a new idea, which is not consistent with classical physics. As we shall see in Chapter 24, an accelerating electron should emit energy by electromagnetic radiation. Thus, according to the continuity equation for energy, the emission of radiation from the atom should result in a decrease in the energy of the atom. Bohr's postulate boldly claims that this simply does not happen.

3. Radiation is emitted by the hydrogen atom when the atom "jumps" from a more energetic initial state to a lower state. The jump cannot be visualized or treated classically. In particular, the frequency f of the radiation emitted in the jump is related to the change in the atom's energy. The frequency of the emitted radiation is found from

$$E_i - E_f = hf \qquad \text{[11.15]}$$

where E_i is the energy of the initial state, E_f is the energy of the final state, h is **Planck's constant** ($h = 6.63 \times 10^{-34}$ J·s; we will see Planck's constant extensively in our studies of modern physics), and $E_i > E_f$. The notion of energy being emitted only when a jump occurs is nonclassical. Given this notion, however, Equation 11.15 is simply the continuity equation for energy, $\Delta E = \Sigma H \rightarrow E_f - E_i = -hf$. On the left is the change in energy of the system—the atom—and on the right is the energy transferred out of the system by electromagnetic radiation.

4. The size of the allowed electron orbits is determined by a condition imposed on the electron's orbital angular momentum. The allowed orbits are those for which the electron's orbital angular momentum about the nucleus is an integral multiple of $\hbar = h/2\pi$:

$$m_e v r = n\hbar \qquad n = 1, 2, 3, \ldots \qquad \text{[11.16]}$$

Niels Bohr (1885–1962)

Bohr, a Danish physicist, was an active participant in the early development of quantum mechanics and provided much of its philosophical framework. During the 1920s and 1930s, Bohr headed the Institute for Advanced Studies in Copenhagen. The institute was a magnet for many of the world's best physicists and provided a forum for the exchange of ideas. Bohr was awarded the 1922 Nobel prize for his investigation of the structure of atoms and of the radiation emanating from them. *(Photo courtesy of AIP Niels Bohr Library, Margarethe Bohr Collection)*

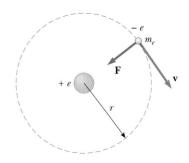

Figure 11.20

A pictorial representation of Bohr's model of the hydrogen atom, in which the electron is in a circular orbit about the proton.

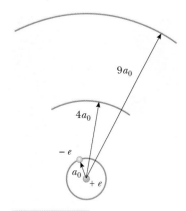

Figure 11.21

The first three Bohr orbits for hydrogen.

This new idea cannot be related to any of the models we have developed so far. It can be related, however, to a model that will be developed in later chapters, and we shall return to this idea at that time to see how it is predicted by the model. This is our first introduction to a notion from **quantum mechanics,** which describes the behavior of microscopic particles.

Using these four assumptions, Bohr built a structural model that explains the emission wavelengths of the hydrogen atom. The electrical potential energy of the system shown in Figure 11.20 is given by Equation 7.22, $U_e = -k_e e^2/r$, where k_e is the electric constant, e is the charge on the electron, and r is the electron–proton separation. Thus, the total energy of the atom, which contains both kinetic and potential energy terms, is

$$E = K + U_e = \tfrac{1}{2}m_e v^2 - k_e \frac{e^2}{r} \qquad [11.17]$$

According to assumption 2, the energy of the system remains constant—the system is isolated because the structural model does not allow for electromagnetic radiation for a given orbit. Applying Newton's second law to this system, we see that the magnitude of the electric attractive force on the electron, $k_e e^2/r^2$ (Eq. 5.15), is equal to the product of its mass and its centripetal acceleration ($a_c = v^2/r$):

$$\frac{k_e e^2}{r^2} = \frac{m_e v^2}{r}$$

From this expression, the kinetic energy is found to be

$$K = \tfrac{1}{2}m_e v^2 = \frac{k_e e^2}{2r} \qquad [11.18]$$

Substituting this value of K into Equation 11.17 gives the following expression for the total energy E of the hydrogen atom:

- *Total energy of the hydrogen atom*

$$E = -\frac{k_e e^2}{2r} \qquad [11.19]$$

Note that the total energy is negative,* indicating a bound electron–proton system. This means that energy in the amount of $k_e e^2/2r$ must be added to the atom just to separate the electron and proton by an infinite distance and make the total energy zero.† An expression for r, the radius of the allowed orbits, can be obtained by eliminating v by substitution between Equations 11.16 and 11.18:

- *Radii of Bohr orbits in hydrogen*

$$r_n = \frac{n^2 \hbar^2}{m_e k_e e^2} \qquad n = 1, 2, 3, \dots \qquad [11.20]$$

This result shows that the radii have discrete values, or are *quantized*. The integer n is called a **quantum number** and specifies the particular allowed **quantum state** of the atomic system.

---

* Compare this expression to Equation 11.8 for a gravitational system.

† This is called *ionizing* the atom. In theory, ionization requires separating the electron and proton by an infinite distance. In reality, however, the electron and proton are in an environment with huge numbers of other particles. Thus, ionization means separating the electron and proton by a distance large enough so that the interaction of these particles with other entities in their environment is larger than the remaining interaction between them.

The orbit for which $n = 1$ has the smallest radius; it is called the **Bohr radius** a_0, and has the value

$$a_0 = \frac{\hbar^2}{m_e k_e e^2} = 0.0529 \text{ nm} \qquad \text{[11.21]}$$

• *The Bohr radius*

The first three Bohr orbits are shown to scale in Figure 11.21.

The quantization of the orbit radii immediately leads to energy quantization. This can be seen by substituting $r_n = n^2 a_0$ into Equation 11.19. The allowed energy levels are

$$E_n = -\frac{k_e e^2}{2a_0}\left(\frac{1}{n^2}\right) \quad n = 1, 2, 3, \ldots \qquad \text{[11.22]}$$

Insertion of numerical values into Equation 11.22 gives

$$E_n = -\frac{13.606 \text{ eV}}{n^2} \quad n = 1, 2, 3, \ldots \qquad \text{[11.23]}$$

• *Energy levels of the hydrogen atom*

(Recall from Section 9.7 that 1 eV = 1.60×10^{-19} J.) The lowest quantum state, corresponding to $n = 1$, is called the **ground state,** and has an energy of $E_1 = -13.606$ eV. The next state, the **first excited state,** has $n = 2$ and an energy of $E_2 = E_1/2^2 = -3.401$ eV. Figure 11.22 is an **energy level diagram** showing the energies of these discrete energy states and the corresponding quantum numbers. This is another semigraphical representation. The vertical axis is linear in energy, but the horizontal axis has no significance. The lines correspond to the allowed energies. The atomic system cannot have any energies other than those represented by the lines. The upper limit of the quantized levels, corresponding to $n \rightarrow \infty$ (or $r \rightarrow \infty$) and $E \rightarrow 0$, represents the state for which the electron is removed from the atom.* Above this energy is a continuum of available states for the ionized atom. The minimum energy required to ionize the atom is called the **ionization energy.** As can be seen

The Rosette Nebula is a region of gas and dust surrounding an open cluster of stars. The red color is due to hydrogen atoms, excited by light from the stars in the cluster, making transitions from the $n = 3$ quantum state to the $n = 2$ state. (© 1987 *Royal Observatory Edinburgh/Anglo-Australian Observatory, photography by David F. Malin from U.K. Schmidt plates*)

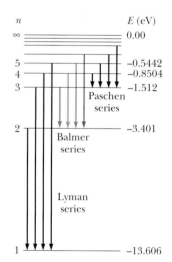

Figure 11.22

An energy level diagram for hydrogen. The discrete allowed energies are plotted on the vertical axis. Nothing is plotted on the horizontal axis, but the horizontal extent of the diagram is made large enough to show allowed transitions. Quantum numbers are given on the left.

---

* The phrase "the electron is removed from the atom" is very commonly used, but, of course, we realize that we mean that the electron and proton are separated *from each other.*

from Figure 11.22, the ionization energy for hydrogen, based on Bohr's calculation, is 13.6 eV. This constituted a major achievement for the Bohr theory because the ionization energy for hydrogen had already been measured to be precisely 13.6 eV.

Figure 11.22 also shows various transitions of the atom from one state to a lower state. These are the jumps referred to in Bohr's assumption 3. As the energy of the atom decreases in a jump, the difference in energy between the states is carried away by electromagnetic radiation. Those transitions ending on $n = 2$ are shown in color, corresponding to the color of the light they represent. The transitions ending on $n = 2$ form the Balmer series of spectral lines, the wavelengths of which are correctly predicted by the Rydberg equation (see Eq. 11.14). Figure 11.22 also shows other spectral series (the Lyman series and the Paschen series) that were found after Balmer's discovery.

Equation 11.22, together with Bohr's third postulate, can be used to calculate the frequency of the radiation that is emitted when the atom makes a transition* from a high-energy state to a low-energy state:

- *Frequency of radiation emitted from hydrogen*

$$f = \frac{E_i - E_f}{h} = \frac{k_e e^2}{2a_0 h}\left(\frac{1}{n_f^2} - \frac{1}{n_i^2}\right)$$ [11.24]

Because the quantity expressed in the Rydberg equation is wavelength, it is convenient to convert frequency to wavelength, using $c = f\lambda$, to obtain

- *Emission wavelengths of hydrogen*

$$\frac{1}{\lambda} = \frac{f}{c} = \frac{k_e e^2}{2a_0 hc}\left(\frac{1}{n_f^2} - \frac{1}{n_i^2}\right)$$ [11.25]

Notice the remarkable fact that the *theoretical* expression, Equation 11.25, is identical to the *empirical* Rydberg equation, Equation 11.14, provided that the combination of constants $k_e e^2/2a_0 hc$ is equal to the experimentally determined Rydberg constant, and that $n_f = 2$. After Bohr demonstrated the agreement of the constants in these two equations to a precision of about 1%, it was soon recognized as the crowning achievement of his structural model of the atom.

One question remains—what is the significance of $n_f = 2$? Its importance is due simply to the fact that those transitions ending on $n_f = 2$ result in radiation that happens to lie in the visible; therefore, they were easily observed! As seen in Figure 11.22, other series of lines end on other final states. These lie in regions of the spectrum not visible to the eye—the infrared and ultraviolet. The generalized Rydberg equation for any initial and final states is

$$\frac{1}{\lambda} = R_H\left(\frac{1}{n_f^2} - \frac{1}{n_i^2}\right)$$ [11.26]

In this equation, different series correspond to different values of n_f, and different lines within a series correspond to varying values of n_i.

Bohr immediately extended his structural model for hydrogen to other elements in which all but one electron had been removed. Ionized elements such as He^+, Li^{2+}, and Be^{3+} were suspected to exist in hot stellar atmospheres, where frequent atomic collisions occur with enough energy to completely remove one or more atomic electrons. Bohr showed that many mysterious lines observed in the Sun and several stars could not be due to hydrogen, but were correctly predicted by his theory if attributed to singly ionized helium.

PITFALL PREVENTION 11.5

The Bohr model is great, but . . .

The Bohr model is successful in some cases, such as predicting the ionization energy for hydrogen, but keep in mind that it is limited in its predictive power. For example, it cannot account for the spectra of more complex atoms and is unable to predict many subtle spectral details of hydrogen and other simple atoms. As a structural model, it is a simple first step toward understanding the atom. The notion of electrons in well-defined orbits around the nucleus is *not* consistent with current models of the atom, despite the cultural popularity of this model. We will see more sophisticated structural models in later chapters of the book.

---

* Another commonly used phrase is, "the electron makes a transition," but we will use, "the atom makes a transition" to emphasize the fact that the energy belongs to the system of the atom, not just to the electron. This is similar to our discussion in Chapter 7 of gravitational potential energy belonging to the system of an object and the Earth, not to the object alone.

Quick Quiz 11.6

A model of the hydrogen atom such as Bohr's is often called a *planetary* model because it is similar in some ways to the structural model of the Solar System. In what ways is the planetary model of the atom similar to the Solar System and in what ways is it different?

Example 11.4 An Electronic Transition in Hydrogen

A hydrogen atom makes a transition from the $n = 2$ energy state to the ground state (corresponding to $n = 1$). Find the wavelength and frequency of the emitted radiation.

Solution   We can use Equation 11.26 directly to obtain λ, with $n_i = 2$ and $n_f = 1$:

$$\frac{1}{\lambda} = R_H \left(\frac{1}{n_f^2} - \frac{1}{n_i^2} \right) = R_H \left(\frac{1}{1^2} - \frac{1}{2^2} \right) = \frac{3R_H}{4}$$

$$\lambda = \frac{4}{3R_H} = \frac{4}{3(1.097 \times 10^{-7}\ \mathrm{m}^{-1})}$$

$$= 1.215 \times 10^{-7}\ \mathrm{m} = \boxed{121.5\ \text{nm (ultraviolet)}}$$

Because $c = f\lambda$, the frequency of the radiation is

$$f = \frac{c}{\lambda} = \frac{3.00 \times 10^8\ \mathrm{m/s}}{1.215 \times 10^{-7}\ \mathrm{m}} = \boxed{2.47 \times 10^{15}\ \mathrm{s}^{-1}}$$

EXERCISE   What is the wavelength of the radiation emitted by hydrogen when the atom makes a transition from the $n = 3$ state to the $n = 1$ state?

Answer $\dfrac{9}{8} R_H = 102.6\ \text{nm}$

---

context connection

11.6 • CHANGING FROM A CIRCULAR TO AN ELLIPTICAL ORBIT

In the Context Connection of Chapter 3, we discussed our spacecraft in a circular parking orbit around the Earth. From our studies of Kepler's laws in this chapter, we are also aware that an elliptical orbit is possible for our spacecraft. Let us investigate how the motion of our spacecraft can be changed from a circular to an elliptical orbit, which will set us up for the Context Conclusion to our *Mission to Mars* Context.

Let us identify the system as the spacecraft and the Earth, *but not the portion of the fuel in the spacecraft that we use to change the orbit*. In a given orbit, the mechanical energy of the spacecraft–Earth system is given by Equation 11.8,

$$E = -\frac{GMm}{2r}$$

This energy includes the kinetic energy of the spacecraft and the potential energy associated with the gravitational force between the spacecraft and the Earth. If the rocket engines are fired, the exhausted fuel can be seen as doing work on the spacecraft–Earth system because the thrust force moves the spacecraft through a displacement. As a result, the mechanical energy of the spacecraft–Earth system increases.

The spacecraft has a new, higher energy but is constrained to be in an orbit that includes the original starting point. It cannot be in a higher energy circular orbit having a larger radius because this orbit would not contain the starting point. The only possibility is that the orbit is elliptical. Figure 11.23 shows the change from the original circular orbit to the new elliptical orbit for our spacecraft.

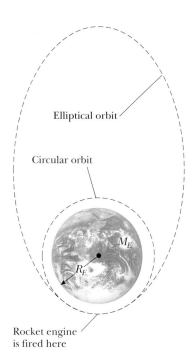

Figure 11.23

A spacecraft, originally in a circular orbit about the Earth, fires its engines and enters an elliptical orbit about the Earth.

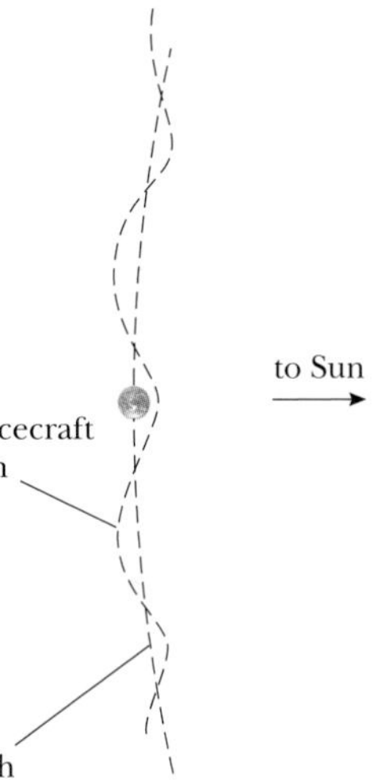

Path of spacecraft around Sun

to Sun

Path of Earth around Sun

Figure 11.24

A spacecraft in orbit about the Earth can be modeled as one in a circular orbit about the Sun, with its orbit about the Earth appearing as small perturbations from the circular orbit.

Equation 11.9 gives the energy of the spacecraft–Earth system for an elliptical orbit. Thus, if we know the new energy of the orbit, we can find the semimajor axis of the elliptical orbit. Conversely, if we know the semimajor axis of an elliptical orbit we would like to achieve, we can calculate how much additional energy is required from the rocket engines. This can then be converted to a required burn time for the rockets.

Larger amounts of energy increase supplied by the rocket engines will move the spacecraft into elliptical orbits with larger semimajor axes. What happens if the burn time of the engines is so long that the total mechanical energy of the space-craft–Earth system becomes *positive?* A positive energy refers to an *unbound* system. Thus, in this case, the spacecraft will *escape* from the Earth, going into a hyperbolic path that would not bring it back to Earth.

This is the essence of what must be done to transfer to Mars. Our rocket engines must be fired to leave the circular parking orbit and escape the Earth. At this point, our thinking must shift to a spacecraft–Sun system rather than a space-craft–Earth system. From this point of view, the spacecraft in orbit around the Earth can also be considered to be in a circular orbit around the Sun, moving along with the Earth, as shown in Figure 11.24. The orbit is not a perfect circle, be-cause there are perturbations corresponding to its extra motion around the Earth, but these perturbations are small compared with the radius of the orbit around the Sun. When our engines are fired to escape from the Earth, our orbit around the Sun changes from a circular orbit (ignoring the perturbations) to an elliptical one with the Sun at one focus. We shall choose the semimajor axis of our elliptical orbit so that it intersects the orbit of Mars! In the Context Conclusion, we shall look at more details of this process.

Example 11.5 How High Do We Go?

Imagine that you are in a spacecraft in circular orbit around the Earth, at a height of 300 km from the surface. You fire your rocket engines and, as a result, the magnitude of the mechanical energy of the spacecraft–Earth system decreases by 10.0%. (Because the mechanical energy is negative, a de-crease in magnitude is an increase in energy.) What is the greatest height of your spacecraft above the surface of the Earth in your new orbit?

Solution We set up a ratio of the energies of the two orbits, using Equations 11.8 and 11.9 for circular and elliptical or-bits:

$$\frac{E_{elliptical}}{E_{circular}} = \frac{\left(-\dfrac{GMm}{2a}\right)}{\left(-\dfrac{GMm}{2r}\right)} = \frac{r}{a}$$

The ratio on the left is 0.900, due to the 10.0% decrease in magnitude of the mechanical energy. Thus,

$$0.900 = \frac{r}{a} \longrightarrow a = \frac{r}{0.900} = 1.11r$$

From this, we can find the semimajor axis for the elliptical orbit:

$$a = 1.11r = 1.11(6.37 \times 10^3 \text{ km} + 3.00 \times 10^2 \text{ km})$$
$$= 7.40 \times 10^3 \text{ km}$$

The maximum distance from the center of the Earth will oc-cur when the spacecraft is at apogee, and is given by

$$r_{max} = 2a - r = 2(7.40 \times 10^3 \text{ km}) - (6.67 \times 10^3 \text{ km})$$
$$= 8.14 \times 10^3 \text{ km}$$

Now, if we subtract the radius of the Earth, we have the maxi-mum height above the surface,

$$h_{max} = r_{max} - R_E = 8.14 \times 10^3 \text{ km} - 6.37 \times 10^3 \text{ km}$$
$$= 1.77 \times 10^3 \text{ km}$$

SUMMARY

Newton's law of universal gravitation states that the gravitational force of attraction between any two particles of masses m_1 and m_2 separated by a distance r has the magnitude

$$F_g = G\frac{m_1 m_2}{r^2} \qquad [11.1]$$

where G is the **universal gravitational constant,** whose value is $6.672 \times 10^{-11} \text{ N} \cdot \text{m}^2/\text{kg}^2$.

Kepler's laws of planetary motion state that

1. Each planet in the Solar System moves in an elliptical orbit with the Sun at one focus.
2. The radius vector drawn from the Sun to any planet sweeps out equal areas in equal time intervals.
3. The square of the orbital period of any planet is proportional to the cube of the semimajor axis of the elliptical orbit.

Kepler's first law is a consequence of the inverse-square nature of the law of universal gravitation. The **semimajor axis** of an ellipse is a, where $2a$ is the longest dimension of the ellipse. The **semiminor axis** of the ellipse is b, where $2b$ is the shortest dimension of the ellipse. The **eccentricity** of the ellipse is $e = c/a$, where $a^2 = b^2 + c^2$.

Kepler's second law is a consequence of the fact that the gravitational force is a central force. For a central force, the angular momentum of the planet is conserved.

Kepler's third law is a consequence of the inverse-square nature of the universal law of gravitation. Newton's second law, together with the force law given by Equation 11.1, verifies that the period T and semimajor axis a of the orbit of a planet about the Sun are related by

$$T^2 = \left(\frac{4\pi^2}{GM_S}\right) a^3 \qquad [11.5]$$

where M_S is the mass of the Sun.

If an isolated system consists of a particle of mass m moving with a speed of v in the vicinity of a massive body of mass M, the *total energy* of the system is constant and is

$$E = \tfrac{1}{2}mv^2 - \frac{GMm}{r} \qquad [11.6]$$

If m moves in an elliptical orbit of major axis $2a$ about M, where $M \gg m$, the total energy of the system is

$$E = -\frac{GMm}{2a} \qquad [11.9]$$

The total energy is negative for any bound system—that is, one in which the orbit is closed, such as a circular or an elliptical orbit.

The Bohr model of the atom successfully describes the spectra of atomic hydrogen and hydrogen-like ions. One of the basic assumptions of this structural model is that the electron can exist only in discrete orbits such that the angular momentum $m_e v r$ is an integral multiple of $h/2\pi = \hbar$. Assuming circular orbits and a simple electrical attraction between the electron and proton, the energies of the quantum states for hydrogen are calculated to be

$$E_n = -\frac{k_e e^2}{2a_0}\left(\frac{1}{n^2}\right) \quad n = 1, 2, 3, \dots \qquad [11.22]$$

where k_e is the Coulomb constant, e is the fundamental electric charge, n is a positive integer called a **quantum number,** and $a_0 = 0.0529$ nm is the **Bohr radius.**

If the hydrogen atom makes a transition from a state whose quantum number is n_i to one whose quantum number is n_f, where $n_f < n_i$, the frequency of the radiation emitted by the atom is

$$f = \frac{k_e e^2}{2a_0 h}\left(\frac{1}{n_f^2} - \frac{1}{n_i^2}\right) \qquad [11.24]$$

Using $E_i - E_f = hf = hc/\lambda$, one can calculate the wavelengths of the radiation for various transitions. The calculated wavelengths are in excellent agreement with those in observed atomic spectra.

QUESTIONS

1. If the gravitational force on an object is directly proportional to its mass, why don't large masses fall with greater acceleration than small ones?
2. At night, you are farther away from the Sun than during the day. What's more, the force from the Sun on you is downward into the Earth at night, and upward into the sky during the day. If you had a sensitive enough bathroom scale, would you appear to weigh more at night than during the day?
3. The gravitational force the Sun exerts on the Moon is about twice as great as the gravitational force that the Earth exerts on the Moon. Why doesn't the Sun pull the Moon away from the Earth during a total eclipse of the Sun?

4. Explain why it takes more fuel for a spacecraft to travel from the Earth to the Moon than for the return trip. Estimate the difference.

5. Explain why no work is done on a planet as it moves in a circular orbit around the Sun, even though a gravitational force is acting on the planet. What is the *net* work done on a planet during each revolution as it moves around the Sun in an elliptical orbit?

6. At what position in its elliptical orbit is the speed of a planet a maximum? At what position is the speed a minimum?

7. Why don't we put a communications satellite in orbit around the 45th parallel? Wouldn't this be more useful in the United States and Canada than one in orbit around the equator?

8. A satellite in orbit is not truly traveling through a vacuum—it is moving through very, very thin air. Does the resulting air friction cause the satellite to slow down?

9. In his 1798 experiment, Cavendish was said to have "weighed the Earth." Explain this statement.

10. If a hole could be dug to the center of the Earth, do you think that the force on an object of mass m would still obey Equation 11.1 there? What do you think the force on the object would be at the center of the Earth?

11. The *Voyager* spacecraft was accelerated toward escape speed from the Sun by the gravitational force exerted on the spacecraft by Jupiter. How is this possible?

12. The *Apollo 13* spacecraft developed trouble in its oxygen system about halfway to the Moon. Why did the mission continue on around the Moon, and then return home, rather than immediately turn back to Earth?

13. How would you explain the fact that Saturn and Jupiter have periods much greater than one year?

14. Suppose the energy of a hydrogen atom is $-E$. What is the kinetic energy of the electron? What is the potential energy of the atom?

15. The Bohr theory of the hydrogen atom is based on several assumptions. Discuss those assumptions and their significance. Do any of them contradict classical physics?

16. Explain the significance behind the fact that the total energy of the atom in the Bohr model is negative.

17. Suppose that the electron in the hydrogen atom obeyed classical mechanics rather than quantum mechanics. Why should such a "hypothetical" atom emit a continuous spectrum rather than the observed line spectrum?

PROBLEMS

1, 2, 3 = straightforward, intermediate, challenging   □ = full solution available in the *Student Solutions Manual and Study Guide*

web = solution posted at **http://www.harcourtcollege.com/physics/**   💻 = computer useful in solving problem

 = Interactive Physics   ▓ = paired numerical/symbolic problems   ▨ = life science application

Section 11.1   Newton's Law of Universal Gravitation Revisited

Note: Problems 33 and 35 in Chapter 5 and Problems 31 through 33 in Chapter 7 can be assigned with this section.

1. Two ocean liners, each with a mass of 40 000 metric tons, are moving on parallel courses, 100 m apart. What is the magnitude of the acceleration of one of the liners toward the other due to their mutual gravitational attraction? Model the ships as particles.

2. A 200-kg object and a 500-kg object are separated by 0.400 m. (a) Find the net gravitational force exerted by these objects on a 50.0-kg object placed midway between them. (b) At what position (other than an infinitely remote one) can the 50.0-kg object be placed so as to experience a net force of zero?

3. Three uniform spheres of mass 2.00 kg, 4.00 kg, and 6.00 kg are placed at the corners of a right triangle, as in Figure P11.3. Calculate the net gravitational force on the

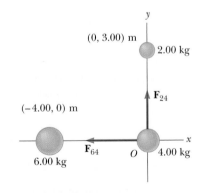

Figure P11.3

4.00-kg sphere, assuming the spheres are isolated from the rest of the Universe.

4. During a solar eclipse, the Moon, Earth, and Sun all lie on the same line, with the Moon between the Earth and the Sun. (a) What force is exerted by the Sun on the Moon?

(b) What force is exerted by the Earth on the Moon?
(c) What force is exerted by the Sun on the Earth?

5. The free-fall acceleration on the surface of the Moon is about one-sixth that on the surface of the Earth. If the radius of the Moon is about $0.250R_E$, find the ratio of their average densities, $\rho_{\text{Moon}}/\rho_{\text{Earth}}$.

6. A student proposes to measure the gravitational constant G by suspending two spherical objects from the ceiling of a tall cathedral and measuring the deflection of the cables from the vertical. Draw a free-body diagram of one of the objects. If two 100.0-kg objects are suspended at the lower ends of cables 45.00 m long, and the cables are attached to the ceiling 1.000 m apart, what is the separation of the objects?

7. On the way to the Moon the *Apollo* astronauts reached a point where the Moon's gravitational pull became stronger than the Earth's. (a) Determine the distance of this point from the center of the Earth. (b) What is the acceleration due to the Earth's gravity at this point?

8. A spacecraft in the shape of a long cylinder has a length of 100 m and its mass with occupants is 1 000 kg. It has strayed too close to a 1.0-m radius black hole having a mass 100 times that of the Sun (Fig. P11.8). The nose of the spacecraft points toward the center of the black hole, and the distance between the nose and the black hole is 10.0 km. (a) Determine the total force on the spacecraft. (b) What is the difference in the gravitational fields acting on the occupants in the nose of the ship and on those in the rear of the ship, farthest from the black hole?

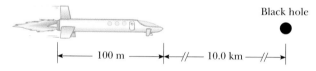

Figure P11.8

9. Compute the magnitude and direction of the gravitational field at a point *P* on the perpendicular bisector of two particles with equal masses separated by a distance $2a$ as shown in Figure P11.9.

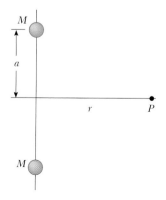

Figure P11.9

10. How much work is done by the Moon's gravitational field as a 1 000-kg meteor comes in from outer space and impacts on the Moon's surface?

11. After our Sun exhausts its nuclear fuel, its ultimate fate may be to collapse to a *white dwarf* state, in which it has approximately the same mass as it has now, but a radius equal to the radius of the Earth. Calculate (a) the average density of the white dwarf, (b) the free-fall acceleration at its surface, and (c) the gravitational potential energy associated with a 1.00-kg object at its surface. (Take $U_g = 0$ at infinity.)

Section 11.3   Kepler's Laws

12. A communication satellite in geosynchronous orbit remains above a single point on the Earth's equator as the planet rotates on its axis. (a) Calculate the radius of its orbit. (b) The satellite relays a radio signal from a transmitter near the North Pole to a receiver, also near the North Pole. Traveling at the speed of light, how long is the radio wave in transit?

13. Plaskett's binary system consists of two stars that revolve in a circular orbit about a center of mass midway between them. This means that the masses of the two stars are equal (Fig. P11.13). If the orbital velocity of each star is 220 km/s and the orbital period of each is 14.4 days, find the mass M of each star. (For comparison, the mass of our Sun is 1.99×10^{30} kg.)

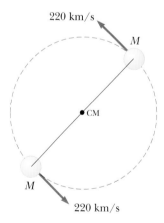

Figure P11.13

14. The *Explorer VIII* satellite, placed into orbit November 3, 1960 to investigate the ionosphere, had the following orbit parameters: perigee, 459 km; apogee, 2 289 km (both distances above the Earth's surface); period, 112.7 min. Find the ratio v_p/v_a of the speed at perigee to that at apogee.

15. Io, a satellite of Jupiter, has an orbital period of 1.77 days and an orbital radius of 4.22×10^5 km. From these data, determine the mass of Jupiter.

16. Comet Halley (Figure P11.16, not to scale) approaches the Sun to within 0.570 AU, and its orbital period is 75.6 years. (AU is the symbol for astronomical unit, where 1 AU = 1.50×10^{11} m is the mean Earth–Sun distance.) How far from the Sun will Comet Halley travel before it starts its return journey?

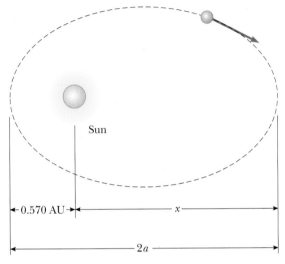

Figure P11.16

17. Two planets X and Y travel counterclockwise in circular orbits about a star as in Figure P11.17. The radii of their orbits are in the ratio 3:1. At some time, they are aligned as in Figure P11.17a, making a straight line with the star. During the next five years, the angular displacement of planet X is 90.0°, as in Figure P11.17b. Where is planet Y at this time?

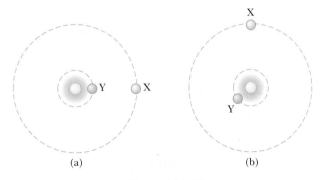

(a)   (b)

Figure P11.17

18. Suppose the Sun's gravity were switched off. The planets would leave their nearly circular orbits and fly away in straight lines, as described by Newton's first law. Would Mercury ever be farther from the Sun than Pluto? If so, find how long it would take for Mercury to achieve this passage. If not, give a convincing argument that Pluto is always farther from the Sun.

19. As thermonuclear fusion proceeds in its core, the Sun loses mass at a rate of 3.64×10^9 kg/s. During the 5 000-yr period of recorded history, by how much has the length of the year changed due to the loss of mass from the Sun? (*Hints:* Assume the Earth's orbit is circular. There is no external torque on the Earth–Sun system, so angular momentum is conserved. If x is small compared with 1, then $(1 + x)^n$ is nearly equal to $1 + nx$.)

Section 11.4   Energy Considerations in Planetary and Satellite Motion

Note: Problems 41 and 42 in Chapter 6 and 34 and 38 in Chapter 7 can be assigned with this section.

20. A 500-kg satellite is in a circular orbit at an altitude of 500 km above the Earth's surface. Because of air friction, the satellite eventually falls to the Earth's surface, where it hits the ground with a speed of 2.00 km/s. How much energy was transformed into internal energy by means of friction?

21. A "treetop satellite" (Fig. P11.21) moves in a circular orbit just above the surface of a planet, assumed to offer no air resistance. Show that its orbital speed v and the escape speed from the planet are related by the expression $v_{esc} = \sqrt{2}v$.

22. A satellite of mass m is placed into Earth orbit at an altitude h. (a) Assuming a circular orbit, how long does the satellite take to complete one orbit? (b) What is the satellite's speed? (c) What is the minimum energy necessary to place this satellite in orbit? Ignore air resistance but include the effect of the planet's daily rotation. At what location on the Earth's surface and in what direction should the satellite be launched to minimize the required energy investment? Represent the mass and radius of the Earth as M_E and R_E.

23. The planet Uranus has a mass about 14 times the Earth's mass, and its radius is equal to about 3.7 Earth radii. (a) By setting up ratios with the corresponding Earth values, find the acceleration due to gravity at the cloud tops of Uranus. (b) Ignoring the rotation of the planet, find the minimum escape speed from Uranus.

24. Derive an expression for the work required to move an Earth satellite of mass m from a circular orbit of radius $2R_E$ to one of radius $3R_E$.

25. A comet of mass 1.20×10^{10} kg moves in an elliptical orbit around the Sun. Its distance from the Sun ranges between 0.500 AU and 50.0 AU. (a) What is the eccentricity of its orbit? (b) What is its period? (c) At aphelion what is the potential energy of the comet–Sun system? (*Note:* 1 AU = one astronomical unit = the average distance from Sun to Earth = 1.496×10^{11} m.)

B.C.   by John Hart

By permission of John Hart and Field Enterprises, Inc.

Figure P11.21

Section 11.5 Atomic Spectra and the Bohr Theory of Hydrogen

26. A hydrogen atom emits light as it undergoes a transition from the $n = 3$ state to the $n = 2$ state. Calculate (a) the energy, (b) the wavelength, and (c) the frequency of the radiation.

27. (a) What value of n_i is associated with the 94.96-nm spectral line in the Lyman series of hydrogen? (b) Could this wavelength be associated with the Paschen or Balmer series?

28. For a hydrogen atom in its ground state, use the Bohr model to compute (a) the orbital speed of the electron, (b) the kinetic energy of the electron, and (c) the electrical potential energy of the atom.

29. A hydrogen atom is in its first excited state ($n = 2$). Using **web** the Bohr theory of the atom, calculate (a) the radius of the orbit, (b) the linear momentum of the electron, (c) the angular momentum of the electron, (d) the kinetic energy, (e) the potential energy, and (f) the total energy.

30. Four possible transitions for a hydrogen atom are as follows:

    i. $n_i = 2$; $n_f = 5$ ii. $n_i = 5$; $n_f = 3$

    iii. $n_i = 7$; $n_f = 4$ iv. $n_i = 4$; $n_f = 7$

(a) In which transition is light of the shortest wavelength emitted? (b) In which transition does the atom gain the most energy? (c) In which transition(s) does the atom lose energy?

31. How much energy is required to ionize hydrogen (a) when it is in the ground state? (b) When it is in the state for which $n = 3$?

32. Show that the speed of the electron in the nth Bohr orbit in hydrogen is given by

$$v_n = \frac{k_e e^2}{n\hbar}$$

33. Two hydrogen atoms collide head-on and end up with zero kinetic energy. Each atom then emits light with a wavelength of 121.6 nm ($n = 2$ to $n = 1$ transition). At what speed were the atoms moving before the collision?

Section 11.6 Context Connection — Changing from a Circular Orbit to an Elliptical Orbit

34. A spacecraft of mass 1.00×10^4 kg is in a circular orbit at an altitude of 500 km above the Earth's surface. Mission Control wants to fire the engines so as to put the spacecraft in an elliptical orbit around the Earth with an apogee of 2.00×10^4 km. How much energy must be used from the fuel to achieve this orbit? (Assume that all of the fuel energy goes into increasing the orbital energy. This will give a lower limit to the required energy because some of the energy from the fuel will appear as internal energy in the hot exhaust gases and engine parts.)

35. A spacecraft is approaching Mars after a long trip from Earth. Its velocity is such that it is traveling along a

parabolic trajectory under the influence of the gravitational force from Mars. The distance of closest approach will be 300 km above the Martian surface. At this point of closest approach, the engines will be fired to slow down the spacecraft and place it in a circular orbit 300 km above the surface. (a) By what percentage must the speed of the spacecraft be reduced to achieve the desired orbit? (b) How would the answer to part (a) change if the distance of closest approach and the desired circular orbit altitude were 600 km instead of 300 km? (*Hint:* The energy of the spacecraft–Mars system for a parabolic orbit is $E = 0$.)

Additional Problems

36. The *Solar and Heliospheric Observatory* (SOHO) spacecraft has a special orbit, chosen so that its view of the Sun is never eclipsed and it is always close enough to Earth to transmit data easily. It moves in a near-circle around the Sun in an orbit that is smaller than the Earth's circular orbit and its period equal to one year. It is always located between Earth and Sun along the line joining them. Both objects exert gravitational forces on the observatory. Show that its distance from the Earth must be between 1.47×10^9 m and 1.48×10^9 m. In 1772, Joseph Louis Lagrange determined theoretically the special location allowing this orbit. The SOHO spacecraft took this position on February 14, 1996. (*Hint:* Use data that are precise to four digits. The mass of the Earth is 5.983×10^{24} kg.)

37. The positron is the antiparticle to the electron. It has the same mass and a positive electric charge of the same magnitude as that of the electron. Positronium is a hydrogen-like atom consisting of a positron and an electron revolving around each other. Using the Bohr model, find the allowed distances between the two particles and the allowed energies of the system.

38. Let Δg_M represent the difference in the gravitational fields produced by the Moon at the points on the Earth's surface nearest to and farthest from the Moon. Find the fraction $\Delta g_M/g$, where g is the Earth's gravitational field. (This difference is responsible for the occurrence of the *lunar tides* on the Earth.)

39. In Larry Niven's science fiction novel *Ringworld*, a rigid ring of material rotates about a star (Fig. P11.39). The tangential speed of the ring is 1.25×10^6 m/s, and its radius is 1.53×10^{11} m. (a) Show that the centripetal acceleration of the inhabitants is 10.2 m/s^2. (b) The inhabitants of this ring world live on the starlit inner surface of the ring. Each person experiences a normal contact force **n**. Acting alone, this normal force would produce an inward acceleration of 9.90 m/s^2. Additionally, the star at the center of the ring exerts a gravitational force on the ring and its inhabitants. The difference between the total accelera-

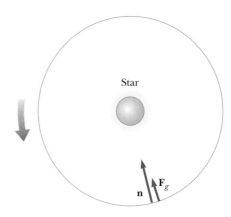

Figure P11.39

tion and the acceleration provided by the normal force is due to the gravitational attraction of the central star. Show that the mass of the star is approximately 10^{32} kg.

40. (a) Show that the rate of change of the free-fall acceleration with distance above the Earth's surface is

$$\frac{dg}{dr} = -\frac{2GM_E}{R_E^3}$$

This rate of change over distance is called a *gradient*. (b) If h is small relative to the radius of the Earth, show that the difference in free-fall acceleration between two points separated by vertical distance h is

$$|\Delta g| = \frac{2GM_E h}{R_E^3}$$

(c) Evaluate this difference for $h = 6.00$ m, a typical height for a two-story building.

41. *Voyagers 1* and *2* surveyed the surface of Jupiter's moon Io and photographed active volcanoes spewing liquid sulfur to heights of 70 km above the surface of this moon. Find the speed with which the liquid sulfur left the volcano. Io's mass is 8.9×10^{22} kg, and its radius is 1820 km.

42. As an astronaut, you observe a small planet to be spherical. After landing on the planet, you set off, walking always straight ahead, and find yourself returning to your spacecraft from the opposite side after completing a lap of 25.0 km. You hold a hammer and a falcon feather at a height of 1.40 m, release them, and observe that they fall together to the surface in 29.2 s. Determine the mass of the planet.

43. In introductory physics laboratories, a typical Cavendish balance for measuring the gravitational constant G uses lead spheres with masses of 1.50 kg and 15.0 g whose centers are separated by about 4.50 cm. Calculate the gravitational force between these spheres, treating each as a point mass located at the center of the sphere.

44. Many people assume that air resistance acting on a moving object will always make the object slow down. It can ac-

tually be responsible for making the object speed up. Consider a 100-kg Earth satellite in a circular orbit at an altitude of 200 km. A small force of air resistance makes the satellite drop slowly into a circular orbit with an altitude of 100 km. (a) Calculate its original speed. (b) Calculate its final speed in this process. (c) Calculate its original energy. (d) Calculate its final energy. (e) Show that it has lost mechanical energy and find the amount of the loss due to friction. (f) What force makes the object's speed increase? You will find a free-body diagram useful in explaining your answer.

45. Two hypothetical planets of masses m_1 and m_2 and radii r_1
web and r_2, respectively, are nearly at rest when they are an infinite distance apart. Because of their gravitational attraction, they head toward each other on a collision course. (a) When their center-to-center separation is d, find expressions for the speed of each planet and for their relative velocity. (b) Find the kinetic energy of each planet just before they collide, assuming $m_1 = 2.00 \times 10^{24}$ kg, $m_2 = 8.00 \times 10^{24}$ kg, $r_1 = 3.00 \times 10^6$ m, and $r_2 = 5.00 \times 10^6$ m. (*Hint:* Both energy and momentum are conserved.)

46. The maximum distance from the Earth to the Sun (at aphelion) is 1.521×10^{11} m, and the distance of closest approach (at perihelion) is 1.471×10^{11} m. If the Earth's orbital speed at perihelion is 3.027×10^4 m/s, determine (a) the Earth's orbital speed at aphelion, (b) the kinetic and potential energies at perihelion, and (c) the kinetic and potential energies at aphelion. Is the total energy constant? (Ignore the effect of the Moon and other planets.)

47. Studies of the relationship of the Sun to its galaxy—the Milky Way—have revealed that the Sun is located near the outer edge of the galactic disc, about 30 000 lightyears from the center. The Sun has an orbital speed of approximately 250 km/s around the galactic center. (a) What is the period of the Sun's galactic motion? (b) What is the order of magnitude of the mass of the Milky Way galaxy? Suppose that the galaxy is made mostly of stars of which the Sun is typical. What is the order of magnitude of the number of stars in the Milky Way?

48. The oldest artificial satellite in orbit is *Vanguard I*, launched March 3, 1958. Its mass is 1.60 kg. In its initial orbit its minimum distance from the center of the Earth was 7.02 Mm and its speed at this perigee point was 8.23 km/s. (a) Find its total energy. (b) Find the magnitude of its angular momentum. (c) Find its speed at apogee and its maximum (apogee) distance from the center of the Earth. (d) Find the semimajor axis of its orbit. (e) Determine its period.

49. A rocket is given an initial speed vertically upward of $v_i = 2\sqrt{Rg}$ at the surface of the Earth, which has radius R

and surface free-fall acceleration g. The rocket motors are quickly cut off, and thereafter the rocket coasts under the action of gravitational forces only. (Ignore atmospheric friction and the Earth's rotation.) Derive an expression for the subsequent speed v as a function of the distance r from the center of the Earth in terms of g, R, and r.

50. Astronomers detect a distant meteoroid moving along a straight line that, if extended, would pass at a distance $3R_E$ from the center of the Earth, where R_E is the radius of the Earth. What minimum speed must the meteoroid have if the Earth's gravity is not to deflect the meteoroid to make it strike the Earth?

51. Two stars of masses M and m, separated by a distance d, revolve in circular orbits about their center of mass (Fig. P11.51). Show that each star has a period given by

$$T^2 = \frac{4\pi^2 d^3}{G}(M + m)$$

(*Hint:* Apply Newton's second law to each star, and note that the center-of-mass condition requires that $Mr_2 = mr_1$, where $r_1 + r_2 = d$.)

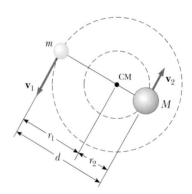

Figure P11.51

52. A spherical planet has uniform density ρ. Show that the minimum period for a satellite in orbit around it is

$$T_{\text{min}} = \sqrt{\frac{3\pi}{G\rho}}$$

independent of the radius of the planet.

53. The acceleration of an object moving in the gravitational field of the Earth is

$$\mathbf{a} = -GM_E\,\frac{\mathbf{r}}{r^3}$$

where $\mathbf{r}$ is the position vector directed from the center of the Earth to the object. Choosing the origin at the center of the Earth and assuming that the small object is moving in the xy plane, we find that the rectangular (cartesian) components of its acceleration are

$$a_x = \frac{-GM_E x}{(x^2 + y^2)^{3/2}} \qquad a_y = \frac{-GM_E y}{(x^2 + y^2)^{3/2}}$$

Use a computer to set up and carry out a numerical prediction of the motion of the object, according to Euler's method. Assume the initial position of the object is $x = 0$ and $y = 2R_E$, where R_E is the radius of the Earth. Give the

object an initial velocity of 5 000 m/s in the x direction. The time increment should be made as small as practical. Try 5 s. Plot the x and y coordinates of the object as time goes on. Does the object hit the Earth? Vary the initial velocity until you find a circular orbit.

ANSWERS TO QUICK QUIZZES

11.1 Kepler's second law is a consequence of conservation of angular momentum. Angular momentum is conserved as long as there is no torque, which is true for any central force, regardless of the particular dependence on separation. Thus, Kepler's second law would be unchanged. The particular mathematical form of Kepler's third law depends on the inverse-square character of the force. If the force were an inverse cube, the relationship between a power of the period and a power of the semimajor axis would still be a proportionality, but the powers would be different from that in the existing law.

11.2 (a). From Kepler's third law and the given period, the major axis of the comet can be calculated. It is found to be 1.2×10^{11} m. Because this is smaller than the Earth–Sun distance, the comet cannot possibly collide with the Earth.

11.3 (b). From conservation of momentum, $m v_p r_p = m v_a r_a$, so that $v_p = (r_a / r_p) v_a = (4D/D) v_a = 4 v_a$.

11.4 (a) Because of conservation of angular momentum, the speed of the comet is highest at its perihelion, its closest position to the Sun. (b) The potential energy of the

comet–Sun system is highest when the comet is at aphelion, its farthest distance from the Sun. (c) The kinetic energy is highest at the point at which the speed is highest, the perihelion. (d) The total energy of the system is the same regardless of where the comet is in its orbit.

11.5 The mass of the asteroid might be so small that you would be able to exceed escape speed by leg power alone. You would jump up, but you would not come down!

11.6 The planetary model of the hydrogen atom is similar to a Solar System model in that the electron is moving about the nucleus in a circular orbit, similar to the way we model the planets in orbit about the Sun. It has several differences, however, aside from the sizes of the systems. The atomic model depends on an electric force, but the Solar System depends on a gravitational force. In the Bohr theory, the allowed radii of the electron orbits are quantized, but there is no detectable quantization in the Solar System. Finally, we see the electron in a Bohr atom make transitions between energy levels, but we do not in the case for planets in a Solar System.

Mission to Mars

A SUCCESSFUL MISSION PLAN

Now that we have explored the physics of classical mechanics, let us return to our central question for the *Mission to Mars* Context:

How can we undertake a successful transfer of a spacecraft from Earth to Mars?

We make use of the physical principles that we now understand and apply them to our journey from Earth to Mars.

Let us start with a more modest proposition. Suppose that our space shuttle is in a circular orbit around the Earth and you are a passenger on the shuttle. If you toss a wrench in the direction of travel, tangent to the circular path, what orbital path will the wrench follow?

Let us adopt a simplification model in which the space shuttle is much more massive than the wrench. Conservation of momentum for the isolated system of the wrench and the spacecraft tells us that the space shuttle must slow down slightly once the wrench is thrown. Because of the mass difference between the wrench and space shuttle, however, we can ignore the small change in the speed of the shuttle. The wrench now enters a new orbit, from its perigee position, with more energy than that for the circular orbit. Because the energy is related to the major axis, the wrench is injected into an elliptical orbit, as discussed in the Context Connection of Chapter 11 and as shown in Figure 1. Thus, the path of the wrench is changed from a circular to an elliptical orbit by providing the wrench–Earth system with extra energy. The energy is provided by the force you apply to the wrench tangent to the circular orbit—you have done work on the system. The elliptical orbit will take the wrench farther from the Earth than the circular orbit. If another shuttle were in a higher circular orbit than the first shuttle, you could throw the wrench so that it transfers from one shuttle to another, as

Elliptical orbit
of wrench   Circular orbit
of shuttle

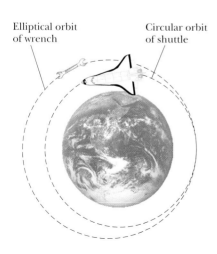

Figure 1

A wrench thrown tangent to the circular orbit of a space shuttle enters an elliptical orbit.

Circular orbit of
second shuttle

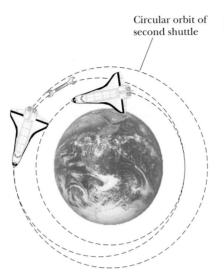

Figure 2

If a second space shuttle were in a higher circular orbit, the wrench could be carefully thrown so as to be transferred from one space shuttle to the other.

shown in Figure 2. For this to occur, the elliptical orbit of the wrench must intersect with the higher shuttle orbit. Furthermore, the wrench and the second shuttle must arrive at the same point at the same time.

This is the essence of our planned mission from Earth to Mars. Rather than transferring a wrench between two shuttles in orbit around the Earth, we will transfer a spacecraft between two planets in orbit around the Sun. Kinetic energy is added to the wrench–Earth system by throwing the wrench. Kinetic energy is added to the spacecraft–Sun system by firing the engines.

What if you were to throw the wrench harder and harder in the previous example? The wrench would be placed in a larger and larger elliptical orbit around the Earth. As you increased the launch velocity, you could inject the wrench into a *hyperbolic* escape orbit, relative to the Earth, and into an *elliptical* orbit around the *Sun*. This is the approach we will take for the trip from Earth to Mars—we will break free from a circular *parking* orbit around the Earth and move into an elliptical *transfer* orbit around the Sun. The spacecraft will then continue on its journey to Mars, where it will enter a new parking orbit.

Now let us focus our attention on the transfer orbit part of the journey. One simple transfer orbit is called a **Hohmann transfer,** the type of transfer imparted to the wrench in the earlier discussion. The Hohmann transfer involves the least energy expenditure and thus requires the smallest amount of fuel. As might be expected for a lowest energy transfer, the transfer time for a Hohmann transfer is longer than for other types of orbits. We shall investigate the Hohmann transfer due to its simplicity and its general usefulness in planetary transfers.

The rocket is fired from the parking orbit such that the spacecraft enters an elliptical orbit around the Sun at its perihelion and encounters the planet at the spacecraft's aphelion. Thus, the spacecraft makes exactly one half of a revolution about its elliptical path during the transfer, as shown in Figure 3.

This is an energy-efficient process, because fuel is expended only at the beginning and at the end. The movement between parking orbits around Earth and Mars is free—the spacecraft simply follows Kepler's laws while in an elliptical orbit around the Sun.

Let us perform a simple numerical example, to see how to apply the mechanical laws to this process. We assume that the spacecraft is in a parking orbit above

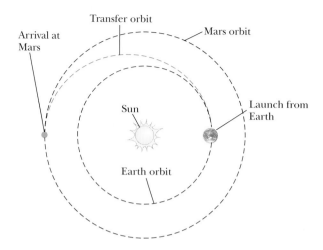

Figure 3

The Hohmann transfer orbit from Earth to Mars. This is similar to transferring the wrench from one shuttle to another in Figure 2—we are transferring a spacecraft from one planet to another.

the Earth's surface. Imagine that the spacecraft is in orbit around the Sun, with a perturbation in its orbit caused by the Earth. Thus, the tangential speed of the Earth about the Sun also represents the average speed of the spacecraft around the Sun. This speed is calculated from Newton's second law for a particle in uniform circular motion:

$$F = ma \longrightarrow G\frac{M_{Sun}m_{Earth}}{r^2} = m_{Earth}\frac{v^2}{r}$$

$$\longrightarrow v = \sqrt{\frac{GM_{Sun}}{r}} = \sqrt{\frac{(6.67 \times 10^{-11}\,\text{N·m}^2/\text{kg}^2)(1.99 \times 10^{30}\,\text{kg})}{1.50 \times 10^{11}\,\text{m}}}$$

$$= 2.97 \times 10^4\,\text{m/s}$$

This is the original speed of the spacecraft, to which we add a change Δv in order to inject the spacecraft into the transfer orbit.

The major axis of the elliptical transfer orbit is found by adding together the orbit radii of Earth and Mars (see Fig. 3):

$$\text{Major axis} = 2a = r_{Earth} + r_{Mars}$$
$$= 1.50 \times 10^{11}\,\text{m} + 2.28 \times 10^{11}\,\text{m} = 3.78 \times 10^{11}\,\text{m}$$

Thus,

$$a = 1.89 \times 10^{11}\,\text{m}$$

From this, Kepler's third law is used to find the travel time, which is one half of the period of the orbit:

$$t_{travel} = \tfrac{1}{2}T = \tfrac{1}{2}\sqrt{\frac{4\pi^2}{GM_{Sun}}\,a^3}$$

$$= \tfrac{1}{2}\sqrt{\frac{4\pi^2}{(6.67 \times 10^{-11}\,\text{N·m}^2/\text{kg}^2)(1.99 \times 10^{30}\,\text{kg})}(1.89 \times 10^{11}\,\text{m})^3}$$

$$= 2.24 \times 10^7\,\text{s} = 0.711\,\text{yr} = 260\,\text{d}$$

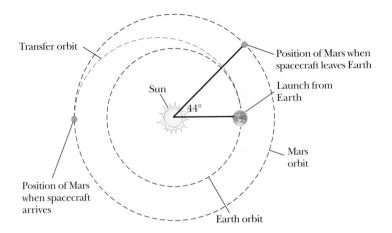

Figure 4

The spacecraft must be launched when Mars is 44° ahead of Earth in its orbit.

Thus, the journey to Mars will require 260 Earth days. We can also determine where in their orbits Earth and Mars must be to ensure the planet will be there when the spacecraft arrives. Mars has an orbital period of 687 Earth days. During the transfer time, the angular position *change* of Mars is

$$\Delta\theta_{\text{Mars}} = \frac{260\ \text{d}}{687\ \text{d}}\ (2\pi) = 2.38\ \text{rad} = 136°$$

Thus, the spacecraft must be launched when Mars is 44° ahead of the Earth in its orbit in order for the spacecraft and Mars to arrive at the same point at the same time. This geometry is shown in Figure 4.

With relatively simple mathematics, this is as far as we can go in describing the details of a trip to Mars—we have found the desired path, the time for the trip, and the position of Mars at launch time. Another important question for the spacecraft captain would be "How much fuel is required for the trip?" This is related to the speed changes necessary to put us into a transfer orbit. These types of calculations involve energy considerations and are explored in Problem 4 below.

Although many considerations necessary for a successful mission to Mars have not been addressed, we have designed a transfer orbit from Earth to Mars that is consistent with the laws of mechanics. We consequently declare success for our endeavor and bring the *Mission to Mars* Context to a close.

PROBLEMS

1. Some science fiction stories describe a twin planet to the Earth. It is exactly 180° ahead of us in the same orbit as the Earth, so we will never see it, because it is on the other side of the Sun. If you are in a spacecraft in orbit around the Earth, describe conceptually how you could visit this planet by altering your orbit.

2. You are in an orbiting space shuttle. Another space shuttle is in exactly the same orbit, but 1 km ahead of you, moving in the same direction around the circle. Through an oversight, your food supplies have been exhausted, and there is more than enough food in the other shuttle. The commander of the other shuttle is going to throw you a picnic basket full of sandwiches, from her shuttle to yours. How should she throw it? (There are no numbers—just give a conceptual response.)

3. Consider a Hohmann transfer from Earth to Venus. (a) How long will this transfer take? (b) Should Venus be ahead of the Earth in its orbit or behind the Earth when the spacecraft leaves the Earth on its way to the rendezvous? How many degrees is Venus ahead or behind the Earth?

4. Investigate what the engine has to do to make a spacecraft follow the Hohmann transfer orbit from the Earth to Mars described in the text. Short-duration burns of our rocket engine are required to change the speed of our spacecraft whenever we alter our orbit. A spacecraft does not have brakes, so fuel is required to both increase and decrease its speed. First ignore the gravitational attraction between the spacecraft and the planets. (a) Calculate the speed change required for switching the craft from a circular orbit around the Sun at the Earth's distance to the transfer orbit to Mars. (b) Calculate the speed change required for switching from the transfer orbit to a circular orbit around the Sun at the distance of Mars. Now consider the effects of the two planets' gravity. (c) Calculate the speed change required for carrying the craft from the Earth's surface to its own independent orbit around the Sun. You may suppose the craft is launched from the Earth's equator toward the east. (d) Model the craft as falling to the surface of Mars from solar orbit. Calculate the magnitude of the speed change required to make a soft landing on Mars at the end of the fall. Mars rotates on its axis with a period of 24.6 h.

Earthquakes

Earthquakes result in massive movement of the ground, as evidenced by the accompanying photograph of railroad tracks in Mexico, damaged severely by an earthquake in 1985. Anyone who has experienced a serious earthquake can attest to the violent shaking it produces. In this Context, we shall focus on earthquakes as an application of our study of the physics of vibrations and waves.

We might expect that the risk of damage in an earthquake decreases as one moves farther from the epicenter. Over long distances, this is correct. For example, structures in Kansas are not affected by earthquakes in California. In regions close to the earthquake, however, the notion of decrease in risk with distance is not consistent. Consider, for example, the following quotations describing damage in two different earthquakes:

After the Northridge, California, earthquake, January 17, 1994:[*]

"Although the city [Santa Monica] sits 25 kilometers from the shock's epicenter, it suffered more [damage] than did other areas less than one-third that distance from the jolt."

After the Michoacán earthquake, September 19, 1985:[†]

" . . . an earthquake rattled the coast of Mexico in the state of Michoacán, about 400 kilometers west of Mexico City. Near the coast, the shaking of the ground was mild and caused little damage. As the seismic waves raced inland, the ground shook even less, and by the time the waves were 100 kilometers from Mexico City, the shaking had nearly subsided. Nevertheless, the seismic waves in-

Severe damage occurred in localized regions of Mexico City in 1985, even though the Michoacán earthquake was hundreds of kilometers away. *(T. Campion/SYGMA)*

The Northridge earthquake in California in 1994 caused billions of dollars in damage. *(Joseph Sohm/ChromoSohm Inc./CORBIS)*

[*] *Science News,* 1994, p. 287.

[†] *American Scientist,* November-December, 1992, p. 566.

duced severe shaking in the city, and some areas continued to shake for several minutes after the seismic waves had passed. Some 300 buildings collapsed and more than 20,000 people died."

It is clear from these quotations that the notion of a simple decrease in risk with distance is misleading. We will use these quotations to motivate us to study the physics of vibrations and waves so we can make a more sophisticated analysis of the risk of damage to structures in an earthquake. Our study here will also be important when we investigate electromagnetic waves in Chapters 24 to 27. In this Context, we shall address the central question

How can we choose locations and build structures to minimize the risk of damage in an earthquake?

chapter 12

The Foucault pendulum at the Franklin Institute in Philadelphia. This type of pendulum was first used by French physicist Jean Foucault to verify the Earth's rotation experimentally. As the pendulum swings, the vertical plane in which it oscillates appears to rotate as the bob successively knocks over the indicators arranged in a circle on the floor. In reality, the plane of oscillation is fixed in space, and the Earth rotating beneath the swinging pendulum moves the indicators into position to be knocked down, one after the other.

(© *Bob Emott, Photographer*)

Oscillatory Motion

You are most likely familiar with several examples of *periodic* motion, such as the oscillations of an object on a spring, the motion of a pendulum, and the vibrations of a stringed musical instrument. Numerous other systems exhibit periodic behavior. For example, the molecules in a solid oscillate about their equilibrium positions; electromagnetic waves, such as light waves, radar, and radio waves, are characterized by oscillating electric and magnetic field vectors; in alternating-current circuits, such as in your household electrical service, voltage and current vary periodically with time. In this chapter, we will investigate mechanical systems that exhibit periodic motion.

We have experienced a number of situations in which the net force on a particle is constant. In these situations, the acceleration of the particle is also constant and we can describe the motion of the particle using the kinematic equations of Chapter 2. If a force acting on a particle varies in time, the acceleration of the particle also changes with time, and the kinematic equations cannot be used.

A special kind of periodic motion occurs when the force that acts on a particle is proportional to the displacement of the particle from the equilibrium position and is always directed toward the equilibrium position. We shall study this special type of varying force in this chapter. When this type of force acts on a particle, the particle exhibits *simple harmonic motion*, which will serve as an analysis model for a large class of oscillation problems.

12.1 • MOTION OF A PARTICLE ATTACHED TO A SPRING

At this point in your study of physics, you have probably started to develop a set of mental models associated with the analysis models that we have developed. By mental model, we mean a typical physical situation that comes to your mind each time you identify the analysis model to be used in a problem. For example, a rock falling in the absence of air resistance is a possible mental model for the analysis model of a particle under constant acceleration. Collisions between billiard balls represent a mental model to help understand the momentum version of the isolated system model. A bicycle wheel might offer a mental model for the rigid body under a net torque model.

In the case of oscillatory motion, a useful mental model is an object of mass m attached to a horizontal spring, as in Figure 12.1. If the spring is unstretched, the object is at rest on a frictionless surface at its **equilibrium position,** which is defined as $x = 0$ (Fig. 12.1b). If the object is pulled to the side to position x and released, it will oscillate back and forth as we discussed in Section 6.4. We will use the particle model to remove effects of the size of the object and analyze a particle on a spring as our system.

Recall from Chapter 6 that when the particle attached to an idealized massless spring is displaced to a position x, the spring exerts a force on it given by **Hooke's law,**

$$F_s = -kx \qquad [12.1]$$

where k is the **force constant** of the spring and F_s is the component of the spring force $\mathbf{F}_s$. We call the force in Equation 12.1 a **linear restoring force** because it is proportional to the displacement from the equilibrium position and always directed toward the equilibrium position, *opposite* the displacement. That is, when the particle is displaced to the right in Figure 12.1a, x is positive and the spring force has a negative component—to the left. When the particle is displaced to the left of $x = 0$ (Fig. 12.1c), x is negative and the spring force has a positive component—to the right. When a particle is under the effect of a linear restoring force, the motion it follows is a special type of oscillatory motion called **simple harmonic motion.** You can test whether or not a particle will undergo simple harmonic motion by seeing if the force on the particle is linear in x. A system undergoing simple harmonic motion is called a **simple harmonic oscillator.**

Let us imagine a particle subject to a linear restoring force such as that given by Equation 12.1. Applying Newton's second law in the x direction to the particle gives us

$$\sum F = F_s = ma \longrightarrow -kx = ma$$

$$a = -\frac{k}{m}x \qquad [12.2]$$

See the *Core Concepts in Physics CD-ROM,* Screens 8.2, 8.3 & 8.10

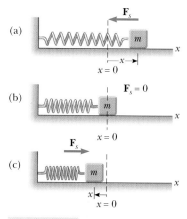

Figure 12.1

A block attached to a spring on a frictionless track moves in simple harmonic motion. (a) When the block is displaced to the right of equilibrium, the position is positive and the force and acceleration are negative. (b) At the equilibrium position $x = 0$, the force and acceleration of the block are zero but the speed is a maximum. (c) When the position is negative, the force and acceleration of the block are positive.

PITFALL PREVENTION 12.1

The orientation of the spring

Figure 12.1 shows the pictorial representation that we will use to study the behavior of systems with springs—a *horizontal* spring, with an attached block sliding on a frictionless surface. Another possible pictorial representation is a block hanging from a *vertical* spring that is attached at the top to a support. All of the results we discuss for the horizontal spring will be the same for the vertical spring, except for one minor difference. When the block is placed on the vertical spring, its weight will cause the spring to extend. If the position of the block at which it hangs at rest on the spring is defined as $x = 0$, the results of this chapter will apply to this system also. We will consider primarily the horizontal spring in this chapter.

- *Acceleration in simple harmonic motion*

That is, the **acceleration is proportional to the displacement of the particle from equilibrium and is in the opposite direction.** If the particle is displaced to a position $x = A$ and released from rest, its *initial* acceleration is $-kA/m$. When the particle passes through the equilibrium position $x = 0$, its acceleration is zero. At this instant, its speed is a maximum because the acceleration changes sign. The particle then continues to travel to the left of equilibrium with a positive acceleration and finally reaches $x = -A$, at which time its acceleration is $+kA/m$ and its speed is again zero. The particle completes a full cycle of its motion by returning to the original position, again passing through $x = 0$ with maximum speed. Thus, we see that the particle oscillates between the turning points $x = \pm A$. In the absence of friction, because the force exerted by the spring is conservative, this motion will continue forever. Real systems are generally subject to friction, and so cannot oscillate forever. We explore the details of the situation with friction in Section 12.6.

PITFALL PREVENTION 12.2

A nonconstant acceleration

Notice that the acceleration of the particle in simple harmonic motion is not constant—Equation 12.2 shows that it varies with position x. Thus, as pointed out in the introduction to the chapter, we *cannot* apply the kinematic equations of Chapter 2 in this situation.

Quick Quiz 12.1

A block on the end of a spring is pulled to position $x = A$ and released. In one full cycle of its motion, through what total distance does it travel? (a) $A/2$ (b) A (c) $2A$ (d) $4A$

See Screens 8.2, 8.3 & 8.10

12.2 • MATHEMATICAL REPRESENTATION OF SIMPLE HARMONIC MOTION

Let us now develop a mathematical representation of the motion we described in the preceding section. Recall that, by definition, $a = dv/dt = d^2x/dt^2$, and so we can express Equation 12.2 as

$$\frac{d^2x}{dt^2} = -\frac{k}{m}x \qquad [12.3]$$

If we denote the ratio k/m with the symbol ω^2 (we choose ω^2 rather than ω to make the solution simpler in form), then

$$\omega^2 = \frac{k}{m} \qquad [12.4]$$

and Equation 12.3 can be written in the form

$$\frac{d^2x}{dt^2} = -\omega^2 x \qquad [12.5]$$

What we now require is a mathematical solution to Equation 12.5—that is, a function $x(t)$ that satisfies this second-order differential equation. This will be a mathematical representation of the position of the particle as a function of time. We seek a function $x(t)$ such that the second derivative of the function is the same as the original function with a negative sign and multiplied by ω^2. The trigonometric functions sine and cosine exhibit this behavior, so we can build a solution around one or both of these. The following cosine function is a solution to the differential equation:

- *Position as a function of time of a particle undergoing simple harmonic motion*

$$x(t) = A\cos(\omega t + \phi) \qquad [12.6]$$

where A, ω, and ϕ are constants.* To see this explicitly, note that

$$\frac{dx}{dt} = A\,\frac{d}{dt}\cos(\omega t + \phi) = -\omega A \sin(\omega t + \phi) \qquad \text{[12.7]}$$

$$\frac{d^2x}{dt^2} = -\omega A\,\frac{d}{dt}\sin(\omega t + \phi) = -\omega^2 A \cos(\omega t + \phi) \qquad \text{[12.8]}$$

Comparing Equations 12.6 and 12.8, we see that $d^2x/dt^2 = -\omega^2 x$ and Equation 12.5 is satisfied.

The parameters A, ω, and ϕ are constants of the motion. To give physical significance to these constants, it is convenient to form a graphical representation of the motion by plotting x as a function of t, as in Figure 12.2a. First, we note that A, called the **amplitude** of the motion, is simply the **maximum value of the position of the particle in either the positive or negative x direction.** The constant ω is called the **angular frequency** and has units of rad/s. From Equation 12.4, the angular frequency is

$$\omega = \sqrt{\frac{k}{m}} \qquad \text{[12.9]}$$

The constant angle ϕ is called the **phase constant** (or phase angle) and, along with the amplitude A, is determined uniquely by the position and velocity of the particle at $t = 0$. If the particle is at its maximum position $x = A$ at $t = 0$, the phase constant is $\phi = 0$ and the graphical representation of the motion is shown in Figure 12.2b. The quantity $(\omega t + \phi)$ is called the **phase** of the motion. Note that the function $x(t)$ is periodic and its value is the same each time ωt increases by 2π radians.

Equations 12.1, 12.5, and 12.6 form the basis for the analysis model of the simple harmonic motion. If we are analyzing a situation and find that the force on a particle is of the mathematical form of Equation 12.1, we know that the motion will be that of a simple harmonic oscillator and that the position is described by Equation 12.6. If we analyze a system and find that it is described by a differential equation of the form of Equation 12.5, the motion will be that of a simple harmonic oscillator. If we analyze a situation and find that the position of a particle is described by Equation 12.6, we know the particle is undergoing simple harmonic motion.

Let us investigate further the mathematical description of the motion. The **period** T of the motion is the time required for the particle to go through one full cycle of its motion (see Fig. 12.2a). That is, the values of x and v for the particle at time t equal the values of x and v at time $t + T$. We can relate the period to the angular frequency by using the fact that the phase increases by 2π radians in a time of T:

$$[\omega(t + T) + \phi] - (\omega t + \phi) = 2\pi$$

---

* In earlier chapters we saw many examples in which we evaluated a trigonometric function of an angle. The argument of a trigonometric function, such as sine or cosine, *must* be a pure number. The radian is a pure number because it is a ratio of lengths. Degrees are pure simply because the degree is a completely artificial "unit"—it is not related to measurements of lengths. The notion of requiring a pure number for a trigonometric function is important in Equation 12.6, where the angle is expressed in terms of other measurements. Thus, ω *must* be expressed in rad/s (and not, for example, in revolutions per second) if t is expressed in seconds. Furthermore, the argument of other types of functions must also be pure numbers, including logarithms and exponential functions.

PITFALL PREVENTION 12.3

Where's the triangle?

Equation 12.6 includes a trigonometric function. We have used trigonometric functions to analyze right triangles as geometric models, but no triangle occurs in this current discussion. This demonstrates that trigonometric functions are just that—mathematical functions that can be used whether they refer to triangles or not. In this case, the cosine function happens to have the correct behavior to provide a mathematical representation of the motion of the particle in simple harmonic motion.

• *Angular frequency*

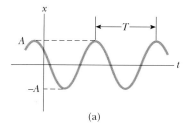

(a)

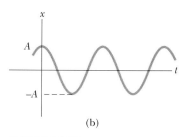

(b)

Figure 12.2

(a) A graphical representation (position versus time) for the system in Figure 12.1—a particle undergoing simple harmonic motion. The amplitude of the motion is A and the period is T. (b) The x-t curve in the special case in which $x = A$ and $v = 0$ at $t = 0$.

Simplifying this expression, we see that $\omega T = 2\pi$, or

- *Period*

$$T = \frac{2\pi}{\omega}$$ [12.10]

The inverse of the period is called the **frequency** f of the motion. Whereas the period is the time interval per oscillation, the frequency represents the **number of oscillations the particle makes per unit time:**

- *Frequency*

$$f = \frac{1}{T} = \frac{\omega}{2\pi}$$ [12.11]

The units of f are cycles per second, or **hertz** (Hz). Rearranging Equation 12.11 gives

- *Relation of angular frequency to frequency and period*

$$\omega = 2\pi f = \frac{2\pi}{T}$$ [12.12]

We can use Equations 12.9, 12.10, and 12.11 to express the period and frequency of the motion for the particle–spring system in terms of the characteristics m and k of the system as

- *Period of a particle–spring system*

$$T = \frac{2\pi}{\omega} = 2\pi\sqrt{\frac{m}{k}}$$ [12.13]

- *Frequency of a particle–spring system*

$$f = \frac{1}{T} = \frac{1}{2\pi}\sqrt{\frac{k}{m}}$$ [12.14]

That is, the period and frequency depend *only* on the mass of the particle and the force constant of the spring, and *not* on the parameters of the motion, such as A or

PITFALL PREVENTION 12.4

Two kinds of frequency

We identify two kinds of frequency for a simple harmonic oscillator: f, called simply the *frequency*, is measured in hertz, and ω, the *angular frequency*, is measured in radians per second. Be sure you are clear about which frequency is being discussed or requested in a given problem. Equations 12.11 and 12.12 show the relationship between the two frequencies.

How does an astronaut measure his weight in orbit when a bathroom scale would be useless? He actually measures mass rather than weight by sitting on a seat of known mass that oscillates due to springs of known force constant. In this photo, astronaut Alan L. Bean sits in the oscillating seat. The measured period of oscillation determines the total oscillating mass, according to Equation 12.13. By subtracting the mass of the seat, the mass of the astronaut is calculated. *(NASA)*

ϕ. As we might expect, the frequency is larger for a stiffer spring (larger value of k) and decreases with increasing mass of the particle.

We can obtain the velocity and acceleration* of a particle undergoing simple harmonic motion from Equations 12.7 and 12.8:

$$v = \frac{dx}{dt} = -\omega A \sin(\omega t + \phi) \qquad [12.15]$$

$$a = \frac{d^2 x}{dt^2} = -\omega^2 A \cos(\omega t + \phi) \qquad [12.16]$$

- *Velocity as a function of time of a particle undergoing simple harmonic motion*
- *Acceleration as a function of time of a particle undergoing simple harmonic motion*

From Equation 12.15 we see that because the sine and cosine functions oscillate between ± 1, the extreme values of v are $\pm \omega A$. Likewise, Equation 12.16 tells us that the extreme values of the acceleration are $\pm \omega^2 A$. Therefore, the *maximum* values of the magnitudes of the velocity and acceleration are

$$v_{\max} = \omega A = \sqrt{\frac{k}{m}}\, A \qquad [12.17]$$

$$a_{\max} = \omega^2 A = \frac{k}{m}\, A \qquad [12.18]$$

- *Maximum values of velocity and acceleration of a particle undergoing simple harmonic motion*

Figure 12.3a plots position versus time for an arbitrary value of the phase constant. The associated velocity–time and acceleration–time curves are illustrated in Figures 12.3b and 12.3c. They show that the phase of the velocity differs from the phase of the position by $\pi/2$ rad, or 90°. That is, when x is a maximum or a minimum, the velocity is zero. Likewise, when x is zero, the speed is a maximum. Furthermore, note that the phase of the acceleration differs from the phase of the position by π radians, or 180°. For example, when x is a maximum, a has a maximum magnitude in the opposite direction.

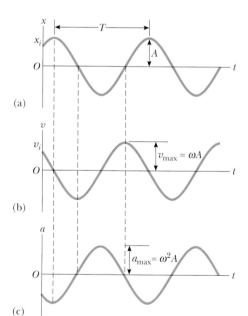

(a)

(b)

(c)

Figure 12.3

Graphical representation of three variables in simple harmonic motion: (a) position versus time, (b) velocity versus time, and (c) acceleration versus time. Note that at any specified time the velocity is 90° out of phase with the position and the acceleration is 180° out of phase with the position.

---

* Because the motion of a simple harmonic oscillator takes place in one dimension, we will denote velocity as v and acceleration as a, with the direction indicated by a positive or negative sign, as in Chapter 2.

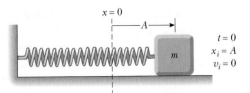

Figure 12.4

$t = 0$
$x_i = A$
$v_i = 0$

A block–spring system that is released from rest at $x_i = A$. In this case, $\phi = 0$, and thus $x = A \cos \omega t$.

Equation 12.6 describes simple harmonic motion of a particle in general. Let us now see how to evaluate the constants of the motion. The angular frequency ω is evaluated using Equation 12.9. The constants A and ϕ are evaluated from the initial conditions, that is, the state of the oscillator at $t = 0$.

Suppose we initiate the motion by pulling the particle from equilibrium by a distance A and releasing it from rest at $t = 0$, as in Figure 12.4. We must then require that our solutions for $x(t)$ and $v(t)$ (Eqs. 12.6 and 12.15) obey the initial conditions that $x(0) = A$ and $v(0) = 0$:

$$x(0) = A \cos \phi = A$$

$$v(0) = -\omega A \sin \phi = 0$$

These conditions are met if we choose $\phi = 0$, giving $x = A \cos \omega t$ as our solution. To check this solution, we note that it satisfies the condition that $x(0) = A$, because $\cos 0 = 1$.

Position, velocity, and acceleration are plotted versus time in Figure 12.5a for this special case. The acceleration reaches extreme values of $\pm \omega A$ when the position has extreme values of $\pm A$. Furthermore, the velocity has extreme values of $\pm \omega A$, which both occur at $x = 0$. Hence, the quantitative solution agrees with our qualitative description of this system.

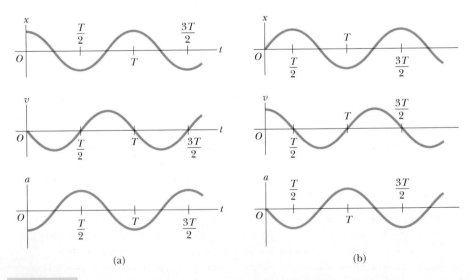

(a)

(b)

Figure 12.5

(a) Position, velocity, and acceleration versus time for a block undergoing simple harmonic motion under the initial conditions that at $t = 0$, $x(0) = A$, and $v(0) = 0$. (b) Position, velocity, and acceleration versus time for a block undergoing simple harmonic motion under the initial conditions that at $t = 0$, $x(0) = 0$, and $v(0) = v_i$.

Let us consider another possibility. Suppose that the system is oscillating and we define $t = 0$ as the instant that the particle passes through the unstretched position of the spring while moving to the right (Fig. 12.6). We must then require that our solutions for $x(t)$ and $v(t)$ obey the initial conditions that $x(0) = 0$ and $v(0) = v_i$:

$$x(0) = A \cos \phi = 0$$

$$v(0) = -\omega A \sin \phi = v_i$$

The first of these conditions tells us that $\phi = -\pi/2$. With this value for ϕ, the second condition tells us that $A = v_i/\omega$. Hence, the solution is given by

$$x = \frac{v_i}{\omega} \cos \left(\omega t - \frac{\pi}{2} \right)$$

Figure 12.5b shows the graphs of position, velocity, and acceleration versus time for this choice of $t = 0$. Note that these curves are the same as those in Figure 12.5a, but shifted to the right by one fourth of a cycle. This is described mathematically by the phase constant $\phi = -\pi/2$, which is one fourth of a full cycle of 2π.

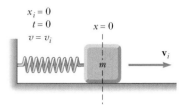

$x_i = 0$
$t = 0$ $x = 0$
$v = v_i$

Figure 12.6

The block–spring system is undergoing oscillation, and $t = 0$ is defined at an instant when the block passes through the equilibrium position $x = 0$ and is moving to the right with speed v_i.

Quick Quiz 12.2

What would be the solution for x if the particle were initially moving to the left in Figure 12.6?

Quick Quiz 12.3

Is a bouncing ball an example of simple harmonic motion? Is the daily movement of a student from home to school and back simple harmonic motion?

THINKING PHYSICS 12.1

We know that the period of oscillation of an object attached to a spring is proportional to the square root of the mass of the object (Eq. 12.13). Thus, if we perform an experiment in which we place objects with a range of masses on the end of a spring and measure the period of oscillation of each mass–spring system, a graph of the square of the period versus the mass will result in a straight line, as suggested in Figure 12.7. But we find that the line does not go through the origin. Why not?

Reasoning   The line does not go through the origin because the spring itself has mass. Thus, the resistance to changes in motion of the system is a combination of the mass of the object on the end of the spring and the mass of the oscillating spring coils. The entire mass of the spring is not oscillating in the same way, however. The coil of the spring attached to the object is oscillating over the same amplitude as the object, but the coil at the opposite end of the spring is not oscillating at all. For a cylindrical spring, energy arguments can be used to show that the effective additional mass representing the oscillations of the spring is one third of the mass of the spring. The square of the period is proportional to the total oscillating mass,

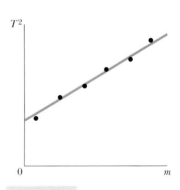

T^2

0 m

Figure 12.7

(Thinking Physics 12.1) A graph of experimental data—the square of the period versus mass of a block in a block–spring system.

but the graph in Figure 12.7 shows the square of the period versus only the mass of the object on the spring. A graph of period squared versus total mass (mass of the object on the spring plus the effective oscillating mass of the spring) would pass through the origin.

THINKING PHYSICS 12.2

Suppose you attach a stapler to a rubber band and allow the stapler to hang in equilibrium from one end of the loop. You now pull the stapler down a short distance and release it, measuring the frequency *f* of its oscillations. After making this measurement, you cut the rubber band in one place and unfold it so that you have a single rubber strip that is twice as long as the length of the original loop. You attach the stapler to one end of the rubber strip, let it hang while you hold the other end, and again measure the frequency of oscillation. How does this frequency compare with the original frequency? (Model the rubber band as obeying Hooke's law.)

Reasoning   First imagine that the cut rubber band is held at the midpoint rather than the end, so the stapler is held from a single rubber strip that has an unstretched length equal to the unstretched length of the original uncut rubber band. Now when the stapler is attached and allowed to hang, the single strip stretches through a distance twice as long as that for the rubber loop, because only one strip is supporting the stapler instead of two in the case of the loop. Now, if the rubber strip is held at the free end and the stapler is hung at the opposite end, each half of the full length of the strip stretches by a distance twice as great as the original stretch of the rubber band. Therefore, the overall stretch will be four times as large as the original rubber band. With four times the stretch for the same hanging weight, the force constant of the rubber strip must be one-fourth as large as that of the original rubber band. Using Equation 12.14, we determine that the frequency of the stapler when oscillating must be $\frac{1}{2}f$.

Example 12.1   A Block–Spring System

A block with a mass of 200 g is connected to a light horizontal spring of force constant 5.00 N/m and is free to oscillate on a horizontal, frictionless surface. (a) If the block is displaced 5.00 cm from equilibrium and released from rest, as in Figure 12.4, find the period of its motion.

Solution   The situation (we assume an ideal spring) tells us to use the simple harmonic motion model. Using Equation 12.13,

$$T = 2\pi\sqrt{\frac{m}{k}} = 2\pi\sqrt{\frac{200 \times 10^{-3}\ \text{kg}}{5.00\ \text{N/m}}} = \boxed{1.26\ \text{s}}$$

(b) Determine the maximum speed and maximum acceleration of the block.

Solution   Using Equations 12.17 and 12.18, with $A = 5.00 \times 10^{-2}$ m, we have

$$v_{\text{max}} = \omega A = \frac{2\pi}{T}A = \left(\frac{2\pi}{1.26\ \text{s}}\right)(5.00 \times 10^{-2}\ \text{m})$$

$$= \boxed{0.250\ \text{m/s}}$$

$$a_{\text{max}} = \omega^2 A = \left(\frac{2\pi}{T}\right)^2 A = \left(\frac{2\pi}{1.26\ \text{s}}\right)^2 (5.00 \times 10^{-2}\ \text{m})$$

$$= \boxed{1.25\ \text{m/s}^2}$$

(c) Express the position, velocity, and acceleration of this object as functions of time, assuming that $\phi = 0$.

Solution   From Equations 12.6, 12.15, and 12.16,

$$x = A\cos\omega t = \boxed{(0.0500\ \text{m})\cos 5.00t}$$

$$v = -\omega A\sin\omega t = \boxed{-(0.250\ \text{m/s})\sin 5.00t}$$

$$a = -\omega^2 A\cos\omega t = \boxed{-(1.25\ \text{m/s}^2)\cos 5.00t}$$

Example 12.2   An Oscillating Particle

A particle oscillates with simple harmonic motion along the x axis. Its position varies with time according to the equation

$$x = (4.00 \text{ m}) \cos\left(\pi t + \frac{\pi}{4}\right)$$

where t is in seconds.

(a) Determine the amplitude, frequency, and period of the motion.

Solution   By comparing this equation with the general equation for simple harmonic motion, $x = A \cos(\omega t + \phi)$, we see that $A = $ 4.00 m and $\omega = \pi$ rad/s; therefore we find $f = \omega/2\pi = \pi/2\pi = $ 0.500 Hz and $T = 1/f = $ 2.00 s .

(b) Calculate the velocity and acceleration of the particle at any time t.

Solution   Using Equations 12.15 and 12.16,

$$v = \frac{dx}{dt} = -(4.00\pi \text{ m/s})\sin\left(\pi t + \frac{\pi}{4}\right)$$

$$a = \frac{dv}{dt} = -(4.00\pi^2 \text{ m/s}^2)\cos\left(\pi t + \frac{\pi}{4}\right)$$

(c) What are the position and the velocity of the particle at time $t = 0$?

Solution   The position function is given in the text of the problem. Evaluating this expression at $t = 0$ gives us

$$x = (4.00 \text{ m})\cos\left(\pi t + \frac{\pi}{4}\right) = (4.00 \text{ m})\cos\left(\frac{\pi}{4}\right) = \boxed{2.83 \text{ m}}$$

From part (b), we evaluate the velocity function at $t = 0$:

$$v = -(4.00\pi \text{ m/s})\sin\left(\pi t + \frac{\pi}{4}\right) = -(4.00\pi \text{ m/s})\sin\left(\frac{\pi}{4}\right)$$
$$= -8.89 \text{ m/s}$$

EXERCISE   Using the results to part (b), determine the position, velocity, and acceleration of the particle at $t = 1.00$ s.

Answer $x = -2.83$ m; $v = 8.89$ m/s; $a = 27.9$ m/s^2

EXERCISE   Determine the maximum velocity and the maximum acceleration of the particle.

Answer $v_{\max} = \pm 12.6$ m/s; $a_{\max} = \pm 39.5$ m/s^2

Example 12.3   Initial Conditions

Suppose that the initial position x_i and initial velocity v_i of a harmonic oscillator of known angular frequency are given; that is, $x(0) = x_i$ and $v(0) = v_i$. Find general expressions for the amplitude and the phase constant in terms of these initial parameters.

Solution   With these initial conditions, Equations 12.6 and 12.15 give us

$$x_i = A \cos \phi \quad \text{and} \quad v_i = -\omega A \sin \phi$$

Dividing these two equations eliminates A, giving $v_i/x_i = -\omega \tan \phi$, or

$$\tan \phi = -\frac{v_i}{\omega x_i}$$

Furthermore, if we take the sum $x_i^2 + (v_i/\omega)^2 = A^2 \cos^2 \phi + A^2 \sin^2 \phi = A^2$ (where we have used Eqs. 12.6 and 12.15) and solve for A, we find that

$$A = \sqrt{x_i^2 + \left(\frac{v_i}{\omega}\right)^2}$$

EXERCISE   A 3.0-kg object is attached to a spring and pulled out horizontally to a maximum displacement from equilibrium of 0.50 m. What force constant must the spring have if the object is to achieve an acceleration equal to the free-fall acceleration?

Answer   59 N/m

12.3 • ENERGY CONSIDERATIONS IN SIMPLE HARMONIC MOTION

If an object attached to a spring slides on a frictionless surface, we can consider the combination of the spring and the attached object to be an isolated system. As a result, we can apply the energy version of the isolated system model to the system. Let us examine the mechanical energy of the system described in Figure 12.1. Because the surface is frictionless, the total mechanical energy of the system is constant. We model the object as a particle. The kinetic energy, in the simplification model in which the spring is massless, is associated only with the motion of

• *Kinetic energy of a simple harmonic oscillator*

the particle of mass m. We use Equation 12.15 to express the kinetic energy as

$$K = \tfrac{1}{2}mv^2 = \tfrac{1}{2}m\omega^2 A^2 \sin^2(\omega t + \phi) \qquad [12.19]$$

Elastic potential energy in this system is associated with the spring, and for any position x of the particle is, as we found in Chapter 7, $U = \tfrac{1}{2}kx^2$. Using Equation 12.6, we have

• *Potential energy of a simple harmonic oscillator*

$$U = \tfrac{1}{2}kx^2 = \tfrac{1}{2}kA^2 \cos^2(\omega t + \phi) \qquad [12.20]$$

We see that K and U are always positive quantities, and each varies with time. We can express the *total energy* of the simple harmonic oscillator as

$$E = K + U = \tfrac{1}{2}m\omega^2 A^2 \sin^2(\omega t + \phi) + \tfrac{1}{2}kA^2 \cos^2(\omega t + \phi)$$

Because $\omega^2 = k/m$, we can write this as

$$E = \tfrac{1}{2}kA^2[\sin^2(\omega t + \phi) + \cos^2(\omega t + \phi)]$$

Because $\sin^2\theta + \cos^2\theta = 1$ for any angle θ, this equation reduces to

• *Total energy of a simple harmonic oscillator*

$$E = \tfrac{1}{2}kA^2 \qquad [12.21]$$

That is, the energy of an isolated simple harmonic oscillator is a constant of the motion and proportional to the square of the amplitude. In fact, the total mechanical energy is just equal to the maximum potential energy stored in the spring when $x = \pm A$. At these points, $v = 0$ and there is no kinetic energy. At the equilibrium position, $x = 0$ and $U = 0$, so that the total energy is all in the form of kinetic energy of the particle,

$$K_{max} = \tfrac{1}{2}mv_{max}^2 = \tfrac{1}{2}kA^2$$

These results are appropriate for the simplification model in which we consider the spring to be massless. In the real situation in which the spring has mass, additional kinetic energy is associated with the motion of the spring. In numerical problems in this book, unless otherwise noted, we will consider only massless springs.

Graphical representations of the kinetic and potential energies versus time for the system of a particle on a massless spring are shown in Figure 12.8a, where $\phi = 0$.

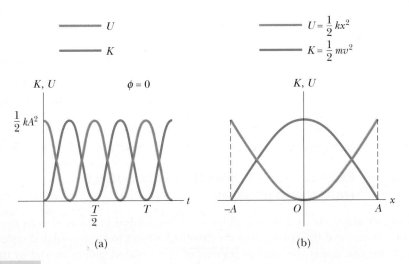

Figure 12.8

(a) Kinetic energy and potential energy versus time for a simple harmonic oscillator with $\phi = 0$. (b) Kinetic energy and potential energy versus position for a simple harmonic oscillator. In either plot, note that $K + U =$ constant.

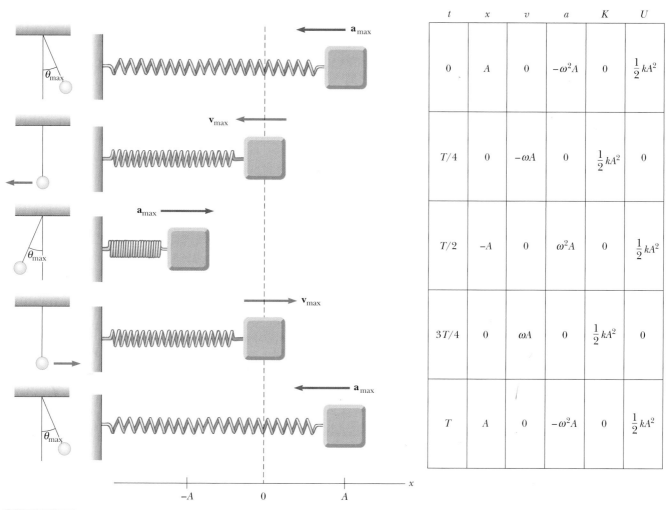

t	x	v	a	K	U
0	A	0	$-\omega^2 A$	0	$\frac{1}{2}kA^2$
$T/4$	0	$-\omega A$	0	$\frac{1}{2}kA^2$	0
$T/2$	$-A$	0	$\omega^2 A$	0	$\frac{1}{2}kA^2$
$3T/4$	0	ωA	0	$\frac{1}{2}kA^2$	0
T	A	0	$-\omega^2 A$	0	$\frac{1}{2}kA^2$

Figure 12.9

Simple harmonic motion for a block–spring system and its analogy to the motion of a simple pendulum (Section 12.4). The parameters in the table at the right refer to the block–spring system, assuming that at $t = 0$, $x = A$ so that $x = A \cos \omega t$.

In this situation, the sum of the kinetic and potential energies at all times is a constant equal to $\frac{1}{2}kA^2$, the total energy of the system. The variations of K and U with position are plotted in Figure 12.8b. Energy in the system is continuously being transformed between potential energy (in the spring) and kinetic energy (of the object attached to the spring). Figure 12.9 illustrates the position, velocity, acceleration, kinetic energy, and potential energy of the particle–spring system for one full period of the motion. Most of the ideas discussed so far for simple harmonic motion are incorporated in this important figure. We suggest that you study it carefully.

Finally, we can use the conservation of mechanical energy expression to obtain the velocity for an arbitrary position x of the particle, expressing the total energy as

$$E = K + U = \tfrac{1}{2}mv^2 + \tfrac{1}{2}kx^2 = \tfrac{1}{2}kA^2$$

$$v = \pm\sqrt{\frac{k}{m}\,(A^2 - x^2)} = \pm\omega\sqrt{A^2 - x^2} \qquad \textbf{[12.22]}$$

• *Velocity as a function of position for a particle undergoing simple harmonic motion*

This expression confirms the facts that the speed is a maximum at $x = 0$ and is zero at the turning points $x = \pm A$.

THINKING PHYSICS 12.3

An object oscillating on the end of a horizontal spring slides back and forth over a frictionless surface. During one oscillation, you set an identical object at the maximum displacement point, with instant-acting glue on its surface. Just as the oscillating object reaches its largest displacement and is momentarily at rest, it adheres to the new object by means of the glue, and the two objects continue the oscillation together. Does the period of the oscillation change? Does the amplitude of oscillation change? Does the energy of the oscillation change?

Reasoning The period of oscillation changes because the period depends on the mass that is oscillating (Eq. 12.13). The amplitude does not change. Because the new object was added while the original object was at rest, the combined objects are at rest at this point also, defining the amplitude as the same as in the original oscillation. The energy does not change, either. At the maximum displacement point, the energy is all potential energy stored in the spring, which depends only on the force constant and the amplitude, not the mass of the object. The object of increased mass will pass through the equilibrium point with lower speed than in the original oscillation, but with the same kinetic energy. Another approach is to think about how energy could be transferred into the oscillating system—no work was done (nor was any other form of energy transferred), so the energy in the system cannot change.

Example 12.4 Oscillations on a Horizontal Surface

A 0.500-kg object connected to a massless spring of force constant 20.0 N/m oscillates on a horizontal, frictionless track. (a) Calculate the total energy of the system and the maximum velocity of the object if the amplitude of the motion is 3.00 cm.

Solution Because the object slides on a frictionless surface, we can use the isolated system model for the system of the object and the spring. Because only conservative forces are acting within the system, the mechanical energy of the system is conserved. Using Equation 12.21, we have

$$E = \tfrac{1}{2}kA^2 = \tfrac{1}{2}(20.0 \text{ N/m})(3.00 \times 10^{-2} \text{ m})^2 = \boxed{9.00 \times 10^{-3} \text{ J}}$$

When the object is at $x = 0$, $U = 0$ and $E = \tfrac{1}{2}mv_{max}^2$; therefore

$$\tfrac{1}{2}mv_{max}^2 = 9.00 \times 10^{-3} \text{ J}$$

$$v_{max} = \sqrt{\frac{18.0 \times 10^{-3} \text{ J}}{0.500 \text{ kg}}} = \boxed{\pm 0.190 \text{ m/s}}$$

The positive and negative signs indicate that the object could be moving to either the right or the left at this instant.

(b) What is the velocity of the object when the position is equal to 2.00 cm?

Solution We apply Equation 12.22 directly:

$$v = \pm\sqrt{\frac{k}{m}(A^2 - x^2)}$$

$$= \pm\sqrt{\frac{20.0 \text{ N/m}}{0.50 \text{ kg}}[(3.00 \times 10^{-2} \text{ m})^2 - (2.00 \times 10^{-2} \text{ m})^2]}$$

$$= \boxed{\pm 0.141 \text{ m/s}}$$

(c) Compute the kinetic and potential energies of the system when the position equals 2.00 cm.

Solution Using the result to part (b), we find

$$K = \tfrac{1}{2}mv^2 = \tfrac{1}{2}(0.500 \text{ kg})(0.141 \text{ m/s})^2 = \boxed{5.00 \times 10^{-3} \text{ J}}$$

$$U = \tfrac{1}{2}kx^2 = \tfrac{1}{2}(20.0 \text{ N/m})(2.00 \times 10^{-2} \text{ m})^2 = \boxed{4.00 \times 10^{-3} \text{ J}}$$

Note that the sum $K + U$ equals the total mechanical energy E found in part (a).

EXERCISE For what values of x does the speed of the object equal 0.100 m/s?

Answer ± 2.55 cm

12.4 • THE SIMPLE PENDULUM

The **simple pendulum** is another mechanical system that exhibits periodic motion. It consists of a point object of mass *m*, suspended by a light string (or rod) of length *L*, where the upper end of the string is fixed, as in Figure 12.10. For a real object, as long as the size of the object is small relative to the length of the string, the pendulum can be modeled as a simple pendulum—this is the particle model. When the object is pulled to the side and released, it oscillates about the lowest point, which is the equilibrium position. The motion occurs in a vertical plane and is driven by the gravitational force.

The forces acting on the object are the force **T** acting along the string and the gravitational force *m***g**. The tangential component of the gravitational force $mg \sin \theta$ always acts toward $\theta = 0$, opposite the displacement. The tangential force is therefore a restoring force, and we can use Newton's second law to write the equation of motion in the tangential direction as

$$F_t = ma_t \longrightarrow -mg \sin \theta = m\frac{d^2s}{dt^2}$$

where *s* is the position measured along the circular arc in Figure 12.10, and the negative sign indicates that F_t acts toward the equilibrium position. Because $s = L\theta$ (Eq. 10.1a) and *L* is constant, this equation reduces to

$$\frac{d^2\theta}{dt^2} = -\frac{g}{L} \sin \theta$$

Compare this equation to Equation 12.5, which is of a similar, but not identical, mathematical form. The right side is proportional to $\sin \theta$ rather than to θ; hence, we conclude that the motion is *not* simple harmonic motion because it is not of the form of Equation 12.5. If we assume that θ is *small* (less than about 10° or 0.2 rad), however, we can use a simplification model called the **small angle approximation,** in which $\sin \theta \approx \theta$, where θ is measured in radians. Table 12.1 shows angles in degrees and radians, and the sines of these angles. As long as θ is less than about 10°, the angle in radians and its sine are the same, at least to within an accuracy of less than 1.0%.

Thus, for small angles, the equation of motion becomes

$$\frac{d^2\theta}{dt^2} = -\frac{g}{L} \theta \qquad\qquad \textbf{[12.23]}$$

 See Screens 8.11 & 8.12

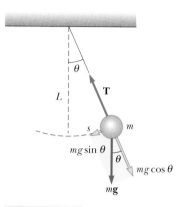

Figure 12.10

When θ is small, the oscillation of the simple pendulum can be modeled as simple harmonic motion about the equilibrium position ($\theta = 0$). The restoring force is $mg \sin \theta$, the component of the gravitational force tangent to the arc.

PITFALL PREVENTION 12.5
Not true simple harmonic motion

 Remember that the pendulum *does not* exhibit true simple harmonic motion for *any* angle. If the angle is less than about 10°, the motion can be *modeled* as simple harmonic.

• *Equation of motion for the simple pendulum (small θ)*

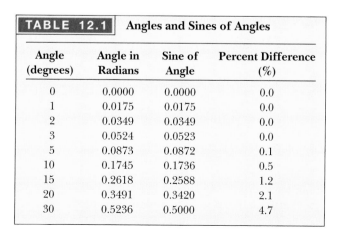

TABLE 12.1	Angles and Sines of Angles		
Angle (degrees)	Angle in Radians	Sine of Angle	Percent Difference (%)
0	0.0000	0.0000	0.0
1	0.0175	0.0175	0.0
2	0.0349	0.0349	0.0
3	0.0524	0.0523	0.0
5	0.0873	0.0872	0.1
10	0.1745	0.1736	0.5
15	0.2618	0.2588	1.2
20	0.3491	0.3420	2.1
30	0.5236	0.5000	4.7

Now we have an expression with exactly the same mathematical form as Equation 12.5, with $\omega^2 = g/L$, and so we conclude that the motion is approximately simple harmonic motion for small amplitudes. Modeling the solution after Equation 12.6, θ can therefore be written as $\theta = \theta_{max} \cos(\omega t + \phi)$, where θ_{max} is the **maximum angular position** and the angular frequency ω is

- *Angular frequency for a simple pendulum*

$$\omega = \sqrt{\frac{g}{L}} \qquad [12.24]$$

The period of the motion is

- *Period for a simple pendulum*

$$T = \frac{2\pi}{\omega} = 2\pi\sqrt{\frac{L}{g}} \qquad [12.25]$$

In other words, the **period and frequency of a simple pendulum oscillating at small angles depend only on the length of the string and the free-fall acceleration.** Because the period is *independent* of the mass, we conclude that *all* simple pendula of equal length at the same location oscillate with equal periods. Experiment shows that this is correct. The analogy between the motion of a simple pendulum and the particle–spring system is illustrated in Figure 12.9.

The simple pendulum can be used as a timekeeper. It is also a convenient device for making precise measurements of the free-fall acceleration.

Quick Quiz 12.4

A simple pendulum is suspended from the ceiling of a stationary elevator, and the period is determined. Describe the changes, if any, in the period when the elevator (a) accelerates upward, (b) accelerates downward, and (c) moves with constant velocity.

Quick Quiz 12.5

A grandfather clock depends on the period of a pendulum to keep correct time. Suppose a grandfather clock is calibrated correctly and then a mischievous child slides the bob of the pendulum downward on the oscillating rod. Does the grandfather clock run (a) slow, (b) fast, or (c) correctly?

THINKING PHYSICS 12.4

You set up two oscillating systems: a simple pendulum and a block hanging from a vertical spring. You carefully adjust the length of the pendulum so that both oscillators have the same period. You now take the two oscillators to the Moon. Will they still have the same period as each other? What happens if you observe the two oscillators in an orbiting space shuttle?

Reasoning The block hanging from the spring will have the same period on the Moon that it had on the Earth because the period depends on the mass of the block and the force constant of the spring, neither of which have changed. The period of the pendulum on the Moon will be different from its period on the Earth because the period of the pendulum depends on the value of g. Because g is smaller on the Moon than on the Earth, the pendulum will oscillate with a longer period.

In the orbiting space shuttle, the block–spring system will oscillate with the same period as on the Earth when it is set into motion, because the period does not depend on gravity. The pendulum will not oscillate at all—if you pull it to the side from a direction you define as "vertical" and release it, it stays there. Because the shuttle is in free-fall while in orbit around the Earth, the effective gravity is zero, and there is no restoring force on the pendulum.

Example 12.5   A Measure of Height

A man enters a tall tower, needing to know its height. He notes that a long pendulum extends from the ceiling almost to the floor and that its period is 12.0 s. How tall is the tower?

Solution   We adopt a simplification model in which the height of the tower is equal to the length of the pendulum. If we use $T = 2\pi\sqrt{L/g}$ and solve for L, we have

$$L = \frac{gT^2}{4\pi^2} = \frac{(9.80 \text{ m/s}^2)(12.0 \text{ s})^2}{4\pi^2} = \boxed{35.7 \text{ m}}$$

EXERCISE   If the pendulum described in this example is taken to the Moon, where the free-fall acceleration is 1.67 m/s^2, what is the period there?

Answer   29.1 s

12.5 • THE PHYSICAL PENDULUM

If a hanging object oscillates about a fixed axis that does not pass through its center of mass, then it must be treated as a **physical, or compound, pendulum.** For the simple pendulum, we generated Equation 12.23 from the particle under a net force model. For the physical pendulum, we will need to use the rigid body under a net torque model from Chapter 10. Consider a rigid body pivoted at a point O that is a distance d from the center of mass (Fig. 12.11). The torque about O is provided by the gravitational force, and its magnitude is $mgd \sin\theta$. Using Newton's second law for rotation, $\sum\tau = I\alpha$ (Eq. 10.27), where I is the moment of inertia about the axis through O, we have

$$-mgd \sin\theta = I\frac{d^2\theta}{dt^2}$$

The negative sign on the left indicates that the torque about O tends to decrease θ. That is, the gravitational force produces a restoring torque.

If we again assume that θ is small, the small angle approximation $\sin\theta \approx \theta$ is valid and the equation of motion reduces to

$$\frac{d^2\theta}{dt^2} = -\left(\frac{mgd}{I}\right)\theta \qquad \qquad \textbf{[12.26]}$$

Note that this equation has the same mathematical form as Equation 12.5, with $\omega^2 = mgd/I$, and so the motion is approximately simple harmonic motion for small amplitudes. That is, the solution of Equation 12.26 is $\theta = \theta_{\max}\cos(\omega t + \phi)$, where $\theta_{\max}$ is the maximum angular position and

$$\omega = \sqrt{\frac{mgd}{I}}$$

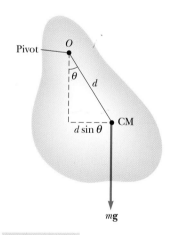

Figure 12.11

The physical pendulum consists of a rigid body pivoted at the point O, which is not at the center of mass.

•   *Angular frequency for a physical pendulum*

The period is

• *Period of motion for a physical pendulum*

$$T = \frac{2\pi}{\omega} = 2\pi \sqrt{\frac{I}{mgd}}$$ [12.27]

One can use this result to measure the moment of inertia of a planar rigid body. If the location of the center of mass and, hence, the distance d are known, the moment of inertia can be obtained through a measurement of the period. Finally, note that Equation 12.27 becomes the equation for the period of a simple pendulum (Eq. 12.25) when $I = md^2$, that is, when all the mass is concentrated at a point, and the physical pendulum reduces to the simple pendulum.

Quick Quiz 12.6

Two students are watching the swinging of a pendulum with a large bob in a museum, such as that in the opening photograph for this chapter. One student says, "I'm going to sneak past the fence and stick some chewing gum on the top of the pendulum bob, to change its period of oscillation." The other student says, "That won't change the period—the period of a pendulum is independent of mass." Which student is correct?

Quick Quiz 12.7

In one of the Moon landings, an astronaut walking on the Moon's surface had a belt hanging from his space suit, and the belt oscillated as a physical pendulum. A scientist on Earth observed this motion on television and from it was able to estimate the free-fall acceleration on the Moon. How do you suppose this calculation was done?

Example 12.6   A Swinging Sign

A circular sign of mass M and radius R is hung on a nail from a small loop located at one edge (Fig. 12.12). After it is placed on the nail, the sign oscillates in a vertical plane. Find the period of oscillation if the amplitude of the motion is small.

Solution   The moment of inertia of a disk about an axis through the center is $\frac{1}{2}MR^2$ (see Table 10.2). The pivot point for the sign is through a point on the rim, so we use the parallel axis theorem (see Eq. 10.43) to find the moment of inertia about the pivot:

$$I_p = \frac{1}{2}MR^2 + MR^2 = \frac{3}{2}MR^2$$

The distance d from the pivot to the center of mass is the radius R. Substituting these quantities into Equation 12.27 gives

$$T = 2\pi \sqrt{\frac{\frac{3}{2}MR^2}{MgR}} = 2\pi \sqrt{\frac{3R}{2g}}$$

Figure 12.12

(Example 12.6) A circular sign oscillating about a pivot as a physical pendulum.

12.6 • DAMPED OSCILLATIONS

The oscillatory motions we have considered so far have occurred under the simplification model of an ideal frictionless system, that is, one that oscillates indefinitely under the action of only a linear restoring force. In realistic systems, resistive forces, such as friction, are present and retard the motion of the system. Consequently, the mechanical energy of the system diminishes in time, and the motion is said to be *damped*.

One common type of resistive force, which we discussed in Chapter 5, is proportional to the velocity and acts in the direction opposite the velocity. This type of force is often observed when an object is oscillating slowly in air, for instance. Because the resistive force can be expressed as $\mathbf{R} = -b\mathbf{v}$, where b is a constant related to the strength of the resistive force, and the restoring force exerted on the system is $-kx$, Newton's second law gives us

$$\sum F_x = -kx - bv = ma_x$$

$$-kx - b\frac{dx}{dt} = m\frac{d^2x}{dt^2} \qquad \textbf{[12.28]}$$

The solution of this differential equation requires mathematics that may not yet be familiar to you, and so it will simply be stated without proof. When the parameters of the system are such that $b < \sqrt{4mk}$, so that the resistive force is small, the solution to Equation 12.28 is

$$x = (Ae^{-(b/2m)t})\cos(\omega t + \phi) \qquad \textbf{[12.29]}$$

where the angular frequency of the motion is

$$\omega = \sqrt{\frac{k}{m} - \left(\frac{b}{2m}\right)^2} \qquad \textbf{[12.30]}$$

This result can be verified by substituting Equation 12.29 into Equation 12.28. Notice that Equation 12.29 is similar to Equation 12.6, with the new feature that the amplitude (in the parentheses before the cosine function) depends on time. Figure 12.13 shows the position as a function of time in this case. We see that, **when the resistive force is relatively small, the oscillatory character of the motion is preserved but the amplitude of vibration decreases in time** and the motion ultimately ceases. This is known as an **underdamped oscillator.** The dashed blue lines in Figure 12.13, which form the *envelope* of the oscillatory curve, represent the exponential factor that appears in Equation 12.29. The exponential factor shows that the *amplitude decays exponentially with time*.

It is convenient to express the angular frequency of vibration of a damped system (Eq. 12.30) in the form

$$\omega = \sqrt{\omega_0^2 - \left(\frac{b}{2m}\right)^2}$$

where $\omega_0 = \sqrt{k/m}$ represents the angular frequency of oscillation in the absence of a resistive force (the undamped oscillator). In other words, when $b = 0$, the resistive force is zero and the system oscillates with angular frequency ω_0, called the **natural frequency.**[*] As the magnitude of the resistive force increases, the oscillations dampen more rapidly. When b reaches a critical value b_c, so that

- *Position as a function of time for a particle undergoing damped oscillations (small damping)*

- *Angular frequency for a particle undergoing damped oscillations (small damping)*

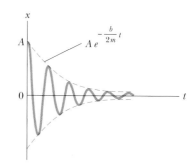

Figure 12.13

Graph of the position versus time for a damped oscillator with small damping. Note the decrease in amplitude with time.

---

[*] In practice, both ω_0 and $f_0 = \omega_0/2\pi$ are described as the natural frequency. The context of the discussion will help you determine which frequency is being discussed.

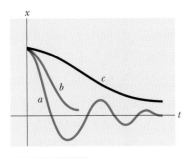

Figure 12.14

Plots of position versus time for (a) an underdamped oscillator, (b) a critically damped oscillator and (c) an overdamped oscillator.

$b_c/2m = \omega_0$, the system does not oscillate and is said to be **critically damped.** In this case it returns to equilibrium in an exponential manner with time, as in Figure 12.14.

If the medium is highly viscous, and the parameters meet the condition that $b/2m > \omega_0$, the system is **overdamped.** Again, the displaced system does not oscillate but simply returns to its equilibrium position. As the damping increases, the time it takes the particle to approach the equilibrium position also increases, as indicated in Figure 12.14. In any case, when a resistive force is present, the mechanical energy of the oscillator eventually falls to zero. The mechanical energy is transformed into internal energy in the oscillating system and the resistive medium.

12.7 • FORCED OSCILLATIONS

We have seen that the mechanical energy of a damped oscillator decreases in time as a result of the resistive force. It is possible to compensate for this energy decrease by applying an external force that does positive work on the system. At any instant, energy can be transferred into the system by an applied force that acts in the direction of motion of the oscillator. For example, a child on a swing can be kept in motion by appropriately timed "pushes." The amplitude of motion remains constant if the energy input per cycle of motion exactly equals the decrease in mechanical energy each cycle that results from resistive forces. This is an example of the non-isolated system in steady state, discussed in Section 7.5.

A common example of a forced oscillator is a damped oscillator driven by an external force that varies periodically, such as $F(t) = F_0 \sin \omega t$, where ω is the angular frequency of the driving force and F_0 is a constant. In general, the frequency ω of the driving force is different from the natural frequency ω_0 of the oscillator. Newton's second law in this situation gives

$$\sum F = ma \longrightarrow F_0 \sin \omega t - b\frac{dx}{dt} - kx = m\frac{d^2x}{dt^2} \qquad \textbf{[12.31]}$$

Again, the solution of this equation is rather lengthy and will not be presented. After the driving force on an initially stationary object begins to act, the amplitude of the oscillation will increase. After a sufficiently long period, when the energy input per cycle from the driving force equals the amount of mechanical energy transformed to internal energy for each cycle, a steady-state condition is reached in which the oscillations proceed with constant amplitude. In this case, Equation 12.31 has the solution

$$x = A\cos(\omega t + \phi) \qquad \textbf{[12.32]}$$

where

- *Amplitude of oscillation for a forced, damped oscillator*

$$A = \frac{F_0/m}{\sqrt{\left(\omega^2 - \omega_0{}^2\right)^2 + \left(\frac{b\omega}{m}\right)^2}} \qquad \textbf{[12.33]}$$

and where $\omega_0 = \sqrt{k/m}$ is the natural frequency of the undamped oscillator ($b = 0$).

Equation 12.33 shows that the amplitude of the forced oscillator is constant for a given driving force because it is being driven in steady state by an external force. For small damping, the amplitude becomes large when the frequency of the driving force is near the natural frequency of oscillation, or when $\omega \approx \omega_0$. The dramatic increase in amplitude near the natural frequency is called **resonance,**

- *Resonance*

and the natural frequency ω_0 is also called the **resonance frequency** of the system.

Figure 12.15 is a graph of amplitude as a function of frequency for the forced oscillator, with varying resistive forces. Note that the amplitude increases with decreasing damping ($b \rightarrow 0$) and that the resonance curve flattens as the damping increases. In the absence of a damping force ($b = 0$), we see from Equation 12.33 that the steady-state amplitude approaches infinity as $\omega \rightarrow \omega_0$. In other words, if there are no resistive forces in the system, and we continue to drive an oscillator with a sinusoidal force at the resonance frequency, the amplitude of motion will build up without limit. This does not occur in practice because some damping is always present in real oscillators.

One experiment that demonstrates a resonance phenomenon is illustrated in Figure 12.16. Several pendula of different lengths are suspended from a stretched string. If one of them, such as P, is set swinging, the others will begin to oscillate, because they are coupled by the stretched string. Of those that are forced into oscillation by this coupling, pendulum Q, which is the same length as P (hence, the two pendula have the same natural frequency), will oscillate with the greatest amplitude.

Resonance appears in other areas of physics. For example, certain electric circuits have resonance frequencies. This fact is exploited in radio tuners, which allow you to select the station you wish to hear. Vibrating strings and columns of air also have resonance frequencies, which allow them to be used for musical instruments, which we shall discuss in Chapter 14.

Quick Quiz 12.8

Some parachutes have holes in them to allow air to move smoothly through the chute. Without the holes, the air gathered under the chute as the parachutist falls is sometimes released from under the edges of the chute periodically from one side and then the other. Why might this periodic release of air cause a problem?

context connection
12.8 • RESONANCE IN STRUCTURES

In the preceding section, we investigated the phenomenon of resonance, in which an oscillating system exhibits its maximum response to a periodic driving force when the frequency of the driving force matches the natural frequency of the oscillator. We now apply this understanding to the interaction between the shaking of the ground during an earthquake and structures attached to the ground. The structure is the oscillator—it has a set of natural frequencies, determined by its stiffness, its mass, and the details of its construction. The periodic driving force is supplied by the shaking of the ground.

A disastrous result can occur if a natural frequency of the building matches a frequency contained in the ground shaking. In this case, the resonance vibrations of the building can build to a very large amplitude, large enough to damage or destroy the building. This can be avoided in two ways. The first involves designing the structure so that natural frequencies of the building lie outside the range of earthquake frequencies. (A typical range of earthquake frequencies is 0–15 Hz.) This can be done by varying the size or mass structure of the building. The second method involves incorporating sufficient damping in the building. This may not change the resonance frequency significantly, but it will lower the response to the

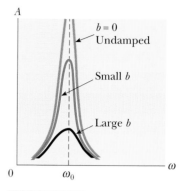

Figure 12.15

Graph of amplitude versus frequency for a damped oscillator when a periodic driving force is present. When the frequency of the driving force equals the natural frequency ω_0, resonance occurs. Note that the shape of the resonance curve depends on the size of the damping coefficient b.

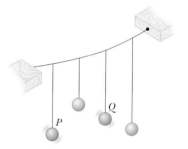

Figure 12.16

If pendulum P is set into oscillation, pendulum Q will eventually oscillate with the greatest amplitude because its length is equal to that of P and so they have the same natural frequency of vibration. The pendula are oscillating in a direction perpendicular to the plane formed by the stationary strings.

(a)   (b)

Figure 12.17

(a) In 1940, steady winds set up vibrations in the Tacoma Narrows Bridge, causing it to oscillate at a frequency near one of the natural frequencies of the bridge structure. (b) Once established, this resonance condition led to the bridge's collapse. *(Special Collections Division, University of Washington Libraries, photo by Farquharson)*

natural frequency, as in Figure 12.15. It will also flatten the resonance curve, so that the building will respond to a wide range of frequencies, but with relatively small amplitude at any given frequency.

We now describe two examples involving resonance excitations in bridge structures. Soldiers are commanded to break step when marching across a bridge. This command takes into account resonance—if the marching frequency of the soldiers matches that of the bridge, the bridge could be set into resonance oscillation. If the amplitude becomes large enough, the bridge could actually collapse. Just such a situation occurred on April 14, 1831, when the Broughton suspension bridge in England collapsed while troops marched over it. Investigations after the accident showed that the bridge was near failure—the resonance vibration induced by the marching soldiers caused it to fail sooner than it otherwise might have.

The second example of such a structural resonance occurred in 1940, when the Tacoma Narrows Bridge in Washington state was destroyed by resonant vibrations (Fig. 12.17). The winds were not particularly strong on that occasion, but the bridge still collapsed because vortices (turbulences) generated by the wind blowing through the bridge occurred at a frequency that matched a natural frequency of the bridge. The flapping of this wind (similar to the flapping of a flag in a strong breeze) provided the periodic driving force that brought the bridge down into the river.

Resonance gives us our first clue to responding to the central question for this Context. Suppose a building is far from the epicenter of an earthquake, so that the ground shaking is small. If the shaking frequency matches a natural frequency of the building, a very effective energy coupling occurs between the ground and the building. Thus, even for relatively small shaking, the ground, by resonance, can feed energy into the building efficiently enough to cause the failure of the structure. The structure must be carefully designed so as to reduce the resonance response.

In the Context introduction, we indicated that buildings in Mexico City were damaged in 1985 even though they were 400 km from the epicenter of the earthquake. How did a release of energy at the epicenter cause oscillations at such a dis-

WEB

Visit a NOVA site about the Tacoma Narrows Bridge at **www.pbs.org/wgbh/nova/ bridge/tacoma3vivisdn.html**

tant location? In the Loma Prieta earthquake during the World Series of 1989, the Nimitz Freeway in Oakland collapsed because of resonance. But only certain portions of it collapsed while other portions survived the shaking. What was special about the portions that collapsed? The answers to these questions involve *seismic waves*, which we will investigate in the next chapter.

WEB

For information on the Loma Prieta earthquake and related links, visit **www.earthquake.org/ lomaprieta.html**

SUMMARY

The simple harmonic motion model is used for a particle experiencing a linear restoring force, expressed by **Hooke's law:**

$$F_s = -kx \qquad \text{[12.1]}$$

where k is the **force constant** of the spring. The motion caused by such a force is called **simple harmonic motion,** and the system is called a **simple harmonic oscillator.** The position of the particle undergoing simple harmonic motion varies periodically in time according to the relation

$$x = A\cos(\omega t + \phi) \qquad \text{[12.6]}$$

where A is the **amplitude** of the motion, ω is the **angular frequency,** and ϕ is the **phase constant.** The values of A and ϕ depend on the initial position and velocity of the oscillator.

The time for one complete oscillation is called the **period** of the motion, defined by

$$T = \frac{1}{f} = \frac{2\pi}{\omega} \qquad \text{[12.10, 12.11]}$$

The inverse of the period is the **frequency** f of the motion, which equals the number of oscillations per second.

The velocity and acceleration of a particle undergoing simple harmonic oscillation are

$$v = \frac{dx}{dt} = -\omega A\sin(\omega t + \phi) \qquad \text{[12.15]}$$

$$a = \frac{d^2x}{dt^2} = -\omega^2 A\cos(\omega t + \phi) \qquad \text{[12.16]}$$

Thus, the maximum speed is ωA, and the maximum acceleration is of magnitude $\omega^2 A$. The speed is zero when the oscillator is at its turning points, $x = \pm A$, and the speed is a maximum at the equilibrium position, $x = 0$. The magnitude of the acceleration is a maximum at the turning points and is zero at the equilibrium position.

A particle–spring system moving on a frictionless track exhibits simple harmonic motion with the period

$$T = \frac{2\pi}{\omega} = 2\pi\sqrt{\frac{m}{k}} \qquad \text{[12.13]}$$

where m is the mass of the particle attached to the spring.

The kinetic energy and potential energy of a simple harmonic oscillator vary with time and are given by

$$K = \tfrac{1}{2}mv^2 = \tfrac{1}{2}m\omega^2 A^2\sin^2(\omega t + \phi) \qquad \text{[12.19]}$$

$$U = \tfrac{1}{2}kx^2 = \tfrac{1}{2}kA^2\cos^2(\omega t + \phi) \qquad \text{[12.20]}$$

The **total energy** of a simple harmonic oscillator is a constant of the motion and is

$$E = \tfrac{1}{2}kA^2 \qquad \text{[12.21]}$$

The potential energy of a simple harmonic oscillator is a maximum when the particle is at its turning points (maximum displacement from equilibrium) and is zero at the equilibrium position. The kinetic energy is zero at the turning points and is a maximum at the equilibrium position.

A **simple pendulum** of length L exhibits simple harmonic motion for small angular displacements from the vertical, with a period of

$$T = 2\pi\sqrt{\frac{L}{g}} \qquad \text{[12.25]}$$

The period of a simple pendulum is independent of the mass of the suspended object.

A **physical pendulum** exhibits simple harmonic motion for small angular displacements from equilibrium about a pivot that does not go through the center of mass. The period of this motion is

$$T = 2\pi\sqrt{\frac{I}{mgd}} \qquad \text{[12.27]}$$

where I is the moment of inertia about an axis through the pivot and d is the distance from the pivot to the center of mass.

Damped oscillations occur in a system in which a resistive force opposes the motion of the oscillating object. If such a system is set in motion and then left to itself, its mechanical energy decreases in time because of the presence of the nonconservative resistive force. It is possible to compensate for this transformation of energy by driving the system with an external periodic force. The oscillator in this case is undergoing **forced oscillations.** When the frequency of the driving force matches the natural frequency of the *undamped* oscillator, energy is efficiently transferred to the oscillator and its steady-state amplitude is a maximum. This situation is called **resonance.**

QUESTIONS

1. Imagine that a pendulum is hanging from the ceiling of a car. As the car coasts freely down a hill, is the equilibrium position of the pendulum vertical? Does the period of oscillation change from that in a stationary car?

2. An object is hung on a spring and the frequency of oscillation f of the system is measured. The object, a second identical object, and the spring are carried in the space shuttle to space. The two objects are attached to the ends of the spring, and the system is taken out into space on a space walk. The spring is extended and the system is released to oscillate while floating in space. What is the frequency of oscillation for this system, in terms of f?

3. You stand on the end of a diving board and bounce to set it in oscillation. You find a maximum response, in terms of the amplitude of oscillation of the end of the board, when you bounce at frequency f. You now move to the middle of the board, and repeat the experiment. Is the resonance frequency for forced oscillations at this point higher, lower, or the same as f? Why?

4. If the position of a particle varies as $x = -A \cos \omega t$, what is the phase constant in Equation 12.6? At what position is the particle at $t = 0$?

5. Does the displacement of an oscillating particle between $t = 0$ and a later time t necessarily equal the position of the particle at time t? Explain.

6. Determine whether the following quantities can have the same sign for a simple harmonic oscillator: (a) position and velocity, (b) velocity and acceleration, (c) position and acceleration.

7. Can the amplitude A and phase constant ϕ be determined for an oscillator if only the position is specified at $t = 0$? Explain.

8. Explain why the kinetic and potential energies of a block–spring system can never be negative.

9. A block–spring system undergoes simple harmonic motion with an amplitude A. Does the total energy change if the mass of the block is doubled but the amplitude is unchanged? Do the kinetic and potential energies depend on the mass? Explain.

10. What happens to the period of a simple pendulum if the pendulum's length is doubled? What happens to the period if the mass of the bob is doubled?

11. A simple pendulum undergoes simple harmonic motion when θ is small. Is the motion periodic when θ is large? How does the period of motion change as θ increases?

12. Will damped oscillations occur for any values of b and k? Explain.

13. Is it possible to have damped oscillations when a system is at resonance? Explain.

14. If a grandfather clock were running slow, how could we adjust the length of the pendulum to correct the time?

PROBLEMS

1, 2, 3 = straightforward, intermediate, challenging   □ = full solution available in the *Student Solutions Manual and Study Guide*

web = solution posted at **http://www.harcourtcollege.com/physics/**   🖥 = computer useful in solving problem

▦ = Interactive Physics   ■ = paired numerical/symbolic problems   ◩ = life science application

Note: Ignore the mass of every spring, except in Problems 47 and 50.

Section 12.1 Motion of a Particle Attached to a Spring

1. An archer pulls his bow string back 0.400 m by exerting a force on the string that increases uniformly from zero to 230 N. (a) What is the equivalent force constant of the bow? (b) How much work does he do in pulling the bow?

Section 12.2 Mathematical Representation of Simple Harmonic Motion

2. A ball dropped from a height of 4.00 m makes a perfectly elastic collision with the ground. Assuming no energy is lost due to air resistance, (a) show that the ensuing motion is periodic and (b) determine the period of the motion. (c) Is the motion simple harmonic? Explain.

3. The position of a particle is given by the expression $x = (4.00 \text{ m}) \cos(3.00\pi t + \pi)$, where x is in meters and t is in seconds. Determine (a) the frequency and period of the motion, (b) the amplitude of the motion, (c) the phase constant, and (d) the position of the particle at $t = 0.250$ s.

4. A particle moves in simple harmonic motion with a frequency of 3.00 Hz and an amplitude of 5.00 cm. (a) Through what total distance does the particle move during one cycle of its motion? (b) What is its maximum speed? Where does this occur? (c) Find the maximum acceleration of the particle. Where in the motion does the maximum acceleration occur?

5. In an engine, a piston oscillates with simple harmonic motion so that its position varies according to the expression

$$x = (5.00 \text{ cm}) \cos(2t + \pi/6)$$

where x is in centimeters and t is in seconds. At $t = 0$, find (a) the position of the particle, (b) its velocity, and (c) its acceleration. (d) Find the period and amplitude of the motion.

6. A spring stretches by 3.90 cm when a 10.0-g object hangs from it at rest. If this object is replaced with a 25.0-g object that is set into simple harmonic motion, calculate the period of the motion.

7. A particle moving along the x axis in simple harmonic motion starts from its equilibrium position, the origin, at $t = 0$ and moves to the right. The amplitude of its motion is 2.00 cm and the frequency is 1.50 Hz. (a) Show that the position of the particle is given by

$$x = (2.00 \text{ cm}) \sin(3.00 \, \pi t)$$

Determine (b) the maximum speed and the earliest time ($t > 0$) at which the particle has this speed, (c) the maximum acceleration and the earliest time ($t > 0$) at which the particle has this acceleration, and (d) the total distance traveled between $t = 0$ and $t = 1.00$ s.

8. A simple harmonic oscillator takes 12.0 s to undergo five complete vibrations. Find (a) the period of its motion, (b) the frequency in hertz, and (c) the angular frequency in radians per second.

9. A 7.00-kg object is hung from the bottom end of a vertical spring fastened to an overhead beam. The object is set into vertical oscillations having a period of 2.60 s. Find the force constant of the spring.

10. The initial position and initial velocity of an object moving in simple harmonic motion are x_i, v_i, and a_i; the angular frequency of oscillation is ω. (a) Show that the position and velocity of the object for all time can be written as

$$x(t) = x_i \cos \omega t + \left(\frac{v_i}{\omega}\right) \sin \omega t$$

$$v(t) = -x_i \omega \sin \omega t + v_i \cos \omega t$$

(b) If the amplitude of the motion is A, show that

$$v^2 - ax = v_i^2 - a_i x_i = \omega^2 A^2$$

11. A 0.500-kg object attached to a spring with a force constant of 8.00 N/m vibrates in simple harmonic motion with an amplitude of 10.0 cm. Calculate (a) the maximum value of its speed and acceleration, (b) the speed and acceleration when the object is 6.00 cm from the equilibrium position, and (c) the time it takes the object to move from $x = 0$ to $x = 8.00$ cm.

12. A piston in a gasoline engine is in simple harmonic motion. If the extremes of its displacement away from its center point are ± 5.00 cm, find the maximum velocity and acceleration of the piston when the engine is running at the rate of 3600 rev/min.

13. After a thrilling plunge, bungee-jumpers bounce freely on the bungee cord through many cycles (Fig. P12.13b). Your little brother can make a pest of himself by figuring out the mass of each person, using a proportion which you set up by solving this problem: An object of mass m is oscillating freely on a vertical spring with a period T (Fig. P12.13a). An object of unknown mass m' on the same spring oscillates with a period T'. Determine (a) the force constant and (b) the unknown mass.

Figure P12.13 (a) Mass–spring system for Problems 13 and 50. (b) Bungee-jumping from a bridge. *(Telegraph Colour Library/FPG International)*

14. A 1.00-kg object attached to a spring with a force constant of 25.0 N/m oscillates on a horizontal, frictionless track. At $t = 0$ the object is released from rest at $x = -3.00$ cm (i.e., the spring is compressed by 3.00 cm). Find (a) the period of its motion; (b) the maximum values of its speed and acceleration; and (c) the displacement, velocity, and acceleration as functions of time.

Section 12.3 Energy Considerations in Simple Harmonic Motion

15. An automobile having a mass of 1 000 kg is driven into a brick wall in a safety test. The bumper behaves like a spring of constant 5.00×10^6 N/m and compresses 3.16 cm as the car is brought to rest. What was the speed of the car before impact, assuming that the mechanical energy of the car remains constant during impact with the wall?

16. A block of unknown mass is attached to a spring with a force constant of 6.50 N/m and undergoes simple harmonic motion with an amplitude of 10.0 cm. When the block is halfway between its equilibrium position and the end point, its speed is measured to be 30.0 cm/s. Calculate (a) the mass of the block, (b) the period of the motion, and (c) the maximum acceleration of the block.

17. A 200-g block is attached to a horizontal spring and executes simple harmonic motion on a frictionless surface with a period of 0.250 s. If the total energy of the system is 2.00 J, find (a) the force constant of the spring and (b) the amplitude of the motion.

18. A block–spring system oscillates with an amplitude of 3.50 cm. If the force constant is 250 N/m and the mass is 0.500 kg, determine (a) the mechanical energy of the system, (b) the maximum speed of the block, and (c) the maximum acceleration.

19. A 50.0-g block connected to a spring with a force constant of 35.0 N/m oscillates on a horizontal, frictionless surface with an amplitude of 4.00 cm. Find (a) the total energy of the system and (b) the speed of the block when the displacement is 1.00 cm. Find (c) the kinetic energy and (d) the potential energy when the displacement is 3.00 cm.

20. A 2.00-kg block is attached to a spring and placed on a horizontal, smooth surface. A horizontal force of 20.0 N is required to hold the block at rest when it is pulled 0.200 m from its equilibrium position. The block is now released from rest from this point, and it subsequently undergoes simple harmonic motion. Find (a) the force constant of the spring, (b) the frequency of the oscillations, and (c) the maximum speed of the block. Where does this maximum speed occur? (d) Find the maximum acceleration of the block. Where does it occur? (e) Find the total energy of the oscillating system. Find (f) the speed and (g) the acceleration when the position equals one third of the maximum value.

21. A particle executes simple harmonic motion with an amplitude of 3.00 cm. At what position does its speed equal one half of its maximum speed?

Section 12.4 The Simple Pendulum

Note: Problem 58 in Chapter 1 involves investigation of the $\sin \theta \approx \theta$ approximation.

22. A "seconds pendulum" is one that moves through its equilibrium position once each second. (The period of the pendulum is 2.000 s.) The length of a seconds pendulum is 0.992 7 m at Tokyo and 0.994 2 m at Cambridge, England. What is the ratio of the free-fall accelerations at these two locations?

23. A simple pendulum has a mass of 0.250 kg and a length of 1.00 m. It is displaced through an angle of 15.0° and then released. What are (a) the maximum speed, (b) the maximum angular acceleration, and (c) the maximum restoring force?

24. The angular position of a simple pendulum is represented by the equation $\theta = (0.320 \text{ rad}) \cos \omega t$, where θ is in radians and $\omega = 4.43$ rad/s. Determine the period and length of the pendulum.

25. A particle of mass m slides without friction inside a hemispherical bowl of radius R. Show that, if it starts from rest with a small displacement from equilibrium, the particle moves in simple harmonic motion with an angular frequency equal to that of a simple pendulum of length R (i.e., $\omega = \sqrt{g/R}$).

Section 12.5 The Physical Pendulum

26. A very light rigid rod with a length of 0.500 m extends straight out from one end of a meter stick. The stick is suspended from a pivot at the far end of the rod and is set into oscillation. (a) Determine the period of oscillation. (*Hint:* Use the parallel-axis theorem from Section 10.11.) (b) By what percentage does the period differ from the period of a simple pendulum 1.00 m long?

27. A physical pendulum in the form of a planar body moves in simple harmonic motion with a frequency of 0.450 Hz. If the pendulum has a mass of 2.20 kg and the pivot is located 0.350 m from the center of mass, determine the moment of inertia of the pendulum about the pivot point.

Section 12.6 Damped Oscillations

28. Show that the time rate of change of mechanical energy for a damped, undriven oscillator is given by $dE/dt = -bv^2$ and hence is always negative. (*Hint:* Differentiate the expression for the mechanical energy of an oscillator, $E = \frac{1}{2}mv^2 + \frac{1}{2}kx^2$, and use Eq. 12.28.)

29. A pendulum with a length of 1.00 m is released from an initial angle of 15.0°. After 1 000 s, its amplitude has been reduced by friction to 5.50°. What is the value of $b/2m$?

30. Show that Equation 12.29 is a solution of Equation 12.28 provided that $b^2 < 4mk$.

Section 12.7 Forced Oscillations

31. The front of her sleeper wet from teething, a baby rejoices in the day by crowing and bouncing up and down in her crib. Her mass is 12.5 kg and the crib mattress can be modeled as a light spring with force constant 4.30 kN/m. (a) The baby soon learns to bounce with maximum amplitude and minimum effort by bending her knees at what frequency? (b) She learns to use the mattress as a trampoline—losing contact with it for part of each cycle—when her amplitude exceeds what value?

32. A 2.00-kg object attached to a spring is driven by an external force given by $F = (3.00 \text{ N}) \cos(2\pi t)$. If the force constant of the spring is 20.0 N/m, determine (a) the period and (b) the amplitude of the motion. (*Hint:* Assume there is no damping—that is, that $b = 0$—and use Eq. 12.33.)

33. Considering an *undamped*, forced oscillator ($b = 0$), show that Equation 12.32 is a solution of Equation 12.31, with an amplitude given by Equation 12.33.

34. Damping is negligible for a 0.150-kg object hanging from a light 6.30-N/m spring. The system is driven by a force oscillating with an amplitude of 1.70 N. At what frequency will the force make the mass vibrate with an amplitude of 0.440 m ?

35. You are a research biologist. Its batteries are getting low, but you take your emergency pager along to a fine restaurant. You switch the small pager to vibrate instead of beep, and you put it into a side pocket of your suit coat. The arm of your chair presses the light cloth against your body at one spot. Fabric with a length of 8.21 cm hangs freely below that spot, with the pager at the bottom. A co-worker urgently needs instructions and calls you from your laboratory. The motion of the pager makes the hanging part of your coat swing back and forth with remarkably large amplitude. The waiter, maître d', wine steward, and nearby diners notice immediately and fall silent. Your daughter pipes up and says, "Daddy, look! Your cockroaches must have gotten out again!" Find the frequency at which your pager vibrates.

Section 12.8 Context Connection — Resonance in Structures

36. Four people, each with a mass of 72.4 kg, are in a car with a mass of 1 130 kg. An earthquake strikes. The driver manages to pull off the road and stop, as the vertical oscillations of the ground surface make the car bounce up and down on its suspension springs. When the frequency of the shaking is 1.80 Hz, the car exhibits a maximum amplitude of vibration. The earthquake ends and the four people leave the car as fast as they can. By what distance does the car's undamaged suspension lift the car body as the people exit the car?

37. People who ride motorcycles and bicycles learn to look out for bumps in the road, and especially for *washboarding*, a condition in which many equally spaced ridges are worn into the road. What is so bad about washboarding? A motorcycle has several springs and shock absorbers in its suspension, but you can model it as a single spring supporting a mass. You can estimate the force constant by thinking about how far the spring compresses when a big biker sits down on the seat. A motorcyclist traveling at highway speed must be particularly careful of washboard bumps that are a certain distance apart. What is the order of magnitude of their separation distance? State the quantities you take as data and the values you measure or estimate for them.

Additional Problems

38. An object with mass $m_1 = 9.00$ kg is in equilibrium while connected to a light spring of constant $k = 100$ N/m that is fastened to a wall as shown in Figure P12.38a. A second object, of mass $m_2 = 7.00$ kg, is slowly pushed up against m_1, compressing the spring by the amount $A = 0.200$ m, (Fig. P12.38b). The system is then released, and both objects start moving to the right on the frictionless surface. (a) When m_1 reaches the equilibrium point, m_2 loses contact with m_1 (Fig. P12.38c) and moves to the right with speed v. Determine the value of v. (b) How far apart are the objects when the spring is fully stretched for the first time (D in Fig. P12.38d)? (*Hint:* First determine the period of oscillation and the amplitude of the m_1–spring system after m_2 loses contact with m_1.)

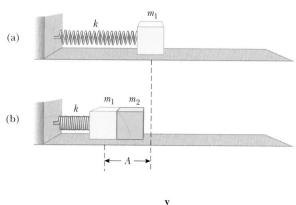

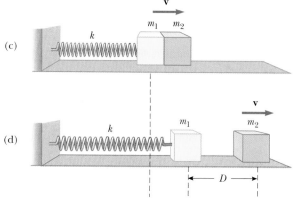

Figure P12.38

39. A large block P executes horizontal simple harmonic motion as it slides across a frictionless surface with a frequency $f = 1.50$ Hz. Block B rests on it, as shown in Figure P12.39, and the coefficient of static friction between the

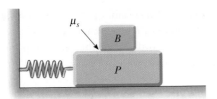

Figure P12.39

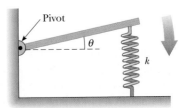

Figure P12.43

two is $\mu_s = 0.600$. What maximum amplitude of oscillation can the system have if block *B* is not to slip?

40. A particle with a mass of 0.500 kg is attached to a spring with a force constant of 50.0 N/m. At time $t = 0$ the particle has its maximum speed of 20.0 m/s and is moving to the left. (a) Determine the particle's equation of motion, specifying its position as a function of time. (b) Where in the motion is the potential energy three times the kinetic energy? (c) Find the length of a simple pendulum with the same period. (d) Find the minimum time required for the particle to move from $x = 0$ to $x = 1.00$ m.

41. A pendulum of length *L* and mass *M* has a spring of force constant *k* connected to it at a distance *h* below its point of suspension (Fig. P12.41). Find the frequency of vibration of the system for small values of the amplitude (small θ). Assume the vertical suspension of length *L* is rigid, but ignore its mass.

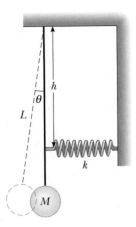

Figure P12.41

42. The mass of the deuterium molecule (D_2) is twice that of the hydrogen molecule (H_2). If the vibrational frequency of H_2 is 1.30×10^{14} Hz, what is the vibrational frequency of D_2? Assume that the "force constant" of attracting forces is the same for the two molecules.

43. A horizontal plank of mass *m* and length *L* is pivoted at one end. The plank's other end is supported by a spring of force constant *k* (Fig P12.43). The moment of inertia of the plank about the pivot is $\frac{1}{3}mL^2$. The plank is displaced by a small angle θ from its horizontal equilibrium position

and released. (a) Show that it moves with simple harmonic motion with an angular frequency $\omega = \sqrt{3k/m}$. (b) Evaluate the frequency if the mass is 5.00 kg and the spring has a force constant of 100 N/m.

44. **Review Problem.** An object of mass 4.00 kg is attached to a spring with a force constant of 100 N/m. It is oscillating on a horizontal frictionless surface with an amplitude of 2.00 m. A 6.00-kg object is dropped vertically on top of the 4.00-kg object as it passes through its equilibrium point. The two objects stick together. (a) By how much does the amplitude of the vibrating system change as a result of the collision? (b) By how much does the period change? (c) By how much does the energy change? (d) Account for the change in energy.

45. One end of a light spring with force constant 100 N/m is attached to a vertical wall. A light string is tied to the other end of the horizontal spring. The string changes from horizontal to vertical as it passes over a solid pulley of diameter 4.00 cm. The pulley is free to turn on a fixed smooth axle. The vertical section of the string supports a 200-g object. The string does not slip at its contact with the pulley. Find the frequency of oscillation of the object if the mass of the pulley is (a) negligible, (b) 250 g, and (c) 750 g.

46. A 2.00-kg block hangs without vibrating at the bottom end of a spring with a force constant of 500 N/m. The top end of the spring is attached to the ceiling of an elevator car. The car is rising with an upward acceleration of $g/3$ when the acceleration suddenly ceases (at $t = 0$). (a) What is the angular frequency of oscillation of the block after the acceleration ceases? (b) By what amount is the spring stretched during the time that the elevator car is accelerating? (c) What are the amplitude of the oscillation and the initial phase angle observed by a rider in the car? Take the upward direction to be positive.

47. A block of mass *M* is connected to a spring of mass *m* and oscillates in simple harmonic motion on a horizontal, frictionless track (Fig. P12.47). The force constant of the spring is *k* and the equilibrium length is ℓ. Find (a) the kinetic energy of the system when the block has a speed *v*, and (b) the period of oscillation. [*Hint:* Assume that all portions of the spring oscillate in phase and that the veloc-

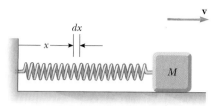

Figure P12.47

ity v_x of a segment dx is proportional to the position x of the segment from the fixed end; that is, $v_x = (x/\ell)v$. Also, note that the mass of a segment of the spring is $dm = (m/\ell)\,dx$.]

48. A simple pendulum with a length of 2.23 m and a mass of 6.74 kg is given an initial speed of 2.06 m/s at its equilibrium position. Assume it undergoes simple harmonic motion, and determine its (a) period, (b) total energy, and (c) maximum angular position.

49. A ball of mass m is connected to two rubber bands of length L, each under tension T, as in Figure P12.49. The ball is displaced by a small distance y perpendicular to the length of the rubber bands. Assuming that the tension does not change, show that (a) the restoring force is $-(2T/L)y$ and (b) the system exhibits simple harmonic motion with an angular frequency $\omega = \sqrt{2T/mL}$.

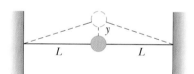

Figure P12.49

50. When an object of mass M, connected to the end of a spring of mass $m_s = 7.40$ g and force constant k, is set into simple harmonic motion, the period of its motion is

$$T = 2\pi\sqrt{\frac{M + (m_s/3)}{k}}$$

A two-part experiment is conducted with the use of various masses suspended vertically from the spring, as shown in Figure P12.13a. (a) Static extensions of 17.0, 29.3, 35.3, 41.3, 47.1, and 49.3 cm are measured for M values of 20.0, 40.0, 50.0, 60.0, 70.0, and 80.0 g, respectively. Construct a graph of Mg versus x, and perform a linear least-squares fit to the data. From the slope of your graph, determine a value for k for this spring. (b) The system is now set into simple harmonic motion, and periods are measured with a stopwatch. With $M = 80.0$ g, the total time for ten oscillations is measured to be 13.41 s. The experiment is re-

peated with M values of 70.0, 60.0, 50.0, 40.0, and 20.0 g, with corresponding times for ten oscillations of 12.52, 11.67, 10.67, 9.62, and 7.03 s. Compute the experimental value for T from each of these measurements. Plot a graph of T^2 versus M and determine a value for k from the slope of the linear least-squares fit through the data points. Compare this value of k with the value obtained in part (a). (c) Obtain a value for m_s from your graph and compare it with the given value of 7.40 g.

51. Imagine that a hole is drilled through the center of the Earth to the other side. It can be shown that an object of mass m at a distance r from the center of the Earth is pulled toward the center of the Earth only by the mass within the sphere of radius r (the reddish region in Fig. P12.51). (a) Write Newton's law of gravitation for an object at the distance r from the center of the Earth, and show that the force on it is of Hooke's law form, $F = -kr$, where the effective force constant is $k = (4/3)\pi\rho Gm$. Here ρ is the density of the Earth, assumed uniform, and G is the gravitational constant. (b) Show that a sack of mail dropped into the hole will execute simple harmonic motion if it moves without friction. How long does it take to arrive at the other side of the Earth?

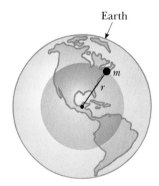

Figure P12.51

52. An object is hung from a spring, set into vertical vibration, and immersed in a beaker of oil. Its motion is graphed in Figure 12.13. Assume the object has mass $m = 375$ g, the spring has force constant $k = 100$ N/m, and $b = 0.100$ N·s/m. (a) How long does it take for the amplitude to drop to half its initial value? (b) How long does it take for the mechanical energy to drop to half its initial value? (c) Show that, in general, the fractional rate at which the amplitude decreases in a damped harmonic oscillator is one-half the fractional rate at which the mechanical energy decreases.

53. A block of mass m is connected to two springs of force constants k_1 and k_2 as shown in Figures P12.53a and P12.53b. In each case, the block moves on a frictionless table after it is displaced from equilibrium and released.

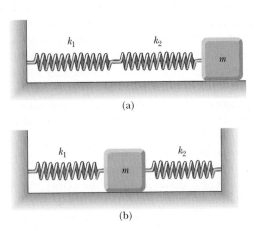

(a)

(b)

Figure P12.53

Show that in the two cases the block exhibits simple harmonic motion with periods

(a) $T = 2\pi\sqrt{\dfrac{m(k_1 + k_2)}{k_1 k_2}}$

(b) $T = 2\pi\sqrt{\dfrac{m}{k_1 + k_2}}$

54. Consider a small object on a light rigid rod, forming a simple pendulum of length $L = 1.20$ m. It is displaced from the vertical by an angle θ_{max} and then released. Predict the subsequent angular positions if θ_{max} is small and also if it is large. Set up and carry out a numerical method to integrate the equation of motion for the simple pendulum:

$$\frac{d^2\theta}{dt^2} = -\frac{g}{L}\sin\theta$$

Take the initial conditions to be $\theta = \theta_{max}$ and $d\theta/dt = 0$ at $t = 0$. On one trial choose $\theta_{max} = 5.00°$, and on another trial take $\theta_{max} = 100°$. In each case find the angular position θ as a function of time. Using the same values of θ_{max}, compare your results for θ with those obtained from the expression $\theta_{max}\cos\omega t$. How does the period for the large value of θ_{max} compare with that for the small value of θ_{max}? *Note:* Using the Euler method to solve this differential equation, you may find that the amplitude tends to increase with time. If you choose Δt small enough, however, the solution that you obtain using Euler's method can still be good.

55. Your thumb squeaks on a plate you have just washed. A hinge creaks. A wheel squeaks. A corkscrew squeals inside the cork. Your sneakers squeak on the gym floor. Car tires squeal when you stop or start abruptly. You can make a goblet "sing" by wiping your moistened finger around its rim. Mortise joints groan in an old table or house. The

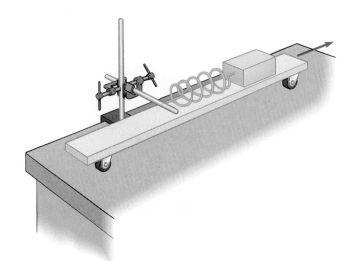

Figure P12.55

ground shudders in an earthquake. As you slide it across a table, a Styrofoam cup may not make much sound, but it makes the surface of some water inside it dance in a complicated resonance vibration.

As these examples suggest, vibration commonly results when friction acts on a moving elastic object. The oscillation is not simple harmonic motion, but is called *stick-and-slip*. When chalk squeaks on a blackboard, you can see that it makes a chain of regularly spaced dashes. This problem models stick-and-slip motion. A block of mass m is attached to a fixed support by a horizontal spring with force constant k and negligible mass (Fig. P12.55). Hooke's law describes the spring both in extension and in compression. The block sits on a long horizontal board, with which it has a coefficient of static friction μ_s and a smaller coefficient of static friction μ_k. The board moves to the right at constant speed v. Assume that the block spends most of its time sticking to the board and moving to the right, so that the speed v is small in comparison to the average speed of the block as it slips back to the left. (a) Show that the maximum extension of the spring from its unstressed position is very nearly given by $\mu_s mg/k$. (b) Show that the block oscillates around an equilibrium position at which the spring is stretched by $\mu_k mg/k$. (c) Graph the block's position versus time. (d) Show that the amplitude of the block's motion is

$$A = \frac{(\mu_s - \mu_k)mg}{k}$$

(e) Show that the period of the block's motion is

$$T = \frac{2(\mu_s - \mu_k)mg}{vk} + \pi\sqrt{\frac{m}{k}}$$

(f) Evaluate the frequency of the motion if $\mu_s = 0.400$, $\mu_k = 0.250$, $m = 0.300$ kg, $k = 12.0$ N/m, and $v = 2.40$ cm/s. (g) What happens to the frequency if the mass increases? (h) If the spring stiffness increases? (i) If the speed of the board increases? (j) If the coefficient of static

friction increases relative to the coefficient of kinetic friction? Observe that the *difference* between static and kinetic friction is important for the vibration. A squeaky wheel is greased because even a viscous fluid cannot exert a force of static friction.

ANSWERS TO QUICK QUIZZES

12.1 (d). In moving from its initial position to the equilibrium position, the block moves through distance A. It then continues to the other side of the equilibrium position, stopping after moving through another distance A, for a total so far of $2A$. It then returns to the initial position, traveling through another distance of $2A$, for a total distance in the full cycle of $4A$.

12.2 $x = -\dfrac{v_i}{\omega} \cos\left(\omega t - \dfrac{\pi}{2}\right)$

12.3 The bouncing ball is not an example of simple harmonic motion. The position of the ball is not described by a sinusoidal function. The daily movement of a student is also not simple harmonic motion because the student stays at a fixed position—school—for a long period. If this motion were sinusoidal, the student would move more and more slowly as she approached her desk, and, as soon as she sat down at the desk, she would start to move back toward home again!

12.4 According to the principle of equivalence (Section 9.9), an acceleration of the elevator is equivalent to a gravitational field. Thus, if the elevator is (a) accelerating upward, this is equivalent to an increased effective gravitational field magnitude g, and, according to Equation 12.25, the period will decrease. Similarly, if the elevator is (b) accelerating downward, the effective value of g is reduced and the period increases. (c) If the elevator moves with constant velocity, the period of the pendulum is the same as that in the stationary elevator.

12.5 (a). With a longer length, the period of the pendulum increases. Thus, it takes longer to execute each swing, so

that each second according to the clock takes longer than an actual second. Thus, the clock runs *slow*.

12.6 The first student is correct. Although changing the mass of a simple pendulum does not change the frequency, the fact that the bob is large means that we must model the pendulum as a physical pendulum rather than a simple pendulum. When the gum is placed on top of the bob, the moment of inertia of the physical pendulum is altered slightly. According to Equation 12.27, this alters the period of the pendulum.

12.7 The period of oscillation T of the swinging belt can be measured by watching the television coverage. Furthermore, the belt can be geometrically modeled as a long thin rod, so that its moment of inertia can be approximated as $\frac{1}{3}mL^2$ from Table 10.2. Finally, because the belt is assumed to be of uniform density, the center of mass is located at $d = L/2$. Thus, Equation 12.27 becomes

$$T = 2\pi\sqrt{\dfrac{\frac{1}{3}mL^2}{mg(L/2)}} = 2\pi\sqrt{\dfrac{2L}{3g}}$$

Upon estimating the length of the swinging belt by comparing it to the height of a typical astronaut this equation can be solved for g.

12.8 If the period of the alternating release of air were to match the swinging period of the parachutist, a resonance response could occur. The swinging of the parachutist could become so energetic that the parachute lines could become fouled and the braking action would disappear.

Drops of water fall from a leaf into a pond. The disturbance caused by the falling water moves away from the drop point as circular ripples on the water's surface.

(Don Bonsey/Stone)

Mechanical Waves

Most of us have experienced waves as children, when we dropped pebbles into a pond. The disturbance created by a pebble manifested itself as ripples that moved outward from the point at which the pebble landed in the water, like the ripples due to the falling water drops in the opening photograph. If you were to carefully examine the motion of a leaf floating near the point where the pebble entered the water, you would see that the leaf moves up and down and back and forth about its original position but does not undergo any net displacement away from or toward the source of the disturbance. The disturbance in the water moves from one place to another, *yet the water is not carried with it*—this is the essence of wave motion.

The world is full of other kinds of waves, including sound waves, waves on strings, seismic waves, radio waves, and x-rays. Most waves can be placed in one of two categories. **Mechanical waves** are waves that disturb and propagate through a medium; the ripple in the water due to the pebble and a sound wave, for which air is the medium, are examples of mechanical waves. **Electromagnetic waves** are a special class of waves that do not require a medium in order to propagate, as discussed with regard to the absence of the ether in Section 9.2; light waves and radio waves are two familiar examples. In this chapter we shall confine our attention to the study of mechanical waves, deferring our study of electromagnetic waves to Chapter 24.

The wave concept is abstract. When we observe a water wave, what we see is a rearrangement of the water's surface. Without the water, there would be no wave. In the case of this or any other mechanical wave, what we interpret as a wave corresponds to the disturbance of a medium. Therefore, we can consider a mechanical wave to be the *propagation of a disturbance in a medium.*

13.1 • PROPAGATION OF A DISTURBANCE

In the introduction, we alluded to the essence of wave motion—the transfer of a *disturbance* through space without the accompanying transfer of *matter.* The propagation of the disturbance also represents a transfer of energy—thus, we can view waves as a means of energy transfer. In the list of energy transfer mechanisms in Section 6.6, we see two entries that depend on waves: mechanical waves and electromagnetic radiation. These are to be contrasted with another entry—matter transfer—in which the energy transfer is accompanied by a movement of matter through space.

All mechanical waves require (1) some source of disturbance, (2) a medium that can be disturbed, and (3) some physical mechanism through which particles of the medium can influence one another. All waves carry energy, but the amount of energy transmitted through a medium and the mechanism responsible for the energy transport differ from case to case. For instance, the power of ocean waves during a storm is much greater than that of sound waves generated by a musical instrument.

One way to demonstrate wave motion is to flip the free end of a long rope that is under tension and has its opposite end fixed, as in Figure 13.1. In this manner, a single **pulse** is formed and travels (to the right in Fig. 13.1) with a definite speed. The rope is the medium through which the pulse travels. Figure 13.1 represents consecutive "snapshots" of the traveling pulse. The shape of the pulse changes very little as it travels along the rope.

 See the *Core Concepts in Physics CD-ROM*, Screens 8.4 & 8.5

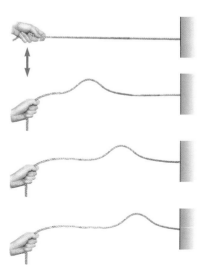

Figure 13.1

A wave pulse traveling down a stretched rope. The shape of the pulse is approximately unchanged as it travels along the rope.

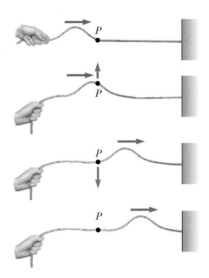

Figure 13.2

A pulse traveling on a stretched rope is a transverse wave. That is, any element *P* on the rope moves *(blue arrows)* in a direction perpendicular to the propagation of the wave *(red arrows)*.

As the pulse travels, each rope segment that is disturbed moves in a direction perpendicular to the direction of propagation. Figure 13.2 illustrates this point for a particular segment, labeled *P*. Note that no motion occurs in any part of the rope in the direction of the wave. A disturbance such as this, in which the particles of the disturbed medium move perpendicularly to the direction of propagation, is called a **transverse wave.**

In another class of waves, called **longitudinal waves,** the particles of the medium undergo displacements *parallel* to the direction of propagation. Sound waves in air, for instance, are longitudinal. Their disturbance corresponds to a series of high- and low-pressure regions that may travel through air or through any material medium with a certain speed. A longitudinal pulse can be easily produced in a stretched spring, as in Figure 13.3. A group of coils at the free end is pushed forward and pulled back. This action produces a pulse in the form of a compressed region of the coil that travels along the spring, parallel to the direction of propagation.

So far we have provided pictorial representations of a traveling pulse and hope you have begun to develop a mental representation of such a pulse. Let us now develop a mathematical representation for the propagation of this pulse. Consider a pulse traveling to the right with constant speed v on a long, stretched string, as in Figure 13.4. The pulse moves along the x axis (the axis of the string), and the transverse (up-and-down) displacement of the string is measured with the position coordinate y.

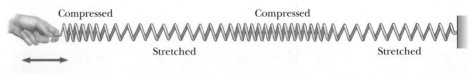

Compressed   Compressed

Stretched   Stretched

Figure 13.3

A longitudinal pulse along a stretched spring. The displacement of the coils is in the direction of the wave motion. For the starting motion described in the text, the compressed region is followed by a stretched region.

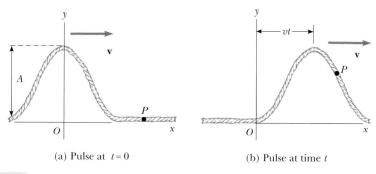

(a) Pulse at $t = 0$ (b) Pulse at time t

Figure 13.4

A one-dimensional wave pulse traveling to the right with a speed v. (a) At $t = 0$, the shape of the pulse is given by $y = f(x)$. (b) At some later time t, the shape remains unchanged and the vertical displacement of any point P of the medium is given by $y = f(x - vt)$.

Figure 13.4a represents the shape and position of the pulse at time $t = 0$. At this time, the shape of the pulse, whatever it may be, can be represented by some mathematical function that we will write as $y(x, 0) = f(x)$. This function describes the vertical position y of the element of the string located at each value of x at time $t = 0$. Because the speed of the pulse is v, the pulse has traveled to the right a distance vt at the time t (Fig. 13.4b). We adopt a simplification model in which the shape of the pulse does not change with time.* Thus, at time t, the shape of the pulse is the same as it was at time $t = 0$, as in Figure 13.4a. Consequently, an element of the string at x at this time has the same y position as an element located at $x - vt$ had at time $t = 0$:

$$y(x, t) = y(x - vt, 0)$$

In general, then, we can represent the displacement y for all positions and times, measured in a stationary frame with the origin at O, as

$$y(x, t) = f(x - vt) \qquad [13.1]$$

• *Pulse traveling to the right*

Similarly, if the wave pulse travels to the left, the displacement of the string is

$$y(x, t) = f(x + vt) \qquad [13.2]$$

• *Pulse traveling to the left*

The function y, sometimes called the **wave function,** depends on the two variables x and t. For this reason, it is often written $y(x, t)$, which is read "y as a function of x and t."

It is important to understand the meaning of y. Consider a point P on the string, identified by a particular value of its x coordinate. As the pulse passes through P, the y coordinate of this point increases, reaches a maximum, and then decreases to zero. **The wave function $y(x, t)$ represents the y coordinate of any point P located at position x at any time t.** Furthermore, if t is fixed (e.g., in the case of taking a snapshot of the pulse), then the wave function y as a function

---

* In reality, the pulse changes its shape and gradually spreads out during the motion. This effect is called *dispersion* and is common to many mechanical waves; however we adopt a simplification model that ignores this effect.

of *x*, sometimes called the **waveform,** defines a curve representing the actual geometric shape of the pulse at that time.

Quick Quiz 13.1

(a) A long line of people is waiting to buy tickets at a movie theater. When the first person leaves, a pulse of motion occurs as people step forward to fill in the gap. As each person steps forward, the gap moves through the line. Is the propagation of this gap transverse or longitudinal? (b) Consider "the wave" at a soccer or American football game: People stand up and shout as the wave arrives at their location, and this pulse moves around the stadium. Is this wave transverse or longitudinal?

THINKING PHYSICS 13.1

Why is it important to be quiet in avalanche country? In the movie *On Her Majesty's Secret Service* (United Artists, 1969), the bad guys try to stop James Bond, who is escaping on skis, by firing a gun and causing an avalanche. Why did this happen?

Reasoning   The essence of a wave is the propagation of a *disturbance* through a medium. An impulsive sound, like the gunshot in the James Bond movie, can cause an acoustic disturbance that propagates through the air and can impact a ledge of unstable snow that is just ready to break free to begin an avalanche. Such a disastrous event occurred in 1916 during World War I, when Austrian soldiers in the Alps were smothered by an avalanche caused by cannon fire.

Example 13.1   A Pulse Moving to the Right

A wave pulse moving to the right along the *x* axis is represented by the wave function

$$y(x, t) = \frac{2.0}{(x - 3.0t)^2 + 1}$$

where *x* and *y* are measured in centimeters and *t* is in seconds. Let us plot the waveform at *t* = 0, *t* = 1.0 s, and *t* = 2.0 s.

Solution   First, note that this function is of the form *y* = *f*(*x* − *vt*). By inspection, we see that the speed of the wave is *v* = 3.0 cm/s. The location of the peak of the pulse occurs at the value of *x* for which the denominator is a minimum, that is, where (*x* − 3.0*t*) = 0. Thus, the peaks occur at *x* = 0.0 cm at *t* = 0, at *x* = 3.0 cm at *t* = 1.0 s, and *x* = 6.0 cm at *t* = 2.0 s.

At times *t* = 0, *t* = 1.0 s, and *t* = 2.0 s, the wave function expressions are

$$y(x, 0) = \frac{2.0}{x^2 + 1} \quad \text{at } t = 0$$

$$y(x, 1.0) = \frac{2.0}{(x - 3.0)^2 + 1} \quad \text{at } t = 1.0 \text{ s}$$

$$y(x, 2.0) = \frac{2.0}{(x - 6.0)^2 + 1} \quad \text{at } t = 2.0 \text{ s}$$

We can now use these expressions to plot the wave function versus *x* at these times. For example, let us evaluate *y*(*x*, 0) at *x* = 0.50 cm:

$$y(0.50, 0) = \frac{2.0}{(0.50)^2 + 1} = 1.6 \text{ cm}$$

Likewise, *y*(1.0, 0) = 1.0 cm, *y*(2.0, 0) = 0.40 cm, and so on. A continuation of this procedure for other values of *x* yields the waveform shown in Figure 13.5a. In a similar manner, one obtains the graphs of *y*(*x*, 1.0) and *y*(*x*, 2.0), shown in Figures 13.5b and 13.5c, respectively. These snapshots show that the wave pulse moves to the right without changing its shape and has a constant speed of 3.0 cm/s.

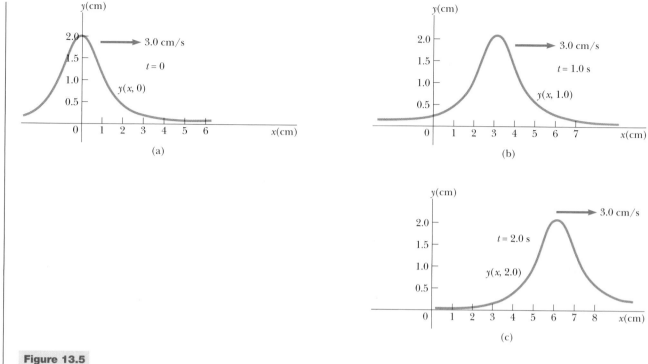

Figure 13.5

(Example 13.1) Graphs of the function $y(x, t) = 2.0/[(x - 3.0t)^2 + 1]$ at (a) $t = 0$, (b) $t = 1.0$ s, and (c) $t = 2.0$ s.

13.2 • THE WAVE MODEL

We have discussed creating a disturbance moving through a medium such as a stretched string by a simple up-and-down displacement of the end of the string. This results in a pulse moving along the medium. A continuous wave is created by shaking the end of the string in simple harmonic motion, which we studied in Chapter 12. If we do this, the string will take on the shape shown by the curve in the graph in Figure 13.6a, with this shape remaining the same but moving toward the right. This is what we call a **sinusoidal** wave because the shape in Figure 13.6 is that of a sine wave. The point with the largest positive displacement of the string is called the **crest** of the wave. The lowest point is called the **trough.** The crest and trough move along with the wave, so that a particular point on the string alternates between locations on a crest and a trough. In idealized wave motion in an idealized medium, each particle of the medium undergoes simple harmonic motion around its equilibrium position.

Three physical characteristics are important in describing a sinusoidal wave: **wavelength, frequency,** and **wave speed. One wavelength is the minimum distance between any two identical points on a wave**—for example, adjacent crests or adjacent troughs, as in Figure 13.6a, which is a graph of displacement versus position for a sinusoidal wave at a specific time. The symbol λ (Greek lambda) is used to denote wavelength.

The frequency of sinusoidal waves is the same as the frequency of simple harmonic motion of a particle of the medium. The **period** T of the wave is the minimum time it takes a particle of the medium to undergo one complete

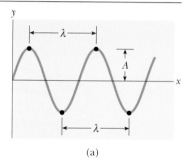

(a)

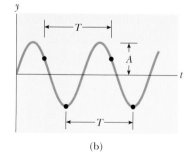

(b)

Figure 13.6

(a) The wavelength λ of a wave is the distance between adjacent crests or adjacent troughs. (b) The period T of a wave is the time it takes the wave to travel one wavelength.

oscillation, and is equal to the inverse of the frequency:

- *Period of a wave*

$$T = \frac{1}{f}$$

Figure 13.6b shows position versus time for a particle of the medium as a sinusoidal wave is passing through its position. The period is the time between instants when the particle has identical displacements and velocities.

Waves travel through the medium with a specific **wave speed,** which depends on the properties of the medium being disturbed. For instance, sound waves travel through air at 20°C with a speed of about 343 m/s, whereas the speed of sound in most solids is higher than 343 m/s. We will learn more about wavelength, frequency, and wave speed in the next section.

Another important parameter for the wave in Figure 13.6 is the **amplitude** of the wave. This is the maximum displacement of a particle of the medium from the equilibrium position. It is denoted by A and is the same as the amplitude of the simple harmonic motion of the particles of the medium.

One method of producing a traveling sinusoidal wave on a very long string is shown in Figure 13.7. One end of the string is connected to a blade that is set vibrating. As the blade oscillates vertically with simple harmonic motion, a traveling wave moving to the right is set up on the string. Figure 13.7 represents snapshots of the wave at intervals of one quarter of a period. **Each particle of the string, such as P, oscillates vertically in the y direction with simple harmonic motion.** Every segment of the string can therefore be treated as a simple harmonic oscillator vibrating with a frequency equal to the frequency of vibration of the blade that drives the string. Although each segment oscillates in the y direction, the wave (or disturbance) travels in the x direction with a speed v. Of course, this

PITFALL PREVENTION 13.1

What's the difference between Figure 13.6a and 13.6b?

Notice the visual similarity between Figures 13.6a and 13.6b. The shapes are the same, but one (a) is a graph of vertical position versus horizontal position and the other (b) is vertical position versus time. Figure 13.6a is a pictorial representation of the wave *for a series of particles of the medium*—this is what you would see at an instant of time. Figure 13.6b is a graphical representation of the position of *one particle of the medium* as a function of time. The fact that both figures have the identical shape represents Equation 13.1—a wave is the *same* function of both x and t.

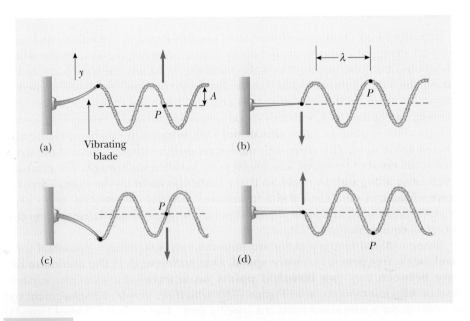

Figure 13.7

One method for producing a sinusoidal wave on a continuous string. The left end of the string is connected to a blade that is set into vibration. Every segment of the string, such as the point P, oscillates with simple harmonic motion in the vertical direction.

is the definition of a transverse wave. In this case, the energy carried by the traveling wave is supplied by the vibrating blade.

In the early chapters of this book, we developed several analysis models based on the particle model. With our introduction to waves, we can develop a new simplification model, the **wave model,** which will allow us to explore more analysis models for solving problems. An ideal particle has zero size. We can build physical objects with nonzero size as combinations of particles. Thus, the particle can be considered a basic building block. An ideal wave has a single frequency and is infinitely long; that is, the wave exists throughout the Universe. (It is beyond the mathematical scope of this text at this point to prove this, but a wave of finite length must necessarily have a mixture of frequencies.) We will find that we can combine ideal waves, just as we combined particles, and thus the ideal wave can be considered as a basic building block.

The wave model starts with an ideal wave having a single frequency, wavelength, wave speed, and amplitude. From this beginning, we describe waves in a variety of situations that serve as analysis models to help us solve problems. In the next section, we further develop our mathematical representation for the wave model.

13.3 • THE TRAVELING WAVE

See Screens 8.6, 8.8, & 8.9

Let us investigate further the mathematics of a sinusoidal wave (Fig. 13.8). The red curve represents a snapshot of a sinusoidal wave at $t = 0$, and the blue curve represents a snapshot of the wave at some later time t. In what follows, we will develop the principal features and mathematical representations of the model of a **traveling wave.** This analysis model is used in situations in which a wave moves through space without interacting with any other waves or particles.

At $t = 0$, the red curve in Figure 13.8 can be described mathematically as

$$y = A \sin \left(\frac{2\pi}{\lambda} x \right) \qquad \text{[13.3]}$$

where the amplitude A, as usual, represents the maximum value of the displacement, and λ is the wavelength as defined in Figure 13.6a. Thus, we see that the value of y is the same when x is increased by an integral multiple of λ. If the wave moves to the right with a speed of v, the wave function at some later time t is

$$y = A \sin \left[\frac{2\pi}{\lambda} (x - vt) \right] \qquad \text{[13.4]}$$

That is, the sinusoidal wave moves to the right a distance of vt in the time t, as in Figure 13.8. Note that the wave function has the form $f(x - vt)$ and represents a wave traveling to the right. If the wave were traveling to the left, the quantity $x - vt$ would be replaced by $x + vt$, just as in the case of the traveling pulse described by Equations 13.1 and 13.2.

Because the period T is the time it takes the wave to travel a distance of one wavelength, the speed, wavelength, and period are related by

$$v = \frac{\lambda}{T} \qquad \text{[13.5]}$$

Substituting Equation 13.5 into Equation 13.4, we find that

$$y = A \sin \left[2\pi \left(\frac{x}{\lambda} - \frac{t}{T} \right) \right] \qquad \text{[13.6]}$$

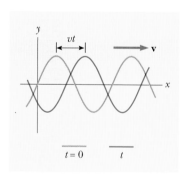

Figure 13.8

A one-dimensional sinusoidal wave traveling to the right with a speed v. The red curve represents a snapshot of the wave at $t = 0$, and the blue curve represents a snapshot at some later time t.

This form of the wave function shows the periodic nature of y in both space and time. That is, at any given time t (a snapshot of the wave), y has the same value at the positions x, $x + \lambda$, $x + 2\lambda$, and so on. Furthermore, at any given position x (at which a single particle of the medium is undergoing simple harmonic motion), the values of y at times t, $t + T$, $t + 2T$, and so on, are the same.

We can express the sinusoidal wave function in a compact form by defining two other quantities: **angular wave number** k (often called simply the **wave number**) and **angular frequency** ω:

•   *Angular wave number*

$$k \equiv \frac{2\pi}{\lambda} \qquad \text{[13.7]}$$

•   *Angular frequency*

$$\omega \equiv \frac{2\pi}{T} = 2\pi f \qquad \text{[13.8]}$$

Note that in Equation 13.8, we use the definition of frequency, $f = 1/T$. Using these definitions, Equation 13.6 can be written in the more compact form

•   *Wave function for a sinusoidal wave*

$$y = A\sin(kx - \omega t) \qquad \text{[13.9]}$$

We shall use this form most frequently.

Using Equations 13.7 and 13.8, we can express the wave speed v (Eq. 13.5) in the alternative forms

$$v = \frac{\omega}{k} \qquad \text{[13.10]}$$

•   *Speed of a traveling sinusoidal wave*

$$v = \lambda f \qquad \text{[13.11]}$$

The wave function given by Equation 13.9 assumes that the displacement y is zero at $x = 0$ and $t = 0$. This need not be the case. If the transverse displacement is not zero at $x = 0$ and $t = 0$, we generally express the wave function in the form

$$y = A\sin(kx - \omega t + \phi) \qquad \text{[13.12]}$$

where ϕ is called the **phase constant** and can be determined from the initial conditions.

Quick Quiz 13.2

Sound waves ($v = 343$ m/s) from a musical performance are encoded onto radio waves for transmission from the radio studio to your home receiver. A radio wave traveling at $c = 3.00 \times 10^8$ m/s (the symbol c, which we used in Chapter 9 for the speed of light, is also used for the speed of all electromagnetic waves traveling through a vacuum) with a wavelength of 3.00 m carries a sound wave that has a wavelength of 3.00 m in air. Which has the higher frequency, the radio wave or the sound wave?

Example 13.2   A Traveling Sinusoidal Wave

A sinusoidal wave traveling in the positive x direction has an amplitude of 15.0 cm, a wavelength of 40.0 cm, and a frequency of 8.00 Hz. The vertical displacement of the medium at $t = 0$ and $x = 0$ is also 15.0 cm, as shown in Figure 13.9. (a) Find the angular wave number, period, angular frequency, and speed of the wave.

Solution   This is a simple problem in which we apply the traveling wave model. Using Equations 13.7, 13.8, and 13.11, we find the following:

$$k = \frac{2\pi}{\lambda} = \frac{2\pi \text{ rad}}{40.0 \text{ cm}} = \boxed{0.157 \text{ rad/cm}}$$

$$T = \frac{1}{f} = \frac{1}{8.00 \text{ s}^{-1}} = \boxed{0.125 \text{ s}}$$

$$\omega = 2\pi f = 2\pi(8.00 \text{ s}^{-1}) = \boxed{50.3 \text{ rad/s}}$$

$$v = f\lambda = (8.00 \text{ s}^{-1})(40.0 \text{ cm}) = \boxed{320 \text{ cm/s}}$$

(b) Determine the phase constant ϕ, and write a general expression for the wave function.

Solution   Because $A = 15.0$ cm and because it is given that $y = 15.0$ cm at $x = 0$ and $t = 0$, substitution into Equation 13.12 gives

$$15.0 = 15.0 \sin \phi \quad \text{or} \quad \sin \phi = 1$$

We see that $\phi = \boxed{\pi/2 \text{ rad}}$ (or 90°). Hence, the wave function is of the form

$$y = A \sin\left(kx - \omega t + \frac{\pi}{2}\right) = A \cos(kx - \omega t)$$

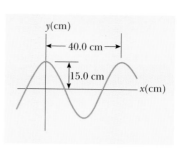

Figure 13.9

(Example 13.2) A sinusoidal wave of wavelength $\lambda = 40.0$ cm and amplitude $A = 15.0$ cm. The wave function can be written in the form $y = A \cos(kx - \omega t)$.

As we can see by inspection, the wave function must have this form because the cosine argument is displaced by 90° from the sine function. Substituting the values for A, k, and ω into this expression gives

$$y = \boxed{(15.0 \text{ cm}) \cos(0.157x - 50.3t)}$$

EXERCISE   When a particular wire is vibrating with a frequency of 4.00 Hz, a transverse wave of wavelength 60.0 cm is produced. Determine the speed of wave pulses along the wire.

Answer   2.40 m/s

The Linear Wave Equation

If the waveform at $t = 0$ is as described in Figure 13.7b, the wave function can be written

$$y = A \sin(kx - \omega t)$$

We can use this expression to describe the motion of any point on the string. The point P (or any other point on the string) moves only vertically, and so its x coordinate *remains constant*. The **transverse velocity** v_y of the point P and its **transverse acceleration** a_y are therefore

$$v_y = \frac{dy}{dt}\bigg]_{x=\text{constant}} = \frac{\partial y}{\partial t} = -\omega A \cos(kx - \omega t) \qquad \text{[13.13]}$$

$$a_y = \frac{dv_y}{dt}\bigg]_{x=\text{constant}} = \frac{\partial v_y}{\partial t} = \frac{\partial^2 y}{\partial t^2} = -\omega^2 A \sin(kx - \omega t) \qquad \text{[13.14]}$$

PITFALL PREVENTION 13.2

Two kinds of speed/velocity

Be sure to differentiate between v, the speed of the wave as it propagates through the medium, and v_y, the transverse velocity of a point on the string. The speed v is constant, whereas v_y varies sinusoidally.

The maximum values of these quantities are simply the absolute values of the coefficients of the cosine and sine functions:

$$v_{y, \text{max}} = \omega A \qquad \text{[13.15]}$$

$$a_{y, \text{max}} = \omega^2 A \qquad \text{[13.16]}$$

You should recognize that the transverse velocity and transverse acceleration of any point on the string do not reach their maximum values simultaneously. In fact, the transverse velocity reaches its maximum value (ωA) when the displacement $y = 0$, whereas the transverse acceleration reaches its maximum magnitude ($\omega^2 A$) when $y = \pm A$. Finally, Equations 13.15 and 13.16 are identical to the corresponding equations for simple harmonic motion (Eq. 12.17 and Eq. 12.18).

Let us take derivatives of our wave function with respect to position at a fixed time, similar to the process by which we took derivatives with respect to time in Equations 13.13 and 13.14:

$$\left.\frac{dy}{dx}\right]_{t=\text{constant}} = \frac{\partial y}{\partial x} = -kA \cos(kx - \omega t) \qquad \text{[13.17]}$$

$$\left.\frac{d^2 y}{dx^2}\right]_{t=\text{constant}} = \frac{\partial^2 y}{\partial x^2} = -k^2 A \sin(kx - \omega t) \qquad \text{[13.18]}$$

Comparing Equations 13.14 and 13.18, we see that

$$A \sin(kx - \omega t) = -\frac{1}{k^2}\frac{\partial^2 y}{\partial x^2} = -\frac{1}{\omega^2}\frac{\partial^2 y}{\partial t^2} \longrightarrow \frac{\partial^2 y}{\partial x^2} = \frac{k^2}{\omega^2}\frac{\partial^2 y}{\partial t^2}$$

Using Equation 13.10, we can rewrite this as

- *Linear wave equation*

$$\frac{\partial^2 y}{\partial x^2} = \frac{1}{v^2}\frac{\partial^2 y}{\partial t^2} \qquad \text{[13.19]}$$

This is known as the **linear wave equation.** If we analyze a situation and find this kind of relationship between derivatives of a function describing the situation, then wave motion is occurring. Equation 13.19 is a differential equation representation of the traveling wave model. The solutions to the equation describe **linear mechanical waves.** We have developed the linear wave equation from a sinusoidal mechanical wave traveling through a medium, but it is much more general. The linear wave equation successfully describes waves on strings, sound waves, and also electromagnetic waves.* What's more, although the sinusoidal wave that we have studied is a solution to Equation 13.19, the general solution to the equation is *any* function of the form $y(x, t) = f(x \pm vt)$, as discussed in Section 13.1.

Example 13.3 A Solution to the Linear Wave Equation

Verify that the wave function presented in Example 13.1 is a solution to the linear wave equation.

Solution The wave function is

$$y(x, t) = \frac{2.0}{(x - 3.0t)^2 + 1}$$

By taking partial derivatives of this function with respect to x and to t, we find

$$\frac{\partial^2 y}{\partial x^2} = \frac{12(x - 3.0t)^2 - 4.0}{[(x - 3.0t)^2 + 1]^3}$$

$$\frac{\partial^2 y}{\partial t^2} = \frac{108(x - 3.0t)^2 - 36}{[(x - 3.0t)^2 + 1]^3}$$

* In the case of electromagnetic waves, y is interpreted to represent an electric field, which we will study in Chapter 24.

Comparing these two expressions, we see that

$$\frac{\partial^2 y}{\partial x^2} = \frac{1}{9.0}\frac{\partial^2 y}{\partial t^2}$$

Comparing this with Equation 13.19, we see that the wave function is a solution to the linear wave equation if the speed

at which the pulse moves is 3.0 cm/s. We have already determined in Example 13.1 that this is indeed the speed of the pulse, so we have proven what we set out to do.

13.4 • THE SPEED OF TRANSVERSE WAVES ON STRINGS

An aspect of the behavior of linear mechanical waves is that the **wave speed depends only on the properties of the medium through which the wave travels.** Waves for which the amplitude A is small relative to the wavelength λ are well represented as linear waves. In this section we determine the speed of a transverse wave traveling on a stretched string.

See Screen 9.2

Let us use a mechanical analysis to derive the expression for the speed of a pulse traveling on a stretched string under tension T. Consider a pulse moving to the right with a uniform speed v, measured relative to a stationary (with respect to the Earth) inertial reference frame. Recall from Chapter 9 that Newton's laws are valid in any inertial reference frame. Thus, let us view this pulse from a different inertial reference frame—one that moves along with the pulse at the same speed, so that the pulse appears to be at rest in the frame, as in Figure 13.10a. In this reference frame, the pulse remains fixed and each point on the string moves to the left through the pulse shape.

A short segment of the string, of length Δs, forms the approximate arc of a circle of radius R, as shown in Figure 13.10a and magnified in Figure 13.10b. We use a simplification model in which this is an arc of a perfect circle. In our moving frame of reference, the segment of the string moves to the left with the speed v through the arc. As it travels through the arc, we can model the segment as a particle in uniform circular motion. This segment has a centripetal acceleration of v^2/R, which is supplied by components of the force $\mathbf{T}$ in the string at each end of the segment. The force $\mathbf{T}$ acts on each side of the segment, tangent to the arc, as

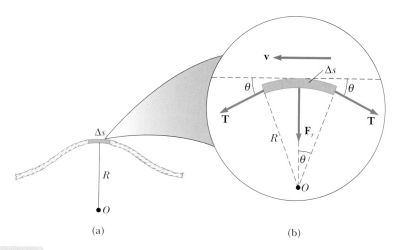

(a)   (b)

Figure 13.10

(a) To obtain the speed v of a wave on a stretched string, it is convenient to describe the motion of a small segment of the string in a moving frame of reference. (b) The net force on a small segment of length Δs is in the radial direction. The horizontal components of the tension force cancel.

• *Speed of a wave on a stretched string*

in Figure 13.10b. The horizontal components of **T** cancel, and each vertical component $T \sin \theta$ acts radially inward toward the center of the arc. Hence, the magnitude of the total radial force is $2T \sin \theta$. Because the segment is small, θ is small and we can use the small-angle approximation $\sin \theta \approx \theta$. Therefore, the magnitude of the total radial force can be expressed as

$$F_r = 2T \sin \theta \approx 2T\theta$$

The segment has mass $m = \mu \, \Delta s$, where μ is the mass per unit length of the string. Because the segment forms part of a circle and subtends an angle of 2θ at the center, $\Delta s = R(2\theta)$, and hence

$$m = \mu \, \Delta s = 2\mu R\theta$$

The radial component of Newton's second law applied to the segment gives

$$F_r = \frac{mv^2}{R} \quad \longrightarrow \quad 2T\theta = \frac{2\mu R\theta v^2}{R} \quad \longrightarrow \quad T = \mu v^2$$

where F_r is the force that supplies the centripetal acceleration of the segment. Solving for v gives

$$v = \sqrt{\frac{T}{\mu}} \qquad \qquad \text{[13.20]}$$

Notice that this derivation is based on the linear wave assumption that the pulse height is small relative to the length of the string. Using this assumption, we were able to use the approximation $\sin \theta \approx \theta$. Furthermore, the model assumes that the tension T is not affected by the presence of the pulse, so that T is the same at all points on the string. Finally, this proof does *not* assume any particular shape for the pulse. We therefore conclude that a pulse of *any shape* will travel on the string with speed $v = \sqrt{T/\mu}$, without changing its shape.

Quick Quiz 13.3

Suppose you create a pulse by moving the free end of a taut string up and down once with your hand. The string is attached at its other end to a distant wall. The pulse reaches the wall in a time interval Δt. Which of the following actions, taken by itself, decreases the time interval required for the pulse to reach the wall? More than one choice may be correct.

(a) Moving your hand more quickly, but still only up and down once by the same amount
(b) Moving your hand more slowly, but still only up and down once by the same amount
(c) Moving your hand a greater distance up and down in the same amount of time
(d) Moving your hand a smaller distance up and down in the same amount of time
(e) Using a heavier string of the same length and under the same tension
(f) Using a lighter string of the same length and under the same tension
(g) Using a string of the same linear mass density but under decreased tension
(h) Using a string of the same linear mass density but under increased tension

Quick Quiz 13.4

In mechanics, a simplification model of massless strings is often adopted. Why is this not a good model when discussing waves on strings?

THINKING PHYSICS 13.2

A secret agent is trapped in a building on top of an elevator car at a lower floor. He attempts to signal a fellow agent on the roof by tapping a message in Morse code on the elevator cable, so that transverse pulses move upward on the cable. As the pulses move up the cable toward the accomplice, does the speed with which they move stay the same, increase, or decrease? If the pulses are sent 1 s apart, are they received 1 s apart by his partner?

Reasoning   The elevator cable can be modeled as a string. The speed of waves on the cable is a function of the tension in the cable. As the waves move higher on the cable, they encounter increased tension because each higher point on the cable must support the weight of all the cable below it (and the elevator). Thus, the speed of the pulses increases as they move higher on the cable. The frequency of the pulses will not be affected because each pulse takes the same total time to reach the top—they will still arrive at the top of the cable at intervals of 1 s.

Example 13.4   The Speed of a Pulse on a Cord

A uniform cord has a mass of 0.300 kg and a total length of 6.00 m. Tension is maintained in the cord by suspending an object of mass 2.00 kg from one end (Fig. 13.11). Find the speed of a pulse on this cord. Assume that the tension is not affected by the mass of the cord.

Solution   Because the suspended object can be modeled as a particle in equilibrium, the tension T in the cord is equal to the weight of the suspended 2.00-kg object:

$$T = mg = (2.00 \text{ kg})(9.80 \text{ m/s}^2) = 19.6 \text{ N}$$

(This calculation of the tension neglects the small mass of the cord. Strictly speaking, the horizontal portion of the cord can never be exactly straight—it will sag slightly—and therefore the tension is not uniform.)

The mass per unit length μ is

$$\mu = \frac{m}{\ell} = \frac{0.300 \text{ kg}}{6.00 \text{ m}} = 0.0500 \text{ kg/m}$$

Therefore, the wave speed is

$$v = \sqrt{\frac{T}{\mu}} = \sqrt{\frac{19.6 \text{ N}}{0.0500 \text{ kg/m}}} = \boxed{19.8 \text{ m/s}}$$

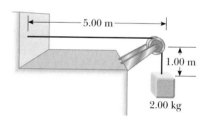

Figure 13.11

(Example 13.4) The tension T in the cord is maintained by the suspended mass. The wave speed is given by the expression $v = \sqrt{T/\mu}$.

EXERCISE   Transverse waves travel with a speed of 20.0 m/s in a string under a tension of 6.00 N. What tension is required for a wave speed of 30.0 m/s in the same string?

Answer   13.5 N

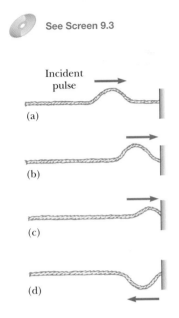

Figure 13.12

The reflection of a traveling wave pulse at the fixed end of a stretched string. The reflected pulse is inverted, but its shape remains the same.

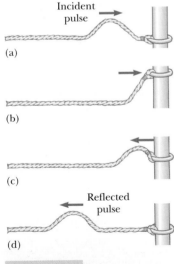

Figure 13.13

The reflection of a traveling wave pulse at the free end of a stretched string. In this case, the reflected pulse is not inverted.

13.5 • REFLECTION AND TRANSMISSION OF WAVES

So far, we have only considered a wave traveling through a medium with no changes in the medium and no interactions with anything other than the particles of the medium—this is the traveling wave model. This situation is similar to a particle traveling through empty space and obeying Newton's first law. Although these situations demonstrate important physics, things become more interesting when particles and waves interact with something. Let us see what happens when a wave encounters a boundary between two media.

For simplicity, consider a single pulse once again. When a traveling pulse reaches a boundary, part or all of the pulse is *reflected*. Any part not reflected is said to be *transmitted* through the boundary. Suppose a pulse travels on a string that is fixed at one end (Fig. 13.12). An upward pulse is initiated by grabbing the end of the string and snapping it upward and then back to the original position. When the pulse reaches the fixed boundary, it is reflected. In the simplification model in which the support attaching the string to the wall is rigid, none of the pulse is transmitted through the wall.

Note that the reflected pulse (Figs. 13.12d and e) has exactly the same amplitude as the incoming pulse but is inverted. The inversion can be explained as follows. The pulse is created initially with an upward and then downward force on the free end of the string. As the pulse arrives at the fixed end of the string, the string first produces an upward force on the support. By Newton's third law, the support exerts an equal and opposite reaction force on the string. Thus, the positive shape of the pulse results in a downward and then upward force on the string as the entirety of the pulse encounters the rigid end. This is equivalent to a person replacing the fixed support and applying a downward and then an upward force to the string. Thus, reflection at a rigid end causes the pulse to invert on reflection.

Now consider a second idealized situation in which reflection is total and transmission is zero. In this simplification model, the pulse arrives at the end of a string that is perfectly free to move vertically, as in Figure 13.13. The tension at the free end is maintained by tying the string to a ring of negligible mass that is free to slide vertically on a frictionless post. Again, the pulse is reflected, but this time it is not inverted. As the pulse reaches the post, it exerts a force on the free end, causing the ring to accelerate upward. In the process, the ring reaches the top of its motion and is then returned to its original position by the downward component of the tension force. Thus, the ring experiences the same motion as if it were raised and lowered by hand. This produces a reflected pulse that is not inverted and whose amplitude is the same as that of the incoming pulse.

Finally, we may have a situation in which the boundary is intermediate between these two extreme cases; that is, it is neither completely rigid nor completely free. In this case, part of the wave is transmitted and part is reflected. For instance, suppose a string is attached to a more dense string as in Figure 13.14. When a pulse traveling on the first string reaches the boundary between the two strings, part of the pulse is reflected and inverted and part is transmitted to the more dense string. Both the reflected and transmitted pulses have smaller amplitude than the incident pulse. The inversion in the reflected wave is similar to the behavior of a pulse meeting a rigid boundary. As the pulse travels from the initial string to the more dense string, the junction acts more like a rigid end than a free end. Thus, the reflected pulse is inverted.

When a pulse traveling on a dense string strikes the boundary of a less dense string, as in Figure 13.15, again part is reflected and part transmitted. This time, however, the reflected pulse is not inverted. As the pulse travels from the dense

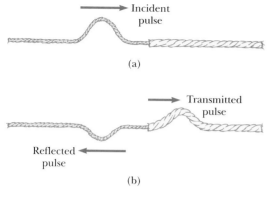

(a)

Incident pulse

Transmitted pulse

Reflected pulse

(b)

(a) A pulse traveling to the right on a light string attached to a heavier string. (b) Part of the incident pulse is reflected (and inverted), and part is transmitted to the heavier string.

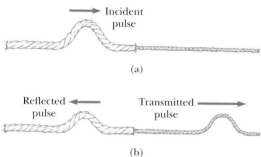

Incident pulse

(a)

Reflected pulse

Transmitted pulse

(b)

(a) A pulse traveling to the right on a heavy string attached to a lighter string. (b) The incident pulse is partially reflected and partially transmitted. In this case, the reflected pulse is not inverted.

string to the less dense one, the junction acts more like a free end than a rigid end.

The limiting value between these two cases would be that in which both strings have the same linear mass density. In this case, no boundary exists between the two media—both strings are identical. As a result, no reflection occurs and transmission is total.

In the preceding section, we found that the speed of a wave on a string increases as the mass per unit length of the string decreases. In other words, a pulse travels more slowly on a dense string than on a less dense one if both are under the same tension. This is illustrated by the lengths of the vectors in Figures 13.14 and 13.15.

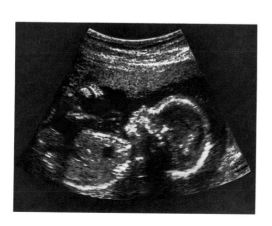

An ultrasound image of a human fetus in the womb after 20 weeks of development, showing the head, body, and arms in profile. Ultrasound refers to sound waves that are higher in frequency than those audible to humans. The sound waves transmit through the body of the mother and reflect from the skin of the fetus. The reflected sound waves are organized by the electronics of the ultrasonic imaging system into a visual image of the fetus. *(U.H.B. Trust/Stone)*

WEB

For a number of ultrasound images of a fetus at various stages of development, visit **www.parenthoodweb.com/ parent_cfmfiles/us.cfm/574**

This discussion has focused on pulses arriving at a boundary. If a sinusoidal wave on a string arrives at a rigid end, the inversion of the waveform is equivalent to shifting the entire wave by half of a wavelength. This equivalence can be seen by looking back at Figure 13.7. The wave in Figure 13.7d is the inversion of the wave in Figure 13.7b. Notice, however, that Figure 13.7b would look like Figure 13.7d if the wave were shifted to the right or left by half of a wavelength. Thus, because a full wavelength can be associated with an angle of 360°, we describe the inversion of a wave at a rigid end as a **180° phase shift.** We will see this effect again in Chapter 27 when we discuss reflection of light waves from materials.

13.6 • RATE OF ENERGY TRANSFER BY SINUSOIDAL WAVES ON STRINGS

See Screen 9.5

As waves propagate through a medium, they transport energy. This fact is easily demonstrated by hanging an object on a stretched string and sending a pulse down the string, as in Figure 13.16. When the pulse meets the suspended object, the object is momentarily displaced, as in Figure 13.16b. In the process, energy is transferred to the object because work must be done in moving it upward. This section examines the rate at which energy is transferred along a string. We shall assume a one-dimensional sinusoidal wave in the calculation of the energy transferred.

Consider a sinusoidal wave traveling on a string (Fig. 13.17). The source of the energy is some external agent at the left end of the string, which does work in producing the oscillations. We can consider the string to be a nonisolated system. As the external agent performs work on the end of the string, moving it up and down, energy enters the system of the string and propagates along its length. Let us focus our attention on an element of the string of length Δx and mass Δm. Each such element moves vertically with simple harmonic motion. Thus, we can model each element of the string as a simple harmonic oscillator, with the oscillation in the y direction. All elements have the same angular frequency ω and the same amplitude A. As we found in Chapter 12, the kinetic energy K associated with a particle in simple harmonic motion is $K = \frac{1}{2}mv^2$, where v varies sinusoidally during the oscillation. If we apply this equation to an element of length Δx, we see that the kinetic energy ΔK of this element is

$$\Delta K = \tfrac{1}{2}(\Delta m)\,v_y^2$$

If μ is the mass per unit length of the string, then the element of length Δx has a mass Δm that is equal to $\mu\,\Delta x$. Hence, we can express the kinetic energy of an ele-

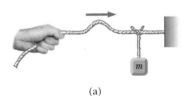

(a)

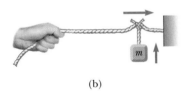

(b)

Figure 13.16

(a) A pulse traveling to the right on a stretched string on which an object has been suspended. (b) Energy is transmitted to the suspended object when the pulse arrives.

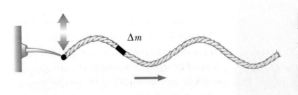

Figure 13.17

A sinusoidal wave traveling along the x axis on a stretched string. Every segment moves vertically, and each segment has the same total energy. The average power transmitted by the wave equals the energy contained in one wavelength divided by the period of the wave.

ment of the string as

$$\Delta K = \tfrac{1}{2}(\mu\,\Delta x)\,v_y{}^2 \qquad\qquad \textbf{[13.21]}$$

As the length of the element of the string shrinks to zero, this becomes a differential relationship:

$$dK = \tfrac{1}{2}(\mu\,dx)\,v_y{}^2$$

We substitute for the general velocity of an element of the string using Equation 13.13:

$$dK = \tfrac{1}{2}\mu\,[\omega A\cos(kx - \omega t)]^2\,dx$$
$$= \tfrac{1}{2}\mu\omega^2 A^2\cos^2(kx - \omega t)\,dx$$

If we take a snapshot of the wave at time $t = 0$, then the kinetic energy in a given element is

$$dK = \tfrac{1}{2}\mu\omega^2 A^2\cos^2 kx\,dx$$

Let us integrate this expression over all the string elements in a wavelength of the wave, which will give us the kinetic energy in one wavelength:

$$K_\lambda = \int_0^\lambda \tfrac{1}{2}\mu\omega^2 A^2\cos^2 kx\,dx = \tfrac{1}{2}\mu\omega^2 A^2 \int_0^\lambda \cos^2 kx\,dx$$

$$= \tfrac{1}{2}\mu\omega^2 A^2\left[\tfrac{1}{2}x + \frac{1}{4k}\sin 2kx\right]_0^\lambda = \tfrac{1}{2}\mu\omega^2 A^2\left[\tfrac{1}{2}\lambda\right] = \tfrac{1}{4}\mu\omega^2 A^2\lambda$$

In addition to this kinetic energy, each element of the string has potential energy, associated with its displacement from the equilibrium position. A similar analysis as that above for the total potential energy in a wavelength gives the same result:

$$U_\lambda = \tfrac{1}{4}\mu\omega^2 A^2\lambda$$

The total energy in one wavelength of the wave is the sum of the kinetic and potential energies:

$$E_\lambda = K_\lambda + U_\lambda = \tfrac{1}{2}\mu\omega^2 A^2\lambda \qquad\qquad \textbf{[13.22]}$$

As the wave moves along the string, this amount of energy passes by a given point on the string during one period of the oscillation. Thus, the power, or rate of energy transfer, associated with the wave is

$$\mathcal{P} = \frac{E_\lambda}{\Delta t} = \frac{\tfrac{1}{2}\mu\omega^2 A^2\lambda}{T} = \tfrac{1}{2}\mu\omega^2 A^2\left(\frac{\lambda}{T}\right) = \tfrac{1}{2}\mu\omega^2 A^2 v \qquad\qquad \textbf{[13.23]}$$

• *Rate of energy transfer for a wave on a string*

This shows that the rate of energy transfer by a sinusoidal wave on a string is proportional to (a) the square of the angular frequency, (b) the square of the amplitude, and (c) the wave speed. In fact, *all* sinusoidal waves have the following general property: **The rate of energy transfer in any sinusoidal wave is proportional to the square of the angular frequency and to the square of the amplitude.**

Example 13.5 Power Supplied to a Vibrating String

A string with linear mass density $\mu = 5.00 \times 10^{-2}$ kg/m is under a tension of 80.0 N. How much power must be supplied to the string to generate sinusoidal waves at a frequency of 60.0 Hz and an amplitude of 6.00 cm?

Solution The wave speed on the string is

$$v = \sqrt{\frac{T}{\mu}} = \left(\frac{80.0 \text{ N}}{5.00 \times 10^{-2} \text{ kg/m}}\right)^{1/2} = 40.0 \text{ m/s}$$

Because $f = 60.0$ Hz, the angular frequency ω of the sinusoidal waves on the string has the value

$$\omega = 2\pi f = 2\pi(60.0 \text{ Hz}) = 377 \text{ s}^{-1}$$

Using these values in Equation 13.23 for the power, with $A = 6.00 \times 10^{-2}$ m, gives

$$\mathcal{P} = \tfrac{1}{2}\mu\omega^2 A^2 v$$
$$= \tfrac{1}{2}(5.00 \times 10^{-2} \text{ kg/m})(377 \text{ s}^{-1})^2$$
$$\times (6.00 \times 10^{-2} \text{ m})^2(40.0 \text{ m/s})$$
$$= \boxed{512 \text{ W}}$$

13.7 • SOUND WAVES

Let us turn our attention from transverse waves to longitudinal waves. As stated in Section 13.2, in longitudinal waves the particles of the medium undergo displacements parallel to the direction of wave motion. Sound waves in air are the most important examples of longitudinal waves. Sound waves can travel through any material medium, however, and their speed depends on the properties of that medium. Table 13.1 provides examples of the speed of sound in different media.

The displacements accompanying a sound wave in air are longitudinal displacements of small elements of the fluid from their equilibrium positions. Such displacements result if the source of the waves, such as the diaphragm of a loudspeaker, oscillates in air. If the oscillation of the diaphragm is described by simple harmonic motion, a sinusoidal sound wave propagates away from the loudspeaker. For instance, one can produce a one-dimensional sound pulse in a long, narrow tube containing a gas by means of a vibrating piston at one end, as in Figure 13.18.

It is difficult to draw a pictorial representation of longitudinal waves because the displacements of the elements of the medium are in the same direction as that of the propagation of the wave. Figure 13.18 is one way to represent these types of waves. The darker color in the figure represents a region where the gas is compressed; consequently, the density and pressure are *above* their equilibrium values. Such a compressed region of gas, called a **compression,** is formed when the piston is being pushed into the tube. The compression moves down the tube as a pulse, continuously compressing the layers in front of it. When the piston is withdrawn from the tube, the gas in front of it expands, and consequently the pressure and density in this region fall below their equilibrium values. These low-pressure regions, called **rarefactions,** are represented by the lighter areas in Figure 13.18. The rarefactions also propagate along the tube, following the compressions. Both regions move with a speed equal to the speed of sound in that medium.

As the piston oscillates back and forth in a sinusoidal fashion, regions of compression and rarefaction are continuously set up. The distance between two successive compressions (or two successive rarefactions) equals the wavelength λ. As these regions travel down the tube, any small element of the medium moves with simple harmonic motion parallel to the direction of the wave (in other words, longitudinally). If $s(x, t)$ is the displacement of a small element measured from its equilibrium position, we can express this displacement function as

$$s(x, t) = s_{max}\sin(kx - \omega t) \qquad [13.24]$$

TABLE 13.1	
Speeds of Sound in Various Media	
Medium	**v (m/s)**
Gases	
Hydrogen (0)	1 286
Helium (0)	972
Air (20)	343
Air (0)	331
Oxygen (0)	317
Liquids (at 25°C)	
Glycerol	1 904
Sea water	1 533
Water	1 493
Mercury	1 450
Kerosene	1 324
Methyl alcohol	1 143
Carbon tetrachloride	926
Solids	
Diamond	12 000
Pyrex glass	5 640
Iron	5 130
Aluminum	5 100
Brass	4 700
Copper	3 560
Gold	3 240
Lucite	2 680
Lead	1 322
Rubber	1 600

where s_{max} is the **maximum displacement from equilibrium** (or the **displacement amplitude**), k is the wave number, and ω is the angular frequency of the piston. The variation ΔP in the pressure* of the gas measured from its equilibrium value is also sinusoidal; it is given by

$$\Delta P = \Delta P_{max} \cos(kx - \omega t) \qquad \text{[13.25]}$$

The pressure amplitude ΔP_{max} is the maximum change in pressure from the equilibrium value. It is proportional to the displacement amplitude s_{max}:

$$\Delta P_{max} = \rho v \omega s_{max} \qquad \text{[13.26]}$$

where ρ is the density of the medium, v is the wave speed, and ωs_{max} is the maximum longitudinal speed of an element of the medium in front of the piston. It is these pressure variations in a sound wave that result in an oscillating force on the eardrum, leading to the sensation of hearing.

Thus, we see that a sound wave may be considered as either a displacement wave or a pressure wave. A comparison of Equations 13.24 and 13.25 shows that **the pressure wave is 90° out of phase with the displacement wave.** Graphs of these functions are shown in Figure 13.19. Note that the change in pressure from equilibrium is a maximum when the displacement is zero, whereas the displacement is a maximum when the pressure change is zero.

Note that Figure 13.19 presents two graphical representations of the longitudinal wave: one for position of the elements of the medium and the other for pressure variation. These are *not* pictorial representations, however, for longitudinal waves. For transverse waves, because the particle displacement is perpendicular to the direction of propagation, the pictorial and graphical representations look the same—the perpendicularity of the oscillations and propagation is matched by the perpendicularity of x and y axes. For longitudinal waves, the oscillations and propagation exhibit no perpendicularity, so that pictorial representations look like Figure 13.18.

The speed of a sound wave in air depends only on the temperature of the air. For a small range of temperatures around room temperature, the speed of sound is described by

$$v = 331 \text{ m/s} + (0.6 \text{ m/s} \cdot °C)\, T_C \qquad \text{[13.27]}$$

where T_C is the temperature in degrees Celsius and the speed of sound at 0°C is 331 m/s.

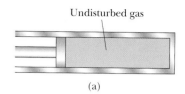

Undisturbed gas

(a)

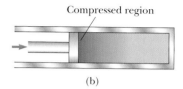

Compressed region

(b)

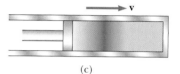

v

(c)

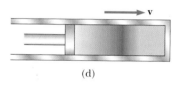

v

(d)

Figure 13.18

A sinusoidal longitudinal wave propagating down a tube filled with a compressible gas. The source of the wave is a vibrating piston at the left. The high- and low-pressure regions are dark and light, respectively.

THINKING PHYSICS 13.3

Garth Brooks had a hit song with "The Thunder Rolls" (Caged Panther Music, Inc., 1990). Why *does* thunder produce an extended "rolling" sound? And how does lightning produce thunder in the first place?

Reasoning Let us assume that we are at ground level and ignore ground reflections. When cloud-to-ground lightning strikes, a channel of ionized air carries a very large electric current from the cloud to the ground. (We will study electric current in Chapter 21.) This results in a very rapid temperature increase of this channel of

---

* We will formally introduce pressure in Chapter 15. In the case of longitudinal waves in a gas, each compressed area is a region of higher-than-average pressure and density, and each stretched region is a region of lower-than-average pressure and density.

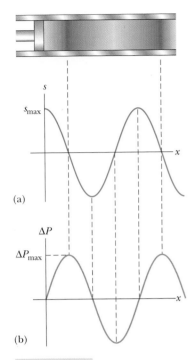

(a)

(b)

Figure 13.19

(a) Displacement versus position and (b) pressure versus position for a sinusoidal longitudinal wave. The displacement wave is 90° out of phase with the pressure wave.

PITFALL PREVENTION 13.4

Doppler effect does not depend on distance

A common misconception about the Doppler effect is that it depends on the distance between the source and the observer. Although the intensity of a sound will vary as the distance changes, the apparent frequency will not—the frequency depends only on the speed. As you listen to an approaching source, you will detect increasing intensity but constant frequency. As the source passes, you will hear the frequency suddenly drop to a new constant value and the intensity begin to decrease.

You can estimate the distance to a lightning strike in kilometers by counting the number of seconds between seeing the strike and hearing the thunder, and then dividing by 3. This works because the speed of sound, about 343 m/s, is approximately one third of a kilometer per second. If you divide the number of seconds by 5, you will have an approximation for the distance in miles. The speed of sound is about 1100 ft/s, which is approximately one fifth of a mile per second. (*Richard Kaylin/Stone*)

air as it carries the current. The temperature increase causes a sudden expansion of the air. This expansion is so sudden and so intense that a tremendous disturbance is produced in the air: thunder. The thunder rolls due to the fact that the lightning channel is a long, extended source—the entire length of the channel produces the sound at essentially the same instant of time. Sound produced at the end of the channel nearest you reaches you first, but sounds from progressively farther portions of the channel reach you shortly thereafter. If the lightning channel were a perfectly straight line, the resulting sound might be a steady roar, but the zigzagged shape of the path results in the rolling variation in loudness.

13.8 • THE DOPPLER EFFECT

When a vehicle sounds its horn as it travels along a highway, the frequency of the sound you hear is higher as the vehicle approaches you than it is as the vehicle moves away from you. This is one example of the **Doppler effect,** named after the Austrian physicist Christian Johann Doppler (1803–1853).

The Doppler effect for sound is experienced whenever there is relative motion between the source of sound and the observer. Motion of the source or observer toward the other results in the observer's hearing a frequency that is higher than the true frequency of the source. Motion of the source or observer away from the other results in the observer's hearing a frequency that is lower than the true frequency of the source.

Although we shall restrict our attention to the Doppler effect for sound waves, it is an effect associated with waves of all types. The Doppler effect for electromagnetic waves is used in police radar systems to measure the speeds of motor vehicles. Likewise, astronomers use the effect to determine the relative motions of stars, galaxies, and other celestial objects. In 1842, Doppler first reported the frequency shift in connection with light emitted by two stars revolving about each other in double-star systems. In the early 20th century, the Doppler effect for light from galaxies was used to argue for the expansion of the Universe. This led to the Big Bang theory, which we discuss in Chapter 31.

(a)

(b)

(c)

Figure 13.20

(a) Waves moving toward a stationary boat. The waves travel to the left, and their source is far to the right of the boat, out of the frame of the drawing. (b) The boat moving toward the wave source. (c) The boat moving away from the wave source.

To see what causes this apparent frequency change, imagine you are in a boat lying at anchor on a gentle sea where the waves have a period of $T = 2.0$ s. This means that every 2.0 s a crest hits your boat. Figure 13.20a shows this situation with the water waves moving toward the left. If you start a stopwatch at $t = 0$ just as one crest hits, the stopwatch reads 2.0 s when the next crest hits, 4.0 s when the third crest hits, and so on. From these observations you conclude that the wave frequency is $f = 1/T = 0.50$ Hz. Now suppose you start your motor and head directly into the oncoming waves, as shown in Figure 13.20b. Again you set your stopwatch to $t = 0$ as a crest hits the front of your boat. Now, however, because you are moving toward the next wave crest as it moves

Figure 13.21

An observer *O* (the cyclist) moving with a speed v_O toward a stationary point source *S*, the horn of a parked car. The observer hears a frequency f' that is greater than the source frequency.

toward you, it hits you less than 2.0 s after the first hit. In other words, the period you observe is shorter than the 2.0-s period you observed when you were stationary. Because $f = 1/T$, you observe a higher wave frequency than when you were at rest.

If you turn around and move in the same direction as the waves (see Fig. 13.20c), you observe the opposite effect. You set your watch to $t = 0$ as a crest hits the back of the boat. Because you are now moving away from the next crest, more than 2.0 s has elapsed on your watch by the time that crest catches you. Thus, you observe a lower frequency than when you were at rest.

These effects occur because the relative speed between your boat and the water depends on the direction of travel and on the speed of your boat. When you are moving toward the right in Figure 13.20b, this relative speed is higher than that of the wave speed, which leads to the observation of an increased frequency. When you turn around and move to the left, the relative speed is lower, as is the observed frequency of the water waves.

Let us now examine an analogous situation with sound waves, in which we replace the water waves with sound waves, the water surface becomes the air, and the person on the boat becomes an observer listening to the sound. In this case, an observer *O* is moving with a speed of v_O and a sound source *S* is stationary. For simplicity, we assume that the air is also stationary and that the observer moves directly toward the source.

The red lines in Figure 13.21 represent circular lines connecting the crests of sound waves moving away from the source. Thus, the radial distance between adjacent red lines is one wavelength. We shall take the frequency of the source to be f, the wavelength to be λ, and the speed of sound to be v. A stationary observer would detect a frequency f, where $f = v/\lambda$ (i.e., when the source and observer are both at rest, the observed frequency must equal the true frequency of the source). If the observer moves toward the source with the speed v_O, however, the relative speed of sound experienced by the observer is higher than the speed of sound in air. Using our relative speed discussion of Section 3.6, if the sound is coming toward the observer at v and the observer is moving toward the sound at v_O, the relative speed of sound as measured by the observer is

$$v_{\text{rel}} = v + v_O$$

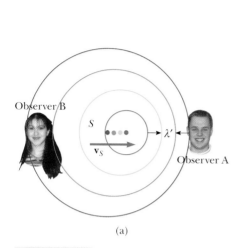

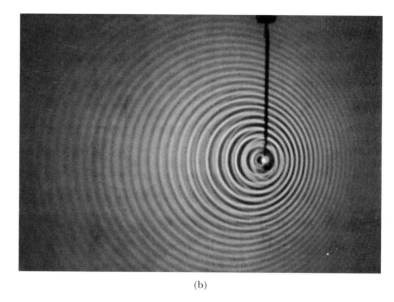

(a)   (b)

Figure 13.22

(a) A source S moving with a speed v_S toward a stationary observer A and away from a stationary observer B. Observer A hears an increased frequency, and observer B hears a decreased frequency.
(b) The Doppler effect in water observed in a ripple tank. The vibrating source is moving to the right. *(Courtesy of the Educational Development Center, Newton, MA)*

Thus, the frequency of sound heard by the observer is based on this apparent speed of sound:

$$f' = \frac{v_{\text{rel}}}{\lambda} = \frac{v + v_O}{\lambda} = f\left(\frac{v + v_O}{v}\right) \quad \text{(observer moving toward source)} \quad \textbf{[13.28]}$$

Now consider the situation in which the source moves with a speed of v_S relative to the medium, and the observer is at rest. Figure 13.22 shows this situation. Because the source is moving, the crest of each new wave is emitted from the source to the right of the position of the emission of the previous crest a distance $v_S T$, where T is the period of the wave being generated by the source. Thus, the center of each colored circle (indicated by the identically colored dot) in Figure 13.22 is shifted to the right by this distance relative to the circle representing the previous crest. If the source moves directly toward observer O_A in Figure 13.22a, the crests detected by the observer along a line between the source and observer are closer to one another than they would be if the source were at rest. As a result, the wavelength λ' measured by observer O_A is shorter than the true wavelength λ of the source. The wavelength is *shortened* by the distance $v_S T$, and the observed wavelength has the value $\lambda' = \lambda - v_S/f$. Because $\lambda = v/f$, the frequency heard by observer O_A is

$$f' = \frac{v}{\lambda'} = f\left(\frac{v}{v - v_S}\right) \quad \text{(source moving toward observer)} \quad \textbf{[13.29]}$$

That is, the frequency is *increased* when the source moves toward the observer. In a similar manner, if the source moves away from observer O_B at rest, the sign of v_S is reversed in Equation 13.29, and the frequency is lower.

In Equation 13.29, notice that the denominator approaches zero when the speed of the source approaches the speed of sound, resulting in the frequency f' approaching infinity. Such a situation results in waves that cannot escape from the source in the direction of motion of the source. This concentration of energy in front of the source results in a *shock wave*. Such a disturbance is noted when a jet aircraft flying at a speed equal to or greater than the speed of sound produces a *sonic boom*.

Finally, if both the source and the observer are in motion, one finds the following general equation for the observed frequency:

- *Frequency heard with observer and source in motion*

$$f' = f\left(\frac{v + v_O}{v - v_S}\right) \qquad [13.30]$$

In this expression, the signs for the values substituted for v_O and v_S depend on the direction of the velocity. A positive value is used for motion of the observer or the source *toward* the other, and a negative sign for motion of one *away from* the other.

When working with any Doppler effect problem, remember the following rule concerning signs: The word *toward* is associated with an *increase* in the observed frequency. The words *away from* are associated with a *decrease* in the observed frequency.

A shock wave due to a jet traveling at the speed of sound is made visible as a fog of water vapor. The large pressure variation in the shock wave causes the water in the air to condense into water droplets. *(U. S. Navy. Photo by Ensign John Gay)*

Quick Quiz 13.5

You are on a cruise and the person next to you at the deck railing falls overboard. During the entire time of fall, the person screams with a constant frequency. How does the apparent frequency of the sound of the scream change with time?

Quick Quiz 13.6

You are driving toward a cliff and you honk your horn. The frequency of the echo you hear will be Doppler shifted (a) up in frequency, equivalent to a moving source; (b) up in frequency, equivalent to a moving observer; (c) up in frequency, equivalent to a moving source *and* a moving observer; (d) down in frequency, equivalent to a moving source; (e) down in frequency, equivalent to a moving observer; (f) down in frequency, equivalent to a moving source *and* a moving observer.

THINKING PHYSICS 13.4

Suppose you run past your stereo speakers, from right to left, or left to right. If you run rapidly enough and have excellent pitch discrimination, you may notice that the music that is playing seems to be out of tune when you are between the speakers. Why?

Reasoning When you are between the speakers, you are running away from one of them and toward the other. Thus, there is a Doppler shift downward for the sound from the speaker behind you and a Doppler shift upward for the sound from the speaker ahead of you. As a result, the sound from the two speakers will not be in tune. A calculation shows that a world-class sprint runner could run fast enough to generate about a semitone difference in the sound from the two speakers.

Example 13.6   The Noisy Siren

An ambulance travels down a highway at a speed of 33.5 m/s (75 mi/h). Its siren emits sound at a frequency of 400 Hz. What is the frequency heard by a passenger in a car traveling at 24.6 m/s (55 mi/h) in the opposite direction (a) as the car approaches the ambulance and (b) as the car moves away from the ambulance?

Solution   (a) Let us take the speed of sound in air to be $v = 343$ m/s. We can use Equation 13.30 in both cases. As the ambulance and car approach each other, the observed apparent frequency is

$$f' = f\left(\frac{v + v_O}{v - v_S}\right) = (400 \text{ Hz})\left(\frac{343 \text{ m/s} + (+24.6 \text{ m/s})}{343 \text{ m/s} - (+33.5 \text{ m/s})}\right)$$

$$= \boxed{475 \text{ Hz}}$$

(b) Similarly, as they recede from each other, a passenger in the car hears a frequency

$$f' = f\left(\frac{v + v_O}{v - v_S}\right) = (400 \text{ Hz})\left(\frac{343 \text{ m/s} + (-24.6 \text{ m/s})}{343 \text{ m/s} - (-33.5 \text{ m/s})}\right)$$

$$= \boxed{338 \text{ Hz}}$$

The *change* in frequency as detected by the passenger in the car is 475 Hz − 338 Hz = 137 Hz, which is more than 30% of the actual frequency emitted.

EXERCISE   Suppose that the passenger car is parked on the side of the highway as the ambulance travels down the highway at the speed of 33.5 m/s. What frequency will the passenger in the car hear as the ambulance (a) approaches the parked car and (b) recedes from the parked car?

Answer   (a) 443 Hz   (b) 364 Hz

EXERCISE   A band is playing on a moving truck. The band strikes the note middle C (262 Hz), but it is heard by spectators ahead of the truck as C# (277 Hz). How fast is the truck moving?

Answer   18.6 m/s

---

context connection

13.9 • SEISMIC WAVES

 See Screen 8.14

When an earthquake occurs, a sudden release of energy takes place at a location called the **focus** or **hypocenter** of the earthquake. (The **epicenter** is the point on the Earth's surface radially above the hypocenter.) This energy will propagate away from the focus of the earthquake by means of **seismic waves.** Seismic waves are like the sound waves that we have studied in the later sections of this chapter—mechanical disturbances moving through a medium.

In discussing mechanical waves in this chapter, we identified two types: transverse and longitudinal. In the case of mechanical waves moving through air, we have only a longitudinal possibility. For mechanical waves moving through a solid, however, both possibilities are available, due to the strong interatomic forces between particles of the solid. Thus, in the case of seismic waves, energy propagates away from the focus both by longitudinal and transverse waves.

In the language used in earthquake studies, these two types of waves are named according to the order of their arrival at a seismograph. The longitudinal wave travels at a higher speed than the transverse wave. As a result, the longitudinal wave arrives at a seismograph first, and is thus called the **P wave,** where *P* stands for *primary.* The slower moving transverse wave arrives next, so it is called the **S wave,** or *secondary* wave.

The wave speed for a seismic wave depends on the medium through which it travels. Typical values are 5 km/s for a *P* wave through granite and 3 km/s for an *S* wave through granite. Figure 13.23 shows a typical seismograph trace of a distant earthquake, with the *S* wave clearly arriving after the *P* wave.

460 CHAPTER 13 *Mechanical Waves*

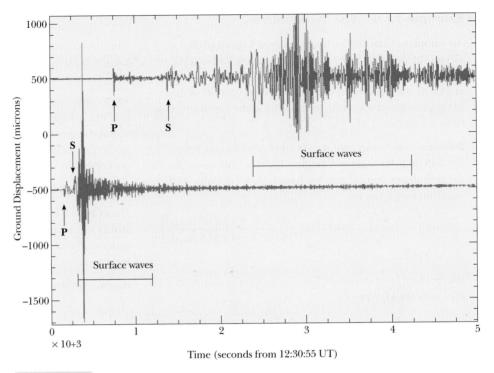

WEB

To access the Princeton Earth
Physics Project seismogram
database, visit
**lasker.princeton.edu/
SeismicDB/select_year.html**

Figure 13.23

A seismograph trace, showing the arrival of *P* and *S* waves from the Northridge earthquake at San
Pablo, Spain (*top trace*) and Albuquerque, New Mexico (*bottom trace*). The *P* wave arrives first, because it
travels the fastest, followed by the slower moving *S* wave. The farther the seismograph station is from
the epicenter, the longer is the time interval between the arrivals of the *P* and *S* waves.

The *P* and *S* waves move through the body of the Earth and can be detected by
seismographs at various locations around the Earth. Once these waves reach the
surface, the energy can propagate by additional types of waves along the surface.
In a *Rayleigh* wave, the motion of the particles at the surface is a combination of
longitudinal and transverse displacements, so that the net motion of a point on
the surface is circular or elliptical. This is similar to the path followed by particles
on the ocean surface as a wave passes by, as in Figure 13.24. The *Love* wave is a

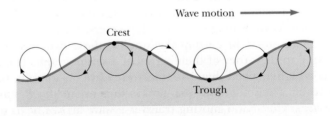

Figure 13.24

The motion of particles on the surface of deep water in which a wave is propagating is a combination
of transverse and longitudinal displacements, with the result that the molecules at the surface move in
nearly circular paths. Each molecule is displaced both horizontally and vertically from its equilibrium
position. This is similar to the motion of the Earth surface for a Rayleigh wave.

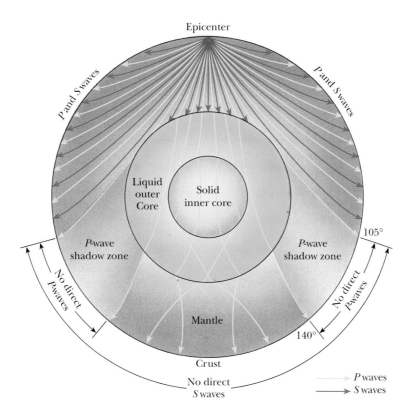

Epicenter

P and S waves

P and S waves

Liquid outer Core

Solid inner core

P-wave shadow zone

P-wave shadow zone

105°

No direct P waves

No direct P waves

Mantle

140°

Crust

No direct S waves

→ P waves
→ S waves

Figure 13.25

Cross-section of the Earth showing paths of waves produced by an earthquake. Only P waves (yellow) can propagate in the liquid core. The S waves (blue) do not enter the liquid core. When the P waves transmit from one region to another, such as from the mantle to the liquid core, they experience *refraction*—a change in the direction of propagation. We will study refraction for light in Chapter 25. Because of the refraction for seismic waves, there is a "shadow" zone between 105° and 140° from the epicenter, in which no waves following a direct path (i.e., a path with no reflections) arrive.

transverse surface wave in which the transverse oscillations are parallel to the surface. Thus, no vertical displacement of the surface occurs in a Love wave.

It is possible to use the P and S waves traveling through the body of the Earth to gain information about the structure of the interior of the Earth. Measurements of a given earthquake by seismographs at various locations on the surface indicate that the Earth has an interior region that allows the passage of P waves, but not S waves. This fact can be understood if this particular region is modeled as having liquid characteristics. Similar to a gas, a liquid cannot sustain a transverse force. For this reason we only have longitudinal sound waves in air. Thus, the transverse S waves cannot pass through this region. This leads to a structural model in which the Earth has a **liquid core** between radii of approximately 1.2×10^3 km and 3.5×10^3 km.

Other measurements of seismic waves allow additional interpretations of layers within the interior of the Earth, including a **solid core** at the center, a rocky region called the **mantle,** and a relatively thin outer layer called the **crust.** Figure 13.25 shows this structure. Using x-rays or ultrasound in medicine to provide information about the interior of the human body is somewhat similar to using seismic waves to provide information about the interior of the Earth.

As P and S waves propagate in the interior of the Earth, they will encounter variations in the medium. At each boundary at which the properties of the medium change, reflection and transmission occur. When the seismic wave arrives at the surface of the Earth, a small amount of the energy is transmitted into the air as low-frequency sound waves. Some of the energy spreads out along the surface in the form of Rayleigh and Love waves. The remaining wave energy is reflected back into the interior. As a result, seismic waves can travel over long distances within the Earth and can be detected at seismographs at many locations around the globe. In addition, because a relatively large fraction of the wave energy continues to be reflected at each encounter with the surface, the wave can propagate for a long time. Data are available showing seismograph activity for several hours after an earthquake, due to the repeated reflections of seismic waves from the surface.

Another example of the reflection of seismic waves is available in the technology of oil exploration. A "thumper truck" applies large impulsive forces to the ground, resulting in low-energy seismic waves propagating into the Earth. Specialized microphones are used to detect the waves reflected from various boundaries between layers under the surface. By using computers to map out the underground structure corresponding to these layers, it is possible to detect layers likely to contain oil.

In this chapter, we have seen how the phenomenon of waves allows energy to be transferred from the location of the earthquake to the location of a structure. In the next chapter, we shall investigate what happens when these waves become trapped in a geological depression, such that the amplitude of oscillation becomes very large, representing even greater risk of damage to structures.

SUMMARY

A **transverse wave** is a wave in which the particles of the medium move in a direction perpendicular to the direction of the wave velocity. An example is a wave moving along a stretched string.

Longitudinal waves are waves in which the elements of the medium move back and forth parallel to the direction of the wave speed. Sound waves in air are longitudinal.

Any one-dimensional wave traveling with a speed of v in the positive x direction can be represented by a **wave function** of the form $y = f(x - vt)$. Likewise, the wave function for a wave traveling in the negative x direction has the form $y = f(x + vt)$.

The wave function for a one-dimensional sinusoidal wave traveling to the right can be expressed as

$$y = A \sin\left[\frac{2\pi}{\lambda}(x - vt)\right] = A \sin(kx - \omega t) \qquad \text{[13.4, 13.9]}$$

where A is the **amplitude,** λ is the **wavelength,** k is the **angular wave number,** and ω is the **angular frequency.** If T is the **period** and f is the **frequency,** then v, k, and ω can be written

$$v = \frac{\lambda}{T} = f\lambda \qquad \text{[13.5, 13.11]}$$

$$k \equiv \frac{2\pi}{\lambda} \qquad \text{[13.7]}$$

$$\omega \equiv \frac{2\pi}{T} = 2\pi f \qquad \text{[13.8]}$$

The speed of a wave traveling on a stretched string of mass per unit length μ and tension T is

$$v = \sqrt{\frac{T}{\mu}} \qquad \text{[13.20]}$$

When a pulse traveling on a string meets a fixed end, the pulse is reflected and inverted. If the pulse reaches a free end, it is reflected but not inverted.

The **power** transmitted by a sinusoidal wave on a stretched string is

$$\mathcal{P} = \tfrac{1}{2}\mu\omega^2 A^2 v \qquad \text{[13.23]}$$

The change in frequency heard by an observer whenever there is relative motion between a wave source and the observer is called the **Doppler effect.** When the source and observer are moving toward each other, the observer hears a higher frequency than the true frequency of the source. When the source and observer are moving away from each other, the observer hears a lower frequency than the true frequency of the source. The following general equation provides the observed frequency:

$$f' = f\left(\frac{v + v_O}{v - v_S}\right) \qquad \text{[13.30]}$$

A positive value is used for v_O and/or v_S for motion of the observer or source *toward* the other, and a negative sign for motion *away from* the other.

QUESTIONS

1. Suppose the wind blows. Does this cause a Doppler effect for a sound moving through the air? Is it like a moving source or a moving observer?

2. How would you set up a longitudinal wave in a stretched spring? Would it be possible to set up a transverse wave in a spring?

3. By what factor would you have to increase the tension in a taut string in order to double the wave speed?

4. When traveling on a taut string, does a wave pulse always invert on reflection? Explain.

5. Can two pulses traveling in opposite directions on the same string reflect from each other? Explain.

6. Does the vertical speed of a segment of a horizontal taut string through which a wave is traveling depend on the wave speed?

7. A vibrating source generates a sinusoidal wave on a string under constant tension. If the power delivered to the string is doubled, by what factor does the amplitude change? Does the wave speed change under these circumstances?

8. Consider a wave traveling on a taut rope. What is the difference, if any, between the speed of the wave and the speed of a small section of the rope?

9. If a long rope is hung from a ceiling and waves are sent up the rope from its lower end, they do not ascend with constant speed. Explain.

10. What happens to the wavelength of a wave on a string when the frequency is doubled? Assume the tension in the string remains the same.

11. What happens to the speed of a wave on a taut string when the frequency is doubled? Assume the tension in the string remains the same.

12. When all the strings on a guitar are stretched to the same tension, will the speed of a wave along the more massive bass strings be faster or slower than the speed of a wave on the lighter strings?

13. If you stretch a rubber hose and pluck it, you can observe a pulse traveling up and down the hose. What happens to the speed if you stretch the hose tighter? If you fill the hose with water?

14. In a longitudinal wave in a spring, the coils move back and forth in the direction of wave motion. Does the speed of the wave depend on the maximum speed of each coil?

15. A solid can transport both longitudinal waves and transverse waves, but a fluid can transport only longitudinal waves. Why?

16. In an earthquake both S (transverse) and P (longitudinal) waves propagate from the focus of the earthquake. The S waves travel through the Earth more slowly than the P waves (≈ 3 km/s versus ≈ 5 km/s). By detecting the time of arrival of the waves, how can one determine how far away the epicenter of the quake was? How many detection centers are necessary to pinpoint the location of the epicenter?

PROBLEMS

1, 2, 3 = straightforward, intermediate, challenging   □ = full solution available in the *Student Solutions Manual and Study Guide*

web = solution posted at **http://www.harcourtcollege.com/physics/**   🖥 = computer useful in solving problem

🌐 = Interactive Physics   ▧ = paired numerical/symbolic problems   ▨ = life science application

Section 13.1   Propagation of a Disturbance

1. At $t = 0$, a transverse wave pulse in a wire is described by the function

$$y = \frac{6}{x^2 + 3}$$

where x and y are in meters. Write the function $y(x, t)$ that describes this wave if it is traveling in the positive x direction with a speed of 4.50 m/s.

2. Ocean waves with a crest-to-crest distance of 10.0 m can be described by the wave function

$$y(x, t) = (0.800 \text{ m}) \sin[0.628(x - vt)]$$

where $v = 1.20$ m/s. (a) Sketch $y(x, t)$ at $t = 0$. (b) Sketch $y(x, t)$ at $t = 2.00$ s. Note how the entire waveform shifts 2.40 m in the positive x direction in this time interval.

Section 13.2   The Wave Model

Section 13.3   The Traveling Wave

3. A sinusoidal wave is traveling along a rope. The oscillator that generates the wave completes 40.0 vibrations in 30.0 s. Also, a given maximum travels 425 cm along the rope in 10.0 s. What is the wavelength?

4. For a certain transverse wave, the distance between two successive crests is 1.20 m, and eight crests pass a given point along the direction of travel every 12.0 s. Calculate the wave speed.

5. The wave function for a traveling wave on a taut string is (in SI units)

$$y(x, t) = (0.350 \text{ m}) \sin(10\pi t - 3\pi x + \pi/4)$$

(a) What are the speed and direction of travel of the wave? (b) What is the vertical displacement of the string at $t = 0$, $x = 0.100$ m? (c) What are the wavelength and frequency of the wave? (d) What is the maximum magnitude of the transverse speed of the string?

6. A wave is described by $y = (2.00 \text{ cm}) \sin(kx - \omega t)$, where $k = 2.11$ rad/m, $\omega = 3.62$ rad/s, x is in meters, and t is in seconds. Determine the amplitude, wavelength, frequency, and speed of the wave.

7. The string shown in Figure 13.7 is driven at a frequency of 5.00 Hz. The amplitude of the motion is 12.0 cm, and the wave speed is 20.0 m/s. Furthermore, the wave is such that $y = 0$ at $x = 0$ and $t = 0$. Determine (a) the angular frequency and (b) wave number for this wave. (c) Write an expression for the wave function. Calculate (d) the maximum transverse speed and (e) the maximum transverse acceleration of a point on the string.

8. A transverse wave on a string is described by the wave function

$$y = (0.120 \text{ m}) \sin(\pi x/8 + 4\pi t)$$

(a) Determine the transverse speed and acceleration at $t = 0.200$ s for the point on the string located at $x = 1.60$ m. (b) What are the wavelength, period, and speed of propagation of this wave?

9. (a) Write the expression for y as a function of x and t for a sinusoidal wave traveling along a rope in the *negative x* direction with the following characteristics: $A = 8.00$ cm, $\lambda = 80.0$ cm, $f = 3.00$ Hz, and $y(0,t) = 0$ at $t = 0$. (b) Write the expression for y as a function of x and t for the wave in part (a) assuming that $y(x, 0) = 0$ at the point $x = 10.0$ cm.

10. A transverse sinusoidal wave on a string has a period $T = 25.0$ ms and travels in the negative x direction with a speed of 30.0 m/s. At $t = 0$, a particle on the string at $x = 0$ has a displacement of 2.00 cm and is traveling downward with a speed of 2.00 m/s. (a) What is the amplitude of the wave? (b) What is the initial phase angle? (c) What is the maximum transverse speed of the string? (d) Write the wave function for the wave.

11. Show that the wave function $y = e^{b(x-vt)}$ is a solution of the linear wave equation (see Eq. 13.19), where b is a constant.

Section 13.4 The Speed of Transverse Waves on Strings

12. A telephone cord is 4.00 m long and has a mass of 0.200 kg. A transverse wave pulse is produced by plucking one end of the taut cord. The pulse makes four trips down and back along the cord in 0.800 s. What is the tension in the cord?

13. A piano string having a mass per unit length equal to 5.00×10^{-3} kg/m is under a tension of 1 350 N. Find the speed with which a wave travels on this string.

14. An astronaut on the Moon wishes to measure the local value of g by timing pulses traveling down a wire, clamped at its top end, that has a large mass suspended from it. Assume a wire has a mass of 4.00 g and a length of 1.60 m, and that a 3.00-kg object is suspended from it. A pulse requires 36.1 ms to traverse the length of the wire. Calculate g_{Moon} from these data. (You may ignore the mass of the wire when calculating the tension in it.)

15. A 30.0-m steel wire and a 20.0-m copper wire, both with 1.00-mm diameters, are connected end to end and stretched to a tension of 150 N. How long does it take a transverse wave to travel the entire length of the two wires?

16. **Review Problem.** A light string with a mass per unit length of 8.00 g/m has its ends tied to two walls separated by a distance equal to three fourths the length of the string (Fig. P13.16). An object of mass m is suspended from the center of the string, creating tension in the string. (a) Find an expression for the transverse wave speed in the string as a function of the hanging mass. (b) How much mass should be suspended from the string to produce a wave speed of 60.0 m/s?

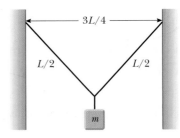

Figure P13.16

Section 13.5 Reflection and Transmission of Waves

17. A series of pulses, each of amplitude 0.150 m, are sent down a string that is attached to a post at one end. The pulses are reflected at the post and travel back along the string without loss of amplitude. When two waves are present on the same string, the net displacement of a given point is the sum of the displacements of the individual waves at that point. What is the net displacement at a

point on the string where two pulses are crossing, (a) if the string is rigidly attached to the post? (b) If the end at which reflection occurs is free to slide up and down?

Section 13.6 Rate of Energy Transfer by Sinusoidal Waves on Strings

18. A taut rope has a mass of 0.180 kg and a length 3.60 m. What power must be supplied to the rope in order to generate sinusoidal waves having an amplitude of 0.100 m and a wavelength of 0.500 m and traveling with a speed of 30.0 m/s?

19. Sinusoidal waves 5.00 cm in amplitude are to be transmitted along a string that has a linear mass density equal to 4.00×10^{-2} kg/m. If the source can deliver a maximum power of 300 W and the string is under a tension of 100 N, what is the highest vibrational frequency at which the source can operate?

web

20. Transverse waves are being generated on a rope under constant tension. By what factor is the required power increased or decreased if (a) the length of the rope is doubled and the angular frequency remains constant, (b) the amplitude is doubled and the angular frequency is halved, (c) both the wavelength and the amplitude are doubled, and (d) both the length of the rope and the wavelength are halved?

21. A 6.00-m segment of a long string has a mass of 180 g. A high-speed photograph shows that the segment contains four complete cycles of a wave. The string is vibrating sinusoidally with a frequency of 50.0 Hz and a peak-to-valley displacement of 15.0 cm. (The "peak-to-valley" displacement is the vertical distance from the farthest positive displacement to the farthest negative displacement.) (a) Write the function that describes this wave traveling in the positive x direction. (b) Determine the power being supplied to the string.

Section 13.7 Sound Waves

> *Note:* Use the following values as needed unless otherwise specified. The equilibrium density of air is $\rho = 1.20$ kg/m^3; the speed of sound in air is $v = 343$ m/s. Problem 52 in Chapter 2 can be assigned with this section.

22. A bat (Fig. P13.22) can detect very small objects, such as an insect whose length is approximately equal to one wavelength of the sound the bat makes. If bats emit a chirp at a frequency of 60.0 kHz, and if the speed of sound in air is 340 m/s, what is the smallest insect a bat can detect?

23. Suppose that you hear a clap of thunder 16.2 s after seeing the associated lightning stroke. The speed of sound waves in air is 343 m/s and the speed of light in air is 3.00×10^8 m/s. How far are you from the lightning stroke?

Figure P13.22 (*Joe McDonald Visuals Unlimited*)

24. A flowerpot is knocked off a balcony 20.0 m above the sidewalk and falls toward an unsuspecting 1.75-m-tall man standing below. How close to the sidewalk can the flowerpot fall before it is too late for a warning shouted from the balcony to reach the man in time? Assume that the man below requires 0.300 s to respond to the warning.

25. A sinusoidal sound wave is described by the displacement

$$s(x, t) = (2.00 \ \mu\text{m}) \cos[(15.7 \ \text{m}^{-1})x - (858 \ \text{s}^{-1})t]$$

(a) Find the amplitude, wavelength, and speed of this wave. (b) Determine the instantaneous displacement of an element of air at the position $x = 0.050\ 0$ m at $t = 3.00$ ms. (c) Determine the maximum speed of an element's oscillatory motion.

26. Calculate the pressure amplitude of a 2.00-kHz sound wave in air if the displacement amplitude is equal to 2.00×10^{-8} m.

27. An experimenter wishes to generate in air a sound wave that has a displacement amplitude of 5.50×10^{-6} m. The pressure amplitude is to be limited to 0.840 N/m^2. What is the minimum wavelength the sound wave can have?

web

28. A sound wave traveling in air has a pressure amplitude of 4.00 N/m^2 and a frequency of 5.00 kHz. Take $\Delta P = 0$ at the point $x = 0$ when $t = 0$. (a) What is ΔP at $x = 0$ when $t = 2.00 \times 10^{-4}$ s? (b) What is ΔP at $x = 0.020\ 0$ m when $t = 0$?

29. Write an expression that describes the pressure variation as a function of position and time for a sinusoidal sound wave in air, if $\lambda = 0.100$ m and $\Delta P_{\text{max}} = 0.200$ N/m^2.

Section 13.8 The Doppler Effect

30. A commuter train passes a passenger platform at a constant speed of 40.0 m/s. The train horn is sounded at its characteristic frequency of 320 Hz. (a) What overall change in frequency is detected by a person on the platform as the train moves from approaching to receding? (b) What wavelength is detected by a person on the platform as the train approaches?

31. Standing at a crosswalk, you hear a frequency of 560 Hz **web** from the siren of an approaching ambulance. After the ambulance passes, the observed frequency of the siren is 480 Hz. Determine the ambulance's speed from these observations.

32. A driver travels northbound on a highway at a speed of 25.0 m/s. A police car, traveling southbound at a speed of 40.0 m/s, approaches with its siren producing sound at a frequency of 2 500 Hz. (a) What frequency does the driver observe as the police car approaches? (b) What frequency does the driver detect after the police car passes him? (c) Repeat parts (a) and (b) for the case when the police car is traveling northbound.

33. A tuning fork vibrating at 512 Hz falls from rest and accelerates at 9.80 m/s². How far below the point of release is the tuning fork when waves of frequency 485 Hz reach the release point? Take the speed of sound in air to be 340 m/s.

34. Expectant parents are thrilled to hear their unborn baby's heartbeat, revealed by an ultrasonic motion detector. Suppose the fetus's ventricular wall moves in simple harmonic motion with an amplitude of 1.80 mm and a frequency of 115/min. (a) Find the maximum linear speed of the heart wall. Suppose the motion detector in contact with the mother's abdomen produces sound at 2 000 000.0 Hz, which travels through tissue at 1.50 km/s. (b) Find the maximum frequency at which sound arrives at the wall of the baby's heart. (c) Find the maximum frequency at which reflected sound is received by the motion detector. By electronically "listening" for echoes at a frequency different from the broadcast frequency, the motion detector can produce beeps of audible sound in synchronization with the fetal heartbeat.

35. A block with a speaker bolted to it is connected to a spring having force constant $k = 20.0$ N/m, as shown in Figure P13.35. The total mass of the block and speaker is 5.00 kg, and the amplitude of this unit's motion is 0.500 m. If the speaker emits sound waves of frequency 440 Hz, determine the highest and lowest frequencies heard by the person to the right of the speaker.

36. At the Winter Olympics, an athlete rides her luge down the track while a bell just above the wall of the chute rings continuously. When her sled passes the bell, she hears the frequency of the bell fall by the musical interval called a minor third. That is, the frequency she hears drops to five sixths of its original value. (a) Find the speed of sound in

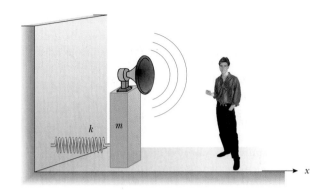

Figure P13.35

air at the ambient temperature $-10.0°C$. (b) Find the speed of the athlete.

Section 13.9   Context Connection — Seismic Waves

37. Two points, A and B, on the surface of the Earth are at the same longitude and 60.0° apart in latitude. Suppose that an earthquake at point A creates a P wave that reaches point B by traveling straight through the body of the Earth at a constant speed of 7.80 km/s. The earthquake also radiates a Rayleigh wave, which travels along the surface of the Earth at 4.50 km/s. (a) Which of these two seismic waves arrives at B first? (b) What is the time difference between the arrivals of the two waves at B? Take the radius of the Earth to be 6 370 km.

38. A seismographic station receives S and P waves from an earthquake, 17.3 s apart. Assume the waves have traveled over the same path at speeds of 4.50 km/s and 7.80 km/s. Find the distance from the seismograph to the hypocenter of the quake.

Additional Problems

39. "The wave" is a particular type of wave pulse that can propagate through a large crowd gathered at a sports arena to watch a soccer or American football match (Figure P13.39). The particles of the medium are the spectators, with zero displacement corresponding to their being seated and maximum displacement corresponding to their standing up and raising their arms. When a large fraction of the spectators participate in the wave motion, a somewhat stable pulse shape can develop. The wave speed depends on people's reaction time, which is typically on the order of 0.1 s. Estimate the order of magnitude, in minutes, of the time required for such a wave pulse to make one circuit around a large sports stadium. State the quantities you measure or estimate and their values.

Figure P13.39 *(Gregg Adams/Stone)*

40. Vocalists sing very high notes most often in ad lib ornaments and cadenzas. The highest note written for a singer in a published score was F-sharp above high C, 1.480 kHz, sung by Zerbinetta in the original version of Richard Strauss's opera *Ariadne auf Naxos*. (a) Find the wavelength of this sound in air. (b) In response to complaints, Strauss later transposed the note down to F above high C, 1.397 kHz. By what increment did the wavelength change?

41. **Review Problem.** A block of mass M, supported by a string, rests on an incline making an angle of θ with the horizontal (Fig. P13.41). The length of the string is L and its mass is $m \ll M$. Derive an expression for the time it takes a transverse wave to travel from one end of the string to the other.

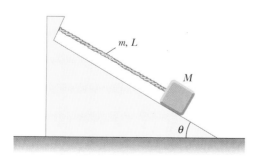

Figure P13.41

42. **Review Problem.** A block of mass M hangs from a rubber cord. The block is supported so that the cord is not stretched. The unstretched length of the cord is L_0 and its mass is m, much less than M. The "force constant" for the cord is k. The block is released and stops at the lowest point. (a) Determine the tension in the string when the block is at this lowest point. (b) What is the length of the cord in this "stretched" position? (c) Find the speed of a transverse wave in the cord if the block is held in this lowest position.

43. A sound wave in a cylinder is described by Equations 13.24 through 13.26. Show that $\Delta P = \pm \rho v \omega \sqrt{s_{max}^2 - s^2}$.

44. A string on a musical instrument is held under tension T and extends from the point $x = 0$ to the point $x = L$. The string is overwound with wire in such a way that its mass per unit length $\mu(x)$ increases linearly from μ_0 at $x = 0$ to μ_L at $x = L$. (a) Find an expression for $\mu(x)$ as a function of x over the range $0 \le x \le L$. (b) Show that the time required for a transverse pulse to travel the length of the string is given by

$$t = \frac{2L\left(\mu_L + \mu_0 + \sqrt{\mu_L \mu_0}\right)}{3\sqrt{T}\left(\sqrt{\mu_L} + \sqrt{\mu_0}\right)}$$

45. A rope of total mass m and length L is suspended vertically. Show that a transverse wave pulse travels the length of the rope in a time $t = 2\sqrt{L/g}$. (*Hint:* First find an expression for the wave speed at any point a distance x from the lower end by considering the tension in the rope as resulting from the weight of the segment below that point.)

46. If an object of mass M is suspended from the bottom of the rope in Problem 45, (a) show that the time for a transverse wave to travel the length of the rope is

$$t = 2\sqrt{\frac{L}{mg}}\left(\sqrt{M + m} - \sqrt{M}\right)$$

(b) Show that this reduces to the result of Problem 45 when $M = 0$. (c) Show that for $m \ll M$, the expression in part (a) reduces to

$$t = \sqrt{\frac{mL}{Mg}}$$

47. (a) Show that the speed of longitudinal waves along a spring of force constant k is $v = \sqrt{kL/\mu}$, where L is the unstretched length of the spring and μ is the mass per unit length. (b) A spring with a mass of 0.400 kg has an unstretched length of 2.00 m and a force constant of 100 N/m. Using the result you obtained in part (a), determine the speed of longitudinal waves along this spring.

48. A wave pulse traveling along a string of linear mass density μ is described by the wave function

$$y = [A_0 e^{-bx}] \sin(kx - \omega t)$$

where the factor in brackets before the sine function is said to be the amplitude. (a) What is the power $\mathcal{P}(x)$ carried by this wave at a point x? (b) What is the power $\mathcal{P}(0)$ carried by this wave at the origin? (c) Compute the ratio $\mathcal{P}(x)/\mathcal{P}(0)$.

49. An earthquake on the ocean floor in the Gulf of Alaska induces a *tsunami* (sometimes called a tidal wave) that reaches Hilo, Hawaii, 4 450 km away, in a time of 9 h 30 min. Tsunamis have enormous wavelengths (100–200 km), and the propagation speed for these waves is $v \approx \sqrt{gd}$, where $\overline{d}$ is the average depth of the water. From the information given, find the average wave speed and the average ocean depth between Alaska and Hawaii. (This method was used in 1856 to estimate the average depth of the Pacific Ocean long before soundings were made to give a direct determination.)

50. On a Saturday morning, pickup trucks and sport utility vehicles carrying garbage to the town dump form a nearly steady procession on a country road, all traveling at 19.7 m/s. From one direction, two trucks arrive at the dump every 3 min. A bicyclist is also traveling this way, at 4.47 m/s. (a) With what frequency do the trucks pass him? (b) A hill does not slow down the trucks, but makes the out-of-shape cyclist's speed drop to 1.56 m/s. How often do the trucks whiz past him now?

51. The ocean floor is underlain by a layer of basalt that constitutes the crust, or uppermost layer, of the Earth in that region. Below this crust is found denser periodotite rock, which forms the Earth's mantle. The boundary between these two layers is called the Mohorovičić discontinuity (Moho for short). If an explosive charge is set off at the surface of the basalt, it generates a seismic wave that is reflected back out at the Moho. If the speed of this wave in basalt is 6.50 km/s and the two-way travel time is 1.85 s, what is the thickness of this oceanic crust?

52. A train whistle ($f = 400$ Hz) appears to be higher or lower in frequency depending on whether it approaches or recedes. (a) Prove that the difference in frequency between the approaching and receding train whistle is

$$\Delta f = \frac{2u/v}{1 - u^2/v^2} f$$

where u is the speed of the train and v is the speed of sound. (b) Calculate this difference for a train moving at a speed of 130 km/h. Take the speed of sound in air to be 340 m/s.

53. A siren mounted on the roof of a firehouse emits sound at a frequency of 900 Hz. A steady wind is blowing with a speed of 15.0 m/s. Taking the speed of sound in calm air to be 343 m/s, find the wavelength of the sound (a) upwind of the siren and (b) downwind of the siren. The volunteer firefighters are approaching the siren from various directions at 15.0 m/s. What frequency does a firefighter hear (c) if he or she is approaching from an upwind position, so that he is moving in the direction in which the wind is blowing? (d) If he or she is approaching from a downwind position and moving against the wind?

54. The Doppler equation presented in the text is valid when the motion between the observer and the source occurs on a straight line, so that the source and observer are moving either directly toward or directly away from each other. If this restriction is relaxed, one must use the more general Doppler equation

$$f' = \left(\frac{v + v_O \cos \theta_O}{v - v_S \cos \theta_S} \right) f$$

where θ_O and θ_S are defined in Figure P13.54a. (a) Show that if the observer and source are moving away from each other, the preceding equation reduces to Equation 13.30 with negative values for both v_O and v_S. (b) Use the preceding equation to solve the following problem. A train moves at a constant speed of 25.0 m/s toward the intersection shown in Figure P13.54b. A car is stopped near the intersection, 30.0 m from the tracks. If the train's horn emits a frequency of 500 Hz, what is the frequency heard by the passengers in the car when the train is 40.0 m from the intersection? Take the speed of sound to be 343 m/s.

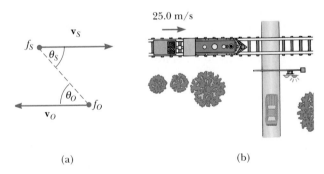

(a)

(b)

Figure P13.54

ANSWERS TO QUICK QUIZZES

13.1 (a) Longitudinal. The movement of the people is parallel to the direction of propagation of the gap. (b) Transverse. The fans stand up vertically, as the wave sweeps past them horizontally.

13.2 The radio wave has the higher frequency. Although both waves have the same wavelength, the speed of the radio wave is much higher than that of the sound wave, so Equation 13.11 gives a higher frequency for the radio wave.

13.3 Only choices (f) and (h) are correct. (a) and (b) affect the transverse speed of a particle of the string but not the wave speed along the string. (c) and (d) change the amplitude. (e) and (g) increase the time by decreasing the wave speed.

13.4 A wave on a massless string would have an infinite speed of propagation, according to Equation 13.20.

13.5 The apparent frequency would be Doppler shifted downward in frequency because the screaming person is moving away from you. Because the speed of the source increases due to the gravitational force, the apparent frequency decreases with time.

13.6 (c). The sound that leaves your horn in the forward direction is Doppler shifted to a higher frequency because it is coming from a moving source. As the sound reflects back and comes toward you, you are a moving observer, so a second Doppler shift occurs to an even higher frequency.

chapter 14

Guitarist Carlos Santana takes advantage of standing waves on strings. He changes to a higher note on the guitar by pushing the strings against the frets on the fingerboard, shortening the lengths of the portions of the strings that vibrate.

(Corbis)

Superposition and Standing Waves

In the previous chapter, we introduced the wave model. We have seen that waves are very different from particles—a particle is of zero size, but an ideal wave is of infinite length. Another important difference between waves and particles is that we can explore the possibility of two or more waves combining at one point in the same medium. We can combine particles to form extended objects, but the particles must be at different locations. In contrast, two waves can both be present at a given location, and the ramifications of this possibility are explored in this chapter.

One of the ramifications of the combination of waves is that only certain allowed frequencies can exist on systems with boundary conditions—that is, the frequencies are *quantized*. In Chapter 11 we learned about quantized energies of the hydrogen atom. Quantization is at the heart of quantum mechanics, a subject that is introduced formally in Chapter 28. We shall see that waves under boundary conditions explain many of the quantum phenomena. For our present purposes in this chapter, quantization enables us to understand the behavior of the wide array of musical instruments that are based on strings and air columns.

14.1 • THE PRINCIPLE OF SUPERPOSITION

Many interesting wave phenomena in nature cannot be described by a single wave. Instead, one must analyze complex waveforms in terms of a combination of many traveling waves. To analyze such wave combinations, we make use of the **principle of superposition,** which states that

See the *Core Concepts in Physics CD-ROM,* Screens 9.6 & 9.7

> If two or more traveling waves are moving through a medium and combine at a given point, the resultant displacement of the medium at that point is the sum of the displacements of the individual waves.

• *Principle of superposition*

This rather striking property is exhibited by many waves in nature, including waves on strings, sound waves, surface water waves, and electromagnetic waves. Waves that obey this principle are called *linear waves.* In general, linear waves have an amplitude that is small relative to their wavelength. Waves that violate the superposition principle are called *nonlinear waves* and are often characterized by large amplitudes. In this book, we shall deal only with linear waves.

A simple pictorial representation of the superposition principle is obtained by considering two pulses traveling in opposite directions on a stretched string, as in Figure 14.1. The wave function for the pulse moving to the right is y_1, and the

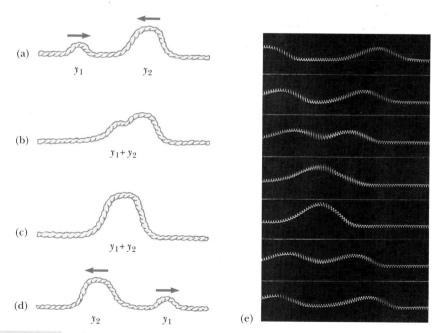

Figure 14.1

(*Left*) Two wave pulses traveling on a stretched string in opposite directions pass through each other. When the pulses overlap, as in (b) and (c), the net displacement of the string equals the sum of the displacements produced by each pulse. Because each pulse produces positive displacements of the string, we refer to their superposition as *constructive interference.* (*Right*) Photograph of the superposition of two equal and symmetric pulses traveling in opposite directions on a stretched spring. (*Photo, Education Development Center, Newton, MA*)

wave function for the pulse moving to the left is y_2. The pulses have the same speed but different shapes. Each pulse is assumed to be symmetric (although this is not a necessary condition), and in both cases the vertical displacements of the string are taken to be positive. When the waves begin to overlap (Fig. 14.1b), the resulting waveform is given by $y_1 + y_2$. When the crests of the pulses exactly coincide (Fig. 14.1c), the resulting waveform is symmetric. The two pulses finally separate and continue moving in their original directions (Fig. 14.1d). Note that the final waveforms remain unchanged, as if the two pulses had never met! The combination of separate waves in the same region of space to produce a resultant wave is called **interference.** Notice that the interference exists only while the waves are in the same region of space, and there is no permanent effect on the pulses after they separate.

One consequence of the superposition principle is that **two traveling waves can pass through each other without being destroyed or even altered.** For instance, when two pebbles are thrown into a pond, the expanding circular surface waves do not destroy one another. In fact, the ripples pass through one another. The complex pattern that is observed can be viewed as a combination of two independent sets of expanding circles.

For the two pulses shown in Figure 14.1, the vertical displacements are in the same direction, and so the resultant waveform (when the pulses overlap) ex-

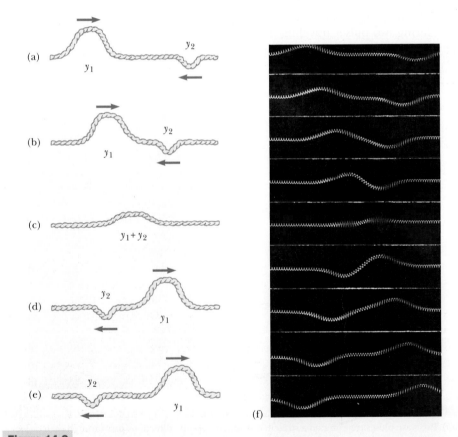

Figure 14.2

(*Left*) Two wave pulses traveling in opposite directions with displacements that are inverted relative to each other. When the two overlap as in (c), their displacements subtract from each other. (*Right*) Photograph of the superposition of two symmetric pulses traveling in opposite directions, where one is inverted relative to the other. (*Photo, Education Development Center, Newton, MA*)

hibits a displacement greater than those of the individual pulses. Now consider two identical pulses, again traveling in opposite directions on a stretched string, but this time one pulse is inverted relative to the other, as in Figure 14.2. In this case, when the pulses begin to overlap, the resultant waveform is the sum of the two separate displacements again, but one of the displacements is negative. Again, the two pulses pass through each other. When they exactly overlap, they partially *cancel* each other. At this time (Fig. 14.2c), the resultant amplitude is small.

14.2 • INTERFERENCE OF WAVES

In this section, we shall investigate the mathematics of waves in interference. Additional applications of this model applied to light waves are presented in Chapter 27.

Let us apply the superposition principle to two sinusoidal waves traveling in the same direction in a medium. If the two waves are traveling to the right and have the same frequency, wavelength, and amplitude but differ in phase, we can express their individual wave functions as

$$y_1 = A \sin(kx - \omega t) \quad \text{and} \quad y_2 = A \sin(kx - \omega t + \phi)$$

where ϕ is the phase difference between the two waves. Let us imagine that these two waves coincide in the medium. For example, these expressions might represent two waves traveling along the same string. In this situation, the resultant wave function y is, according to the principle of superposition

$$y = y_1 + y_2 = A[\sin(kx - \omega t) + \sin(kx - \omega t + \phi)]$$

To simplify this expression, it is convenient to use the trigonometric identity

$$\sin a + \sin b = 2 \cos\left(\frac{a - b}{2}\right) \sin\left(\frac{a + b}{2}\right)$$

If we let $a = kx - \omega t$ and $b = kx - \omega t + \phi$, the resultant wave function y reduces to

$$y = \left(2A \cos \frac{\phi}{2}\right) \sin\left(kx - \omega t + \frac{\phi}{2}\right) \qquad \textbf{[14.1]}$$

This mathematical representation of the resultant wave has several important features. The resultant wave function y is also a sinusoidal wave and has the *same* frequency and wavelength as the individual waves. The amplitude of the resultant wave is $2A \cos(\phi/2)$, and the phase angle is $\phi/2$. If the phase angle ϕ equals 0, then $\cos(\phi/2) = \cos 0 = 1$ and the amplitude of the resultant wave is $2A$. In other words, the amplitude of the resultant wave is twice the amplitude of either individual wave. In this case, the waves are said to be everywhere *in phase* ($\phi = 0$) and to **interfere constructively.** That is, the crests of the individual waves occur at the same positions, as is shown by the blue line in Figure 14.3a. In general, constructive interference occurs when $\cos(\phi/2) = \pm 1$, or when $\phi = 0, 2\pi, 4\pi, \ldots$

On the other hand, if ϕ is equal to π radians, or to any *odd* multiple of π, then $\cos(\phi/2) = \cos(\pi/2) = 0$ and the resultant wave has *zero* amplitude everywhere. In this case, the two waves **interfere destructively.** That is, the crest of one wave coincides with the trough of the second (Fig. 14.3b) and their displacements cancel at every point. Finally, when the phase constant has a value between 0 and π, as in Figure 14.3c, the resultant wave has an amplitude whose value is somewhere between 0 and $2A$.

• *Constructive interference*

• *Destructive interference*

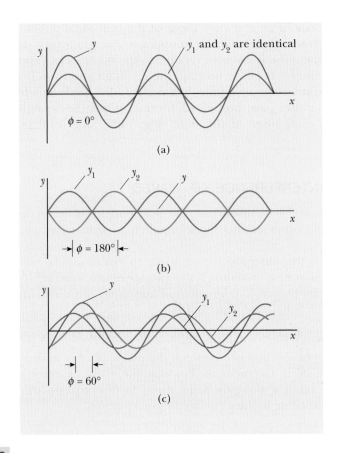

Figure 14.3

The superposition of two identical waves y_1 and y_2. (a) When the two waves are in phase, the result is constructive interference. (b) When the two waves are π rad out of phase, the result is destructive interference. (c) When the phase angle has a value other than 0 or π rad, the resultant wave y falls somewhere between the extremes shown in (a) and (b).

Although we have used waves having the same amplitude in the preceding discussion, waves of differing amplitudes will interfere in a similar way. If the waves are in phase, the combined amplitude is the sum of the individual amplitudes. If they are 180° out of phase, the combined amplitude is the difference between the individual amplitudes.

It is often useful to express the phase difference between the two waves in terms of a fraction of the wavelength λ. Because a difference of one wavelength corresponds to a phase difference of 2π rad, an arbitrary shift Δr in position of one wave with respect to another is related to the phase difference as follows:

- *Relationship between wavelength fraction and phase angle*

$$\frac{\Delta r}{\lambda} = \frac{\phi}{2\pi} \longrightarrow \Delta r = \frac{\phi}{2\pi}\lambda \qquad [14.2]$$

Thus, for example, a phase difference of 180° or π rad corresponds to a shift of $\lambda/2$. Conversely, a 1/4-wavelength shift corresponds to a 90° phase difference.

Figure 14.4 shows a simple device for demonstrating interference of sound waves. Sound from speaker S is sent into a tube at P, where there is a T-shaped junction. Half the sound energy travels in one direction, and half in the opposite

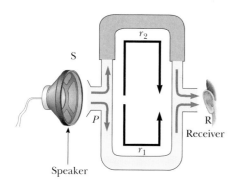

Figure 14.4

An acoustical system for demonstrating interference of sound waves. Sound waves from the speaker propagate into the tube and the energy splits into two parts at point *P*. The waves from the two paths, which superimpose at the opposite side, are detected at the receiver *R*. The upper path length r_2 can be varied by sliding the upper section.

direction. Thus, the sound waves that reach receiver R at the other side can travel along either of two paths. The total distance from speaker to receiver is called the **path length** *r*. The length of the lower path is fixed at r_1. The upper path length r_2 can be varied by sliding the U-shaped tube (similar to that on a slide trombone). When the difference in the path lengths $\Delta r = |r_2 - r_1|$ is either zero or some integral multiple of the wavelength λ, the two waves reaching the receiver are in phase and interfere constructively, as in Figure 14.3a. In this case, a maximum in the sound intensity is detected at the receiver. If path length r_2 is adjusted so that Δr is $\lambda/2, 3\lambda/2, \ldots n\lambda/2$ (for *n* odd), the two waves are exactly 180° out of phase at the receiver and hence cancel each other. In this case of completely destructive interference, no sound is detected at the receiver. This simple experiment is a striking illustration of interference. In addition, it demonstrates the fact that a phase difference may arise between two waves generated by the same source when they travel along paths of unequal lengths.

Nature provides many other examples of interference phenomena. Later in the text we shall describe several interesting interference effects involving light waves.

THINKING PHYSICS 14.1

If stereo speakers are connected to the amplifier "out of phase," one speaker is moving outward when the other is moving inward. This results in a weakness in the bass notes, which can be corrected by reversing the wires on one of the speaker connections. Why are only the bass notes affected in this case, and not the treble notes?

Reasoning Imagine that you are sitting in front of the speakers, midway between them. Then, the sound from each speaker travels the same distance to you, so the phase difference in the sound is not due to a path difference. Because the speakers are connected out of phase, the sound waves are half a wavelength out of phase on leaving the speaker and, consequently, on arriving at your ear. As a result, the sound for all frequencies cancels, in the simplification model of a zero-size head located exactly on the midpoint between the speakers. If the ideal head were moved off the centerline, an additional phase difference is introduced by the path length difference for the sound from the two speakers. In the case of low-frequency, long-wavelength bass notes, the path length differences are a small fraction of a wavelength, so significant cancellation still occurs. For the high-frequency, short-wavelength treble notes, a small movement of the ideal head results in a much larger fraction of a wavelength in path length difference, or even multiple wavelengths. Thus, the treble notes could be in phase with this head movement. If we now add the fact that the head is not of zero size, and the fact that it has two ears, we can see that complete cancellation is not possible, and, with even small movements of the head, one or both ears will be at or near maxima for the treble notes. The size of the head is much smaller than bass wavelengths, however, so the bass notes are significantly weakened over much of the region in front of the speakers.

Example 14.1   Two Speakers Driven by the Same Source

Two speakers placed 3.00 m apart are driven in phase by the same oscillator (Fig. 14.5). A listener is originally at point O, which is located 8.00 m from the center of the line connecting the two speakers. The listener then moves to point P, which is a perpendicular distance 0.350 m from O before reaching the first cancellation of waves, resulting in a minimum in sound intensity. What is the frequency of the oscillator?

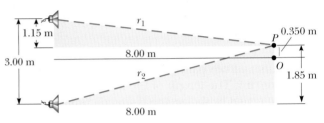

Figure 14.5

(Example 14.1).

Solution   The first cancellation occurs when the two waves reaching the listener at P are 180° out of phase—in other words, when their path difference equals $\lambda/2$. To calculate the path difference, we must first find the path lengths r_1 and r_2. Consider the two geometric model triangles shaded in Figure 14.5. Making use of these triangles, we find the path lengths to be

$$r_1 = \sqrt{(8.00 \text{ m})^2 + (1.15 \text{ m})^2} = 8.08 \text{ m}$$

$$r_2 = \sqrt{(8.00 \text{ m})^2 + (1.85 \text{ m})^2} = 8.21 \text{ m}$$

Hence, the path difference is $r_2 - r_1 = 0.13$ m. Because we require that this path difference be equal to $\lambda/2$ for the first minimum, we find that $\lambda = 0.26$ m.

To obtain the oscillator frequency, we use $v = \lambda f$, where v is the speed of sound in air, 343 m/s:

$$f = \frac{v}{\lambda} = \frac{343 \text{ m/s}}{0.26 \text{ m}} = \boxed{1.3 \text{ kHz}}$$

EXERCISE   If the oscillator frequency is adjusted such that the listener hears the first minimum at a distance of 0.75 m from O, what is the new frequency?

Answer   0.63 kHz

14.3 • STANDING WAVES

In the previous section, we considered two waves moving in the same direction and combining according to the principle of superposition. Now imagine that we have two waves that are identical except that they are traveling in opposite directions in the same medium. This happens, for example, when you shake one end of a rope whose opposite end is attached to a rigid wall. The waves reflected from the wall are identical to the waves approaching the wall, except for the direction of travel and an inversion of the wave on reflection. An interesting pattern emerges from this type of combination of waves.

If we ignore any phase angle ϕ for simplicity, the wave functions for the two waves we have described can be written

$$y_1 = A \sin(kx - \omega t) \qquad \text{and} \qquad y_2 = A \sin(kx + \omega t)$$

where y_1 represents a wave traveling to the right and y_2 represents a wave traveling to the left. According to the principle of superposition, adding these two functions

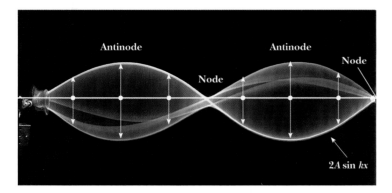

Figure 14.6

Multiflash photograph of a standing wave on a string. The vertical displacement from equilibrium of an individual element of the string is given by cos ωt. That is, each element vibrates at an angular frequency ω. The amplitude of the vertical oscillation of any element on the string depends on the horizontal position of the element. Each element vibrates within the confines of the envelope function $2A \sin kx$. (©1991 Richard Megna/Fundamental Photographs)

gives the resultant wave function y:

$$y = y_1 + y_2 = A \sin(kx - \omega t) + A \sin(kx + \omega t)$$

Using the trigonometric identity $\sin(a \pm b) = \sin a \cos b \pm \cos a \sin b$, this reduces to

$$y = (2A \sin kx) \cos \omega t \qquad [14.3]$$

Notice that this function does not look mathematically like a traveling wave — there is no function of $kx - \omega t$. Equation 14.3 represents the wave function of a **standing wave**, such as that shown in Figure 14.6. A standing wave is an oscillation pattern that results from two waves traveling in opposite directions. Mathematically, this equation looks more like simple harmonic motion than wave motion for traveling waves. Every particle of the string vibrates in simple harmonic motion with the same angular frequency ω (according to the factor cos ωt). The amplitude of motion of a given particle (the factor $2A \sin kx$), however, depends on its position along the string, described by the variable x. This is in contrast to the situation involving a traveling sinusoidal wave, in which all particles oscillate with the same amplitude as well as the same frequency. From this result, we see that the simple harmonic motion of every particle has an angular frequency of ω and a position-dependent amplitude of $2A \sin kx$.

Because the amplitude of the simple harmonic motion at any value of x is equal to $2A \sin kx$, we see that the *maximum* amplitude of the simple harmonic motion has the value $2A$. This is described as the amplitude of the standing wave. This maximum amplitude occurs when the coordinate x satisfies the condition $\sin kx = 1$, or when

$$kx = \frac{\pi}{2}, \frac{3\pi}{2}, \frac{5\pi}{2}, \ldots$$

• *Wave function for a standing wave*

PITFALL PREVENTION 14.2

Three types of amplitude

 We need to distinguish carefully here between the **amplitude of the individual waves,** which is A, and the **amplitude of the simple harmonic motion of the particles of the medium,** which is $2A \sin kx$. A given particle in a standing wave vibrates within the constraints of the *envelope* function $2A \sin kx$, where x is that particle's position in the medium. This is in contrast to traveling sinusoidal waves, in which all particles oscillate with the same amplitude and the same frequency, and the amplitude A of the wave is the same as the amplitude A of the simple harmonic motion of the particles. Furthermore, we can identify the **amplitude of the standing wave** as $2A$.

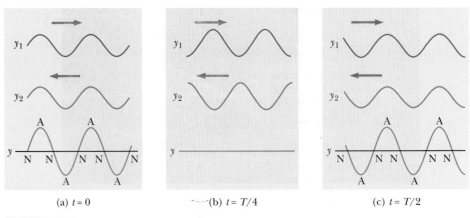

Figure 14.7

Standing wave patterns at various times produced by two waves of equal amplitude traveling in opposite directions. For the resultant wave *y*, the nodes (N) are points of zero displacement and the antinodes (A) are points of maximum displacement.

Because $k = 2\pi/\lambda$, the positions of maximum amplitude, called **antinodes,** are

- *Position of antinodes*

$$x = \frac{\lambda}{4}, \frac{3\lambda}{4}, \frac{5\lambda}{4}, \ldots = \frac{n\lambda}{4} \qquad [14.4]$$

where $n = 1, 3, 5, \ldots$. Note that **adjacent antinodes are separated by a distance of λ/2.**

Similarly, the simple harmonic motion has a *minimum* amplitude of zero when *x* satisfies the condition $\sin kx = 0$, or when $kx = \pi, 2\pi, 3\pi, \ldots$ giving

- *Position of nodes*

$$x = \frac{\lambda}{2}, \lambda, \frac{3\lambda}{2}, \ldots = \frac{n\lambda}{2} \qquad [14.5]$$

where $n = 1, 2, 3, \ldots$. **These points of zero amplitude, called nodes, are also spaced by λ/2.** The distance between a node and an adjacent antinode is $\lambda/4$.

The standing wave patterns produced at various times by two waves traveling in opposite directions are represented graphically in Figure 14.7. The upper part of each figure represents the individual traveling waves, and the lower part represents the standing wave patterns. The nodes of the standing wave are labeled N, and the antinodes are labeled A. At $t = 0$ (Fig. 14.7a), the two waves are in phase, giving a wave pattern with amplitude 2A. One quarter of a period later, at $t = T/4$ (Fig. 14.7b), the individual waves have moved one quarter of a wavelength (one to the right and the other to the left). At this time, the waves are 180° out of phase. The individual displacements are of equal magnitude and opposite direction for all values of *x*, and hence the resultant wave has zero displacement everywhere. At $t = T/2$ (Fig. 14.7c), the individual waves are again in phase, producing a wave pattern that is inverted relative to the $t = 0$ pattern.

Quick Quiz 14.1

For the standing wave described by Equation 14.3, what is the maximum transverse speed of a particle of the medium located at a node? What is the maximum transverse speed of a particle of the medium located at an antinode?

Example 14.2 Formation of a Standing Wave

Two waves traveling in opposite directions produce a standing wave. The individual wave functions are

$$y_1 = (4.0 \text{ cm}) \sin(3.0x - 2.0t)$$

$$y_2 = (4.0 \text{ cm}) \sin(3.0x + 2.0t)$$

where x and y are in centimeters. (a) Find the maximum displacement of a particle of the medium at $x = 2.3$ cm.

Solution When the two waves are summed, the result is a standing wave whose mathematical representation is given by Equation 14.3, with $A = 4.0$ cm and $k = 3.0$ rad/cm:

$$y = (2A \sin kx) \cos \omega t = [(8.0 \text{ cm}) \sin 3.0x] \cos 2.0t$$

Thus, the maximum displacement of a particle at the position $x = 2.3$ cm is

$$y_{max} = [(8.0 \text{ cm}) \sin 3.0x]_{x=2.3 \text{ cm}}$$
$$= (8.0 \text{ cm}) \sin(6.9 \text{ rad}) = \boxed{4.6 \text{ cm}}$$

(b) Find the positions of the nodes and antinodes.

Solution Because $k = 2\pi/\lambda = 3.0$ rad/cm, we see that $\lambda = 2\pi/3$ cm. Therefore, from Equation 14.4 we find that the antinodes are located at

$$x = n \left(\frac{\pi}{6.0} \right) \text{cm} \quad (n = 1, 3, 5, \ldots)$$

and from Equation 14.5 we find that the nodes are located at

$$x = n \frac{\lambda}{2} = n \left(\frac{\pi}{3.0} \right) \text{cm} \quad (n = 1, 2, 3, \ldots)$$

14.4 • STANDING WAVES IN STRINGS

 See Screen 9.9

In the preceding section, we discussed standing waves formed by identical waves moving in opposite directions in the same medium. As you will recall, one way to do this on a string is to combine incoming and reflected waves from a rigid end. If a string is stretched between *two* rigid supports (Fig. 14.8a) and waves are established on the string, standing waves will be set up in the string by the continuous superposition of the waves incident on and reflected from the ends. This physical system is a model for the source of sound in any stringed instrument, such as the guitar, the violin, and the piano. The string has a number of natural patterns of

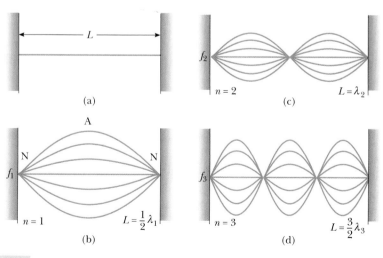

Figure 14.8

(a) A string of length L fixed at both ends. The normal modes of vibration form a harmonic series:
(b) the fundamental frequency, or first harmonic; (c) the second harmonic; and (d) the third harmonic.

vibration, called **normal modes.** Each of these modes has a characteristic frequency.

This discussion is our first introduction to an important analysis model—the **wave under boundary conditions.** When boundary conditions are applied to a wave, we find very interesting behavior that has no analog in the physics of particles. The most prominent aspect of this behavior is **quantization.** We shall find that only certain waves are allowed—those that satisfy the boundary conditions. The notion of quantization was introduced in Chapter 11 when we discussed the Bohr model of the atom. In that model, angular momentum was quantized. As we shall see in Chapter 29, this is just an application of the wave under boundary conditions model.

In the standing wave pattern on a stretched string, the ends of the string must be nodes, because these points are fixed. This is the boundary condition on the waves. The rest of the pattern can be built from this boundary condition along with the requirement that nodes and antinodes are equally spaced and separated by one fourth of a wavelength. The simplest pattern that satisfies these conditions has the required nodes at the ends of the string and an antinode at the center point (Fig. 14.8b). For this normal mode, the length of the string equals $\lambda/2$ (the distance between adjacent nodes):

$$L = \frac{\lambda_1}{2} \quad \text{or} \quad \lambda_1 = 2L$$

The next normal mode, of wavelength λ_2 (Fig. 14.8c), occurs when the length of the string equals one wavelength, that is, when $\lambda_2 = L$. In this mode, the two halves of the string are moving in opposite directions at a given instant—we sometimes say that two *loops* occur. The third normal mode (Fig. 14.8d) corresponds to the case when the length equals $3\lambda/2$; therefore, $\lambda_3 = 2L/3$. In general, the wavelengths of the various normal modes can be conveniently expressed as

• *Wavelengths of normal modes*

$$\lambda_n = \frac{2L}{n} \quad (n = 1, 2, 3, \ldots) \tag{14.6}$$

where the index n refers to the nth mode of vibration. The natural frequencies associated with these modes are obtained from the relationship $f = v/\lambda$, where the wave speed v is determined by the tension T and linear mass density μ of the string and, therefore, is the same for all frequencies. Using Equation 14.6, we find that the frequencies of the normal modes are

• *Frequencies of normal modes as functions of wave speed and length of string*

$$f_n = \frac{v}{\lambda_n} = \frac{n}{2L} v \quad (n = 1, 2, 3, \ldots) \tag{14.7}$$

Because $v = \sqrt{T/\mu}$ (Equation 13.20), we can express the natural frequencies of a stretched string as

• *Frequencies of normal modes as functions of string tension and linear mass density*

$$f_n = \frac{n}{2L} \sqrt{\frac{T}{\mu}} \quad (n = 1, 2, 3, \ldots) \tag{14.8}$$

Equation 14.8 demonstrates the quantization that we mentioned as a feature of the wave under boundary conditions model. The frequencies are quantized—only certain frequencies of waves satisfy the boundary conditions and can exist on the

PITFALL PREVENTION 14.3

Jumping rope

Standing waves are hard to describe via the printed word. Your instructor will likely demonstrate a standing wave to you using a rope or a long spring. If you jumped rope as a child, you set up a pattern on the rope something like that in Figure 14.8b, although the particles of the rope also exhibited a circular motion, rather than a simple up-and-down motion. If you moved your hand faster, you could set up a pattern something like Figure 14.6c, and you could have two friends jumping rope—one in each of the two loops!

string. The lowest frequency, corresponding to $n = 1$, is called the **fundamental frequency** f_1 and is

$$f_1 = \frac{1}{2L}\sqrt{\frac{T}{\mu}}$$

[14.9]

• *Fundamental frequency of a taut string*

Equation 14.8 shows that the frequencies of the higher modes are integral multiples of the fundamental frequency—that is, $2f_1$, $3f_1$, $4f_1$, and so on. These higher natural frequencies, together with the fundamental frequency, form a **harmonic series** and the various frequencies are called **harmonics.** The fundamental f_1 is the first harmonic; the frequency $f_2 = 2f_1$ is the second harmonic; the frequency f_n is the nth harmonic.

We can obtain the foregoing results in an alternative manner. Because the string is fixed at $x = 0$ and $x = L$, we require that the wave function $y(x, t)$ given by Equation 14.3 must be *zero* at these points for *all* times—this is a mathematical representation for the boundary conditions on the waves. Thus, the boundary conditions require that $y(0, t) = 0$ and $y(L, t) = 0$ for all values of t. Because $y = (2A \sin kx) \cos \omega t$, the first condition, $y(0, t) = 0$, is automatically satisfied because $\sin kx = 0$ at $x = 0$. To meet the second condition, $y(L, t) = 0$, we require that $\sin kL = 0$. This condition is satisfied when the angle kL equals an integral multiple of π radians or $180°$. Therefore, the allowed values of k are*

$$k_n L = n\pi \quad (n = 1, 2, 3, \ldots)$$

[14.10]

where the notation k_n indicates that each value of n has an allowed value of k. Because $k_n = 2\pi/\lambda_n$, we find that

$$\left(\frac{2\pi}{\lambda_n}\right) L = n\pi \quad \text{or} \quad \lambda_n = \frac{2L}{n}$$

which is identical to Equation 14.6.

If a stretched string is distorted to a shape that corresponds to any one of its harmonics, after being released it will vibrate at the frequency of that harmonic. If the string is plucked, bowed, or struck, however, as occurs when playing a stringed instrument, the resulting vibration will include frequencies of many modes, including the fundamental. In effect, the string "selects" a mixture of normal-mode frequencies when disturbed by a finger or a bow. The frequency of the combination is that of the fundamental because that is the rate at which the waveform repeats—the frequency associated with the string by a listener is that of the fundamental.

Plucking the bottom string on the guitar in the photograph without pressing it to the fingerboard produces the note E. By pressing the string to the fingerboard at the first fret, the length of the string is reduced and the fundamental frequency of vibration is increased. The first fret is located so that the note being played is F. *(Charles D. Winters)*

* We exclude $n = 0$ because this corresponds to the trivial case when no wave exists ($k = 0$).

The frequency of a given string on a stringed instrument can be changed either by varying the string's tension T or by changing the length L of the vibrating portion of the string. For example, the tension in the strings of guitars and violins is adjusted by a screw mechanism or by tuning pegs on the neck of the instrument. As the tension increases, the frequencies of the normal modes increase according to Equation 14.8. Once the instrument is "tuned," the player varies the frequency by moving his or her fingers along the neck, thereby changing the length of the vibrating portion of the string. As this length is reduced, the frequency increases because the normal-mode frequencies are inversely proportional to the length of the vibrating portion of the string.

Imagine that we have several strings of the same length under the same tension but varying linear mass density μ. The strings will have different wave speeds and, therefore, different fundamental frequencies. The linear mass density can be changed either by varying the diameter of the string or by wrapping extra mass around the string. Both of these possibilities can be seen on the guitar, on which the higher frequency strings vary in diameter and the lower frequency strings have additional wire wrapped around them.

Quick Quiz 14.2

The bass strings on a piano are longer and thicker than the treble strings. Why?

Example 14.3 Give Me a C Note

A middle C string on a piano has a fundamental frequency of 262 Hz, and the A note has a fundamental frequency of 440 Hz. (a) Calculate the frequencies of the next two harmonics of the C string.

Solution Because $f_1 = 262$ Hz for the C string, we can use Equation 14.7 to find the frequencies f_2 and f_3:

$$f_2 = 2f_1 = \boxed{524 \text{ Hz}}$$

$$f_3 = 3f_1 = \boxed{786 \text{ Hz}}$$

(b) If the strings for the A and C notes are assumed to have the same mass per unit length and the same length, determine the ratio of tensions in the two strings.

Solution Using Equation 14.9 for the two strings vibrating at their fundamental frequencies gives

$$f_{1A} = \frac{1}{2L}\sqrt{\frac{T_A}{\mu}}$$
$$f_{1C} = \frac{1}{2L}\sqrt{\frac{T_C}{\mu}} \longrightarrow \frac{f_{1A}}{f_{1C}} = \sqrt{\frac{T_A}{T_C}}$$

$$\frac{T_A}{T_C} = \left(\frac{f_{1A}}{f_{1C}}\right)^2 = \left(\frac{440 \text{ Hz}}{262 \text{ Hz}}\right)^2 = \boxed{2.82}$$

(c) In a real piano, the assumption we made in part (b) is only partially true. The string densities are equal, but the A string is 64% as long as the C string. What is the ratio of their tensions?

Solution We start from the same point as in part (b), but the string lengths do not cancel in the ratio:

$$f_{1A} = \frac{1}{2L_A}\sqrt{\frac{T_A}{\mu}}$$
$$f_{1C} = \frac{1}{2L_C}\sqrt{\frac{T_C}{\mu}} \longrightarrow \frac{f_{1A}}{f_{1C}} = \frac{L_C}{L_A}\sqrt{\frac{T_A}{T_C}}$$

$$\frac{T_A}{T_C} = \left(\frac{L_A}{L_C}\right)^2\left(\frac{f_{1A}}{f_{1C}}\right)^2 = (0.64)^2\left(\frac{440 \text{ Hz}}{262 \text{ Hz}}\right)^2 = \boxed{1.16}$$

EXERCISE A stretched string is 160 cm long and has a linear mass density of 0.0150 g/cm. What tension in the string will result in a second harmonic of 460 Hz?

Answer 813 N

14.5 • STANDING WAVES IN AIR COLUMNS

 See Screen 9.9

We have discussed musical instruments that use strings, which include guitars, violins, and pianos. What about instruments classified as brasses or woodwinds? These produce music using a column of air. Standing longitudinal waves can be set up in an air column, such as an organ pipe, as the result of interference between longitudinal sound waves traveling in opposite directions. Whether a node or an anti-node occurs at the end of an air column depends on whether that end is open or closed. **The closed end of an air column is a displacement node,** just as the fixed end of a vibrating string is a displacement node. Furthermore, because the pressure wave is 90° out of phase with the displacement wave (Section 13.7), the **closed end of an air column corresponds to a pressure antinode** (i.e., a point of maximum pressure variation). On the other hand, the **open end of an air column is approximately a displacement antinode and a pressure node.**

Strictly speaking, the open end of an air column is not exactly an antinode. A compression in the sound wave does not reach full expansion until it passes somewhat beyond the open end. Thus, to accurately calculate frequencies of the normal modes, an **end correction** must be added to the length of the air column at each open end. For a thin-walled tube of circular cross-section, this end correction is about $0.6R$, where R is the tube's radius. Hence, the effective length of the tube is somewhat greater than the true length L.

We can determine the modes of vibration of an air column by applying the appropriate boundary condition at the end of the column, along with the requirement that nodes and antinodes be separated by one fourth of a wavelength. We shall find that the frequency for sound waves in air columns is quantized, similar to the results found for waves on strings under boundary conditions.

The first three modes of vibration of an air column that is open at both ends are shown in Figure 14.9a. Note that the ends are displacement antinodes (approximately). In the fundamental mode, the wavelength is twice the length of the air column, and hence the frequency of the fundamental f_1 is $v/2L$. Similarly, the frequencies of the higher harmonics are $2f_1, 3f_1, \ldots$ Thus, in an air column that is open at both ends, the natural frequencies of vibration form a harmonic series; that is, the higher harmonics are integral multiples of the fundamental frequency. Because all harmonics are present, we can express the natural frequencies of vibration as

$$f_n = n\frac{v}{2L} \quad (n = 1, 2, 3, \ldots) \qquad \text{[14.11]}$$

• *Natural frequencies of an air column open at both ends*

where v is the speed of sound in air.

If an air column is closed at one end and open at the other, the closed end is a displacement node (Fig. 14.9b). In this case, the wavelength for the fundamental mode is four times the length of the column. Hence, the fundamental frequency f_1 is equal to $v/4L$, and the frequencies of the higher harmonics are equal to $3f_1$, $5f_1, \ldots$ That is, **in an air column that is closed at one end, only odd harmonics are present,** and these are

$$f_n = n\frac{v}{4L} \quad (n = 1, 3, 5, \ldots) \qquad \text{[14.12]}$$

• *Natural frequencies of an air column closed at one end and open at the other*

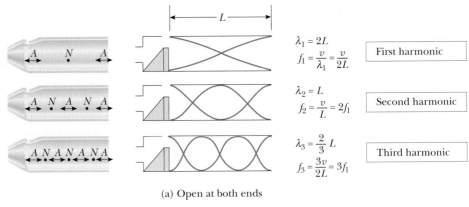

(a) Open at both ends

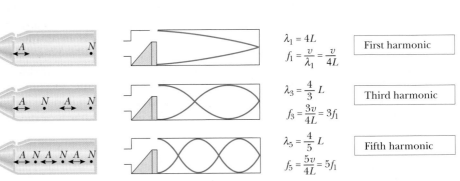

(b) Closed at one end, open at the other

Figure 14.9

Motion of elements of air in standing longitudinal waves in an air column, along with graphical representations of the displacements of the elements. (a) In an air column open at both ends, the harmonic series created consists of all integer multiples of the fundamental frequency: $f_1, 2f_1, 3f_1, \ldots$ (b) In an air column closed at one end and open at the other, the harmonic series consists of only odd-integer multiples of the fundamental frequency: $f_1, 3f_1, 5f_1, \ldots$

PITFALL PREVENTION 14.4

Sound waves in air are not transverse

Note that the standing longitudinal waves are drawn as transverse waves in Figure 14.9. This is because it is difficult to draw longitudinal displacements—they are in the same direction as the propagation. Thus, it is best to interpret the curves in Figure 14.9 as a graphical representation of the waves (our diagrams of string waves are pictorial representations), with the vertical axis representing horizontal displacement of the elements of the medium.

Standing waves in air columns are the primary sources of the sounds produced by wind instruments. In a woodwind instrument, a key is pressed, opening a hole in the side of the column. This hole defines the end of the vibrating column of air (because the hole acts as an open end—pressure can be released), so that the column is effectively shortened and the fundamental frequency rises. In a brass instrument, the length of the air column is changed by an adjustable section, as in a trombone, or by adding segments of tubing, as is done in a trumpet when a valve is pressed.

Quick Quiz 14.3

A pipe open at both ends resonates at a fundamental frequency f_{open}. When one end is closed and the pipe is again made to resonate, the fundamental frequency is f_{closed}. Which of the following expressions describes how these two resonance frequencies compare? (a) $f_{\text{closed}} = f_{\text{open}}$ (b) $f_{\text{closed}} = \frac{1}{2}f_{\text{open}}$ (c) $f_{\text{closed}} = 2f_{\text{open}}$ (d) $f_{\text{closed}} = \frac{3}{2}f_{\text{open}}$

Quick Quiz 14.4

Balboa Park in San Diego, California, has an outdoor organ. Does the fundamental frequency of a particular air column on this organ change on hot and cold days? How about on days with high and low atmospheric pressure?

THINKING PHYSICS 14.2

A bugle has no valves, keys, slides, or finger holes—how can it play a song?

Reasoning   Songs for the bugle are limited to harmonics of the fundamental frequency because the bugle has no control over frequencies by means of valves, keys, slides, or finger holes. The player obtains different notes by changing the tension in the lips as the bugle is played, in order to excite different harmonics. The normal playing range of a bugle is among the third, fourth, fifth, and sixth harmonics of the fundamental. For example, "Reveille" is played with just the three notes D (294 Hz), G (392 Hz), and B (494 Hz), and "Taps" is played with these same three notes and the D one octave above the lower D (588 Hz).

THINKING PHYSICS 14.3

If an orchestra doesn't warm up before a performance, the strings go flat and the wind instruments go sharp during the performance. Why?

Reasoning   Without warming up, all the instruments will be at room temperature at the beginning of the concert. As the wind instruments are played, they fill with warm air from the player's exhalation. The increase in temperature of the air in the instrument causes an increase in the speed of sound, which raises the resonance frequencies of the air columns. As a result, the wind instruments go sharp. The strings on the stringed instruments also increase in temperature due to the friction of rubbing with the bow. This results in thermal expansion, which causes a decrease in the tension in the strings. (We will study thermal expansion in Chapter 16.) With a decrease in tension, the wave speed on the strings drops, and the fundamental frequencies decrease. Thus, the stringed instruments go flat.

Example 14.4   Resonance in a Pipe

A pipe has a length of 1.23 m. (a) Determine the frequencies of the first three harmonics if the pipe is open at each end. Take $v = 343$ m/s as the speed of sound in air.

Solution   The first harmonic of a pipe open at both ends is

$$f_1 = \frac{v}{2L} = \frac{343 \text{ m/s}}{2(1.23 \text{ m})} = \boxed{139 \text{ Hz}}$$

Because all harmonics are possible for a pipe open at both ends, the second and third harmonics are $f_2 = 2f_1 = \boxed{278 \text{ Hz}}$ and $f_3 = 3f_1 = \boxed{417 \text{ Hz}}$.

(b) What are the three frequencies determined in part (a) if the pipe is closed at one end?

Solution   The fundamental frequency of a pipe closed at one end is

$$f_1 = \frac{v}{4L} = \frac{343 \text{ m/s}}{4(1.23 \text{ m})} = \boxed{69.7 \text{ Hz}}$$

In this case, only odd harmonics are present, and so the next two resonances have frequencies $f_3 = 3f_1 = \boxed{209 \text{ Hz}}$ and $f_5 = 5f_1 = \boxed{349 \text{ Hz}}$.

(c) For the pipe open at both ends, how many harmonics are present in the normal human hearing range (20–20 000 Hz)?

Solution   Because all harmonics are present, $f_n = nf_1$. For $f_n = 20\,000$ Hz, we have $n = 20\,000/139 = 144$, so that 144

harmonics are present in the audible range. Actually, only the first few harmonics have sufficient amplitude to be heard.

Example 14.5   Measuring the Frequency of a Tuning Fork

A simple apparatus for demonstrating resonance in a tube is described in Figure 14.10a. A long, vertical tube open at both ends is partially submerged in a beaker of water, and a vibrating tuning fork of unknown frequency is placed near the top. The length L of the air column is adjusted by moving the tube vertically. The sound waves generated by the fork are reinforced when the length of the air column corresponds to one of the resonant frequencies of the tube.

For a certain tube, the smallest value of L for which a peak occurs in the sound intensity is 9.00 cm. From this measurement, determine the frequency of the tuning fork and the value of L for the next two resonant modes.

Reasoning   Although the tube is open at the bottom end to allow the water in, the water surface acts like a rigid barrier at that end. This setup can therefore be modeled as an air column closed at one end, and the fundamental has a frequency of $v/4L$, where L is the length of the tube from the open end to the water surface (Fig. 14.10b).

Solution   Taking $v = 343$ m/s for the speed of sound in air and $L = 0.090\,0$ m, we have

$$f_1 = \frac{v}{4L} = \frac{343 \text{ m/s}}{4(0.0900 \text{ m})} = \boxed{953 \text{ Hz}}$$

From the information about the fundamental mode, we see that the wavelength is $\lambda = 4L = 0.360$ m. Because the frequency of the source is constant, the next two resonance modes (Fig. 14.10b) correspond to lengths of $3\lambda/4 = $ $\boxed{0.270 \text{ m}}$ and $5\lambda/4 = $ $\boxed{0.450 \text{ m}}$.

EXERCISE   The longest pipe on a certain organ is 4.88 m.

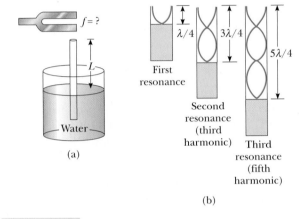

(a)

(b)

Figure 14.10

(Example 14.5) (a) Apparatus for demonstrating the resonance of sound waves in a tube closed at one end. The length L of the air column is varied by moving the tube vertically while it is partially submerged in water. (b) The first three normal modes of the system shown in part (a).

What is the fundamental frequency (at 0.0°C) if the non-driven end of the pipe is (a) closed and (b) open?

Answer   (a) 17.0 Hz   (b) 34.0 Hz

EXERCISE   An air column 2.00 m in length is open at both ends. The frequency of a certain harmonic is 410 Hz, and the frequency of the next higher harmonic is 492 Hz. Determine the speed of sound in the air column.

Answer   328 m/s

14.6 • BEATS: INTERFERENCE IN TIME

The interference phenomena that we have discussed so far involve the superposition of two or more waves with the same frequency. Because the resultant displacement of a particle in the medium in this case depends on the position of the particle, we can refer to the phenomenon as **spatial interference.** Standing waves in strings and air columns are common examples of spatial interference.

We now consider another type of interference effect, one that results from the superposition of two waves with slightly *different* frequencies. In this case, when the two waves of amplitudes A_1 and A_2 are observed at a given point, they are alternately in and out of phase. We refer to this phenomenon as **interference in time** or **temporal interference.** When the waves are in phase, the combined amplitude is $A_1 + A_2$. When they are out of phase, the combined amplitude is $|A_1 - A_2|$. The combination therefore varies between small and large amplitudes, resulting in what we call **beats.**

Although beats occur for all types of waves, they are particularly noticeable for sound waves. For example, if two tuning forks of slightly different frequencies are struck, you hear a sound of pulsating intensity.

The number of beats you hear per second, the **beat frequency,** equals the difference in frequency between the two sources. The maximum beat frequency that the human ear can detect is about 20 beats/s. When the beat frequency exceeds this value, it blends with the sounds producing the beats.

One can use beats to tune a stringed instrument, such as a piano, by beating a note against a reference tone of known frequency. The frequency of the string can then be adjusted to equal the frequency of the reference by changing the string's tension until the beats disappear—then the two frequencies are the same.

Let us look at the mathematical representation of beats. Consider two waves with equal amplitudes, traveling through a medium with slightly different frequencies, f_1 and f_2. We can represent the displacement that each wave would produce at a fixed point, which we choose as $x = 0$, as

$$y_1 = A \cos 2\pi f_1 t \qquad \text{and} \qquad y_2 = A \cos 2\pi f_2 t$$

Using the superposition principle, we find that the resultant displacement at that point is given by

$$y = y_1 + y_2 = A(\cos 2\pi f_1 t + \cos 2\pi f_2 t)$$

It is convenient to write this in a form that uses the trigonometric identity

$$\cos a + \cos b = 2 \cos \left(\frac{a - b}{2} \right) \cos \left(\frac{a + b}{2} \right)$$

Letting $a = 2\pi f_1 t$ and $b = 2\pi f_2 t$, we find that

$$y = \left[2A \cos 2\pi \left(\frac{f_1 - f_2}{2} \right) t \right] \cos 2\pi \left(\frac{f_1 + f_2}{2} \right) t \qquad \textbf{[14.13]}$$

Graphs demonstrating the individual waveforms as well as the resultant wave are shown in Figure 14.11. From the factors in Equation 14.13, we see that the resultant simple harmonic motion at a point has an effective frequency equal to the average frequency $(f_1 + f_2)/2$ and an amplitude of

$$A_{x=0} = 2A \cos 2\pi \left(\frac{f_1 - f_2}{2} \right) t \qquad \textbf{[14.14]}$$

That is, the *amplitude varies in time* with a frequency of $(f_1 - f_2)/2$. When f_1 is close to f_2, this amplitude variation is slow compared with the frequency of the individual waves, as illustrated by the envelope (broken line) of the resultant waveform in Figure 14.11b.

Note that a maximum in amplitude will be detected whenever

$$\cos 2\pi \left(\frac{f_1 - f_2}{2} \right) t = \pm 1$$

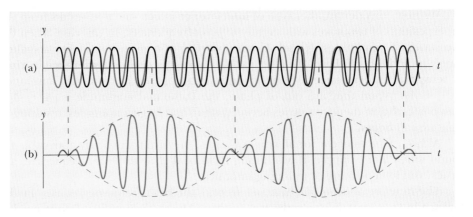

Figure 14.11

Beats are formed by the combination of two waves of slightly different frequencies. (a) The individual waves. (b) The combined wave has an amplitude (broken line) that oscillates in time.

That is, the amplitude maximizes twice in each cycle of the function on the left in the preceding expression. Thus, the number of beats per second, or the beat frequency f_b is twice the frequency of this function:

- *Beat frequency*

$$f_b = |f_1 - f_2| \tag{14.15}$$

For instance, if two tuning forks vibrate individually at frequencies of 438 Hz and 442 Hz, respectively, the resultant sound wave of the combination has a frequency of $(f_1 + f_2)/2 = 440$ Hz (the musical note A) and a beat frequency of $|f_1 - f_2| = 4$ Hz. That is, the listener hears the 440-Hz sound wave go through an intensity maximum four times every second.

Quick Quiz 14.5

You are tuning a guitar by comparing the sound of the string with that of a standard tuning fork. You notice a beat frequency of 5 Hz when both sounds are present. You tighten the guitar string and the beat frequency rises to 8 Hz. Should you continue to tighten the string or should you loosen it to match the frequency of the string to that of the tuning fork?

14.7 • NONSINUSOIDAL WAVE PATTERNS

See Screen 9.6

The sound wave patterns produced by most instruments are not sinusoidal. Some characteristic waveforms produced by a tuning fork, a flute, and a clarinet are shown in Figure 14.12. Although each instrument has its own characteristic pattern, Figure 14.12 shows that all three waveforms are periodic. A struck tuning fork produces primarily one harmonic (the fundamental), whereas the flute and clarinet produce many frequencies, which include the fundamental and various harmonics. The nonsinusoidal waveforms produced by a violin or clarinet, and the corresponding richness of musical tones, are the result of the superposition of various harmonics.

Figure 14.12

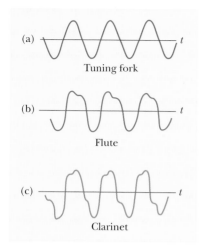

(a) Tuning fork

(b) Flute

(c) Clarinet

Figure 14.12

Waveforms of sound produced by (a) a tuning fork, (b) a flute, and (c) a clarinet, each at approximately the same frequency. (*Adapted from C. A. Culver,* Musical Acoustics, *4th ed., New York, McGraw-Hill, 1956, p. 128.*)

PITFALL PREVENTION 14.5
Pitch versus frequency

A very common mistake made in speech when talking about sound is to use the term *pitch* when one means *frequency*. Frequency is the physical measurement of the number of oscillations per second, as we have defined. Pitch is a psychological reaction of humans to sound that enables a human to place the sound on a scale from high to low, or from treble to bass. Thus, frequency is the stimulus and pitch is the response. Although pitch is related mostly (but not completely) to frequency, they are not the same. A phrase such as "the pitch of the sound" is incorrect because pitch is not a physical property of the sound.

This phenomenon is in contrast to a percussive musical instrument, such as the drum, in which the combination of frequencies do not form a harmonic series. When frequencies that are integer multiples of a fundamental frequency are combined, the result is a *musical* sound. A listener can assign a pitch to the sound, based on the fundamental frequency. Pitch is a psychological reaction to a sound that allows the listener to place the sound on a scale of low to high (bass to treble). Combinations of frequencies that are not integer multiples of a fundamental result in a *noise,* rather than a musical sound. It is much harder for a listener to assign a pitch to a noise than for a musical sound.

Analysis of nonsinusoidal waveforms appears at first sight to be a formidable task. If the waveform is periodic, however, it can be represented with arbitrary precision by the combination of a sufficiently large number of sinusoidal waves that form a harmonic series. In fact, one can represent any periodic function or any function over a finite interval as a series of sine and cosine terms by using a mathematical technique based on *Fourier's theorem.* The corresponding sum of terms that represents the periodic waveform is called a **Fourier series.**

Let $y(t)$ be any function that is periodic in time, with a period of T, so that $y(t + T) = y(t)$. **Fourier's theorem** states that this function can be written

$$y(t) = \sum_n (A_n \sin 2\pi f_n t + B_n \cos 2\pi f_n t) \qquad \text{[14.16]}$$

• *Fourier's theorem*

where the lowest frequency is $f_1 = 1/T$. The higher frequencies are integral multiples of the fundamental, and so $f_n = nf_1$. The coefficients A_n and B_n represent the amplitudes of the various harmonics.

Figure 14.13 represents a harmonic analysis of the waveforms shown in Figure 14.12. Note the variation of relative intensity with harmonic content for the flute and clarinet. In general, any musical sound contains components that are members of a harmonic series with varying relative intensities.

We have discussed the *analysis* of a wave pattern using Fourier's theorem. The analysis involves determining the coefficients of the harmonics in Equation 14.16 from a knowledge of the wave pattern. We can also perform the reverse process, *Fourier synthesis.* In this process, the various harmonics are added together to form a resultant wave pattern. As an example of Fourier synthesis, consider the building

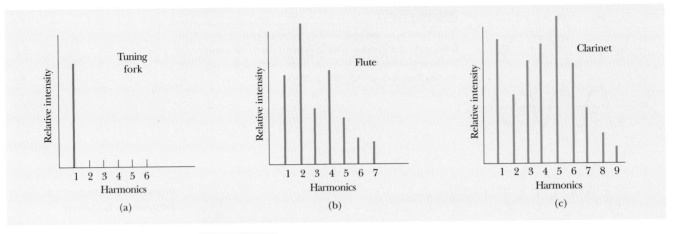

Figure 14.13

Harmonics of the waveforms shown in Figure 14.12. Note the variations in intensity of the various harmonics. *(Adapted from C. A. Culver,* Musical Acoustics, *4th ed., New York, McGraw-Hill, 1956.)*

of a square wave, as shown in Figure 14.14. The symmetry of the square wave results in only odd multiples of the fundamental combining in the synthesis. In Figure 14.14a, the orange curve shows the combination of f and $3f$. In Figure 14.14b, we have added $5f$ to the combination and obtained the green curve. Notice how the general shape of the square wave is approximated, even though the upper and lower portions are not as flat as they should be.

Figure 14.14c shows the result of adding odd frequencies up to $9f$—the purple curve. This approximation to the square wave (black curve) is better than in parts a and b. To approximate the square wave as closely as possible, we would need to add all odd multiples of the fundamental frequency up to infinite frequency.

Using modern technology, one can produce a variety of musical sounds electronically by mixing harmonics with varying amplitudes. Electronic music synthesizers that do this are now widely used.

The physical mixture of harmonics can be described as the **spectrum** of the sound, with the spectrum displayed in a graphical representation such as Figure 14.13. The psychological reaction to changes in the spectrum of a sound is the detection of a change in the **timbre** or the **quality** of the sound. If a clarinet and a trumpet are both playing the same note, then you will assign the same pitch to the two notes. Yet, if only one of the instruments then plays the note, you will likely be able to tell which instrument is playing. The sounds you hear from the two instruments differ in timbre, due to a different physical mixture of harmonics. For example, the timbre due to the sound of a trumpet is different from that of a clarinet. You have probably developed words to describe timbres of various instruments, such as "brassy," "mellow," "tinny," and the like.

Fourier's theorem allows us to understand the excitation process of musical instruments. In a stringed instrument that is plucked, such as a guitar, the string is pulled aside and released. After release, the string oscillates almost freely; a small damping causes the amplitude to eventually decay to zero. The mixture of harmonic frequencies depends on the length of the string, its linear mass density, and the plucking point.

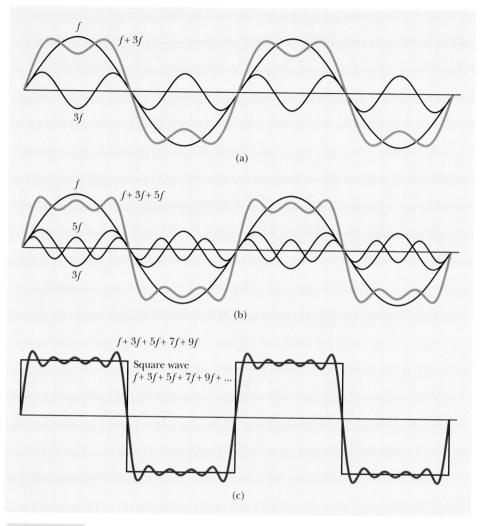

Figure 14.14

Fourier synthesis of a square wave, which is represented by the sum of odd multiples of the first harmonic, which has frequency f. (a) Waves of frequency f and $3f$ are added. (b) One more odd frequency of $5f$ is added. (c) The synthesis curve approaches the square wave when odd frequencies up to $9f$ are combined.

On the other hand, a bowed string instrument, such as a violin, or a wind instrument, is a forced oscillator. In the case of the violin, the alternate sticking and slipping of the bow on the string provides the periodic driving force. In the case of a wind instrument, the vibration of a reed (in a woodwind), of the lips of the player (in a brass), or the blowing of air across an edge (as in a flute) provides the periodic driving force. According to Fourier's theorem, these periodic driving forces contain a mixture of harmonic frequencies. The violin string or the air column in a wind instrument is therefore driven with a wide variety of frequencies. The frequency actually played is determined by *resonance*, which we studied in Chapter 12. The maximum response of the instrument will be to those frequencies that match or are very close to the harmonic frequencies of the instrument. The spectrum of the instrument, then, depends heavily on the strengths of the various harmonics in the initial periodic driving force.

APPLICATION

Effect of helium

THINKING PHYSICS 14.4

A professor performs a demonstration in which he breathes a small amount of helium and then speaks with a comical voice. One student explains, "the velocity of sound in helium is higher than in air, so the fundamental frequency of the standing waves in the mouth is increased." Another student says, "No, the fundamental frequency is determined by the vocal folds and cannot be changed. Only the *quality* of the voice has changed." Which student is correct?

Reasoning   The second student is correct. The fundamental frequency of the complex tone from the voice is determined by the vibration of the vocal folds and is not changed by substituting a different gas in the mouth. The introduction of the helium into the mouth results in higher harmonics being excited more than in the normal voice, but the fundamental frequency of the voice is the same—only the spectrum has changed. The unusual inclusion of the higher frequency harmonics results in a common description of this effect as a "high-pitched" voice, but this is incorrect.

context connection 🌍
14.8 • BUILDING ON ANTINODES

As an example of the application of standing waves to earthquakes, we consider the effects of standing waves in *sedimentary basins*. Many of the world's major cities are built on sedimentary basins, which are topographic depressions that over geologic time have filled with sediment. These areas provide large expanses of flat land, often surrounded by attractive mountains, as in the Los Angeles basin. Flat land for building and attractive scenery attracted early settlers and led to today's cities.

Destruction from an earthquake can increase dramatically if the natural frequencies of buildings or other structures coincide with the resonant frequencies of the underlying basin. These resonant frequencies are associated with three-dimensional standing waves, formed from seismic waves reflecting from the boundaries of the basin.

To understand these standing waves, let us assume a simple model of a basin shaped like a half-ellipsoid, similar to an egg sliced in half along its long diameter. Four possible patterns of vertical ground motion in such a basin are shown in the pictorial representation in Figure 14.15. The long axis of the ellipsoid is designated by *x* and the short axis is *y*. Each of the four patterns is shown as a pair, separated in time by half of a period. In Figure 14.15a, the entire surface of the ground moves up and down except at a nodal curve running around the edge of the basin. The drawing on the left shows the ground level raised above its equilibrium position. Half a period later, the drawing on the right shows the ground forming a depression whose surface is below that of the flat surface.

In Figures 14.15b and 14.15c, half the basin lies above and half lies below the equilibrium position, and each half oscillates up and down on either side of a nodal line. The nodal line is along the *y* axis in Figure 14.15b and along the *x* axis in Figure 14.15c. In Figure 14.15d, nodal lines occur along both the *x* and *y* axes and the surface oscillates in four segments, with two above the equilibrium position at any time and the other two below.

The standing wave patterns in a basin arise from seismic waves traveling vertically between the top surface of the ground and the bottom of the basin, and hori-

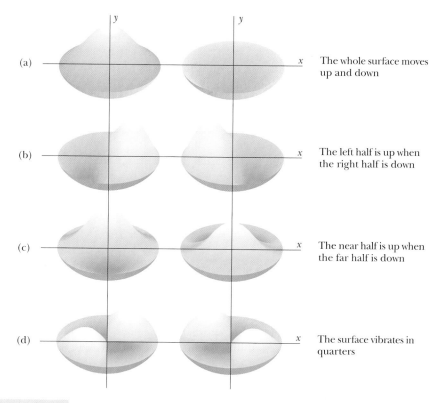

(a) The whole surface moves up and down

(b) The left half is up when the right half is down

(c) The near half is up when the far half is down

(d) The surface vibrates in quarters

Figure 14.15

Standing waves in a basin shaped like a half-ellipsoid. Displacements from equilibrium are exaggerated in order to clearly show the nature of the oscillations. Actual displacements are much smaller than those shown. (a) In the fundamental mode of oscillation, the only nodal line is the rim of the basin and the entire surface moves up and down together. (b) The surface oscillates in two halves, with a nodal line along the *y* axis. (c) The surface oscillates in two halves, with a nodal line along the *x* axis. (d) The surface oscillates in four quarters, with nodal lines along both the *x* and *y* axes.

zontally between the boundaries of the basin. For structures built on sedimentary basins, the degree of seismic risk will depend on the standing wave modes excited by the interference of seismic waves trapped in the basin. It is clear that structures built on regions of maximum ground motion (i.e., the antinodes) will suffer maximum shaking, whereas structures residing near nodes will experience relatively mild ground motion. These considerations appear to have played an important role in the selective destruction that occurred in Mexico City in the Michoacán earthquake in 1985 and, more recently, in the 1989 Loma Prieta earthquake, which caused the collapse of a section of the Nimitz Freeway in Oakland.

A similar effect occurs in bounded bodies of water, such as harbors and bays. A standing wave pattern established in such a body of water is called a **seiche.** This can result in variations in the water level that exhibit a period of several minutes, superposed on the longer period tidal variations. Seiches can be caused by earthquakes, tsunamis, winds, and weather disturbances. You can create a seiche in your bathtub by sliding back and forth at just the right frequency such that the water sloshes back and forth at such a large amplitude that much of it spills out onto the floor.

During the Northridge earthquake of 1994, swimming pools throughout southern California overflowed due to seiches set up by the shaking of the ground.

WEB

For an animation of a seiche, with ability to change parameters, visit **www.coastal.udel.edu/faculty/ rad/seiche.html**

WEB

For information on the Northridge earthquake and related links, visit **www.earthquake.org/ northridge.html**

In a more dramatic example, seismic waves from the 1964 Alaska earthquake caused severe seiches in the bays and bayous of Louisiana, some causing the water level to shift by 2 m.

We have now considered the role of standing waves in the damage caused by an earthquake. In the Context Conclusion, we will gather together the principles of vibrations and waves that we have learned to respond more fully to the central question of this Context.

SUMMARY

The **principle of superposition** states that if two or more traveling waves are moving through a medium and combine at a given point, the resultant displacement of the medium at that point is the sum of the displacements of the individual waves.

When two waves with equal amplitudes and frequencies superpose, the resultant wave has an amplitude that depends on the phase angle ϕ between the two waves. **Constructive interference** occurs when the two waves are *in phase* everywhere, corresponding to $\phi = 0, 2\pi, 4\pi, \ldots$. **Destructive interference** occurs when the two waves are 180° out of phase everywhere, corresponding to $\phi = \pi, 3\pi, 5\pi, \ldots$.

Standing waves are formed from the superposition of two sinusoidal waves that have the same frequency, amplitude, and wavelength but are traveling in *opposite* directions. The resultant standing wave is described by the wave function

$$y = (2A \sin kx) \cos \omega t \qquad [14.3]$$

The maximum amplitude points (called **antinodes**) are separated by a distance $\lambda/2$. Halfway between antinodes are points of zero amplitude (called **nodes**).

One can set up standing waves with quantized frequencies in such systems as stretched strings and air columns. The natural frequencies of vibration of a stretched string of length L, fixed at both ends, are

$$f_n = \frac{n}{2L} \sqrt{\frac{T}{\mu}} \quad (n = 1, 2, 3, \ldots) \qquad [14.8]$$

where T is the tension in the string and μ is its mass per unit length. The natural frequencies of vibration form a **harmonic series,** that is, $f_1, 2f_1, 3f_1, \ldots$

The standing wave patterns for longitudinal waves in an air column depend on whether the ends of the column are open or closed. If the column is open at both ends, the natural frequencies of vibration form a harmonic series. If one end is closed, only odd harmonics of the fundamental are present.

The phenomenon of **beats** occurs as a result of the superposition of two traveling waves of slightly different frequencies. For sound waves at a given point, one hears an alternation in sound intensity with time.

Any periodic waveform can be represented by the combination of sinusoidal waves that form a harmonic series. The process is based on **Fourier's theorem.**

QUESTIONS

1. When two waves interfere constructively or destructively, is there any gain or loss in energy? Explain.

2. Does the phenomenon of wave interference apply only to sinusoidal waves?

3. If you have a series of identical glass bottles, with varying amounts of water in them, you can play musical notes by either striking the bottles with a spoon, or blowing across the open tops of the bottles. When striking the bottles, the frequency of the note decreases as the water level rises. When blowing on the bottles, the frequency of the note increases as the water level rises. Why is the behavior of the frequency different in these two cases?

4. Some singers claim to be able to shatter a wine glass by maintaining a certain vocal pitch over a period of several seconds (Fig. Q14.4). What mechanism causes the glass to break? (The glass must be very clean in order to break.)

5. What limits the amplitude of motion of a real vibrating system that is driven at one of its resonant frequencies?

6. Guitarists sometimes play a "harmonic," by lightly touching a string at the exact center and plucking the string. The result is a clear note one octave higher than the fundamental of the string, even though the string is not pressed to the fingerboard. Why does this happen?

7. An archer shoots an arrow from a bow. Does the string of the bow exhibit standing waves after the arrow leaves? If so, and if the bow is perfectly symmetric, so that the arrow leaves from the center of the string, what harmonics are excited?

8. Explain why all harmonics are present in an organ pipe open at both ends, but only the odd harmonics are present in a pipe closed at one end.

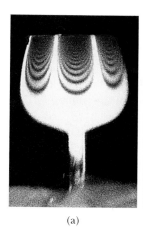 

(a)   (b)

Figure Q14.4   (a) Standing wave pattern in a vibrating wine glass. *(Courtesy of Professor Thomas D. Rossing, Northern Illinois University)* (b) The glass shatters if the amplitude of the vibrations becomes too large. In this photograph, the vibrations are caused by the amplified sound of a human voice. *(©1992 Ben Rose/The Image Bank)*

9. You have a standard tuning fork whose frequency is known and a second tuning fork whose frequency you don't know. When you play them together, you hear a beat frequency of 4 Hz. You know that the frequency of the mystery tuning fork is different from that of the standard fork by 4 Hz, but you can't tell whether it is higher or lower. What could you do to determine this? (*Hint:* You are chewing gum at the time that you perform these measurements.)

10. An airplane mechanic notices that the sound from a twin-engine aircraft rapidly varies in loudness when both engines are running. What could be causing this variation from loud to soft?

11. Why does a vibrating guitar string sound louder when placed on the instrument than it would if allowed to vibrate in the air while off the instrument?

12. When the base of a vibrating tuning fork is placed against a chalk board, the sound becomes louder. How does this affect the length of time for which the fork vibrates? Does this agree with conservation of energy?

13. Stereo speakers are supposed to be "phased" when set up. That is, the waves emitted from them should be in phase with each other. What would the sound be like along the center line of the speakers if one speaker were wired up backwards, that is, out of phase?

14. To keep animals away from their cars, some people mount short, thin pipes on the fenders. The pipes give out a high-pitched wail when the cars are moving. How do they create the sound?

15. If you wet your fingers and lightly run them around the rim of a fine wine glass, a high-pitched sound is heard. Why? How could you produce various musical notes with a set of wine glasses?

16. When a bell is rung, standing waves are set up around the bell's circumference. What boundary conditions must be satisfied by the resonant wavelengths? How does a crack in the bell, such as in the Liberty Bell, affect the satisfying of the boundary conditions and the sound emanating from the bell?

17. Despite a reasonably steady hand, a certain person often spills his coffee when carrying it to his seat. Discuss resonance as a possible cause of this difficulty, and devise a means for solving the problem.

18. Explain why your voice seems to sound better than usual when you sing in the shower.

19. What is the purpose of the slide on a trombone or the valves on a trumpet?

PROBLEMS

$1, 2, 3$ = straightforward, intermediate, challenging   □ = full solution available in the *Student Solutions Manual and Study Guide*

web = solution posted at **http://www.harcourtcollege.com/physics/**   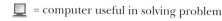 = computer useful in solving problem

 = Interactive Physics   ▨ = paired numerical/symbolic problems    = life science application

Section 14.1   The Principle of Superposition

1. Two waves on one string are described by the wave functions

$$y_1 = 3.0 \cos(4.0x - 1.6t)$$

and

$$y_2 = 4.0 \sin(5.0x - 2.0t)$$

where y and x are in centimeters and t is in seconds. Find the superposition of the waves $y_1 + y_2$ at the points (a) $x = 1.00$, $t = 1.00$, (b) $x = 1.00$, $t = 0.500$, and (c) $x = 0.500$, $t = 0$. (Remember that the arguments of the trigonometric functions are in radians.)

2. Two wave pulses A and B are moving in opposite directions along a taut string with a speed of 2.00 cm/s. The amplitude of A is twice the amplitude of B. The pulses are shown in Figure P14.2 at $t = 0$. Sketch the shape of the string at $t = 1.0, 1.5, 2.0, 2.5$, and 3.0 s.

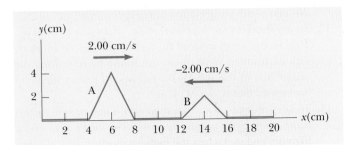

Figure P14.2

3. Two pulses traveling on the same string are described by

$$y_1 = \frac{5}{(3x - 4t)^2 + 2}$$

and

$$y_2 = \frac{-5}{(3x + 4t - 6)^2 + 2}$$

(a) In which direction does each pulse travel? (b) At what time do the two cancel everywhere? (c) At what point do the two waves always cancel?

Section 14.2 Interference of Waves

4. Two waves are traveling in the same direction along a stretched string. The waves are 90.0° out of phase. Each wave has an amplitude of 4.00 cm. Find the amplitude of the resultant wave.

5. Two sinusoidal waves are described by the wave functions

$$y_1 = (5.00 \text{ m}) \sin[\pi(4.00x - 1\,200t)]$$

and

$$y_2 = (5.00 \text{ m}) \sin[\pi(4.00x - 1\,200t - 0.250)]$$

where x, y_1, and y_2 are in meters and t is in seconds. (a) What is the amplitude of the resultant wave? (b) What is the frequency of the resultant wave?

6. Two identical sinusoidal waves with wavelengths of 3.00 m travel in the same direction at a speed of 2.00 m/s. The second wave originates from the same point as the first, but at a later time. Determine the minimum possible time interval between the starting moments of the two waves if the amplitude of the resultant wave is the same as that of each of the two initial waves.

7. A tuning fork generates sound waves with a frequency of 246 Hz. The waves travel in opposite directions along a hallway, are reflected by end walls, and return. The hallway is 47.0 m long, and the tuning fork is located 14.0 m from one end. What is the phase difference between the reflected waves when they meet at the tuning fork? The speed of sound in air is 343 m/s.

8. Two loudspeakers are placed on a wall 2.00 m apart. A listener stands 3.00 m from the wall directly in front of one of the speakers. A single oscillator is driving the speakers in phase at a frequency of 300 Hz. (a) What is the phase difference between the two waves when they reach the observer? (b) What is the frequency closest to 300 Hz to which the oscillator may be adjusted such that the observer hears minimal sound?

9. Two speakers are driven in phase by the same oscillator of frequency f. They are located a distance d from each other on a vertical pole. A man walks straight toward the lower speaker in a direction perpendicular to the pole, as shown in Figure P14.9. (a) How many times will he hear a minimum in sound intensity, and (b) how far is he from the pole at these moments? Let v represent the speed of sound and assume that the ground does not reflect sound.

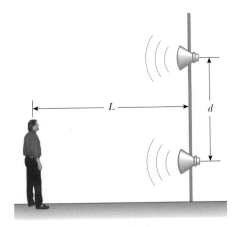

Figure P14.9

Section 14.3 Standing Waves

10. Two sinusoidal waves traveling in opposite directions interfere to produce a standing wave with the wave function

$$y = (1.50 \text{ m}) \sin(0.400x) \cos(200t)$$

where x is in meters and t is in seconds. Determine the wavelength, frequency, and speed of the interfering waves.

11. Two speakers are driven in phase by a common oscillator at 800 Hz and face each other at a distance of 1.25 m. Locate the points along the line between the speakers where

minima of the sound pressure amplitude would be expected. (Use $v = 343$ m/s.)

12. Two waves that set up a standing wave in a long string are given by the wave functions

$$y_1 = A \sin(kx - \omega t + \phi)$$

and

$$y_2 = A \sin(kx + \omega t)$$

Show (a) that the addition of the arbitrary phase constant changes only the position of the nodes, and in particular (b) that the distance between nodes is still one-half the wavelength.

13. Two sinusoidal waves combining in a medium are described by the wave functions

$$y_1 = (3.0 \text{ cm}) \sin \pi(x + 0.60t)$$

and

$$y_2 = (3.0 \text{ cm}) \sin \pi(x - 0.60t)$$

where x is in centimeters and t is in seconds. Determine the *maximum* displacement of the motion at (a) $x = 0.250$ cm, (b) $x = 0.500$ cm, and (c) $x = 1.50$ cm. (d) Find the three smallest values of x corresponding to antinodes.

14. Verify by direct substitution that the wave function for a standing wave given in Equation 14.3,

$$y = (2A \sin kx) \cos \omega t$$

is a solution of the general linear wave equation, Equation 13.19:

$$\frac{\partial^2 y}{\partial x^2} = \frac{1}{v^2} \frac{\partial^2 y}{\partial t^2}$$

Section 14.4   Standing Waves in Strings

15. A student wants to establish a standing wave on a wire that is 1.80 m long and clamped at both ends. The wave speed is 540 m/s. What is the minimum frequency the student should apply to set up a standing wave?

16. One end of a string is attached to a wall. The other end is draped over a small pulley 5.00 m away from the wall and is attached to a hanging mass of 4.00 kg. If the string is plucked, what is the fundamental frequency of vibration? The portion of the string that vibrates transversely has a mass of 8.00 g.

17. Find the fundamental frequency and the next three frequencies that could cause standing wave patterns on a 30.0-m string that has a mass per length of 9.00×10^{-3} kg/m and is stretched to a tension of 20.0 N.

18. A vibrating string with a uniform linear mass density exhibits a single-loop standing wave pattern with a frequency of 800 Hz. (a) If the tension in the string is changed to re-

duce the fundamental frequency to 500 Hz, determine the ratio of the new tension to the old. (b) Alternatively, if the original tension in the string is increased by a factor of 4, determine the new fundamental frequency.

19. A standing wave is established in a 120-cm string fixed at both ends. The string vibrates in four segments when driven at 120 Hz. (a) Determine the wavelength of the wave. (b) What is the fundamental frequency of the string?

20. In the arrangement shown in Figure P14.20, an object can be hung from a string (with linear mass density $\mu = 0.002\ 00$ kg/m) that passes over a light pulley. The string is connected to a vibrator (of constant frequency f), and the length of the string between point P and the pulley is $L = 2.00$ m. When the mass m is either 16.0 kg or 25.0 kg, standing waves are observed; however, no standing waves are observed with any mass between these values. (a) What is the frequency of the vibrator? (*Hint:* The greater the tension in the string, the smaller the number of nodes in the standing wave.) (b) What is the largest mass for which standing waves could be observed?

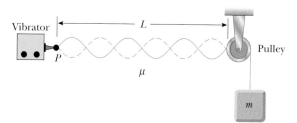

Figure P14.20

21. A cello A-string vibrates in its first normal mode with a frequency of 220 vibrations/s. The vibrating segment is 70.0 cm long and has a mass of 1.20 g. (a) Find the tension in the string. (b) Determine the frequency of vibration when the string vibrates in three segments.

22. A violin string has a length of 0.350 m and is tuned to concert G, with $f_G = 392$ Hz. Where must the violinist place her finger to play concert A, with $f_A = 440$ Hz? If this position is to remain correct to one-half the width of a finger (i.e., to within 0.600 cm), what is the maximum allowable percentage change in the string tension?

23. A standing wave pattern is observed in a thin wire with a length of 3.00 m. The equation of the wave is

$$y = (0.002 \text{ m}) \sin(\pi x) \cos(100\pi t)$$

where x is in meters and t is in seconds. (a) How many loops does this pattern exhibit? (b) What is the fundamental frequency of vibration of the wire? (c) If the original frequency is held constant and the tension in the wire is increased by a factor of 9, how many loops are present in the new pattern?

Section 14.5   Standing Waves in Air Columns

Assume that the speed of sound in air is 343 m/s unless otherwise stated.

24. A glass tube (open at both ends) of length L is positioned near an audio speaker of frequency $f = 680$ Hz. For what values of L will the tube resonate with the speaker?

25. Calculate the length of a pipe that has a fundamental frequency of 240 Hz if the pipe is (a) closed at one end and (b) open at both ends.

26. The overall length of a piccolo is 32.0 cm. The resonating air column vibrates in the same way as a pipe open at both ends. (a) Find the frequency of the lowest note that a piccolo can play, assuming that the speed of sound in air is 340 m/s. (b) Opening holes in the side effectively shortens the length of the resonant column. If the highest note a piccolo can sound is 4 000 Hz, find the distance between adjacent antinodes for this mode of vibration.

27. The fundamental frequency of an open organ pipe corresponds to middle C (261.6 Hz on the chromatic musical scale). The third resonance of a closed organ pipe has the same frequency. What are the lengths of the two pipes?

28. Do not stick anything into your ear! Estimate the length of your ear canal, from its opening at the external ear to the eardrum. If you regard the canal as a tube that is open at one end and closed at the other, at approximately what fundamental frequency would you expect your hearing to be most sensitive? Explain why you can hear especially soft sounds just around this frequency.

29. A pipe that is open at both ends has a fundamental frequency of 300 Hz when the speed of sound in air is 333 m/s. (a) What is the length of the pipe? (b) What is the frequency of the second harmonic when the temperature of the air is increased so that the speed of sound in the pipe is 344 m/s?

30. As shown in Figure P14.30, water is pumped into a tall vertical cylinder at a volume flow rate R. The radius of the cylinder is r, and at the open top of the cylinder a tuning fork vibrates with a frequency f. As the water rises, how much time elapses between successive resonances?

31. A shower stall measures 86.0 cm × 86.0 cm × 210 cm. If you were singing in this shower, which frequencies would sound the richest (because of resonance)? Assume that the stall acts as a pipe closed at both ends, with nodes at opposite sides. Assume that the voices of various singers range from 130 Hz to 2 000 Hz. Let the speed of sound in the hot shower stall be 355 m/s.

32. A tuning fork with a frequency of 512 Hz is placed near the top of the tube shown in Figure 14.10. The water level is lowered so that the length L slowly increases from an initial value of 20.0 cm. Determine the next two values of L that correspond to resonant modes.

Figure P14.30

33. A glass tube is open at one end and closed at the other by a movable piston. The tube is filled with air warmer than that at room temperature, and a 384-Hz tuning fork is held at the open end. Resonance is heard when the piston is 22.8 cm from the open end and again when it is 68.3 cm from the open end. (a) What speed of sound is implied by these data? (b) How far from the open end will the piston be when the next resonance is heard?

34. A student uses an audio oscillator of adjustable frequency to measure the depth of a water well. Two successive resonances are heard at 51.5 Hz and 60.0 Hz. How deep is the well?

Section 14.6   Beats: Interference in Time

35. In certain ranges of a piano keyboard, more than one string is tuned to the same note to provide extra loudness. For example, the note at 110 Hz has two strings at this frequency. If one string slips from its normal tension of 600 N to 540 N, what beat frequency is heard when the hammer strikes the two strings simultaneously?

36. While attempting to tune the note C at 523 Hz, a piano tuner hears 2 beats/s between a reference oscillator and the string. (a) What are the possible frequencies of the string? (b) When she tightens the string slightly, she hears 3 beats/s. What is the frequency of the string now? (c) By what percentage should the piano tuner now change the tension in the string to bring it into tune?

37. A student holds a tuning fork oscillating at 256 Hz. He walks toward a wall at a constant speed of 1.33 m/s. (a) What beat frequency does he observe between the tuning fork and its echo? (b) How fast must he walk away from the wall to observe a beat frequency of 5.00 Hz?

Section 14.7 Nonsinusoidal Wave Patterns

38. Suppose that a flutist plays a 523-Hz C note with first harmonic displacement amplitude $A_1 = 100$ nm. From Figure 14.13b read, by proportion, the displacement amplitudes of harmonics 2 through 7. Take these as the values A_2 through A_7 in the Fourier analysis of the sound, and assume that $B_1 = B_2 = \cdots = B_7 = 0$. Construct a graph of the waveform of the sound. Your waveform will not look exactly like the flute waveform in Figure 14.12b because you simplify by ignoring cosine terms; nevertheless, it produces the same sensation to human hearing.

Section 14.8 Context Connection — Building on Antinodes

39. An earthquake can produce a seiche in a lake, in which the water sloshes back and forth from end to end with remarkably large amplitude and long period. Consider a seiche produced in a rectangular farm pond, as diagrammed in the cross-sectional view of Figure P14.39. (The figure is not drawn to scale.) Suppose that the pond is 9.15 m long and of uniform width and depth. You measure that a wave pulse produced at one end reaches the other end in 2.50 s. (a) What is the wave speed? (b) To produce the seiche, you have several people stand on the bank at one end and paddle together with snow shovels, moving them in simple harmonic motion. What should be the frequency of this motion?

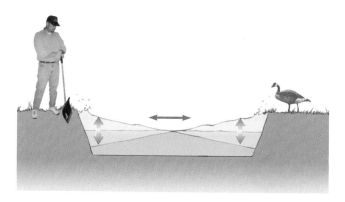

Figure P14.39

40. The Bay of Fundy, Nova Scotia, has the highest tides in the world. Assume that in midocean and at the mouth of the bay, the Moon's gravity gradient and the Earth's rotation make the water surface oscillate with an amplitude of a few centimeters and a period of 12 h 24 min. At the head of the bay, the amplitude is several meters. Argue for or against the proposition that the tide is amplified by standing-wave resonance. Assume the bay has a length of 210 km and a uniform depth of 36.1 m. The speed of long-wavelength water waves is given by $\sqrt{gd}$, where d is the water's depth.

Additional Problems

41. Two identical speakers 10.0 m apart are driven by the same oscillator with a frequency of $f = 21.5$ Hz (Fig. P14.41). (a) Explain why a receiver at point A records a minimum in sound intensity from the two speakers. (b) If the receiver is moved in the plane of the speakers, what path should it take so that the intensity remains at a minimum? That is, determine the relationship between x and y (the coordinates of the receiver) that causes the receiver to record a minimum in sound intensity. Take the speed of sound to be 343 m/s.

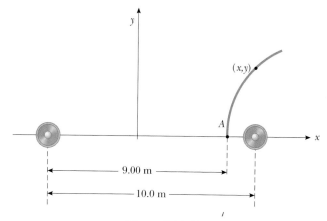

Figure P14.41

42. A major chord on the physical pitch musical scale has frequencies in the ratios 4:5:6:8. A set of pipes, closed at one end, must be cut so that when they are sounded in their first normal mode, they will produce a major chord. (a) What is the ratio of the lengths of the pipes? (b) What are the lengths of the pipes needed if the lowest frequency of the chord is 256 Hz? (c) What are the frequencies of the notes in this chord?

43. Standing-wave vibrations are set up in a crystal goblet with four nodes and four antinodes equally spaced around the 20.0-cm circumference of its rim. If transverse waves move around the glass at 900 m/s, an opera singer would have to produce a high harmonic with what frequency to shatter the glass with a resonant vibration?

44. When a metal pipe is cut into two pieces, the lowest resonance frequency for one piece is 256 Hz and for the other is 440 Hz. (a) What resonant frequency would have been produced by the original length of pipe? (b) How long was the original pipe?

45. On a marimba (Fig. P14.45), the wooden bar that sounds a tone when struck vibrates in a transverse standing wave

Figure P14.45   Marimba players in Mexico City. *(Murray Greenberg)*

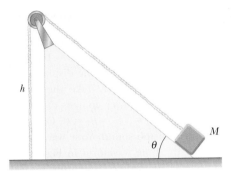

Figure P14.49

having three antinodes and two nodes. The lowest frequency note is 87.0 Hz, produced by a bar 40.0 cm long. (a) Find the speed of transverse waves on the bar. (b) The loudness of the emitted sound is enhanced by a resonant pipe suspended vertically below the center of the bar. If the pipe is open at the top end only and the speed of sound in air is 340 m/s, what is the length of the pipe required to resonate with the bar in part (a)?

46. Two train whistles have identical frequencies of 180 Hz. When one train is at rest in the station and the other is moving nearby, a commuter standing on the station platform hears beats with a frequency of 2.00 beats/s. What are the two possible speeds and directions that the moving train can have?

47. A string fixed at both ends and having a mass of 4.80 g, a length of 2.00 m, and a tension of 48.0 N vibrates in its second ($n = 2$) normal mode. What is the wavelength in air of the sound emitted by this vibrating string?

48. A string 0.400 m long has a mass per unit length of 9.00×10^{-3} kg/m. What must be the tension in the string if its second harmonic has the same frequency as the second resonance mode of a 1.75-m-long pipe open at one end?

49. **Review Problem.** For the arrangement shown in Figure P14.49, $\theta = 30.0°$, the inclined plane and the small pulley are frictionless, the string supports the object of mass M at the bottom of the plane, and the string has a mass m that is small relative to M. The system is in equilibrium and the vertical part of the string has a length h. Standing waves are set up in the vertical section of the string. (a) Find the tension in the string. (b) Model the shape of the string as one leg and the hypotenuse of a right triangle. Find the whole length of the string. (c) Find the mass per unit

length of the string. (d) Find the speed of waves on the string. (e) Find the lowest frequency for a standing wave. (f) Find the period of the standing wave having three nodes. (g) Find the wavelength of the standing wave having three nodes. (h) Find the frequency of the beats resulting from the interference of the sound wave of lowest frequency generated by the string with another sound wave having a frequency that is 2.00% greater.

50. Two wires are welded together end to end. They are made of the same material, but the diameter of one is twice that of the other. They are subjected to a tension of 4.60 N. The thin wire has a length of 40.0 cm and a linear mass density of 2.00 g/m. The combination is fixed at both ends and vibrated in such a way that two antinodes are present, with the node between them being exactly at the location of the weld. (a) What is the frequency of vibration? (b) How long is the thick wire?

51. A standing wave is set up in a string of variable length and tension by a vibrator of variable frequency. Both ends of the string are fixed. When the vibrator has a frequency f, in a string of length L and under tension T, n antinodes are set up in the string. (a) If the length of the string is doubled, by what factor should the frequency be changed so that the same number of antinodes is produced? (b) If the frequency and length are held constant, what tension will produce $n + 1$ antinodes? (c) If the frequency is tripled and the length of the string is halved, by what factor should the tension be changed so that twice as many antinodes are produced?

52. Two identical strings, each fixed at both ends, are arranged near each other. If string A starts oscillating in its first normal mode, it is observed that string B will begin vibrating in its third ($n = 3$) normal mode. Determine the ratio of the tension of string B to the tension of string A.

53. If two adjacent natural frequencies of an organ pipe are determined to be 550 Hz and 650 Hz, calculate the fundamental frequency and length of this pipe. (Use $v = 340$ m/s.)

54. A 0.010 0-kg wire, 2.00 m long, is fixed at both ends and vibrates in its fundamental mode under a tension of 200 N. When a vibrating tuning fork is placed near the wire, a beat frequency of 5.00 Hz is heard. (a) What could be the frequency of the tuning fork? (b) What should the tension in the wire be to make the beats disappear?

55. Two waves are described by the wave functions

$$y_1(x, t) = 5.0 \sin(2.0x - 10t)$$

and

$$y_2(x, t) = 10 \cos(2.0x - 10t)$$

where x is in meters and t is in seconds. Show that the wave resulting from their superposition is sinusoidal. Determine the amplitude and phase of this sinusoidal wave.

56. The wave function for a standing wave is given in Equation 14.3 as $y = 2A \sin kx \cos \omega t$. (a) Rewrite this wave function in terms of the wavelength λ and the wave speed v of the wave. (b) Write the wave function of the simplest standing-wave vibration of a stretched string of length L. (c) Write the wave function for the second harmonic. (d) Generalize these results and write the wave function for the nth resonance vibration.

57. **Review Problem.** A 12.0-kg object hangs in equilibrium from a string with a total length of $L = 5.00$ m and a linear mass density of $\mu = 0.001\,00$ kg/m. The string is wrapped around two light, frictionless pulleys that are separated by a distance of $d = 2.00$ m (Fig. P14.57a). (a) Determine the tension in the string. (b) At what frequency must the string between the pulleys vibrate in order to form the standing wave pattern shown in Figure P14.57b?

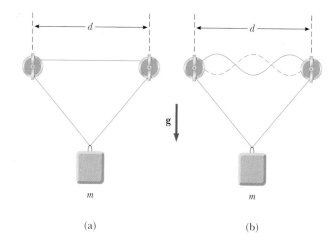

Figure P14.57

58. A quartz watch contains a crystal oscillator in the form of a block of quartz that vibrates by contracting and expanding. Two opposite faces of the block, 7.05 mm apart, are antinodes, moving alternately toward each other and away from each other. The plane halfway between these two faces is a node of the vibration. The speed of sound in quartz is 3.70 km/s. Find the frequency of the vibration. An oscillating electric voltage accompanies the mechanical oscillation—the quartz is described as *piezoelectric*. An electric circuit feeds in energy to maintain the oscillation and also counts the voltage pulses to keep time.

ANSWERS TO QUICK QUIZZES

14.1 At a node, the speed of the particle of the medium at all times is zero. We can find the transverse velocity of a particle at any nonnode point by taking the partial derivative with respect to time of the transverse position:

$$v_y = \frac{\partial y}{\partial t} = \frac{\partial}{\partial t}[(2A \sin kx) \cos \omega t)] = -\omega(2A \sin kx) \sin \omega t$$

At an antinode, the factor in parentheses has its maximum value of $2A$. The maximum speed of the particle is found by allowing the sin ωt term to have its maximum value of 1. Thus, the maximum transverse speed is $2\omega A$.

14.2 The bass strings have low fundamental frequencies. To achieve these low frequencies, we can make the string longer than the treble strings—as indicated in Equation 14.9, if L increases, f_1 decreases. If we depend only on string length to lower the frequency, however, pianos would need to be many meters long to accommodate the lowest strings. The frequency of the bass strings is also lowered by increasing the mass per unit length, which is also indicated in Equation 14.9—as μ increases, f_1 decreases. Thus, the bass strings are thicker. Between these two effects, the frequencies of the bass strings can be achieved in a piano of reasonable size.

14.3 (b). The open tube has a fundamental frequency given by Equation 14.11: $f_{\text{open}} = v/2L$. The closed tube's fundamental frequency is given by Equation 14.12:

$$f_{\text{closed}} = \frac{v}{4L} = \frac{1}{2}\frac{v}{2L} = \frac{1}{2}f_{\text{open}}$$

14.4 A change in temperature will result in a change in the speed of sound and, therefore, in the fundamental frequency of a particular pipe. The speed of sound for audible frequencies in the open atmosphere is only a function of temperature and does not depend on the pressure. Thus, variations in atmospheric pressure will have no effect on the fundamental frequency.

14.5 By tightening the string, you have caused an increase in the beat frequency from 5 to 8 Hz. Thus, the difference between the frequency of the guitar string and that of the tuning fork has increased. To reduce the frequency of the guitar string to make the beats vanish and match the frequency of the tuning fork, you must loosen the guitar string.

Earthquakes

MINIMIZING THE RISK

We have explored the physics of vibrations and waves. Let us now return to our central question for the *Earthquakes* Context:

> *How can we choose locations and build structures to minimize the risk of damage in an earthquake?*

To answer this question, we shall use the physical principles that we now understand more clearly and apply them to our choices of locations and structural design.

In our discussion of simple harmonic oscillation, we learned about resonance. Designs for structures in earthquake-prone areas need to pay careful attention to resonance vibrations from shaking of the ground. The design features to be considered include ensuring that the resonance frequency of the building does not match typical earthquake frequencies. In addition, the structural details should include sufficient damping to ensure that the amplitude of resonance vibration does not destroy the structure.

In Chapter 13, we discussed the role of the medium in the propagation of a wave. For seismic waves moving across the surface of the Earth, the soil on the surface is the medium. Because soil varies from one location to another, the speed of seismic waves will vary at different locations. A particularly dangerous situation exists for structures built on loose soil or mudfill. In these types of media, the interparticle forces are much weaker than in a more solid foundation, such as granite bedrock. As a result, the wave speed is less in loose soil than in bedrock.

Consider Equation 13.23, which provides an expression for the rate of energy transfer by waves. Although this equation was derived for waves on strings, the general dependence on the square of the angular frequency, the square of the amplitude, and the speed is general. To conserve energy, the rate of energy transfer for a wave must remain constant regardless of the medium. Thus, according to Equation 13.23, **if the wave speed decreases, as it does for seismic waves moving through loose soil, the amplitude must increase.** As a result, the shaking of structures built on loose soil is of larger magnitude than for those built on solid bedrock.

This factor contributed to the collapse of the Nimitz Freeway during the Loma Prieta earthquake, near San Francisco, in 1989. Figure 1 shows the results of the earthquake on the freeway. The portion of the freeway that collapsed was built on mudfill, but the surviving portion was built on bedrock. The amplitude of oscillation in the portion built on mudfill was more than five times as large as the amplitude of other portions.

Another danger for structures on loose soil is the possibility of **liquefaction** of the soil. When soil is shaken, the particles can move with respect to one another, and the soil tends to act like a liquid rather than a solid. It is possible for the structure to sink into the soil during an earthquake. If the liquefaction is not uniform over the foundation of the structure, the structure can tip over, as seen in Figure 2. As a result, even if the earthquake is not sufficient to damage the structure, it will be unusable in its tipped-over orientation.

WEB

For a variety of information and photographs about the Loma Prieta earthquake from the Museum of the City of San Francisco, visit **www.sfmuseum.org/1906/89.html**

Figure 1

Portions of the double-decked Nimitz Freeway in Oakland collapsed during the Loma Prieta earthquake of 1989. *(Paul X. Scott/SYGMA)*

WEB

For a report on the Kobe earthquake, visit **mceer.buffalo.edu/publications/sp_pubs/kobereport/KobeReport.asp**

Figure 2

The collapse of this crane during the Kobe, Japan, earthquake of 1995 was caused by liquefaction of the underlying soil. *(Photo by M. Hamada; used by permission from the Multidisciplinary Center for Earthquake Engineering Research, University at Buffalo)*

As discussed in Chapter 14, building structures where standing seismic waves can be established is dangerous. This was a factor in the Michoacán Earthquake of 1985. The shape of the bedrock under Mexico City resulted in standing waves, with severe damage to buildings located at antinodes.

In summary, to minimize risk of damage in an earthquake, architects and engineers must design structures to prevent destructive resonances, avoid building on loose soil, and pay attention to the underground rock formations so as to be aware of possible standing wave patterns. Other precautions can also be taken. For example, buildings can be constructed with **seismic isolation** from the ground. This involves mounting the structure on **isolation dampers**, heavy-duty bearings that dampen the oscillations of the building, resulting in reduced amplitude of vibration.

We have not addressed many other considerations for earthquake safety in structures, but we have been able to apply many of our concepts from oscillations and waves so as to understand some aspects of logical choices in locating and designing structures.

PROBLEMS

1. For seismic waves spreading out from a point (the epicenter) on the surface of the Earth, the intensity of the waves falls off with distance with an inverse dependence on the distance, that is, as $1/r$, where r is the distance from the epicenter to the observation point. This assumes a uniform medium. We have also shown that the energy of vibration of an oscillator is proportional to the square of the amplitude of the vibration of the oscillator. Suppose an earthquake occurs, and, at a distance of 10 km from the epicenter, the *amplitude* of the ground shaking is 5.0 cm. Assuming a uniform medium, what is the *amplitude* of the ground shaking at a point 20 km from the epicenter?

2. As mentioned in the text, the amplitudes of oscillation during the Loma Prieta earthquake of 1989 were five times larger in areas of mudfill as in areas of bedrock. From this information, estimate the percentage by which the seismic wave speed changed as the waves moved from the bedrock to the mudfill.

3. Figure 3 is a graphical representation of the travel time for *P* and *S* waves from the epicenter to a seismograph as a function of the distance between the epicenter and the seismograph. The following table shows the times of day for arrival of *P* waves from an earthquake at three seismograph locations. Fill in the times of day in the last column for the arrival of the *S* waves at the three seismograph locations.

Seismograph Station	Distance from Epicenter (km)	*P* Wave Arrival Time	*S* Wave Arrival Time
#1	170	15:46:05	
#2	160	15:46:03	
#3	105	15:45:54	

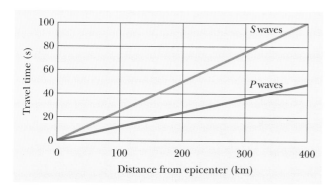

Figure 3

A graph of travel time versus distance from the epicenter for *P* and *S* waves.

Search for the *Titanic*

The *Titanic* and her sister ships, the *Olympic* and the *Britannic*, were designed to be the largest and most luxurious liners sailing the ocean at the time. The *Titanic* was lost in the North Atlantic on its maiden voyage from Southampton, England, to New York, on April 15, 1912. It was claimed by many (although not by the White Star Line, which operated the *Titanic*) that the ship was unsinkable. This was proven to be untrue when the *Titanic* struck an iceberg at 11:40 PM on April 14 and sank less than 3 h later. This event was one of the worst maritime disasters of all time, with over 1500 lives lost due to a severe shortage of lifeboats. Amazingly, British ships of the time were not required to carry enough lifeboats for all passengers on board. It is fortunate that the *Titanic* was not completely full on its maiden voyage, as there would have been only enough lifeboats for a third of its passengers. As it was, there were enough lifeboats for a little over half of the 2200 passengers on board, but only 705 were saved due to the partial filling of the boats, which occurred for a number of reasons.

The mass of the *Titanic* was over 4.2×10^7 kg and it was 269 m long. It was designed to be able to travel at 24 to 25 knots (about 12–13 m/s), and the safety of the ship was ensured by lateral bulkheads

During the sinking process in the early morning of April 15, 1912, the bow of the *Titanic* is underwater and the stern is lifted out of the water. The huge forces necessary to hold the stern aloft caused the *Titanic* to split in the middle before sinking. *(Ken Marschall Collection)*

at several places across the ship with electrically operated watertight doors. It is ironic to note that the design of the bulkheads actually worked against the safety of the ship, and the closing of the watertight doors, according to some experts after the disaster, caused the ship to sink more rapidly than if they were left open.

The accidental sinking of the *Titanic* resulted from a remarkable confluence of bad luck, complacency, and poor policy. Several events occurred during the last day of its voyage that would not have led to the foundering of the ship if they had happened in just a slightly different way.

The *Titanic* has captured the interest of the public for many years. Four theatrical movies related to the ship have been produced: *Titanic* (1953), *A Night to Remember* (1958), *Raise the Titanic* (1980), and *Titanic* (1997), along with a Broadway and movie musical, *The Unsinkable Molly Brown* (1964), and a more recent musical play, *Titanic* (1997). A large number of books have been written on the disaster, many of which were reissued when the *Titanic* became wildly popular in response to the 1997 film.

The *Titanic* on its way from Southampton to New York. *(Ken Marschall Collection)*

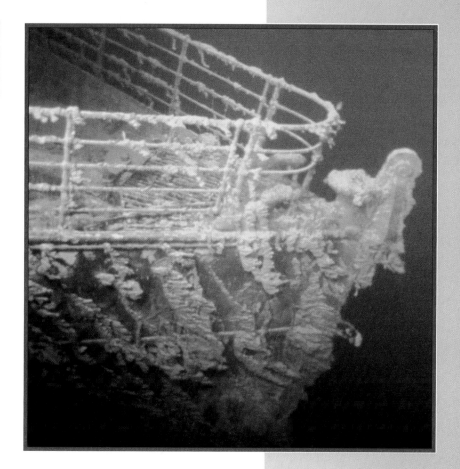

Finding the *Titanic* on the ocean floor was a dream of many individuals ever since she was lost. The wreck of the *Titanic* was discovered in 1985 by a research team from Woods Hole Oceanographic Institute, led by Dr. Robert Ballard. This discovery led to interesting ethical questions as subsequent expeditions to the site salvaged items from the wreck, making them available for sale to the public. We will use the *Titanic*, the search for its wreckage, and the underwater visits to its gravesite in this short Context on fluids, as we address the central question

How can we safely visit the wreck of the **Titanic**?

An iceberg floats in the cold waters of the North Atlantic. Although the visible portion of the iceberg may tower over a passing ship, only about 11% of the iceberg is above water.

(Geraldine Prentice/Stone)

Fluid Mechanics

Matter is normally classified as being in one of three states: solid, liquid, or gas. Everyday experience tells us that a solid has a definite volume and shape. A brick maintains its familiar shape and size over a long time. We also know that a liquid has a definite volume but no definite shape—a cup of liquid water has a fixed volume but assumes the shape of its container. Finally, an unconfined gas has neither definite volume nor definite shape. These definitions help us to picture the states of matter, but they are somewhat artificial. For example, asphalt, glass, and plastics are normally considered solids, but over a long time they tend to flow like liquids. Likewise, most substances can be a solid, liquid, or gas (or combinations of these), depending on the temperature and pressure. In general, the time it takes a particular substance to change its shape in response to an external force determines whether we treat the substance as a solid, liquid, or gas.

A **fluid** is a collection of molecules that are randomly arranged and held together by weak cohesive forces between molecules and forces exerted by the walls of a container. Both liquids and gases are fluids. In our treatment of the mechanics of fluids, we shall see that no new physical principles are needed to explain such effects as the buoyant force on a submerged object and vascular flutter in an artery. In this chapter, we shall apply a number of familiar analysis models to the physics of fluids.

15.1 • PRESSURE

Our first task in understanding the physics of fluids is to define a new quantity to describe fluids. Imagine applying a force to the surface of an object, with the force having components both parallel to and perpendicular to the surface. If the object is a solid at rest on a table, the force component perpendicular to the surface may cause the object to flatten, depending on how hard the object is. Assuming that the object does not slide on the table, the component of the force parallel to the surface of the object will cause the object to distort. As an example, suppose you place your physics book flat on a table and apply a force with your hand parallel to the front cover and perpendicular to the spine. The book will distort, with the bottom pages staying fixed at their original location and the top pages shifting horizontally by some distance. The cross-section of the book changes from a rectangle to a parallelogram. This kind of force parallel to the surface is called a *shearing force*.

We shall adopt a simplification model in which the fluids we study will be nonviscous; that is, no friction exists between adjacent layers of the fluid. Nonviscous fluids do not sustain shearing forces—if you imagine placing your hand on a water surface and pushing parallel to the surface, your hand simply slides over the water—you cannot distort the water as you did the book. This phenomenon occurs because the interatomic forces in a fluid are not strong enough to lock atoms in place with respect to one another. The fluid cannot be modeled as a rigid body, as for rotating objects in Chapter 10. If we try to apply a shearing force, the molecules of the fluid simply slide past one another.

Thus, the only type of force that can exist in a fluid is one that is perpendicular to a surface. For example, the forces exerted by the fluid on the object in Figure 15.1 are everywhere perpendicular to the surfaces of the object.

The force that a fluid exerts on a surface originates in the collisions of molecules of the fluid with the surface. Each collision results in the reversal of the component of the velocity vector of the molecule perpendicular to the surface. By the impulse–momentum theorem and Newton's third law, each collision results in a force on the surface. A huge number of these impulsive forces occur every second, resulting in a constant macroscopic force on the surface. This force is spread out over the area of the surface and is related to a new quantity called **pressure.**

The pressure at a specific point in a fluid can be measured with the device pictured in Figure 15.2. The device consists of an evacuated cylinder enclosing a light piston connected to a spring. As the device is submerged in a fluid, the fluid presses in on the top of the piston and compresses the spring until the inward force of the fluid is balanced by the outward force of the spring. The force exerted on the piston by the fluid can be measured if the spring is calibrated in advance.

If F is the magnitude of the force exerted by the fluid on the piston and A is the surface area of the piston, then the **pressure P of the fluid at the level to which the device has been submerged is defined as the ratio of force to area:**

$$P \equiv \frac{F}{A}$$

[15.1]

Figure 15.1

The force of the fluid on a submerged object at any point is perpendicular to the surface of the object. The force of the fluid on the walls of the container is perpendicular to the walls at all points.

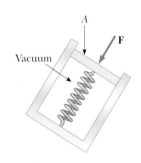

Figure 15.2

A simple device for measuring pressure in a fluid.

PITFALL PREVENTION 15.1

Force and pressure

Equation 15.1 makes a clear distinction between force and pressure. Another important distinction is that *force is a vector* and *pressure is a scalar.* No direction is associated with pressure, but the direction of the force associated with the pressure is perpendicular to the surface of interest.

• *Definition of pressure*

Although pressure is defined in terms of our device in Figure 15.2, the definition is general. Because pressure is force per unit area, it has units of N/m^2 in the SI system. Another name for the SI unit of pressure is the **pascal** (Pa):

- *The pascal*

$$1 \text{ Pa} \equiv 1 \text{ N/m}^2 \qquad \qquad [15.2]$$

Notice that pressure and force are different quantities. We can have a very large pressure from a relatively small force by making the area over which the force is applied small. This is the case with hypodermic needles. The area of the tip of the needle is very small, so a small force pushing on the needle is sufficient to cause a pressure large enough to puncture the skin. We can also create a small pressure from a large force by enlarging the area over which the force acts. This is the principle behind the design of snowshoes. If a person were to walk on deep snow with regular shoes, it is possible for his or her feet to break through the snow and sink. Snowshoes, however, allow the force on the snow due to the weight of the person to spread out over a larger area, reducing the pressure enough so that the snow surface is not broken.

The atmosphere exerts a pressure on the surface of the Earth and all objects at the surface. This pressure is responsible for the action of suction cups, drinking straws, vacuum cleaners, and many other devices. In our calculations and end-of-chapter problems, we usually take atmospheric pressure to be

- *Atmospheric pressure*

$$P_0 = 1.00 \text{ atm} \approx 1.013 \times 10^5 \text{ Pa} \qquad \qquad [15.3]$$

APPLICATION

Hypodermic needles

Quick Quiz 15.1

Suppose someone standing directly in front of you steps back and accidentally stomps on your foot with the heel of one shoe. Would you be better off if that person were a professional basketball player wearing sneakers or a petite woman wearing spike-heeled shoes?

Snowshoes prevent a person from sinking into soft snow because the person's weight is spread over a larger area, reducing the pressure on the snow's surface. *(Earl Young/FPG)*

THINKING PHYSICS 15.1

Suction cups can be used to hold objects onto surfaces. Why don't astronauts use suction cups to hold onto the outside surface of the space shuttle?

Reasoning A suction cup works because air is pushed out from under the cup when it is pressed against a surface. When the cup is released, it tends to spring back a bit, causing the trapped air under the cup to expand. This expansion causes a reduced pressure inside the cup. Thus, the difference between the atmospheric pressure on the outside of the cup and the reduced pressure inside provides a net force pushing the cup against the surface. For astronauts in orbit around the Earth, almost no air exists outside the surface of the spacecraft. Thus, if a suction cup were to be pressed against the outside surface of the spacecraft, the pressure differential needed to press the cup to the surface is not present.

Figure 15.3

(Thinking Physics 15.2) The daring physics professor takes a nap on a bed of nails. (*Jim Lehman*)

THINKING PHYSICS 15.2

The daring physics professor, after a long lecture, stretches out for a nap on a bed of nails as shown in Figure 15.3. How is this possible?

Reasoning If you try to support your entire weight on a single nail, the pressure on your body is your weight divided by the very small area of the nail. This pressure is sufficiently large to penetrate the skin, as in the case of the hypodermic needle discussed earlier. If you distribute your weight over several hundred nails, however, as the professor is doing, the pressure is considerably reduced because the area that supports your weight is the total area of all nails in contact with your body. (Note that lying on a bed of nails is much more comfortable than sitting on a bed of nails. Standing on the bed of nails without shoes is not recommended.)

EXERCISE Estimate the order of magnitude of the pressure on the floor under the chair legs when you sit in a typical chair.

Answer $\approx 10^5$–10^6 Pa

15.2 • VARIATION OF PRESSURE WITH DEPTH

The study of fluid mechanics involves the density of a substance, defined in Equation 1.1 as the mass per unit volume for the substance. Table 15.1 lists the densities of various substances. These values vary slightly with temperature, because the volume of a substance is temperature-dependent (as we shall see in Chapter 16). Note that under standard conditions (0°C and atmospheric pressure) the densities of gases are on the order of 1/1000 the densities of solids and liquids. This difference implies that the average molecular spacing in a gas under these conditions is about ten times greater in each dimension than in a solid or liquid.

As divers know well, the pressure in the sea or a lake increases as they dive to greater depths. Likewise, atmospheric pressure decreases with increasing altitude. For this reason, aircraft flying at high altitudes must have pressurized cabins in order to provide sufficient oxygen for the passengers.

We now show mathematically how the pressure in a liquid increases with depth. Consider a liquid of density ρ at rest as in Figure 15.4. Let us select a sample of the liquid contained within an imaginary cylinder of cross-sectional area A extending from depth d to depth $d + h$. This sample of liquid is at rest. Thus,

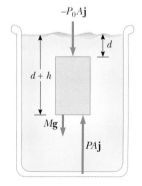

Figure 15.4

The net force on the sample of liquid within the darker region must be zero because the sample is in equilibrium.

TABLE 15.1		Densities of Some Common Substances	
Substance	**ρ (kg/m^3)^a**	**Substance**	**ρ (kg/m^3)^a**
Ice	0.917×10^3	Water	1.00×10^3
Aluminum	2.70×10^3	Sea water	1.03×10^3
Iron	7.86×10^3	Ethyl alcohol	0.806×10^3
Copper	8.92×10^3	Benzene	0.879×10^3
Silver	10.5×10^3	Mercury	13.6×10^3
Lead	11.3×10^3	Air	1.29
Gold	19.3×10^3	Oxygen	1.43
Platinum	21.4×10^3	Hydrogen	8.99×10^{-2}
Glycerin	1.26×10^3	Helium	1.79×10^{-1}

^{a} All values are at standard atmospheric pressure and temperature (STP), that is, atmospheric pressure and 0°C. To convert the density in kg/m^3 to units of g/cm^3, multiply by 10^{-3}.

according to Newton's second law, the net force on the sample must be equal to zero. We will investigate the forces on the sample related to the pressure on it.

The liquid external to our sample exerts forces at all points on the surface of the sample, perpendicular to the surface. The pressure exerted by the liquid on the bottom face of the sample is P, and the pressure on the top face is P_0. Therefore, from Equation 15.1, the magnitude of the upward force exerted by the liquid on the bottom of the sample is PA, and the magnitude of the downward force exerted by the liquid on the top is P_0A. In addition, a gravitational force is exerted on the sample. Because the sample is in equilibrium, the net force in the vertical direction must be zero:

$$\sum F_y = 0 \quad \longrightarrow \quad PA - P_0A - mg = 0$$

Because the mass of liquid in the sample is $m = \rho V = \rho Ah$, the gravitational force on the liquid in the sample is $mg = \rho gAh$. Thus,

$$PA = P_0A + \rho gAh$$

or

- *Variation of pressure with depth*

$$P = P_0 + \rho gh \qquad [15.4]$$

If the top surface of our sample is at $d = 0$, so that it is open to the atmosphere, then P_0 is atmospheric pressure. Equation 15.4 indicates that the pressure in a liquid depends only on the depth h within the liquid. The pressure is therefore the same at all points having the same depth, independent of the shape of the container.

In view of Equation 15.4, any increase in pressure at the surface must be transmitted to every point in the liquid. This was first recognized by the French scientist Blaise Pascal (1623–1662) and is called **Pascal's law: A change in the pressure applied to an enclosed liquid is transmitted undiminished to every point of the fluid and to the walls of the container.**

- *Pascal's law*

An important application of Pascal's law is the hydraulic press illustrated by Figure 15.5. A force $\mathbf{F}_1$ is applied to a small piston of area A_1. The pressure is transmitted through a liquid to a larger piston of area A_2, and force $\mathbf{F}_2$ is exerted

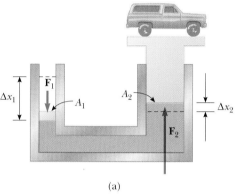

(a)   (b)

Figure 15.5

(a) Diagram of a hydraulic press. Because the increase in pressure is the same at the left and right sides, a small force **F**$_1$ at the left produces a much larger force **F**$_2$ at the right. (b) A vehicle under repair is supported by a hydraulic lift in a garage. *(David Frazier)*

by the liquid on this piston. Because the pressure is the same at both pistons, we see that $P = F_1/A_1 = F_2/A_2$. The force magnitude F_2 is therefore larger than F_1 by the multiplying factor A_2/A_1. Hydraulic brakes, car lifts, hydraulic jacks, and forklifts all make use of this principle.

THINKING PHYSICS 15.3

Blood pressure is normally measured with the cuff of the sphygmomanometer around the arm. Suppose that the blood pressure were measured with the cuff around the calf of the leg of a standing person. Would the reading of the blood pressure be the same here as it is for the arm?

Reasoning The blood pressure measured at the calf would be larger than that measured at the arm. If we imagine the vascular system of the body to be a vessel containing a liquid (blood), the pressure in the liquid will increase with depth. The blood at the calf is deeper in the liquid than that at the arm and is at a higher pressure.

Blood pressures are normally taken at the arm because it is at approximately the same height as the heart. If blood pressures at the calf were used as a standard, adjustments would need to be made for the height of the person, and the blood pressure would be different if the person were lying down.

APPLICATION

Measuring blood pressure

Example 15.1 The Car Lift

In a car lift used in a service station, compressed air exerts a force on a small piston of circular cross-section having a radius of 5.00 cm. This pressure is transmitted by an incompressible liquid to a second piston of radius 15.0 cm. (a) What force must the compressed air exert in order to lift a car weighing 13 300 N? (b) What air pressure will produce this force? (c) Consider the lift as a nonisolated system, and show that the input energy transfer is equal in magnitude to the output energy transfer.

Solution (a) Because the pressure exerted by the compressed air is transmitted undiminished throughout the liquid, we have

$$F_1 = \left(\frac{A_1}{A_2}\right) F_2 = \frac{\pi(5.00 \times 10^{-2} \text{ m})^2}{\pi(15.0 \times 10^{-2} \text{ m})^2} (1.33 \times 10^4 \text{ N})$$

$$= 1.48 \times 10^3 \text{ N}$$

(b) The air pressure that will produce this force is

$$P = \frac{F_1}{A_1} = \frac{1.48 \times 10^3 \text{ N}}{\pi(5.00 \times 10^{-2} \text{ m})^2} = 1.88 \times 10^5 \text{ Pa}$$

This pressure is approximately twice atmospheric pressure.

(c) The energy input and output are by means of work done by the forces as the pistons move. To determine the work done, we must find the magnitude of the displacement through which each force acts. Because the liquid is modeled to be incompressible, the volume of the cylinder through which the input piston moves must equal that through which the output piston moves. The lengths of these cylinders are the magnitudes Δx_1 and Δx_2 of the displacements of the forces (see Fig. 15.5). Setting the volumes equal, we have

$$V_1 = V_2 \longrightarrow A_1 \Delta x_1 = A_2 \Delta x_2$$

$$\frac{A_1}{A_2} = \frac{\Delta x_2}{\Delta x_1}$$

Evaluating the ratio of the input work to the output work, we find

$$\frac{W_1}{W_2} = \frac{F_1 \Delta x_1}{F_2 \Delta x_2} = \left(\frac{F_1}{F_2}\right)\left(\frac{\Delta x_1}{\Delta x_2}\right) = \left(\frac{A_1}{A_2}\right)\left(\frac{A_2}{A_1}\right) = 1$$

which verifies that the work input and output are the same, as they must be to conserve energy.

Example 15.2 The Force on a Dam

Water is filled to a height H behind a dam of width w (Fig. 15.6). Determine the resultant force on the dam.

Reasoning We cannot calculate the force on the dam by simply multiplying the area by the pressure, because the pressure varies with depth. The problem can be solved by finding the force dF on a narrow horizontal strip at depth h, and then integrating the expression to find the total force on the dam.

Solution The pressure at the depth h beneath the surface at the shaded portion in Figure 15.6 is

$$P = \rho g h = \rho g (H - y)$$

(We have not included atmospheric pressure in our calculation because it acts on both sides of the dam, resulting in a net contribution of zero to the total force.) From Equation 15.1, we find the force on the shaded strip of area dA:

$$F = PA \longrightarrow dF = P dA$$

Because $dA = w\, dy$, we have

$$dF = P dA = \rho g (H - y) w\, dy$$

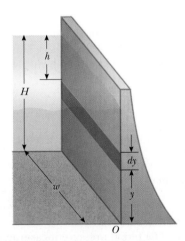

Figure 15.6

(Example 15.2) The total force on a dam is obtained from the expression $F = \int P dA$, where dA is the area of the dark strip.

Therefore, the total force on the dam is

$$F = \int_0^H \rho g (H - y) w\, dy = \tfrac{1}{2}\rho g w H^2$$

Note that because the pressure increases with depth, the dam is designed such that its thickness increases with depth, as in Figure 15.6.

EXERCISE Use the fact that the pressure increases linearly with depth to find the average pressure on the dam.

Answer $\frac{1}{2}\rho gH$

EXERCISE What is the hydrostatic force on the back of

Grand Coulee Dam if the water in the reservoir is 150 m deep and the width of the dam is 1 200 m?

Answer 1.32×10^{11} N

Exercise In some places, the Greenland ice sheet is 1.0 km thick. Estimate the pressure on the ground underneath the ice. ($\rho_{ice} = 920$ kg/m^3)

Answer 9×10^6 Pa

15.3 • PRESSURE MEASUREMENTS

During the weather report on a television news program, the *barometric pressure* is often provided. This is the current pressure of the atmosphere, which varies over a small range from the standard value provided earlier. How is this pressure measured?

One instrument used to measure atmospheric pressure is the common barometer, invented by Evangelista Torricelli (1608–1647). A long tube closed at one end is filled with mercury and then inverted into a dish of mercury (Fig. 15.7a). The closed end of the tube is nearly a vacuum, so the pressure at the top of the mercury column can be taken as zero. In Figure 15.7a, the pressure at point A due to the column of mercury must equal the pressure at point B due to the atmosphere. If this were not the case, a net force would move mercury from one point to the other until equilibrium was established. It therefore follows that $P_0 = \rho_{Hg}gh$, where ρ_{Hg} is the density of the mercury and h is the height of the mercury column. As atmospheric pressure varies, the height of the mercury column varies, so the height can be calibrated to measure atmospheric pressure. Let us determine the height of a mercury column for one atmosphere of pressure, $P_0 =$ 1 atm = 1.013×10^5 Pa:

$$P_0 = \rho_{Hg}gh \quad \longrightarrow \quad h = \frac{P_0}{\rho_{Hg}g} = \frac{1.013 \times 10^5 \text{ Pa}}{(13.6 \times 10^3 \text{ kg/m}^3)(9.80 \text{ m/s}^2)} = 0.760 \text{ m}$$

Based on a calculation such as this, one atmosphere of pressure is defined as the pressure equivalent of a column of mercury that is exactly 0.760 0 m in height at 0°C.

The open-tube manometer illustrated in Figure 15.7b is a device for measuring the pressure of a gas contained in a vessel. One end of a U-shaped tube containing a liquid is open to the atmosphere, and the other end is connected to a system of unknown pressure P. The pressures at points A and B must be the same (otherwise, the curved portion of the liquid would experience a net force and would accelerate), and the pressure at A is the unknown pressure of the gas. Therefore, equating the unknown pressure P to the pressure at point B, we see that $P = P_0 + \rho gh$. The difference in pressure $P - P_0$ is equal to ρgh. The pressure P is called the **absolute pressure,** and the difference $P - P_0$ is called the **gauge pressure.** For example, the pressure you measure in your bicycle tire is gauge pressure.

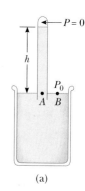

(a)

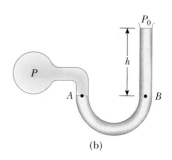

(b)

Figure 15.7

Two devices for measuring pressure: (a) a mercury barometer and (b) an open-tube manometer.

Quick Quiz 15.2

It is possible to build a barometer like that in Figure 15.7a with water rather than mercury. Why is this impractical?

15.4 • BUOYANT FORCES AND ARCHIMEDES'S PRINCIPLE

In this section, we investigate the origin of a **buoyant force,** which is **an upward force exerted on an object by the surrounding fluid.** Buoyant forces are evident in many situations—anyone who has ridden in a boat has experienced a buoyant force. Another common example is the relative ease with which you can lift someone in a swimming pool compared with lifting that same individual on dry land. Evidently, water provides partial support to any object placed in it. According to **Archimedes's principle,**

> Any object completely or partially submerged in a fluid experiences an upward buoyant force whose magnitude is equal to the weight of the fluid displaced by the object.

Archimedes's principle can be verified in the following manner. Suppose we focus our attention on the indicated cube of fluid in the container of Figure 15.8. This cube of fluid is in equilibrium under the action of the forces on it. One of these forces in the vertical direction is the gravitational force. Because the cube is in equilibrium, the net force on it in the vertical direction must be zero. What cancels the downward gravitational force so that the cube remains in equilibrium? Apparently, the rest of the fluid inside the container is applying an upward force— the buoyant force. Thus, the magnitude B of the buoyant force must be exactly equal to the weight of the fluid inside the cube:

$$\sum F_y = 0 \quad \longrightarrow \quad B - F_g = 0 \quad \longrightarrow \quad B = mg$$

where m is the mass of the fluid in the cube.

Now, imagine that the cube of fluid is replaced by a cube of steel of the same dimensions. What is the buoyant force on the steel? The fluid surrounding a cube behaves in the same way whether it is exerting pressure on a cube of fluid or a cube of steel; therefore, the **buoyant force acting on the steel is the same as the buoyant force acting on a cube of fluid of the same dimensions.** This result applies for a submerged object of any shape, size, or density.

Let us now show more explicitly that the magnitude of the buoyant force is equal to the weight of the displaced fluid. Although this is true for both liquids and gases, we will perform the derivation for a liquid. On the sides of the cube of liquid in Figure 15.8, forces due to the pressure act horizontally and cancel in pairs on opposite sides of the cube for a net horizontal force of zero. In a liquid, the pressure at the bottom of the cube is greater than the pressure at the top by an amount $\rho_f gh$, where ρ_f is the density of the liquid and h is the height of the cube. Thus, the upward force F_{bot} on the bottom is greater than the downward force F_{top} on the top of the cube. The net vertical force *exerted by the liquid* (we are ignoring the gravitational force for now) is

$$\sum F_{\text{liquid}} = B = F_{\text{bot}} - F_{\text{top}}$$

Expressing this in terms of pressure gives us

$$B = P_{\text{bot}}A - P_{\text{top}}A = \Delta PA = \rho_f ghA$$

$$B = \rho_f gV \qquad\qquad [15.5]$$

where V is the volume of the cube. Because the mass of the liquid in the cube is

Archimedes (287–212 BC)

Archimedes, a Greek mathematician, physicist, and engineer, was perhaps the greatest scientist of antiquity. He was the first to compute accurately the ratio of a circle's circumference to its diameter, and he showed how to calculate the volume and surface area of spheres, cylinders, and other geometric shapes. He is well known for discovering the nature of the buoyant force.

Figure 15.8

The external forces on the cube of liquid are the gravitational force $\mathbf{F}_g$ and the buoyant force $\mathbf{B}$. Under equilibrium conditions, $B = F_g$.

• *Archimedes's principle*

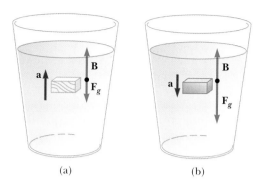

(a)   (b)

Figure 15.9

(a) A totally submerged object that is less dense than the fluid in which it is submerged experiences a net upward force. (b) A totally submerged object that is denser than the fluid sinks.

PITFALL PREVENTION 15.2

Buoyant force is exerted by the fluid

Notice a very important point in this discussion. The buoyant force is a force **exerted by the fluid.** It is not determined by properties of the object, except for the amount of fluid displaced by the object. Thus, if several objects of different densities but the same volume are immersed in a fluid, they will all experience the same buoyant force. Whether they sink or float is determined by the relationship between the buoyant force and their weight. It is not the natural tendency for helium to rise—a helium balloon rises because its overall density is less than that of air.

$M = \rho_f V$, we see that

$$B = Mg$$

which is the weight of the displaced liquid.

Before proceeding with a few examples, it is instructive to compare two common cases: the buoyant force acting on a totally submerged object and that acting on a floating object.

Case I: A Totally Submerged Object   When an object is totally submerged in a liquid of density ρ_f, the magnitude of the upward buoyant force is $B = \rho_f g V_O$, where V_O is the volume of the object. Because the object is totally submerged, the volume of the object and the volume of liquid displaced by the object are the same, $V = V_O$. If the object has a density ρ_O, its weight is $Mg = \rho_O V_O g$. Thus, the net force on it is $\Sigma F = B - Mg = (\rho_f - \rho_O) V_O g$. We see that if the density of the object is less than the density of the liquid as in Figure 15.9a, the net force is positive and the unsupported object accelerates upward. If the density of the object is greater than the density of the liquid as in Figure 15.9b, the net force is negative and the unsupported object sinks.

The same behavior is exhibited by an object immersed in a gas, such as the air in the atmosphere.* If the object is less dense than air, like a helium-filled balloon, the object floats upward. If it is denser, like a rock, it falls downward.

Case II: A Floating Object   Now consider an object in static equilibrium floating on the surface of a fluid, that is, an object that is only partially submerged, such as the ice cube floating in water in Figure 15.10. Because it is only partially submerged, the volume V of fluid displaced by the object is only a fraction of the total volume V_O of the object. The volume of the fluid displaced by the object corresponds to that volume of the object beneath the fluid surface. Because the object is in equilibrium, the upward buoyant force is balanced by the downward gravitational force exerted on the object. The buoyant force has a magnitude $B = \rho_f g V$.

Figure 15.10

An object that is less dense than a fluid floats in the fluid, such as this ice cube in water.

---

* The general behavior is the same, but the buoyant force varies with height in the atmosphere due to the variation in density of the air.

Notice that the sinking or floating behavior of an object is determined by a comparison of the fluid and object *densities*. In the case of aircraft, one often hears descriptive phrases such as "heavier than air" (for an airplane) or "lighter than air" (for a hot-air balloon). These are misguided phrases, because they describe a comparison between the *weight* of an object and that of an unspecified amount of air. Despite the popularity of these phrases, try to train yourself to say "denser than air" and "less dense than air."

Because the weight of the object is $Mg = \rho_O V_{Og}$, and because Newton's second law tells us that $\Sigma F = 0$ in the vertical direction, $Mg = B$, and we see that $\rho_f g V = \rho_O V_{Og}$, or

$$\frac{\rho_O}{\rho_f} = \frac{V}{V_O} \qquad [15.6]$$

Thus, the fraction of the volume of the object under the fluid surface is equal to the ratio of the object density to the fluid density.

Let us consider examples of both cases. Under normal conditions, the average density of a fish is slightly greater than the density of water. This being the case, a fish would sink if it did not have some mechanism for adjusting its density. The fish accomplishes this by internally regulating the size of its swim bladder, a gas-filled cavity within the fish's body. Increasing its size increases the amount of water displaced, which increases the buoyant force. In this manner, fish are able to swim to various depths. Because the fish is totally submerged in the water, this is an example of Case I.

As an example of Case II, imagine a large cargo ship. When the ship is at rest, the upward buoyant force from the water balances the weight so that the ship is in equilibrium. Only part of the volume of the ship is under water. If the ship is loaded with heavy cargo, it sinks deeper into the water. The increased weight of the ship due to the cargo is balanced by the extra buoyant force related to the extra volume of the ship that is now beneath the water surface.

Quick **Quiz** 15.3

Atmospheric pressure varies from day to day. Does a ship float higher in the water on a high-pressure day than on a low-pressure day?

Quick **Quiz** 15.4

A pound of styrofoam and a pound of lead have the same weight. If they are placed on an equal-arm balance surrounded by air, will it balance?

Quick **Quiz** 15.5

What is the origin of the saying, "That's only the tip of the iceberg"?

Hot-air balloons. Because hot air is less dense than cold air, a net upward force acts on the balloons. *(Richard Megna/Fundamental Photographs)*

THINKING PHYSICS 15.4

Suppose an office party is taking place on the top floor of a tall building. Carrying an iced soft drink, you step on the elevator, which begins to accelerate downward. What happens to the ice in the drink? Does it rise farther out of the liquid? Sink deeper into the liquid? Or is it unaffected by the motion?

Reasoning The acceleration of the elevator is equivalent to a change in the gravitational field, according to the principle of equivalence (Section 9.9). If the elevator accelerates downward, one might be tempted to say that the effect is the same as if gravity decreases—the weight of the ice cube decreases, causing it to float higher in the liquid. Recall, however, that the magnitude of the buoyant force is equal to the weight of the liquid displaced by the ice cube. The weight of the liquid also decreases with the effectively decreased gravity. Because both the weight of the ice cube and the buoyant force decrease by the same factor, the level of the ice cube in the liquid is unaffected.

THINKING PHYSICS 15.5

A florist delivery person is delivering a flower basket to a home. The basket includes an attached helium-filled balloon, which suddenly comes loose from the basket and begins to accelerate upward toward the sky. Startled by the release of the balloon, the delivery person drops the flower basket. As the basket falls, the basket–Earth system experiences an increase in kinetic energy and a decrease in gravitational potential energy, consistent with conservation of mechanical energy. The balloon–Earth system, however, experiences an increase in *both* gravitational potential energy and kinetic energy. Is this inconsistent with the principle of conservation of mechanical energy? If not, from where is the extra energy coming?

Reasoning   In the case of the system of the flower basket and the Earth, a good approximation to the motion of the basket can be made by ignoring the effects of the air. Thus, the basket–Earth system can be analyzed with the isolated system model—mechanical energy is conserved. For the balloon–Earth system, we cannot ignore the effects of the air—it is the buoyant force of the air that causes the balloon to rise. Thus, the balloon–Earth system is analyzed with the nonisolated system model. The buoyant force of the air does work across the boundary of the system, and that work results in an increase in both the kinetic and gravitational potential energies of the system.

Example 15.3   A Submerged Object

A piece of aluminum is suspended from a string and then completely immersed in a container of water (Fig. 15.11). The mass of the aluminum is 1.0 kg, and its density is 2.7×10^3 kg/m^3. Calculate the tension in the string before and after the aluminum is immersed.

Solution   When the aluminum is suspended in air as in Figure 15.11a, Newton's second law tells us that the tension T_1 in the string (the reading on the scale) is equal to the weight Mg of the aluminum, assuming that the buoyant force of air can be ignored:

$$T_1 = Mg = (1.0 \text{ kg})(9.80 \text{ m/s}^2) = \boxed{9.8 \text{ N}}$$

When immersed in water, we use Newton's second law again. The aluminum experiences an upward buoyant force **B** as in Figure 15.11b, which reduces the tension in the string. To calculate B, we must first calculate the volume of the aluminum:

$$V_{\text{Al}} = \frac{M}{\rho_{\text{Al}}} = \frac{1.0 \text{ kg}}{2.7 \times 10^3 \text{ kg/m}^3} = 3.7 \times 10^{-4} \text{ m}^3$$

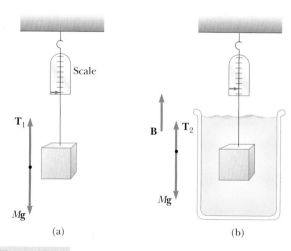

Figure 15.11

(Example 15.3) (a) When the aluminum is suspended in air, the scale reads the true weight Mg (ignoring the buoyancy of air). (b) When the aluminum is immersed in water, the buoyant force **B** reduces the scale reading to $T_2 = Mg - B$.

From Equation 15.5,

$$B = \rho_w V_{Al} g = (1.0 \times 10^3 \text{ kg/m}^3)(3.7 \times 10^{-4} \text{ m}^3)(9.80 \text{ m/s}^2)$$
$$= 3.6 \text{ N}$$

Now we apply Newton's second law to the aluminum,

$$\sum F = T_2 + B - Mg = 0$$
$$T_2 = Mg - B = 9.8 \text{ N} - 3.6 \text{ N} = \boxed{6.2 \text{ N}}$$

Example 15.4 Changing String Vibration with Water

One end of a horizontal string is attached to a vibrating blade, and the other end passes over a pulley as in Figure 15.12a. A sphere of mass 2.00 kg hangs on the end of the string. The string is vibrating in its second harmonic. A container of water is raised under the sphere so that the sphere is completely submerged. After this is done, the string vibrates in its fifth harmonic, as shown in Figure 15.12b. What is the radius of the sphere?

Solution In Figure 15.12a, Newton's second law applied to the sphere tells us that the initial tension T_i in the string is equal to the weight of the sphere:

$$T_i - mg = 0 \longrightarrow T_i = mg$$
$$T_i = (2.00 \text{ kg})(9.80 \text{ m/s}^2) = 19.6 \text{ N}$$

where the subscript i is used to indicate initial variables before we immerse the sphere in water. Once the sphere is immersed in water, the tension in the string will decrease to T_f. Applying Newton's second law to the sphere again in this situation, we have

$$T_f + B - mg = 0 \longrightarrow B = mg - T_f \quad (1)$$

The desired quantity, the radius of the sphere, will appear in the expression for the buoyant force B. Before we can proceed in this direction, however, we need to evaluate T_f. We do this from the standing wave information. We write the equation for the frequency of a standing wave on a string (Equation 14.8) twice: once before we immerse the sphere and once after, and divide the equations:

$$\begin{aligned} f &= \frac{n_i}{2L}\sqrt{\frac{T_i}{\mu}} \\ f &= \frac{n_f}{2L}\sqrt{\frac{T_f}{\mu}} \end{aligned} \longrightarrow 1 = \frac{n_i}{n_f}\sqrt{\frac{T_i}{T_f}} \quad (2)$$

where the frequency f is the same in both cases, because it is determined by the vibrating blade. In addition, the linear mass density μ and the length L of the vibrating portion of the string are the same in both cases. Solving (2) for T_f gives

$$T_f = \left(\frac{n_i}{n_f}\right)^2 T_i = \left(\frac{2}{5}\right)^2 (19.6 \text{ N}) = 3.14 \text{ N}$$

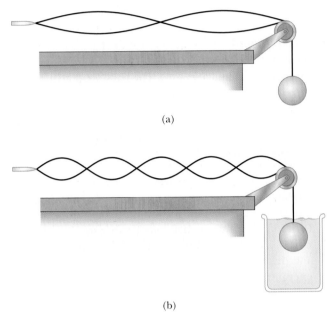

(a)

(b)

Figure 15.12

(Example 15.4) When the sphere hangs in air, the string vibrates in its second harmonic. When the sphere is immersed in water, the string vibrates in its fifth harmonic.

Substituting this into equation (1), we can evaluate the buoyant force on the sphere:

$$B = mg - T_f = 19.6 \text{ N} - 3.14 \text{ N} = 16.5 \text{ N}$$

Finally, expressing the buoyant force in terms of the radius of the sphere, we solve for the radius,

$$B = \rho_{water} g V_{sphere} = \rho_{water} g (\tfrac{4}{3}\pi r^3)$$

$$r = \sqrt[3]{\frac{3B}{4\pi\rho_{water}g}} = \sqrt[3]{\frac{3(16.5 \text{ N})}{4\pi(1\,000 \text{ kg/m}^3)(9.80 \text{ m/s}^2)}}$$

$$= 7.38 \times 10^{-2} \text{ m} = \boxed{7.38 \text{ cm}}$$

15.5 • FLUID DYNAMICS

Thus far, our study of fluids has been restricted to fluids at rest—**fluid statics.** We now turn our attention to **fluid dynamics,** that is, the study of fluids in motion. Instead of trying to study the motion of each particle of the fluid as a function of time, we describe the properties of the fluid as a whole.

Flow Characteristics

When fluid is in motion, its flow is of one of two main types. The flow is said to be **steady,** or **laminar,** if each particle of the fluid follows a smooth path, so that the paths of different particles never cross each other, as in Figure 15.13. Thus, in steady flow, the velocity of the fluid at any point remains constant in time.

Above a certain critical speed, fluid flow becomes **turbulent.** Turbulent flow is an irregular flow characterized by small whirlpool-like regions as in Figure 15.14. As an example, the flow of water in a river becomes turbulent in regions where rocks and other obstructions are encountered, often forming "white water" rapids.

The term **viscosity** is commonly used in fluid flow to characterize the degree of internal friction in the fluid. This internal friction, or viscous force, is associated with the resistance of two adjacent layers of the fluid against moving relative to each other. Because viscosity represents a nonconservative force, part of the kinetic energy of a fluid is converted to internal energy when layers of fluid slide past one another. This is similar to the mechanism by which an object sliding on a rough horizontal surface experiences a transformation of kinetic energy to internal energy.

Because the motion of a real fluid is very complex and not yet fully understood, we adopt a simplification model. As we shall see, many features of real fluids in motion can be understood by considering the behavior of an ideal fluid. In our

Figure 15.14

Turbulent flow: the tip of a rotating blade (the dark region at the top) forms a vortex in air that is being heated by an alcohol lamp (the wick is at the bottom). Note the air turbulence on both sides of the rotating blade. (*Kim Vandiver and Harold Edgerton, Palm Press, Inc.*)

Figure 15.13

An illustration of steady flow around an automobile in a test wind tunnel. The streamlines in the airflow are made visible by smoke particles. (*Andy Sacks/Stone*)

• *Properties of an ideal fluid under ideal flow*

simplification model, we make the following four assumptions:

- **Nonviscous fluid** In a nonviscous fluid, internal friction is ignored. An object moving through the fluid experiences no viscous force.
- **Incompressible fluid** The density of the fluid is assumed to remain constant regardless of the pressure in the fluid.
- **Steady flow** In steady flow, we assume that the velocity of the fluid at each point remains constant in time.
- **Irrotational flow** Fluid flow is irrotational if the fluid about any point has no angular momentum. If a small paddle wheel placed anywhere in the fluid does not rotate about the wheel's center of mass, the flow is irrotational. (If the wheel were to rotate, as it would if turbulence were present, the flow would be rotational.)

The first two assumptions in our simplification model are properties of our ideal fluid. The last two are descriptions of the way that the fluid flows.

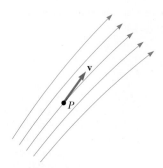

Figure 15.15

This diagram represents a set of streamlines (*blue lines*). A particle at *P* follows one of these streamlines, and its velocity is tangent to the streamline at each point along its path.

15.6 • STREAMLINES AND THE CONTINUITY EQUATION FOR FLUIDS

If you are watering your garden, and your garden hose is too short, you might do one of two things to help you reach the garden with the water (before you look for a longer hose!). You might attach a nozzle to the end of the hose, or, in the absence of a nozzle, you might place your thumb over the end of the hose, allowing the water to come out of a narrower opening. Why does either of these techniques cause the water to come out faster, so that it can be projected over a longer range? We shall see the answer to this question in this section.

The path taken by a particle of the fluid under steady flow is called a **streamline.** The velocity of the particle is always tangent to the streamline, as shown in Figure 15.15. No two streamlines can cross each other, for if they did, a particle could move either way at the crossover point and then the flow would not be steady. A set of streamlines as shown in Figure 15.15 forms what is called a *tube of flow.* Particles of the fluid cannot flow into or out of the sides of this tube, because if they did, the streamlines would be crossing each other.

Consider an ideal fluid flowing through a pipe of nonuniform size as in Figure 15.16. The particles in the fluid move along the streamlines in steady flow. Let us analyze this situation using the nonisolated system in steady-state model. We have seen this model used for energy in Chapter 7, but we were told at that time that the model can be used for any conserved quantity. The volume of an incompressible fluid is a conserved quantity. Assuming no leaks in our pipe, we can neither create nor destroy fluid, just as we could not create nor destroy energy in Chapters 6 and 7.

We choose as our system the region of space in the pipe from point 1 to point 2 in Figure 15.16. Let us assume that this region is filled with fluid at all times. As the fluid flows in the pipe, fluid enters the system at point 1 and leaves the system at point 2. Imagine that the fluid moves through a distance Δx_1 at point 1, and moves a distance Δx_2 at point 2 as it leaves the system. The volume of fluid entering the system at point 1 is $A_1 \Delta x_1$ and the volume leaving at point 2 is $A_2 \Delta x_2$. Because the volume of an incompressible fluid is a conserved quantity, these two volumes must be equal for the system to be in steady state. If this were not true, the volume of fluid in the system would be changing. Thus,

$$A_1 \Delta x_1 = A_2 \Delta x_2$$

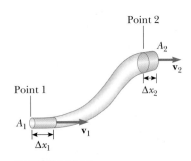

Figure 15.16

A fluid moving with steady flow through a pipe of varying cross-sectional area. The volume of fluid flowing through A_1 in a time interval Δt must equal the volume flowing through A_2 in the same time interval.

Let us divide this equation by the time interval during which the fluid moves:

$$\frac{A_1 \, \Delta x_1}{\Delta t} = \frac{A_2 \, \Delta x_2}{\Delta t}$$

In the limit as the time interval shrinks to zero, the ratio of distance traveled to the time interval is the instantaneous speed of the fluid, so we can write this as

$$A_1 v_1 = A_2 v_2 \qquad \qquad \text{[15.7]}$$

• *Continuity equation for fluids*

This expression, called the **continuity equation for fluids,** says that the **product of the area and the fluid speed at all points along the pipe is a constant.** Therefore, the speed is high where the tube is constricted and low where the tube is wide. The product Av, which has the dimensions of volume per time, is called the **volume flow rate.** This is why a nozzle or your thumb over the garden hose allows you to project the water farther. By reducing the area through which the water flows, you increase its speed. Thus, you project the water from the hose with a high initial velocity, resulting in a large value of the range, as discussed for projectiles in Chapter 3.

Quick **Quiz** 15.6

As water exits a faucet at a low setting so that a streamline flow of water results, the stream becomes narrower as it descends. Explain why this happens.

PITFALL PREVENTION 15.4

A new continuity equation

This is the first time that we have seen a continuity equation for anything other than energy. A continuity equation exists for *any* conserved quantity. Equation 15.7 states that the rate of flow of fluid into the system equals the rate of flow out of the system because fluid cannot be stored in the system once the pipe is full.

Example 15.5   Watering a Garden

A water hose 2.50 cm in diameter is used by a gardener to fill a 30.0-L bucket. The gardener notes that it takes 1.00 min to fill the bucket. A nozzle with an opening of cross-sectional area 0.500 cm^2 is then attached to the hose. The nozzle is held so that water is projected horizontally from a point 1.00 m above the ground. Over what horizontal distance can the water be projected?

Solution   We identify point 1 within the hose and point 2 at the exit of the nozzle. We first find the speed of the water in the hose from the bucket-filling information. The cross-sectional area of the hose is

$$A_1 = \pi r^2 = \pi \frac{d^2}{4} = \pi \left[\frac{(2.50 \text{ cm})^2}{4} \right] = 4.91 \text{ cm}^2$$

According to the data given, the volume flow rate is equal to 30.0 liters per minute:

$$A_1 v_1 = 30.0 \text{ L/min} = \frac{30.0 \times 10^3 \text{ cm}^3}{60.0 \text{ s}} = 500 \text{ cm}^3/\text{s}$$

$$v_1 = \frac{500 \text{ cm}^3/\text{s}}{4.91 \text{ cm}^2} = 102 \text{ cm/s} = 1.02 \text{ m/s}$$

Now we use the continuity equation for fluids to find the speed $v_2 = v_{xi}$ with which the water exits the nozzle. The subscript i anticipates that this will be the *initial* velocity component of the water projected from the hose, and the sub-

script x indicates that the initial velocity vector of the projected water is in the horizontal direction.

$$A_1 v_1 = A_2 v_2 = A_2 v_{ix}$$

$$v_{ix} = \frac{A_1}{A_2} v_1 = \frac{4.91 \text{ cm}^2}{0.500 \text{ cm}^2} (1.02 \text{ m/s}) = 10.0 \text{ m/s}$$

We now shift our thinking away from fluids and to projectile motion because the water is in free-fall once it exits the nozzle. A particle of the water falls through a vertical distance of 1.00 m starting from rest, which requires a time, which we find from Equation 3.13:

$$y_f = y_i + v_{yi}t - \tfrac{1}{2} gt^2$$

$$-1.00 \text{ m} = 0 + 0 - \tfrac{1}{2} (9.80 \text{ m/s}^2) \, t^2$$

$$t = \sqrt{\frac{2(1.00 \text{ m})}{9.80 \text{ m/s}^2}} = 0.452 \text{ s}$$

In the horizontal direction, we apply Equation 3.12 to a particle of water to find the horizontal distance:

$$x_f = x_i + v_{xi}t = 0 + (10.0 \text{ m/s})(0.452 \text{ s}) = 4.52 \text{ m}$$

EXERCISE   The Garfield Thomas water tunnel at Pennsylvania State University has a circular cross-section that constricts from a diameter of 3.6 m to the test section, which is 1.2 m in diameter. If the speed of flow is 3.0 m/s in the larger diameter pipe, determine the speed of flow in the test section.

Answer   27 m/s

Daniel Bernoulli (1700–1782)

Bernoulli, a Swiss physicist and mathematician, made important discoveries in fluid dynamics. His most famous work, *Hydrodynamica,* published in 1738, is both a theoretical and a practical study of equilibrium, pressure, and speed in fluids. In this publication, Bernoulli also attempted the first explanation of the behavior of gases with changing pressure and temperature; this was the beginning of the kinetic theory of gases, which we will study in Chapter 16. *(Corbis-Bettmann)*

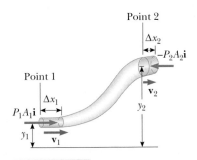

Figure 15.17

A fluid flowing through a constricted pipe with streamline flow. In a time interval Δt, the change in the system is the same as if the fluid in the section of length Δx_1 moves to the position of the section of length Δx_2. The volumes of fluid in the two sections are equal.

15.7 • BERNOULLI'S PRINCIPLE

You have probably had the experience of driving on a highway and having a large truck pass by you at high speed. In that situation, you may have had the frightening feeling that your car was being pulled in toward the truck as it passed. We will see the origin for this effect in this section.

As a fluid moves through a region in which its speed or elevation above the Earth's surface changes, the pressure in the fluid varies with these changes. In 1738, the Swiss physicist Daniel Bernoulli first derived an expression that relates the pressure to fluid speed and elevation.

Let us develop a mathematical representation for **Bernoulli's principle,** which shows explicitly the dependence of pressure on speed and elevation. Consider the flow of an ideal fluid (Section 15.5) through a nonuniform pipe in a time Δt, as illustrated in Figure 15.17. The system is chosen to be the section of fluid that is initially between points 1 and 2 in Figure 15.17 and the Earth. This choice of system is different from that for the derivation of the continuity equation for fluids, in which we chose a region of space as the system. We analyze the current situation with the nonisolated system model. Work is being done on our system by the external fluid that is in contact with the two ends of the fluid in the system, and the kinetic and gravitational potential energy of the system is changing as a result. Thus, the continuity equation for energy (Eq. 6.20) for the system in this situation is

$$\Delta K + \Delta U = W \qquad [15.8]$$

Let us evaluate each of the terms in Equation 15.8. First notice that the shaded elements of fluid of lengths Δx_1 and Δx_2 in Figure 15.17 represent the only change between the initial situation and the final situation. The fluid in between these elements experiences no change in either kinetic or gravitational potential energy. Thus, we can restrict our attention to the shaded elements of fluid. The difference in kinetic energy is that between the element at point 2 and the element at point 1:

$$\Delta K = \tfrac{1}{2}(\Delta m)\, v_2{}^2 - \tfrac{1}{2}(\Delta m)\, v_1{}^2 \qquad [15.9]$$

We have used the same increment of mass Δm for both elements of fluid— both elements have the same volume and, because the fluid is incompressible, the same mass. Now, the change in potential energy of the fluid–Earth system is that associated with moving the element of fluid located at point 1 up to the location at point 2:

$$\Delta U = (\Delta m)\, g y_2 - (\Delta m)\, g y_1 \qquad [15.10]$$

Finally, we evaluate the work done on the section of fluid. The fluid to the left of point 1 does positive work on our section because it applies a force in the same direction as the displacement. The fluid to the right of the element at point 2 does negative work because the force and displacement vectors are in opposite directions. Thus, the net work done on the system is

$$W = F_1\, \Delta x_1 - F_2\, \Delta x_2$$

We replace the force with the product of pressure and cross-sectional area of the pipe:

$$W = P_1\, A_1\, \Delta x_1 - P_2\, A_2\, \Delta x_2$$

Finally, we recognize the product of cross-sectional area and displacement as the volume of the elements of fluid:

$$W = P_1 \, \Delta V - P_2 \, \Delta V \qquad [15.11]$$

where we have not included subscripts on the volume element ΔV because both elements have the same volume for an incompressible fluid. We now use Equations 15.9, 15.10, and 15.11 to substitute for each term in Equation 15.8:

$$\Delta K + \Delta U = W$$

$$\tfrac{1}{2}(\Delta m)\,v_2^{\,2} - \tfrac{1}{2}(\Delta m)\,v_1^{\,2} + (\Delta m)\,gy_2 - (\Delta m)\,gy_1 = P_1 \, \Delta V - P_2 \, \Delta V$$

Dividing each term by the volume element ΔV gives

$$\tfrac{1}{2}\left(\frac{\Delta m}{\Delta V}\right) v_2^{\,2} - \tfrac{1}{2}\left(\frac{\Delta m}{\Delta V}\right) v_1^{\,2} + \left(\frac{\Delta m}{\Delta V}\right) gy_2 - \left(\frac{\Delta m}{\Delta V}\right) gy_1 = P_1 - P_2$$

The ratio of the mass Δm for a fluid element to the volume ΔV of the element is the density ρ of the fluid. Incorporating this fact and rearranging terms, we have

$$P_1 + \tfrac{1}{2}\rho v_1^{\,2} + \rho gy_1 = P_2 + \tfrac{1}{2}\rho v_2^{\,2} + \rho gy_2 \qquad [15.12]$$

● *Bernoulli's equation*

This is **Bernoulli's equation** applied to an ideal fluid. It is often expressed as

$$P + \tfrac{1}{2}\rho v^2 + \rho gy = \text{constant} \qquad [15.13]$$

Bernoulli's equation says that the sum of the pressure P, the kinetic energy per unit volume $\tfrac{1}{2}\rho v^2$, and gravitational potential energy per unit volume ρgy has the same value at all points along a streamline.

When the fluid is at rest, $v_1 = v_2 = 0$ and Equation 15.12 becomes

$$P_1 - P_2 = \rho g(y_2 - y_1) = \rho gh$$

which agrees with Equation 15.4.

PITFALL PREVENTION 15.5
Bernoulli's principle for gases

 For gases, we cannot use the assumption that the fluid is incompressible. Thus, Equation 15.12 is not true for gases. The qualitative behavior is the same, however—as the speed of the gas increases, its pressure decreases.

Quick Quiz 15.7

You observe two helium balloons floating next to each other at the ends of strings secured to a table. The facing surfaces of the balloons are separated by 1 to 2 cm. You blow through the opening between the balloons. What happens to the balloons? (a) They move toward each other; (b) they move away from each other; (c) they are unaffected.

Quick Quiz 15.8

You may have noticed a convertible passing you on a highway with its cloth top bulging upward, as if the interior of the car is pressurized. Why does this happen?

THINKING PHYSICS 15.6

During a tornado, the windows of a house sometimes explode outward. At other times, the roof of a house is lifted off the house. What causes these events to happen? How does opening the windows in a tornado (which may seem counterintuitive!) help?

WEB

For information on tornadoes and tornado safety, visit **www.usatoday.com/weather/wtwisto.htm**

Reasoning The wind speed in a tornado is very high. According to Bernoulli's principle, such high wind speeds will result in low pressures. When these winds pass by a closed window, the pressure outside the window is reduced well below that of the pressure on the inside surface, which is due to the still air inside the house. As a result, the net outward force can burst the window. When these high-speed winds pass over the roof of a house, the pressure above the roof is much lower than the pressure below the roof, again due to the still air inside the house. The pressure difference can be enough to lift the roof off the house.

Example 15.6 Sinking the Cruise Ship

A scuba diver is hunting for fish with a spear gun. He accidentally fires the gun so that a spear punctures the side of a cruise ship. The hole is located at a depth of 10.0 m below the water surface. With what speed does the water enter the cruise ship through the hole?

Reasoning We identify point 1 as the water surface outside the ship, which we will assign as $y = 0$. At this point, the water is static, so $v_1 = 0$. We identify point 2 as a point just inside the hole in the interior of the ship because that is the point at which we wish to evaluate the speed of the water.

This point is at a depth $y = -h = -10.0$ m below the water surface. We use Bernoulli's equation to compare these two points. At both points, the water is open to atmospheric pressure, so that $P_1 = P_2 = P_0$.

Solution Based on the arguments made in the Reasoning, Bernoulli's equation becomes

$$P_0 + \tfrac{1}{2}\rho(0)^2 + \rho g(0) = P_0 + \tfrac{1}{2}\rho v_2^2 + \rho g(-h)$$

$$\longrightarrow v_2 = \sqrt{2gh} = \sqrt{2(9.80 \text{ m/s}^2)(10.0 \text{ m})} = \boxed{14 \text{ m/s}}$$

Example 15.7 The Venturi Tube

The horizontal constricted pipe illustrated in Figure 15.18, known as a *Venturi tube,* can be used to measure the flow speed of an incompressible fluid. Let us determine the flow speed at point 2 if the pressure difference $P_1 - P_2$ is known.

Solution Because the pipe is horizontal, $y_1 = y_2$ and Equation 15.12 applied to points 1 and 2 gives

$$P_1 + \tfrac{1}{2}\rho v_1^2 = P_2 + \tfrac{1}{2}\rho v_2^2$$

From the continuity equation for fluids (Eq. 15.7), we see that $A_1 v_1 = A_2 v_2$ or

$$v_1 = \frac{A_2}{A_1} v_2$$

Substituting this expression into the previous equation gives

$$P_1 + \tfrac{1}{2}\rho\left(\frac{A_2}{A_1}\right)^2 v_2^2 = P_2 + \tfrac{1}{2}\rho v_2^2$$

$$v_2 = \boxed{A_1 \sqrt{\frac{2(P_1 - P_2)}{\rho(A_1{}^2 - A_2{}^2)}}}$$

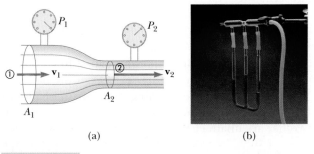

(a) (b)

Figure 15.18

(a) (Example 15.7) The pressure P_1 is greater that the pressure P_2, because $v_1 < v_2$. This device can be used to measure the speed of fluid flow. (b) A Venturi tube. The higher level of fluid in the middle column shows that the pressure at the top of the column, which is in the constricted region of the Venturi tube, is lower. *(Courtesy of Central Scientific Company)*

We can also obtain an expression for v_1 using this result and the continuity equation for fluids. Note that because $A_2 < A_1$, it follows that $P_1 > P_2$. In other words, the pressure is reduced in the constricted part of the pipe.

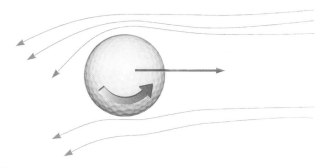

Figure 15.20

A spinning golf ball experiences a lifting force that allows it to travel much farther than it would if it were not spinning.

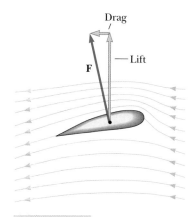

Figure 15.19

Streamline flow around a moving airplane wing. The air approaching from the right is deflected downward by the wing. Because the airstream is deflected, it exerts a force **F** on the wing.

15.8 • OTHER APPLICATIONS OF FLUID DYNAMICS

Consider the streamlines that flow around an airplane wing as shown in Figure 15.19. Let us assume that the airstream approaches the wing horizontally from the right. The tilt of the wing causes the airstream to be deflected downward. Because the airstream is deflected by the wing, the wing must exert a force on the airstream. According to Newton's third law, the airstream must exert an equal and opposite force **F** on the wing. This force has a vertical component called the **lift** (or aerodynamic lift) and a horizontal component called **drag.** The lift depends on several factors, such as the speed of the airplane, the area of the wing, its curvature, and the angle between the wing and the horizontal. As this angle increases, turbulent flow can set in above the wing to reduce the lift.

In general, an object experiences lift by any effect that causes the fluid to change its direction as it flows past the object. Some factors that influence lift are the shape of the object, its orientation with respect to the fluid flow, spinning motion (for example, a spinning golf ball as in Figure 15.20), and the texture of the object's surface.

A number of devices operate in a manner similar to the *atomizer* in Figure 15.21. A stream of air passing over an open tube reduces the pressure above the tube. This reduction in pressure causes the liquid to rise into the air stream. The liquid is then dispersed into a fine spray of droplets. This type of system is used in perfume bottles and paint sprayers.

Bernoulli's principle explains one symptom of advanced arteriosclerosis called *vascular flutter.* The artery is constricted as a result of an accumulation of plaque on its inner walls (Fig. 15.22). Plaque is a combination of fat, cell debris, connective tissue, and sometimes calcium that forms a flat patch inside a blood vessel. The blood speed through the constriction is higher than elsewhere according to the continuity equation for fluids. According to Bernoulli's principle, the pressure in the constriction is lower than elsewhere. If the blood speed is sufficiently high in the constricted region, the artery may collapse under the larger external pressure, causing a momentary interruption in blood flow. At this point, the speed of the blood goes to zero, its pressure rises again, and the vessel reopens. As the blood rushes through the constricted artery, the internal pressure drops and again the artery closes. Such variations in blood flow can be heard with a stethoscope.

APPLICATION

Vascular flutter

Figure 15.21

A stream of air passing over a tube dipped into a liquid will cause the liquid to rise in the tube.

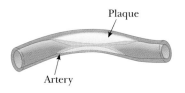

Figure 15.22

Blood must travel faster than normal through a constricted region of an artery.

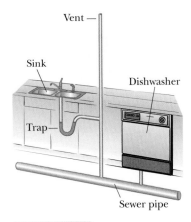

Figure 15.23

(Thinking Physics 15.7) What is the purpose of the vent—the vertical pipe open to the atmosphere?

THINKING PHYSICS 15.7

Consider the portion of a home plumbing system shown in Figure 15.23. The water trap in the pipe below the sink captures a plug of water that prevents sewer gas from finding its way from the sewer pipe, up the sink drain, and into the home. Suppose the dishwasher is draining, so that water is moving to the left in the sewer pipe. What is the purpose of the vent, which is open to the air above the roof of the house? In which direction is air moving at the opening of the vent, upward or downward?

Reasoning Let us imagine that the vent is not present, so that the drain pipe for the sink is simply connected through the trap to the sewer pipe. As water from the dishwasher moves to the left in the sewer pipe, the pressure in the sewer pipe is reduced below atmospheric pressure, according to Bernoulli's principle. The pressure at the drain in the sink is still at atmospheric pressure. Thus, this pressure differential pushes the plug of water in the water trap of the sink down the drain pipe into the sewer pipe, removing it as a barrier to sewer gas. With the addition of the vent to the roof, the reduced pressure of the dishwasher water causes air to enter the vent pipe at the roof. This keeps the pressure in the vent pipe and the right-hand side of the sink drain pipe close to atmospheric pressure, so that the plug of water in the water trap remains in the trap and prevents sewer gas from entering the home.

context connection
15.9 • A NEAR MISS EVEN BEFORE LEAVING SOUTHAMPTON

As the *Titanic* began its maiden voyage, it experienced a potentially disastrous incident before it left Southampton Harbor. (If this disaster had occurred, however, it is highly likely that it would have been much less disastrous in terms of lives lost than the incident that actually occurred later in the voyage.) The *Titanic* passed closely by the *New York*, which was tied securely next to the *Oceanic* at the dock, with the keels of the two ships parallel. As the *Titanic* passed by, the *New York* was forced toward it, the *New York*'s mooring ropes snapped, and its stern swung out toward the *Titanic*. It was only quick thinking by the harbor pilot on the *Titanic*, who reversed the engines, causing the *Titanic* to slow and allow the *New York* to pass by safely, that saved the two ships from a collision. As it was, a collision was averted by only a few feet, and the *Titanic* was delayed by over an hour in her departure. Figure 15.24 is a photograph taken from the *Titanic*, showing how close the ships came to colliding.

It is ironic that the captain of the *Titanic*, E. J. Smith, who watched the near miss from the bridge, was captain on one of the *Titanic*'s sister ships, *Olympic*, when a similar incident occurred seven months before the *New York* incident. In this case, the cruiser *Hawke* was pulled toward the *Olympic* and a collision was not averted. The *Hawke*'s bow was seriously damaged in the collision, and the hull of the *Olympic* was punctured above and below the waterline. Both ships were able to return to port but needed extensive repairs.

Why did these events occur? The answer lies in Bernoulli's principle. As ships move through the water, they push water out of the way, and the water moves around the sides of the ship. Imagine now that a ship such as the *Titanic* passes near another ship, such as the *New York*, with their keels parallel. The water moving around the side of the *Titanic* toward the *New York* is forced into a

Figure 15.24

While leaving Southampton harbor, the *Titanic (left)* experienced a near miss with the *New York (right),* due to Bernoulli's principle. If this accident had actually occurred, it may have changed the timing enough that the *Titanic,* once underway, might not have been sunk by an iceberg.

narrow channel between the ships. Because water is incompressible, its volume remains constant when it is squeezed into a narrow channel. The initial tendency is for the compressed water to rise into the air between the ships because the air above the water offers little resistance to being compressed. As soon as the water level between the ships rises, however, the water will begin to flow in a direction parallel to the keels toward the lower level water near the bow and stern of the ships. Thus, the *water between the ships is moving at a higher speed than the water on the opposite sides of the ships.* According to Bernoulli's principle, this rapidly moving water exerts less pressure on the sides of the ships than the slower moving water on the outer sides. The result is a *net force pushing the two ships toward each other.*

Thus, captains of boats and ships are advised not to pass too close by other boats in a parallel direction. If this occurs, then the boats could be pushed into each other. This effect occurs for air, explaining the effect of the passing truck at the beginning of Section 15.7. As the truck passes, air is squeezed into the channel between the truck and your car. The high-speed air exerts less pressure on your car than the slower moving air on the opposite side, resulting in a net force toward the truck.

In this Context Connection section, we investigated an application of Bernoulli's principle. In the Context Conclusion, we shall explore the difficulties in visiting the *Titanic* due to its great depth under the ocean surface.

SUMMARY

The **pressure** P in a fluid is the force per unit area that the fluid exerts on a surface:

$$P \equiv \frac{F}{A} \qquad [15.1]$$

In the SI system, pressure has units of N/m², and 1 N/m² = 1 pascal (Pa).

The pressure in a liquid varies with depth h according to the expression

$$P = P_0 + \rho gh \qquad [15.4]$$

where P_0 is atmospheric pressure ($= 1.013 \times 10^5$ N/m²) and ρ is the density of the liquid, assumed uniform.

Pascal's law states that when a change in pressure is applied to a fluid, the change in pressure is transmitted undiminished to every point in the fluid and to every point on the walls of the container.

When an object is partially or fully submerged in a fluid, the fluid exerts an upward force on the object called the **buoyant force**. According to **Archimedes's principle,** the buoyant

force is equal to the weight of the fluid displaced by the object.

Various aspects of fluid dynamics can be understood by adopting a simplification model in which the fluid is nonviscous and incompressible and the fluid motion is a steady flow with no turbulence.

Using this model, two important results regarding fluid flow through a pipe of nonuniform size can be obtained:

• The flow rate through the pipe is a constant, which is equivalent to stating that the product of the cross-sectional area A and the speed v at any point is a constant. This gives the **continuity equation for fluids:**

$$A_1 v_1 = A_2 v_2 = \text{constant} \qquad [15.7]$$

• The sum of the pressure, kinetic energy per unit volume, and gravitational potential energy per unit volume has the same value at all points along a streamline. This corresponds to **Bernoulli's equation:**

$$P_1 + \tfrac{1}{2}\rho v_1^2 + \rho g y_1 = P_2 + \tfrac{1}{2}\rho v_2^2 + \rho g y_2 \qquad [15.12]$$

QUESTIONS

1. Suppose a damaged ship can just barely float in the ocean. It is pulled toward shore and into a river, heading toward a dry dock for repair. As it is pulled up the river, it sinks. Why?

2. Figure Q15.2 shows aerial views from directly above two dams. Both dams are equally long (the vertical dimension in the diagram) and equally deep (into the page in the diagram). The dam on the left holds back a very large lake, while the dam on the right holds back a narrow river. Which dam has to be built more strongly?

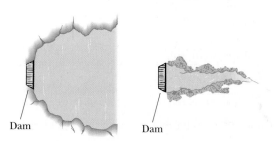

Dam   Dam

Figure Q15.2

3. Some physics students attach a long tube to the opening of a balloon. Leaving the balloon on the ground, the other end of the tube is hoisted to the roof of a multistory campus building. Students at the top of the building start pouring water into the tube. The students on the ground watch the balloon fill with water. On the roof, the students

are surprised to see that the tube never seems to fill up—more and more water can continue to be poured into the tube. On the ground, the balloon bursts and drenches the students there. Why did the balloon–tube system never fill up, and why did the balloon burst?

4. A barge is carrying a load of gravel along a river. It approaches a low bridge, and the captain realizes that the top of the pile of gravel is not going to make it under the bridge. The captain orders the crew to quickly shovel gravel from the pile into the water. Is this a good decision?

5. A person in a boat floating in a small pond throws an anchor overboard. Does the level of the pond rise, fall, or remain the same?

6. Two drinking glasses having equal weights but different shapes and different cross-sectional areas are filled to the same level with water. According to the expression $P = P_0 + \rho gh$, the pressure is the same at the bottom of both glasses. In view of this, why does one weigh more than the other?

7. A helium-filled balloon rises until its density becomes the same as that of the air. If a sealed submarine begins to sink, will it go all the way to the bottom of the ocean or will it stop when its density becomes the same as that of the surrounding water?

8. A fish rests on the bottom of a bucket of water while the bucket is being weighed. When the fish begins to swim around, does the weight change?

9. Will a ship ride higher in the water of an inland lake or in the ocean? Why?

10. Lead has a greater density than iron, and both are denser than water. Is the buoyant force on a lead object greater than, less than, or equal to the buoyant force on an iron object of the same volume?

11. An ice cube is placed in a glass of water. What happens to the level of the water as the ice melts?

12. The water supply for a city is often provided from reservoirs built on high ground. Water flows from the reservoir, through pipes, and into your home when you turn the tap on your faucet. Why is the water flow faster out of a faucet on the first floor of a building than in an apartment on a higher floor?

13. If air from a hair dryer is blown upward under a Ping-Pong ball, the ball can be suspended in air. Explain.

14. When ski-jumpers are airborne, why do they bend their bodies forward and keep their hands at their sides?

15. Explain why a sealed bottle partially filled with a liquid can float.

16. When is the buoyant force on a swimmer greater—after exhaling or after inhaling?

17. A piece of unpainted wood is partially submerged in a container filled with water. If the container is sealed and pressurized above atmospheric pressure, does the wood rise, fall, or remain at the same level? (*Hint:* Wood is porous.)

18. Because atmospheric pressure is about 10^5 N/m^2 and the area of a person's chest is about 0.13 m^2, the force of the atmosphere on one's chest is approximately 13 000 N. In view of this enormous force, why don't our bodies collapse?

19. If you release a ball while inside a freely falling elevator, the ball remains in front of you rather than falling to the floor, because the ball, the elevator, and you all experience the same downward acceleration g. What happens if you repeat this experiment with a helium-filled balloon? (This one is tricky.)

20. Two identical ships set out to sea. One is loaded with a cargo of Styrofoam, and the other is empty. Which ship is more submerged?

21. A small piece of steel is tied to a block of wood. When the wood is placed in a tub of water with the steel on top, half of the block is submerged. If the block is inverted so that the steel is under water, does the amount of the block submerged increase, decrease, or remain the same? What happens to the water level in the tub when the block is inverted?

22. Prairie dogs ventilate their burrows by building a mound over one entrance, which is open to a stream of air. A second entrance at ground level is open to almost stagnant air. How does this construction create an air flow through the burrow?

23. An unopened can of diet cola floats when placed in a tank of water, whereas a can of regular cola of the same brand sinks in the tank. What do you suppose could explain this behavior?

PROBLEMS

1, 2, 3 = straightforward, intermediate, challenging   ☐ = full solution available in the *Student Solutions Manual and Study Guide*

web = solution posted at **http://www.harcourtcollege.com/physics/**   ☐ = computer useful in solving problem

☐ = Interactive Physics   ☐ = paired numerical/symbolic problems   ☐ = life science application

Section 15.1   Pressure

1. Calculate the mass of a solid iron sphere that has a diameter of 3.00 cm.

2. A king orders a gold crown having a mass of 0.500 kg. When it arrives from the metalsmith, the volume of the crown is found to be 185 cm^3. Is the crown made of solid gold?

3. A 50.0-kg woman balances on one heel of a pair of high-heeled shoes. If the heel is circular and has a radius of 0.500 cm, what pressure does she exert on the floor?

4. The four tires of an automobile are inflated to a gauge pressure of 200 kPa. Each tire has an area of 0.024 0 m^2 in contact with the ground. Determine the weight of the automobile.

5. What is the total mass of the Earth's atmosphere? (The radius of the Earth is 6.37×10^6 m, and atmospheric pressure at the surface is 1.013×10^5 N/m^2.)

Section 15.2   Variation of Pressure with Depth

6. (a) Calculate the absolute pressure at an ocean depth of 1 000 m. Assume the density of sea water is 1 024 kg/m^3 and that the air above exerts a pressure of 101.3 kPa. (b) At this depth, what force must the frame around a circular submarine porthole having a diameter of 30.0 cm exert to counterbalance the force exerted by the water?

7. The spring of the pressure gauge shown in Figure 15.2 has a force constant of 1 000 N/m, and the piston has a

(a)

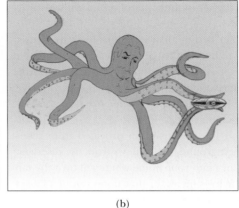
(b)

Figure P15.10

diameter of 2.00 cm. As the gauge is lowered into water, what change in depth causes the piston to move in by 0.500 cm?

8. The small piston of a hydraulic lift has a cross-sectional area of 3.00 cm^2, and its large piston has a cross-sectional area of 200 cm^2 (see Figure 15.5). What force must be applied to the small piston for the lift to raise a load of 15.0 kN? (In service stations, this force is usually exerted by compressed air.)

9. What must be the contact area between a suction cup (completely exhausted) and a ceiling if the cup is to support the weight of an 80.0-kg student?

web

10. (a) A very powerful vacuum cleaner has a hose 2.86 cm in diameter. With no nozzle on the hose, what is the weight of the heaviest brick that the cleaner can lift? (Fig. P15.10) (b) A very powerful octopus uses one sucker of diameter 2.86 cm on each of the two shells of a clam in an attempt to pull the shells apart. Find the greatest force the octopus can exert in salt water 32.3 m deep. (*Caution:* Experimental verification can be interesting, but do not drop a brick on your foot. Do not overheat the motor of a vacuum cleaner. Do not get an octopus mad at you.)

11. A swimming pool has dimensions 30.0 m × 10.0 m and a flat bottom. When the pool is filled to a depth of 2.00 m with fresh water, what is the force caused by the water on the bottom? On each end? On each side?

Section 15.3 Pressure Measurements

12. Figure P15.12 shows Superman attempting to drink water through a very long straw. With his great strength he achieves maximum possible suction. The walls of the tubular straw do not collapse. (a) Find the maximum height through which he can lift the water. (b) Still thirsty, the Man of Steel repeats his attempt on the Moon, which has

Figure P15.12

no atmosphere. Find the difference between the water levels inside and outside the straw.

13. Blaise Pascal duplicated Torricelli's barometer using a red Bordeaux wine, of density 984 kg/m^3, as the working liquid (Fig. P15.13). What was the height h of the wine column for normal atmospheric pressure? Would you expect the vacuum above the column to be as good as for mercury?

web

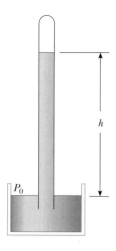

Figure P15.13

14. Mercury is poured into a U-tube as in Figure P15.14a. The left arm of the tube has cross-sectional area A_1 of 10.0 cm^2, and the right arm has a cross-sectional area A_2 of 5.00 cm^2. One hundred grams of water is then poured into the right arm as in Figure P15.14b. (a) Determine the length of the water column in the right arm of the U-tube. (b) The density of mercury is 13.6 g/cm^3. Through what distance h does the mercury rise in the left arm?

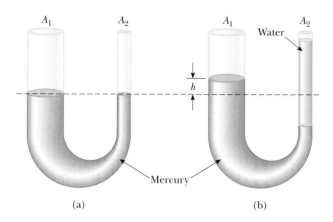

Figure P15.14

15. Normal atmospheric pressure is 1.013×10^5 Pa. The approach of a storm causes the height of a mercury barometer to drop by 20.0 mm from the normal height. What is the atmospheric pressure? (The density of mercury is 13.59 g/cm^3.)

16. The human brain and spinal cord are immersed in the cerebrospinal fluid. The fluid is normally continuous between the cranial and spinal cavities. It normally exerts a pressure of 100 to 200 mm of H$_2$O above the prevailing at-

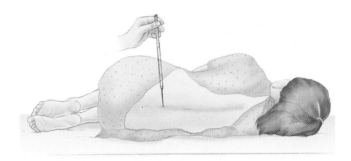

Figure P15.16

mospheric pressure. In medical work pressures are often measured in units of millimeters of H$_2$O because body fluids, including the cerebrospinal fluid, typically have the same density as water. The pressure of the cerebrospinal fluid can be measured be means of a *spinal tap,* as illustrated in Figure P15.16. A hollow tube is inserted into the spinal column, and the height to which the fluid rises is observed. If the fluid rises to a height of 160 mm, its gauge pressure is recorded as 160 mm H$_2$O. (a) Express this pressure in pascals, in atmospheres, and in millimeters of mercury. (b) Sometimes it is necessary to determine if an accident victim has suffered a crushed vertebra that is blocking flow of the cerebrospinal fluid in the spinal column. In other cases a physician may suspect a tumor or other growth is blocking the spinal column and inhibiting flow of cerebrospinal fluid. Such conditions can be investigated by means of the *Queckenstedt test.* In this procedure, the veins in the patient's neck are compressed to make the blood pressure rise in the brain. The increase in pressure in the blood vessels is transmitted to the cerebrospinal fluid. What should be the normal effect on the height of the fluid in the spinal tap? (c) Suppose that compressing the veins had no effect on the fluid level. What might account for this?

Section 15.4 Buoyant Forces and Archimedes's Principle

17. A Ping-Pong ball has a diameter of 3.80 cm and average density of 0.084 0 g/cm^3. What force is required to hold it completely submerged under water?

18. A swimmer of mass m rests on top of a Styrofoam slab, which has thickness h and density ρ_s. What is the area of the slab if it floats in fresh water with its upper surface just awash?

19. A 10.0-kg block of metal measuring 12.0 cm $\times$ 10.0 cm $\times$ 10.0 cm is suspended from a scale and immersed in water as in Figure P15.19. The 12.0-cm dimension is vertical, and the top of the block is 5.00 cm below the

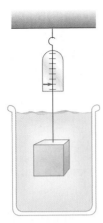

Figure P15.19

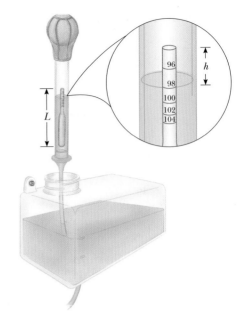

Figure P15.22

surface of the water. (a) What are the forces acting on the top and on the bottom of the block? (Take atmospheric pressure to be $1.013\ 0 \times 10^5\ \text{N/m}^2$.) (b) What is the reading of the spring scale? (c) Show that the buoyant force equals the difference between the forces at the top and bottom of the block.

20. To an order of magnitude, how many helium-filled toy balloons would be required to lift you? Because helium is an irreplaceable resource, develop a theoretical answer rather than an experimental one. In your solution state what physical quantities you take as data and the values you measure or estimate for them.

21. A cube of wood having an edge dimension of 20.0 cm and
web a density of 650 kg/m³ floats on water. (a) What is the distance from the horizontal top surface of the cube to the water level? (b) How much lead weight has to be placed on top of the cube so that its top is just level with the water?

22. Determination of the density of a fluid has many important applications. A car battery contains sulfuric acid, and the battery will not function properly if the acid density is too low. Similarly, the effectiveness of antifreeze in your car's engine coolant depends on the density of the mixture (usually ethylene glycol and water). When you donate blood to a blood bank, its screening includes determination of the density of the blood, because higher density indicates higher hemoglobin content. A *hydrometer* is an instrument used to determine liquid density. A simple one is sketched in Figure P15.22. The bulb of a syringe is squeezed and released to lift a sample of the liquid of interest into a tube containing a calibrated rod of known density. The rod, of length L and average density ρ_0, floats partially immersed in the liquid of density ρ. A length h of the rod protrudes above the surface of the liquid. Show that the density of the liquid is given by

$$\rho = \frac{\rho_0 L}{L - h}$$

23. How many cubic meters of helium are required to lift a balloon with a 400-kg payload to a height of 8 000 m? (Take $\rho_{\text{He}} = 0.180\ \text{kg/m}^3$.) Assume that the balloon maintains a constant volume and that the density of air decreases with the altitude z according to the expression $\rho_{\text{air}} = \rho_0 e^{-z/8000}$, where z is in meters and $\rho_0 = 1.25\ \text{kg/m}^3$ is the density of air at sea level.

24. A bathysphere used for deep-sea exploration has a radius of 1.50 m and a mass of 1.20×10^4 kg. To dive, this submarine takes on mass in the form of sea water. Determine the amount of mass the submarine must take on if it is to descend at a constant speed of 1.20 m/s, when the resistive force on it is 1 100 N in the upward direction. The density of sea water is $1.03 \times 10^3\ \text{kg/m}^3$.

25. A plastic sphere floats in water with 50.0% of its volume submerged. This same sphere floats in glycerin with 40.0% of its volume submerged. Determine the densities of the glycerin and the sphere.

26. **Review Problem.** A long cylindrical rod of radius r is weighted on one end so that it floats upright in a fluid having a density ρ. It is pushed down a distance x from its equilibrium position and released. Show that the rod will execute simple harmonic motion if the resistive effects of the fluid are ignored, and determine the period of the oscillations.

Section 15.5   Fluid Dynamics

Section 15.6   Streamlines and the Continuity Equation for Fluids

Section 15.7   Bernoulli's Principle

27. (a) A water hose 2.00 cm in diameter is used to fill a 20.0-L bucket. If it takes 1.00 min to fill the bucket, what is the speed v at which water moves through the hose? (*Note:* 1 L = 1 000 cm^3.) (b) If the hose has a nozzle 1.00 cm in diameter, find the speed of the water at the nozzle.

28. A horizontal pipe 10.0 cm in diameter has a smooth reduction to a pipe 5.00 cm in diameter. If the pressure of the water in the larger pipe is 8.00×10^4 Pa and the pressure in the smaller pipe is 6.00×10^4 Pa, at what rate does water flow through the pipes?

29. Water falls over a dam of height h with a mass flow rate of R, in units of kg/s. (a) Show that the power available from the water is

$$\mathscr{P} = Rgh$$

where g is the acceleration due to gravity. (b) Each hydroelectric unit at the Grand Coulee Dam takes in water at a rate of 8.50×10^5 kg/s from a height of 87.0 m. The power developed by the falling water is converted to electric power with an efficiency of 85.0%. How much electric power is produced by each hydroelectric unit?

30. *Torricelli's Law:* A tank with a cover, containing a liquid of density ρ, has a hole in its side at a distance y_1 from the bottom (Fig. P15.30). The diameter of the hole is small relative to the diameter of the tank. The air above the liquid is maintained at a pressure P. Assume steady frictionless flow. Show that the speed at which the fluid leaves the hole when the liquid level is a distance h above the hole is

$$v_1 = \sqrt{\frac{2(P - P_0)}{\rho} + 2gh}$$

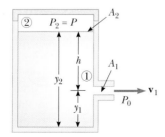

Figure P15.30

31. A large storage tank with an open top is filled to a height h_0. The tank is punctured at a height h above the bottom of the tank (Fig. P15.31). Find an expression for how far from the tank the exiting stream lands.

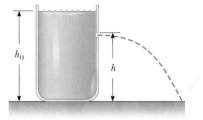

Figure P15.31   Problems 31 and 32.

32. A hole is punched at height h in the side of an open tank of height h_0. The container stands on level ground and is full of water, as shown in Figure P15.31. If the water is to shoot as far as possible horizontally, (a) how far from the bottom of the container should the hole be punched? (b) Ignoring friction losses, how far (initially) from the side of the container will the water land?

33. A legendary Dutch boy saved Holland by plugging a hole in a dike with his finger, 1.20 cm in diameter. If the hole was located 2.00 m below the surface of the North Sea (density 1 030 kg/m^3), (a) what was the force on his finger? (b) If he pulled his finger out of the hole, how long would it take the released water to fill 1 acre of land to a depth of 1 foot, assuming the hole remained constant in size? (A typical U.S. family of four uses 1 acre-foot of water, 1 234 m^3, in 1 year.)

34. Water is pumped through a pipe 15.0 cm in diameter, from the Colorado River up to Grand Canyon Village, located on the rim of the canyon. The river is at an elevation of 564 m and the village is at an elevation of 2 096 m. (a) What is the minimum pressure at which the water must be pumped if it is to arrive at the village? (b) If 4 500 m^3 is pumped per day, what is the speed of the water in the pipe? (c) What additional pressure is necessary to deliver this flow? (*Note:* You may assume that the acceleration due to gravity and the density of air are constant over this range of elevations.)

35. Old Faithful Geyser in Yellowstone Park erupts at approximately 1-h intervals, and the height of the water column reaches 40.0 m. (Fig. P15.35). (a) Consider the rising stream as a series of separate drops. Analyze the free-fall motion of one of the drops to determine the speed at which the water leaves the ground. (b) Treat the rising stream as an ideal fluid in streamline flow. Use Bernoulli's equation to determine the speed of the water as it leaves ground level. (c) What is the pressure (above

atmospheric) in the heated underground chamber if its depth is 175 m? You may assume that the chamber is large compared with the geyser's vent.

Figure P15.35 *(Stan Osolinski/Dembinsky Photo Associates)*

36. A Venturi tube may be used as a fluid flow meter (see Fig. 15.18a). The radius of the outlet tube is 1.00 cm, the radius of the inlet tube is 2.00 cm, and the fluid is gasoline ($\rho = 700$ kg/m^3). If the difference in pressure is $P_1 - P_2 =$ 21.0 kPa, find the fluid flow rate in cubic meters per second.

Section 15.8 Other Applications of Fluid Dynamics

37. An airplane has a mass of 1.60×10^4 kg, and each wing has an area of 40.0 m^2. During level flight, the pressure on the lower wing surface is 7.00×10^4 Pa. Determine the pressure on the upper wing surface.

38. A siphon is used to drain water from a tank, as illustrated in Figure P15.38. The siphon has a uniform diameter. Assume steady flow without friction. (a) If the distance $h = 1.00$ m, find the speed of outflow at the end of the siphon. (b) What is the limitation on the height of the top of the siphon above the water surface? (For the flow of the liquid to be continuous, the pressure must not drop below the vapor pressure of the liquid.)

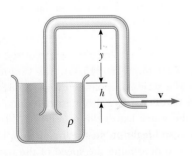

Figure P15.38

39. Bernoulli's principle can have important consequences for the design of buildings. For example, wind can move around a skyscraper at remarkably high speed, creating low pressure. The higher atmospheric pressure in the still air inside the buildings can cause windows to pop out. As originally constructed, the John Hancock building in Boston popped window panes, which fell many stories to the sidewalk below. (a) Suppose that a horizontal wind moves with a speed of 11.2 m/s outside a large pane of plate glass with dimensions 4.00 m $\times$ 1.50 m. Assume the density of the air to be 1.30 kg/m^3. The air inside the building is at atmospheric pressure. What is the total force exerted by air on the window pane? (b) If a second skyscraper is built nearby, the air speed can be especially high where wind passes through the narrow separation between the buildings. Solve part (a) again if the wind speed is 22.4 m/s, twice as high.

40. A hypodermic syringe contains a medicine with the density of water (Figure P15.40). The barrel of the syringe has a cross-sectional area $A = 2.50 \times 10^{-5}$ m^2, and the needle has a cross-sectional area 1.00×10^{-8} m^2. In the absence of a force on the plunger, the pressure everywhere is 1 atm. A force **F** of magnitude 2.00 N acts on the plunger, making medicine squirt horizontally from the needle. Determine the speed of the medicine as it leaves the tip of the needle.

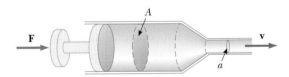

Figure P15.40

Section 15.9 Context Connection—A Near Miss Even Before Leaving Southampton

41. According to the caption of the chapter-opening photograph, about 11% of an iceberg is above water. (a) Confirm this value mathematically. (b) Suppose an iceberg were floating in fresh water rather than sea water. Would a larger or smaller percentage be above the water? Calculate this percentage.

42. The *Titanic* is docked in Southampton harbor just before boarding. You, as a ticket agent for White Star Lines, notice where the water level is on a scale of numbers marked on the side of the vessel. Because your ticket-collecting job is so boring that you need something to occupy your mind, you make a note of the water level with the intent to check it after everyone boards. During the next 2 h, 2 205 passengers, of average mass 75.0 kg, board the *Titanic*. You

notice that the ship has sunk 1.00 cm deeper in the water with the passengers on board. What is the cross-sectional area of the hull of the *Titanic* right at the waterline?

43. **Review Problem.** The *Titanic* is drifting in Southampton harbor before its fateful journey, and the captain wishes to stop the drift by dropping an anchor. The anchor is made of iron and has mass 2 000 kg. It is attached to a massless rope. The rope is wrapped around a reel in the form of a solid disk of radius 0.250 m and mass 300 kg, which rotates on a frictionless axle. (a) Find the angular displacement of the reel when the anchor moves down 15.0 m. (b) Find the acceleration of the anchor as it falls through the air, which offers negligible resistance. (c) While the anchor continues to drop through the water, the water exerts a friction force of 2 500 N on it. With what acceleration does the anchor move through the water? (d) While the anchor drops through the water, what torque is exerted on the reel?

Additional Problems

44. Water is forced out of a fire extinguisher by air pressure, as shown in Figure P15.44. How much gauge air pressure in the tank (above atmospheric) is required for the water jet to have a speed of 30.0 m/s when the water level is 0.500 m below the nozzle?

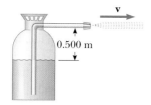

Figure P15.44

45. The true weight of an object can be measured in a vacuum, where buoyant forces are absent. A body of volume V is weighed in air on a balance with the use of weights of density ρ. If the density of air is ρ_{air} and the balance reads F'_g, show that the true weight F_g of the object is

$$F_g = F'_g + \left(V - \frac{F'_g}{\rho g} \right) \rho_{air} g$$

46. A light spring with force constant $k = 90.0$ N/m is attached vertically to a table (Fig. P15.46a). A 2.00-g balloon is filled with helium (density $= 0.180$ kg/m^3) to a volume of 5.00 m^3 and is then connected to the spring, causing it to stretch as in Figure P15.46b. Determine the extension distance L when the balloon is in equilibrium.

47. **Review Problem.** With reference to Figure 15.6, show that
web the total torque exerted by the water behind the dam about a horizontal axis through O is $\frac{1}{6}\rho gwH^3$. Show that

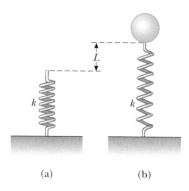

(a)   (b)

Figure P15.46

the effective line of action of the total force exerted by the water is at a distance $\frac{1}{3}H$ above O.

48. Evangelista Torricelli was the first person to realize that we live at the bottom of an ocean of air. He correctly surmised that the pressure of our atmosphere is attributable to the weight of the air. The density of air at 0°C at the Earth's surface is 1.29 kg/m^3. The density decreases with increasing altitude (as the atmosphere thins). On the other hand, if we assume that the density is constant at 1.29 kg/m^3 up to some altitude h, and zero above that altitude, then h would represent the depth of the ocean of air. Use this model to determine the value of h that gives a pressure of 1.00 atm at the surface of the Earth. Would the peak of Mount Everest rise above the surface of such an atmosphere?

49. A beaker of mass m_b containing oil of mass m_o (density = ρ_o) rests on a scale. A block of iron of mass m_i is suspended from a spring scale and completely submerged in the oil as in Figure P15.49. Determine the equilibrium readings of both scales.

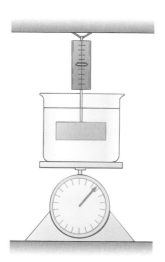

Figure P15.49

50. **Review Problem.** A copper cylinder hangs at the bottom of a steel wire of negligible mass. The top end of the wire is fixed. When the wire is struck, it emits sound with a fundamental frequency of 300 Hz. If the copper cylinder is then submerged in water so that half its volume is below the water line, determine the new fundamental frequency.

51. In about 1657, Otto von Guericke, inventor of the air pump, evacuated a sphere made of two brass hemispheres. Two teams of eight horses each could pull the hemispheres apart only on some trials, and then "with greatest difficulty," with the resulting sound likened to a cannon firing (Fig. P15.51). (a) Show that the force F required to pull the evacuated hemispheres apart is $\pi R^2(P_0 - P)$, where R is the radius of the hemispheres and P is the pressure inside the hemispheres, which is much less than P_0. (b) Determine the force if $P = 0.100P_0$ and $R = 0.300$ m.

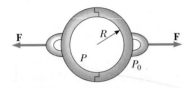

Figure P15.51   The colored engraving, dated 1672, illustrates Otto von Guericke's demonstration of the force due to air pressure as performed before Emperor Ferdinand III in 1657. *(The Granger Collection)*

52. Show that the variation of atmospheric pressure with altitude is given by $P = P_0e^{-\alpha h}$, where $\alpha = \rho_0 g/P_0$, P_0 is atmospheric pressure at some reference level $y = 0$, and ρ_0 is the atmospheric density at this level. Assume that the decrease in atmospheric pressure over an infinitesimal change in altitude (so that the density is approximately uniform) is given by $dP = -\rho g\, dy$, and that the density of air is proportional to the pressure.

53. An incompressible, nonviscous fluid is initially at rest in the vertical portion of the pipe shown in Figure P15.53a, where $L = 2.00$ m. When the valve is opened, the fluid flows into the horizontal section of the pipe. What is the speed of the fluid when all of it is in the horizontal section, as in Figure P15.53b? Assume the cross-sectional area of the entire pipe is constant.

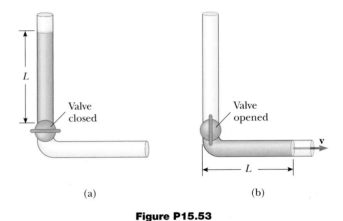

Figure P15.53

54. A cube of ice whose edges measure 20.0 mm is floating in a glass of ice-cold water with one of its faces parallel to the water's surface. (a) How far below the water surface is the bottom face of the block? (b) Ice-cold ethyl alcohol is gently poured onto the water surface to form a layer 5.00 mm thick above the water. The alcohol does not mix with the water. When the ice cube again attains hydrostatic equilibrium, what is the distance from the top of the water to the bottom face of the block? (c) Additional cold ethyl alcohol is poured onto the water's surface until the top surface of the alcohol coincides with the top surface of the ice cube (in hydrostatic equilibrium). How thick is the required layer of ethyl alcohol?

55. A U-tube open at both ends is partially filled with water (Fig. P15.55a). Oil having a density of 750 kg/m³ is then poured into the right arm and forms a column of height $L = 5.00$ cm (Fig. P 15.55b). (a) Determine the difference h in the heights of the two liquid surfaces. (b) The right

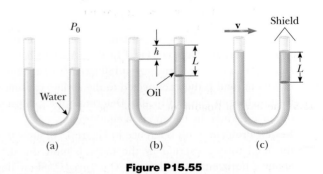

Figure P15.55

arm is then shielded from any air motion while air is blown across the top of the left arm until the surfaces of the two liquids are at the same height (Fig. P15.55c). Determine the speed of the air being blown across the left arm. Take the density of air as 1.29 kg/m^3.

56. The water supply of a building is fed through a main pipe 6.00 cm in diameter. A 2.00-cm-diameter faucet tap, located 2.00 m above the main pipe, is observed to fill a 25.0-L container in 30.0 s. (a) What is the speed at which the water leaves the faucet? (b) What is the gauge pressure in the 6.00-cm main pipe? (Assume the faucet is the only "leak" in the building.)

57. The *spirit-in-glass thermometer,* invented in Florence, Italy, around 1654, consists of a tube of liquid (the spirit) containing a number of submerged glass spheres with slightly different masses (Fig. P15.57). At sufficiently low temperatures all the spheres float, but as the temperature rises, the spheres sink one after another. The device is a crude but interesting tool for measuring temperature. Suppose that the tube is filled with ethyl alcohol, whose density is $0.789\,45 \text{ g/cm}^3$ at $20.0°C$ and decreases to $0.780\,97 \text{ g/cm}^3$ at $30.0°C$. (a) If one of the spheres has a radius of 1.000 cm and is in equilibrium halfway up the tube at $20.0°C$, determine its mass. (b) When the temperature increases to $30.0°C$, what mass must a second sphere of the same radius have to be in equilibrium at the halfway point? (c) At $30.0°C$ the first sphere has fallen to the bottom of the tube. What upward force does the bottom of the tube exert on this sphere?

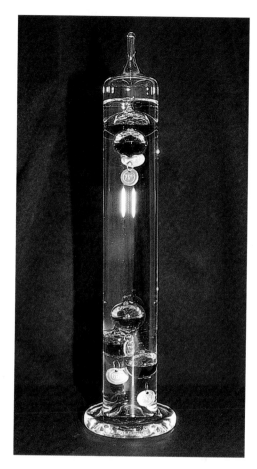

Figure P15.57 *(Courtesy of Jeanne Maier)*

ANSWERS TO QUICK QUIZZES

15.1 You would be better off with the basketball player. The basketball player has a larger weight, but it is distributed over a large area of the bottom of the sneaker, so that the pressure, which is force per unit area, is relatively small. The woman's smaller weight is distributed over the very small area of the spiked heel, resulting in a large pressure. Some museums require women in high-heeled shoes to wear slippers or special heel attachments so that they do not damage the wood floors.

15.2 Because water is so much less dense than mercury, the column for a water barometer would have to be $h = P_0/\rho g = 10.3$ m high, making it inconveniently tall.

15.3 The level of floating of a ship in the water is unaffected by the atmospheric pressure. The buoyant force results from the pressure *differential* in the fluid. On a high-pressure day, the pressure at all points in the water is higher than on a low-pressure day. Because water is almost incompressible, however, the rate of change of pressure with depth is the same, resulting in no change in the buoyant force.

15.4 The balance will not be in equilibrium—the lead side will be lower. Despite the fact that the weights on both sides of the balance are the same, the Styrofoam, due to its larger volume, experiences a larger buoyant force from the surrounding air. Thus, the net force of gravity plus the buoyant force is larger, in the downward direction, for the lead than for the Styrofoam.

15.5 Because the density of ice is about 89% that of sea water, 89% of a floating iceberg is underwater. Thus, only 11% is visible above water—this is the "tip of the iceberg."

15.6 As the water falls, its speed increases. Because the flow rate Av must remain constant at all points (Eq. 15.7), the stream must become narrower as the speed increases.

15.7 (a). The air between the balloons is moving at high speed. Therefore, it is at lower pressure than the stagnant area on the outer edges of the balloons, according to Bernoulli's principle. Thus, the balloons move toward each other.

15.8 The air passing over the roof of the car has a high speed relative to the car. The air inside the car is moving with the car and thus has a speed of zero relative to the car. As a result, the air inside the car has a higher pressure than the air above the roof and the flexible top is pushed upward.

Search for the *Titanic*

FINDING AND VISITING THE *TITANIC*

We have now investigated the physics of fluids and can respond to our central question for the *Search for the Titanic* Context:

How can we safely visit the wreck of the Titanic?

Many individuals felt that the *Titanic* was unsinkable. One of the factors in this belief was the series of watertight bulkheads that divided the hull of the ship into several watertight compartments. Even if the hull were breached so that a compartment became flooded, the incoming water could be isolated to that compartment by closing watertight doors in the bulkhead.

According to the design of the ship, the *Titanic* could be kept afloat if her four forwardmost compartments were flooded. Unfortunately, the collision with the iceberg caused a breach in the first five compartments. As these forward compartments filled, the extra weight of the water in the bow of the ship resulted in the bow sinking into the water and the stern lifting out of the water (Figure 1).

Figure 1

The *Titanic* struck the iceberg near the bow, so that the forward compartments filled with water and sank, lifting the stern of the ship above the water. *(Painting by Ken Marschall)*

541

Despite the shipbuilders' pride in their watertight compartments, they were not watertight at the top. The bulkheads only went up to a certain height in the ship and then ended. Thus, as the *Titanic* tilted forward, water from one compartment simply spilled over the top of the bulkhead into the next compartment, and the compartments filled one by one.

Some experts after the disaster claimed that opening the watertight doors in the bulkheads would have kept the *Titanic* afloat longer, with an increased possibility of another ship arriving in time to save those who were not able to leave in the lifeboats. According to this hypothesis, if the water entering the forward compartments had been allowed to distribute evenly along the ship by passing through the doors in the bulkheads, the ship would not have tilted so that water could spill over the tops of the bulkheads. The sinking of a ship is a complicated event, however, and this hypothesis is not universally accepted.

In the region of the sinking, the depth of the ocean is about 4 km. When the *Titanic* was located and visited, the depth was measured to be 3 784 m. Let us use Equation 15.3 to calculate the pressure at this depth of sea water:

$$P = P_0 + \rho g h$$
$$= 1.013 \times 10^5 \text{ Pa} + (1.03 \times 10^3 \text{ kg/m}^3)(9.80 \text{ m/s}^2)(3\,784 \text{ m})$$
$$= 3.83 \times 10^7 \text{ Pa} = 378 \text{ atm}$$

Thus, the pressure is 378 times that at the surface! A human being could not survive at this pressure.

Plans for finding and possibly salvaging the *Titanic* began immediately after she sank. Families of some of the wealthy victims contacted salvage companies with requests for a salvage operation. One plan suggested filling the *Titanic* with Ping-Pong balls, so that its overall density would be less than that of water, and she would float to the surface! This plan, of course, ignores the obvious problems of the Ping-Pong balls' failure to withstand the tremendous pressure at that depth.

An early expedition to find the *Titanic* occurred in 1980 and met with failure. After unsuccessful searches were carried out by a number of teams, Dr. Robert Ballard, of Woods Hole Oceanographic Institute, discovered the wreck in 1985, in cooperation with a team from IFREMER, the French National Institute of Oceanography. The search began by towing a sonar device, which emitted sound waves through the water and analyzed the reflection of the waves from solid objects such as the hull of the *Titanic*. The search pattern was a tedious back-and-forth sweeping of the area near the reported sinking of the ship, looking for sonar reflections, and checking possible sites with a magnetometer for the presence of an iron hull.

After failing to find the *Titanic* with the sonar system, Ballard switched to a visual search using an underwater video system called *Argo*. After three more grueling weeks with no reward, the searchers saw one of the *Titanic's* boilers in the early morning of September 1, 1985. This first evidence of the wreck having been found, the remainder was quickly located.

The visual evidence indicated clearly that the *Titanic* had split in two, as had been reported by some of the survivors in 1912. As it tilted steeply in the water due to the sinking of the bow, the midsection was subjected to forces that it was not designed to sustain. After the break occurred, but while the two sections were still connected, the stern section settled back into the water, with the bow section hanging from it underwater. As more water entered the bow section, it pulled the stern section into a vertical orientation and then broke free, beginning its trip to the bottom. The stern bobbed for a while as it filled with water and then sank into the ocean.

WEB

For information on Woods Hole Oceanographic Institute, visit **www.whoi.edu/**
For information on IFREMER, visit **www.ifremer.fr/anglais/**

Figure 2

The bow section of the *Titanic* rests on the ocean floor relatively intact. *(Painting by Ken Marschall)*

The two sections of the *Titanic* lie about 600 m apart on the ocean floor. The bow section (Figure 2) is fairly intact, but the stern section (Figure 3) is tremendously damaged. As the bow section sank, it was already filled with water. As the pressure of the water outside the bow section increased during the plummet to the bottom, the pressure inside the section increased. On the other hand, the stern section spent most of the time in the air before sinking. Thus, as it sank, a significant volume of air was still trapped inside the stern section. As the water pressure increased while the stern section sank, the air pressure inside could not increase along with the external water pressure because many air pockets existed in the relatively sealed sections of the structure. Thus, some areas of the hull of the stern section experienced very large pressure on the outside surface, with relatively low pressure on the inside surface. This extreme imbalance in pressures possibly caused an implosion of the stern section at some depth, causing severe destruction of the structure. Further damage was caused by the sudden impact of hitting the bottom. With little structural integrity left, the decks pancaked downward as the stern hit the ocean floor.

The *Titanic* has been visited by a number of teams, for purposes of research, salvage, and even filmmaking, by James Cameron, the director of the 1997 version of the film *Titanic*. What is necessary to travel to such depths? Our calculation of the pressure at the location of the *Titanic* indicates that special submarines must be used that can withstand such high pressure while maintaining normal atmospheric pressure inside for the human occupants. This was first done by Ballard in the summer of 1986, using a deep-sea submersible called *Alvin*, with a remote-controlled robot named *Jason Junior*. Figure 4 shows the structure of *Alvin*.

Alvin has a titanium alloy hull that can withstand the pressure at the depth of the *Titanic*. The submersible has room for three occupants although they are quite cramped. A number of air tanks on the craft can be flooded with water. Blocks of

Figure 3

The stern section of the *Titanic* is heavily damaged and lies in pieces on the ocean floor. *(Painting by Ken Marschall)*

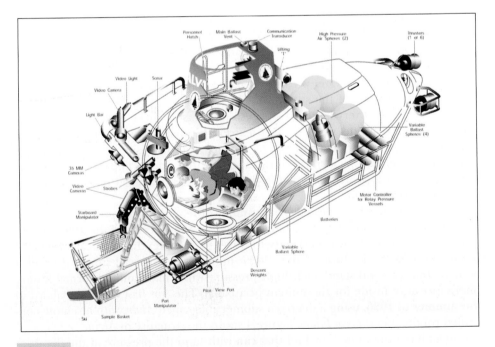

Figure 4

The submersible *Alvin*, which can carry three scientists to the great depths at which the *Titanic* currently lies. *(E. Paul Oberlander/Woods Hole Oceanographic Institution)*

iron can also be jettisoned. When the tanks are filled with air and the iron blocks are attached, *Alvin* floats on water. When the air tanks are flooded with water, the submersible sinks. In visiting the *Titanic*, the first step is to fill the air tanks with water, and then wait for 2.5 h to sink to the bottom. After visiting the wreckage, the iron blocks are jettisoned. After doing this, the buoyant force on *Alvin* is larger than its weight, and it starts upward on another long journey to the surface. The descent and ascent of *Alvin* are effective examples of applications of the physics described in Case I in Section 15.4.

Once *Alvin* reaches the *Titanic*, the only means of viewing it are by visual inspection through the portholes or by video, using *Jason Junior*. It is impossible to don scuba gear and exit the submersible because of the tremendous pressure.

There is much more to the story of the *Titanic*, but we need to return to our investigations into physics.

PROBLEMS

1. When the *Titanic* is in its normal sailing position, the torque about the midpoint of the ship is zero. Imagine now that the bow of the ship is under water and the stern is in the air above the water during the sinking process, as in Figure 1. The waterline is at the midpoint of the ship, and the keel of the ship makes a 45° angle with the horizontal. The torque is zero here also, because the *Titanic* is in equilibrium. However, the torque counteracting the weight of the stern section must be applied by the structure of the ship at the midpoint. Calculate the torque about the midpoint of the *Titanic* required to hold the stern section in the air. Model the *Titanic* as a uniform rod of length 269 m and mass 4.2×10^7 kg. This torque caused the *Titanic* to split near the middle of the ship during the sinking process.

2. The *Titanic* had two almost identical sister ships: the *Britannic* and the *Olympic*. The *Britannic* sank in 1916 off the coast of Athens, Greece, possibly due to a mine planted during World War I. It now sits below 119 m of sea water. (a) What is the pressure at the location of the *Britannic*? (b) Can you visit the *Britannic* by scuba diving?

3. The submersible *Alvin* requires 2.5 h to sink to the location of the *Titanic*. (a) What is the average speed during descent? (b) Suppose the speed of *Alvin* remains constant during the entire descent. Is the average density of *Alvin* greater, less than, or equal to that of sea water? (*Hint:* Include the resistive force on *Alvin* as it moves through the water in your analysis.)

4. The *Titanic* is only one of many maritime disasters. In 1956, despite the presence of radar, which was not invented at the time of the *Titanic*, a collision occurred between the Italian luxury liner *Andrea Doria* and the Swedish liner *Stockholm*. Deaths were few because of the long time interval between the collision and the sinking of the *Andrea Doria*, but a remarkable event occurred. A 14-year-old girl was sleeping in her bed on the *Andrea Doria* before the collision and awoke on the bow of the *Stockholm* after the collision. The bow of the *Stockholm* pierced the hull of the *Andrea Doria* at the point where the girl was sleeping and she was miraculously transferred from one ship to the other! Her only injuries were two broken kneecaps.

   Let us imagine that the collision between the *Andrea Doria* and the *Stockholm* is perfectly inelastic. (This was not the case, as the *Stockholm* pulled away after the collision, but it is a reasonable model.) The weight of the *Andrea Doria* is 29 100 tons, and it is traveling at full speed of 23 knots at the time of the collision, 15° south of west. The *Stockholm* has a weight of 12 165 tons and is traveling at 18 knots 30° east of south at the time of the collision. (a) Just after the collision, what is the velocity, in knots, of the combined wreckage? (b) What fraction of the initial kinetic energy was transformed or transferred away in the collision?

Tables

TABLE A.1	**Conversion Factors**

Length

	m	cm	km	in.	ft	mi
1 meter	1	10^2	10^{-3}	39.37	3.281	6.214×10^{-4}
1 centimeter	10^{-2}	1	10^{-5}	0.393 7	3.281×10^{-2}	6.214×10^{-6}
1 kilometer	10^3	10^5	1	3.937×10^4	3.281×10^3	0.621 4
1 inch	2.540×10^{-2}	2.540	2.540×10^{-5}	1	8.333×10^{-2}	1.578×10^{-5}
1 foot	0.304 8	30.48	3.048×10^{-4}	12	1	1.894×10^{-4}
1 mile	1 609	1.609×10^5	1.609	6.336×10^4	5 280	1

Mass

	kg	g	slug	u
1 kilogram	1	10^3	6.852×10^{-2}	6.024×10^{26}
1 gram	10^{-3}	1	6.852×10^{-5}	6.024×10^{23}
1 slug	14.59	1.459×10^4	1	8.789×10^{27}
1 atomic mass unit	1.660×10^{-27}	1.660×10^{-24}	1.137×10^{-28}	1

Note: 1 metric ton = 1 000 kg.

Time

	s	min	h	day	yr
1 second	1	1.667×10^{-2}	2.778×10^{-4}	1.157×10^{-5}	3.169×10^{-8}
1 minute	60	1	1.667×10^{-2}	6.994×10^{-4}	1.901×10^{-6}
1 hour	3 600	60	1	4.167×10^{-2}	1.141×10^{-4}
1 day	8.640×10^4	1 440	24	1	2.738×10^{-5}
1 year	3.156×10^7	5.259×10^5	8.766×10^3	365.2	1

Speed

	m/s	cm/s	ft/s	mi/h
1 meter per second	1	10^2	3.281	2.237
1 centimeter per second	10^{-2}	1	3.281×10^{-2}	2.237×10^{-2}
1 foot per second	0.304 8	30.48	1	0.681 8
1 mile per hour	0.447 0	44.70	1.467	1

Note: 1 mi/min = 60 mi/h = 88 ft/s.

continued

TABLE A.1 *Continued*

Force

	N	lb
1 newton	1	0.224 8
1 pound	4.448	1

Work, Energy, Heat

	J	ft·lb	eV
1 joule	1	0.737 6	6.242×10^{18}
1 ft·lb	1.356	1	8.464×10^{18}
1 eV	1.602×10^{-19}	1.182×10^{-19}	1
1 cal	4.186	3.087	2.613×10^{19}
1 Btu	1.055×10^{3}	7.779×10^{2}	6.585×10^{21}
1 kWh	3.600×10^{6}	2.655×10^{6}	2.247×10^{25}

	cal	Btu	kWh
1 joule	0.238 9	9.481×10^{-4}	2.778×10^{-7}
1 ft·lb	0.323 9	1.285×10^{-3}	3.766×10^{-7}
1 eV	3.827×10^{-20}	1.519×10^{-22}	4.450×10^{-26}
1 cal	1	3.968×10^{-3}	1.163×10^{-6}
1 Btu	2.520×10^{2}	1	2.930×10^{-4}
1 kWh	8.601×10^{5}	3.413×10^{2}	1

Pressure

	Pa	atm
1 pascal	1	9.869×10^{-6}
1 atmosphere	1.013×10^{5}	1
1 centimeter mercury[a]	1.333×10^{3}	1.316×10^{-2}
1 pound per inch^2	6.895×10^{3}	6.805×10^{-2}
1 pound per foot^2	47.88	4.725×10^{-4}

	cm Hg	lb/in.^{2}	lb/ft^2
1 pascal	7.501×10^{-4}	1.450×10^{-4}	2.089×10^{-2}
1 atmosphere	76	14.70	2.116×10^{3}
1 centimeter mercury[a]	1	0.194 3	27.85
1 pound per inch^2	5.171	1	144
1 pound per foot^2	3.591×10^{-2}	6.944×10^{-3}	1

[a] At 0°C and at a location where the acceleration due to gravity has its "standard" value, 9.806 65 m/s^2.

TABLE A.2 Symbols, Dimensions, and Units of Physical Quantities

Quantity	Common Symbol	Unit[a]	Dimensions[b]	Unit in Terms of Base SI Units
Acceleration	**a**	m/s^2	L/T^2	m/s^2
Amount of substance	n	mole		mol
Angle	θ, ϕ	radian (rad)		
Angular acceleration	α	rad/s^2	T^{-2}	s^{-2}
Angular frequency	ω	rad/s	T^{-1}	s^{-1}
Angular momentum	**L**	$kg \cdot m^2/s$	ML^2/T	$kg \cdot m^2/s$
Angular speed	ω	rad/s	T^{-1}	s^{-1}
Area	A	m^2	L^2	m^2
Atomic number	Z			
Capacitance	C	farad (F)	Q^2T^2/ML^2	$A^2 \cdot s^4/kg \cdot m^2$
Charge	q, Q, e	coulomb (C)	Q	$A \cdot s$
Charge density				
Line	λ	C/m	Q/L	$A \cdot s/m$
Surface	σ	C/m^2	Q/L^2	$A \cdot s/m^2$
Volume	ρ	C/m^3	Q/L^3	$A \cdot s/m^3$
Conductivity	σ	$1/\Omega \cdot m$	Q^2T/ML^3	$A^2 \cdot s^3/kg \cdot m^3$
Current	I	AMPERE	Q/T	A
Current density	**J**	A/m^2	Q/T^2	A/m^2
Density	ρ	kg/m^3	M/L^3	kg/m^3
Dielectric constant	κ			
Displacement	**r, s**	METER	L	m
Distance	d, h			
Length	ℓ, L			
Electric dipole moment	**p**	$C \cdot m$	QL	$A \cdot s \cdot m$
Electric field	**E**	V/m	ML/QT^2	$kg \cdot m/A \cdot s^3$
Electric flux	Φ_E	$V \cdot m$	ML^3/QT^3	$kg \cdot m^3/A \cdot s^3$
Electromotive force	ε	volt (V)	ML^2/QT^2	$kg \cdot m^2/A \cdot s^3$
Energy	E, U, K	joule (J)	ML^2/T^2	$kg \cdot m^2/s^2$
Entropy	S	J/K	$ML^2/T^2 \cdot K$	$kg \cdot m^2/s^2 \cdot K$
Force	**F**	newton (N)	ML/T^2	$kg \cdot m/s^2$
Frequency	f	hertz (Hz)	T^{-1}	s^{-1}
Heat	Q	joule (J)	ML^2/T^2	$kg \cdot m^2/s^2$
Inductance	L	henry (H)	ML^2/Q^2	$kg \cdot m^2/A^2 \cdot s^2$
Magnetic dipole moment	$\boldsymbol{\mu}$	$N \cdot m/T$	QL^2/T	$A \cdot m^2$
Magnetic field	**B**	tesla (T) ($= Wb/m^2$)	M/QT	$kg/A \cdot s^2$
Magnetic flux	Φ_B	weber (Wb)	ML^2/QT	$kg \cdot m^2/A \cdot s^2$
Mass	m, M	KILOGRAM	M	kg
Molar specific heat	C	$J/mol \cdot K$		$kg \cdot m^2/s^2 \cdot mol \cdot K$
Moment of inertia	I	$kg \cdot m^2$	ML^2	$kg \cdot m^2$
Momentum	**p**	$kg \cdot m/s$	ML/T	$kg \cdot m/s$
Period	T	s	T	s
Permeability of space	μ_0	$N/A^2 (= H/m)$	ML/Q^2T	$kg \cdot m/A^2 \cdot s^2$
Permittivity of space	ϵ_0	$C^2/N \cdot m^2 (= F/m)$	Q^2T^2/ML^3	$A^2 \cdot s^4/kg \cdot m^3$
Potential	V	volt (V) ($= J/C$)	ML^2/QT^2	$kg \cdot m^2/A \cdot s^3$
Power	$\mathscr{P}$	watt (W) ($= J/s$)	ML^2/T^3	$kg \cdot m^2/s^3$

continued

| TABLE A.2 | *Continued* | | | |

Quantity	Common Symbol	Unit[a]	Dimensions[b]	Unit in Terms of Base SI Units
Pressure	P	pascal (Pa) $= (N/m^2)$	M/LT^2	$kg/m \cdot s^2$
Resistance	R	ohm $(\Omega) (= V/A)$	ML^2/Q^2T	$kg \cdot m^2/A^2 \cdot s^3$
Specific heat	c	$J/kg \cdot K$	$L^2/T^2 \cdot K$	$m^2/s^2 \cdot K$
Speed	v	m/s	L/T	m/s
Temperature	T	KELVIN	K	K
Time	t	SECOND	T	s
Torque	τ	$N \cdot m$	ML^2/T^2	$kg \cdot m^2/s^2$
Volume	V	m^3	L^3	m^3
Wavelength	λ	m	L	m
Work	W	joule $(J) (= N \cdot m)$	ML^2/T^2	$kg \cdot m^2/s^2$

[a] The base SI units are given in uppercase letters.

[b] The symbols M, L, T, Q, and K denote mass, length, time, charge, and temperature, respectively.

| TABLE A.3 | Table of Atomic Masses[a] |

Atomic Number Z	Element	Symbol	Chemical Atomic Mass (u)	Mass Number (*Indicates Radioactive) A	Atomic Mass (u)	Percent Abundance	Half-Life (If Radioactive) $T_{1/2}$
0	(Neutron)	n		1*	1.008 665		10.4 min
1	Hydrogen	H	1.007 9	1	1.007 825	99.985	
	Deuterium	D		2	2.014 102	0.015	
	Tritium	T		3*	3.016 049		12.33 yr
2	Helium	He	4.002 60	3	3.016 029	0.000 14	
				4	4.002 602	99.999 86	
				6*	6.018 886		0.81 s
3	Lithium	Li	6.941	6	6.015 121	7.5	
				7	7.016 003	92.5	
				8*	8.022 486		0.84 s
4	Beryllium	Be	9.012 2	7*	7.016 928		53.3 days
				9	9.012 174	100	
				10*	10.013 534		1.5×10^6 yr
5	Boron	B	10.81	10	10.012 936	19.9	
				11	11.009 305	80.1	
				12*	12.014 352		0.020 2 s
6	Carbon	C	12.011	10*	10.016 854		19.3 s
				11*	11.011 433		20.4 min
				12	12.000 000	98.90	
				13	13.003 355	1.10	
				14*	14.003 242		5 730 yr
				15*	15.010 599		2.45 s
7	Nitrogen	N	14.006 7	12*	12.018 613		0.011 0 s
				13*	13.005 738		9.96 min
				14	14.003 074	99.63	
				15	15.000 108	0.37	
				16*	16.006 100		7.13 s
				17*	17.008 450		4.17 s

continued

TABLE A.3 *Continued*

Atomic Number Z	Element	Symbol	Chemical Atomic Mass (u)	Mass Number (*Indicates Radioactive) A	Atomic Mass (u)	Percent Abundance	Half-Life (If Radioactive) $T_{1/2}$
8	Oxygen	O	15.999 4	14*	14.008 595		70.6 s
				15*	15.003 065		122 s
				16	15.994 915	99.761	
				17	16.999 132	0.039	
				18	17.999 160	0.20	
				19*	19.003 577		26.9 s
9	Fluorine	F	18.998 40	17*	17.002 094		64.5 s
				18*	18.000 937		109.8 min
				19	18.998 404	100	
				20*	19.999 982		11.0 s
				21*	20.999 950		4.2 s
10	Neon	Ne	20.180	18*	18.005 710		1.67 s
				19*	19.001 880		17.2 s
				20	19.992 435	90.48	
				21	20.993 841	0.27	
				22	21.991 383	9.25	
				23*	22.994 465		37.2 s
11	Sodium	Na	22.989 87	21*	20.997 650		22.5 s
				22*	21.994 434		2.61 yr
				23	22.989 770	100	
				24*	23.990 961		14.96 h
12	Magnesium	Mg	24.305	23*	22.994 124		11.3 s
				24	23.985 042	78.99	
				25	24.985 838	10.00	
				26	25.982 594	11.01	
				27*	26.984 341		9.46 min
13	Aluminum	Al	26.981 54	26*	25.986 892		7.4×10^5 yr
				27	26.981 538	100	
				28*	27.981 910		2.24 min
14	Silicon	Si	28.086	28	27.976 927	92.23	
				29	28.976 495	4.67	
				30	29.973 770	3.10	
				31*	30.975 362		2.62 h
				32*	31.974 148		172 yr
15	Phosphorus	P	30.973 76	30*	29.978 307		2.50 min
				31	30.973 762	100	
				32*	31.973 908		14.26 days
				33*	32.971 725		25.3 days
16	Sulfur	S	32.066	32	31.972 071	95.02	
				33	32.971 459	0.75	
				34	33.967 867	4.21	
				35*	34.969 033		87.5 days
				36	35.967 081	0.02	
17	Chlorine	Cl	35.453	35	34.968 853	75.77	
				36*	35.968 307		3.0×10^5 yr
				37	36.965 903	24.23	

continued

TABLE A.3						

Continued

Atomic Number Z	Element	Symbol	Chemical Atomic Mass (u)	Mass Number (*Indicates Radioactive) A	Atomic Mass (u)	Percent Abundance	Half-Life (If Radioactive) $T_{1/2}$
18	Argon	Ar	39.948	36	35.967 547	0.337	
				37*	36.966 776		35.04 days
				38	37.962 732	0.063	
				39*	38.964 314		269 yr
				40	39.962 384	99.600	
				42*	41.963 049		33 yr
19	Potassium	K	39.098 3	39	38.963 708	93.258 1	
				40*	39.964 000	0.011 7	1.28×10^9 yr
				41	40.961 827	6.730 2	
20	Calcium	Ca	40.08	40	39.962 591	96.941	
				41*	40.962 279		1.0×10^5 yr
				42	41.958 618	0.647	
				43	42.958 767	0.135	
				44	43.955 481	2.086	
				46	45.953 687	0.004	
				48	47.952 534	0.187	
21	Scandium	Sc	44.955 9	41*	40.969 250		0.596 s
				45	44.955 911	100	
22	Titanium	Ti	47.88	44*	43.959 691		49 yr
				46	45.952 630	8.0	
				47	46.951 765	7.3	
				48	47.947 947	73.8	
				49	48.947 871	5.5	
				50	49.944 792	5.4	
23	Vanadium	V	50.941 5	48*	47.952 255		15.97 days
				50*	49.947 161	0.25	1.5×10^{17} yr
				51	50.943 962	99.75	
24	Chromium	Cr	51.996	48*	47.954 033		21.6 h
				50	49.946 047	4.345	
				52	51.940 511	83.79	
				53	52.940 652	9.50	
				54	53.938 883	2.365	
25	Manganese	Mn	54.938 05	54*	53.940 361		312.1 days
				55	54.938 048	100	
26	Iron	Fe	55.847	54	53.939 613	5.9	
				55*	54.938 297		2.7 yr
				56	55.934 940	91.72	
				57	56.935 396	2.1	
				58	57.933 278	0.28	
				60*	59.934 078		1.5×10^6 yr
27	Cobalt	Co	58.933 20	59	58.933 198	100	
				60*	59.933 820		5.27 yr
28	Nickel	Ni	58.693	58	57.935 346	68.077	
				59*	58.934 350		7.5×10^4 yr
				60	59.930 789	26.223	
				61	60.931 058	1.140	
				62	61.928 346	3.634	
				63*	62.929 670		100 yr
				64	63.927 967	0.926	

continued

TABLE A.3 *Continued*

Atomic Number Z	Element	Symbol	Chemical Atomic Mass (u)	Mass Number (*Indicates Radioactive) A	Atomic Mass (u)	Percent Abundance	Half-Life (If Radioactive) $T_{1/2}$
29	Copper	Cu	63.54	63	62.929 599	69.17	
				65	64.927 791	30.83	
30	Zinc	Zn	65.39	64	63.929 144	48.6	
				66	65.926 035	27.9	
				67	66.927 129	4.1	
				68	67.924 845	18.8	
				70	69.925 323	0.6	
31	Gallium	Ga	69.723	69	68.925 580	60.108	
				71	70.924 703	39.892	
32	Germanium	Ge	72.61	70	69.924 250	21.23	
				72	71.922 079	27.66	
				73	72.923 462	7.73	
				74	73.921 177	35.94	
				76	75.921 402	7.44	
33	Arsenic	As	74.921 6	75	74.921 594	100	
34	Selenium	Se	78.96	74	73.922 474	0.89	
				76	75.919 212	9.36	
				77	76.919 913	7.63	
				78	77.917 307	23.78	
				79*	78.918 497		$\leqslant 6.5 \times 10^4$ yr
				80	79.916 519	49.61	
				82*	81.916 697	8.73	1.4×10^{20} yr
35	Bromine	Br	79.904	79	78.918 336	50.69	
				81	80.916 287	49.31	
36	Krypton	Kr	83.80	78	77.920 400	0.35	
				80	79.916 377	2.25	
				81*	80.916 589		2.1×10^5 yr
				82	81.913 481	11.6	
				83	82.914 136	11.5	
				84	83.911 508	57.0	
				85*	84.912 531		10.76 yr
				86	85.910 615	17.3	
37	Rubidium	Rb	85.468	85	84.911 793	72.17	
				87*	86.909 186	27.83	4.75×10^{10} yr
38	Strontium	Sr	87.62	84	83.913 428	0.56	
				86	85.909 266	9.86	
				87	86.908 883	7.00	
				88	87.905 618	82.58	
				90*	89.907 737		29.1 yr
39	Yttrium	Y	88.905 8	89	88.905 847	100	
40	Zirconium	Zr	91.224	90	89.904 702	51.45	
				91	90.905 643	11.22	
				92	91.905 038	17.15	
				93*	92.906 473		1.5×10^6 yr
				94	93.906 314	17.38	
				96	95.908 274	2.80	

continued

TABLE A.3   *Continued*

Atomic Number Z	Element	Symbol	Chemical Atomic Mass (u)	Mass Number (*Indicates Radioactive) A	Atomic Mass (u)	Percent Abundance	Half-Life (If Radioactive) $T_{1/2}$
41	Niobium	Nb	92.906 4	91*	90.906 988		6.8×10^2 yr
				92*	91.907 191		3.5×10^7 yr
				93	92.906 376	100	
				94*	93.907 280		2×10^4 yr
42	Molybdenum	Mo	95.94	92	91.906 807	14.84	
				93*	92.906 811		3.5×10^3 yr
				94	93.905 085	9.25	
				95	94.905 841	15.92	
				96	95.904 678	16.68	
				97	96.906 020	9.55	
				98	97.905 407	24.13	
				100	99.907 476	9.63	
43	Technetium	Tc		97*	96.906 363		2.6×10^6 yr
				98*	97.907 215		4.2×10^6 yr
				99*	98.906 254		2.1×10^5 yr
44	Ruthenium	Ru	101.07	96	95.907 597	5.54	
				98	97.905 287	1.86	
				99	98.905 939	12.7	
				100	99.904 219	12.6	
				101	100.905 558	17.1	
				102	101.904 348	31.6	
				104	103.905 428	18.6	
45	Rhodium	Rh	102.905 5	103	102.905 502	100	
46	Palladium	Pd	106.42	102	101.905 616	1.02	
				104	103.904 033	11.14	
				105	104.905 082	22.33	
				106	105.903 481	27.33	
				107*	106.905 126		6.5×10^6 yr
				108	107.903 893	26.46	
				110	109.905 158	11.72	
47	Silver	Ag	107.868	107	106.905 091	51.84	
				109	108.904 754	48.16	
48	Cadmium	Cd	112.41	106	105.906 457	1.25	
				108	107.904 183	0.89	
				109*	108.904 984		462 days
				110	109.903 004	12.49	
				111	110.904 182	12.80	
				112	111.902 760	24.13	
				113*	112.904 401	12.22	9.3×10^{15} yr
				114	113.903 359	28.73	
				116	115.904 755	7.49	
49	Indium	In	114.82	113	112.904 060	4.3	
				115*	114.903 876	95.7	4.4×10^{14} yr
50	Tin	Sn	118.71	112	111.904 822	0.97	
				114	113.902 780	0.65	
				115	114.903 345	0.36	
				116	115.901 743	14.53	
				117	116.902 953	7.68	

continued

TABLE A.3 *Continued*

Atomic Number Z	Element	Symbol	Chemical Atomic Mass (u)	Mass Number (*Indicates Radioactive) A	Atomic Mass (u)	Percent Abundance	Half-Life (If Radioactive) $T_{1/2}$
(50)	(Tin)			118	117.901 605	24.22	
				119	118.903 308	8.58	
				120	119.902 197	32.59	
				121*	120.904 237		55 yr
				122	121.903 439	4.63	
				124	123.905 274	5.79	
51	Antimony	Sb	121.76	121	120.903 820	57.36	
				123	122.904 215	42.64	
				125*	124.905 251		2.7 yr
52	Tellurium	Te	127.60	120	119.904 040	0.095	
				122	121.903 052	2.59	
				123*	122.904 271	0.905	1.3×10^{13} yr
				124	123.902 817	4.79	
				125	124.904 429	7.12	
				126	125.903 309	18.93	
				128*	127.904 463	31.70	$> 8 \times 10^{24}$ yr
				130*	129.906 228	33.87	$\leqslant 1.25 \times 10^{21}$ yr
53	Iodine	I	126.904 5	127	126.904 474	100	
				129*	128.904 984		1.6×10^7 yr
54	Xenon	Xe	131.29	124	123.905 894	0.10	
				126	125.904 268	0.09	
				128	127.903 531	1.91	
				129	128.904 779	26.4	
				130	129.903 509	4.1	
				131	130.905 069	21.2	
				132	131.904 141	26.9	
				134	133.905 394	10.4	
				136*	135.907 215	8.9	$\geqslant 2.36 \times 10^{21}$ yr
55	Cesium	Cs	132.905 4	133	132.905 436	100	
				134*	133.906 703		2.1 yr
				135*	134.905 891		2×10^6 yr
				137*	136.907 078		30 yr
56	Barium	Ba	137.33	130	129.906 289	0.106	
				132	131.905 048	0.101	
				133*	132.905 990		10.5 yr
				134	133.904 492	2.42	
				135	134.905 671	6.593	
				136	135.904 559	7.85	
				137	136.905 816	11.23	
				138	137.905 236	21.70	
57	Lanthanum	La	138.905	137*	136.906 462		6×10^4 yr
				138*	137.907 105	0.090 2	1.05×10^{11} yr
				139	138.906 346	99.909 8	
58	Cerium	Ce	140.12	136	135.907 139	0.19	
				138	137.905 986	0.25	
				140	139.905 434	88.43	
				142*	141.909 241	11.13	$> 5 \times 10^{16}$ yr
59	Praseodymium	Pr	140.907 6	141	140.907 647	100	

continued

| TABLE A.3 | *Continued* | | | | | | |

Atomic Number Z	Element	Symbol	Chemical Atomic Mass (u)	Mass Number (*Indicates Radioactive) A	Atomic Mass (u)	Percent Abundance	Half-Life (If Radioactive) $T_{1/2}$
60	Neodymium	Nd	144.24	142	141.907 718	27.13	
				143	142.909 809	12.18	
				144*	143.910 082	23.80	2.3×10^{15} yr
				145	144.912 568	8.30	
				146	145.913 113	17.19	
				148	147.916 888	5.76	
				150*	149.920 887	5.64	$>1 \times 10^{18}$ yr
61	Promethium	Pm		143*	142.910 928		265 days
				145*	144.912 745		17.7 yr
				146*	145.914 698		5.5 yr
				147*	146.915 134		2.623 yr
62	Samarium	Sm	150.36	144	143.911 996	3.1	
				146*	145.913 043		1.0×10^{8} yr
				147*	146.914 894	15.0	1.06×10^{11} yr
				148*	147.914 819	11.3	7×10^{15} yr
				149*	148.917 180	13.8	$>2 \times 10^{15}$ yr
				150	149.917 273	7.4	
				151*	150.919 928		90 yr
				152	151.919 728	26.7	
				154	153.922 206	22.7	
63	Europium	Eu	151.96	151	150.919 846	47.8	
				152*	151.921 740		13.5 yr
				153	152.921 226	52.2	
				154*	153.922 975		8.59 yr
				155*	154.922 888		4.7 yr
64	Gadolinium	Gd	157.25	148*	147.918 112		75 yr
				150*	149.918 657		1.8×10^{6} yr
				152*	151.919 787	0.20	1.1×10^{14} yr
				154	153.920 862	2.18	
				155	154.922 618	14.80	
				156	155.922 119	20.47	
				157	156.923 957	15.65	
				158	157.924 099	24.84	
				160	159.927 050	21.86	
65	Terbium	Tb	158.925 3	159	158.925 345	100	
66	Dysprosium	Dy	162.50	156	155.924 277	0.06	
				158	157.924 403	0.10	
				160	159.925 193	2.34	
				161	160.926 930	18.9	
				162	161.926 796	25.5	
				163	162.928 729	24.9	
				164	163.929 172	28.2	
67	Holmium	Ho	164.930 3	165	164.930 316	100	
				166*	165.932 282		1.2×10^{3} yr
68	Erbium	Er	167.26	162	161.928 775	0.14	
				164	163.929 198	1.61	
				166	165.930 292	33.6	

continued

TABLE A.3 *Continued*

Atomic Number Z	Element	Symbol	Chemical Atomic Mass (u)	Mass Number (*Indicates Radioactive) A	Atomic Mass (u)	Percent Abundance	Half-Life (If Radioactive) $T_{1/2}$
(68)	(Erbium)			167	166.932 047	22.95	
				168	167.932 369	27.8	
				170	169.935 462	14.9	
69	Thulium	Tm	168.934 2	169	168.934 213	100	
				171*	170.936 428		1.92 yr
70	Ytterbium	Yb	173.04	168	167.933 897	0.13	
				170	169.934 761	3.05	
				171	170.936 324	14.3	
				172	171.936 380	21.9	
				173	172.938 209	16.12	
				174	173.938 861	31.8	
				176	175.942 564	12.7	
71	Lutecium	Lu	174.967	173*	172.938 930		1.37 yr
				175	174.940 772	97.41	
				176*	175.942 679	2.59	3.78×10^{10} yr
72	Hafnium	Hf	178.49	174*	173.940 042	0.162	2.0×10^{15} yr
				176	175.941 404	5.206	
				177	176.943 218	18.606	
				178	177.943 697	27.297	
				179	178.945 813	13.629	
				180	179.946 547	35.100	
73	Tantalum	Ta	180.947 9	180	179.947 542	0.012	
				181	180.947 993	99.988	
74	Tungsten (Wolfram)	W	183.85	180	179.946 702	0.12	
				182	181.948 202	26.3	
				183	182.950 221	14.28	
				184	183.950 929	30.7	
				186	185.954 358	28.6	
75	Rhenium	Re	186.207	185	184.952 951	37.40	
				187*	186.955 746	62.60	4.4×10^{10} yr
76	Osmium	Os	190.2	184	183.952 486	0.02	
				186*	185.953 834	1.58	2.0×10^{15} yr
				187	186.955 744	1.6	
				188	187.955 832	13.3	
				189	188.958 139	16.1	
				190	189.958 439	26.4	
				192	191.961 468	41.0	
				194*	193.965 172		6.0 yr
77	Iridium	Ir	192.2	191	190.960 585	37.3	
				193	192.962 916	62.7	
78	Platinum	Pt	195.08	190*	189.959 926	0.01	6.5×10^{11} yr
				192	191.961 027	0.79	
				194	193.962 655	32.9	
				195	194.964 765	33.8	
				196	195.964 926	25.3	
				198	197.967 867	7.2	
79	Gold	Au	196.966 5	197	196.966 543	100	

continued

TABLE A.3 *Continued*

Atomic Number Z	Element	Symbol	Chemical Atomic Mass (u)	Mass Number (*Indicates Radioactive) A	Atomic Mass (u)	Percent Abundance	Half-Life (If Radioactive) $T_{1/2}$
80	Mercury	Hg	200.59	196	195.965 806	0.15	
				198	197.966 743	9.97	
				199	198.968 253	16.87	
				200	199.968 299	23.10	
				201	200.970 276	13.10	
				202	201.970 617	29.86	
				204	203.973 466	6.87	
81	Thallium	Tl	204.383	203	202.972 320	29.524	
				204*	203.973 839		3.78 yr
				205	204.974 400	70.476	
		(Ra E″)		206*	205.976 084		4.2 min
		(Ac C″)		207*	206.977 403		4.77 min
		(Th C″)		208*	207.981 992		3.053 min
		(Ra C″)		210*	209.990 057		1.30 min
82	Lead	Pb	207.2	202*	201.972 134		5×10^4 yr
				204*	203.973 020	1.4	$\geqslant 1.4 \times 10^{17}$ yr
				205*	204.974 457		1.5×10^7 yr
				206	205.974 440	24.1	
				207	206.975 871	22.1	
				208	207.976 627	52.4	
		(Ra D)		210*	209.984 163		22.3 yr
		(Ac B)		211*	210.988 734		36.1 min
		(Th B)		212*	211.991 872		10.64 h
		(Ra B)		214*	213.999 798		26.8 min
83	Bismuth	Bi	208.980 3	207*	206.978 444		32.2 yr
				208*	207.979 717		3.7×10^5 yr
				209	208.980 374	100	
		(Ra E)		210*	209.984 096		5.01 days
		(Th C)		211*	210.987 254		2.14 min
				212*	211.991 259		60.6 min
		(Ra C)		214*	213.998 692		19.9 min
				215*	215.001 836		7.4 min
84	Polonium	Po		209*	208.982 405		102 yr
		(Ra F)		210*	209.982 848		138.38 days
		(Ac C′)		211*	210.986 627		0.52 s
		(Th C′)		212*	211.988 842		0.30 μs
		(Ra C′)		214*	213.995 177		164 μs
		(Ac A)		215*	214.999 418		0.001 8 s
		(Th A)		216*	216.001 889		0.145 s
		(Ra A)		218*	218.008 965		3.10 min
85	Astatine	At		215*	214.998 638		≈ 100 μs
				218*	218.008 685		1.6 s
				219*	219.011 294		0.9 min
86	Radon	Rn					
		(An)		219*	219.009 477		3.96 s
		(Tn)		220*	220.011 369		55.6 s
		(Rn)		222*	222.017 571		3.823 days
87	Francium	Fr					
		(Ac K)		223*	223.019 733		22 min

continued

TABLE A.3   *Continued*

Atomic Number Z	Element	Symbol	Chemical Atomic Mass (u)	Mass Number (*Indicates Radioactive) A	Atomic Mass (u)	Percent Abundance	Half-Life (If Radioactive) $T_{1/2}$
88	Radium	Ra					
		(Ac X)		223*	223.018 499		11.43 days
		(Th X)		224*	224.020 187		3.66 days
		(Ra)		226*	226.025 402		1 600 yr
		(Ms Th$_1$)		228*	228.031 064		5.75 yr
89	Actinium	Ac		227*	227.027 749		21.77 yr
		(Ms Th$_2$)		228*	228.031 015		6.15 h
90	Thorium	Th	232.038 1				
		(Rd Ac)		227*	227.027 701		18.72 days
		(Rd Th)		228*	228.028 716		1.913 yr
				229*	229.031 757		7 300 yr
		(Io)		230*	230.033 127		75.000 yr
		(UY)		231*	231.036 299		25.52 h
		(Th)		232*	232.038 051	100	1.40×10^{10} yr
		(UX1)		234*	234.043 593		24.1 days
91	Protactinium	Pa		231*	231.035 880		32.760 yr
		(Uz)		234*	234.043 300		6.7 h
92	Uranium	U	238.028 9	232*	232.037 131		69 yr
				233*	233.039 630		1.59×10^5 yr
				234*	234.040 946	0.005 5	2.45×10^5 yr
		(Ac U)		235*	235.043 924	0.720	7.04×10^8 yr
				236*	236.045 562		2.34×10^7 yr
		(UI)		238*	238.050 784	99.274 5	4.47×10^9 yr
93	Neptunium	Np		235*	235.044 057		396 days
				236*	236.046 560		1.15×10^5 yr
				237*	237.048 168		2.14×10^6 yr
94	Plutonium	Pu		236*	236.046 033		2.87 yr
				238*	238.049 555		87.7 yr
				239*	239.052 157		2.412×10^4 yr
				240*	240.053 808		6 560 yr
				241*	241.056 846		14.4 yr
				242*	242.058 737		3.73×10^6 yr
				244*	244.064 200		8.1×10^7 yr

[a] The masses in the sixth column are atomic masses, which include the mass of Z electrons. Data are from the National Nuclear Data Center, Brookhaven National Laboratory, prepared by Jagdish K. Tuli, July 1990. The data are based on experimental results reported in *Nuclear Data Sheets* and *Nuclear Physics* and also from *Chart of the Nuclides*, 14th ed. Atomic masses are based on those by A. H. Wapstra, G. Audi, and R. Hoekstra. Isotopic abundances are based on those by N. E. Holden.

Mathematics Review

This appendix in mathematics is intended as a brief review of operations and methods. Early in this course, you should be totally familiar with basic algebraic techniques, analytic geometry, and trigonometry. The sections on differential and integral calculus are more detailed and are intended for those students who have difficulty applying calculus concepts to physical situations.

B.1 • SCIENTIFIC NOTATION

Many quantities that scientists deal with often have very large or very small values. For example, the speed of light is about 300 000 000 m/s, and the ink required to make the dot over an i in this textbook has a mass of about 0.000 000 001 kg. Obviously, it is very cumbersome to read, write, and keep track of numbers such as these. We avoid this problem by using a method dealing with powers of the number 10:

$$10^0 = 1$$
$$10^1 = 10$$
$$10^2 = 10 \times 10 = 100$$
$$10^3 = 10 \times 10 \times 10 = 1000$$
$$10^4 = 10 \times 10 \times 10 \times 10 = 10\ 000$$
$$10^5 = 10 \times 10 \times 10 \times 10 \times 10 = 100\ 000$$

and so on. The number of zeros corresponds to the power to which 10 is raised, called the **exponent** of 10. For example, the speed of light, 300 000 000 m/s, can be expressed as 3×10^8 m/s.

In this method, some representative numbers smaller than unity are

$$10^{-1} = \frac{1}{10} = 0.1$$

$$10^{-2} = \frac{1}{10 \times 10} = 0.01$$

$$10^{-3} = \frac{1}{10 \times 10 \times 10} = 0.001$$

$$10^{-4} = \frac{1}{10 \times 10 \times 10 \times 10} = 0.000\ 1$$

$$10^{-5} = \frac{1}{10 \times 10 \times 10 \times 10 \times 10} = 0.000\ 01$$

In these cases, the number of places the decimal point is to the left of the digit 1 equals the value of the (negative) exponent. Numbers expressed as some power of 10 multiplied by another number between 1 and 10 are said to be in **scientific notation.** For example, the scientific notation for 5 943 000 000 is 5.943×10^9 and that for 0.000 083 2 is 8.32×10^{-5}.

When numbers expressed in scientific notation are being multiplied, the following general rule is very useful:

$$10^n \times 10^m = 10^{n+m} \tag{B.1}$$

where n and m can be *any* numbers (not necessarily integers). For example, $10^2 \times 10^5 = 10^7$. The rule also applies if one of the exponents is negative: $10^3 \times 10^{-8} = 10^{-5}$.

When dividing numbers expressed in scientific notation, note that

$$\frac{10^n}{10^m} = 10^n \times 10^{-m} = 10^{n-m} \tag{B.2}$$

EXERCISES

With help from the preceding rules, verify the answers to the following:

1. $86\ 400 = 8.64 \times 10^4$
2. $9\ 816\ 762.5 = 9.816\ 762\ 5 \times 10^6$
3. $0.000\ 000\ 039\ 8 = 3.98 \times 10^{-8}$
4. $(4.0 \times 10^8)(9.0 \times 10^9) = 3.6 \times 10^{18}$
5. $(3.0 \times 10^7)(6.0 \times 10^{-12}) = 1.8 \times 10^{-4}$
6. $\dfrac{75 \times 10^{-11}}{5.0 \times 10^{-3}} = 1.5 \times 10^{-7}$
7. $\dfrac{(3 \times 10^6)(8 \times 10^{-2})}{(2 \times 10^{17})(6 \times 10^5)} = 2 \times 10^{-18}$

B.2 • ALGEBRA

Some Basic Rules

When algebraic operations are performed, the laws of arithmetic apply. Symbols such as x, y, and z are usually used to represent unspecified quantities, called the **unknowns.**

First, consider the equation

$$8x = 32$$

If we wish to solve for x, we can divide (or multiply) each side of the equation by the same factor without destroying the equality. In this case, if we divide both sides by 8, we have

$$\frac{8x}{8} = \frac{32}{8}$$

$$x = 4$$

Next consider the equation

$$x + 2 = 8$$

In this type of expression, we can add or subtract the same quantity from each side. If we subtract 2 from each side, we have

$$x + 2 - 2 = 8 - 2$$

$$x = 6$$

In general, if $x + a = b$, then $x = b - a$.

Now consider the equation

$$\frac{x}{5} = 9$$

If we multiply each side by 5, we are left with x on the left by itself and 45 on the right:

$$\left(\frac{x}{5}\right)(5) = 9 \times 5$$

$$x = 45$$

In all cases, *whatever operation is performed on the left side of the equality must also be performed on the right side.*

The following rules for multiplying, dividing, adding, and subtracting fractions should be recalled, where a, b, and c are three numbers:

	Rule	Example
Multiplying	$\left(\dfrac{a}{b}\right)\left(\dfrac{c}{d}\right) = \dfrac{ac}{bd}$	$\left(\dfrac{2}{3}\right)\left(\dfrac{4}{5}\right) = \dfrac{8}{15}$
Dividing	$\dfrac{(a/b)}{(c/d)} = \dfrac{ad}{bc}$	$\dfrac{2/3}{4/5} = \dfrac{(2)(5)}{(4)(3)} = \dfrac{10}{12}$
Adding	$\dfrac{a}{b} \pm \dfrac{c}{d} = \dfrac{ad \pm bc}{bd}$	$\dfrac{2}{3} - \dfrac{4}{5} = \dfrac{(2)(5) - (4)(3)}{(3)(5)} = -\dfrac{2}{15}$

EXERCISES

In the following exercises, solve for x:

Answers

1. $a = \dfrac{1}{1 + x}$ $x = \dfrac{1 - a}{a}$

2. $3x - 5 = 13$ $x = 6$

3. $ax - 5 = bx + 2$ $x = \dfrac{7}{a - b}$

4. $\dfrac{5}{2x + 6} = \dfrac{3}{4x + 8}$ $x = -\dfrac{11}{7}$

Powers

When powers of a given quantity x are multiplied, the following rule applies:

$$x^n x^m = x^{n+m} \tag{B.3}$$

For example, $x^2 x^4 = x^{2+4} = x^6$.

When dividing the powers of a given quantity, the rule is

$$\frac{x^n}{x^m} = x^{n-m} \tag{B.4}$$

For example, $x^8 / x^2 = x^{8-2} = x^6$.

A power that is a fraction, such as $\frac{1}{3}$, corresponds to a root as follows:

$$x^{1/n} = \sqrt[n]{x} \tag{B.5}$$

For example, $4^{1/3} = \sqrt[3]{4} = 1.5874$. (A scientific calculator is useful for such calculations.)

Finally, any quantity x^n raised to the mth power is

$$(x^n)^m = x^{nm} \tag{B.6}$$

Table B.1 summarizes the rules of exponents.

TABLE B.1
Rules of Exponents
$x^0 = 1$
$x^1 = x$
$x^n x^m = x^{n+m}$
$x^n / x^m = x^{n-m}$
$x^{1/n} = \sqrt[n]{x}$
$(x^n)^m = x^{nm}$

EXERCISES

Verify the following:

1. $3^2 \times 3^3 = 243$
2. $x^5 x^{-8} = x^{-3}$
3. $x^{10} / x^{-5} = x^{15}$
4. $5^{1/3} = 1.709\ 975$ (Use your calculator.)
5. $60^{1/4} = 2.783\ 158$ (Use your calculator.)
6. $(x^4)^3 = x^{12}$

Factoring

Some useful formulas for factoring an equation are

$$ax + ay + az = a(x + y + x) \qquad \text{common factor}$$

$$a^2 + 2ab + b^2 = (a + b)^2 \qquad \text{perfect square}$$

$$a^2 - b^2 = (a + b)(a - b) \qquad \text{differences of squares}$$

Quadratic Equations

The general form of a quadratic equation is

$$ax^2 + bx + c = 0 \qquad \text{(B.7)}$$

where x is the unknown quantity and a, b, and c are numerical factors referred to as **coefficients** of the equation. This equation has two roots, given by

$$x = \frac{-b \pm \sqrt{b^2 - 4ac}}{2a} \qquad \text{(B.8)}$$

If $b^2 \geq 4ac$, the roots are real.

Example 1   The equation $x^2 + 5x + 4 = 0$ has the following roots corresponding to the two signs of the square-root term:

$$x = \frac{-5 \pm \sqrt{5^2 - (4)(1)(4)}}{2(1)} = \frac{-5 \pm \sqrt{9}}{2} = \frac{-5 \pm 3}{2}$$

$$x_+ = \frac{-5 + 3}{2} = \boxed{-1} \qquad x_- = \frac{-5 - 3}{2} = \boxed{-4}$$

where x_+ refers to the root corresponding to the positive sign and x_- refers to the root corresponding to the negative sign.

EXERCISES

Solve the following quadratic equations:

Answers

1. $x^2 + 2x - 3 = 0$ $x_+ = 1$ $x_- = -3$
2. $2x^2 - 5x + 2 = 0$ $x_+ = 2$ $x_- = \frac{1}{2}$
3. $2x^2 - 4x - 9 = 0$ $x_+ = 1 + \sqrt{22}/2$ $x_- = 1 - \sqrt{22}/2$

Linear Equations

A linear equation has the general form

$$y = mx + b \qquad \text{(B.9)}$$

where m and b are constants. This equation is referred to as being linear because the graph of y versus x is a straight line, as shown in Figure B.1. The constant b, called the **y-intercept,** represents the value of y at which the straight line intersects the y axis. The constant m is equal to the **slope** of the straight line and is also equal to the tangent of the angle that the line makes with the x axis. If any two points on

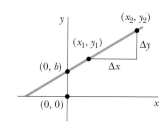

Figure B.1

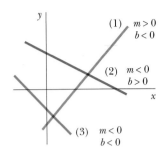

(1) $m > 0$
$b < 0$

(2) $m < 0$
$b > 0$

(3) $m < 0$
$b < 0$

Figure B.2

the straight line are specified by the coordinates (x_1, y_1) and (x_2, y_2), as in Figure B.1, then the slope of the straight line can be expressed as

$$\text{Slope} = \frac{y_2 - y_1}{x_2 - x_1} = \frac{\Delta y}{\Delta x} \tag{B.10}$$

Note that m and b can have either positive or negative values. If $m > 0$, the straight line has a *positive* slope, as in Figure B.1. If $m < 0$, the straight line has a *negative* slope. In Figure B.1, both m and b are positive. Three other possible situations are shown in Figure B.2.

EXERCISES

1. Draw graphs of the following straight lines:
   (a) $y = 5x + 3$ (b) $y = -2x + 4$ (c) $y = -3x - 6$
2. Find the slopes of the straight lines described in Exercise 1.

Answers   (a) 5   (b) -2 (c) -3

3. Find the slopes of the straight lines that pass through the following sets of points:
   (a) $(0, -4)$ and $(4, 2)$ (b) $(0, 0)$ and $(2, -5)$ (c) $(-5, 2)$ and $(4, -2)$

Answers   (a) $3/2$ (b) $-5/2$ (c) $-4/9$

Solving Simultaneous Linear Equations

Consider the equation $3x + 5y = 15$, which has two unknowns, x and y. Such an equation does not have a unique solution. For example, note that $(x = 0, y = 3)$, $(x = 5, y = 0)$, and $(x = 2, y = 9/5)$ are all solutions to this equation.

If a problem has two unknowns, a unique solution is possible only if we have *two* equations. In general, if a problem has n unknowns, its solution requires n equations. To solve two simultaneous equations involving two unknowns, x and y, we solve one of the equations for x in terms of y and substitute this expression into the other equation.

Example 2   Solve the following two simultaneous equations:

(1) $5x + y = -8$

(2) $2x - 2y = 4$

Solution   From (2), $x = y + 2$. Substitution of this into (1) gives

$$5(y + 2) + y = -8$$

$$6y = -18$$

$$y = -3$$

$$x = y + 2 = \boxed{-1}$$

Alternative Solution   Multiply each term in (1) by the factor 2 and add the result to (2):

$$10x + 2y = -16$$

$$\underline{2x - 2y = 4}$$

$$12x = -12$$

$$x = -1$$

$$y = x - 2 = \boxed{-3}$$

Two linear equations containing two unknowns can also be solved by a graphical method. If the straight lines corresponding to the two equations are plotted in a conventional coordinate system, the intersection of the two lines represents the solution. For example, consider the two equations

$$x - y = 2$$
$$x - 2y = -1$$

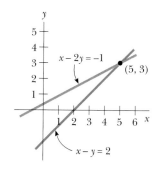

Figure B.3

These are plotted in Figure B.3. The intersection of the two lines has the coordinates $x = 5$, $y = 3$. This represents the solution to the equations. You should check this solution by the analytical technique discussed earlier.

EXERCISES

Solve the following pairs of simultaneous equations involving two unknowns:

Answers

1. $x + y = 8$ $x = 5$, $y = 3$
   $x - y = 2$
2. $98 - T = 10a$ $T = 65$, $a = 3.27$
   $T - 49 = 5a$
3. $6x + 2y = 6$ $x = 2$, $y = -3$
   $8x - 4y = 28$

Logarithms

Suppose that a quantity x is expressed as a power of some quantity a:

$$x = a^y \tag{B.11}$$

The number a is called the **base** number. The **logarithm** of x with respect to the base a is equal to the exponent to which the base must be raised to satisfy the expression $x = a^y$:

$$y = \log_a x \tag{B.12}$$

Conversely, the **antilogarithm** of y is the number x:

$$x = \text{antilog}_a y \tag{B.13}$$

In practice, the two bases most often used are base 10, called the *common* logarithm base, and base $e = 2.718 \ldots$, called Euler's constant or the *natural* logarithm base. When common logarithms are used,

$$y = \log_{10} x \quad (\text{or } x = 10^y) \tag{B.14}$$

When natural logarithms are used,

$$y = \ln x \quad (\text{or } x = e^y) \tag{B.15}$$

For example, $\log_{10} 52 = 1.716$, so that $\text{antilog}_{10} 1.716 = 10^{1.716} = 52$. Likewise, $\ln 52 = 3.951$, so $\text{antiln } 3.951 = e^{3.951} = 52$.

In general, note that you can convert between base 10 and base e with the equality

$$\ln x = (2.302\ 585) \log_{10} x \tag{B.16}$$

Finally, some useful properties of logarithms are

$$\log(ab) = \log a + \log b$$
$$\log(a/b) = \log a - \log b$$
$$\log(a^n) = n \log a$$
$$\ln e = 1$$
$$\ln e^a = a$$
$$\ln\left(\frac{1}{a}\right) = -\ln a$$

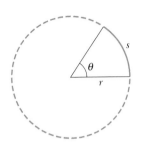

Figure B.4

B.3 • GEOMETRY

The **distance** d between two points having coordinates (x_1, y_1) and (x_2, y_2) is

$$d = \sqrt{(x_2 - x_1)^2 + (y_2 - y_1)^2} \tag{B.17}$$

Radian measure: The arc length s of a circular arc (Fig. B.4) is proportional to the radius r for a fixed value of θ (in radians):

$$s = r\theta$$
$$\theta = \frac{s}{r} \tag{B.18}$$

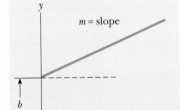

Figure B.5

The equation of a **straight line** (Fig. B.5) is

$$y = mx + b \tag{B.19}$$

where b is the y-intercept and m is the slope of the line.

The equation of a **circle** of radius R centered at the origin is

$$x^2 + y^2 = R^2 \tag{B.20}$$

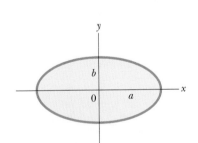

Figure B.6

The equation of an **ellipse** having the origin at its center (Fig. B.6) is

$$\frac{x^2}{a^2} + \frac{y^2}{b^2} = 1 \tag{B.21}$$

where a is the length of the semimajor axis (the longer one) and b is the length of the semiminor axis (the shorter one).

The equation of a **parabola** the vertex of which is at $y = b$ (Fig. B.7) is

$$y = ax^2 + b \tag{B.22}$$

The equation of a **rectangular hyperbola** (Fig. B.8) is

$$xy = \text{constant} \tag{B.23}$$

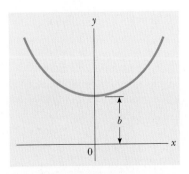

Figure B.7

| TABLE B.2 | Useful Information for Geometry | | |

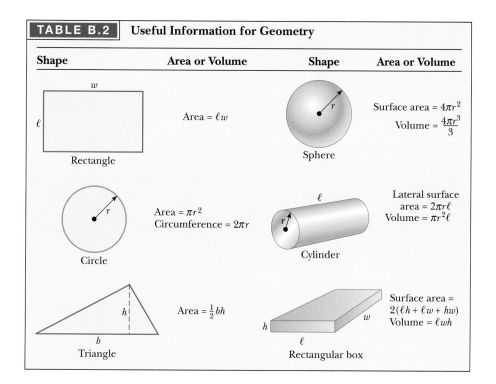

Shape	Area or Volume	Shape	Area or Volume
Rectangle	Area = ℓw	Sphere	Surface area = $4\pi r^2$ Volume = $\frac{4\pi r^3}{3}$
Circle	Area = πr^2 Circumference = $2\pi r$	Cylinder	Lateral surface area = $2\pi r \ell$ Volume = $\pi r^2 \ell$
Triangle	Area = $\frac{1}{2}bh$	Rectangular box	Surface area = $2(\ell h + \ell w + hw)$ Volume = ℓwh

Table B.2 gives the areas and volumes for several geometric shapes used throughout this text.

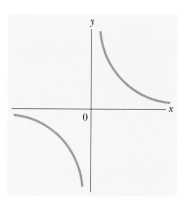

Figure B.8

B.4 • TRIGONOMETRY

That portion of mathematics based on the special properties of the right triangle is called trigonometry. By definition, a right triangle is one containing a 90° angle. Consider the right triangle shown in Figure B.9, where side a is opposite the angle θ, side b is adjacent to the angle θ, and side c is the hypotenuse of the triangle. The three basic trigonometric functions defined by such a triangle are the sine (sin), cosine (cos), and tangent (tan) functions. In terms of the angle θ, these functions are defined by

$$\sin \theta \equiv \frac{\text{side opposite } \theta}{\text{hypotenuse}} = \frac{a}{c} \qquad \textbf{(B.24)}$$

$$\cos \theta \equiv \frac{\text{side adjacent to } \theta}{\text{hypotenuse}} = \frac{b}{c} \qquad \textbf{(B.25)}$$

$$\tan \theta \equiv \frac{\text{side opposite } \theta}{\text{side adjacent to } \theta} = \frac{a}{b} \qquad \textbf{(B.26)}$$

The Pythagorean theorem provides the following relationship between the sides of a right triangle:

$$c^2 = a^2 + b^2 \qquad \textbf{(B.27)}$$

a = opposite side
b = adjacent side
c = hypotenuse

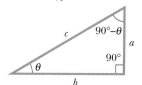

Figure B.9

From the preceding definitions and the Pythagorean theorem, it follows that

$$\sin^2 \theta + \cos^2 \theta = 1$$

$$\tan \theta = \frac{\sin \theta}{\cos \theta}$$

The cosecant, secant, and cotangent functions are defined by

$$\csc \theta \equiv \frac{1}{\sin \theta} \qquad \sec \theta \equiv \frac{1}{\cos \theta} \qquad \cot \theta \equiv \frac{1}{\tan \theta}$$

The following relationships are derived directly from the right triangle shown in Figure B.9:

$$\sin \theta = \cos(90° - \theta)$$

$$\cos \theta = \sin(90° - \theta)$$

$$\cot \theta = \tan(90° - \theta)$$

Some properties of trigonometric functions are

$$\sin(-\theta) = -\sin \theta$$

$$\cos(-\theta) = \cos \theta$$

$$\tan(-\theta) = -\tan \theta$$

The following relationships apply to *any* triangle, as shown in Figure B.10:

$$\alpha + \beta + \gamma = 180°$$

Law of cosines
$$a^2 = b^2 + c^2 - 2bc \cos \alpha$$
$$b^2 = a^2 + c^2 - 2ac \cos \beta$$
$$c^2 = a^2 + b^2 - 2ab \cos \gamma$$

Law of sines
$$\frac{a}{\sin \alpha} = \frac{b}{\sin \beta} = \frac{c}{\sin \gamma}$$

Table B.3 lists a number of useful trigonometric identities.

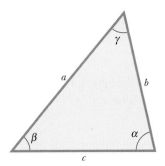

Figure B.10

TABLE B.3	Some Trigonometric Identities
$\sin^2 \theta + \cos^2 \theta = 1$	$\csc^2 \theta = 1 + \cot^2 \theta$
$\sec^2 \theta = 1 + \tan^2 \theta$	$\sin^2 \dfrac{\theta}{2} = \frac{1}{2}(1 - \cos \theta)$
$\sin 2\theta = 2 \sin \theta \cos \theta$	$\cos^2 \dfrac{\theta}{2} = \frac{1}{2}(1 + \cos \theta)$
$\cos 2\theta = \cos^2 \theta - \sin^2 \theta$	$1 - \cos \theta = 2 \sin^2 \dfrac{\theta}{2}$
$\tan 2\theta = \dfrac{2 \tan \theta}{1 - \tan^2 \theta}$	$\tan \dfrac{\theta}{2} = \sqrt{\dfrac{1 - \cos \theta}{1 + \cos \theta}}$

$$\sin(A \pm B) = \sin A \cos B \pm \cos A \sin B$$
$$\cos(A \pm B) = \cos A \cos B \mp \sin A \sin B$$
$$\sin A \pm \sin B = 2 \sin[\tfrac{1}{2}(A \pm B)]\cos[\tfrac{1}{2}(A \mp B)]$$
$$\cos A + \cos B = 2 \cos[\tfrac{1}{2}(A + B)]\cos[\tfrac{1}{2}(A - B)]$$
$$\cos A - \cos B = 2 \sin[\tfrac{1}{2}(A + B)]\sin[\tfrac{1}{2}(B - A)]$$

Example 3   Consider the right triangle in Figure B.11, in which $a = 2$, $b = 5$, and c is unknown. From the Pythagorean theorem, we have

$$c^2 = a^2 + b^2 = 2^2 + 5^2 = 4 + 25 = 29$$

$$c = \sqrt{29} = 5.39$$

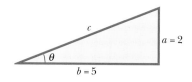

Figure B.11

To find the angle θ, note that

$$\tan \theta = \frac{a}{b} = \frac{2}{5} = 0.400$$

From a table of functions or from a calculator, we have

$$\theta = \tan^{-1}(0.400) = 21.8°$$

where $\tan^{-1}(0.400)$ is the notation for "angle whose tangent is 0.400," sometimes written as arctan (0.400).

EXERCISES

1. In Figure B.12, identify (a) the side opposite θ and (b) the side adjacent to ϕ and then find (c) $\cos \theta$, (d) $\sin \phi$, and (e) $\tan \phi$.

Answers   (a) 3   (b) 3   (c) $\frac{4}{5}$ (d) $\frac{4}{5}$ (e) $\frac{4}{3}$

2. In a certain right triangle, the two sides that are perpendicular to each other are 5m and 7m long. What is the length of the third side?

Answer   8.60 m

3. A right triangle has a hypotenuse of length 3 m, and one of its angles is 30°. What is the length of (a) the side opposite the 30° angle and (b) the side adjacent to the 30° angle?

Answers   (a) 1.5 m   (b) 2.60 m

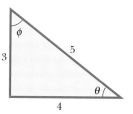

Figure B.12

B.5 • SERIES EXPANSIONS

$$(a + b)^n = a^n + \frac{n}{1!} a^{n-1}b + \frac{n(n-1)}{2!} a^{n-2}b^2 + \cdots$$

$$(1 + x)^n = 1 + nx + \frac{n(n-1)}{2!} x^2 + \cdots$$

$$e^x = 1 + x + \frac{x^2}{2!} + \frac{x^3}{3!} + \cdots$$

$$\ln(1 \pm x) = \pm x - \tfrac{1}{2}x^2 \pm \tfrac{1}{3}x^3 - \cdots$$

$$\left. \begin{aligned} \sin x &= x - \frac{x^3}{3!} + \frac{x^5}{5!} - \cdots \\[4pt] \cos x &= 1 - \frac{x^2}{2!} + \frac{x^4}{4!} - \cdots \\[4pt] \tan x &= x + \frac{x^3}{3} + \frac{2x^5}{15} + \cdots \qquad |x| < \pi/2 \end{aligned} \right\} \quad x \text{ in radians}$$

For $x \ll 1$, the following approximations can be used*:

$$(1 + x)^n \approx 1 + nx \qquad \sin x \approx x$$

$$e^x \approx 1 + x \qquad \cos x \approx 1$$

$$\ln(1 \pm x) \approx \pm x \qquad \tan x \approx x$$

B.6 • DIFFERENTIAL CALCULUS

In various branches of science, it is sometimes necessary to use the basic tools of calculus, invented by Newton, to describe physical phenomena. The use of calculus is fundamental in the treatment of various problems in Newtonian mechanics, electricity, and magnetism. In this section, we simply state some basic properties and "rules of thumb" that should be a useful review to the student.

First, a **function** must be specified that relates one variable to another (e.g., a coordinate as a function of time). Suppose one of the variables is called y (the dependent variable), the other x (the independent variable). We might have a function relationship such as

$$y(x) = ax^3 + bx^2 + cx + d$$

If a, b, c, and d are specified constants, then y can be calculated for any value of x. We usually deal with continuous functions, that is, those for which y varies "smoothly" with x.

The **derivative** of y with respect to x is defined as the limit, as Δx approaches zero, of the slopes of chords drawn between two points on the y versus x curve. Mathematically, we write this definition as

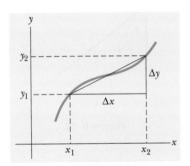

Figure B.13

$$\frac{dy}{dx} = \lim_{\Delta x \to 0} \frac{\Delta y}{\Delta x} = \lim_{\Delta x \to 0} \frac{y(x + \Delta x) - y(x)}{\Delta x} \tag{B.28}$$

where Δy and Δx are defined as $\Delta x = x_2 - x_1$ and $\Delta y = y_2 - y_1$ (Fig. B.13). It is important to note that dy/dx *does not* mean dy divided by dx, but is simply a notation of the limiting process of the derivative as defined by Equation B.28.

A useful expression to remember when $y(x) = ax^n$, where a is a *constant* and n is *any* positive or negative number (integer or fraction), is

$$\frac{dy}{dx} = nax^{n-1} \tag{B.29}$$

If $y(x)$ is a polynomial or algebraic function of x, we apply Equation B.29 to *each* term in the polynomial and take $d[\text{constant}]/dx = 0$. In Examples 4 through 7, we evaluate the derivatives of several functions.

Example 4 Suppose $y(x)$ (that is, y as a function of x) is given by

$$y(x) = ax^3 + bx + c$$

where a and b are constants. Then it follows that

$$y(x + \Delta x) = a(x + \Delta x)^3 + b(x + \Delta x) + c$$

$$y(x + \Delta x) = a(x^3 + 3x^2\Delta x + 3x\Delta x^2 + \Delta x^3) + b(x + \Delta x) + c$$

so

$$\Delta y = y(x + \Delta x) - y(x) = a(3x^2\Delta x + 3x\Delta x^2 + \Delta x^3) + b\Delta x$$

Substituting this into Equation B.28 gives

$$\frac{dy}{dx} = \lim_{\Delta x \to 0} \frac{\Delta y}{\Delta x} = \lim_{\Delta x \to 0} a[3x^2 + 3x\Delta x + \Delta x^2] + b$$

$$\frac{dy}{dx} = 3ax^2 + b$$

* The approximations for the functions $\sin x$, $\cos x$, and $\tan x$ are for $x \leq 0.1$ rad.

Example 5 Find the derivative of

$$y(x) = 8x^5 + 4x^3 + 2x + 7$$

Solution Applying Equation B.29 to each term independently, and remembering that d/dx (constant) $= 0$, we have

$$\frac{dy}{dx} = 8(5)x^4 + 4(3)x^2 + 2(1)x^0 + 0$$

$$\frac{dy}{dx} = 40x^4 + 12x^2 + 2$$

Special Properties of the Derivative

A. Derivative of the product of two functions If a function $f(x)$ is given by the product of two functions, say, $g(x)$ and $h(x)$, then the derivative of $f(x)$ is defined as

$$\frac{d}{dx} f(x) = \frac{d}{dx} [g(x)h(x)] = g\frac{dh}{dx} + h\frac{dg}{dx} \qquad \textbf{(B.30)}$$

B. Derivative of the sum of two functions If a function $f(x)$ is equal to the sum of two functions, then the derivative of the sum is equal to the sum of the derivatives:

$$\frac{d}{dx} f(x) = \frac{d}{dx} [g(x) + h(x)] = \frac{dg}{dx} + \frac{dh}{dx} \qquad \textbf{(B.31)}$$

C. Chain rule of differential calculus If $y = f(x)$ and $x = g(z)$, then dy/dz can be written as the product of two derivatives:

$$\frac{dy}{dz} = \frac{dy}{dx}\frac{dx}{dz} \qquad \textbf{(B.32)}$$

D. The second derivative The second derivative of y with respect to x is defined as the derivative of the function dy/dx (the derivative of the derivative). It is usually written

$$\frac{d^2y}{dx^2} = \frac{d}{dx}\left(\frac{dy}{dx}\right) \qquad \textbf{(B.33)}$$

Example 6 Find the derivative of $y(x) = x^3/(x+1)^2$ with respect to x.

Solution We can rewrite this function as $y(x) = x^3(x+1)^{-2}$ and apply Equation B.30:

$$\frac{dy}{dx} = (x+1)^{-2}\frac{d}{dx}(x^3) + x^3\frac{d}{dx}(x+1)^{-2}$$

$$= (x+1)^{-2}3x^2 + x^3(-2)(x+1)^{-3}$$

$$\frac{dy}{dx} = \frac{3x^2}{(x+1)^2} - \frac{2x^3}{(x+1)^3}$$

Example 7 A useful formula that follows from Equation B.30 is the derivative of the quotient of two functions. Show that

$$\frac{d}{dx}\left[\frac{g(x)}{h(x)}\right] = \frac{h\dfrac{dg}{dx} - g\dfrac{dh}{dx}}{h^2}$$

Solution We can write the quotient as gh^{-1} and then apply Equations B.29 and B.30:

$$\frac{d}{dx}\left(\frac{g}{h}\right) = \frac{d}{dx}(gh^{-1}) = g\frac{d}{dx}(h^{-1}) + h^{-1}\frac{d}{dx}(g)$$

$$= -gh^{-2}\frac{dh}{dx} + h^{-1}\frac{dg}{dx}$$

$$= \frac{h\dfrac{dg}{dx} - g\dfrac{dh}{dx}}{h^2}$$

TABLE B.4

Derivatives for Several Functions

$$\frac{d}{dx}(a) = 0$$

$$\frac{d}{dx}(ax^n) = nax^{n-1}$$

$$\frac{d}{dx}(e^{ax}) = ae^{ax}$$

$$\frac{d}{dx}(\sin ax) = a\cos ax$$

$$\frac{d}{dx}(\cos ax) = -a\sin ax$$

$$\frac{d}{dx}(\tan ax) = a\sec^2 ax$$

$$\frac{d}{dx}(\cot ax) = -a\csc^2 ax$$

$$\frac{d}{dx}(\sec x) = \tan x \sec x$$

$$\frac{d}{dx}(\csc x) = -\cot x \csc x$$

$$\frac{d}{dx}(\ln ax) = \frac{1}{x}$$

Note: The letters a and n are constants.

Some of the more commonly used derivatives of functions are listed in Table B.4.

B.7 • INTEGRAL CALCULUS

We think of integration as the inverse of differentiation. As an example, consider the expression

$$f(x) = \frac{dy}{dx} = 3ax^2 + b \tag{B.34}$$

which was the result of differentiating the function

$$y(x) = ax^3 + bx + c$$

in Example 4. We can write Equation B.34 as $dy = f(x)\,dx = (3ax^2 + b)\,dx$ and obtain $y(x)$ by "summing" over all values of x. Mathematically, we write this inverse operation

$$y(x) = \int f(x)\,dx$$

For the function $f(x)$ given by Equation B.34, we have

$$y(x) = \int (3ax^2 + b)\,dx = ax^3 + bx + c$$

where c is a constant of the integration. This type of integral is called an *indefinite integral* because its value depends on the choice of c.

A general **indefinite integral** $I(x)$ is defined as

$$I(x) = \int f(x)\,dx \tag{B.35}$$

where $f(x)$ is called the *integrand* and $f(x) = dI(x)/dx$.

For a *general continuous* function $f(x)$, the integral can be described as the area under the curve bounded by $f(x)$ and the x axis, between two specified values of x, say, x_1 and x_2, as in Figure B.14.

The area of the blue element is approximately $f(x_i)\Delta x_i$. If we sum all these area elements from x_1 and x_2 and take the limit of this sum as $\Delta x_i \rightarrow 0$, we obtain

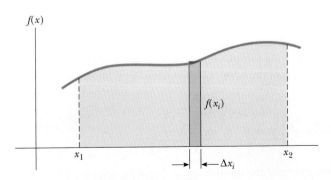

Figure B.14

the *true* area under the curve bounded by $f(x)$ and x, between the limits x_1 and x_2:

$$\text{Area} = \lim_{\Delta x_i \to 0} \sum_i f(x_i)\,\Delta x_i = \int_{x_1}^{x_2} f(x)\,dx \tag{B.36}$$

Integrals of the type defined by Equation B.36 are called **definite integrals.**
One common integral that arises in practical situations has the form

$$\int x^n\,dx = \frac{x^{n+1}}{n+1} + c \quad (n \neq -1) \tag{B.37}$$

This result is obvious, being that differentiation of the right-hand side with respect to x gives $f(x) = x^n$ directly. If the limits of the integration are known, this integral becomes a *definite integral* and is written

$$\int_{x_1}^{x_2} x^n\,dx = \frac{x_2^{n+1} - x_1^{n+1}}{n+1} \quad (n \neq -1) \tag{B.38}$$

EXAMPLES

1. $\displaystyle\int_0^a x^2\,dx = \frac{x^3}{3}\bigg]_0^a = \frac{a^3}{3}$

2. $\displaystyle\int_0^b x^{3/2}\,dx = \frac{x^{5/2}}{5/2}\bigg]_0^b = \tfrac{2}{5} b^{5/2}$

3. $\displaystyle\int_3^5 x\,dx = \frac{x^2}{2}\bigg]_3^5 = \frac{5^2 - 3^2}{2} = 8$

Partial Integration

Sometimes it is useful to apply the method of *partial integration* (also called "integrating by parts") to evaluate certain integrals. The method uses the property that

$$\int u\,dv = uv - \int v\,du \tag{B.39}$$

where u and v are *carefully* chosen so as to reduce a complex integral to a simpler one. In many cases, several reductions have to be made. Consider the function

$$I(x) = \int x^2 e^x\,dx$$

This can be evaluated by integrating by parts twice. First, if we choose $u = x^2$, $v = e^x$, we get

$$\int x^2 e^x\,dx = \int x^2\,d(e^x) = x^2 e^x - 2\int e^x x\,dx + c_1$$

Now, in the second term, choose $u = x$, $v = e^x$, which gives

$$\int x^2 e^x\,dx = x^2 e^x - 2x e^x + 2\int e^x\,dx + c_1$$

or

$$\int x^2 e^x\,dx = x^2 e^x - 2x e^x + 2e^x + c_2$$

The Perfect Differential

Another useful method to remember is the use of the *perfect differential*, in which we look for a change of variable such that the differential of the function is the differential of the independent variable appearing in the integrand. For example, consider the integral

$$I(x) = \int \cos^2 x \sin x \, dx$$

TABLE B.5 | **Some Indefinite Integrals (An arbitrary constant should be added to each of these integrals.)**

$$\int x^n \, dx = \frac{x^{n+1}}{n+1} \quad \text{(provided } n \neq -1\text{)}$$

$$\int \frac{dx}{x} = \int x^{-1} \, dx = \ln x$$

$$\int \frac{dx}{a+bx} = \frac{1}{b} \ln(a+bx)$$

$$\int \frac{x \, dx}{a+bx} = \frac{x}{b} - \frac{a}{b^2} \ln(a+bx)$$

$$\int \frac{dx}{x(x+a)} = -\frac{1}{a} \ln \frac{x+a}{x}$$

$$\int \frac{dx}{(a+bx)^2} = -\frac{1}{b(a+bx)}$$

$$\int \frac{dx}{a^2+x^2} = \frac{1}{a} \tan^{-1} \frac{x}{a}$$

$$\int \frac{dx}{a^2-x^2} = \frac{1}{2a} \ln \frac{a+x}{a-x} \quad (a^2-x^2 > 0)$$

$$\int \frac{dx}{x^2-a^2} = \frac{1}{2a} \ln \frac{x-a}{x+a} \quad (x^2-a^2 > 0)$$

$$\int \frac{x \, dx}{a^2 \pm x^2} = \pm \frac{1}{2} \ln(a^2 \pm x^2)$$

$$\int \frac{dx}{\sqrt{a^2-x^2}} = \sin^{-1} \frac{x}{a} = -\cos^{-1} \frac{x}{a} \quad (a^2-x^2 > 0)$$

$$\int \frac{dx}{\sqrt{x^2 \pm a^2}} = \ln(x + \sqrt{x^2 \pm a^2})$$

$$\int \frac{x \, dx}{\sqrt{a^2-x^2}} = -\sqrt{a^2-x^2}$$

$$\int \frac{x \, dx}{\sqrt{x^2 \pm a^2}} = \sqrt{x^2 \pm a^2}$$

$$\int \sqrt{a^2-x^2} \, dx = \frac{1}{2}\left(x\sqrt{a^2-x^2} + a^2 \sin^{-1} \frac{x}{a}\right)$$

$$\int x\sqrt{a^2-x^2} \, dx = -\frac{1}{3}(a^2-x^2)^{3/2}$$

$$\int \sqrt{x^2 \pm a^2} \, dx = \frac{1}{2}[x\sqrt{x^2 \pm a^2} \pm a^2 \ln(x + \sqrt{x^2 \pm a^2})]$$

$$\int x(\sqrt{x^2 \pm a^2}) \, dx = \frac{1}{3}(x^2 \pm a^2)^{3/2}$$

$$\int e^{ax} \, dx = \frac{1}{a} e^{ax}$$

$$\int \ln ax \, dx = (x \ln ax) - x$$

$$\int xe^{ax} \, dx = \frac{e^{ax}}{a^2}(ax - 1)$$

$$\int \frac{dx}{a+be^{cx}} = \frac{x}{a} - \frac{1}{ac} \ln(a+be^{cx})$$

$$\int \sin ax \, dx = -\frac{1}{a} \cos ax$$

$$\int \cos ax \, dx = \frac{1}{a} \sin ax$$

$$\int \tan ax \, dx = -\frac{1}{a} \ln(\cos ax) = \frac{1}{a} \ln(\sec ax)$$

$$\int \cot ax \, dx = \frac{1}{a} \ln(\sin ax)$$

$$\int \sec ax \, dx = \frac{1}{a} \ln(\sec ax + \tan ax) = \frac{1}{a} \ln\left[\tan\left(\frac{ax}{2} + \frac{\pi}{4}\right)\right]$$

$$\int \csc ax \, dx = \frac{1}{a} \ln(\csc ax - \cot ax) = \frac{1}{a} \ln\left(\tan \frac{ax}{2}\right)$$

$$\int \sin^2 ax \, dx = \frac{x}{2} - \frac{\sin 2ax}{4a}$$

$$\int \cos^2 ax \, dx = \frac{x}{2} + \frac{\sin 2ax}{4a}$$

$$\int \frac{dx}{\sin^2 ax} = -\frac{1}{a} \cot ax$$

$$\int \frac{dx}{\cos^2 ax} = \frac{1}{a} \tan ax$$

$$\int \tan^2 ax \, dx = \frac{1}{a}(\tan ax) - x$$

$$\int \cot^2 ax \, dx = -\frac{1}{a}(\cot ax) - x$$

$$\int \sin^{-1} ax \, dx = x(\sin^{-1} ax) + \frac{\sqrt{1-a^2x^2}}{a}$$

$$\int \cos^{-1} ax \, dx = x(\cos^{-1} ax) - \frac{\sqrt{1-a^2x^2}}{a}$$

$$\int \frac{dx}{(x^2+a^2)^{3/2}} = \frac{x}{a^2\sqrt{x^2+a^2}}$$

$$\int \frac{x \, dx}{(x^2+a^2)^{3/2}} = -\frac{1}{\sqrt{x^2+a^2}}$$

TABLE B.6	**Gauss's Probability Integral and Other Definite Integrals**

$$\int_0^\infty x^n e^{-ax}\, dx = \frac{n!}{a^{n+1}}$$

$$I_0 = \int_0^\infty e^{-ax^2}\, dx = \frac{1}{2}\sqrt{\frac{\pi}{a}} \quad \text{(Gauss's probability integral)}$$

$$I_1 = \int_0^\infty x e^{-ax^2}\, dx = \frac{1}{2a}$$

$$I_2 = \int_0^\infty x^2 e^{-ax^2}\, dx = -\frac{dI_0}{da} = \frac{1}{4}\sqrt{\frac{\pi}{a^3}}$$

$$I_3 = \int_0^\infty x^3 e^{-ax^2}\, dx = -\frac{dI_1}{da} = \frac{1}{2a^2}$$

$$I_4 = \int_0^\infty x^4 e^{-ax^2}\, dx = \frac{d^2 I_0}{da^2} = \frac{3}{8}\sqrt{\frac{\pi}{a^5}}$$

$$I_5 = \int_0^\infty x^5 e^{-ax^2}\, dx = \frac{d^2 I_1}{da^2} = \frac{1}{a^3}$$

$$\vdots$$

$$I_{2n} = (-1)^n \frac{d^n}{da^n} I_0$$

$$I_{2n+1} = (-1)^n \frac{d^n}{da^n} I_1$$

This becomes easy to evaluate if we rewrite the differential as $d(\cos x) = -\sin x\, dx$. The integral then becomes

$$\int \cos^2 x \sin x\, dx = -\int \cos^2 x\, d(\cos x)$$

If we now change variables, letting $y = \cos x$, we obtain

$$\int \cos^2 x \sin x\, dx = -\int y^2 dy = -\frac{y^3}{3} + c = -\frac{\cos^3 x}{3} + c$$

Table B.5 lists some useful indefinite integrals. Table B.6 gives Gauss's probability integral and other definite integrals. A more complete list can be found in various handbooks, such as *The Handbook of Chemistry and Physics*, CRC Press.

Periodic Table of the Elements

Group I	Group II	Transition elements							
H 1 1.008 0 $1s^1$									
Li 3 6.94 $2s^1$	**Be** 4 9.012 $2s^2$								
Na 11 22.99 $3s^1$	**Mg** 12 24.31 $3s^2$								

Symbol — **Ca** 20 — Atomic number
Atomic mass † — 40.08
$4s^2$ — Electron configuration

Group I	Group II								
K 19 39.102 $4s^1$	**Ca** 20 40.08 $4s^2$	**Sc** 21 44.96 $3d^14s^2$	**Ti** 22 47.90 $3d^24s^2$	**V** 23 50.94 $3d^34s^2$	**Cr** 24 51.996 $3d^54s^1$	**Mn** 25 54.94 $3d^54s^2$	**Fe** 26 55.85 $3d^64s^2$	**Co** 27 58.93 $3d^74s^2$
Rb 37 85.47 $5s^1$	**Sr** 38 87.62 $5s^2$	**Y** 39 88.906 $4d^15s^2$	**Zr** 40 91.22 $4d^25s^2$	**Nb** 41 92.91 $4d^45s^1$	**Mo** 42 95.94 $4d^55s^1$	**Tc** 43 (99) $4d^55s^2$	**Ru** 44 101.1 $4d^75s^1$	**Rh** 45 102.91 $4d^85s^1$
Cs 55 132.91 $6s^1$	**Ba** 56 137.34 $6s^2$	57-71*	**Hf** 72 178.49 $5d^26s^2$	**Ta** 73 180.95 $5d^36s^2$	**W** 74 183.85 $5d^46s^2$	**Re** 75 186.2 $5d^56s^2$	**Os** 76 190.2 $5d^66s^2$	**Ir** 77 192.2 $5d^76s^2$
Fr 87 (223) $7s^1$	**Ra** 88 (226) $7s^2$	89-103**	**Rf** 104 (261) $6d^27s^2$	**Db** 105 (262) $6d^37s^2$	**Sg** 106 (263)	**Bh** 107 (262)	**Hs** 108 (265)	**Mt** 109 (266)

*Lanthanide series

La 57 138.91 $5d^16s^2$	**Ce** 58 140.12 $5d^14f^16s^2$	**Pr** 59 140.91 $4f^36s^2$	**Nd** 60 144.24 $4f^46s^2$	**Pm** 61 (147) $4f^56s^2$	**Sm** 62 150.4 $4f^66s^2$
Ac 89 (227) $6d^17s^2$	**Th** 90 (232) $6d^27s^2$	**Pa** 91 (231) $5f^26d^17s^2$	**U** 92 (238) $5f^36d^17s^2$	**Np** 93 (239) $5f^46d^17s^2$	**Pu** 94 (239) $5f^66d^07s^2$

**Actinide series

WEB

For a description of the atomic data, visit **physics.nist.gov/atomic**

† Atomic mass values given are averaged over isotopes in the percentages in which they exist in nature; for an unstable element, mass number of the most stable known isotope is given in parentheses.
†† Elements 110, 111, 112, and 114 have not yet been named.

	Group III	Group IV	Group V	Group VI	Group VII	Group 0
					H 1 $1.008\ 0$ $1s^1$	**He** 2 $4.002\ 6$ $1s^2$
	B 5 10.81 $2p^1$	**C** 6 12.011 $2p^2$	**N** 7 14.007 $2p^3$	**O** 8 15.999 $2p^4$	**F** 9 18.998 $2p^5$	**Ne** 10 20.18 $2p^6$
	Al 13 26.98 $3p^1$	**Si** 14 28.09 $3p^2$	**P** 15 30.97 $3p^3$	**S** 16 32.06 $3p^4$	**Cl** 17 35.453 $3p^5$	**Ar** 18 39.948 $3p^6$

Ni 28 58.71 $3d^84s^2$	**Cu** 29 63.54 $3d^{10}4s^1$	**Zn** 30 65.37 $3d^{10}4s^2$	**Ga** 31 69.72 $4p^1$	**Ge** 32 72.59 $4p^2$	**As** 33 74.92 $4p^3$	**Se** 34 78.96 $4p^4$	**Br** 35 79.91 $4p^5$	**Kr** 36 83.80 $4p^6$
Pd 46 106.4 $4d^{10}$	**Ag** 47 107.87 $4d^{10}5s^1$	**Cd** 48 112.40 $4d^{10}5s^2$	**In** 49 114.82 $5p^1$	**Sn** 50 118.69 $5p^2$	**Sb** 51 121.75 $5p^3$	**Te** 52 127.60 $5p^4$	**I** 53 126.90 $5p^5$	**Xe** 54 131.30 $5p^6$
Pt 78 195.09 $5d^96s^1$	**Au** 79 196.97 $5d^{10}6s^1$	**Hg** 80 200.59 $5d^{10}6s^2$	**Tl** 81 204.37 $6p^1$	**Pb** 82 207.2 $6p^2$	**Bi** 83 208.98 $6p^3$	**Po** 84 (210) $6p^4$	**At** 85 (218) $6p^5$	**Rn** 86 (222) $6p^6$
110†† (269)	111†† (272)	112†† (277)		114†† (289)				

Eu 63 152.0 $4f^76s^2$	**Gd** 64 157.25 $5d^14f^76s^2$	**Tb** 65 158.92 $5d^14f^86s^2$	**Dy** 66 162.50 $4f^{10}6s^2$	**Ho** 67 164.93 $4f^{11}6s^2$	**Er** 68 167.26 $4f^{12}6s^2$	**Tm** 69 168.93 $4f^{13}6s^2$	**Yb** 70 173.04 $4f^{14}6s^2$	**Lu** 71 174.97 $5d^14f^{14}6s^2$
Am 95 (243) $5f^76d^07s^2$	**Cm** 96 (245) $5f^76d^17s^2$	**Bk** 97 (247) $5f^86d^17s^2$	**Cf** 98 (249) $5f^{10}6d^07s^2$	**Es** 99 (254) $5f^{11}6d^07s^2$	**Fm** 100 (253) $5f^{12}6d^07s^2$	**Md** 101 (255) $5f^{13}6d^07s^2$	**No** 102 (255) $6d^07s^2$	**Lr** 103 (257) $6d^17s^2$

SI Units

TABLE D.1 SI Units

| | SI Base Unit | |
Base Quantity	Name	Symbol
Length	Meter	m
Mass	Kilogram	kg
Time	Second	s
Electric current	Ampere	A
Temperature	Kelvin	K
Amount of substance	Mole	mol
Luminous intensity	Candela	cd

TABLE D.2 Some Derived SI Units

Quantity	Name	Symbol	Expression in Terms of Base Units	Expression in Terms of Other SI Units
Plane angle	radian	rad	m/m	
Frequency	hertz	Hz	s^{-1}	
Force	newton	N	$kg \cdot m/s^2$	J/m
Pressure	pascal	Pa	$kg/m \cdot s^2$	N/m^2
Energy; work	joule	J	$kg \cdot m^2/s^2$	$N \cdot m$
Power	watt	W	$kg \cdot m^2/s^3$	J/s
Electric charge	coulomb	C	$A \cdot s$	
Electric potential	volt	V	$kg \cdot m^2/A \cdot s^3$	W/A
Capacitance	farad	F	$A^2 \cdot s^4/kg \cdot m^2$	C/V
Electric resistance	ohm	Ω	$kg \cdot m^2/A^2 \cdot s^3$	V/A
Magnetic flux	weber	Wb	$kg \cdot m^2/A \cdot s^2$	$V \cdot s$
Magnetic field	tesla	T	$kg/A \cdot s^2$	
Inductance	henry	H	$kg \cdot m^2/A^2 \cdot s^2$	$T \cdot m^2/A$

Nobel Prizes

All Nobel Prizes in physics are listed (and marked with a P), as well as relevant Nobel Prizes in Chemistry (C). The key dates for some of the scientific work are supplied; they often antedate the prize considerably.

1901   (P) *Wilhelm Roentgen* for discovering x-rays (1895).

1902   (P) *Hendrik A. Lorentz* for predicting the Zeeman effect and *Pieter Zeeman* for discovering the Zeeman effect, the splitting of spectral lines in magnetic fields.

1903   (P) *Antoine-Henri Becquerel* for discovering radioactivity (1896) and *Pierre* and *Marie Curie* for studying radioactivity.

1904   (P) *John William Strutt, Lord Rayleigh* for studying the density of gases and discovering argon.

  (C) *William Ramsay* for discovering the inert gas elements helium, neon, xenon, and krypton, and placing them in the periodic table.

1905   (P) *Philipp Lenard* for studying cathode rays, electrons (1898–1899).

1906   (P) *J. J. Thomson* for studying electrical discharge through gases and discovering the electron (1897).

1907   (P) *Albert A. Michelson* for inventing optical instruments and measuring the speed of light (1880s).

1908   (P) *Gabriel Lippmann* for making the first color photographic plate, using interference methods (1891).

  (C) *Ernest Rutherford* for discovering that atoms can be broken apart by alpha rays and for studying radioactivity.

1909   (P) *Guglielmo Marconi* and *Carl Ferdinand Braun* for developing wireless telegraphy.

1910   (P) *Johannes D. van der Waals* for studying the equation of state for gases and liquids (1881).

1911   (P) *Wilhelm Wien* for discovering Wien's law giving the peak of a blackbody radiation spectrum (1893).

  (C) *Marie Curie* for discovering radium and polonium (1898) and isolating radium.

1912   (P) *Nils Dalén* for inventing automatic gas regulators for lighthouses.

1913   (P) *Heike Kamerlingh-Onnes* for the discovery of superconductivity and liquefying helium (1908).

1914   (P) *Max T. F. von Laue* for studying x-rays from their diffraction by crystals, showing that x-rays are electromagnetic waves (1912).

  (C) *Theodore W. Richards* for determining the atomic weights of 60 elements, indicating the existence of isotopes.

1915   (P) *William Henry Bragg* and *William Lawrence Bragg*, his son, for studying the diffraction of x-rays in crystals.

1917   (P) *Charles Barkla* for studying atoms by x-ray scattering (1906).

1918   (P) *Max Planck* for discovering energy quanta (1900).

1919   (P) *Johannes Stark*, for discovering the Stark effect, the splitting of spectral lines in electric fields (1913).

1920   (P) *Charles-Édouard Guillaume* for discovering invar, a nickel–steel alloy with low coefficient of expansion.

(C) *Walther Nernst* for studying energy changes in chemical reactions and formulating the third law of thermodynamics (1918).

1921   (P) *Albert Einstein* for explaining the photoelectric effect and for his services to theoretical physics (1905).

(C) *Frederick Soddy* for studying the chemistry of radioactive substances and discovering isotopes (1912).

1922   (P) *Niels Bohr* for his model of the atom and its radiation (1913).

(C) *Francis W. Aston* for using the mass spectrograph to study atomic weights, thus discovering 212 of the 287 naturally occurring isotopes.

1923   (P) *Robert A. Millikan* for measuring the charge on an electron (1911) and for studying the photoelectric effect experimentally (1914).

1924   (P) *Karl M. G. Siegbahn* for his work in x-ray spectroscopy.

1925   (P) *James Franck* and *Gustav Hertz* for discovering the Franck–Hertz effect in electron–atom collisions.

1926   (P) *Jean-Baptiste Perrin* for studying Brownian motion to validate the discontinuous structure of matter and measure the size of atoms.

1927   (P) *Arthur Holly Compton* for discovering the Compton effect on x-rays, their change in wavelength when they collide with matter (1922), and *Charles T. R. Wilson* for inventing the cloud chamber, used to study charged particles (1906).

1928   (P) *Owen W. Richardson* for studying the thermionic effect and electrons emitted by hot metals (1911).

1929   (P) *Louis-Victor de Broglie* for discovering the wave nature of electrons (1923).

1930   (P) *Chandrasekhara Venkata Raman* for studying Raman scattering, the scattering of light by atoms and molecules with a change in wavelength (1928).

1932   (P) *Werner Heisenberg* for creating quantum mechanics (1925).

1933   (P) *Erwin Schrödinger* and *Paul A. M. Dirac* for developing wave mechanics (1925) and relativistic quantum mechanics (1927).

1934   (C) *Harold Urey* for discovering heavy hydrogen, deuterium (1931).

1935   (P) *James Chadwick* for discovering the neutron (1932).

(C) *Irène* and *Frédéric Joliot-Curie* for synthesizing new radioactive elements.

1936   (P) *Carl D. Anderson* for discovering the positron in particular and antimatter in general (1932) and *Victor F. Hess* for discovering cosmic rays.

(C) *Peter J. W. Debye* for studying dipole moments and diffraction of x-rays and electrons in gases.

1937   (P) *Clinton Davisson* and *George Thomson* for discovering the diffraction of electrons by crystals, confirming de Broglie's hypothesis (1927).

1938   (P) *Enrico Fermi* for producing the transuranic radioactive elements by neutron irradiation (1934–1937).

1939   (P) *Ernest O. Lawrence* for inventing the cyclotron.

1943   (P) *Otto Stern* for developing molecular-beam studies (1923), and using them to discover the magnetic moment of the proton (1933).

1944   (P) *Isidor I. Rabi* for discovering nuclear magnetic resonance in atomic and molecular beams.

(C) *Otto Hahn* for discovering nuclear fission (1938).

1945 (P) *Wolfgang Pauli* for discovering the exclusion principle (1924).

1946 (P) *Percy W. Bridgman* for studying physics at high pressures.

1947 (P) *Edward V. Appleton* for studying the ionosphere.

1948 (P) *Patrick M. S. Blackett* for studying nuclear physics with cloud-chamber photographs of cosmic-ray interactions.

1949 (P) *Hideki Yukawa* for predicting the existence of mesons (1935).

1950 (P) *Cecil F. Powell* for developing the method of studying cosmic rays with photographic emulsions and discovering new mesons.

1951 (P) *John D. Cockcroft* and *Ernest T. S. Walton* for transmuting nuclei in an accelerator (1932).

(C) *Edwin M. McMillan* for producing neptunium (1940) and *Glenn T. Seaborg* for producing plutonium (1941) and further transuranic elements.

1952 (P) *Felix Bloch* and *Edward Mills Purcell* for discovering nuclear magnetic resonance in liquids and gases (1946).

1953 (P) *Frits Zernike* for inventing the phase-contrast microscope, which uses interference to provide high contrast.

1954 (P) *Max Born* for interpreting the wave function as a probability (1926) and other quantum-mechanical discoveries and *Walther Bothe* for developing the coincidence method to study subatomic particles (1930–1931), producing, in particular, the particle interpreted by Chadwick as the neutron.

1955 (P) *Willis E. Lamb, Jr.,* for discovering the Lamb shift in the hydrogen spectrum (1947) and *Polykarp Kusch* for determining the magnetic moment of the electron (1947).

1956 (P) *John Bardeen, Walter H. Brattain,* and *William Shockley* for inventing the transistor (1956).

1957 (P) *T.-D. Lee* and *C.-N. Yang* for predicting that parity is not conserved in beta decay (1956).

1958 (P) *Pavel A. Cerenkov* for discovering Cerenkov radiation (1935) and *Ilya M. Frank* and *Igor Tamm* for interpreting it (1937).

1959 (P) *Emilio G. Segrè* and *Owen Chamberlain* for discovering the antiproton (1955).

1960 (P) *Donald A. Glaser* for inventing the bubble chamber to study elementary particles (1952).

(C) *Willard Libby* for developing radiocarbon dating (1947).

1961 (P) *Robert Hofstadter* for discovering internal structure in protons and neutrons and *Rudolf L. Mössbauer* for discovering the Mössbauer effect of recoilless gamma-ray emission (1957).

1962 (P) *Lev Davidovich Landau* for studying liquid helium and other condensed matter theoretically.

1963 (P) *Eugene P. Wigner* for applying symmetry principles to elementary-particle theory and *Maria Goeppert-Mayer* and *J. Hans D. Jensen* for studying the shell model of nuclei (1947).

1964 (P) *Charles H. Townes, Nikolai G. Basov,* and *Alexandr M. Prokhorov* for developing masers (1951–1952) and lasers.

1965 (P) *Sin-Itiro Tomonaga, Julian S. Schwinger,* and *Richard P. Feynman* for developing quantum electrodynamics (1948).

1966 (P) *Alfred Kastler* for his optical methods of studying atomic energy levels.

1967 (P) *Hans Albrecht Bethe* for discovering the routes of energy production in stars (1939).

1968 (P) *Luis W. Alvarez* for discovering resonance states of elementary particles.

1969 (P) *Murray Gell-Mann* for classifying elementary particles (1963).

1970 (P) *Hannes Alfvén* for developing magnetohydrodynamic theory and *Louis Eugène Félix Néel* for discovering antiferromagnetism and ferrimagnetism (1930s).

1971 (P) *Dennis Gabor* for developing holography (1947).

(C) *Gerhard Herzberg* for studying the structure of molecules spectroscopically.

1972 (P) *John Bardeen, Leon N. Cooper,* and *John Robert Schrieffer* for explaining superconductivity (1957).

1973 (P) *Leo Esaki* for discovering tunneling in semiconductors, *Ivar Giaever* for discovering tunneling in superconductors, and *Brian D. Josephson* for predicting the Josephson effect, which involves tunneling of paired electrons (1958–1962).

1974 (P) *Anthony Hewish* for discovering pulsars and *Martin Ryle* for developing radio interferometry.

1975 (P) *Aage N. Bohr, Ben R. Mottelson,* and *James Rainwater* for discovering why some nuclei take asymmetric shapes.

1976 (P) *Burton Richter* and *Samuel C. C. Ting* for discovering the J/psi particle, the first charmed particle (1974).

1977 (P) *John H. Van Vleck, Nevill F. Mott,* and *Philip W. Anderson* for studying solids quantum-mechanically.

(C) *Ilya Prigogine* for extending thermodynamics to show how life could arise in the face of the second law.

1978 (P) *Arno A. Penzias* and *Robert W. Wilson* for discovering the cosmic background radiation (1965) and *Pyotr Kapitsa* for his studies of liquid helium.

1979 (P) *Sheldon L. Glashow, Abdus Salam,* and *Steven Weinberg* for developing the theory that unified the weak and electromagnetic forces (1958–1971).

1980 (P) *Val Fitch* and *James W. Cronin* for discovering CP (charge-parity) violation (1964), which possibly explains the cosmological dominance of matter over antimatter.

1981 (P) *Nicolaas Bloembergen* and *Arthur L. Schawlow* for developing laser spectroscopy and *Kai M. Siegbahn* for developing high-resolution electron spectroscopy (1958).

1982 (P) *Kenneth G. Wilson* for developing a method of constructing theories of phase transitions to analyze critical phenomena.

1983 (P) *William A. Fowler* for theoretical studies of astrophysical nucleosynthesis and *Subramanyan Chandrasekhar* for studying physical processes of importance to stellar structure and evolution, including the prediction of white dwarf stars (1930).

1984 (P) *Carlo Rubbia* for discovering the W and Z particles, verifying the electroweak unification, and *Simon van der Meer,* for developing the method of stochastic cooling of the CERN beam that allowed the discovery (1982–1983).

1985 (P) *Klaus von Klitzing* for the quantized Hall effect, relating to conductivity in the presence of a magnetic field (1980).

1986 (P) *Ernst Ruska* for inventing the electron microscope (1931), and *Gerd Binnig* and *Heinrich Rohrer* for inventing the scanning-tunneling microscope (1981).

1987 (P) *J. Georg Bednorz* and *Karl Alex Müller* for the discovery of high-temperature superconductivity (1986).

1988 (P) *Leon M. Lederman, Melvin Schwartz,* and *Jack Steinberger* for a collaborative experiment that led to the development of a new tool for studying the weak nuclear force, which affects the radioactive decay of atoms.

1989 (P) *Norman Ramsey* for various techniques in atomic physics; and *Hans Dehmelt* and *Wolfgang Paul* for the development of techniques for trapping single charge particles.

1990 (P) *Jerome Friedman, Henry Kendall,* and *Richard Taylor* for experiments important to the development of the quark model.

1991 (P) *Pierre-Gilles de Gennes* for discovering that methods developed for studying order phenomena in simple systems can be generalized to more complex forms of matter, in particular to liquid crystals and polymers.

1992 (P) *Georges Charpak* for developing detectors that trace the paths of evanescent subatomic particles produced in particle accelerators.

1993   (P) *Russell Hulse* and *Joseph Taylor* for discovering evidence of gravitational waves.

1994   (P) *Bertram N. Brockhouse* and *Clifford G. Shull* for pioneering work in neutron scattering.

1995   (P) *Martin L. Perl* and *Frederick Reines* for discovering the tau particle and the neutrino, respectively.

1996   (P) *David M. Lee, Douglas C. Osheroff,* and *Robert C. Richardson* for developing a superfluid using helium-3.

1997   (P) *Steven Chu, Claude Cohen-Tannoudji,* and *William D. Phillips* for developing methods to cool and trap atoms with laser light.

1998   (P) *Robert B. Laughlin, Horst L. Störmer,* and *Daniel C. Tsui* for discovering a new form of quantum fluid with fractionally charged excitations.

1999   (P) *Gerardus 'T Hooft* and *Martinus J. G. Veltman* for elucidating the quantum structure of electroweak interactions in physics.

2000   (P) *Zhores I. Alferov* and *Herbert Kroemer* for developing semiconductor heterostructures used in high-speed electronics and opto-electronics, and *Jack S. Kilby* for his part in the invention of the integrated circuit.

Answers to Odd-Numbered Problems

Chapter 1

1. $2.14 \times 10^4 \, \text{kg/m}^3$
3. $23.0 \, \text{kg}$
5. $8.72 \times 10^{11} \, \text{atom/s}$
7. $2.57 \times 10^{-10} \, \text{m}$
9. No.
11. $0.579 \, t \, \text{ft}^3/\text{s} + 1.19 \times 10^{-9} \, t^2 \, \text{ft}^3/\text{s}^2$
13. (a) $0.071 \, 4 \, \text{gal/s}$ (b) $2.70 \times 10^{-4} \, \text{m}^3/\text{s}$ (c) $1.03 \, \text{h}$
15. $4.05 \times 10^3 \, \text{m}^2$
17. $151 \, \mu\text{m}$
19. $2.86 \, \text{cm}$
21. $\sim 10^6 \, \text{balls}$
23. $\sim 10^2 \, \text{kg}; \sim 10^3 \, \text{kg}$
25. (a) 797 (b) 1.1 (c) 17.66
27. (a) 3 (b) 4 (c) 3 (d) 2
29. $5.2 \, \text{m}^3, 3\%$
31. (a) $2.24 \, \text{m}$ (b) $2.24 \, \text{m}$ at $26.6°$
33. (a) $r, 180° - \theta$ (b) $2r, 180° + \theta$ (c) $3r, -\theta$
35. (a) $10.0 \, \text{m}$ (b) $15.7 \, \text{m}$ (c) 0
37. Approximately $420 \, \text{ft}$ at $-3°$
39. $47.2 \, \text{units}$ at $122°$
41. $240 \, \text{m}$ at $237°$
43. (a) $49.5\mathbf{i} + 27.1\mathbf{j}$ (b) $56.4 \, \text{units}$ at $28.7°$
45. (a) $2.00\mathbf{i} - 6.00\mathbf{j}$ (b) $4.00\mathbf{i} + 2.00\mathbf{j}$ (c) 6.32
    (d) 4.47 (e) $288°; 26.6°$
47. $\mathbf{A} + \mathbf{B} = (2.60\mathbf{i} + 4.50\mathbf{j}) \, \text{m}$
49. (a) $8.00\mathbf{i} + 12.0\mathbf{j} - 4.00\mathbf{k}$ (b) $2.00\mathbf{i} + 3.00\mathbf{j} - 1.00\mathbf{k}$
    (c) $-24.0\mathbf{i} - 36.0\mathbf{j} + 12.0\mathbf{k}$
51. $|\mathbf{B}| = 7.81, \alpha = 59.2°, \beta = 39.8°, \gamma = 67.4°$
53. $106°$
55. $0.141 \, \text{nm}$
57. $4.50 \, \text{m}^2$
59. 0.449%
61. $\sim 10^{11} \, \text{stars}$
63. $2 \tan^{-1}(1/n)$
65. (a) zero (b) zero

Chapter 2

1. (a) $2.30 \, \text{m/s}$ (b) $16.1 \, \text{m/s}$ (c) $11.5 \, \text{m/s}$
3. (a) $5 \, \text{m/s}$ (b) $1.2 \, \text{m/s}$ (c) $-2.5 \, \text{m/s}$
   (d) $-3.3 \, \text{m/s}$ (e) 0
5. (a) $2v_1 v_2/(v_1 + v_2)$ (b) 0
7. (a) $-2.4 \, \text{m/s}$ (b) $-3.8 \, \text{m/s}$ (c) $4.0 \, \text{s}$
9. (a) $5.0 \, \text{m/s}$ (b) $-2.5 \, \text{m/s}$ (c) 0 (d) $5.0 \, \text{m/s}$
11. $5.00 \, \text{m}$
13. (a) $1.3 \, \text{m/s}^2$ (b) $2 \, \text{m/s}^2$ at $3 \, \text{s}$
    (c) at $t = 6 \, \text{s}$ and for $t > 10 \, \text{s}$ (d) $-1.5 \, \text{m/s}^2$ at $8 \, \text{s}$

15. (a) $2.00 \, \text{m}$ (b) $-3.00 \, \text{m/s}$ (c) $-2.00 \, \text{m/s}^2$
17. (a) $20.0 \, \text{m/s}, 5.00 \, \text{m/s}$ (b) $262 \, \text{m}$
19. (a) $4.53 \, \text{s}$ (b) $14.1 \, \text{m/s}$
21. $-16.0 \, \text{cm/s}^2$
23. (a) $12.7 \, \text{m/s}$ (b) $-2.30 \, \text{m/s}$
25. (a) $20.0 \, \text{s}$ (b) no
27. $3.10 \, \text{m/s}$
29. (a) $-202 \, \text{m/s}^2$ (b) $198 \, \text{m}$
31. (a) $35.0 \, \text{s}$ (b) $15.7 \, \text{m/s}$
33. They collide $212 \, \text{m}$ into the tunnel after $11.4 \, \text{s}$.
35. (a) $10.0 \, \text{m/s}$ up (b) $4.68 \, \text{m/s}$ down
37. (a) $29.4 \, \text{m/s}$ (b) $44.1 \, \text{m}$
39. (a) $7.82 \, \text{m}$ (b) $0.782 \, \text{s}$
41. (a) $1.53 \, \text{s}$ (b) $11.5 \, \text{m}$ (c) $-4.60 \, \text{m/s}, -9.80 \, \text{m/s}^2$
43. $2.74 \times 10^5 \, \text{m/s}^2$, which is $2.79 \times 10^4 \, g$
45. (a) $41.0 \, \text{s}$ (b) $1.73 \, \text{km}$ (c) $-184 \, \text{m/s}$
47. (a) $5.43 \, \text{m/s}^2$ and $3.83 \, \text{m/s}^2$
    (b) $10.9 \, \text{m/s}$ and $11.5 \, \text{m/s}$ (c) Maggie by $2.62 \, \text{m}$
49. $\sim 10^3 \, \text{m/s}^2$
51. (a) $2.99 \, \text{s}$ (b) $-15.4 \, \text{m/s}$
    (c) $31.3 \, \text{m/s}$ down and $34.9 \, \text{m/s}$ down
53. (c) $v_{\text{boy}}^2/h, 0$ (d) $v_{\text{boy}}, 0$
55. $1.60 \, \text{m/s}^2$
57. $0.577 \, v$

Chapter 3

1. (a) $4.87 \, \text{km}$ at $209°$ from east (b) $23.3 \, \text{m/s}$
   (c) $13.5 \, \text{m/s}$ at $209°$
3. (a) $18.0 \, t\mathbf{i} + (4.00 \, t - 4.90 \, t^2)\mathbf{j}$
   (b) $18.0 \, \mathbf{i} + (4.00 - 9.80 \, t)\mathbf{j}$ (c) $-9.80 \, \mathbf{j} \, \text{m/s}^2$
   (d) $(54.0 \, \mathbf{i} - 32.1 \, \mathbf{j}) \, \text{m}$ (e) $(18.0 \, \mathbf{i} - 25.4 \, \mathbf{j}) \, \text{m/s}$
   (f) $-9.80 \, \mathbf{j} \, \text{m/s}^2$
5. (a) $(0.800 \, \mathbf{i} - 0.300 \, \mathbf{j}) \, \text{m/s}^2$ (b) $339°$
   (c) $(360 \, \mathbf{i} - 72.7 \, \mathbf{j}) \, \text{m}, -15.2°$
7. (a) $x = 0.010 \, 0 \, \text{m}, y = 2.41 \times 10^{-4} \, \text{m}$
   (b) $\mathbf{v} = (1.84 \times 10^7 \, \mathbf{i} + 8.78 \times 10^5 \, \mathbf{j}) \, \text{m/s}$
   (c) $v = 1.85 \times 10^7 \, \text{m/s}$ (d) $\theta = 2.73°$
9. (a) $3.34 \, \mathbf{i} \, \text{m/s}$ (b) $-50.9°$
11. $48.6 \, \text{m/s}$
13. (a) $22.6 \, \text{m}$ (b) $52.3 \, \text{m}$ (c) $1.18 \, \text{s}$
15. (a) The ball clears by $0.889 \, \text{m}$ while (b) descending
17. $67.8°$
19. $d \tan \theta_i - gd^2/(2v_i^2 \cos^2 \theta_i)$
21. (a) $18.1 \, \text{m/s}$ (b) $1.13 \, \text{m}$ (c) $2.79 \, \text{m}$
23. $\tan^{-1}[(2gh)^{1/2}/v]$
25. $377 \, \text{m/s}^2$
27. $10.5 \, \text{m/s}, 219 \, \text{m/s}^2$ inward

29. 1.48 m/s^2 inward and $29.9°$ backward
31. (a) 13.0 m/s^2 (b) 5.70 m/s (c) 7.50 m/s^2
33. 2.02×10^3 s; 21.0% longer
35. 153 km/h at $11.3°$ north of west
37. 15.3 m
39. $7.58 \times 10^3 \text{ m/s}, 5.80 \times 10^3$ s
41. (a) $26.6°$ (b) 0.949
43. (b) $45° + \phi/2$; $v_i^2 (1 - \sin \phi)/g \cos^2 \phi$
45. (a) 41.7 m/s (b) 3.81 s
(c) $(34.1 \mathbf{i} - 13.4 \mathbf{j})$ m/s; 36.6 m/s
47. 10.7 m/s
49. (a) $v_i > (gR)^{1/2}$ (b) $(\sqrt{2} - 1)R$
51. 7.50 m/s in the direction the ball was thrown
53. (a) 20.0 m/s, 5.00 s (b) $(16.0 \mathbf{i} - 27.1 \mathbf{j})$ m/s
(c) 6.54 s (d) $24.6 \mathbf{i}$ m
55. (a) 43.2 m (b) $(9.66 \mathbf{i} - 25.6 \mathbf{j})$ m/s
57. Imagine you have a sick child and are shaking down the mercury in a fever thermometer. Starting with your hand at the level of your shoulder, move your hand down as fast as you can and snap it around an arc at the bottom. $\sim 100 \text{ m/s}^2 \sim 10 \, g$

Chapter 4

1. (a) 1/3 (b) 0.750 m/s^2
3. $(6.00 \mathbf{i} + 15.0 \mathbf{j})$ N; 16.2 N
5. (a) $(-45.0 \mathbf{i} + 15.0 \mathbf{j})$ m/s (b) $162°$ from the $+ x$ axis
(c) $(-225 \mathbf{i} + 75.0 \mathbf{j})$ m (d) $(-227, 79.0)$ m
7. (a) 5.00 m/s^2 at $36.9°$ (b) 6.08 m/s^2 at $25.3°$
9. 112 N
11. 2.38 kN
13. (a) 3.64×10^{-18} N
(b) 8.93×10^{-30} N is 408 billion times smaller
15. 2.55 N for an 88.7-kg person
17. (a) $\sim 10^{-22} \text{ m/s}^2$ (b) $\sim 10^{-23}$ m
19. (a) 15.0 lb up (b) 5.00 lb up (c) 0
23. 8.66 N east
25. (a) 49.0 N (b) 98.0 N (c) 24.5 N
27. $a = F/(m_1 + m_2)$; $T = F m_1/(m_1 + m_2)$
29. 3.73 m
31. (a) 36.8 N (b) 2.45 m/s^2 (c) 1.23 m
33. (a) $F_x > 19.6$ N (b) $F_x \leq -78.4$ N
(c) Figure at top of next column.

F_x (N)	a_x (m/s^2)
-100	-12.5
-78.4	-9.80
-50	-6.96
0	-1.96
50	3.04
100	8.04

35. (a) 706 N (b) 814 N (c) 706 N (d) 648 N
37. (a) $(2.50 \mathbf{i} + 5.00 \mathbf{j})$ N (b) 5.59 N
39. (a) Figure in middle of next column. (b) 0.408 m/s^2
(c) 83.3 N
41. 1.18 kN
43. (a) 2.20 m/s^2 (b) 27.4 N

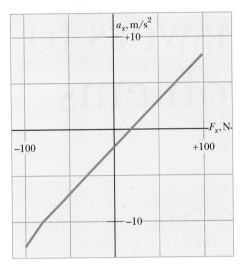

Chapter 4, Problem 33(c)

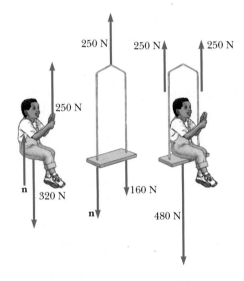

Chapter 4, Problem 39(a)

45. 173 lb
47. (a) $Mg/2, Mg/2, Mg/2, 3Mg/2, Mg$ (b) $Mg/2$
49. (a) 4.90 m/s^2 (b) 3.13 m/s (c) 1.35 m
(d) 1.14 s (e) no
51. $(M + m_1 + m_2)(m_2 g/ m_1)$
53. (a) $30.7°$ (b) 0.843 N
55. (a) 0.515 s, 5.05 m/s (b) 182 s, 0.014 3 m/s

Chapter 5

1. $\mu_s = 0.306$; $\mu_k = 0.245$
3. (a) 1.11 s (b) 0.875 s
5. (a) 14.7 m (b) Neither mass is necessary.
7. (a) 1.78 m/s^2 (b) 0.368 (c) 9.37 N
(d) 2.67 m/s
9. (a) Figure at top of next page. (b) 27.2 N, 1.29 m/s^2

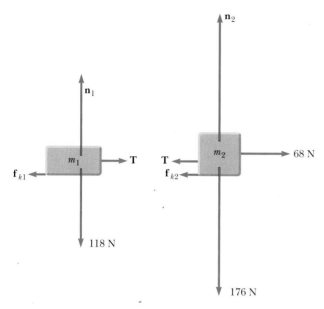

Chapter 5, Problem 9(a)

11. Any value between 31.7 N and 48.6 N
13. Any speed up to 8.08 m/s
15. $v \le 14.3$ m/s
17. No. The jungle-lord needs a vine of tensile strength 1.38 kN.
19. 3.13 m/s
21. (a) 8.62 m   (b) Mg downward   (c) 8.45 m/s^2
    Unless they have belts, the riders will fall from the cars.
23. (a) 1.47 N·s/m   (b) 2.04×10^{-3} s
    (c) 2.94×10^{-2} N
25. (a) 0.034 7 s^{-1} (b) 2.50 m/s   (c) $a = -cv$
27. (a) 13.7 m/s down   (b)

t (s)	x (m)	v (m/s)
0	0	0
0.2	0	-1.96
0.4	-0.392	-3.88
. . .	. . .	. . .
1.0	-3.77	-8.71
. . . 2.0	-14.4	-12.56
. . . 4.0	-41.0	-13.67

29. (a) 7.70×10^{-4} kg/m   (b) 0.998 N
    (c) approximately 49 m, 6.3 s, 27 m/s
31. 2.97 nN
33. $\sim 10^{-7}$ N toward you
35. 0.613 m/s^2 toward the Earth
37. (a) 1.52 m/s^2 (b) 1.66 km/s   (c) 6.82 ks
39. (b)

θ	P (N)
0	40.0
15°	46.4
30°	60.1
45°	94.3
60°	260

41. They do not; 29.4 N
43. (a) 2.13 s   (b) 1.67 m
45. 6.84 m
47. (a) m_2g (b) m_2g (c) $(m_2gR/m_1)^{1/2}$
49. (b) 732 N down at the equator and 735 N down at the poles
51. 2.14 rev/min
53. (b) 2.54 s; 23.6 rev/min
55. 12.8 N
57. $\Sigma \mathbf{F} = -km\mathbf{v}$

Chapter 6

1. (a) 31.9 J   (b) 0   (c) 0   (d) 31.9 J
3. -4.70 kJ
5. 28.9
7. (a) 16.0 J   (b) 36.9°
9. (a) 11.3°   (b) 156°   (c) 82.3°
11. (a) 7.50 J   (b) 15.0 J   (c) 7.50 J   (d) 30.0 J
13. (a) 0.938 cm   (b) 1.25 J
15. 0.299 m/s
17. 50.0 J
19. (a) 1.20 J   (b) 5.00 m/s   (c) 6.30 J
21. (a) 60.0 J   (b) 60.0 J
23. 878 kN up
25. 0.116 m
27. 1.25 m/s
29. (a) 650 J   (b) $\Delta E_{int} = 588$ J   (c) 0   (d) 0
    (e) 62.0 J   (f) 1.76 m/s
31. 2.04 m
33. (a) -168 J   (b) $\Delta E_{int} = 184$ J   (c) 500 J
    (d) 148 J   (e) 5.65 m/s
35. 875 W
37. (a) 20.6 kJ   (b) 686 W
39. (a) 2.91×10^3 times; impractical
    (b) 90.5 W = 0.121 hp
41. 1.66×10^4 m/s
43. 90.0 J
45. (a) $(2 + 24t^2 + 72t^4)$ J   (b) $12t$ m/s^2; $48t$ N
    (c) $(48t + 288t^3)$ W   (d) 1 250 J
47. -0.047 5 J
49. (b) 240 W
51. (a) 4.12 m   (b) 3.35 m
53. (a) $\mathbf{F}_1 = (20.5\,\mathbf{i} + 14.3\,\mathbf{j})$ N $\mathbf{F}_2 = (-36.4\,\mathbf{i} + 21.0\,\mathbf{j})$ N   (b) $(-15.9\,\mathbf{i} + 35.3\,\mathbf{j})$ N   (c) $(-3.18\,\mathbf{i} + 7.07\,\mathbf{j})$ m/s^2
    (d) $(-5.54\,\mathbf{i} + 23.7\,\mathbf{j})$ m/s   (e) $(-2.30\,\mathbf{i} + 39.3\,\mathbf{j})$ m
    (f) 1.48 kJ (g) 1.48 kJ
55. 1.68 m/s
57. (a) 14.5 m/s   (b) 1.75 kg   (c) 0.350 kg
59. 0.799 J

Chapter 7

1. (a) 259 kJ, 0, -259 kJ   (b) 0, -259 kJ, -259 kJ
3. 22.0 kW
5. (a) $v = (3\,g\,R)^{1/2}$ (b) 0.098 0 N down
7. (a) 2.29 m/s   (b) 1.98 m/s
9. (a) 4.43 m/s   (b) 5.00 m

11. (a) 22.0 J, 40.0 J   (b) Yes, $\Delta K + \Delta U \neq 0$
13. (a) 125 J   (b) 50.0 J   (c) 66.7 J
    (d) Nonconservative. The results differ.
15. 7.04 m/s
17. $d = kx^2/(2mg \sin \theta) - x$
19. 10.2 m
21. 3.74 m/s
23. (a) -160 J   (b) 73.5 J   (c) 28.8 N   (d) 0.679
25. (a) 0.381 m   (b) 0.143 m   (c) 0.371 m
27. (a) 40.0 J   (b) -40.0 J   (c) 62.5 J
29. (A/r^2) away from the other particle
31. (a) -4.77×10^9 J   (b) 569 N   (c) 569 N up
33. 2.52×10^7 m
35. (a) + at Ⓑ, − at Ⓓ, 0 at Ⓐ, Ⓒ, and Ⓔ
    (b) Ⓒ stable; Ⓐ and Ⓔ unstable
    (c)

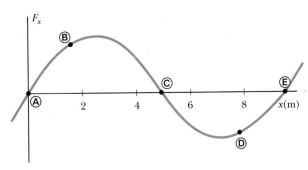

37. (b) Equilibrium at $x = 0$

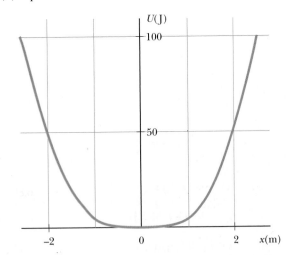

    (c) 0.823 m/s
39. 1.78 km
41. $\sim 10^3$ W peak or $\sim 10^2$ W sustainable
43. (a) 0.225 J   (b) $\Delta E_f = -0.363$ J
    (c) No; the normal force changes in a complicated way.
45. 48.2°
47. 0.328
49. 1.24 m/s
51. (a) 0.400 m   (b) 4.10 m/s
    (c) The block stays on the track.

53. $(h/5)(4 \sin^2 \theta + 1)$
55. 0.923 m/s
57. (a) $2.5R$

Chapter 8

1. (a) $(9.00\mathbf{i} - 12.0\mathbf{j})$ kg·m/s   (b) 15.0 kg·m/s at 307°
3. $\sim 10^{-23}$ m/s
5. (b) $p = \sqrt{2mK}$
7. (a) 13.5 N·s   (b) 9.00 kN   (c) 18.0 kN
9. 260 N normal to the wall
11. (a) $(9.05\mathbf{i} + 6.12\mathbf{j})$ N·s   (b) $(377\mathbf{i} + 255\mathbf{j})$ N
13. (b) Small   (d) large   (e) no difference
15. (a) $v_{gx} = 1.15$ m/s   (b) $v_{px} = -0.346$ m/s
17. (a) 2.50 m/s   (b) $\Delta K = -37.5$ kJ
19. (a) 0.284   (b) 115 fJ and 45.4 fJ
21. 91.2 m/s
23. $(4M/m)\sqrt{g\ell}$
25. (a) 2.88 m/s at 32.3° north of east
    (b) 783 J into internal energy
27. 2.50 m/s at $-60.0°$
29. $(3.00\mathbf{i} - 1.20\mathbf{j})$ m/s
31. (a) $(-9.33\mathbf{i} - 8.33\mathbf{j})$ Mm/s   (b) 439 fJ
33. $\mathbf{r}_{CM} = (0\mathbf{i} + 1.00\mathbf{j})$ m
35. $\mathbf{r}_{CM} = (11.7\mathbf{i} + 13.3\mathbf{j})$ cm
37. (a) $(1.40\mathbf{i} + 2.40\mathbf{j})$ m/s   (b) $(7.00\mathbf{i} + 12.0\mathbf{j})$ kg·m/s
39. 0.700 m
41. (a) 39.0 MN   (b) 3.20 m/s² up
43. (a) 442 metric tons   (b) 19.2 metric tons
45. (a) 3.7 km/s   (b) 153 km
47. 0.980 m
49. 3.20×10^4 N, 7.13 MW
51. (a) 3.54 m/s   (b) 1.77 m
    (c) 3.54×10^4 N to the right
    (d) No; the normal force of the rails contributes upward momentum to the system.
53. (a) $\sqrt{2}\, v_i$ and $\sqrt{2/3}\, v_i$ (b) 35.3°
55. (a) $(20.0\mathbf{i} + 7.00\mathbf{j})$ m/s   (b) 4.00i m/s²
    (c) 4.00i m/s²   (d) $(50.0\mathbf{i} + 35.0\mathbf{j})$ m   (e) 600 J
    (f) 674 J   (g) 674 J
57. $(3Mgx/L)\mathbf{j}$

Chapter 9

5. $0.866c$
7. (a) 64.9/min   (b) 10.6/min
9. 1.54 ns
11. $0.800c$
13. $0.140c$
15. (a) 21.0 yr   (b) 14.7 ly   (c) 10.5 ly   (d) 35.7 yr
17. $0.696c$
19. $0.960c$
21. (a) 2.50×10^8 m/s   (b) 4.97 m   (c) -1.33×10^{-8} s
23. (a) 2.73×10^{-24} kg·m/s   (b) 1.58×10^{-22} kg·m/s
    (c) 5.64×10^{-22} kg·m/s
25. 4.50×10^{-14}
27. $0.285c$
29. 1.63×10^3 MeV/c

31. (a) 938 MeV   (b) 3.00 GeV   (c) 2.07 GeV

33. 18.4 g/cm^3

37. 3.88 MeV and 28.8 MeV

39. 4.19 × 10^9 kg/s

41. 1.02 MeV

43. (a) $v/c = 1 - 1.12 \times 10^{-10}$ (b) 6.00 × 10^{27} J
   (c) \$2.17 × 10^{20}

45. (a) A few hundred seconds   (b) ∼ 10^8 km

47. (a) The charged battery has a mass greater by 4.00 × 10^{-14} kg   (b) 1.60 × 10^{-12}

51. Yes, with 18.8 m to spare

53. (b) For u small compared with c, the relativistic expression agrees with the classical expression. As u approaches c, the acceleration approaches zero, so that the object can never reach or surpass the speed of light.
   (c) Perform $\int (1 - u^2/c^2)^{-3/2}\, du = (qE/m)\int dt$ to obtain $u = qEct(m^2c^2 + q^2E^2t^2)^{-1/2}$ and then $\int dx = \int qEct(m^2c^2 + q^2E^2t^2)^{-1/2}\, dt$ to obtain $x = (c/qE)[(m^2c^2 + q^2E^2t^2)^{1/2} - mc]$

57. (a) The refugees conclude that Tau Ceti exploded 16.0 yr before the Sun.   (b) An observer at the midpoint and at rest with respect to the Sun and Tau Ceti concludes that they exploded simultaneously.

Chapter 10

1. (a) 5.24 s   (b) 27.4 rad

3. (a) 822 rad/s^2 (b) 4.21 × 10^3 rad

5. 50.0 rev

7. (a) 7.27 × 10^{-5} rad/s   (b) 428 min

9. (a) 126 rad/s   (b) 3.77 m/s   (c) 1.26 km/s^2
   (d) 20.1 m

11. (a) 54.3 rev   (b) 12.1 rev/s

13. (a) 92.0 kg·m^2, 184 J
   (b) 6.00 m/s, 4.00 m/s, 8.00 m/s, 184 J

17. − 3.55 N·m

19. − 7.00**i** + 16.0**j** − 10.0**k**

21. (a) 2.00 N·m   (b) **k**

23. $\dfrac{[(m_1 + m_b)d + m_1\ell/2]}{m_2}$

25. (a) $\left[\dfrac{m_1 g}{2} + \dfrac{xm_2 g}{L}\right]\cot\theta$, $(m_1 + m_2)g$
   (b) $\dfrac{m_1/2 + dm_2/L}{(m_1 + m_2)\tan\theta}$

27. (a) $T = \dfrac{F_g(L + d)}{\sin\theta\,(2L + d)}$
   (b) $R_x = \dfrac{F_g\,(L + d)\cot\theta}{(2L + d)}$; $R_y = \dfrac{F_g L}{(2L + d)}$

29. (a) 24.0 N·m   (b) 0.035 6 rad/s^2 (c) 1.07 m/s^2

31. 21.5 N

33. (a) 118 N and 156 N   (b) 1.17 kg·m^2

35. (a) 11.4 N, 7.57 m/s^2, 9.53 m/s down   (b) 9.53 m/s

37. $3g/2L$; $-(3g/2)$**j**

39. (60.0**k**) kg·m^2/s

41. (a) 0.433 kg·m^2/s   (b) 1.73 kg·m^2/s

43. (a) $\omega_f = \omega_i I_1/(I_1 + I_2)$ (b) $I_1/(I_1 + I_2)$

45. 7.14 rev/min

47. 12.3 m/s^2

49. (a) 500 J   (b) 250 J   (c) 750 J

51. 1.21 × 10^{-4} kg·m^2; height is unnecessary.

53. 131 s

55. (a) 0.360 rad/s counterclockwise   (b) 99.9 J

57. (a) $(3g/L)^{1/2}$ (b) $3g/2L$ (c) $-(3/2)g$**i** $-(3/4)g$**j**
   (d) $-(3/2)Mg$**i** $+(1/4)Mg$**j**

61. (a) 0.309 m/s^2 (b) 7.67 N and 9.22 N

63. (a) $\mu_k = 0.571$; the normal force acts 20.1 cm to the left of the front edge of the sliding cabinet.   (b) 0.501 m

65. $(3/8)F_g$

67. $T = 2.71$ kN; $R_x = 2.65$ kN

71. (a) − 794 N·m; − 2 510 N·m; 0; − 1 160 N·m; − 2 940 N·m
   (b) At the following times:

12:00:00	12:30:55	12:58:19	1:32:31	1:57:01
2:33:25	2:56:29	3:33:22	3:56:55	4:32:24
4:58:14	5:30:52	6:00:00	6:29:08	7:01:46
7:27:36	8:03:05	8:26:38	9:03:31	9:26:35
10:02:59	10:27:29	11:01:41	11:29:05	

Chapter 11

1. 2.67 × 10^{-7} m/s^2

3. (− 100**i** + 59.3**j**) pN

5. $\rho_M/\rho_E = 2/3$

7. (a) 3.46 × 10^8 m   (b) 3.34 × 10^{-3} m/s^2 toward the Earth

9. **g** $= 2MGr(r^2 + a^2)^{-3/2}$ toward the center of mass

11. (a) 1.84 × 10^9 kg/m^3 (b) 3.27 × 10^6 m/s^2
   (c) − 2.08 × 10^{13} J

13. 12.6 × 10^{31} kg

15. 1.90 × 10^{27} kg

17. Planet Y has gone through 1.30 revolutions.

19. 18.2 ms

23. (a) 10.0 m/s^2 (b) 21.8 km/s

25. (a) 0.980   (b) 127 yr   (c) − 2.13 × 10^{17} J

27. (a) 5   (b) No; no

29. (a) 0.212 nm   (b) 9.95 × 10^{-25} kg·m/s
   (c) 2.11 × 10^{-34} kg·m^2/s   (d) 3.40 eV   (e) − 6.80 eV
   (f) − 3.40 eV

31. (a) 13.6 eV   (b) 1.51 eV

33. 4.42 × 10^4 m/s

35. (a) 29.3%   (b) no change

37. $r_n = (0.106$ nm$)n^2$, $E_n = -6.80$ eV/n^2,
   for $n = 1, 2, 3, \ldots$

39. (b) 1.10 × 10^{32} kg

41. $v = 492$ m/s

43. 7.41 × 10^{-10} N

45. (a) $m_2(2G/d)^{1/2}(m_1 + m_2)^{-1/2}$ and
   $m_1(2G/d)^{1/2}(m_1 + m_2)^{-1/2}$ and
   $(2G/d)^{1/2}(m_1 + m_2)^{1/2}$
   (b) 1.07 × 10^{32} J and 2.67 × 10^{31} J

47. (a) 200 Myr   (b) ∼ 10^{41} kg; ∼ 10^{11} stars

49. $v = \sqrt{2R^2g\left(\dfrac{1}{R} + \dfrac{1}{r}\right)}$

53.

t (s)	x (m)	y (m)	v_x (m/s)	v_y (m/s)
0	0	12 740 000	5 000	0
10	50 000	12 740 000	4 999.9	−24.6
20	99 999	12 739 754	4 999.7	−49.1
30	149 996	12 739 263	4 999.4	−73.7 . . .

The object does not hit the Earth; its minimum radius is $1.33R_E$. Its period is 1.09×10^4 s. A circular orbit would require speed 5.60 km/s.

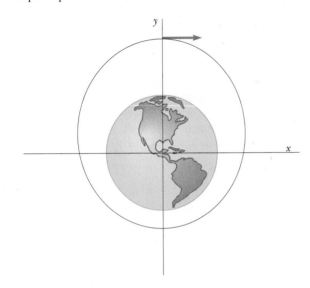

Context 1 Conclusion

3. (a) 146 d   (b) Venus 53.9° behind the Earth
4. (a) 2.95 km/s   (b) 2.65 km/s   (c) 10.7 km/s
     (d) 4.80 km/s

Chapter 12

1. (a) 575 N/m   (b) 46.0 J
3. (a) 1.50 Hz, 0.667 s   (b) 4.00 m   (c) π rad   (d) 2.83 m
5. (a) 4.33 cm   (b) −5.00 cm/s   (c) −17.3 cm/s²
     (d) 3.14 s, 5.00 cm
7. (b) 18.8 cm/s, 0.333 s   (c) 178 cm/s², 0.500 s
     (d) 12.0 cm
9. 40.9 N/m
11. (a) 40.0 cm/s, 160 cm/s²   (b) 32.0 cm/s, −96.0 cm/s²
     (c) 0.232 s
13. (a) $4\pi^2m/T^2$ (b) $m\,(T'/T)^2$
15. 2.23 m/s
17. (a) 126 N/m   (b) 0.178 m
19. (a) 28.0 mJ   (b) 1.02 m/s   (c) 12.2 mJ   (d) 15.8 mJ
21. 2.60 cm and −2.60 cm
23. (a) 0.817 m/s   (b) 2.54 rad/s²   (c) 0.634 N
27. 0.944 kg·m²
29. 1.00×10^{-3} s⁻¹

31. (a) 2.95 Hz   (b) 2.85 cm
35. 1.74 Hz
37. $\sim 10^1$ m
39. 6.62 cm
41. $f = (2\pi L)^{-1}\,(gL + kh^2/M)^{1/2}$
43. (b) 1.23 Hz
45. (a) 3.56 Hz   (b) 2.79 Hz   (c) 2.10 Hz
47. (a) $\frac{1}{2}\,(M + m/3)\,v^2$ (b) $2\pi(M + m/3)^{1/2}k^{-1/2}$
51. (b) After 42.1 min
55. (c)

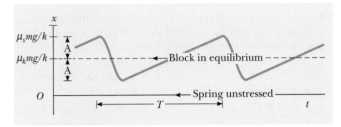

(f) 0.281 Hz   **(g)** frequency decreases
(h) frequency increases   **(i)** frequency increases
(j) frequency decreases

Chapter 13

1. $y = 6\,[(x - 4.5t)^2 + 3]^{-1}$
3. 0.319 m
5. (a) 3.33**i** m/s   (b) −5.48 cm   (c) 0.667m, 5.00 Hz
     (d) 11.0 m/s
7. (a) 31.4 rad/s   (b) 1.57 rad/m
     (c) $y = (0.120 \text{ m}) \sin(1.57x - 31.4t)$ (d) 3.77 m/s
     (e) 118 m/s²
9. (a) $y = (8.00 \text{ cm}) \sin(7.85x + 6\pi t)$
     (b) $y = (8.00 \text{ cm}) \sin(7.85x + 6\pi t - 0.785)$
13. 520 m/s
15. 0.329 s
17. (a) Zero   (b) 0.300 m
19. 55.1 Hz
21. (a) $y = (7.50 \text{ cm}) \sin(4.19x - 314t)$ (b) 625 W
23. 5.56 km
25. (a) 2.00 μm, 40.0 cm, 54.6 m/s
     (b) −0.433 μm (c) 1.72 mm/s
27. 5.81 m
29. $\Delta P = (0.200 \text{ N/m}^2) \sin(62.8x/\text{m} - 2.16 \times 10^4 t/\text{s})$
31. 26.4 m/s
33. 19.3 m
35. 439 Hz and 441 Hz
37. (a) The P wave   (b) 665 s
39. $\sim$1 min
41. $\left(\dfrac{Lm}{Mg\sin\theta}\right)^{1/2}$

47. (b) 31.6 m/s
49. 130 m/s, 1.73 km
51. 6.01 km
53. (a) 0.364 m   (b) 0.398 m   (c) 941 Hz   (d) 938 Hz

Chapter 14

1. (a) -1.65 cm   (b) -6.02 cm   (c) 1.15 cm
3. (a) y_1 in the $+x$ direction, y_2 in the $-x$ direction
   (b) 0.750 s   (c) 1.00 m
5. (a) 9.24 m   (b) 600 Hz
7. $91.3°$
9. (a) The number of minima is the greatest integer less than or equal to $df/v + \frac{1}{2}$
   (b) $L_n = \dfrac{d^2 - (n - 1/2)^2(v/f)^2}{2(n - 1/2)(v/f)}$,
   where $n = 1, 2, 3, \ldots$, number of minima
11. At $0.089\,1$ m, 0.303 m, 0.518 m, 0.732 m, 0.947 m, 1.16 m from one speaker
13. (a) 4.24 cm   (b) 6.00 cm   (c) 6.00 cm
    (d) 0.500 cm, 1.50 cm, 2.50 cm
15. 150 Hz
17. 0.786 Hz, 1.57 Hz, 2.36 Hz, 3.14 Hz
19. (a) 0.600 m   (b) 30.0 Hz
21. (a) 163 N   (b) 660 Hz
23. (a) Three loops   (b) 16.7 Hz   (c) One loop
25. (a) 0.357 m   (b) 0.715 m
27. 0.656 m and 1.64 m
29. (a) 0.555 m   (b) 620 Hz
31. $n(206\text{ Hz})$ for $n = 1$ to 9 and $n(84.5\text{ Hz})$ for $n = 2$ to 23
33. (a) 350 m/s   (b) 1.14 m
35. 5.64 beats/s
37. (a) 1.99 beats/s   (b) 3.38 m/s
39. (a) 3.66 m/s   (b) 0.200 Hz
41. (a) $\Delta r = \lambda/2$ (b) $9.00\,x^2 - 16.0\,y^2 = 144$
43. 9.00 kHz
45. (a) 34.8 m/s   (b) 0.977 m
47. 4.85 m
49. (a) $\frac{1}{2}Mg$ (b) $3h$ (c) $m/3h$ (d) $\sqrt{\dfrac{3Mgh}{2m}}$ (e) $\sqrt{\dfrac{3Mg}{8mh}}$
    (f) $\sqrt{\dfrac{2mh}{3Mg}}$ (g) h (h) $(2.00 \times 10^{-2})\sqrt{\dfrac{3Mg}{8mh}}$
51. (a) $\frac{1}{2}$ (b) $\left[\dfrac{n}{n+1}\right]^2 T$ (c) $\frac{9}{16}$
53. 50.0 Hz, 1.70 m
55. $y_1 + y_2 = 11.2\sin(2.00x - 10.0t + 63.4°)$
57. (a) 78.9 Hz   (b) 211 Hz

Context 2 Conclusion

1. 3.5 cm
2. -96%
3. Station 1: 15:46:26
   Station 2: 15:46:23
   Station 3: 15:46:08

Chapter 15

1. 0.111 kg
3. 6.24 MPa
5. 5.27×10^{18} kg
7. 1.62 m
9. 7.76×10^{-3} m^2
11. 5.88 MN, 196 kN, 588 kN
13. 10.5 m; no; some alcohol and water evaporate.
15. 98.6 kPa
17. 0.258 N
19. (a) $1.017\,9 \times 10^3$ N down, $1.029\,7 \times 10^3$ N up
    (b) 86.2 N   (c) 11.8 N
21. (a) 7.00 cm   (b) 2.80 kg
23. $1\,430$ m^3
25. $\rho_{\text{glycerin}} = 1\,250$ kg/m^3 and $\rho_{\text{sphere}} = 500$ kg/m^3
27. (a) 1.06 m/s   (b) 4.24 m/s
29. 616 MW
31. $2\sqrt{h(h_0 - h)}$
33. (a) 2.28 N   (b) 1.74×10^6 s
35. (a) 28.0 m/s   (b) 28.0 m/s   (c) 2.11 MPa
37. 68.0 kPa
39. (a) 489 N outward   (b) 1.96 kN outward
41. (b) Smaller; 8.30%
43. (a) 60.0 rad   (b) 9.12 m/s^2 (c) 6.79 m/s^2
    (d) 255 N·m
49. The top scale reads $(1 - \rho_0/\rho_{\text{Fe}})m_{\text{Fe}}g$. The bottom scale reads $[m_b + m_0 + \rho_0 m_{\text{Fe}}/\rho_{\text{Fe}}]g$.
51. (b) 2.58×10^4 N
53. 4.43 m/s
55. (a) 1.25 cm   (b) 13.8 m/s
57. (a) 3.307 g   (b) 3.271 g   (c) 3.48×10^{-4} N

Context 3 Conclusion

1. 2.0×10^{10} N·m
2. (a) 1.30×10^6 Pa   (b) Yes, but only with specialized equipment and techniques
3. (a) 0.42 m/s   (b) greater
4. (a) 16 knots, $56°$ west of south   (b) 47%

Index

Page numbers in *italics* indicate figures; page numbers followed by "n" indicate footnotes; page numbers followed by "t" indicate tables.

Standard Abbreviations and Symbols for Units

Symbol	Unit	Symbol	Unit
A	ampere	K	kelvin
u	atomic mass unit	kg	kilogram
atm	atmosphere	kmol	kilomole
Btu	British thermal unit	L	liter
C	coulomb	lb	pound
°C	degree Celsius	ly	lightyear
cal	calorie	m	meter
d	day	min	minute
eV	electron volt	mol	mole
°F	degree Fahrenheit	N	newton
F	farad	Pa	pascal
ft	foot	rad	radian
G	gauss	rev	revolution
g	gram	s	second
H	henry	T	tesla
h	hour	V	volt
hp	horsepower	W	watt
Hz	hertz	Wb	weber
in.	inch	yr	year
J	joule	Ω	ohm

Mathematical Symbols Used in the Text and Their Meaning

Symbol	Meaning
$=$	is equal to
$\equiv$	is defined as
$\neq$	is not equal to
$\propto$	is proportional to
$\sim$	is on the order of
$>$	is greater than
$<$	is less than
$>>(<<)$	is much greater (less) than
$\approx$	is approximately equal to
Δx	the change in x
$\displaystyle\sum_{i=1}^{N} x_i$	the sum of all quantities x_i from $i = 1$ to $i = N$
$|x|$	the magnitude of x (always a nonnegative quantity)
$\Delta x \to 0$	Δx approaches zero
$\dfrac{dx}{dt}$	the derivative of x with respect to t
$\dfrac{\partial x}{\partial t}$	the partial derivative of x with respect to t
$\displaystyle\int$	integral

Conversions[a]

Length
1 in. = 2.54 cm (exact)
1 m = 39.37 in. = 3.281 ft
1 ft = 0.304 8 m
12 in. = 1 ft
3 ft = 1 yd
1 yd = 0.914 4 m
1 km = 0.621 mi
1 mi = 1.609 km
1 mi = 5 280 ft
$1 \mu m = 10^{-6}$ m $= 10^3$ nm
1 lightyear $= 9.461 \times 10^{15}$ m

Area
$1 \text{ m}^2 = 10^4 \text{ cm}^2 = 10.76 \text{ ft}^2$
$1 \text{ ft}^2 = 0.092 \, 9 \text{ m}^2 = 144 \text{ in.}^2$
$1 \text{ in.}^2 = 6.452 \text{ cm}^2$

Volume
$1 \text{ m}^3 = 10^6 \text{ cm}^3 = 6.102 \times 10^4 \text{ in.}^3$
$1 \text{ ft}^3 = 1 \, 728 \text{ in.}^3 = 2.83 \times 10^{-2} \text{ m}^3$
$1 \text{ L} = 1 \, 000 \text{ cm}^3 = 1.057 \, 6 \text{ qt} = 0.035 \, 3 \text{ ft}^3$
$1 \text{ ft}^3 = 7.481 \text{ gal} = 28.32 \text{ L} = 2.832 \times 10^{-2} \text{ m}^3$
$1 \text{ gal} = 3.786 \text{ L} = 231 \text{ in.}^3$

Mass
1 000 kg = 1 t (metric ton)
1 slug = 14.59 kg
$1 \text{ u} = 1.66 \times 10^{-27} \text{ kg} \doteq 931.5 \text{ MeV}/c^2$

Force
1 N = 0.224 8 lb
1 lb = 4.448 N

Velocity
1 mi/h = 1.47 ft/s = 0.447 m/s = 1.61 km/h
1 m/s = 100 cm/s = 3.281 ft/s
1 mi/min = 60 mi/h = 88 ft/s

Acceleration
$1 \text{ m/s}^2 = 3.28 \text{ ft/s}^2 = 100 \text{ cm/s}^2$
$1 \text{ ft/s}^2 = 0.304 \, 8 \text{ m/s}^2 = 30.48 \text{ cm/s}^2$

Pressure
$1 \text{ bar} = 10^5 \text{ N/m}^2 = 14.50 \text{ lb/in.}^2$
1 atm = 760 mm Hg = 76.0 cm Hg
$1 \text{ atm} = 14.7 \text{ lb/in.}^2 = 1.013 \times 10^5 \text{ N/m}^2$
$1 \text{ Pa} = 1 \text{ N/m}^2 = 1.45 \times 10^{-4} \text{ lb/in.}^2$

Time
$1 \text{ yr} = 365 \text{ days} = 3.16 \times 10^7 \text{ s}$
$1 \text{ day} = 24 \text{ h} = 1.44 \times 10^3 \text{ min} = 8.64 \times 10^4 \text{ s}$

Energy
1 J = 0.738 ft·lb
1 cal = 4.186 J
$1 \text{ Btu} = 252 \text{ cal} = 1.054 \times 10^3 \text{ J}$
$1 \text{ eV} = 1.602 \times 10^{-19} \text{ J}$
$1 \text{ kWh} = 3.60 \times 10^6 \text{ J}$

Power
1 hp = 550 ft·lb/s = 0.746 kW
1 W = 1 J/s = 0.738 ft·lb/s
1 Btu/h = 0.293 W

Some Approximations Useful for Estimation Problems
1 m ≈ 1 yd
1 kg ≈ 2 lb
$1 \text{ N} \approx \frac{1}{4} \text{ lb}$
$1 \text{ L} \approx \frac{1}{4} \text{ gal}$

1 m/s ≈ 2 mi/h
$1 \text{ yr} \approx \pi \times 10^7 \text{ s}$
60 mi/h ≈ 100 ft/s
$1 \text{ km} \approx \frac{1}{2} \text{ mi}$

[a] See Table A.1 of Appendix A for a more complete list.

The Greek Alphabet

Alpha	A	α	Iota	I	ι	Rho	P	ρ
Beta	B	β	Kappa	K	κ	Sigma	Σ	σ
Gamma	Γ	γ	Lambda	Λ	λ	Tau	T	τ
Delta	Δ	δ	Mu	M	μ	Upsilon	Υ	υ
Epsilon	E	ϵ	Nu	N	ν	Phi	Φ	ϕ
Zeta	Z	ζ	Xi	Ξ	ξ	Chi	X	χ
Eta	H	η	Omicron	O	o	Psi	Ψ	ψ
Theta	Θ	θ	Pi	Π	π	Omega	Ω	ω